ATLAS OF

CARBON-13 NMR DATA

VOLUME 2
Compounds 1000–3017

ATLAS OF

CARBON-13 NMR DATA

E. BREITMAIER

Institut für Organische Chemie
und Biochemie der Universität Bonn

G. HAAS

Institut für Organische Chemie
der Universität Bremen

W. VOELTER

Institut für Organische Chemie
der Universität Tübingen

VOLUME 2

Compounds 1000–3017

London — Philadelphia — Rheine    HEYDEN

Heyden & Son Ltd., Spectrum House, Hillview Gardens, London NW4 2JQ, U.K.
Heyden & Son Inc., 247 South 41st Street, Philadelphia, Pennsylvania 19104, U.S.A.
Heyden & Son GmbH, Münsterstrasse 22, 4440 Rheine, West Germany.

ISBN 0 85501 482 2 (Volume 2)
ISBN 0 85501 480 6 (Complete Set)

Printed in Great Britain by J. W. Arrowsmith Ltd., Winterstoke Road, Bristol BS3 2NT.

CONTENTS

Carbon-13 NMR Data: Compounds 1000–3017

FOREWORD

A few years ago the practice and application of carbon-13 nuclear magnetic resonance spectroscopy was restricted to a handful of laboratories, and a search of the ^{13}C n.m.r. literature was a relatively simple and straightforward matter. Then, with the development of commercial Fourier transform n.m.r. spectrometers, the technique became generally available and ^{13}C n.m.r. applications rapidly became routine, with the result that the ^{13}C literature seems to be following an exponential growth pattern. Consequently a search of the literature has become a formidable task. Although during this period several monographs and reviews of the chemical applications of ^{13}C n.m.r. have appeared and some of these present extensive tabulations of data, these data tend to be illustrative rather than exhaustive. There is a real need for a systematic, regularly appearing collection of data to enable one to locate the results for specific compounds quickly and easily.

This is a continuing series that will serve to fill this need. It should prove extremely valuable as the ^{13}C n.m.r. literature continues its explosive growth, since ^{13}C spectra now rival proton spectra as a primary means for the characterization and identification of a wide variety of compounds. A particularly attractive feature of the Atlas is that all the information is stored on computer files, and these plus search programs are also available from the publishers. The Atlas should be welcomed both by newcomers to the field as well as those currently active in ^{13}C n.m.r.

J. B. STOTHERS

Department of Chemistry
University of Western Ontario
London
Ontario
Canada

PREFACE

It is the aim of the Atlas to present all reasonably verified carbon-13 n.m.r. results, predominantly selected from the published literature. Therefore the literature has been researched from the earliest publication of reasonably accurate results. The data reported in the earlier literature are largely chemical shifts. Only more recently have couplings and spin-lattice relaxation been reported. The first volumes (about 3000 compounds) therefore report shifts and assignments. Shifts are all referred to TMS. When the original reference standard was not TMS the value referred to TMS has been calculated.

It would be in one way logical to present the data in the Atlas in order of publication. However, this would have tended to leave an emphasis on certain chemical classes so the data have been sorted and early volumes contain a representative selection. When concentrations were available in the original literature they are listed in weight per cent. If reported as coupling constants or reduced couplings from proton off-resonance decoupled spectra, carbon–proton multiplicities are given in the Atlas and denoted below the shift values as S, D, T, and Q for singlet, doublet, triplet and quartet, respectively. If only carbon shieldings are available, the same multiplicities are derived from the formula and the assignment and then denoted as 1, 2, 3, and 4, respectively.

Very full indexing is a special feature because of the necessity for ready access to the data, bearing in mind the use of this Atlas in conjunction with other data. Therefore cumulative molecular formula, molecular weight, chemical class, alphabetical and shift indexes are provided and are renewed with each volume. At present the chemical class index (chemical classes according to *Chemischer Informationsdienst*) will offer reasonable access to substructures. Changes in indexing will be considered as the Atlas grows.

The Compilers welcome comment from users and will be pleased to receive suitable carbon-13 data for inclusion in subsequent volumes.

We are especially grateful to Dr. Hanswolf Kilian and his co-workers for their help in producing the computer printout at Hoechst Aktiengesellschaft.

We are indebted to J. Brun, A. R. Hernanto, R. Kimmich, M. Tikhomirov, and H. Wollmann for assistance. Especially the help of Mrs. Erika Ochterbeck, Bonn, is greatly acknowledged.

W. Voelter
E. Breitmaier
G. Haas

GUIDE TO THE PRESENTATION OF THE DATA

This collection of carbon-13 chemical shift data presents the following information for each compound.

1. Name
2. Structural formula
3. Molecular formula
4. Molecular weight
5. Solvent
6. Original standard

7. pH Value (when available)
8. Temperature
9. Carbon-13 chemical shifts
10. Assignments of shifts to carbons
11. Multiplicities of signals
12. Original literature
13. Chemical class code

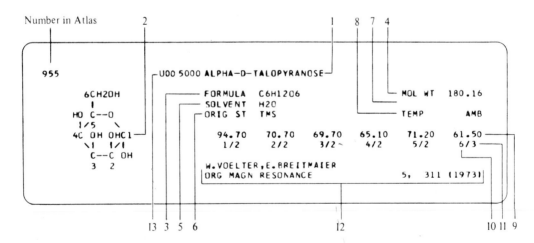

All compounds are numbered and listed in the Atlas in this reference number order. To facilitate fullest possible use of the Atlas five indexes are provided.

1. Alphabetical Index
2. Chemical Class Index
3. Molecular Formula Index
4. Molecular Weight Index
5. Chemical Shift Index

Specially written programs provide computer controlled acquisition, correction and printing of all the parameters and other information listed in the Atlas. The data are stored on magnetic tape which, together with a retrieval program, may be obtained on application to Heyden and Son.

1. Compound Name

Whenever possible and useful, IUPAC names are given. Otherwise the names given in the original literature are cited.

2. Structural Formula

The structural formulae are printed using a special printing chain ("QN" chain) containing the usual characters for single and multiple bonds. The arrangements of the structural formulae largely correspond to normal organic chemistry practice. Coded formulae, such as the Wiswesser line notation, although useful for computer retrieval, are not employed in this Atlas because data are acquired visually and there would be no justification for the reader to learn and code.

Hydrogen atoms attached to ring carbons are not printed. However, those linked to open-chain carbons are given throughout the Atlas.

The numbering of the carbon atoms (for chemical shift assignment) does not always follow IUPAC rules. Frequently, the numbering used in the original publication has been adopted.

3. Molecular Formula

The molecular formula is given as well as the molecular weight to facilitate co-ordination with mass spectral data collections. The symbols and the numbers of atoms in the molecular formulae are printed on the same line.

4. Molecular Weight

The molecular weight, as computed from the molecular formula, is printed to two decimal places.

5. Solvent

The solvent is printed as its molecular formula and, where feasible, additionally by its name. The molecular formulae of some common solvents are:

```
CHCL3      CHLOROFORM            C4H10O    DIETHYL ETHER
CDCL3      DEUTERIOCHLOROFORM    CH3NO2    NITROMETHANE
CH4O       METHANOL              C4H8O2    DIOXANE
CD4O       TETRADEUTERIOMETHANOL C2H6OS    DIMETHYLSULFOXIDE
C5H5N      PYRIDINE              C2D6OS    HEXADEUTERIODIMETHYLSULFOXIDE
CCL4       CARBON TETRACHLORIDE  C6H12     CYCLOHEXANE
C6H6       BENZENE               C6D12     DODECADEUTERIOCYCLOHEXANE
C6D6       HEXADEUTERIOBENZENE   CS2       CARBON DISULFIDE
C3H6O      ACETONE               H2O       WATER
C3D6O      HEXADEUTERIOACETONE   D2O       DEUTERIUMOXIDE
```

Mixtures of solvents are indicated by a sequence of the corresponding molecular formulae. The names of less commonly used solvents are printed in the next line, e.g.

SOLVENT C2H4CL2/C6H12

If neat or almost neat liquid samples are used, the formula of the measured compound is given as the solvent. If known, the mixture ratio is added in parentheses, e.g.

SOLVENT NEAT/CYCLOHEXANE (9/1)

6. Original Standard

The standard to which the carbon-13 shifts were referred in the original literature is given, e.g. TMS,

`C4H12SI  TETRAMETHYLSILANE  CS2  CARBON  DISULFIDE.`

7. pH Value

If the carbon-13 shifts reported were obtained at a certain pH value, this value is presented in the Atlas.

8. Temperature

The temperature of measurement is given in degrees Kelvin. If the sample temperature is not mentioned in the publication, the abbreviation " **AMB** " for ambient is printed.

9. Carbon-13 Chemical Shifts

All carbon-13 chemical shifts tabulated in this collection refer to tetramethylsilane (TMS) irrespective of the standard used by the original researchers. They are given to two decimal places with an accuracy of ± 0.1 ppm but not better than ± 0.05 ppm. If reference compounds other than TMS were used originally, the values have been converted to the δ scale relative to TMS using the following shift conversions.

Cyclohexane	26.55	Carbon tetrachloride	96.10
Dimethylsulfoxide	40.60	Benzene	128.50
Methanol	49.90	Acetic acid (carbonyl)	177.05
1,4-Dioxane	66.50	Carbon disulfide	192.50
Chloroform	77.20		

10. Signal Assignments

The assignments of the carbon-13 shifts are given using the numbering of the carbon atoms in the structural formula. These numbers also define the sequence in which the shift values are listed. The identification number of the carbon atom is printed underneath the corresponding shift value and to the left of the decimal point, e.g.

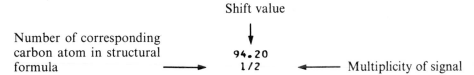

11. Multiplicities of Signals

The multiplicities of signals (1 for a quaternary carbon, 2 for a CH, 3 for a CH2 and 4 for a CH3 carbon, respectively) refer to those obtained or obtainable from proton off-resonance decoupling experiments. If no experimental multiplicity data were given in the original literature, the multiplicities were derived from the structural formula. The multiplicity is printed underneath the corresponding shift and to the right of the decimal point (see the example above). Couplings other than carbon-proton one-bond splittings have not been considered.

12. Original Literature

The full literature source from which the carbon chemical shifts and the assignments were taken is the final item in the box of data for each compound. The abbreviations of the journals are those used in *Chemical Abstracts*.

13. Chemical Class Code

The classification follows the system of *Chemischer Informationsdienst*, Verlag Chemie, Weinheim, 1973. Tables listing the code numbers, the chemical classes and the compounds in this Atlas are given at the beginning of the Cumulative Indexes Volume.

CARBON-13 NMR DATA

Compounds 1000–3017

1000   U 005000 3,4-DI-O-METHYL-D-GALACTITOL

```
    1CH2-OH
      |
    2CH-OH
  7   |
H3CO-HC3
  8   |
H3CO-HC4
      |
    5CH-OH
      |
    6CH2-OH
```

FORMULA C8H18O6   MOL WT 210.23
SOLVENT D2O
ORIG ST TMS   TEMP   AMB

62.90	70.65	78.85	60.30
1/3	2/3	3/3	7/4

W.VOELTER,E.BREITMAIER,E.B.RATHBONE,E.M.STEPHEN
TETRAHEDRON   29, 3845 (1973)

1001   U 013000 PROSTAGLANDIN E1

```
      7   5   3   1
     CH2 CH2 CH2 COOH
 O     .  \ / \ / \ /
  \\      .  CH2 CH2 CH2
 9C---C8  6   4   2
 10C 12C  14  16  18  20
   \ / \  CH  CH2 CH2 CH3
  11C  \ / \ / \ / \ /
    .   CH  CH  CH2 CH2
    OH 13 15.  17  19
         OH
```

FORMULA C20H34O5   MOL WT 354.49
SOLVENT CHCL3/CH3OH
ORIG ST TMS   TEMP   AMB

176.70	33.80	24.50	29.00	28.60	26.30
1/1	2/3	3/3	4/3	5/3	6/3
27.40	54.60	215.20	45.90	71.60	54.20
7/3	8/2	9/1	10/3	11/2	12/2
131.90	136.60	72.90	36.90	25.00	31.50
13/2	14/2	15/2	16/3	17/3	18/3
22.50	13.80				
19/3	20/4				

G.LUKACS,F.PIRIOU,S.D.GERO,D.A.VAN DORP,
E.W.HAGAMAN,E.WENKERT
TETRAHEDRON LETTERS   1973, 515 (1973)

1002   U 013000 PROSTAGLANDIN E2

```
      7   5   3   1
     CH2 CH  CH2 COOH
 O     .  \ / \ / \ /
  \\      CH  CH2 CH2
 9C---C8  6   4   2
 10C 12C  14  16  18  20
   \ / \  CH  CH2 CH2 CH3
  11C  \ / \ / \ / \ /
    .   CH  CH  CH2 CH2
    OH 13 15.  17  19
         OH
```

FORMULA C20H32O5   MOL WT 352.48
SOLVENT CHCL3/CH3OH
ORIG ST TMS   TEMP   AMB

177.20	33.40	24.60	26.50	131.00	126.70
1/1	2/3	3/3	4/3	5/2	6/2
25.20	54.60	215.40	46.20	72.00	53.60
7/3	8/2	9/1	10/3	11/2	12/2
131.70	136.70	73.20	37.00	25.20	31.80
13/2	14/2	15/2	16/3	17/3	18/3
22.70	14.10				
19/3	20/4				

G.LUKACS,F.PIRIOU,S.D.GERO,D.A.VAN DORP,
E.W.HAGAMAN,E.WENKERT
TETRAHEDRON LETTERS   1973, 515 (1973)

1003   U 013000 PROSTAGLANDIN F2ALPHA

```
      7   5   3   1
     CH2 CH  CH2 COOH
 HO    .  \ / \ / \ /
  .       .  CH  CH2 CH2
 9C---C8  6   4   2
  |   |
 10C 12C  14  16  18  20
   \ / \  CH  CH2 CH2 CH3
  11C  \ / \ / \ / \ /
    .   CH  C   CH2 CH2
    OH 13 15.  17  19
         OH
```

FORMULA C20H34O5   MOL WT 354.49
SOLVENT CHCL3/CH3OH
ORIG ST TMS   TEMP   AMB

176.60	33.20	24.50	26.30	132.80	128.90
1/1	2/3	3/3	4/3	5/2	6/2
25.10	49.90	71.80	42.60	77.20	55.00
7/3	8/2	9/2	10/3	11/2	12/2
129.10	135.00	72.90	36.80	25.10	31.50
13/2	14/2	15/2	16/3	17/3	18/3
22.40	13.90				
19/3	20/4				

G.LUKACS,F.PIRIOU,S.D.GERO,D.A.VAN DORP,
E.W.HAGAMAN,E.WENKERT
TETRAHEDRON LETTERS   1973, 515 (1973)

1004　　　　P 000200 ALLENE

```
                    FORMULA  C3H4              MOL WT   40.07
        3 2 1       SOLVENT
       H2C=C=CH2    ORIG ST  CS2              TEMP     AMB

                    72.30  211.40
                     1/3    2/1

                    J.K.CRANDALL,S.A.SOJKA
                    J AM CHEM SOC              94, 5084 (1972)
```

1005　　　　P 000200 DIMETHYLCARBENIUM ICN

```
        3           FORMULA  C3H7              MOL WT   43.09
       H3C          SOLVENT  CLFO2S/F5SB
         \ +        ORIG ST  CS2              TEMP     198
        H-C
         /1         317.50   59.70   59.70
       H3C           1/2      2/4     3/4
        2
                    G.A.OLAH,P.W.WESTERMAN
                    J AM CHEM SOC              95, 7530 (1973)
```

1006　　　　P 000200 1,2-BUTADIENE

```
                    FORMULA  C4H6              MOL WT   54.09
      4 3   2 1     SOLVENT
     H3C-CH=C=CH2   ORIG ST  CS2              TEMP     AMB

                    72.20  208.20   83.00   12.00
                     1/3    2/1      3/2      4/4

                    J.K.CRANDALL,S.A.SOJKA
                    J AM CHEM SOC              94, 5084 (1972)
```

1007　　　　P 000200 TRIMETHYLCARBENIUM ION

```
                    FORMULA  C4H9              MOL WT   57.12
         4          SOLVENT  CLFO2S/F5SB
       H3C          ORIG ST  CS2              TEMP     198
         \ +
       H3C-C        327.90   47.20   47.20   47.20
        3/1          1/1      2/4     3/4      4/4
       H3C
        2           G.A.OLAH,P.W.WESTERMAN
                    J AM CHEM SOC              95, 7530 (1973)
```

1008　　　　P 000200 3-METHYL-1,2-BUTADIENE

```
                    FORMULA  C5H8              MOL WT   68.12
      4   3 2 1     SOLVENT
     CH3-C=C=CH2    ORIG ST  CS2              TEMP     AMB
         |
        CH3         70.50  205.70   91.80   17.90
         5           1/3    2/1      3/1      4/4

                    J.K.CRANDALL,S.A.SOJKA
                    J AM CHEM SOC              94, 5084 (1972)
```

1009   P 000200 1,2-PENTADIENE

```
 4   3  2 1
 CH2-CH=C=CH2
5/
CH3
```

	FORMULA	C5H8		MOL WT	68.12
	SOLVENT				
	ORIG ST	CS2		TEMP	AMB

73.50	207.60	90.40	20.40	12.00
1/3	2/1	3/2	4/3	5/4

J.K.CRANDALL,S.A.SOJKA
J AM CHEM SOC   94, 5084 (1972)

1010   P 000200 2,3-PENTADIENE

```
 5   4  3 2  1
 CH3-CH=C=CH-CH3
```

	FORMULA	C5H8		MOL WT	68.12
	SOLVENT				
	ORIG ST	CS2		TEMP	AMB

13.40	83.70	205.30
1/4	2/2	3/1

J.K.CRANDALL,S.A.SOJKA
J AM CHEM SOC   94, 5084 (1972)

1011   P 000200 2-METHYL-2-BUTENE

```
 H3C        CH3
   5\    /4
     C=C
    /2  3\
 H3C        H
  1
```

	FORMULA	C5H10		MOL WT	77.06
	SOLVENT	C4H8O2			
	ORIG ST	C4H12SI		TEMP	AMB

25.50	131.70	118.70	13.30	17.10
1/4	2/1	3/2	4/4	5/4

L.F.JOHNSON,W.C.JANKOWSKI
CARBON-13 NMR SPECTRA,JOHN
WILEY AND SONS,NEW YORK   119 (1972)

1012   P 000200 1,2-HEXADIENE

```
 4   3  2 1
 CH2-CH=C=CH2
 /
5CH2
 \
 CH3
 6
```

	FORMULA	C6H10		MOL WT	82.15
	SOLVENT				
	ORIG ST	CS2		TEMP	AMB

73.40	208.30	88.70	29.30	18.10	12.50
1/3	2/1	3/2	4/3	5/3	6/4

J.K.CRANDALL,S.A.SOJKA
J AM CHEM SOC   94, 5084 (1972)

1013   P 000200 HEXYNE-1

	FORMULA	C6H10		MOL WT	82.15
	SOLVENT	CDCL3			
	ORIG ST	C4H12SI		TEMP	AMB

68.10	84.50	18.10	30.70	21.90	13.50
1/2	2/1	3/3	4/3	5/3	6/4

```
HC≡C-CH2-CH2-CH2-CH3
 1 2 3   4   5   6
```

L.F.JOHNSON,W.C.JANKOWSKI
CARBON-13 NMR SPECTRA,JOHN
WILEY AND SONS,NEW YORK   175 (1972)

1014　　　　　P 000200　2-METHYL-2,3-PENTADIENE

```
                              FORMULA   C6H10              MOL WT   82.15
    5   4  3 2 1              SOLVENT
    CH3-CH=C=C-CH3            ORIG ST   CS2                TEMP     AMB
            |
            CH3                  19.50   92.90  202.30   82.30   13.80
            6                     1/4     2/1     3/1      4/2     5/4
```

J.K.CRANDALL,S.A.SOJKA
J AM CHEM SOC　　　　　　　　　　94, 5084 (1972)

1015　　　　　P 000200　4-METHYL-1,2-PENTADIENE

```
    5
    CH3                       FORMULA   C6H10              MOL WT   82.15
      \   3  2 1              SOLVENT
      4CH-CH=C=CH2            ORIG ST   CS2                TEMP     AMB
      /
    CH3                          75.00  206.60   96.60   26.70   21.70
    6                             1/3     2/1     3/2      4/2     5/4
```

J.K.CRANDALL,S.A.SOJKA
J AM CHEM SOC　　　　　　　　　　94, 5084 (1972)

1016　　　　　P 000200　CIS-HEXENE-2

```
                              FORMULA   C6H12              MOL WT   84.16
       H H                    SOLVENT   CDCL3
       | |                    ORIG ST   C4H12SI            TEMP     AMB
       C=C
      /2 3\                      12.70  123.80  130.70   29.10   22.90   13.70
    H3C     CH2-CH2-CH3           1/4     2/2     3/2      4/3     5/3     6/4
     1      4   5   6
```

L.F.JOHNSON,W.C.JANKOWSKI
CARBON-13 NMR SPECTRA,JOHN
　　WILEY AND SONS,NEW YORK　　　　188 (1972)

1017　　　　　P 000200　NEOHEXENE

```
            4
            CH3               FORMULA   C6H12              MOL WT   84.16
            /5                SOLVENT   C4H8O2
     H    3C-CH3              ORIG ST   C4H12SI            TEMP     AMB
      \1 2/ \
       C=C   CH3                 109.00  149.50   38.70   34.10   34.10   34.10
      /   \  6                     1/3     2/2     3/1      4/4     5/4     6/4
     H     H
```

L.F.JOHNSON,W.C.JANKOWSKI
CARBON-13 NMR SPECTRA,JOHN
　　WILEY AND SONS,NEW YORK　　　　187 (1972)

1018　　　　　P 000200　2,2-DIMETHYLBUTANE

```
        5
        CH3                   FORMULA   C6H14              MOL WT   86.18
        |2                    SOLVENT   CDCL3
    H3C-C-CH2-CH3             ORIG ST   C4H12SI            TEMP     AMB
     1  |3   4
        CH3                      28.90   30.40   36.50    8.80    28.90   28.90
        6                         1/4     2/1     3/3      4/4     5/4     6/4
```

L.F.JOHNSON,W.C.JANKOWSKI
CARBON-13 NMR SPECTRA,JOHN
　　WILEY AND SONS,NEW YORK　　　　206 (1972)

1019   P 000200 2,3-DIMETHYLBUTANE

```
      5        6
    H3C      CH3        FORMULA   C6H14              MOL WT   86.18
      \      /          SOLVENT   CDCL3
       CH-CH            ORIG ST   C4H12SI            TEMP       AMB
      /2  3\
    H3C      CH3          19.50    34.00    34.00    19.50    19.50    19[50
      1       4            1/4      2/2      3/2      4/4      5/4      6/4
```

L.F.JOHNSON,W.C.JANKOWSKI
CARBON-13 NMR SPECTRA,JOHN
WILEY AND SONS,NEW YORK   208 (1972)

1020   P 000200 2-METHYLPENTANE

```
                     FORMULA   C6H14              MOL WT   86.18
             1       SOLVENT   CDCL3
            CH3      ORIG ST   C4H12SI            TEMP       AMB
     5 4   3  2/
    H3C-CH2-CH2-CH       22.60    27.90    41.60    20.60    14.30    22.60
               \          1/4      2/2      3/3      4/3      5/4      6/4
               CH3
               6     L.F.JOHNSON,W.C.JANKOWSKI
                     CARBON-13 NMR SPECTRA,JOHN
                     WILEY AND SONS,NEW YORK        209 (1972)
```

1021   P 000200 3-METHYLPENTANE

```
                     FORMULA   C6H14              MOL WT   86.18
        4            SOLVENT   CDCL3
       CH3           ORIG ST   C4H12SI            TEMP       AMB
        |
    H3C-CH2-CH-CH2-CH3   11.40    29.30    36.40    18.80    29.30    11.40
     1  2   3  5   6      1/4      2/3      3/2      4/4      5/3      6/4


                     L.F.JOHNSON,W.C.JANKOWSKI
                     CARBON-13 NMR SPECTRA,JOHN
                     WILEY AND SONS,NEW YORK        207 (1972)
```

1022   P 000200 2,4-DIMETHYL-2,3-PENTADIENE

```
                          FORMULA   C7H12              MOL WT   96.17
    5    4 3 2 1          SOLVENT
    CH3-C=C=C-CH3         ORIG ST   CS2                TEMP       AMB
        |   |
       CH3 CH3              18.80    90.90   198.80
        7   6                1/4      2/1      3/1

                     J.K.CRANDALL,S.A.SOJKA
                     J AM CHEM SOC                    94, 5084 (1972)
```

1023   P 000200 4,4-DIMETHYL-1,2-PENTADIENE

```
      6
     CH3
    5  \4  3 2 1          FORMULA   C7H12              MOL WT   96.17
    CH3-C-CH=C=CH2        SOLVENT
        /                 ORIG ST   CS2                TEMP       AMB
     CH3
                           75.80   205.80   101.20    30.50    29.60
                            1/2      2/1      3/2      4/1      5/4

                     J.K.CRANDALL,S.A.SOJKA
                     J AM CHEM SOC                    94, 5084 (1972)
```

1024 P 000200 1,2-HEPTADIENE

```
   4   3  2 1
  CH2-CH=C=CH2
  /
5CH2
  \
   CH2-CH3
   6   7
```

FORMULA	C7H12			MOL WT	96.17
SOLVENT					
ORIG ST	CS2			TEMP	AMB

72.60	207.80	88.40	29.90	26.40	20.60
1/3	2/1	3/2	4/3	5/3	6/3
12.00					
7/4					

J.K.CRANDALL,S.A.SOJKA
J AM CHEM SOC   94, 5084 (1972)

1025 P 000200 CIS-HEPTENE-2

```
  1    4   5   6   7
 H3C      CH2-CH2-CH2-CH3
   \2 3/
    C=C
   /    \
  H      H
```

FORMULA	C7H14			MOL WT	98.19
SOLVENT	CDCL3				
ORIG ST	C4H12SI			TEMP	AMB

12.60	123.50	130.30	26.70	32.00	22.40
1/4	2/2	3/2	4/3	5/3	6/3
13.90					
7/4					

L.F.JOHNSON,W.C.JANKOWSKI
CARBON-13 NMR STECTRA,JOHN
WILEY AND SONS,NEW YORK   264 (1972)

1026 P 000200 TRANS-HEPTENE-2

```
  1
 H3C      H
   \2 3/
    C=C3
   /    \4   5   6   7
  H      CH2-CH2-CH2-CH3
```

FORMULA	C7H14			MOL WT	98.19
SOLVENT	CDCL3				
ORIG ST	C4H12SI			TEMP	AMB

17.80	124.50	131.70	32.40	32.00	22.40
1/4	2/2	3/2	4/3	5/3	6/3
13.90					
7/4					

L.F.JOHNSON,W.C.JANKOWSKI
CARBON-13 NMR STECTRA,JOHN
WILEY AND SONS,NEW YORK   264 (1972)

1027 P 000200 N-HEPTANE

```
    CH2-CH2-CH3
   /3  2  1
   CH2
   4\
    CH2-CH2-CH3
    5   6   7
```

FORMULA	C7H16			MOL WT	100.21
SOLVENT	C4H8O2				
ORIG ST	C4H12SI			TEMP	AMB

14.20	23.20	32.50	29.60	32.50	23.20
1/4	2/3	3/3	4/3	5/3	6/3
14.20					
7/4					

L.F.JOHNSON,W.C.JANKOWSKI
CARBON-13 NMR SPECTRA,JOHN
WILEY AND SONS,NEW YORK   277 (1972)

1028 P 000200 2,4,6-OCTATRIYNE

```
    1 2 3 4 5 6 7 8
  H3C-C≡C-C≡C-C≡C-CH3
```

FORMULA	C8H6		MOL WT	102.14
SOLVENT	CDCL3		
ORIG ST	C4H12SI		TEMP	AMB

4.40	74.80	65.00	60.00
1/Q	2/S	3/S	4/S

R.ZEISBERG,F.BOHLMANN
CHEM BER   107, 3800 (1974)

```
1029                P 000200  1,2-OCTADIENE

    4   3  2 1                 FORMULA   C8H14                MOL WT  110.20
    CH2-CH=C=CH2               SOLVENT
   /                           ORIG ST   CS2                  TEMP       AMB
  5CH2
    \                            73.80   208.20    89.20    30.70    28.30    27.60
    CH2-CH2                        1/3      2/1      3/2      4/3      5/3      6/3
    6   7\                        22.00    13.40
         CH3                       7/3      8/4
         8
                               J.K.CRANDALL,S.A.SOJKA
                               J AM CHEM SOC                  94, 5084 (1972)
```

```
1030                P 000200  TRANS-2,5-DIMETHYL-3-HEXENE

      8                        FORMULA   C8H16                MOL WT  112.22
     H3C                       SOLVENT   CDCL3
       \                       ORIG ST   C4H12SI              TEMP       AMB
       6CH      H
      / \5 1/  3               134.50    31.00    22.80    22.80   134.50    31.00
   H3C    C=C    CH3             1/2      2/2      3/4      4/4      5/2      6/2
    7   /    \ /                 22.80    22.80
        H      2CH               7/4      8/4
               \
               CH3             L.F.JOHNSON,W.C.JANKOWSKI
               4               CARBON-13 NMR SPECTRA,JOHN
                               WILEY AND SONS,NEW YORK          315 (1972)
```

```
1031                P 000200  OCTENE-1

                    7          FORMULA   C8H16                MOL WT  112.22
   CH2-CH2-CH2-CH2-CH2         SOLVENT   CDCL3
   /3   4   5   6  |           ORIG ST   C4H12SI              TEMP       AMB
  H2C=C             CH3
   1 2\             8           139.00   114.10    34.00    29.00    29.10    31.90
      H                          1/3      2/2      3/3      4/3      5/3      6/3
                                22.70    14.10
                                 7/3      8/4

                               L.F.JOHNSON,W.C.JANKOWSKI
                               CARBON-13 NMR SPECTRA,JOHN
                               WILEY AND SONS,NEW YORK          316 (1972)
```

```
1032                P 000200  CIS-OCTENE-2

                               FORMULA   C8H16                MOL WT  112.22
                               SOLVENT   CDCL3
       5   7                   ORIG ST   C4H12SI              TEMP       AMB
       CH2 CH2
   1   / \ / \                  12.70   123.60   130.90    27.00    29.50    31.70
  H3C    CH2 CH2 CH3             1/4      2/2      3/2      4/3      5/3      6/3
    \2 3/4   6   8              22.80    14.10
     C=C                         7/3      8/4
    /   \
   H     H                     L.F.JOHNSON,W.C.JANKOWSKI
                               CARBON-13 NMR SPECTRA,JOHN
                               WILEY AND SONS,NEW YORK          313 (1972)
```

```
1033                P 000200  TRANS-OCTENE-2

                               FORMULA   C8H16                MOL WT  112.22
     1                         SOLVENT   CDCL3
   H3C    H                    ORIG ST   C4H12SI              TEMP       AMB
     \2 3/
       C=C                      17.80   124.50   131.70    32.70    29.90    31.70
      /  \4   6   8              1/4      2/2      3/2      4/3      5/3      6/3
    H    CH2 CH2 CH3            22.80    14.10
          \ / \ /                7/3      8/4
          CH2 CH2
           5   7               L.F.JOHNSON,W.C.JANKOWSKI
                               CARBON-13 NMR SPECTRA,JOHN
                               WILEY AND SONS,NEW YORK          313 (1972)
```

1034   P 000200   2,4,4-TRIMETHYL-2-PENTENE

```
      H3C    CH3
        8\ /7
   H3C    4C        FORMULA   C8H16              MOL WT  112.22
    1\   / \        SOLVENT   C4H8O2
     C=C   CH3      ORIG ST   C4H12SI            TEMP      AMB
    /2 3\  6
   H3C    H          18.70   129.80  135.10  32.20   27.80   31.20
    5                 1/4      2/1     3/2     4/1     5/4     6/4
                      31.20    31.20
                      7/4      8/4
```

L.F.JOHNSON,W.C.JANKOWSKI
CARBON-13 NMR SPECTRA,JOHN
WILEY AND SONS,NEW YORK   314 (1972)

1035   P 000200   2,2,4-TRIMETHYLPENTANE

```
                     FORMULA   C8H18              MOL WT  114.23
          5          SOLVENT   C4H8O2
   1CH3    CH3       ORIG ST   C4H12SI            TEMP      AMB
   2| 3  4/
   H3C-C-CH2-CH       30.20    31.20   53.40  24.90   25.50   25.50
   8 |      \         1/4      2/1     3/3     4/2     5/4     6/4
     CH3     CH3      30.20    30.20
     7       6        7/4      8/4
```

L.F.JOHNSON,W.C.JANKOWSKI
CARBON-13 NMR SPECTRA,JOHN
WILEY AND SONS,NEW YORK   321 (1972)

1036   P 000200   N-UNDECANE

```
   1 2   3   4   5   FORMULA   C11H24             MOL WT  156.31
   H3C-CH2-CH2-CH2-CH2  SOLVENT   CDCL3
                |    ORIG ST   C4H12SI            TEMP      AMB
              6CH2
                |     14.10    22.80   32.10  29.50   29.90   29.90
   H3C-CH2-CH2-CH2-CH2  1/4     2/3     3/3     4/3     5/3     6/3
   11 10  9   8   7   29.90    29.50   32.10  22.80   14.10
                      7/3      8/3     9/3    10/3    11/4
```

L.F.JOHNSON,W.C.JANKOWSKI
CARBON-13 NMR SPECTRA,JOHN
WILEY AND SONS,NEW YORK   428 (1972)

1037   P 000200   7,9-DIMETHYLDECENE-1

```
        12          11   FORMULA   C12H24         MOL WT  168.33
        CH3         CH3  SOLVENT   C4H8O2
     3   5   |        /  ORIG ST   C4H12SI        TEMP      AMB
     CH2 CH2 CH-CH2-CH
   1 2/ \ / \ /7  8   9\  114.20  138.90   34.10  29.60   26.80   37.60
   H2C=C   CH2 CH2      CH3  1/3     2/2     3/3    4/3     5/3     6/3
       \  4   6        10   30.60    47.20   25.60  23.50   22.40   19.10
       H                    7/2      8/3     9/2   10/4    11/4    12/4
```

L.F.JOHNSON,W.C.JANKOWSKI
CARBON-13 NMR SPECTRA,JOHN
WILEY AND SONS,NEW YORK   445 (1972)

1038   P 000200   TRIDECANE

```
   13  11   9   7   5   FORMULA   C13H28          MOL WT  184.37
   H3C  CH2 CH2 CH2 CH2 SOLVENT   C4H8O2
    \ / \ / \ / \ / \ 4 ORIG ST   C4H12SI         TEMP      AMB
   CH2 CH2 CH2 CH2 CH2
   12  10   8   6   |   14.20    23.00   29.80  32.40   30.10   30.10
                  3CH2   1/4      2/3     3/3     4/3     5/3     6/3
                    /    30.10    30.10   30.10  32.40   29.80   23.00
               H3C-CH2   7/3      8/3     9/3    10/3    11/3    12/3
               1   2     14.20
                         13/4
```

L.F.JOHNSON,W.C.JANKOWSKI
CARBON-13 NMR SPECTRA,JOHN
WILEY AND SONS,NEW YORK   455 (1972)

```
1039          P 000200  DI-TERT-BUTYLOCTATETRAYNE

 13                    15   FORMULA   C16H18              MOL WT  210.32
H3C                   CH3   SOLVENT   CDCL3
   1\2 3 4 5 6 7 8 9 11/12  ORIG ST   C4H12SI             TEMP       AMB
H3C-C-C≡C-C≡C-C≡C-C≡C-C-CH3
  14/              10  \16     30.30    28.70    88.60    64.70    62.20    61.90
H3C                   CH3      1/4      2/1      3/1      4/1      5/1      6/1

                             R.ZEISBERG,F.BOHLMANN
                             CHEM BER                     107, 3800 (1974)
```

```
1040          P 000200  DI-TERT-BUTYLDECAPENTAYNE

                             FORMULA   C18H18              MOL WT  234.34
    8                        SOLVENT   CDCL3
   H3C                       ORIG ST   C4H12SI             TEMP       AMB
    1\2 3 4 5 6 7
  (H3C-C-C≡C-C≡C-C≡)           30.30    28.30    88.50    64.60    62.30    62.15
    9/             2            1/4      2/1      3/1      4/1      5/1      6/1
   H3C                         61.80
                               7/1

                             R.ZEISBERG,F.BOHLMANN
                             CHEM BER                     107, 3800 (1974)
```

```
1041          P 000300  TETRACHLOROETHYLENE

                             FORMULA   C2CL4               MOL WT  155.83
   CL     CL                 SOLVENT   CHCL3
    \1 2/                     ORIG ST   C4H12SI             TEMP       AMB
     C=C
    /    \                    117.40
   CL     CL                   1/1

                             G.E.HAWKES,R.A.SMITH,J.D.ROBERTS
                             J ORG CHEM                   39, 1276 (1974)
```

```
1042          P 000300  1,1,2,2-TETRACHLORO-1,2-DIFLUOROETHANE

                             FORMULA   C2CL4F2             MOL WT  203.83
   CL     CL                 SOLVENT   C6H12
    \1 2/                     ORIG ST   C4H12SI             TEMP       AMB
   F-C-C-F
    /    \                    120.20
   CL     CL                   1/D

                             G.E.HAWKES,R.A.SMITH,J.D.ROBERTS
                             J ORG CHEM                   39, 1276 (1974)
```

```
1043          P 000300  HEXACHLOROETHANE

                             FORMULA   C2CL6               MOL WT  236.74
    CL     CL                SOLVENT   CHCL3
     \1 2/                    ORIG ST   C4H12SI             TEMP       AMB
   CL-C-C-CL
     /    \                   105.30
    CL     CL                  1/1

                             G.E.HAWKES,R.A.SMITH,J.D.ROBERTS
                             J ORG CHEM                   39, 1276 (1974)
```

1044   P 000300   1,2,2-TRICHLOROETHENE

```
                           FORMULA   C2HCL3            MOL WT   131.39
                           SOLVENT   CHCL3
   CL     CL               ORIG ST   C4H12SI           TEMP        AMB
    \1  2/
     C=C                      116.80   125.40
    /    \                     1/D       2/S
   H      CL
                           G.E.HAWKES,R.A.SMITH,J.D.ROBERTS
                           J ORG CHEM                  39, 1276 (1974)
```

1045   P 000300   (E)-1,2-DICHLOROETHENE

```
                           FORMULA   C2H2CL2           MOL WT   96.94
                           SOLVENT   CHCL3
   CL     H                ORIG ST   C4H12SI           TEMP        AMB
    \1  2/
     C=C                      120.20
    /    \                     1/D
   H      CL
                           G.E.HAWKES,R.A.SMITH,J.D.ROBERTS
                           J ORG CHEM                  39, 1276 (1974)
```

1046   P 000300   (Z)-1,2-DICHLOROETHENE

```
                           FORMULA   C2H2CL2           MOL WT   96.94
                           SOLVENT   CHCL3
   H      H                ORIG ST   C4H12SI           TEMP        AMB
    \1  2/
     C=C                      120.50
    /    \                     1/D
   CL     CL
                           G.E.HAWKES,R.A.SMITH,J.D.ROBERTS
                           J ORG CHEM                  39, 1276 (1974)
```

1047   P 000300   1,1,2-TRIBROMOETHANE

```
   BR                      FORMULA   C2H3BR3           MOL WT   266.76
    |                      SOLVENT   CDCL3
  2CH-CH2-BR               ORIG ST   C4H12SI           TEMP        AMB
    |  1
   BR                         37.70    40.30
                              1/3       2/2

                           L.F.JOHNSON,W.C.JANKOWSKI
                           CARBON-13 NMR SPECTRA,JOHN
                           WILEY AND SONS,NEW YORK          2 (1972)
```

1048   P 000300   1,1,1-TRICHLOROETHANE

```
                           FORMULA   C2H3CL3           MOL WT   133.41
                           SOLVENT   CHCL3
   CL                      ORIG ST   C4H12SI           TEMP        AMB
   1 |2
  H3C-C-CL                    60.00   109.90
    |                         1/Q       2/S
   CL
                           G.E.HAWKES,R.A.SMITH,J.D.ROBERTS
                           J ORG CHEM                  39, 1276 (1974)
```

1049   P 000300 1,1,2-TRICHLOROETHANE

```
CL
 \
  CH-CH2-CL
 /1    2
CL
```

FORMULA	C2H3CL3		MOL WT 133.41
SOLVENT	CDCL3		
ORIG ST	C4H12SI		TEMP AMB

70.40	50.10
1/2	2/3

L.F.JOHNSON,W.C.JANKOWSKI
CARBON-13 NMR SPECTRA,JOHN
WILEY AND SONS,NEW YORK   3 (1972)

1050   P 000300 ETHYL IODIDE

```
J-CH2-CH3
  1    2
```

FORMULA	C2H5I	MOL WT 155.97
SOLVENT	CDCL3	
ORIG ST	C4H12SI	TEMP AMB

-1.20	20.50
1/3	2/4

L.F.JOHNSON,W.C.JANKOWSKI
CARBON-13 NMR SPECTRA,JOHN
WILEY AND SONS,NEW YORK   9 (1972)

1051   P 000300 2,2,3,3,3-PENTACHLORO-1,1,1-TRIFLUOROPROPANE

```
F   CL  CL
 \1  12 /
F-C-C-C-CL
 /  | 3\
F   CL  CL
```

FORMULA	C3CL5F3		MOL WT 270.29
SOLVENT	CHCL3		
ORIG ST	C4H12SI		TEMP AMB

121.20	92.50	99.50
1/Q	2/S	3/S

G.E.HAWKES,R.A.SMITH,J.D.ROBERTS
J ORG CHEM   39, 1276 (1974)

1052   P 000300 HEXACHLOROPROPENE

```
CL       CL
 \1 2 3/
  C=C-C-CL
 /  |  \
CL  CL  CL
```

FORMULA	C3CL6		MOL WT 248.75
SOLVENT	CHCL3		
ORIG ST	C4H12SI		TEMP AMB

127.30	132.10	92.90
1/S	2/S	3/S

G.E.HAWKES,R.A.SMITH,J.D.ROBERTS
J ORG CHEM   39, 1276 (1974)

1053   P 000300 1,2,2,3,3,3-HEXACHLORO-1,1-DIFLUOROPROPANE

```
F   CL  CL
 \1  12 /
CL-C-C-C-CL
 /  | 3\
F   CL  CL
```

FORMULA	C3CL6F2		MOL WT 286.75
SOLVENT	CHCL3		
ORIG ST	C4H12SI		TEMP AMB

128.10	97.00	100.20
1/T	2/S	3/S

G.E.HAWKES,R.A.SMITH,J.D.ROBERTS
J ORG CHEM   39, 1276 (1974)

1054   P 000300 OCTACHLOROPROPANE

```
    CL  CL  CL
     \1 |2 /
    CL-C-C-C-CL
     / | 3\
    CL  CL  CL
```

FORMULA   C3CL8   MOL WT   319.66
SOLVENT   C4H8O2
ORIG ST   C4H12SI   TEMP   AMB

101.40   103.00
 1/S   2/S

G.E.HAWKES,R.A.SMITH,J.D.ROBERTS
J ORG CHEM   39, 1276 (1974)

1055   P 000300 1,1,1,2,3,3,3—HEPTACHLOROPROPANE

```
    CL  H   CL
     \1 |2 /
    CL-C-C-C-CL
     / | 3\
    CL  CL  CL
```

FORMULA   C3HCL7   MOL WT   285.21
SOLVENT   C4H8O2
ORIG ST   C4H12SI   TEMP   AMB

95.90   79.00
 1/S   2/D

G.E.HAWKES,R.A.SMITH,J.D.ROBERTS
J ORG CHEM   39, 1276 (1974)

1056   P 000300 1,1,2,2,3,3,3—HEPTACHLOROPROPANE

```
    CL  CL  CL
     \  |2 /
    H-C-C-C-CL
     /1 | 3\
    CL  CL  CL
```

FORMULA   C3HCL7   MOL WT   285.21
SOLVENT   C4H8O2
ORIG ST   C4H12SI   TEMP   AMB

75.00   99.30   101.50
 1/D   2/S   3/S

G.E.HAWKES,R.A.SMITH,J.D.ROBERTS
J ORG CHEM   39, 1276 (1974)

1057   P 000300 ALLYL BROMIDE

```
          3
    H    CH2-BR
     \1 2/
      C=C
     /    \
    H      H
```

FORMULA   C3H5BR   MOL WT   120.98
SOLVENT   CDCL3
ORIG ST   C4H12SI   TEMP   AMB

118.80   134.20   32.60
 1/3   2/2   3/3

L.F.JOHNSON,W.C.JANKOWSKI
CARBON-13 NMR SPECTRA,JOHN
WILEY AND SONS,NEW YORK   21 (1972)

1058   P 000300 CIS-1-BROMO-1-PROPENE

```
    BR     CH3
     \    /3
      C=C
     /1 2\
    H      H
```

FORMULA   C3H5BR   MOL WT   120.98
SOLVENT   CDCL3
ORIG ST   C4H12SI   TEMP   AMB

108.90   129.40   15.30
 1/2   2/2   3/4

L.F.JOHNSON,W.C.JANKOWSKI
CARBON-13 NMR SPECTRA,JOHN
WILEY AND SONS,NEW YORK   20 (1972)

1059   P 000300   TRANS-1-BROMO-1-PROPENE

FORMULA   C3H5BR   MOL WT   120.98
SOLVENT   CDCL3
ORIG ST   C4H12SI   TEMP   AMB

```
BR      H
  \    /
   C=C
  /1  2\
 H    CH3
      3
```

104.70   132.70   18.10
  1/2   2/2   3/4

L.F.JOHNSON,W.C.JANKOWSKI
CARBON-13 NMR SPECTRA,JOHN
WILEY AND SONS,NEW YORK   20 (1972)

1060   P 000300   3-CHLORO-1-PROPENE

FORMULA   C3H5CL   MOL WT   76.53
SOLVENT   C4H8O2
ORIG ST   C4H12SI   TEMP   AMB

CH2=CH-CH2-CL
 1   2   3

118.10   134.60   45.10
  1/3   2/2   3/3

L.F.JOHNSON,W.C.JANKOWSKI
CARBON-13 NMR SPECTRA,JOHN
WILEY AND SONS,NEW YORK   22 (1972)

1061   P 000300   1,2,3-TRICHLOROPROPANE

FORMULA   C3H5CL3   MOL WT   147.43
SOLVENT   C4H8O2
ORIG ST   C4H12SI   TEMP   AMB

```
      CL
      |
CL-CH2-CH-CH2-CL
  1    2    3
```

45.30   59.00   45.30
  1/3   2/2   3/3

L.F.JOHNSON,W.C.JANKOWSKI
CARBON-13 NMR SPECTRA,JOHN
WILEY AND SONS,NEW YORK   23 (1972)

1062   P 000300   1,2-DIBROMOPROPANE

FORMULA   C3H6BR2   MOL WT   201.89
SOLVENT   CDCL3
ORIG ST   C4H12SI   TEMP   AMB

```
      BR
      |
CH3-CH-CH2-BR
 3   2   1
```

37.60   45.70   24.10
  1/3   2/2   3/4

L.F.JOHNSON,W.C.JANKOWSKI
CARBON-13 NMR SPECTRA,JOHN
WILEY AND SONS,NEW YORK   26 (1972)

1063   P 000300   1,2-DICHLOROPROPANE

FORMULA   C3H6CL2   MOL WT   112.99
SOLVENT   CDCL3
ORIG ST   C4H12SI   TEMP   AMB

```
      CL
      |
CL-CH2-CH-CH3
  1    2   3
```

49.50   55.80   22.40
  1/3   2/2   3/4

L.F.JOHNSON,W.C.JANKOWSKI
CARBON-13 NMR SPECTRA,JOHN
WILEY AND SONS,NEW YORK   27 (1972)

1064   P 000300 1-IODOPROPANE

                      FORMULA   C3H7I   MOL WT   169.99
                      SOLVENT   CDCL3
                      ORIG ST   C4H12SI   TEMP   AMB
    J-CH2-CH2-CH3
       1   2   3   9.20   26.80   15.30
                            1/3   2/3   3/4

                      L.F.JOHNSON,W.C.JANKOWSKI
                      CARBON-13 NMR SPECTRA,JOHN
                      WILEY AND SONS,NEW YORK   33 (1972)

1065   P 000300 HEXACHLOROBUTADIENE

                      FORMULA   C4CL6   MOL WT   260.76
    CL   CL CL   SOLVENT   CDCL3
     \1 2 1 /   ORIG ST   C4H12SI   TEMP   AMB
      C=C-C=C
     / 1 3 4\   123.60 126.40 126.40 123.60
    CL   CL   CL   1/1   2/1   3/1   4/1

                      L.F.JOHNSON,W.C.JANKOWSKI
                      CARBON-13 NMR SPECTRA,JOHN
                      WILEY AND SONS,NEW YORK   43 (1972)

1066   P 000300 HEXACHLORO-1,3-BUTADIENE

                      FORMULA   C4CL6   MOL WT   260.76
    CL   CL CL   SOLVENT   CHCL3
     \1 2 1 /   ORIG ST   C4H12SI   TEMP   AMB
      C=C-C=C4
     / 1 3 \   127.70 126.70
    CL CL   CL   1/S   2/S

                      G.E.HAWKES,R.A.SMITH,J.D.ROBERTS
                      J ORG CHEM   39, 1276 (1974)

1067   P 000300 DECACHLOROBUTANE

                      FORMULA   C4CL10   MOL WT   402.57
    CL CL CL CL   SOLVENT   CHCL3
     \1 12 13 /   ORIG ST   C4H12SI   TEMP   AMB
    CL-C-C--C-C-CL
     / 1   1 4\   102.80 103.50
    CL CL CL CL   1/S   2/S

                      G.E.HAWKES,R.A.SMITH,J.D.ROBERTS
                      J ORG CHEM   39, 1276 (1974)

1068   P 000300 1,1,3,4,4-PENTACHLORO-1,3-BUTADIENE

                      FORMULA   C4HCL5   MOL WT   226.32
    CL H   CL   SOLVENT   CHCL3
     \1 1 3 4/   ORIG ST   C4H12SI   TEMP   AMB
      C=C-C=C
     / 2 1 \   123.40 120.10 123.70 129.80
    CL   CL CL   1/S   2/D   3/S   4/S

                      G.E.HAWKES,R.A.SMITH,J.D.ROBERTS
                      J ORG CHEM   39, 1276 (1974)

1069   P 000300 1,2-DIBROMOBUTANE

```
                         FORMULA   C4H8BR2              MOL WT  215.92
                         SOLVENT   CDCL3
        BR               ORIG ST   C4H12SI              TEMP       AMB
        |
   BR-CH2-CH-CH2-CH3       35.50    54.30    29.00    10.90
    1   2   3   4           1/3      2/2      3/3      4/4
```

L.F.JOHNSON,W.C.JANKOWSKI
CARBON-13 NMR SPECTRA,JOHN
WILEY AND SONS,NEW YORK   69 (1972)

1070   P 000300 1,2-DIBROMO-2-METHYLPROPANE

```
                         FORMULA   C4H8BR2              MOL WT  215.92
                         SOLVENT   CDCL3
        BR               ORIG ST   C4H12SI              TEMP       AMB
        |
   BR-CH2-C-CH3            44.60    61.70    31.80    31.80
    1   2| 3               1/3      2/1      3/4      4/4
       4CH3
```

L.F.JOHNSON,W.C.JANKOWSKI
CARBON-13 NMR SPECTRA,JOHN
WILEY AND SONS,NEW YORK   70 (1972)

1071   P 000300 2-BROMOBUTANE

```
                         FORMULA   C4H9BR               MOL WT  137.02
                         SOLVENT   CDCL3
        BR               ORIG ST   C4H12SI              TEMP       AMB
        |
   H3C-CH-CH2-CH3          26.00    53.10    34.20    12.10
    1   2   3    4          1/4      2/2      3/3      4/4
```

L.F.JOHNSON,W.C.JANKOWSKI
CARBON-13 NMR SPECTRA,JOHN
WILEY AND SONS,NEW YORK   80 (1972)

1072   P 000300 1-BROMO-2-METHYLPROPANE

```
                         FORMULA   C4H9BR               MOL WT  137.02
        3                SOLVENT   CDCL3
        CH3              ORIG ST   C4H12SI              TEMP       AMB
    1    2/
   BR-CH2-CH               42.20    30.70    20.90    20.90
        \                 1/3      2/2      3/4      4/4
        CH3
        4
```

L.F.JOHNSON,W.C.JANKOWSKI
CARBON-13 NMR SPECTRA,JOHN
WILEY AND SONS,NEW YORK   79 (1972)

1073   P 000300 2-BROMO-2-METHYLPROPANE

```
        2                FORMULA   C4H9BR               MOL WT  137.02
        CH3              SOLVENT   CDCL3
       1| 3              ORIG ST   C4H12SI              TEMP       AMB
   BR-C-CH3
        |                 62.10    36.40    36.40    36.40
       4CH3               1/1      2/4      3/4      4/4
```

L.F.JOHNSON,W.C.JANKOWSKI
CARBON-13 NMR SPECTRA,JOHN
WILEY AND SONS,NEW YORK   81 (1972)

1074   P 000300 2-CHLOROBUTANE

FORMULA   C4H9CL   MOL WT   92.57
SOLVENT   C4H8O2
CL   ORIG ST   C4H12SI   TEMP   AMB
|
H3C-CH-CH2-CH3   25.00   60.10   33.70   11.10
1 2   3   4   1/4   2/2   3/3   4/4

L.F.JOHNSON,W.C.JANKOWSKI
CARBON-13 NMR SPECTRA,JOHN
WILEY AND SONS,NEW YORK   83 (1972)

1075   P 000300 2-CHLORO-2-METHYLPROPANE

FORMULA   C4H9CL   MOL WT   92.57
3CH3   SOLVENT   CDCL3
| ORIG ST   C4H12SI   TEMP   AMB
CL-C-CH3
1| 2   66.80   34.40   34.40   34.40
4CH3   1/1   2/4   3/4   4/4

L.F.JOHNSON,W.C.JANKOWSKI
CARBON-13 NMR SPECTRA,JOHN
WILEY AND SONS,NEW YORK   82 (1972)

1076   P 000300 1-BROMO-3-METHYLBUTANE

FORMULA   C5H11BR   MOL WT   151.05
5   SOLVENT   CDCL3
CH3   ORIG ST   C4H12SI   TEMP   AMB
/
BR-CH2-CH2-CH   31.70   41.70   26.80   21.80   21.80
1   2   3\   1/3   2/3   3/2   4/4   5/4
CH3
4
L.F.JOHNSON,W.C.JANKOWSKI
CARBON-13 NMR SPECTRA,JOHN
WILEY AND SONS,NEW YORK   129 (1972)

1077   P 000300 1-CHLOROPENTANE

FORMULA   C5H11CL   MOL WT   106.60
SOLVENT   CDCL3
CL-CH2-CH2-CH2-CH2-CH3   ORIG ST   C4H12SI   TEMP   AMB
1   2   3   4   5
44.90   29.20   32.50   22.10   13.90
1/3   2/3   3/3   4/3   5/4

L.F.JOHNSON,W.C.JANKOWSKI
CARBON-13 NMR SPECTRA,JOHN
WILEY AND SONS,NEW YORK   132 (1972)

1078   P 000300 2-CHLORO-2-METHYLBUTANE

FORMULA   C5H11CL   MOL WT   106.60
CL   SOLVENT   CDCL3
| 1   ORIG ST   C4H12SI   TEMP   AMB
H3C-CH2-C-CH3
4 3 21 32.00   71.10   38.80   9.40   32.00
CH3   1/4   2/1   3/3   4/4   5/4
5
L.F.JOHNSON,W.C.JANKOWSKI
CARBON-13 NMR SPECTRA,JOHN
WILEY AND SONS,NEW YORK   130 (1972)

1079   P 000300 1-CHLORO-3-METHYLBUTANE

```
                5
               CH3
                /
   CL-CH2-CH2-CH
    1   2   3 \
               CH3
               4
```

FORMULA	C5H11CL			MOL WT	106.60
SOLVENT	CDCL3				
ORIG ST	C4H12SI			TEMP	AMB

43.10	41.60	25.70	22.00	22.00
1/3	2/3	3/2	4/4	4/4

L.F.JOHNSON,W.C.JANKOWSKI
CARBON-13 NMR SPECTRA,JOHN
WILEY AND SONS,NEW YORK   131 (1972)

1080   P 000300 1-BROMODECANE

```
                        6
  BR-CH2-CH2-CH2-CH2-CH2-CH2
    1   2   3   4   5  |
                     7CH2
                      |
                     8CH2
                      |
                     9CH2
                      |
                    10CH3
```

FORMULA	C10H21BR			MOL WT	221.18
SOLVENT	CDCL3				
ORIG ST	C4H12SI			TEMP	AMB

33.50	32.90	28.20	28.80	29.50	29.50
1/3	2/3	3/3	4/3	5/3	6/3
29.30	31.90	22.70	14.10		
7/3	8/3	9/3	10/4		

L.F.JOHNSON,W.C.JANKOWSKI
CARBON-13 NMR SPECTRA,JOHN
WILEY AND SONS,NEW YORK   370 (1972)

1081   P 000400 NITROETHANE

```
   CH3-CH2-NO2
    1   2
```

FORMULA	C2H5NO2	MOL WT	75.07
SOLVENT	CDCL3		
ORIG ST	C4H12SI	TEMP	AMB

12.30	70.70
1/4	2/3

L.F.JOHNSON,W.C.JANKOWSKI
CARBON-13 NMR SPECTRA,JOHN
WILEY AND SONS,NEW YORK   10 (1972)

1082   P 000450 N-NITROSODIMETHYLAMINE

```
            1
    O     CH3
     \\   /
      N-N
         \
          CH3
          2
```

FORMULA	C2H6N2O	MOL WT	74.08
SOLVENT	C4H8O2		
ORIG ST	C4H12SI	TEMP	AMB

32.10	39.90
1/4	2/4

L.F.JOHNSON,W.C.JANKOWSKI
CARBON-13 NMR SPECTRA,JOHN
WILEY AND SONS,NEW YORK   11 (1972)

1083   P 000500 ACETONE OXIME

```
        CH3
        /1
    N=C2
   /    \3
  HO    CH3
```

FORMULA	C3H7NO	MOL WT	73.10
SOLVENT	CDCL3		
ORIG ST	C4H12SI	TEMP	AMB

21.50	155.20	14.90
1/4	2/1	3/4

L.F.JOHNSON,W.C.JANKOWSKI
CARBON-13 NMR SPECTRA,JOHN
WILEY AND SONS,NEW YORK   35 (1972)

1084   P 000500 2,3-BUTANEDIONE MONOOXIME

```
    O      N-OH
     \    //
      C-C
     /2 3\
 H3C1    4CH3
```

FORMULA C4H7NO2			MOL WT 101.11
SOLVENT CDCL3			
ORIG ST C4H12SI			TEMP AMB

25.00	198.80	157.00	8.00
1/4	2/1	3/1	4/4

L.F.JOHNSON,W.C.JANKOWSKI
CARBON-13 NMR SPECTRA,JOHN
WILEY AND SONS,NEW YORK   68 (1972)

1085   P 000600 ISOPROPYLAMINE

```
     2
   H3C
     \1
      CH-NH2
    3/
   H3C
```

FORMULA C3H9N			MOL WT 59.11
SOLVENT C6D6			
ORIG ST C4H12SI			TEMP AMB

42.96	26.45	26.45
1/2	2/4	3/4

H.EGGERT,C.DJERASSI
J AM CHEM SOC   95, 3710 (1973)

1086   P 000600 PROPYLAMINE

FORMULA C3H9N			MOL WT 59.11
SOLVENT C6D6			

```
  3 2   1
 H3C-CH2-CH2-NH2
```

ORIG ST C4H12SI   TEMP   AMB

44.58	27.40	11.54
1/3	2/3	3/4

H.EGGERT,C.DJERASSI
J AM CHEM SOC   95, 3710 (1973)

1087   P 000600 TRIMETHYLAMINE

```
      3
     CH3
      |
      N
    / \
 H3C   CH3
  2     1
```

FORMULA C3H9N			MOL WT 59.11
SOLVENT C6D6			
ORIG ST C4H12SI			TEMP AMB

47.56	47.56	47.56
1/4	2/4	3/4

H.EGGERT,C.DJERASSI
J AM CHEM SOC   95, 3710 (1973)

1088   P 000600 BUTYLAMINE

FORMULA C4H11N			MOL WT 73.14
SOLVENT C6D6			

```
  4 3   2   1
 H3C-CH2-CH2-CH2-NH2
```

ORIG ST C4H12SI   TEMP   AMB

42.33	36.75	20.47	14.16
1/3	2/3	3/3	4/4

H.EGGERT,C.DJERASSI
J AM CHEM SOC   95, 3710 (1973)

1089   P 000600   SEC—BUTYLAMINE

```
 4  3   2  1
H3C-CH2-CH-CH3
          |
         NH2
```

FORMULA	C4H11N			MOL WT	73.14
SOLVENT	C6D6				
ORIG ST	C4H12SI			TEMP	AMB

23.97	48.79	33.42	10.81
1/4	2/2	3/3	4/4

H.EGGERT,C.DJERASSI
J AM CHEM SOC   95, 3710 (1973)

1090   P 000600   TERT—BUTYLAMINE

```
    4CH3
   2 | 1
H3C-C-NH2
    |
   3CH3
```

FORMULA	C4H11N			MOL WT	73.14
SOLVENT	C6D6				
ORIG ST	C4H12SI			TEMP	AMB

47.20	32.87	32.87	32.87
1/1	2/4	3/4	4/4

H.EGGERT,C.DJERASSI
J AM CHEM SOC   95, 3710 (1973)

1091   P 000600   DIETHYLAMINE

```
 1   2
CH2-CH3
|
NH
|
CH2-CH3
 3   4
```

FORMULA	C4H11N			MOL WT	73.14
SOLVENT	CDCL3				
ORIG ST	C4H12SI			TEMP	AMB

44.10	15.40	44.10	15.40
1/3	2/4	3/3	4/4

L.F.JOHNSON,W.C.JANKOWSKI
CARBON-13 NMR SPECTRA,JOHN
WILEY AND SONS,NEW YORK   98 (1972)

1092   P 000600   DIETHLAMINE

```
 1   2
CH2-CH3
|
NH
|
CH2-CH3
 3   4
```

FORMULA	C4H11N			MOL WT	73.14
SOLVENT	C6D6				
ORIG ST	C4H12SI			TEMP	AMB

44.43	15.72	44.43	15.72
1/3	2/4	3/3	4/4

H.EGGERT,C.DJERASSI
J AM CHEM SOC   95, 3710 (1973)

1093   P 000600   ISOPROPYLMETHYLAMINE

```
       4
  H   CH3
  |  /
H3C-N-CH
 1   2\
      CH3
       3
```

FORMULA	C4H11N			MOL WT	73.14
SOLVENT	CDCL3				
ORIG ST	C4H12SI			TEMP	AMB

33.90	50.50	22.50	22.50
1/4	2/2	3/4	4/4

L.F.JOHNSON,W.C.JANKOWSKI
CARBON-13 NMR SPECTRA,JOHN
WILEY AND SONS,NEW YORK   97 (1972)

1094   P 000600   ISOBUTYLAMINE

```
      4                FORMULA   C4H11N                MOL WT   73.14
    H3C                SOLVENT   C6D6
      \2  1            ORIG ST   C4H12SI               TEMP       AMB
       CH-CH2-NH2
      3/                  50.62    32.05    20.20    20.20
    H3C                    1/3      2/2      3/4      4/4

                       H.EGGERT,C.DJERASSI
                       J AM CHEM SOC                    95, 3710 (1973)
```

1095   P 000600   SYM—DIMETHYLETHYLENEDIAMINE

```
                       FORMULA   C4H12N2               MOL WT   88.15
              1        SOLVENT   C4H8O2
     H        CH3      ORIG ST   C4H12SI               TEMP       AMB
      \   3  2  /
       N-CH2-CH2-N        52.00    36.40    52.00    36.40
     4/         \          2/3      1/4      3/3      4/4
    H3C          H

                       L.F.JOHNSON,W.C.JANKOWSKI
                       CARBON-13 NMR SPECTRA,JOHN
                       WILEY AND SONS,NEW YORK          101 (1972)
```

1096   P 000600   1—METHYLBUTYLAMINE

```
                       FORMULA   C5H13N                MOL WT   87.17
    4  3   2   1        SOLVENT   C6D6
   H3C-CH2-CH2-CH-NH2   ORIG ST   C4H12SI               TEMP       AMB
                |
              5CH3         46.85    42.78    19.85    14.32    24.02
                           1/2      2/3      3/3      4/4      5/4

                       H.EGGERT,C.DJERASSI
                       J AM CHEM SOC                    95, 3710 (1973)
```

1097   P 000600   2—METHYLBUTYLAMINE

```
                       FORMULA   C5H13N                MOL WT   87.17
                       SOLVENT   C6D6
    4  3   2  1        ORIG ST   C4H12SI               TEMP       AMB
   H3C-CH2-CH-CH2-NH2
             |            48.36    38.50    27.20    11.55    17.20
           5CH3           1/3      2/2      3/3      4/4      5/4

                       H.EGGERT,C.DJERASSI
                       J AM CHEM SOC                    95, 3710 (1973)
```

1098   P 000600   3—METHYLBUTYLAMINE

```
                       FORMULA   C5H13N                MOL WT   87.17
                       SOLVENT   C6D6
    4  3  2   1        ORIG ST   C4H12SI               TEMP       AMB
   H3C-CH-CH2-CH2-NH2
         |                40.58    43.71    25.89    22.89    22.89
       5CH3               1/3      2/3      3/2      4/4      5/4

                       H.EGGERT,C.DJERASSI
                       J AM CHEM SOC                    95, 3710 (1973)
```

1099   P 000600   N—METHYL—SEC—BUTYLAMINE

```
        5
    H   CH3                FORMULA   C5H13N                    MOL WT   87.17
     \ /                   SOLVENT   CDCL3
      N                    ORIG ST   C4H12SI                   TEMP     AMB
      |
   H3C-CH-CH2-CH3            19.30    56.40    29.30   10.20    33.80
    1  2  3   4              1/4      2/2      3/3     4/4      5/4
```

L.F.JOHNSON,W.C.JANKOWSKI
CARBON—13 NMR SPECTRA,JOHN
WILEY AND SONS,NEW YORK   141 (1972)

1100   P 000600   N—METHYL—SEC—BUTYLAMINE

```
        5
       CH3                  FORMULA   C5H13N                    MOL WT   87.17
        |                   SOLVENT   C6D6
   H3C-NH-CH-CH2-CH3        ORIG ST   C4H12SI                   TEMP     AMB
    1    2  3   4
                              33.92    56.64    29.63   10.29    19.55
                              1/4      2/2      3/3     4/4      5/4
```

H.EGGERT,C.DJERASSI
J AM CHEM SOC   95, 3710 (1973)

1101   P 000600   N—METHYL—TERT—BUTYLAMINE

```
        5
       CH3                  FORMULA   C5H13N                    MOL WT   87.17
       |2                   SOLVENT   C6D6
   H3C-NH-C-CH3             ORIG ST   C4H12SI                   TEMP     AMB
    1   | 3
       H3C                    28.52    50.40    28.17   28.17    28.17
        4                     1/4      2/1      3/4     4/4      5/4
```

H.EGGERT,C.DJERASSI
J AM CHEM SOC   95, 3710 (1973)

1102   P 000600   N—METHYLDIETHYLAMINE

```
        3
       CH3                  FORMULA   C5H13N                    MOL WT   87.17
        |                   SOLVENT   C6D6
        N                   ORIG ST   C4H12SI                   TEMP     AMB
      / \
   H3C-CH2 CH2-CH3           51.42    12.83    40.97   51.42    12.83
    5  4   1   2             1/3      2/4      3/4     4/3      5/4
```

H.EGGERT,C.DJERASSI
J AM CHEM SOC   95, 3710 (1973)

1103   P 000600   2,2-DIMETHYLPROPYLAMINE

```
                            FORMULA   C5H13N                    MOL WT   87.17
        4                   SOLVENT   C6H6
       CH3                  ORIG ST   C4H12SI                   TEMP     AMB
     3 |2 1
   H3C-C-CH2-NH2              54.45    32.08    27.00   27.00    27.00
       |                      1/3      2/1      3/4     4/4      5/4
       CH3
        5                   H.EGGERT,C.DJERASSI
                            J AM CHEM SOC                       95, 3710 (1973)
```

1104   P 000600   PENTYLAMINE

FORMULA   C5H13N   MOL WT   87.17
SOLVENT   C6D6
ORIG ST   C4H12SI   TEMP   AMB

```
 5  4    3    2    1
H3C-CH2-CH2-CH2-CH2-NH2
```

42.65	34.30	29.69	23.10	14.28
1/3	2/3	3/3	4/3	5/4

H.EGGERT,C.DJERASSI
J AM CHEM SOC   95, 3710 (1973)

1105   P 000600   1,5-DIAMINOPENTANE

FORMULA   C5H14N2   MOL WT   102.18
SOLVENT   CDCL3

```
H2N-CH2-CH2-CH2-CH2-CH2-NH2
  1   2   3   4   5
```
ORIG ST   C4H12SI   TEMP   AMB

42.10	33.80	24.20	33.80	42.10
1/3	2/3	3/3	4/3	5/3

L.F.JOHNSON,W.C.JANKOWSKI
CARBON-13 NMR SPECTRA,JOHN
WILEY AND SONS,NEW YORK   144 (1972)

1106   P 000600   N,N-DIMETHYL-1,3-PROPANEDIAMINE

FORMULA   C5H14N2   MOL WT   102.18
SOLVENT   C4H8O2
ORIG ST   C4H12SI   TEMP   AMB

```
                5
               CH3
                 \
                  /
H2N-CH2-CH2-CH2-N
  1   2   3       \
                   CH3
                    4
```

40.80	32.20	57.80	45.40	45.40
1/3	2/3	3/3	4/4	5/4

L.F.JOHNSON,W.C.JANKOWSKI
CARBON-13 NMR SPECTRA,JOHN
WILEY AND SONS,NEW YORK   143 (1972)

1107   P 000600   DIISOPROPYLAMINE

FORMULA   C6H15N   MOL WT   101.19
SOLVENT   C6D6
ORIG ST   C4H12SI   TEMP   AMB

```
  5      2
 CH3    CH3
  I      I I
4CH-NH-CH
  I      I
 CH3    CH3
  6      3
```

45.30	23.72	23.72	45.30	23.72	23.72
1/2	2/4	3/4	4/2	5/4	6/4

H.EGGERT,C.DJERASSI
J AM CHEM SOC   95, 3710 (1973)

1108   P 000600   1,3-DIMETHYLBUTYLAMINE

FORMULA   C6H15N   MOL WT   101.19
SOLVENT   C6D6
ORIG ST   C4H12SI   TEMP   AMB

```
  4   3   2   1
H3C-CH-CH2-CH-NH2
    I       I
   CH3     CH3
    6       5
```

44.71	50.15	25.14	22.49	25.14	23.37
1/2	2/3	3/2	4/4	5/4	6/4

H.EGGERT,C.DJERASSI
J AM CHEM SOC   95, 3710 (1973)

1109   P 000600   2,2-DIMETHYLBUTYLAMINE

```
        6
        CH3
   4 3  21 1
 H3C-CH2-C-CH2-NH2
        |
        CH3
        5
```

FORMULA	C6H15N			MOL WT	101.19
SOLVENT	C6D6				
ORIG ST	C4H12SI			TEMP	AMB

52.46	34.63	31.84	8.36	24.32	24.32
1/3	2/1	3/3	4/4	5/4	6/4

H.EGGERT,C.DJERASSI
J AM CHEM SOC   95, 3710 (1973)

1110   P 000600   N,N-DIMETHYL-SEC-BUTYLAMINE

```
   6       4
   H3C     CH3
     \    /
      N-CH
     /  1\
   H3C    CH2-CH3
   5      2   3
```

FORMULA	C6H15N			MOL WT	101.19
SOLVENT	C6D6				
ORIG ST	C4H12SI			TEMP	AMB

60.92	26.94	11.36	13.15	40.60	40.60
1/2	2/3	3/4	4/4	5/4	6/4

H.EGGERT,C.DJERASSI
J AM CHEM SOC   95, 3710 (1973)

1111   P 000600   N,N-DIMETHYL-TERT-BUTYLAMINE

```
   5       4
   H3C     CH3
     \    /3
      N-C-CH3
     /  1\
   H3C    CH3
   6      2
```

FORMULA	C6H15N			MOL WT	101.19
SOLVENT	C6D6				
ORIG ST	C4H12SI			TEMP	AMB

53.15	25.62	25.62	25.62	38.74	38.74
1/1	2/4	3/4	4/4	5/4	6/4

H.EGGERT,C.DJERASSI
J AM CHEM SOC   95, 3710 (1973)

1112   P 000600   DIPROPYLAMINE

```
   1   2   3
   CH2-CH2-CH3
  /
HN
  \
   CH2-CH2-CH3
   4   5   6
```

FORMULA	C6H15N			MOL WT	101.19
SOLVENT	C6D6				
ORIG ST	C4H12SI			TEMP	AMB

52.34	23.94	11.98	52.34	23.94	11.98
1/3	2/3	3/4	4/3	5/3	6/4

H.EGGERT,C.DJERASSI
J AM CHEM SOC   95, 3710 (1973)

1113   P 000600   N-ETHYLBUTYLAMINE

```
   1  2
   CH2-CH3
  /
NH
  \3  4   5   6
   CH2-CH2-CH2-CH3
```

FORMULA	C6H15N			MOL WT	101.19
SOLVENT	C6D6				
ORIG ST	C4H12SI			TEMP	AMB

44.58	15.73	49.98	33.14	20.97	14.23
1/3	2/4	3/3	4/3	5/3	6/4

H.EGGERT,C.DJERASSI
J AM CHEM SOC   95, 3710 (1973)

1114   P 000600 N-ETHYL-SEC-BUTYLAMINE

```
    1    2
    CH2-CH3              FORMULA   C6H15N              MOL WT   101.19
   /                    SOLVENT   C6D6
  NH                    ORIG ST   C4H12SI             TEMP       AMB
   \3   4   5
    CH-CH2-CH3           41.71    16.01   54.68   30.17   10.18   20.31
    |                    1/3      2/4      3/2     4/3     5/4     6/4
    6CH3
                        H.EGGERT,C.DJERASSI
                        J AM CHEM SOC                   95, 3710 (1973)
```

1115   P 000600 N-HEXYLAMINE

```
                        FORMULA   C6H15N              MOL WT   101.19
                        SOLVENT   CDCL3
H2N-CH2-CH2-CH2-CH2-CH2-CH3 ORIG ST  C4H12SI          TEMP       AMB
    1   2   3   4   5   6
                         42.30    34.00   26.70   31.90   22.70   14.00
                         1/3      2/3      3/3     4/3     5/3     6/4

                        L.F.JOHNSON,W.C.JANKOWSKI
                        CARBON-13 NMR STECTRA,JOHN
                        WILEY AND SONS,NEW YORK         218 (1972)
```

1116   P 000600 HEXYLAMINE

```
                        FORMULA   C6H15N              MOL WT   101.19
                        SOLVENT   C6D6
   6 5   4   3   2   1  ORIG ST   C4H12SI             TEMP       AMB
  H3C-CH2-CH2-CH2-CH2-CH2
                  |      42.66    34.59   27.12   32.34   23.15   14.22
                  NH2    1/3      2/3      3/3     4/3     5/3     6/1

                        H.EGGERT,C.DJERASSI
                        J AM CHEM SOC                   95, 3710 (1973)
```

1117   P 000600 TRIETHYLAMINE

```
                        FORMULA   C6H15N              MOL WT   101.19
     1    2             SOLVENT   C6D6
     CH2-CH3            ORIG ST   C4H12SI             TEMP       AMB
       |
       N                 46.92    12.60   46.92   12.60   46.92   12.60
     /   \               1/3      2/4      3/3     4/4     5/3     6/4
  H3C-CH2 CH2-CH3
    6 5   3   4          H.EGGERT,C.DJERASSI
                        J AM CHEM SOC                   95, 3710 (1973)
```

1118   P 000600 1,2,2-TRIMETHYLPROPYLAMINE

```
       6                FORMULA   C6H15N              MOL WT   101.19
       CH3              SOLVENT   C6D6
     3 | 1              ORIG ST   C4H12SI             TEMP       AMB
   H3C-C-CH-NH2
     | 2|               55.62    34.35   25.97   18.71   25.97   25.97
    H3C CH3             1/2      2/1      3/4     4/4     5/4     6/4
     5  4
                        H.EGGERT,C.DJERASSI
                        J AM CHEM SOC                   95, 3710 (1973)
```

1119   P 000600 1,3-DIMETHYLPENTYLAMINE

```
   5 4    3  2   1
 H3C-CH2-CH-CH2-CH-NH2
         |       |
         CH3     CH3
         7       6
```

FORMULA	C7H17N			MOL WT	115.22
SOLVENT	C6D6				
ORIG ST	C4H12SI			TEMP	AMB

44.57	48.02	31.67	30.06	11.34	25.14
1/2	2/3	3/2	4/3	5/4	6/4
19.43					
7/4					

H.EGGERT,C.DJERASSI
J AM CHEM SOC   95, 3710 (1973)

1120   P 000600 N-ETHYLPENTYLAMINE

```
   1  2
   CH2-CH3
  /
 NH
  \3   4   5   6   7
   CH2-CH2-CH2-CH2-CH3
```

FORMULA	C7H17N			MOL WT	115.22
SOLVENT	C6D6				
ORIG ST	C4H12SI			TEMP	AMB

44.35	15.57	50.04	30.34	29.96	23.00
1/3	2/4	3/3	4/3	5/3	6/3
14.24					
7/4					

H.EGGERT,C.DJERASSI
J AM CHEM SOC   95, 3710 (1973)

1121   P 000600 1-ETHYLPENTYLAMINE

```
   5 4   3   2   1
 H3C-CH2-CH2-CH2-CH-NH2
                 |
             H3C-CH2
              7  6
```

FORMULA	C7H17N			MOL WT	115.22
SOLVENT	C6D6				
ORIG ST	C4H12SI			TEMP	AMB

53.00	38.13	28.88	23.30	14.00	31.45
1/2	2/3	3/3	4/3	5/4	6/3
10.83					
7/4					

H.EGGERT,C.DJERASSI
J AM CHEM SOC   95, 3710 (1973)

1122   P 000600 HEPTYLAMINE

```
   7 6   5   4   3   2   1
 H3C-CH2-CH2-CH2-CH2-CH2-CH2
                         |
                         NH2
```

FORMULA	C7H17N			MOL WT	115.22
SOLVENT	C6D6				
ORIG ST	C4H12SI			TEMP	AMB

42.53	34.45	27.36	29.71	32.37	23.05
1/3	2/3	3/3	4/3	5/3	6/3
14.20					
7/4					

H.EGGERT,C.DJERASSI
J AM CHEM SOC   95, 3710 (1973)

1123   P 000600 N-ISOPROPYLBUTYLAMINE

```
  2
  CH3
  | 1  4   5   6   7
  CH-NH-CH2-CH2-CH2-CH3
  |
  CH3
  3
```

FORMULA	C7H17N			MOL WT	115.22
SOLVENT	C6D6				
ORIG ST	C4H12SI			TEMP	AMB

48.94	23.41	23.41	47.42	33.34	20.87
1/2	2/4	3/4	4/3	5/3	6/3
14.17					
7/4					

H.EGGERT,C.DJERASSI
J AM CHEM SOC   95, 3710 (1973)

1124   P 000600 N-ISOPROPYL-SEC-BUTYLAMINE

```
 2      7
CH3    CH3
 | |    |
CH-NH-CH-CH2-CH3
 |    4   5   6
CH3
 3
```

FORMULA	C7H17N			MOL WT	115.22
SOLVENT	C6D6				
ORIG ST	C4H12SI			TEMP	AMB

45.33	23.98	23.28	51.23	30.39	10.33
1/2	2/4	3/4	4/2	5/3	6/4
20.56					
7/4					

H.EGGERT,C.DJERASSI
J AM CHEM SOC   95, 3710 (1973)

1125   P 000600 1-METHYLHEXYLAMINE

```
 6  5    4    3    2    1
H3C-CH2-CH2-CH2-CH2-CH-NH2
                      |
                     CH3
                      7
```

FORMULA	C7H17N			MOL WT	115.22
SOLVENT	C6D6				
ORIG ST	C4H12SI			TEMP	AMB

47.14	40.81	26.52	32.46	23.05	14.20
1/2	2/3	3/3	4/3	5/3	6/4
24.45					
7/4					

H.EGGERT,C.DJERASSI
J AM CHEM SOC   95, 3710 (1973)

1126   P 000600 N-METHYLHEXYLAMINE

```
 1
CH3                    |
 /
NH
 \2   3   4   5   6   7
 CH2-CH2-CH2-CH2-CH2-CH3
```

FORMULA	C7H17N			MOL WT	115.22
SOLVENT	C6D6				
ORIG ST	C4H12SI			TEMP	AMB

36.67	52.36	30.54	27.56	32.41	23.15
1/4	2/3	3/3	4/3	5/3	6/3
14.23					
7/4					

H.EGGERT,C.DJERASSI
J AM CHEM SOC   95, 3710 (1973)

1127   P 000600 DIBUTYLAMINE

```
         H
         |
         N
   7  5/  \1   3
  CH2 CH2 CH2 CH2
  / \ / \   \ / \
 H3C  CH2    CH2 CH3
 8    6      2   4
```

FORMULA	C8H19N			MOL WT	129.25
SOLVENT	C6D6				
ORIG ST	C4H12SI			TEMP	AMB

50.11	33.09	20.89	14.20	50.11	33.09
1/3	2/3	3/3	4/4	5/3	6/3
20.89	14.20				
7/3	8/4				

H.EGGERT,C.DJERASSI
J AM CHEM SOC   95, 3710 (1973)

1128   P 000600 N,N-DIETHYLBUTYLAMINE

```
    7   8
   CH2-CH3
    |
    N
   / \1   3
 H3C-CH2 CH2 CH2
  6  5   \ / \
       CH2 CH3
        2   4
```

FORMULA	C8H19N			MOL WT	129.25
SOLVENT	C6D6				
ORIG ST	C4H12SI			TEMP	AMB

53.21	30.37	20.94	14.22	47.43	12.57
1/3	2/3	3/3	4/4	5/3	6/4
47.43	12.57				
7/3	8/4				

H.EGGERT,C.DJERASSI
J AM CHEM SOC   95, 3710 (1973)

1129   P 000600 N,N-DIETHYL-SEC-BUTYLAMINE

```
    5    6
   CH2-CH3            FORMULA    C8H19N              MOL WT   129.25
    |                SOLVENT    C6D6
    N   4CH3          ORIG ST    C4H12SI             TEMP        AMB
   /\ /
H3C-CH2 CH              56.54   27.52  11.78  13.93  43.44   15.12
  8  7   1\             1/2     2/3    3/4    4/4    5/3     6/4
         CH2-CH3        43.44   15.12
          2   3         7/3     8/4

                    H.EGGERT,C.DJERASSI
                    J AM CHEM SOC                    95, 3710 (1973)
```

1130   P 000600 DIISOBUTYLAMINE

```
       H
       |               FORMULA    C8H19N              MOL WT   129.25
  8    N    4          SOLVENT    C6D6
H3C    |    CH3        ORIG ST    C4H12SI             TEMP        AMB
  \  / \  /
   HC-CH2 CH2-CH          58.59   28.97  20.87  20.87  58.59   28.97
  /6 5   1  2\            1/3     2/2    3/4    4/4    5/3     6/2
H3C          CH3         20.87   20.87
  7            3         7/4     8/4

                    H.EGGERT,C.DJERASSI
                    J AM CHEM SOC                    95, 3710 (1973)
```

1131   P 000600 DIISOBUTYLAMINE

```
                        FORMULA    C8H19N              MOL WT   129.25
                        SOLVENT    CDCL3
                        ORIG ST    C4H12SI             TEMP        AMB
   7          3
 H3C   NH    CH3           58.30   28.40  20.60  20.60  58.30   28.40
  \  / \    /             1/3     2/2    3/4    4/4    5/3     6/2
   HC-CH2 CH2-CH          20.60   20.60
  8/6 5   1  2\4          7/4     8/4
H3C          CH3
                        L.F.JOHNSON,W.C.JANKOWSKI
                        CARBON-13 NMR SPECTRA,JOHN
                        WILEY AND SONS,NEW YORK         327 (1972)
```

1132   P 000600 N,N-DIMETHYLHEXYLAMINE

```
      8
      CH3               FORMULA    C8H19N              MOL WT   129.25
      |                 SOLVENT    C6D6
      N-CH3             ORIG ST    C4H12SI             TEMP        AMB
  5   3  1/  7
 CH2 CH2 CH2               60.11   28.34  27.54  32.34  23.16   14.33
 /  \ / \ /              1/3     2/3    3/3    4/3    5/3     6/4
H3C  CH2 CH2             45.64   45.64
  6   4   2              7/4     8/4

                    H.EGGERT,C.DJERASSI
                    J AM CHEM SOC                    95, 3710 (1973)
```

1133   P 000600 N-ETHYLDIISOPROPYLAMINE

```
   4    5
  CH2-CH3              FORMULA    C8H19N              MOL WT   129.25
 8  |                 SOLVENT    C6D6
H3C  N 3CH3           ORIG ST    C4H12SI             TEMP        AMB
  \ / \ /
   CH  CH                48.27   21.06  21.06  38.67  17.28   48.27
  /6  1\                 1/2     2/4    3/4    4/3    5/4     6/2
H3C       CH3           21.06   21.06
  7        2           7/4     8/4

                    H.EGGERT,C.DJERASSI
                    J AM CHEM SOC                    95, 3710 (1973)
```

1134   P 000600 N-ETHYLHEXYLAMINE

```
              H
              |
              N
     7   5   3/ \
    CH2 CH2 CH2 CH2-CH3
    / \ / \ /   1   2
  H3C   CH2 CH2
   8    6   4
```

FORMULA	C8H19N			MOL WT	129.25
SOLVENT	C6D6				
ORIG ST	C4H12SI			TEMP	AMB

44.39	15.60	50.12	30.68	27.49	32.25
1/3	2/4	3/3	4/3	5/3	6/3
23.05	14.22				
7/3	8/4				

H.EGGERT,C.DJERASSI
J AM CHEM SOC   95, 3710 (1973)

1135   P 000600 OCTYLAMINE

```
  8  7  6  5  4  3  2
H3C-CH2-CH2-CH2-CH2-CH2-CH2
                      |
                     1CH2
                      |
                     NH2
```

FORMULA	C8H19N			MOL WT	129.25
SOLVENT	C6D6				
ORIG ST	C4H12SI			TEMP	AMB

42.54	34.47	27.39	30.02	29.83	32.33
1/3	2/3	3/3	4/3	5/3	6/3
23.10	14.22				
7/3	8/4				

H.EGGERT,C.DJERASSI
J AM CHEM SOC   95, 3710 (1973)

1136   P 000600 TERT-OCTYLAMINE

```
             6
   NH2      CH3
  5 |        |
  H3C-C-CH2-C-CH3
   4| 3  2| 1
   CH3    CH3
    8      7
```

FORMULA	C8H19N			MOL WT	129.25
SOLVENT	CDCL3				
ORIG ST	C4H12SI			TEMP	AMB

31.60	31.60	57.10	51.10	32.80	31.60
1/4	2/1	3/3	4/1	5/4	6/4
31.60	32.80				

L.F.JOHNSON,W.C.JANKOWSKI
CARBON-13 NMR SPECTRA,JOHN
WILEY AND SONS,NEW YORK   328 (1972)

1137   P 000600 N-SEC-BUTYL-3-METHYLBUTYLAMINE

```
            H
   8   6    |   4
  H3C   CH2 N   CH3
   \ / \ / \1/
   7CH  CH2 CH
   /   5    \
  H3C        CH2-CH3
   9          2   3
```

FORMULA	C9H21N			MOL WT	143.27
SOLVENT	C6D6				
ORIG ST	C4H12SI			TEMP	AMB

54.77	30.07	10.15	20.33	45.52	40.30
1/2	2/3	3/4	4/4	5/3	6/3
26.24	22.84	22.84			
7/2	8/4	9/4			

H.EGGERT,C.DJERASSI
J AM CHEM SOC   95, 3710 (1973)

1138   P 000600 N-SEC-BUTYLPENTYLAMINE

```
             H
             |   4
             N  CH3
    9   7   5/ \1/
   H3C   CH2 CH2 CH
    \ / \ /       \
    CH2 CH2       CH2-CH3
     8   6         2   3
```

FORMULA	C9H21N			MOL WT	143.27
SOLVENT	C6D6				
ORIG ST	C4H12SI			TEMP	AMB

54.75	30.09	10.24	20.34	47.55	30.90
1/2	2/3	3/4	4/4	5/3	6/3
30.09	23.06	14.26			
7/3	8/3	9/4			

H.EGGERT,C.DJERASSI
J AM CHEM SOC   95, 3710 (1973)

1139   P 000600 N—TERT—BUTYL—3—METHYLBUTYLAMINE

```
              H
    5   2   |   7
  H3C   CH2 N   CH3
     \ / \ / /8
      3CH  CH2 C—CH3
     /    1  6\
   H3C         CH3
   4           9
```

FORMULA	C9H21N			MOL WT	143.27
SOLVENT	C6D6				
ORIG ST	C4H12SI			TEMP	AMB

40.36	40.80	26.05	22.81	22.81	49.30
1/3	2/3	3/2	4/4	5/4	6/1
29.28	29.28	29.28			
7/4	8/4	9/4			

H.EGGERT,C.DJERASSI
J AM CHEM SOC   95, 3710 (1973)

1140   P 000600 1—ISOPROPYLHEXYLAMINE

```
        9   8
      H3C   CH3
        \ /
       7CH
        |
H3C—CH2—CH2—CH2—CH2—CH—NH2
 6  5   4   3   2   1
```

FORMULA	C9H21N			MOL WT	143.27
SOLVENT	C6D6				
ORIG ST	C4H12SI			TEMP	AMB

56.64	35.31	26.70	32.45	23.07	14.28
1/2	2/3	3/3	4/3	5/3	6/4
33.79	19.51	17.08			
7/2	8/4	9/4			

H.EGGERT,C.DJERASSI
J AM CHEM SOC   95, 3710 (1973)

1141   P 000600 N—METHYL—1,1,3,3—TETRAMETHYLBUTYLAMINE

```
              H
    5   6   |
  H3C   CH3 N—CH3
    \4  \ / 1
  H3C—C   C2
   8/ \ / \
  H3C   CH2 CH3
   9   3   7
```

FORMULA	C9H21N			MOL WT	143.27
SOLVENT	C6D6				
ORIG ST	C4H12SI			TEMP	AMB

28.80	53.88	52.66	31.80	31.80	28.67
1/4	2/1	3/3	4/1	5/4	6/4
28.67	31.80	31.80			
7/4	8/4	9/4			

H.EGGERT,C.DJERASSI
J AM CHEM SOC   95, 3710 (1973)

1142   O 000600 NONYLAMINE

```
   9 8   7   6   5
  H3C—CH2—CH2—CH2—CH2
                  |
  H2N—CH2—CH2—CH2—CH2
   1   2   3   4
```

FORMULA	C9H21N			MOL WT	143.27
SOLVENT	C6D6				
ORIG ST	C4H12SI			TEMP	AMB

42.61	34.57	27.42	30.13	30.13	29.28
1/3	2/3	3/3	4/3	5/3	6/3
32.38	23.08	14.25			
7/3	8/3	9/4			

H.EGGERT,C.DJERASSI
J AM CHEM SOC   95, 3710 (1973)

1143   P 000600 N—PROPYLHEXYLAMINE

```
    3 2   1
  H3C—CH2—CH2
             \
              N—H
   8  6   4/
   CH2 CH2 CH2
  / \ / \ /
 H3C   CH2 CH2
  9   7   5
```

FORMULA	C9H21N			MOL WT	143.27
SOLVENT	C6D6				
ORIG ST	C4H12SI			TEMP	AMB

52.23	23.76	12.00	50.29	30.73	27.48
1/3	2/3	3/4	4/3	5/3	6/3
32.26	23.06	14.25			
7/3	8/3	9/4			

H.EGGERT,C.DJERASSI
J AM CHEM SOC   95, 3710 (1973)

1144   P 000600   TRIPROPYLAMINE

```
        1    2    3        FORMULA   C9H21N                    MOL WT   143.27
       CH2-CH2-CH3         SOLVENT   C6D6
        |                  ORIG ST   C4H12SI                   TEMP        AMB
        N
       / \                  56.76    21.24    12.01    56.76    21.24    12.01
 H3C-CH2-CH2 CH2-CH2-CH3    1/3      2/3      3/4      4/3      5/3      6/4
  9 8   7    4   5   6      56.76    21.24    12.01
                            7/3      8/3      9/4
```

H.EGGERT,C.DJERASSI
J AM CHEM SOC   95, 3710 (1973)

1145   P 000600   N-BENZYLISOPROPYLAMINE

```
   3 2
   C-C            H          FORMULA   C10H15N                 MOL WT   149.24
  4/   \1  7    /  9         SOLVENT   C4H8O2
 C       C-CH2-N   CH3       ORIG ST   C4H12SI                 TEMP        AMB
  \    /       \8/
   C=C          CH           141.70   128.10   128.10   126.50   128.10   128.10
   5 6          |            1/1      2/2      3/2      4/2      5/2      6/2
              10CH3          51.60    48.20    23.10    23.10
                             7/3      8/2      9/4      10/4
```

L.F.JOHNSON,W.C.JANKOWSKI
CARBON-13 NMR SPECTRA,JOHN
WILEY AND SONS,NEW YORK   388 (1972)

1146   P 000600   N-BUTYLHEXYLAMINE

```
             H              FORMULA   C10H23N                 MOL WT   157.30
             |              SOLVENT   C6D6
             N              ORIG ST   C4H12SI                 TEMP        AMB
   5    3  1/ \7   9
  CH2  CH2 CH2 CH2 CH2       50.37    30.73    27.48    32.25    23.05    14.25
  / \ / \ /     \ / \        1/3      2/3      3/3      4/3      5/3      6/4
 H3C   CH2 CH2    CH2 CH3    50.02    32.90    20.83    14.23
  6    4   2      8   10     7/3      8/3      9/3      10/4
```

H.EGGERT,C.DJERASSI
J AM CHEM SOC   95, 3710 (1973)

1147   P 000600   N-SEC-BUTYL-3,3-DIMETHYLBUTYLAMINE

```
             H              FORMULA   C10H23N                 MOL WT   157.30
    8        |    4         SOLVENT   C6D6
   H3C       N    CH3       ORIG ST   C4H12SI                 TEMP        AMB
    9\      5/ \  /
   H3C-C7  CH2  CH           54.79    30.04    10.15    20.30    43.59    45.13
    / \ /      1\            1/2      2/3      3/4      4/4      5/3      6/3
  H3C   CH2    CH2-CH3       29.83    29.83    29.83    29.83
  10    6      2   3         7/1      8/4      9/4      10/4
```

H.EGGERT,C.DJERASSI
J AM CHEM SOC   95, 3710 (1973)

1148   P 000600   N-TERT-BUTYLHEXYLAMINE

```
                            FORMULA   C10H23N                 MOL WT   157.30
         H                  SOLVENT   C6D6
         |   10             ORIG ST   C4H12SI                 TEMP        AMB
         N   CH3
   7   5  3/ \ /9
  CH2 CH2 CH2 C-CH3          49.94    29.20    42.55    31.58    27.50    32.24
  / \ / \ /    1\            1/1      2/4      3/3      4/3      5/3      6/3
 H3C   CH2 CH2   CH3         22.99    14.20    29.20    29.20
  8    6   4     2           7/3      8/4      9/4      10/4
```

H.EGGERT,C.DJERASSI
J AM CHEM SOC   95, 3710 (1973)

1149   P 000600 DECYLAMINE

```
                              FORMULA   C10H23N              MOL WT   157.30
                              SOLVENT   C6D6
  10 9   8   7   6   5        ORIG ST   C4H12SI              TEMP       AMB
H3C-CH2-CH2-CH2-CH2-CH2
                     |           42.59    34.53    27.37    30.08    30.08    30.08
    H2N-CH2-CH2-CH2-CH2          1/3      2/3      3/3      4/3      5/3      6/3
         1    2    3    4        29.82    32.35    23.05    14.23
                                 7/3      8/3      9/3      10/4
```

H.EGGERT,C.DJERASSI
J AM CHEM SOC   95, 3710 (1973)

1150   P 000600 N,N-DIISOPROPYLBUTYLAMINE

```
      1    2    3    4    FORMULA   C10H23N              MOL WT   157.30
     CH2-CH2-CH2-CH3 SOLVENT   C6D6
  10   |                 ORIG ST   C4H12SI              TEMP       AMB
  H3C     N  7CH3
     \  /  \  /             44.53    33.75    20.71    14.30    48.09    21.00
      8CH  5CH              1/3      2/3      3/3      4/4      5/2      6/4
     /        \             21.00    48.09    21.00    21.00
   H3C        CH3           7/4      8/2      9/4      10/4
    9          6
                           H.EGGERT,C.DJERASSI
                           J AM CHEM SOC                95, 3710 (1973)
```

1151   P 000600 N-ETHYLDIBUTYLAMINE

```
      1    2          FORMULA   C10H23N              MOL WT   157.30
     CH2-CH3          SOLVENT   C6D6
       |             ORIG ST   C4H12SI              TEMP       AMB
       N
   9  7/ \3   5         47.89    12.49    53.76    30.34    20.92    14.23
  CH2 CH2 CH2 CH2       1/3      2/4      3/3      4/3      5/3      6/4
  / \ /     \ / \       53.76    30.34    20.92    14.23
 H3C   CH2    CH2 CH3   7/3      8/3      9/3      10/4
 10    8      4   6
                       H.EGGERT,C.DJERASSI
                       J AM CHEM SOC                95, 3710 (1973)
```

1152   P 000600 DI(3-METHYLBUTYL)AMINE

```
          H              FORMULA   C10H23N              MOL WT   157.30
   9      |      5       SOLVENT   C6D6
  H3C     N     CH3      ORIG ST   C4H12SI              TEMP       AMB
    \8   / \   3/
     CH CH2 CH2 CH         48.46    39.93    26.27    22.87    22.87    48.46
  10/ \ /6  1\ / \         1/3      2/3      3/2      4/4      5/4      6/3
  H3C   CH2    CH2 CH3     39.93    26.27    22.87    22.87
   7     2      4          7/3      8/2      9/4      10/4
                          H.EGGERT,C.DJERASSI
                          J AM CHEM SOC                95, 3710 (1973)
```

1153   P 000600 DIPENTYLAMINE

```
          H              FORMULA   C10H23N              MOL WT   157.30
          |              SOLVENT   C6D6
          N              ORIG ST   C4H12SI              TEMP       AMB
 10   8  6/ \1   3    5
 H3C  CH2 CH2 CH2 CH2 CH3    50.44    30.58    30.09    23.10    14.27    50.44
   \ / \ /     \ / \ /       1/3      2/3      3/3      4/3      5/4      6/3
   CH2 CH2     CH2 CH2       30.58    30.09    23.10    14.27
    9   7       2   4        7/3      8/3      9/3      10/4
                            H.EGGERT,C.DJERASSI
                            J AM CHEM SOC                95, 3710 (1973)
```

1154   P 000600   N—BUTYL—1—ETHYLPENTYLAMINE

```
              H
     11  10  |
    H3C—H2C  N
      |/   \1    3
       5CH   CH2 CH2
   9   7   /     \ /  \
  H3C  CH2—CH2    CH2  CH3
    \  /    6      2    4
     CH2
      8
```

FORMULA	C11H25N			MOL WT	171.33
SOLVENT	C6D6				
ORIG ST	C4H12SI			TEMP	AMB

46.96	33.48	20.88	14.24	58.95	33.99
1/3	2/3	3/3	4/4	5/2	6/3
26.93	23.45	14.24	28.39	9.85	
7/3	8/3	9/4	10/3	11/4	

H.EGGERT,C.DJERASSI
J AM CHEM SOC   95, 3710 (1973)

1155   P 000600   N—BUTYL—1—METHYLHEXYLAMINE

```
     11 10  9   8
    H3C—CH2—CH2—CH2
                   \
                    N—H
  6  5   4   3   2  1/
 H3C—CH2—CH2—CH2—CH2—CH
                    |
                    CH3
                    7
```

FORMULA	C11H25N			MOL WT	171.33
SOLVENT	C6D6				
ORIG ST	C4H12SI			TEMP	AMB

53.39	38.18	25.98	32.62	23.05	14.20
1/2	2/3	3/3	4/3	5/3	6/4
20.84	47.16	33.41	20.84	14.20	
7/4	8/3	9/3	10/3	11/4	

H.EGGERT,C.DJERASSI
J AM CHEM SOC   95, 3710 (1973)

1156   P 000600   N—(3,3—DIMETHYLBUTYL)PENTYLAMINE

```
             H
   9         |
  H3C        N
   10\   6/  \1   3   5
   H3C—C9  CH2 CH2 CH2 CH3
    / \ /      \ / \ /
  H3C  CH2    CH2 CH2
   11   7      2   4
```

FORMULA	C11H25N			MOL WT	171.33
SOLVENT	C6D6				
ORIG ST	C4H12SI			TEMP	AMB

50.43	30.45	29.82	22.94	14.19	46.44
1/3	2/3	3/3	4/3	5/4	6/3
44.71	29.82	29.82	29.82	29.82	
7/3	8/1	9/4	10/4	11/4	

H.EGGERT,C.DJERASSI
J AM CHEM SOC   95, 3710 (1973)

1157   P 000600   N—ISOPROPYLDIBUTYLAMINE

```
    H3C    CH3
     6\   /7
      5CH
       |
       N
  10  8/ \1    3
  CH2 CH2 CH2 CH2
   / \ /     \ / \
 H3C  CH2    CH2 CH3
  11   9      2   4
```

FORMULA	C11H25N			MOL WT	171.33
SOLVENT	C6D6				
ORIG ST	C4H12SI			TEMP	AMB

49.81	32.01	20.83	14.34	50.09	18.15
1/3	2/3	3/3	4/4	5/2	6/4
18.15	49.81	32.01	20.83	14.34	
7/4	8/3	9/3	10/3	11/4	

H.EGGERT,C.DJERASSI
J AM CHEM SOC   95, 3710 (1973)

1158   P 000600   N—METHYL—N—BUTYLHEXYLAMINE

```
         11
        CH3
         |
         N
  5   3  1/ \7   9
 CH2 CH2 CH2 CH2 CH2
  / \ / \     \ / \
 H3C  CH2 CH2  CH2 CH3
  6   4   2     8   10
```

FORMULA	C11H25N			MOL WT	171.33
SOLVENT	C6D6				
ORIG ST	C4H12SI			TEMP	AMB

58.19	27.99	27.43	32.25	23.04	14.18
1/3	2/3	3/3	4/3	5/3	6/4
57.85	30.20	20.78	14.18	42.07	
7/3	8/3	9/3	10/4	11/4	

H.EGGERT,C.DJERASSI
J AM CHEM SOC   95, 3710 (1973)

1159   P 000600 N—PENTYLHEXYLAMINE

```
                                FORMULA   C11H25N              MOL WT  171.33
                H               SOLVENT   C6D6
                I               ORIG ST   C4H12SI              TEMP      AMB
                N
     10   8   6/ \1   3    5     50.37    30.45    30.00    23.05    14.25    50.37
     CH2  CH2 CH2 CH2 CH2 CH3    1/3      2/3      3/3      4/3      5/4      6/3
    / \ ./ \ /      \ / \ /      30.74    27.50    32.27    23.05    14.25
   H3C   CH2 CH2    CH2 CH2      7/3      8/3      9/3      10/3     11/4
   11    9   7      2   4
                                H.EGGERT,C.DJERASSI
                                J AM CHEM SOC              95, 3710 (1973)
```

1160   P 000600 N—PENTYL—1,3—DIMETHYLBUTYLAMINE

```
                                FORMULA   C11H25N              MOL WT  171.33
                H               SOLVENT   C6D6
    9    10    I               ORIG ST   C4H12SI              TEMP      AMB
   H3C    CH3 N
     \    \ / \1   3    5        47.70    30.86    30.00    23.17    14.18    51.14
    8CH 6CH  CH2 CH2 CH3         1/3      2/3      3/3      4/3      5/4      6/2
   / \ /     \ / \ /             47.30    25.10    23.17    21.21    23.17
   H3C  CH2   CH2 CH2            7/3      8/2      9/4      10/4     11/4
   11   7     2   4
                                H.EGGERT,C.DJERASSI
                                J AM CHEM SOC              95, 3710 (1973)
```

1161   P 000600 N—PENTYL—1,2,2—TRIMETHYLPROPYLAMINE

```
                                FORMULA   C11H25N              MOL WT  171.33
                H               SOLVENT   C6D6
    9    I                      ORIG ST   C4H12SI              TEMP      AMB
   CH3 N
    10   \ / \1   3    5         49.15    30.89    29.99    22.98    14.27    62.32
   H3C  6CH  CH2 CH2 CH3         1/3      2/3      3/3      4/3      5/4      6/2
    \ /      \ / \ /             34.74    26.57    15.12    26.57    26.57
    7C       CH2 CH2             7/1      8/4      9/4      10/4     11/4
   / \       2   4
  H3C   CH3                     H.EGGERT,C.DJERASSI
   8    11                      J AM CHEM SOC              95, 3710 (1973)
```

1162   P 000600 N—PROPYLDIBUTYLAMINE

```
                                FORMULA   C11H25N              MOL WT  171.33
    5    6    7                 SOLVENT   C6D6
   CH2—CH2—CH3                  ORIG ST   C4H12SI              TEMP      AMB
        I
        N                       54.35    30.32    20.90    14.25    56.67    21.17
    10  8/ \1    3              1/3      2/3      3/3      4/4      5/3      6/3
   CH2 CH2 CH2 CH2              12.04    54.35    30.32    20.90    14.25
   / \ /     \ / \              7/4      8/3      9/3      10/3     11/4
  H3C  CH2   CH2 CH3
   11   9    2   4             H.EGGERT,C.DJERASSI
                               J AM CHEM SOC              95, 3710 (1973)
```

1163   P 000600 N—TERT—BUTYLDIBUTYLAMINE

```
    7CH3
   6 I 8                        FORMULA   C12H27N              MOL WT  185.36
   H3C—C—CH3                    SOLVENT   C6D6
      I5                        ORIG ST   C4H12SI              TEMP      AMB
      N
   11  9/ \1    3               50.68    34.60    20.77    14.32    54.49    27.43
   CH2 CH2 CH2 CH2              1/3      2/3      3/3      4/4      5/1      6/4
   / \ /     \ / \              27.43    27.43    50.68    34.60    20.77    14.32
  H3C  CH2   CH2 CH3            7/4      8/4      9/3      10/3     11/3     12/4
   12   10    2   4
                               H.EGGERT,C.DJERASSI
                               J AM CHEM SOC              95, 3710 (1973)
```

1164   P 000600 DIHEXYLAMINE

```
              H
              |
              N
  11    9    / \1    3    5
    CH2 CH2 CH2 CH2 CH2 CH2
12/ \ / \ /7   \ / \ / \
H3C   CH2 CH2      CH2 CH2 CH3
     10   8        2   4   6
```

FORMULA	C12H27N			MOL WT	185.36
SOLVENT	C6D6				
ORIG ST	C4H12SI			TEMP	AMB

50.42	30.34	27.53	32.33	23.08	14.21
1/3	2/3	3/3	4/3	5/3	6/4
50.42	30.34	27.53	32.33	23.08	14.21
7/3	8/3	9/3	10/3	11/3	12/4

H.EGGERT,C.DJERASSI
J AM CHEM SOC   95, 3710 (1973)

1165   P 000600 N-(1,3-DIMETHYLBUTYL)HEXYLAMINE

FORMULA	C12H27N			MOL WT	185.36
SOLVENT	C6D6				
ORIG ST	C4H12SI			TEMP	AMB

```
              H
              |    11
              N    CH3
   5    3    1/ \ /        12
  CH2 CH2 CH2 CH       CH3
 / \ / \ /     7\    9/
H3C   CH2 CH2    CH2-CH
  6    4    2       8   \10
                 CH3
```

47.66	31.13	27.45	32.21	23.05	14.17
1/3	2/3	3/3	4/3	5/3	6/4
51.10	47.33	25.08	23.05	21.19	23.05
7/2	8/3	9/2	10/4	11/4	12/4

H.EGGERT,C.DJERASSI
J AM CHEM SOC   95, 3710 (1973)

1166   P 000600 N-ETHYLDIPENTYLAMINE

FORMULA	C12H27N			MOL WT	185.36
SOLVENT	C6D6				
ORIG ST	C4H12SI			TEMP	AMB

```
     6    7
     CH2-CH3
        |
        N
  12    10  8/ \1    3    5
  H3C   CH2 CH2 CH2 CH2 CH3
   \ / \ /     \ / \ /
    CH2 CH2      CH2 CH2
    11   9       2    4
```

54.01	27.75	30.09	23.03	14.26	47.86
1/3	2/3	3/3	4/3	5/4	6/3
12.46	54.01	27.75	30.09	23.03	14.26
7/4	8/3	9/3	10/3	11/3	12/4

H.EGGERT,C.DJERASSI
J AM CHEM SOC   95, 3710 (1973)

1167   P 000600 TRIBUTYLAMINE

```
 11  10   9      1    2    3
 CH2-CH2-CH2    CH2-CH2-CH2
  |        \ /              |
 CH3        N            4CH3
 12         |    7
         CH2-CH2-CH2
          5    6    |
                   CH3
                    8
```

FORMULA	C12H27N			MOL WT	185.36
SOLVENT	CDCL3				
ORIG ST	C4H12SI			TEMP	AMB

54.10	29.50	20.80	14.10	54.10	29.50
1/3	2/3	3/3	4/4	5/3	6/3
20.80	14.10	54.10	29.50	20.80	14.10
7/3	8/4	9/3	10/3	11/3	12/4

L.F.JOHNSON,W.C.JANKOWSKI
CARBON-13 NMR SPECTRA,JOHN
WILEY AND SONS,NEW YORK   449 (1972)

1168   P 000600 TRIBUTYLAMINE

```
       10CH2 CH3
        / \ /12
       9CH2 CH2
        |    11
        N
    7    5/ \1    3
   CH2 CH2 CH2 CH2
  / \ /     \ / \
 H3C   CH2    CH2 CH3
  8    6      2    4
```

FORMULA	C12H27N			MOL WT	185.36
SOLVENT	C6D6				
ORIG ST	C4H12SI			TEMP	AMB

54.28	30.28	20.96	14.24	54.28	30.28
1/3	2/3	3/3	4/4	5/3	6/3
20.96	14.24	54.28	30.28	20.96	14.24
7/3	8/4	9/3	10/3	11/3	12/4

H.EGGERT,C.DJERASSI
J AM CHEM SOC   95, 3710 (1973)

1169 P 000600 N,N-DIBUTYL-3-METHYLBUTYLAMINE

```
              11   13
              CH2 CH3          FORMULA   C13H29N              MOL WT   199.38
               / \ /           SOLVENT   C6D6
              10CH2 CH2        ORIG ST   C4H12SI              TEMP     AMB
      4        |   12
     H3C       N              52.54   36.98   26.31   22.92   22.92   54.18
       \3   1/ \6   8          1/3     2/3     3/2     4/4     5/4     6/3
        CH  CH2 CH2 CH2       30.21   20.84   14.22   54.18   30.21   20.84
       / \ /       \ /          7/3     8/3     9/4    10/3    11/3    12/3
     H3C   CH2    CH2 CH3      14.22
      5     2      7   9        13/4
```

H.EGGERT,C.DJERASSI
J AM CHEM SOC   95, 3710 (1973)

1170 P 000600 N-PENTYL-1,1,3,3-TETRAMETHYLBUTYLAMINE

```
                                FORMULA   C13H29N              MOL WT   199.38
                                SOLVENT   C6D6
                                ORIG ST   C4H12SI              TEMP     AMB
              H
    4    2    |   10      9    41.94   31.28   29.96   22.90   14.18   53.82
   CH2 CH2 N   CH3    CH3       1/3     2/3     3/3     4/3     5/4     6/1
   / \ / \ / \6/7   8/12      53.47   31.86   31.86   29.52   29.52   31.86
  H3C  CH2 CH2 C-CH2-C-CH3      7/3     8/1     9/4    10/4    11/4    12/4
   5   3   1   |       \       31.86
              CH3     CH3       13/4
              11       13
```

H.EGGERT,C.DJERASSI
J AM CHEM SOC   95, 3710 (1973)

1171 P 000600 N-SEC-BUTYLDIPENTYLAMINE

```
      13     11
      H2C    CH2            FORMULA   C14H31N              MOL WT   213.41
       / \  / \             SOLVENT   C6D6
     H3C H2C 10CH2          ORIG ST   C4H12SI              TEMP     AMB
     14   12   |
              N   9CH3     50.07   29.51   30.04   23.09   14.31   56.84
     5    3   1/ \ /         1/3     2/3     3/3     4/3     5/4     6/2
    H3C   CH2 CH2 CH       27.40   11.96   13.70   50.07   29.51   30.04
      \ / \ /    6\          7/3     8/4     9/4    10/3    11/3    12/3
     CH2  CH2     CH2-CH3   23.09   14.31
      4    2       7   8     13/3    14/4
```

H.EGGERT,C.DJERASSI
J AM CHEM SOC   95, 3710 (1973)

1172 P 000600 N,N-DIBUTYLHEXYLAMINE

```
              8    10
              CH2 CH3        FORMULA   C14H31N              MOL WT   213.41
               / \ /         SOLVENT   C6D6
             7CH2 CH2        ORIG ST   C4H12SI              TEMP     AMB
               |   9
               N            54.54   27.96   27.48   32.26   23.05   14.20
     5    3   1/ \11   13     1/3     2/3     3/3     4/3     5/3     6/4
   CH2 CH2 CH2 CH2 CH2      54.24   30.24   20.63   14.20   54.24   30.24
    / \ / \ /      \ / \      7/3     8/3     9/3    10/4    11/3    12/3
  H3C   CH2 CH2    CH2 CH3  20.63   14.20
   6     4   2      12  14    13/3    14/4
```

H.EGGERT,C.DJERASSI
J AM CHEM SOC   95, 3710 (1973)

1173 P 000600 N,N-DIBUTYL-3,3-DIMETHYLBUTYLAMINE

```
              12   14
              CH2 CH3        FORMULA   C14H31N              MOL WT   213.41
               / \ /         SOLVENT   C6D6
            11CH2 CH2        ORIG ST   C4H12SI              TEMP     AMB
     5    6   |   13
    H3C   CH3 N             50.22   41.08   29.69   30.04   30.04   30.04
      \ / \ / \7    9        1/3     2/3     3/1     4/4     5/4     6/4
      3C  1CH2 CH2 CH2      54.27   30.28   20.94   14.28   54.27   30.28
      / \ /       \ / \       7/3     8/3     9/3    10/4    11/3    12/3
    H3C   CH2    CH2 CH3    20.94   14.28
     4     2      8   10      13/3    14/4
```

H.EGGERT,C.DJERASSI
J AM CHEM SOC   95, 3710 (1973)

```
1174              P 000600  N,N-DIBUTYL-1-METHYLPENTYLAMINE
         6    2    4
       H3C   CH2 CH2      FORMULA   C14H31N              MOL WT  213.41
         \1/ \ / \        SOLVENT   C6D6
          CH  CH2 CH3     ORIG ST   C4H12SI              TEMP      AMB
          |   3   5
          N               54.85   34.31   29.72   23.25   14.35   14.10
      13 11/ \7    9       1/2     2/3     3/3     4/3     5/4     6/4
      CH2 CH2 CH2 CH2     49.81   32.10   20.89   14.35   49.81   32.10
     / \ /     \ / \       7/3     8/3     9/3    10/4    11/3    12/3
    H3C  CH2    CH2 CH3   20.89   14.35
    14   12     8   10     13/3    14/4

                          H.EGGERT,C.DJERASSI
                          J AM CHEM SOC                95, 3710 (1973)
```

```
1175              P 000600  N-(1-ETHYLPENTYL)-1-PROPYLBUTYLAMINE
              12  13  14
         H    CH2-CH2-CH3  FORMULA   C14H31N              MOL WT  213.41
         |   /            SOLVENT   C6D6
         N-CH             ORIG ST   C4H12SI              TEMP      AMB
   5   3  1/  8\   11
  H3C  CH2 CH  9CH2 CH3   55.85   34.40   27.34   23.47   14.33   28.26
   \ / \ / \    \ /        1/2     2/3     3/3     4/3     5/4     6/3
   CH2 CH2 CH2   CH2      9.84    54.02   37.74   19.20   14.67   37.58
   4   2   6\    10        7/4     8/2     9/3    10/3    11/4    12/3
          7CH3            19.07   14.67
                          13/2    14/4

                          H.EGGERT,C.DJERASSI
                          J AM CHEM SOC                95, 3710 (1973)
```

```
1176              P 000600  N-BUTYL-N-(1,2,2-TRIMETHYLPROPYL)PENTYLAMINE
         7   9
       CH2 CH3           FORMULA   C15H33N              MOL WT  227.44
      / \ /              SOLVENT   C6D6
    6CH2 CH2             ORIG ST   C4H12SI              TEMP      AMB
  13   |   8
  H3C   N               52.79   29.30   29.99   23.10   14.28   52.56
   \ / \1   3   5        1/3     2/3     3/3     4/3     5/4     6/3
  H3C  CH CH2 CH2 CH3   31.88   20.95   14.28   64.41   35.74   27.74
  14\ /10  \ / \ /       7/3     8/3     9/4    10/2    11/1    12/4
   C11      CH2 CH2     7.74    27.74   27.74
   / \       2   4      13/4    14/4    15/4
 H3C   CH3
 12    15                 H.EGGERT,C.DJERASSI
                          J AM CHEM SOC                95, 3710 (1973)
```

```
1177              P 000600  N,N-DIBUTYLHEPTYLAMINE
         13  15
       CH2 CH3           FORMULA   C15H33N              MOL WT  227.44
      / \ /              SOLVENT   C6D6
    12CH2 CH2            ORIG ST   C4H12SI              TEMP      AMB
       |   14
       N                54.56   28.01   27.79   29.72   32.34   23.01
   7   5   3  1/ \8   10  1/3     2/3     3/3     4/3     5/3     6/3
  H3C  CH2 CH2 CH2 CH2 CH2  14.23   54.26   30.24   20.87   14.23   54.26
   \ / \ / \ /     \ / \     7/4     8/3     9/3    10/3    11/4    12/3
   CH2 CH2 CH2     CH2 CH3  30.24   20.87   14.23
   6   4   2       9   11   13/3    14/3    15/4

                          H.EGGERT,C.DJERASSI
                          J AM CHEM SOC                95, 3710 (1973)
```

```
1178              P 000600  TRIPENTYLAMINE

                          FORMULA   C15H33N              MOL WT  227.44
         12  14           SOLVENT   C6D6
       CH2 CH2            ORIG ST   C4H12SI              TEMP      AMB
      / \ / \
    11CH2 CH2 CH3         54.50   27.63   30.05   22.98   14.24   54.50
       |  13  15           1/3     2/3     3/3     4/3     5/4     6/3
       N                  27.63   30.05   22.98   14.24   54.50   27.63
  10   8   6/ \1   3   5   7/3     8/3     9/3    10/4    11/3    12/3
 H3C  CH2 CH2 CH2 CH2 CH3  30.05   22.98   14.24
  \ / \ / \     \ / \ /    13/3    14/3    15/4
  CH2 CH2       CH2 CH2
  9   7         2   4      H.EGGERT,C.DJERASSI
                          J AM CHEM SOC                95, 3710 (1973)
```

1179   P 000600 DI(2-ETHYLHEXYL)AMINE

```
                 H
                 |
                 N
    13   11   9/ \1   3    5
    CH2  CH2  CH2 CH2 CH2 CH2
   / \ / \ /       \ / \ / \
 H3C H2C 10CH     2CH  CH2 CH3
  14   12  |       |   4   6
        15CH2    7CH2
          |        |
        16CH3    8CH3
```

FORMULA	C16H35N			MOL WT	241.46
SOLVENT	C6D6				
ORIG ST	C4H12SI			TEMP	AMB
53.72	40.21	31.88	29.51	23.50	14.26
1/3	2/2	3/3	4/3	5/3	6/4
25.02	11.22	53.72	40.21	31.88	29.51
7/3	8/4	9/3	10/2	11/3	12/3
23.50	14.26	25.02	11.22		
13/3	14/4	15/3	16/4		

H.EGGERT,C.DJERASSI
J AM CHEM SOC   95, 3710 (1973)

1180   P 000600 DIOCTYLAMINE

```
   16    14   12   10
   H3C   CH2  CH2  CH2
    \ / \ / \ / \
    CH2 CH2 CH2 CH2
    15  13   11   9\
                    N-H
     7    5    3   1/
    CH2  CH2  CH2 CH2
   / \ / \ / \ /
  H3C  CH2 CH2 CH2
   8    6    4    2
```

FORMULA	C16H35N			MOL WT	241.46
SOLVENT	C6D6				
ORIG ST	C4H12SI			TEMP	AMB
50.40	30.06	27.84	29.92	29.80	32.30
1/3	2/3	3/3	4/3	5/3	6/3
23.03	14.23	50.40	30.06	27.84	29.92
7/3	8/4	9/3	10/3	11/3	12/3
29.80	32.30	23.03	14.23		
13/3	14/3	15/3	16/4		

H.EGGERT,C.DJERASSI
J AM CHEM SOC   95, 3710 (1973)

1181   P 000600 TRIHEXYLAMINE

```
 1    2    3    4    5    6
CH2-CH2-CH2-CH2-CH2-CH3
 | 7   8    9   10   11   12
N-CH2-CH2-CH2-CH2-CH2-CH3
 |13  14   15   16   17   18
CH2-CH2-CH2-CH2-CH2-CH3
```

FORMULA	C18H39N			MOL WT	269.52
SOLVENT	C6D6				
ORIG ST	C4H12SI			TEMP	AMB
54.54	27.96	27.51	32.27	23.07	14.22
1/3	2/3	3/3	4/3	5/3	6/4
54.54	27.96	27.51	32.27	23.07	14.22
7/3	8/3	9/3	10/3	11/3	12/4
54.54	27.96	27.51	32.27	23.07	14.22
13/3	14/3	15/3	16/3	17/3	18/4

H.EGGERT,C.DJERASSI
J AM CHEM SOC   95, 3710 (1973)

1182   P 000600 N-BUTYLDI-2-ETHYLHEXYLAMINE

```
    19   20
    CH2-CH3
 13  |  15   16   17   18
CH2-CH-CH2-CH2-CH2-CH3
 |  14 2    3    4
 N-CH2-CH2-CH2-CH3
 |  1 6    8    9   10
CH2-CH-CH2-CH2-CH2-CH3
 5   | 7
    CH2-CH3
    11   12
```

FORMULA	C20H43N			MOL WT	297.57
SOLVENT	C6D6				
ORIG ST	C4H12SI			TEMP	AMB
55.14	29.83	21.00	14.25	60.19	37.93
1/3	2/3	3/3	4/4	5/3	6/2
31.65	29.48	23.55	14.25	24.77	10.98
7/3	8/3	9/3	10/4	11/3	12/4
60.19	37.93	31.65	29.48	23.55	14.25
13/3	14/2	15/3	16/3	17/3	18/4
24.77	10.98				
19/3	20/4				

H.EGGERT,C.DJERASSI
J AM CHEM SOC   95, 3710 (1973)

1183   P 000600   TRIOCTYLAMINE

```
                    8CH3
1   2   3   4   5   6   |7    FORMULA   C24H51N              MOL WT  353.68
CH2-CH2-CH2-CH2-CH2-CH2-CH2  SOLVENT   C6D6
| 9  10  11  12  13  14       ORIG ST   C4H12SI             TEMP      AMB
N-CH2-CH2-CH2-CH2-CH2-CH2
|17                      |        54.50   27.87   27.87   30.00   29.75   32.25
CH2                   15CH2        1/3     2/3     3/3     4/3     5/3     6/3
|18                      |        22.98   14.28   54.50   27.87   27.87   30.00
CH2                   16CH3         7/3     8/4     9/3    10/3    11/3    12/3
|19 20  21  22  23  24            29.75   32.25   22.98   14.28   54.50   27.87
CH2-CH2-CH2-CH2-CH2-CH3           13/3    14/3    15/3    16/4    17/3    18/3
                                  27.87   30.00   29.75   32.25   22.98   14.28
                                  19/3    20/3    21/3    22/3    23/3    24/4
```

H.EGGERT,C.DJERASSI
J AM CHEM SOC   95, 3710 (1973)

1184   P 000600   TRIDODECYLAMINE

```
    12  11  10   9   8        FORMULA   C36H75N              MOL WT  522.01
    CH3-CH2-CH2-CH2-CH2       SOLVENT   C6D6
1   2   3   4   5   6   |7    ORIG ST   C4H12SI             TEMP      AMB
CH2-CH2-CH2-CH2-CH2-CH2-CH2
| 13 14  15  16  17  19  19       54.48   27.96   27.79   30.05   30.05   30.05
N-CH2-CH2-CH2-CH2-CH2-CH2-CH2      1/3     2/3     3/3     4/3     5/3     6/3
|25      24  23  22  21  |20      30.05   30.05   29.72   32.36   22.94   14.18
CH2      CH3-CH2-CH2-CH2-CH2       7/3     8/3     9/3    10/3    11/3    12/4
|26 27  28  29  30  31  32        54.48   27.96   27.79   30.05   30.05   30.05
CH2-CH2-CH2-CH2-CH2-CH2-CH2       13/3    14/3    15/3    16/3    17/3    18/3
        36  35  34      33        30.05   30.05   29.72   32.36   22.94   14.18
        CH3-CH2-CH2-CH2           19/3    20/3    21/3    22/3    23/3    24/4
                                  54.48   27.96   27.79   30.05   30.05   30.05
                                  25/3    26/3    27/3    28/3    29/3    30/3
                                  30.05   30.05   29.72   32.36   22.94   14.18
                                  31/3    32/3    33/3    34/3    35/3    36/4
```

H.EGGERT,C.DJERASSI
J AM CHEM SOC   95, 3710 (1973)

1185   P 000700   TRIMETHYLVINYLAMMONIUM BROMIDE

```
                    FORMULA   C5H12BRN             MOL WT  166.06
                    SOLVENT   H2O
  1 2               ORIG ST   C4H12SI             TEMP      AMB
H2C=CH
  3 |+  -              112.70  143.10   55.30   55.30   55.30
H3C-N  BR               1/3     2/2     3/4     4/4     5/4
   /|
H3C  CH3            L.F.JOHNSON,W.C.JANKOWSKI
  4  5              CARBON-13 NMR SPECTRA,JOHN
                    WILEY AND SONS,NEW YORK            136 (1972)
```

1186   P 000700   CHOLINE CHLORIDE

```
   4                FORMULA   C5H14CLNO            MOL WT  139.63
   CH3              SOLVENT   H2O
   |+         -     ORIG ST   C4H12SI             TEMP      AMB
H3C-N-CH2-CH2-OH  CL
  3 | 2   1           56.60   68.30   54.80   54.80   54.80
  5CH3                 1/3     2/3     3/4     4/4     5/4
```

L.F.JOHNSON,W.C.JANKOWSKI
CARBON-13 NMR SPECTRA,JOHN
WILEY AND SONS,NEW YORK   142 (1972)

1187   P 001100 2-PROPYNOL

```
                        FORMULA   C3H4O              MOL WT   56.06
       1 2 3           SOLVENT   CDCL3
     HC≡C-CH2-OH       ORIG ST   C4H12SI            TEMP       AMB


                        73.80     82.00    50.40
                        1/2       2/1      3/3


                        M.T.W.HEARN
                        TETRAHEDRON                 32,  115 (1976)
```

1188   P 001100 2-PROPYN-1-OL

```
                        FORMULA   C3H4O              MOL WT   56.06
                        SOLVENT   C4H8O2
       1 2 3           ORIG ST   C4H12SI            TEMP       AMB
     HC≡C-CH2-OH
                        73.80     83.00    50.00
                        1/2       2/1      3/3


                        L.F.JOHNSON,W.C.JANKOWSKI
                        CARBON-13 NMR SPECTRA,JOHN
                        WILEY AND SONS,NEW YORK         19 (1972)
```

1189   P 001100 3-HYDROXY-1-PROPENE

```
                        FORMULA   C3H6O              MOL WT   58.08
                        SOLVENT   CDCL3
                        ORIG ST   C4H12SI            TEMP       AMB
     CH2=CH-CH2-OH
      1    2   3        114.90   137.50    63.40
                        1/3      2/2       3/3


                        L.F.JOHNSON,W.C.JANKOWSKI
                        CARBON-13 NMR SPECTRA,JOHN
                        WILEY AND SONS,NEW YORK         30 (1972)
```

1190   P 001100 2-PROPEN-1-OL   ALLYL ALCOHOL

```
          1
     H      CH2-OH      FORMULA   C3H6O              MOL WT   58.08
       \   /            SOLVENT   C4H8O2
        3C=C2           ORIG ST   C4H12SI            TEMP       313
       /   \
     H       H          63.30    139.10   113.70
                        1/3      2/2       3/3


                        H.BROUWER,J.B.STOTHERS
                        CAN J CHEM                   50, 1361 (1972)
```

1191   P 001100 HYDROXYDIMETHYLCARBENIUM ION

```
          3
         H3C           FORMULA   C3H7O              MOL WT   59.09
           \  +        SOLVENT   CLFO2S/F5SB
            C1         ORIG ST   CS2                TEMP       198
           / \
         H3C   OH       248.20    30.50    29.20
          2             1/1       2/4      3/4


                        G.A.OLAH,P.W.WESTERMAN
                        J AM CHEM SOC                95, 7530 (1973)
```

1192   P 001100   N-PROPANOL

	FORMULA	C3H8O		MOL WT	60.10
	SOLVENT	C4H8O2			
	ORIG ST	C4H12SI		TEMP	AMB

HO-CH2-CH2-CH3   64.00   26.30   10.50
   1   2   3   1/3   2/3   3/4

L.F.JOHNSON,W.C.JANKOWSKI
CARBON-13 NMR SPECTRA,JOHN
WILEY AND SONS,NEW YORK   39 (1972)

1193   P 001100   PROPYLENE GLYCOL

	FORMULA	C3H8O2		MOL WT	76.10
	SOLVENT	CDCL3			
	ORIG ST	C4H12SI		TEMP	AMB

      OH
      |
HO-CH2-CH-CH3   67.70   68.20   18.70
  1   2   3   1/3   2/2   3/4

L.F.JOHNSON,W.C.JANKOWSKI
CARBON-13 NMR SPECTRA,JOHN
WILEY AND SONS,NEW YORK   40 (1972)

1194   P 001100   2-BUTYN-1-OL

	FORMULA	C4H6O		MOL WT	70.09
	SOLVENT	C4H8O2			
	ORIG ST	C4H12SI		TEMP	AMB

HO-CH2-C≡C-CH3   50.50   80.00   78.90   3.20
  1   2 3 4   1/3   2/1   3/1   4/4

L.F.JOHNSON,W.C.JANKOWSKI
CARBON-13 NMR SPECTRA,JOHN
WILEY AND SONS,NEW YORK   58 (1972)

1195   P 001100   2-BUTYN-1-OL

	FORMULA	C4H6O		MOL WT	70.09
	SOLVENT	CDCL3			
1 2 3 4	ORIG ST	C4H12SI		TEMP	AMB

H3C-C≡C-CH2-OH   3.60   81.60   77.70   50.40
               1/4   2/1   3/1   4/3

M.T.W.HEARN
TETRAHEDRON   32,   115 (1976)

1196   P 001100   3-BUTYN-1-OL

	FORMULA	C4H6O		MOL WT	70.09
	SOLVENT	CDCL3			
1 2 3 4	ORIG ST	C4H12SI		TEMP	AMB

HC≡C-CH2-CH2-OH   70.50   80.70   22.90   60.70
               1/2   2/1   3/3   4/3

M.T.W.HEARN
TETRAHEDRON   32,   115 (1976)

1197 P 001100 3-BUTYN-2-OL

```
        OH
  1 2 3/
 HC≡C-CH
      \4
      CH3
```

FORMULA	C4H6O			MOL WT	70.09
SOLVENT	CDCL3				
ORIG ST	C4H12SI			TEMP	AMB

72.00	85.80	57.70	24.00
1/2	2/1	3/2	4/4

M.T.W.HEARN
TETRAHEDRON   32, 115 (1976)

1198 P 001100 2-BUTYNE-1,4-DIOL

```
  1    2 3 4
 HO-CH2-C≡C-CH2-OH
```

FORMULA	C4H6O2	MOL WT	86.09
SOLVENT	CDCL3		
ORIG ST	C4H12SI	TEMP	AMB

50.30	83.70
1/3	2/1

M.T.W.HEARN
TETRAHEDRON   32, 115 (1976)

1199 P 001100 E-2-BUTEN-1-OL

```
        1
 H      CH2-OH
  \    /
   3C=C2
  /    \
 H3C    H
  4
```

FORMULA	C4H8O			MOL WT	72.11
SOLVENT	C4H8O2				
ORIG ST	C4H12SI			TEMP	313

62.90	132.10	126.00	17.30
1/3	2/2	3/2	4/4

H.BROUWER,J.B.STOTHERS
CAN J CHEM   50, 1361 (1972)

1200 P 001100 Z-2-BUTEN-1-OL

```
  4     1
 H3C    CH2-OH
  \    /
   3C=C2
  /    \
 H      H
```

FORMULA	C4H8O			MOL WT	72.11
SOLVENT	C4H8O2				
ORIG ST	C4H12SI			TEMP	313

57.90	131.40	125.30	12.70
1/3	2/2	3/2	4/4

H.BROUWER,J.B.STOTHERS
CAN J CHEM   50, 1361 (1972)

1201 P 001100 2-METHYL-2-PROPEN-1-OL

```
        1
 H      CH2-OH
  \    /
   3C=C2
  /    \
 H      CH3
        4
```

FORMULA	C4H8O			MOL WT	72.11
SOLVENT	C4H8O2				
ORIG ST	C4H12SI			TEMP	313

66.20	146.20	108.70	18.90
1/3	2/1	3/3	4/4

H.BROUWER,J.B.STOTHERS
CAN J CHEM   50, 1361 (1972)

1202   P 001100 1,3-BUTANEDIOL

```
                          FORMULA   C4H10O              MOL WT   74.12
                          SOLVENT   CDCL3
            OH            ORIG ST   C4H12SI             TEMP       AMB
            |
   HO-CH2-CH2-CH-CH3        23.40    66.30    40.60    60.00
    4    3   2  1            1/4      2/2      3/3      4/3
```

L.F.JOHNSON,W.C.JANKOWSKI
CARBON-13 NMR SPECTRA,JOHN
WILEY AND SONS,NEW YORK   91 (1972)

1203   P 001100 SEC-BUTYL ALCOHOL

```
                          FORMULA   C4H10O              MOL WT   74.12
                          SOLVENT   CDCL3
          OH              ORIG ST   C4H12SI             TEMP       AMB
          |
   H3C-CH2-CH-CH3           22.70    69.20    32.00    10.00
    4   3   2  1            1/4      2/2      3/3      4/4
```

L.F.JOHNSON,W.C.JANKOWSKI
CARBON-13 NMR SPECTRA,JOHN
WILEY AND SONS,NEW YORK   89 (1972)

1204   P 001100 TERT-BUTYL ALCOHOL

```
        3                 FORMULA   C4H10O              MOL WT   74.12
       CH3                SOLVENT   CDCL3
        |                 ORIG ST   C4H12SI             TEMP       AMB
   HO-C-CH3
    1|  2                  68.90    31.20    31.20    31.20
    4CH3                   1/1      2/4      3/4      4/4
```

L.F.JOHNSON,W.C.JANKOWSKI
CARBON-13 NMR SPECTRA,JOHN
WILEY AND SONS,NEW YORK   88 (1972)

1205   P 001100 ISOBUTYL ALCOHOL

```
       3CH3               FORMULA   C4H10O              MOL WT   74.12
        |                 SOLVENT   CDCL3
   HO-CH2-CH               ORIG ST   C4H12SI            TEMP       AMB
    1   21
       4CH3                 69.40    30.80    18.90    18.90
                            1/3      2/2      3/4      4/4
```

L.F.JOHNSON,W.C.JANKOWSKI
CARBON-13 NMR SPECTRA,JOHN
WILEY AND SONS,NEW YORK   90 (1972)

1206   P 001100 BUTANE-1,4-DIOL

```
                          FORMULA   C4H10O2             MOL WT   90.12
                          SOLVENT   D2O
    1   2   3   4         ORIG ST   C4H12SI             TEMP       AMB
   HO-CH2-CH2-CH2-CH2-OH
                            62.60    29.20
                            1/3      2/3
```

R.G.S.RITCHIE,N.CYR,B.KORSCH,H.J.KOCH,
A.S.PERLIN
CAN J CHEM   53, 1424 (1975)

1207   P 001100 MESO-ERYTHRITOL

```
        HO  OH
   1   2| |  4
HO-CH2-C-C-CH2-OH
     |  |3
     H  H
```

FORMULA	C4H10O4		MOL WT	122.12
SOLVENT	D2O			
ORIG ST	C4H12SI		TEMP	AMB

63.90	73.20
1/3	2/2

R.G.S.RITCHIE,N.CYR,B.KORSCH,H.J.KOCH,
A.S.PERLIN
CAN J CHEM   53, 1424 (1975)

1208   P 001100 D-THREITOL

```
        HO  H
   1   2| |  4
HO-CH2-C-C-CH2-OH
     |  |3
     H  OH
```

FORMULA	C4H10O4		MOL WT	122.12
SOLVENT	D2O			
ORIG ST	C4H12SI		TEMP	AMB

63.90	72.80
1/3	2/2

R.G.S.RITCHIE,N.CYR,B.KORSCH,H.J.KOCH,
A.S.PERLIN
CAN J CHEM   53, 1424 (1975)

1209   P 001100 2,4-PENTADIYN-1-OL

```
   1    2 3 4 5
HO-CH2-C≡C-C≡CH
```

FORMULA	C5H4O				MOL WT	80.09
SOLVENT						
ORIG ST	C4H12SI				TEMP	AMB

50.80	74.70	69.70	67.50	68.80
1/T	2/S	3/S	4/S	5/D

R.ZEISBERG,F.BOHLMANN
CHEM BER   107, 3800 (1974)

1210   P 001100 PENTA-1,4-DIYN-3-OL

```
        OH
   1 2  |  4 5
HC≡C-CH-C≡CH
      3
```

FORMULA	C5H4O		MOL WT	80.09
SOLVENT	CDCL3			
ORIG ST	C4H12SI		TEMP	AMB

72.90	80.40	50.90
1/2	2/1	3/2

M.T.W.HEARN
TETRAHEDRON   32, 115 (1976)

1211   P 001100 1-HYDROXY-2-PENTEN-4-YNE

FORMULA	C5H6O				MOL WT	82.10
SOLVENT						
ORIG ST	C4H12SI				TEMP	AMB

```
   1   2  3  4 5
HO-CH2-CH=CH-C≡CH
```

60.70	137.90	118.90	82.00	82.50
1/T	2/D	3/D	4/S	5/D

R.ZEISBERG,F.BOHLMANN
CHEM BER   107, 3800 (1974)

1212   P 001100 1-PENTEN-4-YN-3-OL

```
        ЭH
   1 2 I  4   5
  HC≡C-CH-CH=CH2
        3
```

FORMULA	C5H6O			MOL WT	82.10
SOLVENT	CDCL3				
ORIG ST	C4H12SI			TEMP	AMB

74.60	82.80	62.60	136.60	116.70
1/2	2/1	3/2	4/2	5/3

M.T.W.HEARN
TETRAHEDRON   32, 115 (1976)

1213   P 001100 2-PENTEN-4-YN-1-OL

```
  1 2 3  4   5
  HC≡C-CH=CH-CH2-OH
```

FORMULA	C5H6O			MOL WT	82.10
SOLVENT	CDCL3				
ORIG ST	C4H12SI			TEMP	AMB

78.20	81.80	108.90	143.70	62.10
1/2	2/1	3/2	4/2	5/3

M.T.W.HEARN
TETRAHEDRON   32, 115 (1976)

1214   P 001100 3-HYDROXY-3-METHYLBUTYNE-1

```
    4CH3
    I 3
  HC≡C-C-ЭH
  1 2 I
    CH3
    5
```

FORMULA	C5H8O			MOL WT	84.12
SOLVENT	C4H8O2				
ORIG ST	C4H12SI			TEMP	AMB

70.00	89.60	64.00	31.30	31.30
1/2	2/1	3/1	4/4	5/4

L.F.JOHNSON,W.C.JANKOWSKI
CARBON-13 NMR SPECTRA,JOHN
WILEY AND SONS,NEW YORK   112 (1972)

1215   P 001100 2-METHYL-3-BUTYN-2-OL

```
      4
     CH3
  1 2 3/
  HC≡C-C-OH
      \
     5CH3
```

FORMULA	C5H8O			MOL WT	84.12
SOLVENT	CDCL3				
ORIG ST	C4H12SI			TEMP	AMB

70.20	88.80	64.90	31.20
1/2	2/1	3/1	4/4

M.T.W.HEARN
TETRAHEDRON   32, 115 (1976)

1216   P 001100 1-PENTYN-3-OL

```
        ЭH
   1 2 I  4   5
  HC≡C-CH-CH2-CH3
        3
```

FORMULA	C5H8O			MOL WT	84.12
SOLVENT	CDCL3				
ORIG ST	C4H12SI			TEMP	AMB

72.90	84.90	63.30	30.60	9.40
1/2	2/1	3/2	4/3	5/4

M.T.W.HEARN
TETRAHEDRON   32, 115 (1976)

1217   P 001100   3-METHYL-2-BUTEN-1-OL

```
    4       1
   H3C    CH2-OH         FORMULA   C5H10O              MOL WT    86.13
     \    /              SOLVENT   C4H8O2
      3C=C2              ORIG ST   C4H12SI             TEMP        313
     /    \
   H3C     H                58.80  125.70  133.70   17.60   25.40
    5                       1/3     2/2     3/1      4/4     5/4

                         H.BROUWER,J.B.STOTHERS
                         CAN J CHEM                    50, 1361 (1972)
```

1218   P 001100   E-2-METHYL-2-BUTEN-1-OL

```
           1
    H     CH2-OH         FORMULA   C5H10O              MOL WT    86.13
     \    /              SOLVENT   C4H8O2
      3C=C2              ORIG ST   C4H12SI             TEMP        313
     /    \
   H3C     CH3              68.10  136.70  118.80   12.90   12.90
    4       5               1/3     2/1     3/2      4/4     5/4

                         H.BROUWER,J.B.STOTHERS
                         CAN J CHEM                    50, 1361 (1972)
```

1219   P 001100   Z-2-METHYL-2-BUTEN-1-OL

```
    4       1
   H3C    CH2-OH         FORMULA   C5H10O              MOL WT    86.13
     \    /              SOLVENT   C4H8O2
      3C=C2              ORIG ST   C4H12SI             TEMP        313
     /    \
    H       CH3             60.50  136.50  120.80   12.80   21.90
            5               1/3     2/1     3/2      4/4     5/4

                         H.BROUWER,J.B.STOTHERS
                         CAN J CHEM                    50, 1361 (1972)
```

1220   P 001100   ISOPENTYL ALCOHOL

```
                         FORMULA   C5H12O              MOL WT    88.15
          5              SOLVENT   CDCL3
          CH3            ORIG ST   C4H12SI             TEMP         AMB
          /
  HO-CH2-CH2-CH            60.70   41.70   24.80   22.60   22.60
   1   2   3  \            1/3     2/3     3/2      4/4     5/4
          CH3
          4              L.F.JOHNSON,W.C.JANKOWSKI
                         CARBON-13 NMR SPECTRA,JOHN
                         WILEY AND SONS,NEW YORK           139 (1972)
```

1221   P 001100   3-METHOXY-1-BUTANOL

```
          5              FORMULA   C5H12O2             MOL WT   104.15
          O-CH3          SOLVENT   CDCL3
          |              ORIG ST   C4H12SI             TEMP         AMB
  HO-CH2-CH2-CH-CH3
   1   2   3  4             59.60   39.10   75.30   18.90   55.80
                           1/3     2/3     3/2      4/4     5/4

                         L.F.JOHNSON,W.C.JANKOWSKI
                         CARBON-13 NMR SPECTRA,JOHN
                         WILEY AND SONS,NEW YORK           140 (1972)
```

1222   P 001100   4-HEXEN-1-YN-3-OL

```
        OH
 1 2 I   4  5  6
HC≡C-CH-CH=CH-CH3
        3
```

FORMULA	C6H8O			MOL WT	96.13
SOLVENT	CDCL3				
ORIG ST	C4H12SI			TEMP	AMB

| 74.00 | 83.60 | 62.20 | 129.90 | 128.60 | 17.40 |
| 1/2 | 2/1 | 3/2 | 4/2 | 5/2 | 6/4 |

M.T.W.HEARN
TETRAHEDRON   32, 115 (1976)

1223   P 001100   3-METHYL-1-PENTYN-3-OL

```
         OH
 1 2 3/4   5
HC≡C-C-CH2-CH3
       \
        CH3
        6
```

FORMULA	C6H10O			MOL WT	98.15
SOLVENT	CDCL3				
ORIG ST	C4H12SI			TEMP	AMB

| 71.30 | 87.70 | 68.50 | 36.30 | 9.00 | 29.10 |
| 1/2 | 2/1 | 3/1 | 4/3 | 5/4 | 6/4 |

M.T.W.HEARN
TETRAHEDRON   32, 115 (1976)

1224   P 001100   2-METHYL-3-PENTYN-2-OL

```
      5
      CH3
 1 2 3 4/
H3C-C≡C-C-OH
       \
        CH3
        6
```

FORMULA	C6H10O			MOL WT	98.15
SOLVENT	CDCL3				
ORIG ST	C4H12SI			TEMP	AMB

| 3.50 | 77.90 | 84.40 | 65.20 | 31.60 | |
| 1/4 | 2/1 | 3/1 | 4/1 | 5/4 | |

M.T.W.HEARN
TETRAHEDRON   32, 115 (1976)

1225   P 001100   4-METHYL-1-PENTYN-3-OL

```
      5
   OH   CH3
 1 2 I  4/
HC≡C-CH-CH
    3   \6
        CH3
```

FORMULA	C6H10O			MOL WT	98.15
SOLVENT	CDCL3				
ORIG ST	C4H12SI			TEMP	AMB

| 73.50 | 83.70 | 67.40 | 34.20 | 18.20 | 17.30 |
| 1/2 | 2/1 | 3/2 | 4/2 | 5/4 | 6/4 |

M.T.W.HEARN
TETRAHEDRON   32, 115 (1976)

1226   P 001100   3-HEXYNE-2,5-DIOL

```
   OH     OH
 1 I  3 4 I  6
H3C-CH-C≡C-CH-CH3
    2      5
```

FORMULA	C6H10O2		MOL WT	114.15
SOLVENT	CDCL3			
ORIG ST	C4H12SI		TEMP	AMB

| 24.10 | 57.80 | 85.60 |
| 1/4 | 2/2 | 3/1 |

M.T.W.HEARN
TETRAHEDRON   32, 115 (1976)

1227   P 001100 2-ETHYL-1-BUTANOL

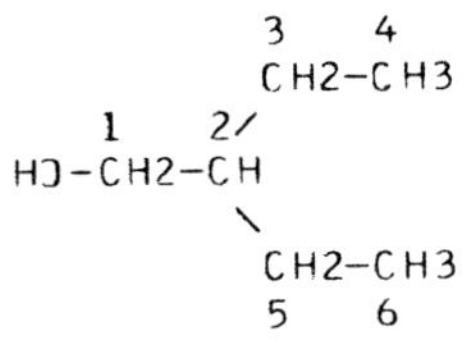

```
        3   4
        CH2-CH3
    1    2/
   HO-CH2-CH
             \
             CH2-CH3
             5   6
```

FORMULA	C6H14O			MOL WT	102.18
SOLVENT	CDCL3				
ORIG ST	C4H12SI			TEMP	AMB

64.60	43.60	23.00	11.10	23.00	11.10
1/3	2/2	3/3	4/4	5/3	6/4

L.F.JOHNSON,W.C.JANKOWSKI
CARBON-13 NMR SPECTRA,JOHN
WILEY AND SONS,NEW YORK   215 (1972)

1228   P 001100 4-METHYL-2-PENTANOL

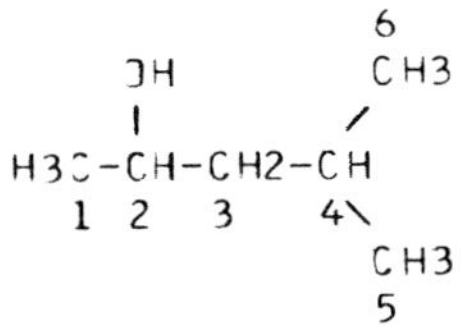

```
            6
    OH      CH3
    |      /
   H3C-CH-CH2-CH
    1  2  3   4\
               CH3
               5
```

FORMULA	C6H14O			MOL WT	102.18
SOLVENT	CDCL3				
ORIG ST	C4H12SI			TEMP	AMB

23.90	65.80	48.70	24.80	23.10	22.40
1/4	2/2	3/3	4/2	5/4	6/4

L.F.JOHNSON,W.C.JANKOWSKI
CARBON-13 NMR SPECTRA,JOHN
WILEY AND SONS,NEW YORK   214 (1972)

1229   P 001100 3-HEPTYN-2-OL

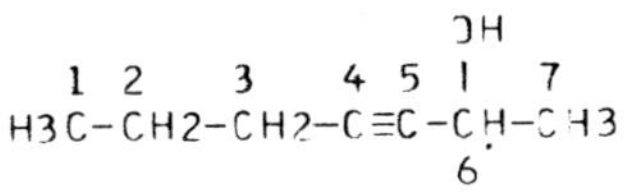

```
                    OH
   1 2  3   4 5 |  7
  H3C-CH2-CH2-C≡C-CH-CH3
                    6
```

FORMULA	C7H12O			MOL WT	112.17
SOLVENT	CDCL3				
ORIG ST	C4H12SI			TEMP	AMB

13.50	22.10	20.70	82.60	84.30	58.40
1/4	2/3	3/3	4/1	5/1	6/2
24.70					
7/4					

M.T.W.HEARN
TETRAHEDRON   32, 115 (1976)

1230   P 001100 4-ETHOXY-2-METHYL-3-BUTYN-2-OL

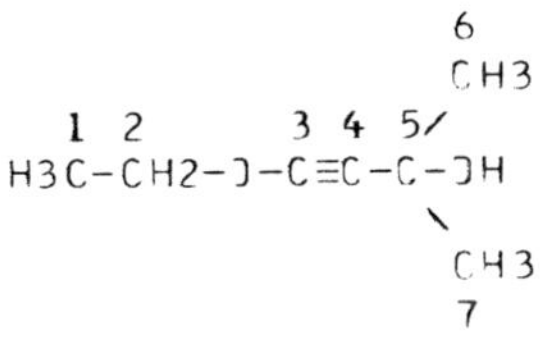

```
            6
            CH3
   1 2    3 4 5/
  H3C-CH2-O-C≡C-C-OH
                  \
                  CH3
                  7
```

FORMULA	C7H12O2			MOL WT	128.17
SOLVENT	CDCL3				
ORIG ST	C4H12SI			TEMP	AMB

14.30	74.30	91.30	43.30	65.10	32.10
1/4	2/3	3/1	4/1	5/1	6/4

M.T.W.HEARN
TETRAHEDRON   32, 115 (1976)

1231   P 001100 4-TRIMETHYLSILYL-3-BUTYNOL

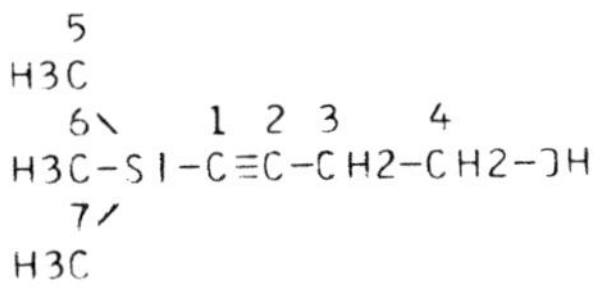

```
     5
    H3C
     6\   1 2 3   4
   H3C-SI-C≡C-CH2-CH2-OH
     7/
    H3C
```

FORMULA	C7H14OSI			MOL WT	142.28
SOLVENT	CDCL3				
ORIG ST	C4H12SI			TEMP	AMB

84.00	103.90	24.50	61.40	0.00	
1/1	2/1	3/3	4/3	5/4	

M.T.W.HEARN
TETRAHEDRON   32, 115 (1976)

1232 P 001100 2,2-DIETHYL-1,3-PROPANEDIOL

```
      1   6   7
  HO-H2C   CH2-CH3
       \ /
       2C
       / \
  HO-H2C   CH2-CH3
    5   3   4
```

FORMULA C7H16O2				MOL WT 132.20	
SOLVENT CDCL3					
ORIG ST C4H12SI			TEMP		AMB
67.60	40.90	22.20	7.10	67.60	22.20
1/3	2/1	3/3	4/4	5/3	6/3
7.10					
7/4					

L.F.JOHNSON,W.C.JANKOWSKI
CARBON-13 NMR SPECTRA,JOHN
WILEY AND SONS,NEW YORK   278 (1972)

1233 P 001100 OCTA-1,4,7-TRIYNE-3,6-DIOL

```
       OH      OH
 1 2 |  4 5 |  7 8
 HC≡C-CH-C≡C-CH-C≡CH
       3       6
```

FORMULA C8H6O2			MOL WT 134.14
SOLVENT CDCL3			
ORIG ST C4H12SI		TEMP	AMB
72.80	80.50	51.20	81.00
1/2	2/1	3/2	4/1

M.T.W.HEARN
TETRAHEDRON   32, 115 (1976)

1234 P 001100 2-ETHYL-1-HEXANOL

```
    7   8
   CH2-CH3
    |
HO-CH2-CH-CH2-CH2-CH2-CH3
 1   2   3   4   5   6
```

FORMULA C8H18O				MOL WT 130.23	
SOLVENT CDCL3					
ORIG ST C4H12SI			TEMP		AMB
65.10	42.10	30.30	29.30	23.20	14.10
1/3	2/2	3/3	4/3	5/3	6/4
23.50	11.10				
7/3	8/4				

L.F.JOHNSON,W.C.JANKOWSKI
CARBON-13 NMR SPECTRA,JOHN
WILEY AND SONS,NEW YORK   322 (1972)

1235 P 001100 2-OCTANOL

```
    OH
    |   4   6   8
 H3C-CH  CH2 CH2 CH3
  1 2 \ / \ / \ /
     CH2 CH2 CH2
      3   5   7
```

FORMULA C8H18O				MOL WT 130.23	
SOLVENT CDCL3					
ORIG ST C4H12SI			TEMP		AMB
23.40	67.80	39.40	25.90	29.50	32.00
1/4	2/2	3/3	4/3	5/3	6/3
22.77	14.10				
7/3	8/4				

L.F.JOHNSON,W.C.JANKOWSKI
CARBON-13 NMR SPECTRA,JOHN
WILEY AND SONS,NEW YORK   323 (1972)

1236 P 001100 2,2,4-TRIMETHYL-1,3-PENTANEDIOL

```
  5      7
 H3C OH CH3
  |  |  |
 4CH-CH-C-CH2-OH
  |  3 21 1
 H3C    CH3
  6      8
```

FORMULA C8H18O				MOL WT 130.23	
SOLVENT CDCL3					
ORIG ST C4H12SI			TEMP		AMB
73.10	39.10	83.00	29.20	23.30	23.30
1/3	2/1	3/2	4/2	5/4	6/4
16.70	19.70				
7/4	8/4				

L.F.JOHNSON,W.C.JANKOWSKI
CARBON-13 NMR SPECTRA,JOHN
WILEY AND SONS,NEW YORK   324 (1972)

1237   P 001100 1-PHENYL-2-PROPYNOL

```
        5 6
   OH   C=C
 1 2 |  4/   \7
HC≡C-CH-C    C
   3     \   //
         C-C
         9 8
```

FORMULA	C9H8O			MOL WT	132.16
SOLVENT	CDCL3				
ORIG ST	C4H12SI			TEMP	AMB

74.90	83.60	63.60	139.90	126.60	128.30
1/2	2/1	3/2	4/1	5/2	6/2
128.30					
7/2					

M.T.W.HEARN
TETRAHEDRON   32, 115 (1976)

1238   P 001100 2-PHENYL-3-BUTYN-2-OL

```
       OH
   1 2 3/4
HC≡C-C-CH3
      \
       5C=C6
    10/    \
    C       C7
    \       //
     C-C
     9 8
```

FORMULA	C10H10O			MOL WT	146.19
SOLVENT	CDCL3				
ORIG ST	C4H12SI			TEMP	AMB

73.10	87.20	69.70	33.10	144.90	124.90
1/2	2/1	3/1	4/4	5/1	6/2
128.20	127.60				
7/2	8/2				

M.T.W.HEARN
TETRAHEDRON   32, 115 (1976)

1239   P 001100 1-DECANOL

```
  1    2    3    4    5    6
HO-CH2-CH2-CH2-CH2-CH2-CH2
                         |
     H3C-CH2-CH2-CH2
     10  9    8    7
```

FORMULA	C10H22O			MOL WT	158.29
SOLVENT	CDCL3				
ORIG ST	C4H12SI			TEMP	AMB

62.60	32.80	25.90	29.70	29.70	29.70
1/3	2/3	3/3	4/3	5/3	6/3
29.40	32.00	22.70	14.10		
7/3	8/3	9/3	10/3		

L.F.JOHNSON,W.C.JANKOWSKI
CARBON-13 NMR SPECTRA,JOHN
WILEY AND SONS,NEW YORK   412 (1972)

1240   P 001100 1-PHENYL-1-PENTEN-4-YN-3-OL

```
           7 8
   OH      C=C
 1 2 |  4 5 6/   \9
HC≡C-CH-CH=CH-C    C
   3            \   //
                C-C
                11 10
```

FORMULA	C11H10O			MOL WT	158.20
SOLVENT	CDCL3				
ORIG ST	C4H12SI			TEMP	AMB

74.70	82.90	62.40	128.60	127.50	135.80
1/2	2/1	3/2	4/2	5/2	6/1
126.70	128.40	131.90			
7/2	8/2	9/2			

M.T.W.HEARN
TETRAHEDRON   32, 115 (1976)

1241   P 001100 1,1-DIPHENYL-3-BUTYN-2-OL

```
      7C=C8
     /    \
    6C      C9
    \\      //
   OH  5C-C10
 1 2 |  4/
HC≡C-CH-CH
   3    \
       11C-C12
       //    \
      16C      C13
       \      /
       15C=C14
```

FORMULA	C16H14O			MOL WT	222.29
SOLVENT	CDCL3				
ORIG ST	C4H12SI			TEMP	AMB

75.10	83.40	64.70	57.30	140.00	128.30
1/2	2/1	3/2	4/2	5/1	6/2
128.80	126.80	140.50			
7/2	8/2	11/1			

M.T.W.HEARN
TETRAHEDRON   32, 115 (1976)

1242   P 001200 3-BROMO-2-PROPYNOL

                                    FORMULA   C3H3BRO   MOL WT   134.96
                                    SOLVENT   CDCL3
        1 2 3   ORIG ST   C4H12SI   TEMP   A8B
     BR-C≡C-CH2-OH
                                       45.70   78.20   51.50
                                       1/1   2/1   3/3

                                    M.T.W.HEARN
                                    TETRAHEDRON   32, 115 (1976)

1243   P 001200 E-3-BROMO-2-PROPEN-1-OL

        H   CH2-OH   FORMULA   C3H5BRO   MOL WT   136.98
         \   /   SOLVENT   C4H8O2
          C=C   ORIG ST   C4H12SI   TEMP   313
         /   \
       BR   H   62.50   138.40   106.80
                                       1/3   2/2   3/2

                                    H.BROUWER,J.B.STOTHERS
                                    CAN J CHEM   50, 1361 (1972)

1244   P 001200 Z-3-BROMO-2-PROPEN-1-OL
              1
       BR   CH2-OH   FORMULA   C3H5BRO   MOL WT   136.98
         \   /   SOLVENT   C4H8O2
        3C=C2   ORIG ST   C4H12SI   TEMP   313
         /   \
        H   H   60.70   135.90   107.50
                                       1/3   2/2   3/2

                                    H.BROUWER,J.B.STOTHERS
                                    CAN J CHEM   50, 1361 (1972)

1245   P 001200 E-3-CHLORO-2-PROPEN-1-OL
              1
       H   CH2-OH   FORMULA   C3H5CLO   MOL WT   92.53
        \   /   SOLVENT   C4H8O2
       3C=C2   ORIG ST   C4H12SI   TEMP   313
        /   \
       CL   H   60.90   134.20   119.30
                                       1/3   2/2   3/2

                                    H.BROUWER,J.B.STOTHERS
                                    CAN J CHEM   50, 1361 (1972)

1246   P 001200 Z-3-CHLORO-2-PROPEN-1-OL
              1
       CL   CH2-OH   FORMULA   C3H5CLO   MOL WT   92.53
         \   /   SOLVENT   C4H8O2
        3C=C2   ORIG ST   C4H12SI   TEMP   313
         /   \
        H   H   58.10   132.90   118.30
                                       1/3   2/2   3/2

                                    H.BROUWER,J.B.STOTHERS
                                    CAN J CHEM   50, 1361 (1972)

1247   P 001200 E-3-IODO-2-PROPEN-1-OL

```
        1
  H   CH2-OH
   \   /
   3C=C2
   /   \
  J     H
```

FORMULA	C3H5IO		MOL WT	183.98
SOLVENT	C4H8O2			
ORIG ST	C4H12SI		TEMP	313

64.70	146.80	76.80
1/3	2/2	3/2

H.BROUWER,J.B.STOTHERS
CAN J CHEM   50, 1361 (1972)

1248   P 001200 Z-3-IODO-2-PROPEN-1-OL

```
        1
  J   CH2-OH
   \   /
   3C=C2
   /   \
  H     H
```

FORMULA	C3H5IO		MOL WT	183.98
SOLVENT	C4H8O2			
ORIG ST	C4H12SI		TEMP	313

65.70	142.00	81.20
1/3	2/2	3/2

H.BROUWER,J.B.STOTHERS
CAN J CHEM   50, 1361 (1972)

1249   P 001200 3-CHLORO-1,2-PROPANEDIOL

```
       OH
        |
 HO-CH2-CH-CH2-CL
    1   2   3
```

FORMULA	C3H7CLO2		MOL WT	110.54
SOLVENT	H2O			
ORIG ST	C4H12SI		TEMP	AMB

63.50	72.00	46.80
1/3	2/2	3/3

L.F.JOHNSON,W.C.JANKOWSKI
CARBON-13 NMR SPECTRA,JOHN
WILEY AND SONS,NEW YORK   32 (1972)

1250   P 001200 E-3-BROMO-2-BUTEN-1-OL

```
   4      1
 H3C    CH2-OH
    \   /
    3C=C2
    /   \
  BR     H
```

FORMULA	C4H7BRO		MOL WT	151.00
SOLVENT	C4H8O2			
ORIG ST	C4H12SI		TEMP	313

59.20	132.80	122.60	23.30
1/3	2/2	3/1	4/4

H.BROUWER,J.B.STOTHERS
CAN J CHEM   50, 1361 (1972)

1251   P 001200 Z-3-BROMO-2-BUTEN-1-OL

```
        1
  BR   CH2-OH
   \   /
   3C=C2
   /   \
 H3C    H
  4
```

FORMULA	C4H7BRO		MOL WT	151.00
SOLVENT	C4H8O2			
ORIG ST	C4H12SI		TEMP	313

62.20	130.10	122.20	28.40
1/3	2/2	3/1	4/4

H.BROUWER,J.B.STOTHERS
CAN J CHEM   50, 1361 (1972)

1252   P 001200 E-3-CHLORO-2-BUTEN-1-OL

```
   4      1
  H3C    CH2-OH
    \    /
    3C=C2
    /    \
  CL      H
```

FORMULA   C4H7CLO   MOL WT 106.55
SOLVENT   C4H8O2
ORIG ST   C4H12SI   TEMP   313

56.60   128.30   132.50   20.70
 1/3   2/2   3/1   4/4

H.BROUWER,J.B.STOTHERS
CAN J CHEM   50, 1361 (1972)

1253   P 001200 Z-3-CHLORO-2-BUTEN-1-OL

```
         1
  CL    CH2-OH
    \    /
    3C=C2
    /    \
  H3C     H
   4
```

FORMULA   C4H7CLO   MOL WT 106.55
SOLVENT   C4H8O2
ORIG ST   C4H12SI   TEMP   313

60.10   127.00   130.70   25.60
 1/3   2/2   3/1   4/4

H.BROUWER,J.B.STOTHERS
CAN J CHEM   50, 1361 (1972)

1254   P 001200 E-3-CHLORO-2-METHYL-2-PROPEN-1-OL

```
         1
  H     CH2-OH
    \    /
    3C=C2
    /    \
  CL     CH3
          4
```

FORMULA   C4H7CLO   MOL WT 106.55
SOLVENT   C4H8O2
ORIG ST   C4H12SI   TEMP   313

65.50   139.50   113.90   17.50
 1/3   2/1   3/2   4/4

H.BROUWER,J.B.STOTHERS
CAN J CHEM   50, 1361 (1972)

1255   P 001200 Z-3-CHLORO-2-METHYL-2-PROPEN-1-OL

```
         1
  CL    CH2-OH
    \    /
    3C=C2
    /    \
  H      CH3
          4
```

FORMULA   C4H7CLO   MOL WT 106.55
SOLVENT   C4H8O2
ORIG ST   C4H12SI   TEMP   313

60.60   139.40   111.80   23.20
 1/3   2/1   3/2   4/4

H.BROUWER,J.B.STOTHERS
CAN J CHEM   50, 1361 (1972)

1256   P 001200 E-3-IODO-2-BUTEN-1-OL

```
   4      1
  H3C    CH2-OH
    \    /
    3C=C2
    /    \
  J       H
```

FORMULA   C4H7IO   MOL WT 198.00
SOLVENT   C4H8O2
ORIG ST   C4H12SI   TEMP   313

59.70   141.70   96.90   28.00
 1/3   2/2   3/1   4/4

H.BROUWER,J.B.STOTHERS
CAN J CHEM   50, 1361 (1972)

1257   P 001200 Z-3-IODO-2-BUTEN-1-OL

```
        1
     CH2-OH          FORMULA  C4H7IO              MOL WT  198.00
 J     /            SOLVENT  C4H8O2
  \   /             ORIG ST  C4H12SI              TEMP       313
   3C=C2
  /     \
H3C     H              67.00  136.30   99.70   33.30
 4                      1/3     2/2     3/1     4/4
```

H.BROUWER,J.B.STOTHERS
CAN J CHEM   50, 1361 (1972)

1258   P 001300 2-METHYLAMINOETHANOL

```
                    FORMULA  C3H9O                MOL WT   61.10
                    SOLVENT  C4H8O2
    H               ORIG ST  C4H12SI              TEMP       AMB
    |
H3C-N-CH2-CH2-OH       36.00   54.30   60.30
 1   2    3             1/4     2/3     3/3
```

L.F.JOHNSON,W.C.JANKOWSKI
CARBON-13 NMR SPECTRA,JOHN
WILEY AND SONS,NEW YORK   41 (1972)

1259   P 001300 2-AMINO-1-BUTANOL

```
                    FORMULA  C4H11NO              MOL WT   89.14
   NH2              SOLVENT  CDCL3
    |               ORIG ST  C4H12SI              TEMP       AMB
HO-CH2-CH-CH2-CH3
  1   2   3   4        65.80   54.30   26.50   10.40
                       1/3     2/2     3/3     4/4
```

L.F.JOHNSON,W.C.JANKOWSKI
CARBON-13 NMR SPECTRA,JOHN
WILEY AND SONS,NEW YORK   100 (1972)

1260   P 001300 2-AMINO-2-METHYL-1-PROPANOL

```
                    FORMULA  C4H11NO              MOL WT   89.14
   4                SOLVENT  CDCL3
  H3C NH2           ORIG ST  C4H12SI              TEMP       AMB
    \|
     C-CH2-OH          71.10   50.60   26.70   26.70
    /2 1               1/3     2/1     3/4     4/4
  H3C
   3                L.F.JOHNSON,W.C.JANKOWSKI
                    CARBON-13 NMR SPECTRA,JOHN
                    WILEY AND SONS,NEW YORK          99 (1972)
```

1261   P 001300 N-(1-METHYLHEPTYL)-ETHANOLAMINE

```
    8CH3            FORMULA  C10H23NO             MOL WT  173.30
   9  |  3  5  7    SOLVENT  CDCL3
 HO H2C   CH CH2 CH2 CH3ORIG ST  C4H12SI          TEMP       AMB
   \ / \ /1\ / \ / \ /
    H2C   N   CH2 CH2 CH2    53.20   37.00   26.10   29.50   31.90   22.60
    10    |  2   4   6        1/2     2/3     3/3     4/3     5/3     6/3
          H                  14.00   20.10   48.90   60.70
                              7/4     8/4     9/3    10/3
```

L.F.JOHNSON,W.C.JANKOWSKI
CARBON-13 NMR SPECTRA,JOHN
WILEY AND SONS,NEW YORK   413 (1972)

1262　　　　　　　P 001400　DIETHYLSULFITE

```
            ]
            ‖
 H3C-CH2-]-S-]-CH2-CH3
  4   3        1   2
```

FORMULA	C4H10O3S			MOL WT	138.19
SOLVENT	C4H8O2				
ORIG ST	C4H12SI			TEMP	AMB

58.30	15.40	58.30	15.40
1/3	2/4	3/3	4/4

L.F.JOHNSON,W.C.JANKOWSKI
CARBON-13 NMR SPECTRA,JOHN
WILEY AND SONS,NEW YORK　　　　94 (1972)

1263　　　　　　　P 001400　ETHYL SULFATE

```
        1  2
  ]   ]-CH2-CH3
   \ /
    S
   / \  4  3
  ]   ]-CH2-CH3
```

FORMULA	C4H10O4S			MOL WT	154.19
SOLVENT	CDCL3				
ORIG ST	C4H12SI			TEMP	AMB

69.60	14.50	14.50	69.60
1/3	2/4	3/4	4/3

L.F.JOHNSON,W.C.JANKOWSKI
CARBON-13 NMR SPECTRA,JOHN
WILEY AND SONS,NEW YORK　　　　13 (1972)

1264　　　　　　　P 001400　OCTYL NITRITE

```
                    6
 ]-CH2-CH2-CH2-CH2-CH2-CH2
 /  1   2   3   4   5   |
 √                     7CH2
  \                     |
   ]                   8CH3
```

FORMULA	C8H17NO2				MOL WT	159.23
SOLVENT	CDCL3					
ORIG ST	C4H12SI				TEMP	AMB

68.30	29.20	26.00	29.30	29.30	31.90
1/3	2/3	3/3	4/3	5/3	6/3
22.7	14.00				
7/3	8/4				

L.F.JOHNSON,W.C.JANKOWSKI
CARBON-13 NMR SPECTRA,JOHN
WILEY AND SONS,NEW YORK　　　320 (1972)

1265　　　　　　　P 001400　TRIALLYPHOSPHATE

```
           H
            \
    ]      C=CH2
   7 ‖    /2  3
  H2C-]-P-]-CH2
  9 /   |   1
 H2C=C8  ]
     \   |4
      H  CH2
          \  6
          5C=CH2
          /
          H
```

FORMULA	C9H15O4P				MOL WT	218.19
SOLVENT	CDCL3					
ORIG ST	C4H12SI				TEMP	AMB

68.10	132.50	118.10	68.10	132.50	118.10
1/3	2/2	3/3	4/3	5/2	6/3
68.10	132.50	118.10			
7/3	8/2	9/3			

L.F.JOHNSON,W.C.JANKOWSKI
CARBON-13 NMR SPECTRA,JOHN
WILEY AND SONS,NEW YORK　　　358 (1972)

1266　　　　　　　P 001400　TRIBUTYL PHOSPHITE

```
             5CH3
      8   7   I6
     ]-CH2-CH2-CH2
 1]  11  12   /
 CH2-CH2-CH2-]-P
 |        \  4  3  2
 CH3       ]-CH2-CH2-CH2
 9                   |
                    1CH3
```

FORMULA	C12H27O3P				MOL WT	250.32
SOLVENT	CDCL3					
ORIG ST	C4H12SI				TEMP	AMB

13.70	19.10	33.40	61.90	13.70	19.10
1/4	2/3	3/3	4/3	5/4	6/3
33.40	61.90	13.70	19.10	33.40	61.90
7/3	8/3	9/4	10/3	11/3	12/3

L.F.JOHNSON,W.C.JANKOWSKI
CARBON-13 NMR SPECTRA,JOHN
WILEY AND SONS,NEW YORK　　　450 (1972)

1267   P 001400 TRIBUTYL PHOSPHATE

```
   4     2              6    8      FORMULA  C12H27O4P          MOL WT  266.32
 H3C   CH2    O      CH2 CH3         SOLVENT  CDCL3
    \ / \     ||    5/ \ /           ORIG ST  C4H12SI            TEMP      AMB
   H2C H2C-O-P-O-CH2 CH2
    3   1      O        7            67.30   32.50   18.70   13.50   67.30   32.50
                                     1/3     2/3     3/3     4/4     5/3     6/3
              \   11                 18.70   13.50   67.30   32.50   18.70   13.50
           9CH2 CH2                  7/3     8/4     9/3    10/3    11/3    12/4
            \ / \
            CH2 CH3                  L.F.JOHNSON,W.C.JANKOWSKI
            10    12                 CARBON-13 NMR SPECTRA,JOHN
                                     WILEY AND SONS,NEW YORK           451 (1972)
```

1268   P 001500 METHOXYETHYLENE

```
    3                               FORMULA  C3H6O              MOL WT  58.08
  H3C-O     H                       SOLVENT  C6D6
     \1 2/                          ORIG ST  C4H12SI            TEMP      AMB
      C=C
     /   \                          153.30   85.20   54.30
    H     H                          1/D      2/T     3/Q

                                    M.HERBERHOLD,G.O.WIEDERSATZ,C.G.KREITER
                                    Z NATURFORSCH              31B,  35 (1976)
```

1269   P 001500 1,1-DIMETHOXYETHYLENE

```
    3                               FORMULA  C4H8O2             MOL WT  88.11
  H3C-O     H                       SOLVENT  C6D6
     \1 2/                          ORIG ST  C4H12SI            TEMP      AMB
      C=C
    4 /   \                         167.90   54.70   55.00
  H3C-O     H                        1/S      2/T     3/Q

                                    M.HERBERHOLD,G.O.WIEDERSATZ,C.G.KREITER
                                    Z NATURFORSCH              31B,  35 (1976)
```

1270   P 001500 (E)-1,2-DIMETHOXYETHYLENE

```
    3                               FORMULA  C4H8O2             MOL WT  88.11
  H3C-O     H                       SOLVENT  C6D6
     \1 2/                          ORIG ST  C4H12SI            TEMP      AMB
      C=C
     /   \ 4                        135.20   57.90
    H     O-CH3                       1/D      3/Q

                                    M.HERBERHOLD,G.O.WIEDERSATZ,C.G.KREITER
                                    Z NATURFORSCH              31B,  35 (1976)
```

1271   P 001500 (Z)-1,2-DIMETHOXYETHYLENE

```
    3          4                    FORMULA  C4H8O2             MOL WT  88.11
  H3C-O     O-CH3                   SOLVENT  C6D6
     \1 2/                          ORIG ST  C4H12SI            TEMP      AMB
      C=C
     /   \                          130.30   59.60
    H     H                          1/D      3/Q

                                    M.HERBERHOLD,G.O.WIEDERSATZ,C.G.KREITER
                                    Z NATURFORSCH              31B,  35 (1976)
```

1272   P 001500 DIETHYLETHER

FORMULA   C4H10O   MOL WT   74.12
SOLVENT   CDCL3
ORIG ST   C4H12SI   TEMP   AMB

```
 4 3     1 2
H3C-CH2-O-CH2-CH3
```

67.40	17.10	67.40	17.10
1/3	2/4	3/3	4/4

N.R.EASTON,JR.,F.A.L.ANET,P.A.BURNS,C.S.FOOTE
J AM CHEM SOC   96, 3945 (1974)

1273   P 001500 1,2-DIMETHOXYETHANE

FORMULA   C4H10O2   MOL WT   90.12
SOLVENT   C4H8O2
ORIG ST   C4H12SI   TEMP   AMB

```
  4   3   2   1
H3C-O-CH2-CH2-O-CH3
```

72.30	58.60	72.30	58.60
2/3	1/4	3/3	4/4

L.F.JOHNSON,W.C.JANKOWSKI
CARBON-13 NMR SPECTRA,JOHN
WILEY AND SONS,NEW YORK   93 (1972)

1274   P 001500 1,2-DIMETHOXYETHANE

FORMULA   C4H10O2   MOL WT   90.12
SOLVENT   CDCL3
ORIG ST   C4H12SI   TEMP   AMB

```
  4   3   1   2
H3C-O-CH2-CH2-O-CH3
```

72.00	59.00	72.00	59.00
1/3	2/4	3/3	4/4

N.R.EASTON,JR.,F.A.L.ANET,P.A.BURNS,C.S.FOOTE
J AM CHEM SOC   96, 3945 (1974)

1275   P 001500 ETHYLENE GLYCOL MONOETHYL ETHER

FORMULA   C4H10O2   MOL WT   90.12
SOLVENT   CDCL3
ORIG ST   C4H12SI   TEMP   AMB

```
HO-CH2-CH2-O-CH2-CH3
   2   1       3   4
```

61.50	72.00	66.50	15.00
1/3	2/3	3/3	4/4

L.F.JOHNSON,W.C.JANKOWSKI
CARBON-13 NMR SPECTRA,JOHN
WILEY AND SONS,NEW YORK   92 (1972)

1276   P 001500 1-AMINO-3-METHOXYPROPANE

FORMULA   C4H11NO   MOL WT   89.14
SOLVENT   CDCL3
ORIG ST   C4H12SI   TEMP   AMB

```
H2N-CH2-CH2-CH2-O-CH3
    1   2   3     4
```

39.60	33.60	70.90	58.40
1/3	2/3	3/3	4/4

L.F.JOHNSON,W.C.JANKOWSKI
CARBON-13 NMR SPECTRA,JOHN
WILEY AND SONS,NEW YORK   96 (1972)

1277   P 001500   1-METHOXY-1-BUTEN-3-YNE

FORMULA C5H6O   MOL WT 82.10
SOLVENT
ORIG ST C4H12SI   TEMP   AMB

```
      5
    H3C
      \  1 2  3 4
       O-CH=CH-C≡CH
```

158.30	84.20	78.60	80.90	60.60
1/D	2/D	3/S	4/D	5/Q

R.ZEISBERG,F.BOHLMANN
CHEM BER   107, 3800 (1974)

1278   P 001500   1-METHOXY-1-PENTEN-3-YNE

FORMULA C6H8O   MOL WT 96.13
SOLVENT
ORIG ST C4H12SI   TEMP   AMB

```
      6
    H3C
      \  1  2  3 4 5
       O-CH=CH-C≡C-CH3
```

155.60	86.10	74.40	88.80	4.30	60.30
1/D	2/D	3/S	4/S	5/Q	6/Q

R.ZEISBERG,F.BOHLMANN
CHEM BER   107, 3800 (1974)

1279   P 001500   TETRAMETHOXYETHYLENE

FORMULA C6H12O4   MOL WT 148.16
SOLVENT C6D6
ORIG ST C4H12SI   TEMP   AMB

```
   3           5
 H3C-O     O-CH3
    \1 2/
     C=C
  4 /    \ 6
 H3C-O     O-CH3
```

141.90	57.70
1/S	3/Q

M.HERBERHOLD,G.O.WIEDERSATZ,C.G.KREITER
Z NATURFORSCH   31B, 35 (1976)

1280   P 001500   ETHYLENE GLYCOL MONOBUTYL ETHER

FORMULA C6H14O2   MOL WT 118.18
SOLVENT CDCL3
ORIG ST C4H12SI   TEMP   AMB

```
HO-CH2-CH2-O-CH2-CH2-CH2-CH3
  6   5    1   2   3   4
```

71.10	31.80	19.30	13.90	61.60	72.20
1/3	2/3	3/3	4/4	5/3	6/3

L.F.JOHNSON,W.C.JANKOWSKI
CARBON-13 NMR STECTRA,JOHN
WILEY AND SONS,NEW YORK   216 (1972)

1281   P 001500   1,1-DIETHOXY-2-PROPYNE

FORMULA C7H12O2   MOL WT 128.17
SOLVENT CDCL3
ORIG ST C4H12SI   TEMP   AMB

```
        4  5
      O-CH2-CH3
   1 2 3/
   HC≡C-CH
      \  6  7
       O-CH2-CH3
```

73.50	79.20	91.00	60.90	15.10
1/2	2/1	3/2	4/3	5/4

M.T.W.HEARN
TETRAHEDRON   32, 115 (1976)

1282   P 001500. 2,2,3-BIS(HYDROXYMETHYL)BUTYL ALLYL ETHER

```
       H
      /          3    4
H2C=C          CH2-CH3
  9 8\            |
      CH2-O-CH2-C-CH2-OH
      7    1  21 5
             CH2-OH
              6
```

FORMULA	C9H18O3			MOL WT	174.24
SOLVENT	C4H8O2				
ORIG ST	C4H12SI			TEMP	AMB
71.60	43.50	22.60	7.60	64.20	64.20
1/3	2/1	3/3	4/4	5/3	6/3
72.30	135.50	115.90			
7/3	8/2	9/3			

L.F.JOHNSON,W.C.JANKOWSKI
CARBON-13 NMR SPECTRA,JOHN
WILEY AND SONS,NEW YORK   364 (1972)

1283   P 001500 HEXYL ETHER

```
 6 5   4   3   2   1
H3C-CH2-CH2-CH2-CH2-CH2
                      \
                       O
                      /
H3C-CH2-CH2-CH2-CH2-CH2
 12 11  10  9   8   7
```

FORMULA	C12H26O			MOL WT	186.34
SOLVENT	CDCL3				
ORIG ST	C4H12SI			TEMP	AMB
71.00	29.90	26.00	31.90	22.70	14.00
1/3	2/3	3/3	4/3	5/3	6/4
71.00	29.90	26.00	31.90	22.70	14.00
7/3	8/3	9/3	10/3	11/3	12/4

L.F.JOHNSON,W.C.JANKOWSKI
CARBON-13 NMR SPECTRA,JOHN
WILEY AND SONS,NEW YORK   447 (1972)

1284   P 001700 ACETALDEHYDE

```
    O
    ||
  H-C-CH3
    1 2
```

FORMULA	C2H4O	MOL WT	44.05
SOLVENT	CDCL3		
ORIG ST	C4H12SI	TEMP	AMB
199.70	30.70		
1/2	2/4		

L.F.JOHNSON,W.C.JANKOWSKI
CARBON-13 NMR SPECTRA,JOHN
WILEY AND SONS,NEW YORK   5 (1972)

1285   P 001700 ACETALDEHYDE

```
      O
   2 1/
  H3C-C
      \
       H
```

FORMULA	C2H4O	MOL WT	44.05
SOLVENT	CDCL3		
ORIG ST	C4H12SI	TEMP	308
199.80	30.90		
1/2	2/4		

G.E.HAWKES,K.HERWIG,J.D.ROBERTS
J ORG CHEM   39, 1017 (1974)

1286   P 001700 PROPANAL

```
        O
   3 2 1/
  H3C-CH2-C
        \
         H
```

FORMULA	C3H6O	MOL WT	58.08
SOLVENT	CDCL3		
ORIG ST	C4H12SI	TEMP	308
202.80	37.30	6.00	
1/2	2/3	3/4	

G.E.HAWKES,K.HERWIG,J.D.ROBERTS
J ORG CHEM   39, 1017 (1974)

1287   P 001700 CROTONALDEHYDE

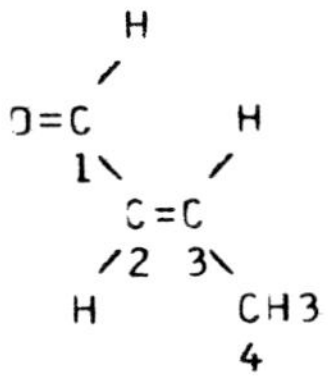

```
        H
        /
    O=C     H
    1\   /
      C=C
     /2 3\
    H     CH3
          4
```

FORMULA   C4H6O   MOL WT   70.09
SOLVENT   C4H8O2
ORIG ST   C4H12SI   TEMP   AMB

193.40   134.90   153.70   18.20
1/2   2/2   3/2   4/4

L.F.JOHNSON,W.C.JANKOWSKI
CARBON-13 NMR SPECTRA,JOHN
WILEY AND SONS,NEW YORK   57 (1972)

1288   P 001700 BUTANAL

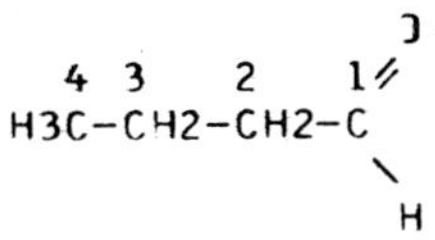

```
               O
  4  3   2   1/
H3C-CH2-CH2-C
             \
              H
```

FORMULA   C4H8O   MOL WT   72.11
SOLVENT   CDCL3
ORIG ST   C4H12SI   TEMP   308

202.50   45.90   15.80   13.80
1/2   2/3   3/3   4/4

G.E.HAWKES,K.HERWIG,J.D.ROBERTS
J ORG CHEM   39, 1017 (1974)

1289   P 001700 BUTYRALDEHYDE

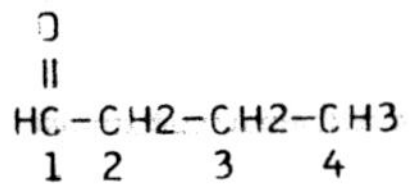

```
  O
  ||
HC-CH2-CH2-CH3
1  2   3   4
```

FORMULA   C4H8O   MOL WT   72.11
SOLVENT   C4H8O2
ORIG ST   C4H12SI   TEMP   AMB

201.90   45.90   16.00   13.80
1/2   2/3   3/3   4/4

L.F.JOHNSON,W.C.JANKOWSKI
CARBON-13 NMR SPECTRA,JOHN
WILEY AND SONS,NEW YORK   74 (1972)

1290   P 001700 2-METHYLPROPANAL

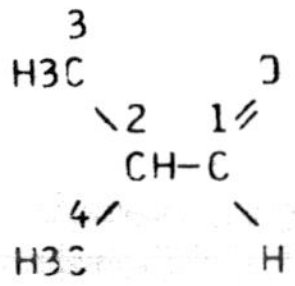

```
     3
   H3C      O
     \2  1/
      CH-C
    4/     \
   H3C      H
```

FORMULA   C4H8O   MOL WT   72.11
SOLVENT   CDCL3
ORIG ST   C4H12SI   TEMP   308

204.70   41.20   15.50   15.50
1/2   2/2   3/4   4/4

G.E.HAWKES,K.HERWIG,J.D.ROBERTS
J ORG CHEM   39, 1017 (1974)

1291   P 001700 3-N,N-DIMETHYLAMINO-2-PROPENAL

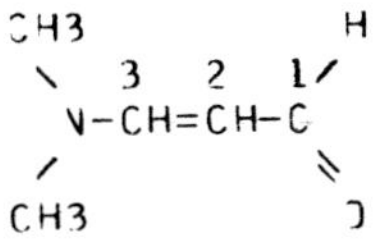

```
 CH3        H
   \ 3  2  1/
    N-CH=CH-C
   /        \\
 CH3         O
```

FORMULA   C5H9NO   MOL WT   99.13
SOLVENT   C2H6OS
ORIG ST   C4H12SI   TEMP   AMB

188.40   101.30   161.60   45.00   37.50
1/2   2/2   3/2   4/4   5/4

R.RADEGLIA,G.ENGELHARDT,E.LIPPMAA,T.PEHK,
K.-D.NOLTE,S.DAEHNE
ORG MAGN RESON   4, 571 (1972)

1292   P 001700 2-METHYLBUTANAL

```
        5
      CH3  O
   4 3  |  1/
  H3C-CH2-CH-C
      2     \
            H
```

FORMULA   C5H10O   MOL WT   86.13
SOLVENT   CDCL3
ORIG ST   C4H12SI   TEMP   308

204.90   47.90   23.20   11.40   12.90
 1/2   2/2   3/3   4/4   5/4

G.E.HAWKES,K.HERWIG,J.D.ROBERTS
J ORG CHEM   39, 1017 (1974)

1293   P 001700 PENTANAL

```
                    O
  5 4   3   2   1/
 H3C-CH2-CH2-CH2-C
                 \
                 H
```

FORMULA   C5H10O   MOL WT   86.13
SOLVENT   CDCL3
ORIG ST   C4H12SI   TEMP   308

202.30   43.70   24.30   22.30   13.80
 1/2   2/3   3/3   4/3   5/4

G.E.HAWKES,K.HERWIG,J.D.ROBERTS
J ORG CHEM   39, 1017 (1974)

1294   P 001700 2-ETHYLBUTANAL

```
   4 3
  H3C-CH2   O
      \2 1/
       CH-C
  6 5/    \
  H3C-CH2  H
```

FORMULA   C6H12O   MOL WT   100.16
SOLVENT   CDCL3
ORIG ST   C4H12SI   TEMP   308

205.00   55.00   21.50   11.40   21.50   11.40
 1/2   2/2   3/3   4/4   5/3   6/4

G.E.HAWKES,K.HERWIG,J.D.ROBERTS
J ORG CHEM   39, 1017 (1974)

1295   P 001700 3-METHYLPENTANAL

```
        6
      CH3       O
  5 4  |  2   1/
 H3C-CH2-CH-CH2-C
        3       \
                H
```

FORMULA   C6H12O   MOL WT   100.16
SOLVENT   CDCL3
ORIG ST   C4H12SI   TEMP   308

202.40   50.90   29.80   29.80   11.40   19.60
 1/2   2/3   3/2   4/3   5/4   6/4

G.E.HAWKES,K.HERWIG,J.D.ROBERTS
J ORG CHEM   39, 1017 (1974)

1296   P 001700 5-N,N-DIMETHYLAMINO-2,4-PENTADIENAL

```
  6
 CH3              H
   \ 5 4 3 2 1/
    N-CH=CH-CH=CH-C
   /              \\
 CH3              O
  7
```

FORMULA   C7H11NO   MOL WT   125.17
SOLVENT   C2H6OS
ORIG ST   C4H12SI   TEMP   AMB

191.10   118.60   154.60   97.30   157.70   41.30
 1/2   2/2   3/2   4/2   5/2   6/4

R.RADEGLIA,G.ENGELHARDT,E.LIPPMAA,T.PEHK,
K.-D.NOLTE,S.DAEHNE
ORG MAGN RESON   4, 571 (1972)

1297   P 001700 3-DIMETHYLAMINO-2,2-DIMETHYLPROPIONALDEHYDE

```
      4        5
 O   CH3      CH3          FORMULA   C7H15NO              MOL WT   129.20
  \   |  2    /            SOLVENT   C4H8O2
   C-C-CH2-N               ORIG ST   C4H12SI              TEMP       AMB
  /1 | 3    \
 H   CH3      CH3            205.10   47.20   67.10   20.40   47.20   20.40
     6        7                1/2     2/1     3/3     4/4     5/4     6/4
                             47.20
                               7/4
```

L.F.JOHNSON,W.C.JANKOWSKI
CARBON-13 NMR SPECTRA,JOHN
WILEY AND SONS,NEW YORK   276 (1972)

1298   P 001900 ANTI-ACETALDOXIME

```
        OH                FORMULA   C2H5NO               MOL WT    59.07
       /                  SOLVENT   CDCL3
      N                   ORIG ST   C4H12SI              TEMP        308
   2  ‖
  H3C-CH                    148.20   15.00
   1                          1/2     2/4
```

G.E.HAWKES,K.HERWIG,J.D.ROBERTS
J ORG CHEM   39, 1017 (1974)

1299   P 001900 SYN-ACETALDOXIME

```
   HO                     FORMULA   C2H5NO               MOL WT    59.07
     \                    SOLVENT   CDCL3
      N                   ORIG ST   C4H12SI              TEMP        308
   2  ‖
  H3C-CH                    147.80   11.20
   1                          1/2     2/4
```

G.E.HAWKES,K.HERWIG,J.D.ROBERTS
J ORG CHEM   39, 1017 (1974)

1300   P 001900 ANTI-PROPIONALDOXIME

```
         OH               FORMULA   C3H7NO               MOL WT    73.10
        /                 SOLVENT   CDCL3
       N                  ORIG ST   C4H12SI              TEMP        308
   3 2  ‖
  H3C-CH2-CH                153.10   23.10   10.90
   1                          1/2     2/3     3/4
```

G.E.HAWKES,K.HERWIG,J.D.ROBERTS
J ORG CHEM   39, 1017 (1974)

1301   P 001900 SYN-PROPIONALDOXIME

```
   HO                     FORMULA   C3H7NO               MOL WT    73.10
     \                    SOLVENT   CDCL3
      N                   ORIG ST   C4H12SI              TEMP        308
   3 2  ‖
  H3C-CH2-CH                153.70   18.60   10.40
   1                          1/2     2/3     3/4
```

G.E.HAWKES,K.HERWIG,J.D.ROBERTS
J ORG CHEM   39, 1017 (1974)

1302   P C01900 ANTI-BUTANAL OXIME

```
          OH        FORMULA   C4H9NO              MOL WT   87.12
         /          SOLVENT   CDCL3
        N           ORIG ST   C4H12SI             TEMP       308
  4 3  2 ||
 H3C-CH2-CH2-CH     152.10    31.50    20.10    13.60
        1            1/2       2/3      3/3      4/4
```

G.E.HAWKES,K.HERWIG,J.D.ROBERTS
J ORG CHEM   39, 1017 (1974)

1303   P 001900 SYN-BUTANAL OXIME

```
       HO          FORMULA   C4H9NO              MOL WT   87.12
        \          SOLVENT   CDCL3
         N         ORIG ST   C4H12SI             TEMP       308
  4 3  2 ||
 H3C-CH2-CH2-CH     152.60    27.00    19.50    13.90
        1            1/2       2/3      3/3      4/4
```

G.E.HAWKES,K.HERWIG,J.D.ROBERTS
J ORG CHEM   39, 1017 (1974)

1304   P 001900 ANTI-2-METHYLPROPANAL OXIME

```
        OH         FORMULA   C4H9NO              MOL WT   87.12
   3   /           SOLVENT   CDCL3
 H3C   N           ORIG ST   C4H12SI             TEMP       308
   \2 ||
    CH-CH          156.90    29.40    20.00    20.00
   4/  1            1/2       2/2      3/4      4/4
 H3C
```

G.E.HAWKES,K.HERWIG,J.D.ROBERTS
J ORG CHEM   39, 1017 (1974)

1305   P 001900 SYN-2-METHYLPROPANAL OXIME

```
      HO           FORMULA   C4H9NO              MOL WT   87.12
   3   \           SOLVENT   CDCL3
 H3C    N          ORIG ST   C4H12SI             TEMP       308
   \2 ||
    CH-CH          157.80    24.50    19.70    19.70
   4/  1            1/2       2/2      3/4      4/4
 H3C
```

G.E.HAWKES,K.HERWIG,J.D.ROBERTS
J ORG CHEM   39, 1017 (1974)

1306   P 001900 ANTI-PENTANAL OXIME

```
           OH        FORMULA   C5H11NO            MOL WT   101.15
          /          SOLVENT   CDCL3
         N           ORIG ST   C4H12SI            TEMP       308
  5 4  3  2 ||
 H3C-CH2-CH2-CH2-CH   152.30   29.20   28.80   22.30   13.70
           1           1/2      2/3     3/3     4/3     5/4
```

G.E.HAWKES,K.HERWIG,J.D.ROBERTS
J ORG CHEM   39, 1017 (1974)

1307   P 001900 SYN-PENTANAL OXIME

```
        HO
          \
            N
 5  4   3   2  ‖
H3C-CH2-CH2-CH2-CH
              1
```

FORMULA   C5H11NO   MOL WT 101.15
SOLVENT   CDCL3
ORIG ST   C4H12SI   TEMP   308

152.80   24.90   28.40   22.60   13.70
 1/2   2/3   3/3   4/3   5/4

G.E.HAWKES,K.HERWIG,J.D.ROBERTS
J ORG CHEM   39, 1017 (1974)

1308   P 001900 ANTI-2-METHYLBUTANAL OXIME

```
        OH
     5  /
   H3C  N
 4  3  |  ‖
H3C-CH2-CH-CH
      2   1
```

FORMULA   C5H11NO   MOL WT 101.15
SOLVENT   CDCL3
ORIG ST   C4H12SI   TEMP   308

156.40   36.10   27.70   11.40   17.70
 1/2   2/2   3/3   4/4   5/4

G.E.HAWKES,K.HERWIG,J.D.ROBERTS
J ORG CHEM   39, 1017 (1974)

1309   P 001900 SYN-2-METHYLBUTANAL OXIME

```
   HO
     \
   H3C5 N
 4  3  |  ‖
H3C-CH2-CH-CH
      2   1
```

FORMULA   C5H11NO   MOL WT 101.15
SOLVENT   CDCL3
ORIG ST   C4H12SI   TEMP   308

157.00   31.20   27.70   11.40   17.10
 1/2   2/2   3/3   4/4   5/4

G.E.HAWKES,K.HERWIG,J.D.ROBERTS
J ORG CHEM   39, 1017 (1974)

1310   P 001900 ANTI-2-ETHYLBUTANAL OXIME

```
          OH
   4  3   /
 H3C-CH2  N
      \2  ‖
       CH-CH
   6  5/   1
 H3C-CH2
```

FORMULA   C6H13NO   MOL WT 115.18
SOLVENT   CDCL3
ORIG ST   C4H12SI   TEMP   308

155.60   43.20   25.70   11.50   25.70   11.50
 1/2   2/2   3/3   4/4   5/3   6/4

G.E.HAWKES,K.HERWIG,J.D.ROBERTS
J ORG CHEM   39, 1017 (1974)

1311   P 001900 SYN-2-ETHYLBUTANAL OXIME

```
      HO
   4  3  \
 H3C-CH2  N
      \2  ‖
       CH-CH
   6  5/   1
 H3C-CH2
```

FORMULA   C6H13NO   MOL WT 115.18
SOLVENT   CDCL3
ORIG ST   C4H12SI   TEMP   308

156.40   38.30   25.40   11.60   25.40   11.60
 1/2   2/2   3/3   4/4   5/3   6/4

G.E.HAWKES,K.HERWIG,J.D.ROBERTS
J ORG CHEM   39, 1017 (1974)

1312   P 001900   ANTI-3-METHYLPENTANAL OXIME

```
                OH
      6        /
     CH3      N
   5 4  |  2  ||
 H3C-CH2-CH-CH2-CH
      3        1
```

FORMULA	C6H13NO			MOL WT	115.18
SOLVENT	CDCL3				
ORIG ST	C4H12SI			TEMP	308

151.60	36.30	33.20	29.30	11.40	19.20
1/2	2/3	3/2	4/3	5/4	6/4

G.E.HAWKES,K.HERWIG,J.D.ROBERTS
J ORG CHEM   39, 1017 (1974)

1313   P 001900   SYN-3-METHYLPENTANAL OXIME

```
             HO
      6       \
     CH3       N
   5 4  |  2  ||
 H3C-CH2-CH-CH2-CH
      3       1
```

FORMULA	C6H13NO			MOL WT	115.18
SOLVENT	CDCL3				
ORIG ST	C4H12SI			TEMP	308

152.00	31.90	32.80	29.50	11.40	19.60
1/2	2/3	3/2	4/3	5/4	6/4

G.E.HAWKES,K.HERWIG,J.D.ROBERTS
J ORG CHEM   39, 1017 (1974)

1314   P 001900   ACETAL

```
   6 5
  H3C-CH2-O \
             \
              CH-CH3
             /1  2
  H3C-CH2-O /
   4  3
```

FORMULA	C6H14O2			MOL WT	118.18
SOLVENT	CDCL3				
ORIG ST	C4H12SI			TEMP	AMB

99.50	19.90	60.60	15.30	60.60	15.30
1/2	2/4	3/3	4/4	5/3	6/4

L.F.JOHNSON,W.C.JANKOWSKI
CARBON-13 NMR SPECTRA,JOHN
WILEY AND SONS,NEW YORK   217 (1972)

1315   P 001900   3-N,N-DIMETHYLAMINO-2-PROPENAL-N,N-DIMETHYL-
                             IMMONIUM PERCHLORATE

```
 CH3           CH3
   \ 3  2  1 +/
    N-CH=CH-CH=N       CLO4
   /           \
 CH3           CH3
 5             4
```

FORMULA	C7H15CLN2O4		MOL WT	226.66
SOLVENT	C2H6OS			
ORIG ST	C4H12SI		TEMP	AMB

163.70	90.80	46.30	38.50
1/2	2/2	4/4	5/4

R.RADEGLIA,G.ENGELHARDT,E.LIPPMAA,T.PEHK,
K.-D.NOLTE,S.DAEHNE
ORG MAGN RESON   4, 571 (1972)

1316   P 001900   5-N,N-DIMETHYLAMINO-2,4-PENTADIENAL-N,N-
                             DIMETHYLIMMONIUM PERCHLORATE

```
 CH3                  CH3
   \ 5  4  3  2  1 +/
    N-CH=CH-CH=CH-CH=N CLO4
   /                  \
 CH3                  CH3
 7                    6
```

FORMULA	C9H17CLN2O4		MOL WT	252.70
SOLVENT	C2H6OS			
ORIG ST	C4H12SI		TEMP	AMB

162.10	103.70	162.60	46.10	38.60
1/2	2/2	3/2	6/4	7/4

R.RADEGLIA,G.ENGELHARDT,E.LIPPMAA,T.PEHK,
K.-D.NOLTE,S.DAEHNE
ORG MAGN RESON   4, 571 (1972)

1317   P 002000 ACETONE

```
                    O
                    ||
            CH3-C-CH3
             1    2 3
```

FORMULA C3H6O   MOL WT 58.08
SOLVENT CDCL3
ORIG ST C4H12SI   TEMP AMB

  30.60 206.00 30.60
  1/4 2/1 3/4

L.F.JOHNSON,W.C.JANKOWSKI
CARBON-13 NMR SPECTRA,JOHN
WILEY AND SONS,NEW YORK 28 (1972)

1318   P 002000 PROPANONE

```
                    O
                1  ||  3
            H3C-C-CH3
                    2
```

FORMULA C3H6O   MOL WT 58.08
SOLVENT CDCL3
ORIG ST C4H12SI   TEMP 308

  30.70 206.20 30.70
  1/4 2/1 3/4

G.E.HAWKES,K.HERWIG,J.D.ROBERTS
J ORG CHEM 39, 1017 (1974)

1319   P 002000 2-BUTANONE

```
                    O
                1  ||  3    4
            H3C-C-CH2-CH3
                    2
```

FORMULA C4H8O   MOL WT 72.11
SOLVENT CDCL3
ORIG ST C4H12SI   TEMP 308

  28.90 209.80 36.40 7.40
  1/4 2/1 3/3 4/4

G.E.HAWKES,K.HERWIG,J.D.ROBERTS
J ORG CHEM 39, 1017 (1974)

1320   P 002000 ACETOIN

```
                    O   3
                    ||
            H3C-C-CH-CH3
             1  2  |  4
                   OH
```

FORMULA C4H8O2   MOL WT 88.11
SOLVENT CDCL3
ORIG ST C4H12SI   TEMP AMB

  24.90 211.20 73.10 19.40
  1/4 2/1 3/2 4/4

L.F.JOHNSON,W.C.JANKOWSKI
CARBON-13 NMR SPECTRA,JOHN
WILEY AND SONS,NEW YORK 76 (1972)

1321   P 002000 METHYL ETHYL KETONE

```
                    O
                    ||
            H3C-C-CH2-CH3
             1  2  3    4
```

FORMULA C4H8O   MOL WT 72.11
SOLVENT C4H8O2
ORIG ST C4H12SI   TEMP AMB

  29.00 207.60 36.50 8.00
  1/4 2/1 3/3 4/4

L.F.JOHNSON,W.C.JANKOWSKI
CARBON-13 NMR SPECTRA,JOHN
WILEY AND SONS,NEW YORK 75 (1972)

1322   P 002000 ACETYLACETONE   ENOL FORM

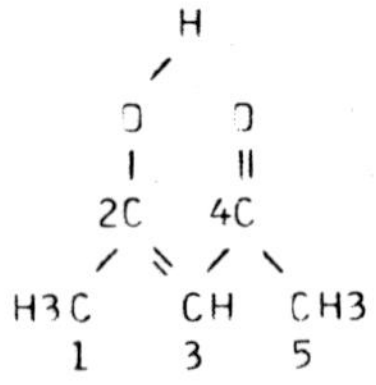

```
            H
           /
        O   O
        |   ||
       2C  4C
       / \ / \
    H3C    CH  CH3
     1      3   5
```

	FORMULA	C5H8O2			MOL WT	100.12
	SOLVENT	C4H8O2				
	ORIG ST	C4H12SI			TEMP	AMB

24.30	191.40	100.30	191.40	24.30
1/4	2/1	3/2	4/1	5/4

L.F.JOHNSON,W.C.JANKOWSKI
CARBON-13 NMR SPECTRA,JOHN
WILEY AND SONS,NEW YORK   115 (1972)

1323   P 002000 ACETYLACETONE   KETO FORM

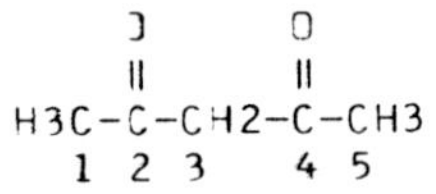

```
    O      O
    ||     ||
  H3C-C-CH2-C-CH3
   1  2  3  4  5
```

	FORMULA	C5H8O2			MOL WT	100.12
	SOLVENT	C4H8O2				
	ORIG ST	C4H12SI			TEMP	AMB

201.90	58.20	30.20	201.90	30.20
2/1	3/3	1/4	4/1	5/4

L.F.JOHNSON,W.C.JANKOWSKI
CARBON-13 NMR SPECTRA,JOHN
WILEY AND SONS,NEW YORK   115 (1972)

1324   P 002000 3-METHYL-2-BUTANONE

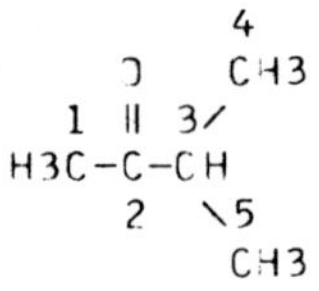

```
         4
    O   CH3
  1 ||  3/
  H3C-C-CH
    2  \5
         CH3
```

	FORMULA	C5H10O			MOL WT	86.13
	SOLVENT	CDCL3				
	ORIG ST	C4H12SI			TEMP	308

27.10	212.10	41.30	17.80	17.80
1/4	2/1	3/2	4/4	5/4

G.E.HAWKES,K.HERWIG,J.D.ROBERTS
J ORG CHEM   39, 1017 (1974)

1325   P 002000 METHYL ISOPROPYL KETONE

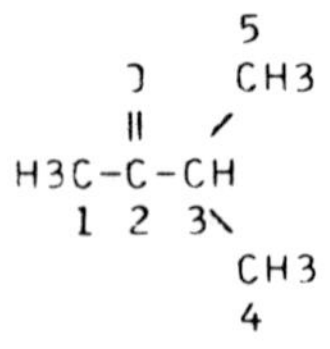

```
      5
    O   CH3
    ||  /
  H3C-C-CH
   1 2 3\
         CH3
          4
```

	FORMULA	C5H10O			MOL WT	86.13
	SOLVENT	CDCL3				
	ORIG ST	C4H12SI			TEMP	AMB

27.30	211.80	41.50	18.10	18.10
1/4	2/1	3/2	4/4	5/4

L.F.JOHNSON,W.C.JANKOWSKI
CARBON-13 NMR SPECTRA,JOHN
WILEY AND SONS,NEW YORK   121 (1972)

1326   P 002000 3-PENTANONE

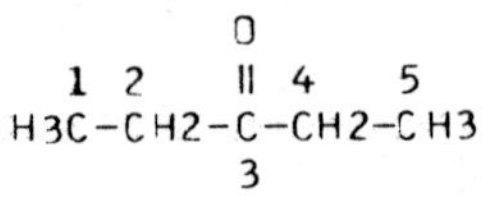

```
          O
   1 2    ||  4   5
  H3C-CH2-C-CH2-CH3
          3
```

	FORMULA	C5H10O			MOL WT	86.13
	SOLVENT	CDCL3				
	ORIG ST	C4H12SI			TEMP	308

8.00	35.60	212.10	35.60	8.00
1/4	2/3	3/1	4/3	5/4

G.E.HAWKES,K.HERWIG,J.D.ROBERTS
J ORG CHEM   39, 1017 (1974)

1327 — P 002000 3-PENTANONE

```
        O
   5    ||
 H3C    C
   \  /3\
  H2C    CH2-CH3
   4    2    1
```

	FORMULA	$C_5H_{10}O$		MOL WT	86.13
	SOLVENT	CDCL3			
	ORIG ST	C4H12SI		TEMP	AMB

7.90	35.40	211.40	35.40	7.90
1/4	2/3	3/1	4/3	5/4

L.F.JOHNSON,W.C.JANKOWSKI
CARBON-13 NMR SPECTRA,JOHN
WILEY AND SONS,NEW YORK   122 (1972)

1328 — P 002000 3,3-DIMETHYL-2-BUTANONE

```
       4
  O    CH3
 1 || 3/5
 H3C-C-C-CH3
  2   \6
       CH3
```

	FORMULA	$C_6H_{12}O$		MOL WT	100.16
	SOLVENT	CDCL3			
	ORIG ST	C4H12SI		TEMP	308

24.50	213.90	44.30	26.40	26.40	26.40
1/4	2/1	3/1	4/4	5/4	6/4

G.E.HAWKES,K.HERWIG,J.D.ROBERTS
J ORG CHEM   39, 1017 (1974)

1329 — P 002000 2-METHYL-3-PENTANONE

```
      5
  O   CH3
1 2 || 4/
H3C-CH2-C-CH
  3   \6
      CH3
```

	FORMULA	$C_6H_{12}O$		MOL WT	100.16
	SOLVENT	CDCL3			
	ORIG ST	C4H12SI		TEMP	308

7.90	33.50	215.20	40.70	18.40	18.40
1/4	2/3	3/1	4/2	5/4	6/4

G.E.HAWKES,K.HERWIG,J.D.ROBERTS
J ORG CHEM   39, 1017 (1974)

1330 — P 002000 METHYL ISOBUTYL KETONE

```
        6
  O    CH3
  ||    |
 H3C-C-CH2-CH
 1 2 3  4|
        CH3
         5
```

	FORMULA	$C_6H_{12}O$		MOL WT	100.16
	SOLVENT	CDCL3			
	ORIG ST	C4H12SI		TEMP	AMB

30.10	208.00	52.70	24.50	22.50	22.50
1/4	2/1	3/3	4/2	5/4	6/4

L.F.JOHNSON,W.C.JANKOWSKI
CARBON-13 NMR SPECTRA,JOHN
WILEY AND SONS,NEW YORK   190 (1972)

1331 — P 002000 HEPTA-3,5-DIEN-2-ONE

```
   1
  CH3
   |   4   6
  C2  CH  CH
  / \ / \ / \
 O  CH  CH  CH3
   3   5   7
```

	FORMULA	$C_7H_{10}O$		MOL WT	110.16
	SOLVENT	CDCL3			
	ORIG ST	C4H12SI		TEMP	305

26.70	197.70	128.30	143.30	130.00	139.70
1/4	2/1	3/2	4/2	5/2	6/2
18.30					
7/4					

R.HOLLENSTEIN,W.VON PHILIPSBORN
HELV CHIM ACTA   55, 2030 (1972)

1332 P 002000 2,2-DIMETHYL-3-PENTANONE

```
                5
        O     CH3
    1 2   II  4/6
  H3C-CH2-C-C-CH3
        3   \7
            CH3
```

FORMULA	C7H14O			MOL WT	114.19
SOLVENT	CDCL3				
ORIG ST	C4H12SI			TEMP	308

8.20	29.60	216.30	44.00	26.50	26.50
1/4	2/3	3/1	4/1	5/4	6/4
26.50					
7/4					

G.E.HAWKES,K.HERWIG,J.D.ROBERTS
J ORG CHEM   39, 1017 (1974)

1333 P 002000 2,4-DIMETHYL-3-PENTANONE

```
   1        5
  H3C   O   CH3
   \2  II  4/
    CH-C-CH
  7/    3  \6
  H3C       CH3
```

FORMULA	C7H14O			MOL WT	114.19
SOLVENT	CDCL3				
ORIG ST	C4H12SI			TEMP	308

18.60	38.90	218.40	38.90	18.60	18.60
1/4	2/2	3/1	4/2	5/4	6/4
18.60					
7/4					

G.E.HAWKES,K.HERWIG,J.D.ROBERTS
J ORG CHEM   39, 1017 (1974)

1334 P 002000 4,4-DIMETHYL-2-PENTANONE

```
              5
      O      CH3
  1 II 3   4/6
  H3C-C-CH2-C-CH3
      2      \7
             CH3
```

FORMULA	C7H14O			MOL WT	114.19
SOLVENT	CDCL3				
ORIG ST	C4H12SI			TEMP	308

32.30	208.50	55.80	30.90	29.60	29.60
1/4	2/1	3/3	4/1	5/4	6/4
29.60					
7/4					

G.E.HAWKES,K.HERWIG,J.D.ROBERTS
J ORG CHEM   39, 1017 (1974)

1335 P 002000 2-HEPTANONE

```
  O
  II
  H3C-C-CH2-CH2-CH2-CH2-CH3
  1 2 3   4   5   6   7
```

FORMULA	C7H14O			MOL WT	114.19
SOLVENT	CDCL3				
ORIG ST	C4H12SI			TEMP	AMB

29.60	208.40	43.70	23.60	31.50	22.60
1/4	2/1	3/3	4/3	5/3	6/3
13.90					
7/4					

L.F.JOHNSON,W.C.JANKOWSKI
CARBON-13 NMR SPECTRA,JOHN
WILEY AND SONS,NEW YORK   270 (1972)

1336 P 002000 3-HEPTANONE

```
   3 2
  H3C-CH2
       \
        1C=O
       /
  H3C-CH2-CH2-CH2
   7  6   5   4
```

FORMULA	C7H14O			MOL WT	114.19
SOLVENT	CDCL3				
ORIG ST	C4H12SI			TEMP	AMB

211.20	35.80	7.80	42.10	26.20	22.50
1/1	2/3	3/4	4/3	5/3	6/3
13.80					
7/4					

L.F.JOHNSON,W.C.JANKOWSKI
CARBON-13 NMR SPECTRA,JOHN
WILEY AND SONS,NEW YORK   269 (1972)

1337   P 002000 4-HEPTANONE

```
        O
        ||
H3C-CH2-CH2-C-CH2-CH2-CH3
 7 6   5   4 3   2   1
```

FORMULA	C7H14O			MOL WT	114.19
SOLVENT	CDCL3				
ORIG ST	C4H12SI			TEMP	AMB

13.70	17.40	44.70	210.60	44.70	17.40
1/4	2/3	3/3	4/1	5/3	6/3
13.70					
7/4					

L.F.JOHNSON,W.C.JANKOWSKI
CARBON-13 NMR SPECTRA,JOHN
WILEY AND SONS,NEW YORK   268 (1972)

1338   P 002000 2-METHYLHEPTA-2,5-DIEN-4-ONE

```
      8
   H3C     O
    |   || 6
    C2  C4  CH
  1/ \ / \ / \
  H3C   CH CH CH3
        3  5  7
```

FORMULA	C8H12O			MOL WT	124.18
SOLVENT	CDCL3				
ORIG ST	C4H12SI			TEMP	305

27.50	154.90	122.50	189.80	133.30	141.10
1/4	2/1	3/2	4/1	5/2	6/2
20.60	17.90				
7/4	8/4				

R.HOLLENSTEIN,W.VON PHILIPSBORN
HELV CHIM ACTA   55, 2030 (1972)

1339   P 002000 6-METHYLHEPTA-3,5-DIEN-2-ONE

```
  1        8
  CH3      CH3
   |   4   |
   C2  CH  C6
  // \ // \ // \
  O   CH  CH CH3
      3   5  7
```

FORMULA	C8H12O			MOL WT	124.18
SOLVENT	CDCL3				
ORIG ST	C4H12SI			TEMP	305

27.40	198.50	128.00	139.40	124.10	147.50
1/4	2/1	3/2	4/2	5/2	6/1
26.60	19.00				
7/4	8/4				

R.HOLLENSTEIN,W.VON PHILIPSBORN
HELV CHIM ACTA   55, 2030 (1972)

1340   P 002000 DIISOBUTYL KETONE

FORMULA	C9H18O			MOL WT	142.24
SOLVENT	CDCL3				
ORIG ST	C4H12SI			TEMP	AMB

```
        O
  8     ||     4
  H3C   C1    CH3
   \7 6 / \2 3/
   HC-CH2  CH2-CH
   /          \
  H3C          CH3
   9            5
```

210.00	52.30	24.40	22.60	22.60	52.30
1/1	2/3	3/2	4/4	5/4	6/3
24.40	22.60	22.60			
7/2	8/4	9/4			

L.F.JOHNSON,W.C.JANKOWSKI
CARBON-13 NMR SPECTRA,JOHN
WILEY AND SONS,NEW YORK   361 (1972)

1341   P 002000 3-UNDECANONE

FORMULA	C11H22O			MOL WT	170.30
SOLVENT	C4H8O2				
ORIG ST	C4H12SI			TEMP	AMB

```
  4    5   6   7   8   9
  CH2-CH2-CH2-CH2-CH2-CH2
  /                    |
O=C3                10CH2
  \                    |
   CH2-CH3            CH3
   2    1             11
```

7.80	35.50	209.00	42.10	24.20	29.80
1/4	2/3	3/1	4/3	5/3	6/3
29.80	29.80	32.20	23.00	14.10	
7/3	8/3	9/3	10/3	11/4	

L.F.JOHNSON,W.C.JANKOWSKI
CARBON-13 NMR SPECTRA,JOHN
WILEY AND SONS,NEW YORK   426 (1972)

1342   P 002200 PROPANONE OXIME

```
        OH
       /
      N                FORMULA   C3H7NO              MOL WT   73.10
    3 ‖ 1              SOLVENT   CDCL3
   H3C-C-CH3           ORIG ST   C4H12SI             TEMP        308
      2
                         15.00    155.40    21.70
                          1/4      2/1       3/4

                      G.E.HAWKES,K.HERWIG,J.D.ROBERTS
                      J ORG CHEM                    39, 1017 (1974)
```

1343   P 002200 HEXACHLOROPROPANONE

```
                       FORMULA   C3CL6O              MOL WT  264.75
                       SOLVENT   CHCL3
   CL  O  CL           ORIG ST   C4H12SI             TEMP        AMB
     \1 ‖ 3/
   CL-C-C-C-CL           90.20    175.50
    / 2  \                1/S      2/S
   CL      CL
                      G.E.HAWKES,R.A.SMITH,J.D.ROBERTS
                      J ORG CHEM                    39, 1276 (1974)
```

1344   P 002200 DIMETHYLGLYOXIME

```
     4CH3
    /                  FORMULA   C4H8N2O2            MOL WT  116.12
   HON=C3              SOLVENT   C2D6OS
    \                  ORIG ST   C4H12SI             TEMP        AMB
     2C=NOH
     |                   153.10     9.20    153.10    9.20
     1CH3                 2/1       1/4      3/1      4/4

                      L.F.JOHNSON,W.C.JANKOWSKI
                      CARBON-13 NMR SPECTRA,JOHN
                      WILEY AND SONS,NEW YORK           71 (1972)
```

1345   P 002200 ANTI-2-BUTANONE OXIME

```
     HO
       \               FORMULA   C4H9NO              MOL WT   87.12
        N              SOLVENT   CDCL3
    1 ‖ 3   4          ORIG ST   C4H12SI             TEMP        308
   H3C-C-CH2-CH3
      2                  13.00    159.10    28.90    10.70
                         1/4      2/1       3/3      4/4

                      G.E.HAWKES,K.HERWIG,J.D.ROBERTS
                      J ORG CHEM                    39, 1017 (1974)
```

1346   P 002200 SYN-2-BUTANONE OXIME

```
       OH
      /                FORMULA   C4H9NO              MOL WT   87.12
     N                 SOLVENT   CDCL3
    1 ‖ 3   4          ORIG ST   C4H12SI             TEMP        308
   H3C-C-CH2-CH3
      2                  18.90    159.50    21.70    9.60
                         1/4      2/1       3/3      4/4

                      G.E.HAWKES,K.HERWIG,J.D.ROBERTS
                      J ORG CHEM                    39, 1017 (1974)
```

1347   P 002200 ANTI-3-METHYL-2-BUTANONE OXIME

```
       HO                FORMULA  C5H11NO              MOL WT  101.15
         \    4          SOLVENT  CDCL3
          N   CH3        ORIG ST  C4H12SI              TEMP        308
     1  II 3/
     H3C-C-CH                 10.80  162.20  34.30  19.50  19.50
      2   \5                   1/4    2/1    3/2    4/4    5/4
          CH3
                           G.E.HAWKES,K.HERWIG,J.D.ROBERTS
                           J ORG CHEM                   39, 1017 (1974)
```

1348   P 002200 SYN-3-METHYL-2-BUTANONE OXIME

```
        OH               FORMULA  C5H11NO              MOL WT  101.15
       /   4             SOLVENT  CDCL3
      N   CH3            ORIG ST  C4H12SI              TEMP        308
     1  II 3/
     H3C-C-CH                 15.10  162.70  25.70  18.70  18.70
      2   \5                   1/4    2/1    3/2    4/4    5/4
          CH3
                           G.E.HAWKES,K.HERWIG,J.D.ROBERTS
                           J ORG CHEM                   39, 1017 (1974)
```

1349   P 002200 3-PENTANONE OXIME

```
       HO                FORMULA  C5H11NO              MOL WT  101.15
         \               SOLVENT  CDCL3
          N              ORIG ST  C4H12SI              TEMP        308
     1  2 II 4   5
     H3C-CH2-C-CH2-CH3         10.10  21.00  163.50  27.10  10.70
         3                      1/4    2/3    3/1    4/3    5/4
                           G.E.HAWKES,K.HERWIG,J.D.ROBERTS
                           J ORG CHEM                   39, 1017 (1974)
```

1350   P 002200 3,3-DIMETHYL-2-BUTANONE OXIME

```
       HO                FORMULA  C6H13NO              MOL WT  115.18
         \    4          SOLVENT  CDCL3
          N   CH3        ORIG ST  C4H12SI              TEMP        308
     1  II 3/5
     H3C-C-C-CH3              10.00  164.10  37.30  27.50  27.50  27.50
      2   \6                   1/4    2/1    3/1    4/4    5/4    6/4
          CH3
                           G.E.HAWKES,K.HERWIG,J.D.ROBERTS
                           J ORG CHEM                   39, 1017 (1974)
```

1351   P 002200 ANTI-2-METHYL-3-PENTANONE OXIME

```
       HO                FORMULA  C6H13NO              MOL WT  115.18
         \    5          SOLVENT  CDCL3
          N   CH3        ORIG ST  C4H12SI              TEMP        308
     1  2 II 4/
     H3C-CH2-C-CH              10.70  20.00  166.40  33.60  20.00  20.00
         3   \6                1/4    2/3    3/1    4/2    5/4    6/4
             CH3
                           G.E.HAWKES,K.HERWIG,J.D.ROBERTS
                           J ORG CHEM                   39, 1017 (1974)
```

1352　　　　　P 002200 SYN-2-METHYL-3-PENTANONE OXIME

```
        OH              FORMULA   C6H13NO              MOL WT  115.18
       / 5             SOLVENT   CDCL3
      N   CH3          ORIG ST   C4H12SI              TEMP       308
  1 2  || 4/
  H3C-CH2-C-CH              10.80   23.30  165.80   26.50   19.00   19.00
      3  \6                1/4     2/3     3/1      4/2     5/4     6/4
        CH3
```

G.E.HAWKES,K.HERWIG,J.D.ROBERTS
J ORG CHEM　　　　　　　　　　　39, 1017 (1974)

1353　　　　　P 002200 2,4-DIMETHYL-3-PENTANONE OXIME

```
      HO              FORMULA   C7H15NO              MOL WT  129.20
   1   \   5          SOLVENT   CDCL3
  H3C   N   CH3       ORIG ST   C4H12SI              TEMP       308
     \2 || 4/
       CH-C-CH             18.90   27.70  168.60   30.80   21.30   21.30
   7/   3  \6             1/4     2/2     3/1      4/2     5/4     6/4
  H3C        CH3           18.90
                          7/4
```

G.E.HAWKES,K.HERWIG,J.D.ROBERTS
J ORG CHEM　　　　　　　　　　　39, 1017 (1974)

1354　　　　　P 002200 ANTI-2,2-DIMETHYL-3-PENTANONE OXIME

```
      HO              FORMULA   C7H15NO              MOL WT  129.20
     \   5            SOLVENT   CDCL3
      N   CH3         ORIG ST   C4H12SI              TEMP       308
  1 2  || 4/6
  H3C-CH2-C-C-CH3          11.10   19.00  168.10   37.60   27.60   27.60
      3   \7             1/4     2/3     3/1      4/1     5/4     6/4
        CH3               27.60
                          7/4
```

G.E.HAWKES,K.HERWIG,J.D.ROBERTS
J ORG CHEM　　　　　　　　　　　39, 1017 (1974)

1355　　　　　P 002200 ANTI-4,4-DIMETHYL-2-PENTANONE OXIME

```
      HO              FORMULA   C7H15NO              MOL WT  129.20
     \   5            SOLVENT   CDCL3
      N   CH3         ORIG ST   C4H12SI              TEMP       308
  1 || 3  4/6
  H3C-C-CH2-C-CH3          16.40  157.00   49.20   31.60   29.90   29.90
    2    \7              1/4     2/1     3/3      4/1     5/4     6/4
        CH3               29.90
                          7/4
```

G.E.HAWKES,K.HERWIG,J.D.ROBERTS
J ORG CHEM　　　　　　　　　　　39, 1017 (1974)

1356　　　　　P 002200 SYN-4,4-DIMETHYL-2-PENTANONE OXIME

```
      OH              FORMULA   C7H15NO              MOL WT  129.20
     / 5              SOLVENT   CDCL3
      N   CH3         ORIG ST   C4H12SI              TEMP       308
  1 || 3  4/6
  H3C-C-CH2-C-CH3          22.70  157.00   42.00   32.20   30.70   30.70
    2    \7              1/4     2/1     3/3      4/1     5/4     6/4
        CH3               30.70
                          7/4
```

G.E.HAWKES,K.HERWIG,J.D.ROBERTS
J ORG CHEM　　　　　　　　　　　39, 1017 (1974)

1357   P 002500 ACETIC ACID

```
    O
     \
      C-CH3
     /1 2
   HO
```

FORMULA	C2H4O2		MOL WT	60.05
SOLVENT	CDCL3			
ORIG ST	C4H12SI		TEMP	AMB

| 178.10 | 20.60 |
| 1/1 | 2/4 |

L.F.JOHNSON,W.C.JANKOWSKI
CARBON-13 NMR SPECTRA,JOHN
WILEY AND SONS,NEW YORK   7 (1972)

1358   P 002500 3—MERCAPTOPROPIONIC ACID

```
    O
    ||
 HO-C-CH2-CH2-SH
    1 2   3
```

FORMULA	C3H6O2S		MOL WT	106.14
SOLVENT	CDCL3			
ORIG ST	C4H12SI		TEMP	AMB

| 177.90 | 38.20 | 19.30 |
| 1/1 | 2/3 | 3/3 |

L.F.JOHNSON,W.C.JANKOWSKI
CARBON-13 NMR SPECTRA,JOHN
WILEY AND SONS,NEW YORK   31 (1972)

1359   P 002500 CITRIC ACID

```
       4
    O  O=C-OH  O
    ||   |     ||
 HO-C-CH2-C-CH2-C-OH
    6 5  3| 2    1
         OH
```

FORMULA	C6H8O7		MOL WT	192.13
SOLVENT	H2O			
ORIG ST	C4H12SI		TEMP	AMB

| 174.20 | 44.10 | 74.20 | 177.50 | 44.10 | 174.20 |
| 1/1 | 2/3 | 3/1 | 4/1 | 5/3 | 6/1 |

L.F.JOHNSON,W.C.JANKOWSKI
CARBON-13 NMR SPECTRA,JOHN
WILEY AND SONS,NEW YORK   172 (1972)

1360   P 002500 CAPROIC ACID

```
    O
    ||
    1C
   /  \
  HO   CH2-CH2-CH2-CH2-CH3
       2   3   4   5   6
```

FORMULA	C6H12O2		MOL WT	116.16
SOLVENT	CDCL3			
ORIG ST	C4H12SI		TEMP	AMB

| 180.60 | 34.20 | 24.50 | 31.40 | 22.40 | 13.80 |
| 1/1 | 2/3 | 3/3 | 4/3 | 5/3 | 6/4 |

L.F.JOHNSON,W.C.JANKOWSKI
CARBON-13 NMR SPECTRA,JOHN
WILEY AND SONS,NEW YORK   193 (1972)

1361   P 002500 2—ETHYLHEXANOIC ACID

```
    O
    ||   3   5
    1C  CH2 CH2
   / \2/ \ / \
  HO  CH 4CH2 CH3
      |        6
     CH2-CH3
      7   8
```

FORMULA	C8H16O2		MOL WT	144.22
SOLVENT	CDCL3			
ORIG ST	C4H12SI		TEMP	AMB

183.20	47.30	31.50	29.60	22.70	13.90
1/1	2/2	3/3	4/3	5/3	6/4
25.30	11.70
7/3	8/4

L.F.JOHNSON,W.C.JANKOWSKI
CARBON-13 NMR SPECTRA,JOHN
WILEY AND SONS,NEW YORK   317 (1972)

1362   P 002500 NONANOIC ACID

```
              O
              ‖                    7
HO-C-CH2-CH2-CH2-CH2-CH2-CH2
   1  2   3   4   5   6   ‖
                         8CH2
                          ‖
                         9CH3
```

FORMULA	C9H18O2			MOL WT	158.24
SOLVENT	CDCL3				
ORIG ST	C4H12SI			TEMP	AMB

180.60	34.20	24.80	29.30	29.20	29.20
1/1	2/3	3/3	4/3	5/3	6/3
31.90	22.70	14.10			
7/3	8/3	9/4			

L.F.JOHNSON,W.C.JANKOWSKI
CARBON-13 NMR SPECTRA,JOHN
WILEY AND SONS,NEW YORK   362 (1972)

1363   P 002500 OLEIC ACID

```
            O
          1‖ 2
       HO-C-CH2
  8  7  6  5  4  13
H   CH2-CH2-CH2-CH2-CH2-CH2
 \ /
  C9
  ‖
  C10
 / \11  12  13  14  15  16
H   CH2-CH2-CH2-CH2-CH2-CH2
        18 I17
        H3C-CH2
```

FORMULA	C18H34O2			MOL WT	282.47
SOLVENT	CDCL3				
ORIG ST	C4H12SI			TEMP	AMB

180.50	34.10	24.70	27.20	129.60	129.90
1/1	2/3	3/3	8/3	9/2	10/2
27.20	32.00	22.70	14.10	29.10	29.40
11/3	16/3	17/3	18/4		
29.60	29.70				

L.F.JOHNSON,W.C.JANKOWSKI
CARBON-13 NMR SPECTRA,JOHN
WILEY AND SONS,NEW YORK   480 (1972)

1364   P 002530 THIOMALIC ACID

```
      O       O
      ‖ 2  3  ‖
  HO-C-CH-CH2-C-OH
    1  |       4
       SH
```

FORMULA	C4H6O4S		MOL WT	150.15
SOLVENT	H2O			
ORIG ST	C4H12SI		TEMP	AMB

175.30	36.90	40.20	177.00
1/1	2/2	3/3	4/1

L.F.JOHNSON,W.C.JANKOWSKI
CARBON-13 NMR SPECTRA,JOHN
WILEY AND SONS,NEW YORK   63 (1972)

1365   P 002600 TRIFLUOROACETIC ACID

```
    O     F
     \   /
      C-C-F
     /1 2\
   HO     F
```

FORMULA	C2HF3O2	MOL WT	114.02
SOLVENT	CDCL3		
ORIG ST	C4H12SI	TEMP	AMB

163.00	115.00
1/1	2/1

L.F.JOHNSON,W.C.JANKOWSKI
CARBON-13 NMR SPECTRA,JOHN
WILEY AND SONS,NEW YORK   1 (1972)

1366   P 002600 2,3-DIBROMOPROPIONIC ACID

```
      O
      ‖ 2  3
  HO-C-CH-CH2-BR
    1  |
       BR
```

FORMULA	C3H4BR2O2	MOL WT	231.87
SOLVENT	CDCL3		
ORIG ST	C4H12SI	TEMP	AMB

173.60	40.40	28.80
1/1	2/2	3/3

L.F.JOHNSON,W.C.JANKOWSKI
CARBON-13 NMR SPECTRA,JOHN
WILEY AND SONS,NEW YORK   18 (1972)

1367   P 002600 ETHYL TRIFLUOROACETATE

```
      O
      ||
F3C-C-O-CH2-CH3
 1  2   3   4
```

FORMULA C4H5F3O2   MOL WT 142.08
SOLVENT CDCL3
ORIG ST C4H12SI   TEMP   AMB

115.30   158.10   64.70   13.80
1/1   2/1   3/3   4/4

L.F.JOHNSON,W.C.JANKOWSKI
CARBON-13 NMR SPECTRA,JOHN
WILEY AND SONS,NEW YORK   52 (1972)

1368   P 002600 2-BROMOBUTYRIC ACID

```
    O BR
    || |
HO-C-CH-CH2-CH3
 1  2   3   4
```

FORMULA C4H7BRO2   MOL WT 167.00
SOLVENT CDCL3
ORIG ST C4H12SI   TEMP   AMB

175.90   47.00   28.10   11.70
1/1   2/2   3/3   4/4

L.F.JOHNSON,W.C.JANKOWSKI
CARBON-13 NMR SPECTRA,JOHN
WILEY AND SONS,NEW YORK   65 (1972)

1369   P 002700 L-VALINAMIDE HYDROBROMIDE

```
   HBR     O
   . 2   1//
  H2N-CH-C
     |    \
   3CH   NH2
   5/  \4
  H3C   CH3
```

FORMULA C5H12N2O   MOL WT 116.16
SOLVENT D2O
ORIG ST C4H12SI   TEMP   303

168.55   57.25   29.65   17.80   18.60
1/1   2/2   3/2   4/4   5/4

W.VOELTER,G.JUNG,E.BREITMAIER,E.BAYER
Z NATURFORSCH   26B, 213 (1971)

1370   P 002800 4-BUTYROLACTONE

```
  3    2
  C---C
  |    |
  C    C1
 4\   /\\
    O   O
```

FORMULA C4H6O2   MOL WT 86.09
SOLVENT CDCL3
ORIG ST C4H12SI   TEMP   AMB

177.90   27.70   22.20   68.60
1/1   2/3   3/3   4/3

L.F.JOHNSON,W.C.JANKOWSKI
CARBON-13 NMR SPECTRA,JOHN
WILEY AND SONS,NEW YORK   60 (1972)

1371   P 002800 L-TARTARIC ACID

```
   O    OH
   || 3  |  1
 HO-C-CH-CH-C-OH
  4  |  2  ||
    OH     O
```

FORMULA C4H6O6   MOL WT 150.09
SOLVENT H2O
ORIG ST C4H12SI   TEMP   AMB

175.30   72.80   72.80   175.30
1/1   2/2   3/2   4/1

L.F.JOHNSON,W.C.JANKOWSKI
CARBON-13 NMR SPECTRA,JOHN
WILEY AND SONS,NEW YORK   64 (1972)

1372   P 003000 MALEIC ANHYDRIDE

```
                        FORMULA   C4H2O3                    MOL WT   98.06
        3C===C2         SOLVENT   CDCL3
         |   |          ORIG ST   C4H12SI                   TEMP      AMB
        4C   C1
       ⁄ \ ⁄ ⧵          164.30   136.60   136.60   164.30
       O   O   O          1/1      2/2      3/2      4/1
```

L.F.JOHNSON,W.C.JANKOWSKI
CARBON-13 NMR SPECTRA,JOHN
WILEY AND SONS,NEW YORK   45 (1972)

1373   P 003000 SUCCINIC ANHYDRIDE

```
                        FORMULA   C4H4O3                    MOL WT   100.07
        3C---C2         SOLVENT   C2D6OS
         |   |          ORIG ST   C4H12SI                   TEMP      AMB
        4C   C1
       ⁄ \ ⁄ ⧵          172.90    20.60    20.60   172.90
       O   O   O          1/1      2/3      3/3      4/1
```

L.F.JOHNSON,W.C.JANKOWSKI
CARBON-13 NMR SPECTRA,JOHN
WILEY AND SONS,NEW YORK   49 (1972)

1374   P 003000 CITRACONIC ANHYDRIDE

```
            3 5         FORMULA   C5H4O3                    MOL WT   112.09
       2C===C-CH3       SOLVENT   CDCL3
         |   |          ORIG ST   C4H12SI                   TEMP      AMB
        1C   C4
       ⁄ \ ⁄ ⧵          164.30   129.70   149.50   166.50    11.30
       O   O   O          1/1      2/2      3/1      4/1      5/4
```

L.F.JOHNSON,W.C.JANKOWSKI
CARBON-13 NMR SPECTRA,JOHN
WILEY AND SONS,NEW YORK   103 (1972)

1375   P 003000 PROPIONIC ANHYDRIDE

```
                        FORMULA   C6H10O3                   MOL WT   130.14
         O   O          SOLVENT   CDCL3
         ||  ||         ORIG ST   C4H12SI                   TEMP      AMB
  H3C-CH2-C-O-C-CH2-CH3
    6 5   4  1 2   3    170.30    28.70     8.40   170.30    28.70     8.40
                          1/1      2/3      3/4      4/1      5/3      6/4
```

L.F.JOHNSON,W.C.JANKOWSKI
CARBON-13 NMR SPECTRA,JOHN
WILEY AND SONS,NEW YORK   182 (1972)

1376   P 003100 ADIPYL CHLORIDE

```
                        FORMULA   C6H8CL2O2                 MOL WT   183.04
                        SOLVENT   CDCL3
         O           O  ORIG ST   C4H12SI                   TEMP      AMB
         ||          ||
  C-CH2-CH2-CH2-CH2-C    173.20    46.40    23.70    23.70    46.40   173.20
  ⁄6 5   4   3   2  1⧵     1/1      2/3      3/3      4/3      5/3      6/1
CL                    CL  L.F.JOHNSON,W.C.JANKOWSKI
                        CARBON-13 NMR SPECTRA,JOHN
                        WILEY AND SONS,NEW YORK            167 (1972)
```

1377   P 003200 DIMETHYLFORMAMIDE

```
        2
    O   CH3
    ‖   |
    CH-N
    |   |
        CH3
        3
```

FORMULA	C3H7NO			MOL WT	73.10
SOLVENT	CDCL3				
ORIG ST	C4H12SI			TEMP	AMB

162.40	31.10	36.20
1/2	2/4	3/4

L.F.JOHNSON,W.C.JANKOWSKI
CARBON-13 NMR SPECTRA,JOHN
WILEY AND SONS,NEW YORK   34 (1972)

1378   P 003200 SUCCINIMIDE

```
    3C---C2
    |     |
    4C    C1
    ⫽ \ / ⫽
    O   N   O
```

FORMULA	C4H5NO2			MOL WT	99.09
SOLVENT	H2O				
ORIG ST	C4H12SI			TEMP	AMB

183.60	30.30	30.30	183.60
1/1	2/3	3/3	4/1

L.F.JOHNSON,W.C.JANKOWSKI
CARBON-13 NMR SPECTRA,JOHN
WILEY AND SONS,NEW YORK   55 (1972)

1379   P 003200 N,N-DIMETHYL ACETAMIDE

```
            4
    O       CH3
     \2    /
      C-N
     /    \
   H3C      CH3
   1        3
```

FORMULA	C4H9NO			MOL WT	87.12
SOLVENT	C4H8O2				
ORIG ST	C4H12SI			TEMP	AMB

21.30	169.60	37.50	34.50
1/4	2/1	3/4	4/4

L.F.JOHNSON,W.C.JANKOWSKI
CARBON-13 NMR SPECTRA,JOHN
WILEY AND SONS,NEW YORK   85 (1972)

1380   P 003200 N,N-DIETHYLDODECANAMIDE

```
         O
  14 13  ‖   3   5   7
  H3C-CH2 C1  CH2 CH2 CH2
   \ / \ / \ / \8
      N  CH2 CH2 CH2 CH2
     /    2   4   6   1
  H3C-CH2          9CH2
  16 15            /
         H3C-CH2-CH2
          12 11  10
```

FORMULA	C16H33NO			MOL WT	255.45
SOLVENT	CDCL3				
ORIG ST	C4H12SI			TEMP	AMB

172.00	33.10	25.50	29.50	29.50	29.50
1/1	2/3	3/3	4/3	5/3	6/3
29.50	29.50	29.40	31.90	22.70	14.10
7/3	8/3	9/3	10/3	11/3	12/4
40.00	13.10	41.90	14.40		
13/3	14/4	15/3	16/4		

L.F.JOHNSON,W.C.JANKOWSKI
CARBON-13 NMR SPECTRA,JOHN
WILEY AND SONS,NEW YORK   473 (1972)

1381   P 003230 MALEIC ACID HYDRAZIDE

```
      N-N
     /   \
  O=C     C=O
   1\    /4
    C=C
    2 3
```

FORMULA	C4H4N2O2			MOL WT	112.09
SOLVENT	C2D6OS				
ORIG ST	C4H12SI			TEMP	AMB

156.20	130.30	130.30	156.20
1/1	2/2	3/2	4/1

L.F.JOHNSON,W.C.JANKOWSKI
CARBON-13 NMR SPECTRA,JOHN
WILEY AND SONS,NEW YORK   47 (1972)

1382   P 003400   TRIETHYLORTHOFORMATE

```
        4    5
     O-CH2-CH3
        |
       1CH
   7 6  / \
 H3C-CH2-O   O-CH2-CH3
        2    3
```

FORMULA	C7H16O3			MOL WT	148.20
SOLVENT	CDCL3				
ORIG ST	C4H12SI			TEMP	AMB

112.50	59.50	15.00	59.50	15.00	59.50
1/2	2/3	3/4	4/3	5/4	6/3
15.00					
7/4					

L.F.JOHNSON,W.C.JANKOWSKI
CARBON-13 NMR SPECTRA,JOHN
WILEY AND SONS,NEW YORK   279 (1972)

1383   P 003400   2,2,4-TRIMETHYL-3-HYDROXY-3-PENTENOIC
                         ACID BETA-LACTONE

```
      7    8
    H3C   CH3
   5   \ /
  H3C    C2
   \4 3/ \1
    C=C   C=O
   /   \ /
  H3C    O
   6
```

FORMULA	C8H12O2			MOL WT	140.18
SOLVENT	CDCL3				
ORIG ST	C4H12SI			TEMP	AMB

173.50	54.00	145.80	103.40	15.90	16.20
1/1	2/1	3/1	4/1	5/4	6/4
20.20	20.20				
7/4	8/4				

L.F.JOHNSON,W.C.JANKOWSKI
CARBON-13 NMR SPECTRA,JOHN
WILEY AND SONS,NEW YORK   307 (1972)

1384   P 003500   ACETONITRILE

```
   2 1
 H3C-C≡N
```

FORMULA	C2H3N	MOL WT	41.05
SOLVENT	C4H8O2		
ORIG ST	C4H12SI	TEMP	AMB

117.70	1.30
1/1	2/4

L.F.JOHNSON,W.C.JANKOWSKI
CARBON-13 NMR SPECTRA,JOHN
WILEY AND SONS,NEW YORK   4 (1972)

1385   P 003500   2-CHLOROACRYLONITRILE

```
  CL
   \1 2
    C=CH2
   3/
 N≡C
```

FORMULA	C3H2CLN	MOL WT	87.51
SOLVENT	CDCL3		
ORIG ST	C4H12SI	TEMP	AMB

110.80	131.70	114.40
1/1	2/3	3/1

L.F.JOHNSON,W.C.JANKOWSKI
CARBON-13 NMR SPECTRA,JOHN
WILEY AND SONS,NEW YORK   15 (1972)

1386   P 003500   ACRYLONITRILE

```
   3CN
    |
  CH2=CH
  1    2
```

FORMULA	C3H3N	MOL WT	53.06
SOLVENT	C4H8O2		
ORIG ST	C4H12SI	TEMP	AMB

137.50	108.10	117.50
1/3	2/2	3/1

L.F.JOHNSON,W.C.JANKOWSKI
CARBON-13 NMR SPECTRA,JOHN
WILEY AND SONS,NEW YORK   16 (1972)

1387   P 003500 PROPIONITRILE

FORMULA   C3H5N   MOL WT   55.08
SOLVENT   COCL3
NC-CH2-CH3   ORIG ST   C4H12SI   TEMP   AMB
3   2   1

10.60   10.80   120.80
1/4   2/3   3/1

L.F.JOHNSON,W.C.JANKOWSKI
CARBON-13 NMR SPECTRA,JOHN
WILEY AND SONS,NEW YORK   24 (1972)

1388   P 003500 CIS-CROTONONITRILE

FORMULA   C4H5N   MOL WT   67.09
SOLVENT   C4H8O2
H   H   ORIG ST   C4H12SI   TEMP   AMB
\2 3/
C=C   17.30   150.20   100.90   116.00
/   \   1/4   2/2   3/2   4/1
H3C   C≡N
1   4   L.F.JOHNSON,W.C.JANKOWSKI
CARBON-13 NMR SPECTRA,JOHN
WILEY AND SONS,NEW YORK   53 (1972)

1389   P 003500 TRANS-CROTONONITRILE

FORMULA   C4H5N   MOL WT   67.09
SOLVENT   C4H8O2
H3C   H   ORIG ST   C4H12SI   TEMP   AMB
1\2 3/
C=C   18.80   151.60   101.20   117.60
/   \   1/4   2/2   3/2   4/1
H   C≡N
4   L.F.JOHNSON,W.C.JANKOWSKI
CARBON-13 NMR SPECTRA,JOHN
WILEY AND SONS,NEW YORK   53 (1972)

1390   P 003500 3-ETHOXYPROPIONITRILE

FORMULA   C5H9NO   MOL WT   99.13
SOLVENT   CDCL3
NC-CH2-CH2-O-CH2-CH3 ORIG ST   C4H12SI   TEMP   AMB
1 2   3   4   5

118.20   18.90   65.10   66.50   14.90
1/1   2/3   3/3   4/3   5/4

L.F.JOHNSON,W.C.JANKOWSKI
CARBON-13 NMR SPECTRA,JOHN
WILEY AND SONS,NEW YORK   116 (1972)

1391   P 003500 ADIPONITRILE

FORMULA   C6H8N2   MOL WT   108.14
SOLVENT   CDCL3
ORIG ST   C4H12SI   TEMP   AMB
1 2   3   4   5   6
NC-CH2-CH2-CH2-CH2-CN   119.30   16.40   24.30   24.30   16.40   119.30
1/1   2/3   3/3   4/3   5/3   6/1

L.F.JOHNSON,W.C.JANKOWSKI
CARBON-13 NMR SPECTRA,JOHN
WILEY AND SONS,NEW YORK   169 (1972)

1392   P 003500 ISOBUTYRONITRILE

```
                     FORMULA  C4H7N              MOL WT   69.11
        3CH3         SOLVENT  CDCL3
         I 2         ORIG ST  C4H12SI            TEMP       AMB
        NC-CH
         1 I            123.70   19.80   19.90   19.90
           CH3           1/1      2/2     3/4     4/4
           4
```

L.F.JOHNSON,W.C.JANKOWSKI
CARBON-13 NMR SPECTRA,JOHN
WILEY AND SONS,NEW YORK   66 (1972)

1393   P 003600 ETHYL TRICHLOROACETATE

```
                        FORMULA  C4H5CL3O2          MOL WT  191.44
                        SOLVENT  CDCL3
  CL    O                ORIG ST  C4H12SI           TEMP       AMB
    \   II
  CL-C-C-O-CH2-CH3          89.90   161.10   65.40   13.70
    /1  2   3   4            1/1      2/1     3/3     4/4
  CL
```

L.F.JOHNSON,W.C.JANKOWSKI
CARBON-13 NMR SPECTRA,JOHN
WILEY AND SONS,NEW YORK   51 (1972)

1394   P 003600 VINYL ACETATE

```
                        FORMULA  C4H6O2             MOL WT   86.09
                        SOLVENT  C4H8O2
      O                 ORIG ST  C4H12SI            TEMP       AMB
      II
  H3C-C-O-CH=CH2           20.20   167.60   141.80   96.80
   1  2  3   4              1/4      2/1     3/2     4/3
```

L.F.JOHNSON,W.C.JANKOWSKI
CARBON-13 NMR SPECTRA,JOHN
WILEY AND SONS,NEW YORK   61 (1972)

1395   P 003600 ISOPROPYLACETATE

```
      O                 FORMULA  C5H10O2            MOL WT  102.13
      II  3             SOLVENT  C4H8O2
      C   CH3           ORIG ST  C4H12SI            TEMP       AMB
    5/4\  I
  H3C   O-CH              21.80   67.40   21.80   169.60   20.80
        21                 1/4     2/2     3/4     4/1      5/4
        CH3
         1
```

L.F.JOHNSON,W.C.JANKOWSKI
CARBON-13 NMR SPECTRA,JOHN
WILEY AND SONS,NEW YORK   126 (1972)

1396   P 003600 METHYL METHACRYLATE

```
  O   O-CH3             FORMULA  C5H8O2             MOL WT  100.12
   \3/  5               SOLVENT  C4H8O2
    C                   ORIG ST  C4H12SI            TEMP       AMB
    I
    C                    124.70   136.90   167.30   18.30   51.50
  1/2\4                   1/3      2/1      3/1     4/4     5/4
  H2C   CH3
```

L.F.JOHNSON,W.C.JANKOWSKI
CARBON-13 NMR SPECTRA,JOHN
WILEY AND SONS,NEW YORK   114 (1972)

1397   P 003600 METHOXYETHYL THIOGLYCOLATE

FORMULA C5H10O3S   MOL WT 150.20
SOLVENT CDCL3
ORIG ST C4H12SI   TEMP   AMB

```
      O
      ||
HS-CH2-C-O-CH2-CH2-O-CH3
  1   2  3   4   5
```

26.30	170.70	64.40	70.20	58.70
1/3	2/1	3/3	4/3	5/4

L.F.JOHNSON,W.C.JANKOWSKI
CARBON-13 NMR SPECTRA,JOHN
WILEY AND SONS,NEW YORK   127 (1972)

1398   P 003600 ETHYL ACETOACETATE

FORMULA C6H10O3   MOL WT 130.14
SOLVENT CDCL3
ORIG ST C4H12SI   TEMP   AMB

```
    O   O
    ||  ||
   2C  4C
   /\  /\
H3C   CH2 O-CH2-CH3
 1    3    5   6
```

29.90	200.50	50.00	167.20	61.10	14.10
1/4	2/1	3/3	4/1	5/3	6/4

L.F.JOHNSON,W.C.JANKOWSKI
CARBON-13 NMR SPECTRA,JOHN
WILEY AND SONS,NEW YORK   181 (1972)

1399   P 003600 DIETHYLOXALATE

FORMULA C6H10O4   MOL WT 146.14
SOLVENT CDCL3
ORIG ST C4H12SI   TEMP   AMB

```
        O
        ||
       1C
  3 2  /\4  5   6
H3C-CH2-O  C-O-CH2-CH3
        ||
        O
```

158.00	62.90	13.90	158.00	62.90	13.90
1/1	2/3	3/4	4/1	5/3	6/4

L.F.JOHNSON,W.C.JANKOWSKI
CARBON-13 NMR SPECTRA,JOHN
WILEY AND SONS,NEW YORK   183 (1972)

1400   P 003600 TERT-BUTYL ACETATE

FORMULA C6H12O2   MOL WT 116.16
SOLVENT CDCL3
ORIG ST C4H12SI   TEMP   AMB

```
         3
   O    CH3
   ||   |
H3C-C-O-C-CH3
 6 5 11 2
      CH3
       4
```

79.90	28.10	28.10	28.10	170.20	22.30
1/1	2/4	3/4	4/4	5/1	6/4

L.F.JOHNSON,W.C.JANKOWSKI
CARBON-13 NMR SPECTRA,JOHN
WILEY AND SONS,NEW YORK   194 (1972)

1401   P 003600 DIETHYLMALONATE

FORMULA C7H12O4   MOL WT 160.17
SOLVENT C4H8O2
ORIG ST C4H12SI   TEMP   AMB

```
    O   O
    ||  ||
   5C  2C
   /\  /\
H3C-CH2-O  CH2 O-CH2-CH3
 7 6      1   3   4
```

41.60	166.70	61.30	14.20	166.70	61.30
1/3	2/1	3/3	4/4	5/1	6/3

14.20
7/4

L.F.JOHNSON,W.C.JANKOWSKI
CARBON-13 NMR SPECTRA,JOHN
WILEY AND SONS,NEW YORK   263 (1972)

1402   P 003600 MALONIC ACID DIETHYL ESTER

```
                              FORMULA  C7H12O4              MOL WT  160.17
   7   6                4   5 SOLVENT  CDCL3
 H3C-H2C-O        O-CH2-CH3   ORIG ST  C4H12SI              TEMP      AMB
           \1  2  3/
             C-CH2-C          166.50   41.70   61.30   14.10
            //       \          1/1     2/3     4/3     5/4
           O          O
```

R.RADEGLIA,S.SCHEITHAUER
Z CHEM   14,   20 (1974)

1403   P 003600 N-AMYL ACETATE

```
                              FORMULA  C7H14O2              MOL WT  130.19
   O                          SOLVENT  C4H8O2
   ||                         ORIG ST  C4H12SI              TEMP      AMB
   C1
  / \                         170.10   20.50   64.30   28.60   28.90   22.80
 H3C    O-CH2-CH2-CH2-CH2-CH3   1/1     2/4     3/3     4/3     5/3     6/3
  2       3    4    5    6   7 14.10
                               7/4
```

L.F.JOHNSON,W.C.JANKOWSKI
CARBON-13 NMR SPECTRA,JOHN
WILEY AND SONS,NEW YORK   273 (1972)

1404   P 003600 METHYLHEXANOATE

```
                              FORMULA  C7H14O2              MOL WT  130.19
        O                     SOLVENT  CDCL3
        ||                    ORIG ST  C4H12SI              TEMP      AMB
 H3C-O-C-CH2-CH2-CH2-CH2-CH3
  1   2 3  4    5    6    7   51.20   174.00   34.10   24.80   31.50   22.40
                               1/4     2/1     3/3     4/3     5/3     6/3
                              13.90
                               7/4
```

L.F.JOHNSON,W.C.JANKOWSKI
CARBON-13 NMR SPECTRA,JOHN
WILEY AND SONS,NEW YORK   274 (1972)

1405   P 003600 2-BUTYNE-1,4-DIACETATE

```
                              FORMULA  C8H10O4              MOL WT  170.17
                              SOLVENT  CDCL3
   O              O           ORIG ST  C4H12SI              TEMP      AMB
   ||             ||
 H3C-C-O-CH2-C≡C-CH2-O-C-CH3  20.50   170.00   52.00   80.80   80.80   52.00
  1 2   3   4 5 6     7 8       1/4     2/1     3/3     4/1     5/1     6/3
                              170.00   20.50
                               7/1     8/4
```

L.F.JOHNSON,W.C.JANKOWSKI
CARBON-13 NMR SPECTRA,JOHN
WILEY AND SONS,NEW YORK   302 (1972)

1406   P 003600 CIS-BUTENE-2-1,4-DIACETATE

```
                              FORMULA  C8H12O4              MOL WT  172.18
   O              O           SOLVENT  CDCL3
   ||             ||          ORIG ST  C4H12SI              TEMP      AMB
  7C             3C
 8/ \ 6   5 1 2    /\4        128.10   59.90   170.40   20.70   128.10   59.90
 H3C    O-CH2-C=C-CH2-O   CH3   1/2     2/3     3/1     4/4     5/2     6/3
                |  |          170.40   20.70
                H  H           7/1     8/4
```

L.F.JOHNSON,W.C.JANKOWSKI
CARBON-13 NMR SPECTRA,JOHN
WILEY AND SONS,NEW YORK   309 (1972)

1407   P 003600   N-BUTYL METHACRYLATE

```
   3   4   5   6   7
O=C-O-CH2-CH2-CH2-CH3
  |
 H2C=C-CH3
  1 2 8
```

FORMULA C8H14O2				MOL WT	142.20
SOLVENT C4H8O2					
ORIG ST C4H12SI				TEMP	AMB

124.40	137.20	166.70	64.40	31.30	19.70
1/3	2/1	3/1	4/3	5/3	6/3
13.90	18.30				
7/4	8/4				

L.F.JOHNSON,W.C.JANKOWSKI
CARBON-13 NMR SPECTRA,JOHN
WILEY AND SONS,NEW YORK   310 (1972)

1408   P 003600   ETHYLHEXANOATE

```
   CH2-CH2-CH2-CH2-CH3
   /2   3   4   5   6
O=C1
   \
    O-CH2-CH3
    7   8
```

FORMULA C8H16O2				MOL WT	144.22
SOLVENT CDCL3					
ORIG ST C4H12SI				TEMP	AMB

173.50	34.40	24.80	31.50	22.40	14.30
1/1	2/3	3/3	4/3	5/3	6/4
60.00	13.90				
7/3	8/4				

L.F.JOHNSON,W.C.JANKOWSKI
CARBON-13 NMR SPECTRA,JOHN
WILEY AND SONS,NEW YORK   318 (1972)

1409   P 003600   DI(3-ACETOXY-1-PROPYNYL)

```
     3 4 5 6 7 8
O    H2C-C≡C-C≡C-CH2    O
 \2 /              \ 9/
  C-O                O-C
 1/                    \10
H3C                    CH3
```

FORMULA C10H10O4				MOL WT	194.19
SOLVENT CDCL3					
ORIG ST C4H12SI				TEMP	AMB

20.20	169.60	51.90	73.90	69.80
1/Q	2/S	3/T	4/S	5/S

R.ZEISBERG,F.BOHLMANN
CHEM BER   107, 3800 (1974)

1410   P 003600   2-ETHYLHEXYLACETATE

```
           H3C
  7   5     1\
  CH2 CH2    2C=O
  / \ / \  3  /
H3C   CH2 CH-CH2-O
 8   6  41
        CH2-CH3
        9   10
```

FORMULA C10H20O2				MOL WT	172.27
SOLVENT CDCL3					
ORIG ST C4H12SI				TEMP	AMB

20.80	170.90	66.90	39.00	30.90	29.10
1/4	2/1	3/3	4/2	5/3	6/3
23.00	14.00	23.90	11.00		
7/3	8/4	9/3	10/4		

L.F.JOHNSON,W.C.JANKOWSKI
CARBON-13 NMR SPECTRA,JOHN
WILEY AND SONS,NEW YORK   411 (1972)

1411   P 003600   DEHYDROMATRICARIA ACID METHYL ESTER

```
O
 \1 2 3  4 5 6 7 8 9 10
  C-CH=CH-C≡C-C≡C-C≡C-CH3
 /
O
|
CH3
11
```

FORMULA C11H8O2				MOL WT	172.19
SOLVENT C6D6					
ORIG ST C4H12SI				TEMP	AMB

163.80	120.60	132.80	71.80	85.90	59.10
1/S	2/D	3/D	4/S	5/S	6/S
72.30	65.20	80.90	3.60	50.90	
7/S	8/S	9/S	10/Q	11/Q	

R.ZEISBERG,F.BOHLMANN
CHEM BER   107, 3800 (1974)

1412   P 003600 LACHNOPHYLLUMIC ACID METHYL ESTER

```
                              FORMULA   C11H12O2              MOL WT   176.22
                    10 9      SOLVENT   CDCL3
        ]           H3C-CH2   ORIG ST   C4H12SI              TEMP       AMB
      \1 2 3  45      1
      C-CH=CH-C≡C-C≡C-CH2   164.90  122.70  131.00   71.00   66.70   65.30
   11  /            6 7 8     1/S     2/D     3/D     4/S     5/S     6/S
   H3C-]                      90.00   21.80   21.70   13.50   51.80
                              7/S     8/T     9/T    10/Q    11/Q
```

          R.ZEISBERG,F.BOHLMANN
          CHEM BER   107, 3800 (1974)

1413   P 003600 DI(4-ACETOXY-1-BUTYNYL)

```
                              FORMULA   C12H14O4             MOL WT   222.24
                              SOLVENT   CDCL3
   ]                      O   ORIG ST   C4H12SI             TEMP       AMB
   \2    456789           //
   C-] H2C-C≡C-C≡C-CH2 ]-C11   20.80  170.60   61.80   19.70   73.60   66.50
  1/   \ /          \ /  \     1/Q     2/S     3/T     4/T     5/S     6/S
  CH3   CH2         H2C  H3C
   3     10         12 R.ZEISBERG,F.BOHLMANN
                      CHEM BER                              107, 3800 (1974)
```

1414   P 003600 DI-N-BUTYLSUCCINATE

```
     4    3    2    1
   ]-CH2-CH2-CH2-CH3     FORMULA   C12H22O4             MOL WT   230.31
   5/                    SOLVENT   CDCL3
  ]=C                    ORIG ST   C4H12SI             TEMP       AMB
     \6
     CH2                   13.70   19.20   30.80   64.40  172.10   29.20
      |                    1/4     2/3     3/3     4/3     5/1     6/3
     7CH2                  29.20  172.10   64.40   30.80   19.20   13.70
      /                    7/3     8/1     9/3    10/3    11/3    12/4
  ]=C8
     \  9   10   11   12  L.F.JOHNSON,W.C.JANKOWSKI
      ]-CH2-CH2-CH2-CH3   CARBON-13 NMR SPECTRA,JOHN
                          WILEY AND SONS,NEW YORK          442 (1972)
```

1415   P 003600 TRIBUTYRIN

```
                    ]      FORMULA   C15H26O6             MOL WT   302.37
                    ||  13 SOLVENT   CDCL3
          ]        8C  CH2 ORIG ST   C4H12SI             TEMP       AMB
     6    ||   1   |\ / \
    CH2 C4  CH2-] CH2 CH3    62.10   69.10   62.10  172.40   36.10   18.40
   / \  / \    2|  9   11    1/3     2/2     3/3     4/1     5/3     6/3
  H3C   CH2 ]-CH              13.50  172.80   35.90   18.40   13.50  172.80
   7    5    |     13  15    7/4     8/1     9/3    10/3    11/4    12/1
            CH2-] CH2 CH3    35.90   18.40   13.50
             3  1/ \ /       13/3    14/3    15/4
            12C    CH2
             ||    14 L.F.JOHNSON,W.C.JANKOWSKI
              ]      CARBON-13 NMR SPECTRA,JOHN
                     WILEY AND SONS,NEW YORK               467 (1972)
```

1416   P 003600 METHYL MYRISTATE

```
  15  14   13   12   11   10 9
  H3C-CH2-CH2-CH2-CH2-CH2-CH2 FORMULA  C15H30O2           MOL WT   242.41
                           |  SOLVENT   CDCL3
    CH2-CH2-CH2-CH2-CH2-CH2 ORIG ST   C4H12SI             TEMP       AMB
    /3   4    5    6    7    8
  ]=C                          51.20  174.00   34.10   25.00   29.40   29.70
    2\                         1/4     2/1     3/3     4/3     5/3     6/3
     ]-CH3                     29.70   29.70   29.70   29.70   29.70   29.50
       1                       7/3     8/3     9/3    10/3    11/3    12/3
                               32.00   22.80   14.10
                               13/3    14/3    15/4
                        L.F.JOHNSON,W.C.JANKOWSKI
                        CARBON-13 NMR SPECTRA,JOHN
                        WILEY AND SONS,NEW YORK            468 (1972)
```

1417   P 004000   ETHANESULFONYL CHLORIDE

```
        O
        ‖
H3C-CH2-S-CL
 1   2  ‖
        O
```

FORMULA	C2H5CLO2S
SOLVENT	CDCL3
ORIG ST	C4H12SI

MOL WT   128.58

TEMP   AMB

9.10   60.20
1/4   2/3

L.F.JOHNSON,W.C.JANKOWSKI
CARBON-13 NMR SPECTRA,JOHN
WILEY AND SONS,NEW YORK   8 (1972)

1418   P 004300   1,2-BUTANEDITHIOL

FORMULA	C4H10S2
SOLVENT	C4H8O2
ORIG ST	C4H12SI

MOL WT   122.25

TEMP   AMB

```
     SH
     |
HS-CH2-CH-CH2-CH3
 1    2   3    4
```

33.20   45.20   29.40   11.40
1/3   2/2   3/3   4/4

L.F.JOHNSON,W.C.JANKOWSKI
CARBON-13 NMR SPECTRA,JOHN
WILEY AND SONS,NEW YORK   95 (1972)

1419   P 004300   2-MERCAPTOETHANOL

FORMULA	C2H6OS
SOLVENT	C4H8O2
ORIG ST	C4H12SI

MOL WT   78.13

TEMP   AMB

```
HS-CH2-CH2-OH
 1    2
```

27.10   64.00
1/3   2/3

L.F.JOHNSON,W.C.JANKOWSKI
CARBON-13 NMR SPECTRA,JOHN
WILEY AND SONS,NEW YORK   12 (1972)

1420   P 004300   1-OCTANETHIOL

FORMULA	C8H18S
SOLVENT	CDCL3
ORIG ST	C4H12SI

MOL WT   146.30

TEMP   AMB

```
                        6
HS-CH2-CH2-CH2-CH2-CH2-CH2
 1   2   3   4   5   |
                    7CH2
                     |
                    8CH3
```

24.60   34.20   28.40   29.20   29.10   31.90
1/3   2/3   3/3   4/3   5/3   6/3
22.70   14.10
7/3   8/4

L.F.JOHNSON,W.C.JANKOWSKI
CARBON-13 NMR SPECTRA,JOHN
WILEY AND SONS,NEW YORK   326 (1972)

1421   P 004400   THIOACETIC ACID

```
      O
      ‖
      C2
     ╱ ╲
   HS    CH3
          1
```

FORMULA	C2H4OS
SOLVENT	C4H8O2
ORIG ST	C4H12SI

MOL WT   76.12

TEMP   AMB

32.60   194.50
1/4   2/1

L.F.JOHNSON,W.C.JANKOWSKI
CARBON-13 NMR SPECTRA,JOHN
WILEY AND SONS,NEW YORK   6 (1972)

1422   P 004400   ETHYLTHIOCYANATE

```
  1   2   3
N≡C-S-CH2-CH3
```

FORMULA C3H5NS   MOL WT 87.14
SOLVENT CDCL3
ORIG ST C4H12SI   TEMP   AMB

111.90	28.60	15.40
1/1	2/3	3/4

L.F.JOHNSON,W.C.JANKOWSKI
CARBON-13 NMR SPECTRA,JOHN
WILEY AND SONS,NEW YORK   25 (1972)

1423   P 004400   N-BUTYLTHIOACETATE

```
      O
      ||
H3C-C-S-CH2-CH2-CH2-CH3
  6  5   1    2    3   4
```

FORMULA C6H12OS   MOL WT 132.23
SOLVENT C4H8O2
ORIG ST C4H12SI   TEMP   AMB

28.70	32.10	22.20	13.60	194.10	30.10
1/3	2/3	3/3	4/4	5/1	6/4

L.F.JOHNSON,W.C.JANKOWSKI
CARBON-13 NMR SPECTRA,JOHN
WILEY AND SONS,NEW YORK   192 (1972)

1424   P 004400 DITHIONOMALONIC ACID DIETHYL ESTER

```
 7   6              4   5
H3C-H2C-O        O-CH2-CH3
      \1  2   3/
        C-CH2-C
       //       \\
      S          S
```

FORMULA C7H12O2S2   MOL WT 192.30
SOLVENT CDCL3
ORIG ST C4H12SI   TEMP   AMB

213.10	63.20	68.60	13.40
1/1	2/3	4/3	5/4

R.RADEGLIA,S.SCHEITHAUER
Z CHEM   14,   20 (1974)

1425   P 004400 DITHIOLOMALONIC ACID DIETHYL ESTER

```
 7   6              4   5
H3C-H2C-S        S-CH2-CH3
      \1  2   3/
        C-CH2-C
       //       \\
      O          O
```

FORMULA C7H12O2S2   MOL WT 192.30
SOLVENT CDCL3
ORIG ST C4H12SI   TEMP   AMB

190.00	57.80	24.00	14.40
1/1	2/3	4/3	5/4

R.RADEGLIA,S.SCHEITHAUER
Z CHEM   14,   20 (1974)

1426   P 004400 1-THIONO-1-THIOLOMALONIC ACID DIETHYL ESTER

```
 7   6              4   5
H3C-H2C-S        O-CH2-CH3
      \1  2   3/
        C-CH2-C
       //       \\
      S          O
```

FORMULA C7H12O2S2   MOL WT 192.30
SOLVENT CDCL3
ORIG ST C4H12SI   TEMP   AMB

225.10	56.10	166.70	61.20	14.00	31.40
1/1	2/3	3/1	4/3	5/4	6/3
12.00					
7/4					

R.RADEGLIA,S.SCHEITHAUER
Z CHEM   14,   20 (1974)

1427   P 004400 3-THIONO-1-THIOLOMALONIC ACID DIETHYL ESTER

FORMULA C7H12O2S2   MOL WT 192.30

7   6   4   5   SOLVENT CDCL3
H3C-H2C-S   O-CH2-CH3   ORIG ST C4H12SI   TEMP   AMB
     \1 2 3/
      C-CH2-C   190.70   61.20   211.30   68.80   13.40   23.80
     //   \\   1/1   2/3   3/1   4/3   5/4   6/3
    O   S   14.50
                            7/4

                          R.RADEGLIA,S.SCHEITHAUER
                          Z CHEM   14,   20 (1974)

1428   P 004400 THIOLOMALONIC ACID DIETHYL ESTER

FORMULA C7H12O3S   MOL WT 176.24

7   6   4   5   SOLVENT CDCL3
H3C-H2C-S   O-CH2-CH3   ORIG ST C4H12SI   TEMP   AMB
     \1 2 3/
      C-CH2-C   190.60   49.70   165.80   61.40   14.10   23.90
     //   \\   1/1   2/3   3/1   4/3   5/4   6/3
    O   O   14.50
                            7/4

                          R.RADEGLIA,S.SCHEITHAUER
                          Z CHEM   14,   20 (1974)

1429   P 004400 THIONOMALONIC ACID DIETHYL ESTER

FORMULA C7H12O3S   MOL WT 176.24

7   6   4   5   SOLVENT CDCL3
H3C-H2C-O   O-CH2-CH3   ORIG ST C4H12SI   TEMP   AMB
     \1 2 3/
      C-CH2-C   212.40   52.70   166.80   61.20   14.10   68.90
     //   \\   1/1   2/3   3/1   4/3   5/4   6/3
    S   O   13.40
                            7/4

                          R.RADEGLIA,S.SCHEITHAUER
                          Z CHEM   14,   20 (1974)

1430   P 004500 DIVINYLSULFONE
   2 1
   H2C=CH
        \   FORMULA C4H6O2S   MOL WT 118.16
         SO2   SOLVENT C4H6O2S
        /   ORIG ST C4H12SI   TEMP   AMB
   H2C=CH
                        137.10   129.80
                         1/2   2/3

                        G.C.LEVY,D.C.DITTMER
                        ORG MAGN RESON   4, 107 (1972)

1431   P 004700 PROPYLENE CARBONATE

                        FORMULA C4H6O3   MOL WT 102.09
    O   SOLVENT CDCL3
    || ORIG ST C4H12SI   TEMP   AMB
    C2
   / \   73.90   155.20   70.80   19.10
  O   O   1/2   2/1   3/3   4/4
  | |
  C---C   L.F.JOHNSON,W.C.JANKOWSKI
  3   1\4   CARBON-13 NMR SPECTRA,JOHN
      CH3   WILEY AND SONS,NEW YORK   62 (1972)

1432   P 004700 DIBUTYL CARBONATE

	FORMULA	C9H18O3			MOL WT	174.24
	SOLVENT	CDCL3				
H3C-CH2-CH2-CH2-O	ORIG ST	C4H12SI			TEMP	AMB
5 4 3 2 \1						
C=O	155.50	67.60	30.90	19.10	13.60	67.60
/	1/1	2/3	3/3	4/3	5/4	6/3
H3C-CH2-CH2-CH2-O	30.90	19.10	13.60			
9 8 7 6	7/3	8/3	9/4			

L.F.JOHNSON,W.C.JANKOWSKI
CARBON-13 NMR SPECTRA,JOHN
WILEY AND SONS,NEW YORK   363 (1972)

1433   P 004800 ETHYL CARBAMATE

	FORMULA	C3H7NO2	MOL WT	89.09
	SOLVENT	CDCL3		
O	ORIG ST	C4H12SI	TEMP	AMB
II				
H2N-C-O-CH2-CH3	157.80	60.90	14.50	
1 2 3	1/1	2/3	3/4	

L.F.JOHNSON,W.C.JANKOWSKI
CARBON-13 NMR SPECTRA,JOHN
WILEY AND SONS,NEW YORK   37 (1972)

1434   P 004800 ETHYL METHYLCARBAMATE

	FORMULA	C4H9NO2	MOL WT	103.12
O	SOLVENT	CDCL3		
II	ORIG ST	C4H12SI	TEMP	AMB
H3C-N-C-O-CH2-CH3				
2 I 1 3 4	157.80	27.40	60.70	14.70
H	1/1	2/4	3/3	4/4

L.F.JOHNSON,W.C.JANKOWSKI
CARBON-13 NMR SPECTRA,JOHN
WILEY AND SONS,NEW YORK   86 (1972)

1435   P 004900 TETRAMETHYLUREA

	FORMULA	C5H12N2O	MOL WT	116.16
4 3	SOLVENT	CDCL3		
H3C O CH3	ORIG ST	C4H12SI	TEMP	AMB
\ II /				
N-C-N	165.40	38.50	38.50	38.50
/ 1 \	1/1	2/4	3/4	4/4
H3C CH3	38.50			
5 2	5/4			

L.F.JOHNSON,W.C.JANKOWSKI
CARBON-13 NMR SPECTRA,JOHN
WILEY AND SONS,NEW YORK   137 (1972)

1436   P 005000 SODIUM DIETHYLDITHIOCARBAMATE

S	FORMULA	C5H10NNAS2	MOL WT	171.26
5 4 II	SOLVENT	H2O		
H3C-CH2 C	ORIG ST	C4H12SI	TEMP	AMB
\ /1\ - +				
N S NA	206.40	49.50	12.30	49.50
/	1/1	2/3	3/4	4/3
H3C-CH2	12.30			
3 2	5/4			

L.F.JOHNSON,W.C.JANKOWSKI
CARBON-13 NMR SPECTRA,JOHN
WILEY AND SONS,NEW YORK   120 (1972)

1437   P 005000 TETRAMETHYLTHIOUREA

```
         S
    5    ||    2
  H3C    C    CH3
     \  /1\  /
      N     N
   4/    \3
  H3C        CH3
```

FORMULA C5H12N2S			MOL WT	132.23
SOLVENT CDCL3				
ORIG ST C4H12SI			TEMP	AMB

193.90	43.00	43.00	43.00	43.00
1/1	2/4	3/4	4/4	5/4

L.F.JOHNSON,W.C.JANKOWSKI
CARBON-13 NMR SPECTRA,JOHN
WILEY AND SONS,NEW YORK   138 (1972)

1438   P 005200 TRICHLOROACETYLISOCYANATE

```
   CL  O
   ||  ||
CL-C--C-N=C=O
   |  2    3
   CL
```

FORMULA C3CL3NO2		MOL WT 188.40
SOLVENT CDCL3		
ORIG ST C4H12SI		TEMP AMB

92.40	158.60	130.20
1/1	2/1	3/1

L.F.JOHNSON,W.C.JANKOWSKI
CARBON-13 NMR SPECTRA,JOHN
WILEY AND SONS,NEW YORK   14 (1972)

1439   Q 000210 TETRACHLOROCYCLOPROPENE

```
      CL
      |
      C2
    1/ \3
CL-C---C-CL
      |
      CL
```

FORMULA C3CL4	MOL WT 177.85
SOLVENT CHCL3	
ORIG ST C4H12SI	TEMP AMB

122.80	62.40
1/S	3/S

G.E.HAWKES,R.A.SMITH,J.D.ROBERTS
J ORG CHEM   39, 1276 (1974)

1440   Q 000210 CYCLOPROPYLCARBINYL CATION

```
3C
 |\   +
 | C-CH2
1/1 4
2C
```

FORMULA C4H7			MOL WT 55.10
SOLVENT			
ORIG ST CS2			TEMP 195

107.80	54.60	54.60	54.60
1/2	2/3	3/3	4/3

G.A.OLAH,G.LIANG
J AM CHEM SOC   95, 3792 (1973)

1441   Q 000210 METHYLCYCLOPROPYLCARBENIUM ION

```
3C
 |\2 1+
 | C-C-CH3
1/  | 5
4C   H
```

FORMULA C5H9			MOL WT	69.13
SOLVENT CLFO2S/F5SB				
ORIG ST CS2			TEMP	198

251.60	66.00	56.20	56.20	32.50
1/2	2/2	3/3	4/3	5/4

G.A.OLAH,P.W.WESTERMAN
J AM CHEM SOC   95, 7530 (1973)

1442   Q 000210 CYCLOPROPYLDIMETHYLCARBINYL CATION

```
                              FORMULA   C6H11                 MOL WT   83.15
                              SOLVENT
      3C                      ORIG ST  C S2                   TEMP      195
     I\2 1+
      I C-C-CH3               279.30    58.70    52.10    52.10    29.80    38.60
     I/  I 6                   1/1       2/2      3/3      4/3      5/4      6/4
   4C    CH3
        5                     G.A.OLAH,G.LIANG
                              J AM CHEM SOC                    95, 3792 (1973)
```

1443   Q 000210 DIMETHYLCYCLOPROPYLCARBENIUM ION

```
                              FORMULA   C6H11                 MOL WT   83.15
      C6                      SOLVENT   CLF02S/F5SB
     /I                       ORIG ST  C S2                   TEMP      198
    / I +
   C--C-C1                    279.30    38.60    29.80    58.70    52.10    52.10
   5  4/ \                     1/1       2/4      3/4      4/2      5/3      6/3
       / CH3
   H3C    2                   G.A.OLAH,P.W.WESTERMAN
     3                        J AM CHEM SOC                    95, 7530 (1973)
```

1444   Q 000210 TRANS-1,2-DIVINYLCYCLOPROPANE

```
    4   5
   CH=CH2                     FORMULA   C7H10                 MOL WT   94.16
   /                          SOLVENT   CCL4/CDCL3
   C1                         ORIG ST  C4H12SI                TEMP      AMB
  /I
 3C I                         24.20    14.70   139.80   112.20
  \I                           1/2      3/3      4/2      5/3
   C2
    \6  7                     H.GUENTHER,G.JIKELI
   CH=CH2                     CHEM BER                   106, 1863 (1973)
```

1445   Q 000210 DICYCLOPROPYLCARBENIUM ION

```
   3C                         FORMULA   C7H11                 MOL WT   95.17
  I\2 1+                      SOLVENT   CLF02S/F5SB
  I C-C-H                     ORIG ST  C S2                   TEMP      198
  I/  I
 4C  5C                       252.40    44.40    37.40    37.40    44.40    37.40
   / \                         1/2       2/2      3/3      4/3      5/2      6/3
  C---C                       37.40
  6   7                        7/3

                              G.A.OLAH,P.W.WESTERMAN
                              J AM CHEM SOC                    95, 7530 (1973)
```

1446   Q 000210 DICYCLOPROPYL KETOXIME

```
                              FORMULA   C7H11NO               MOL WT  125.17
     OH                       SOLVENT   CDCL3
    /                         ORIG ST  C4H12SI                TEMP      AMB
  7C   N   C4
  I\  II  /I                  161.50     9.10     5.00     5.00     9.20     5.30
  I C-C-C I                    1/1       2/2      3/3      4/3      5/2      6/3
  I/5 1 2\I                   5.30
  6C       C3                  7/3

                              L.F.JOHNSON,W.C.JANKOWSKI
                              CARBON-13 NMR SPECTRA,JOHN
                              WILEY AND SONS,NEW YORK          261 (1972)
```

1447   Q 000210 HYDROXYDICYCLOPROPYLCARBENIUM ION

```
     C7
     /|
    / |
  C--C
  6  5\ +
       C1
      / \
   C2  OH
   /|
  / |
4C--C3
```

FORMULA	C7H11O			MOL WT	111.17
SOLVENT	CLFO2S/F5SB				
ORIG ST	CS2			TEMP	198

236.10	18.50	25.70	25.70	27.80	25.10
1/1	2/2	3/3	4/3	5/2	6/3
25.10					
7/3					

G.A.OLAH,P.W.WESTERMAN
J AM CHEM SOC   95, 7530 (1973)

1448   Q 000210 CYCLOPROPYLMETHYLETHYLCARBINYL CATION

FORMULA	C7H13			MOL WT	97.18
SOLVENT					
ORIG ST	CS2			TEMP	195

```
 3C
  |\2 1+
  | C-C-CH2-CH3
  |/  | 5  6
4C  CH3
    7
```

280.40	59.10	52.50	52.50	27.30	11.00
1/1	2/2	3/3	4/3	5/3	6/4
50.00					
7/4					

G.A.OLAH,G.LIANG
J AM CHEM SOC   95, 3792 (1973)

1449   Q 000210 DICYCLOPROPYLMETHYLCARBENIUM ION

```
    C7
    /|
   / |
  C--C6
  8   \ +
      1C
  5  / \
  C--C3  CH3
   \ |    2
    \|
     C4
```

FORMULA	C8H13			MOL WT	109.19
SOLVENT	CLFO2S/F5SB				
ORIG ST	CS2			TEMP	198

274.10	37.10	43.90	37.10	37.10	43.90
1/1	2/4	3/2	4/3	5/3	6/2
37.10	37.10				
7/3	8/3				

G.A.OLAH,P.W.WESTERMAN
J AM CHEM SOC   95, 7530 (1973)

1450   Q 000210 TRICYCLOPROPYLCARBENIUM ION

```
    10C
    /|
   / |
  9C--C8
     \ +
 7C--C-C1
  \ |5 \
   \|  2C--C4
   C   | /
   6   |/
       3C
```

FORMULA	C10H15			MOL WT	135.23
SOLVENT	CLFO2S/F5SB				
ORIG ST	CS2			TEMP	198

270.30	31.30	29.60	29.60	31.30	29.60
1/1	2/2	3/3	4/3	5/2	6/3
29.60	31.30	29.60	29.60		
7/3	8/2	9/3	10/3		

G.A.OLAH,P.W.WESTERMAN
J AM CHEM SOC   95, 7530 (1973)

1451   Q 000220 CYCLOBUTANONE

```
      O
  2 1/
  C-C
  | |
  C-C
  3 4
```

FORMULA	C4H6O			MOL WT	70.09
SOLVENT	CDCL3				
ORIG ST	C4H12SI			TEMP	308

208.90	47.80	9.90	47.80
1/1	2/3	3/3	4/3

G.E.HAWKES,K.HERWIG,J.D.ROBERTS
J ORG CHEM   39, 1017 (1974)

1452   Q 000220 CYCLOBUTANONE OXIME

```
        N
   4 1⁄ ＼          FORMULA   C4H7NO              MOL WT   85.11
   C-C    OH        SOLVENT   CDCL3
   I I              ORIG ST   C4H12SI            TEMP       308
   C-C
   3 2                 159.70    30.70    14.60    31.60
                        1/1       2/3      3/3      4/3

                     G.E.HAWKES,K.HERWIG,J.D.ROBERTS
                     J ORG CHEM                    39, 1017 (1974)
```

1453   Q 000220 3,3,4,4-TETRACHLORO-1,1,2,2-TETRAFLUOROCYCLO-
                              BUTANE

```
    F CL
   1I I4             FORMULA   C4CL4F4            MOL WT   265.85
 F-C-C-CL            SOLVENT   C4H8O2
   I I              ORIG ST   C4H12SI            TEMP       AMB
 F-C-C-CL
   2I I3               114.00    88.40
    F CL               1/T       3/S

                     G.E.HAWKES,R.A.SMITH,J.D.ROBERTS
                     J ORG CHEM                    39, 1276 (1974)
```

1454   Q 000220 TETRACHLOROCYCLOBUTENE-1-ONE

```
   ⊃    CL
    ＼1 2⁄           FORMULA   C4CL4O             MOL WT   205.86
    C-C             SOLVENT   CHCL3
    I II            ORIG ST   C4H12SI            TEMP       AMB
 CL-C-C3
   4I  ＼               174.00   134.10   166.80   189.80
   CL  CL              1/S      2/S      3/S      4/S

                     G.E.HAWKES,R.A.SMITH,J.D.ROBERTS
                     J ORG CHEM                    39, 1276 (1974)
```

1455   Q 000220 HEXACHLOROCYCLOBUTENE

```
   CL  CL
   4I  ⁄             FORMULA   C4CL6              MOL WT   260.76
 CL-C-C1             SOLVENT   CHCL3
   I II             ORIG ST   C4H12SI            TEMP       AMB
 CL-C-C2
   3I  ＼               134.10    92.10
   CL  CL              1/S       3/S

                     G.E.HAWKES,R.A.SMITH,J.D.ROBERTS
                     J ORG CHEM                    39, 1276 (1974)
```

1456   Q 000220 CYCLOBUTYLNITRILE

```
    5CN
   2 II              FORMULA   C5H6N              MOL WT   80.11
   C-C               SOLVENT   CDCL3
   I I              ORIG ST   C4H12SI            TEMP       AMB
   C-C
   3 4                 22.00     27.10    19.90    27.10   122.40
                       1/2       2/3      3/3      4/3      5/1

                     L.F.JOHNSON,W.C.JANKOWSKI
                     CARBON-13 NMR SPECTRA,JOHN
                     WILEY AND SONS,NEW YORK              56 (1972)
```

1457   Q 000220 METHYLENE CYCLOBUTANE

```
        5
       CH2
       //
   4C—C1
   |  |
   C—C2
   3
```

FORMULA	C5H8			MOL WT	68.12
SOLVENT	CDCL3				
ORIG ST	C4H12SI			TEMP	AMB

150.40	32.10	16.80	32.10	104.80
1/1	2/3	3/3	4/3	5/3

L.F.JOHNSON,W.C.JANKOWSKI
CARBON-13 NMR SPECTRA,JOHN
WILEY AND SONS,NEW YORK   110 (1972)

1458   Q 000220 3,4-DIMETHYLENECYCLOBUTENE

```
    6C
    \\
   4C—C1
   |  ||
   3C—C2
    //
    5C
```

FORMULA	C6H6		MOL WT	78.11
SOLVENT	CDCL3			
ORIG ST	C4H12SI		TEMP	AMB

145.87	149.76	94.56
1/2	3/1	5/3

A.J.JONES,P.J.GARRATT,K.P.C.VOLLHARDT
ANGEW CHEM   85, 260 (1973)

1459   Q 000220 1,4-BIS(DICHLOROMETHYLENE)-2,3-DICHLORO-CYCLOBUTENE

```
   CL
   |
   C5
  / \1 2
 CL  C—C—CL
     |  ||
 CL  C—C—CL
  \  //4 3
   C6
   |
   CL
```

FORMULA	C6CL6		MOL WT	284.78
SOLVENT	CHCL3/C4H8O2			
ORIG ST	C4H12SI		TEMP	AMB

136.10	131.50	109.50
1/S	2/S	5/S

G.E.HAWKES,R.A.SMITH,J.D.ROBERTS
J ORG CHEM   39, 1276 (1974)

1460   Q 000220 1,2-BIS(DICHLOROMETHYLENE)-TETRACHLORO-CYCLOBUTANE

```
   CL
   |
   C5  CL
  / \1 |2
 CL  C—C—CL
     |  |
 CL  C—C—CL
  \  //4 |3
   C6  CL
   |
   CL
```

FORMULA	C6CL8		MOL WT	355.69
SOLVENT	CHCL3			
ORIG ST	C4H12SI		TEMP	AMB

133.00	89.50	127.30
1/S	2/S	5/S

G.E.HAWKES,R.A.SMITH,J.D.ROBERTS
J ORG CHEM   39, 1276 (1974)

1461   Q 000220 BENZOCYCLOBUTENE-5,6-DIONE

```
         1
    O    C
    \\  / \
   6C—C7  C2
   |  ||  |
   5C—C8  C3
    //  \ //
    O    C
         4
```

FORMULA	C8H4O2			MOL WT	132.12
SOLVENT	CDCL3				
ORIG ST	C4H12SI			TEMP	AMB

135.45	122.37	194.28	173.20
1/2	2/2	5/1	7/1

A.J.JONES,P.J.GARRATT,K.P.C.VOLLHARDT
ANGEW CHEM   85, 260 (1973)

1462   Ɔ 000220 BENZOCYCLOBUTENE

```
        1
        C
      ╱ ╲
   6C-C7  C2
    |  ‖   |
   5C-C8  C3
      ╲ ╱
        C4
```

FORMULA   C8H8   MOL WT   104.15
SOLVENT   CDCL3
ORIG ST   C4H12SI   TEMP   AMB

122.29   126.50   133.07   146.31
 1/2   2/2   5/3   7/1

A.J.JONES,P.J.GARRATT,K.P.C.VOLLHARDT
ANGEW CHEM   85, 260 (1973)

1463   Q 000220 2,2,4,4-TETRAMETHYLCYCLOBUTANEDIONE

```
        5
    Ɔ  CH3
     ╲  12
   1C-C—CH3
   4|  | 6
  H3C-C-C3
   8 |  ╲
  H3C   O
    7
```

FORMULA   C8H12O2   MOL WT   140.18
SOLVENT   CDCL3
ORIG ST   C4H12SI   TEMP   AMB

215.00   70.40   215.00   70.40   18.80   18.80
 1/1   2/1   3/1   4/1   5/4   6/4
18.80   18.80
 7/4   8/4

L.F.JOHNSON,W.C.JANKOWSKI
CARBON-13 NMR STECTRA,JOHN
WILEY AND SONS,NEW YORK   308 (1972)

1464   Ɔ 000220 2-THIANORBIPHENYLENE DIOXIDE

```
   7      1
   C      C
  ╱ ╲8  ╱ ╲
 6╱   C-C10 C2
Ɔ2S   |  ‖   |
  ╲   C-C11 C3
   ╲ ╱9  ╲ ╱
   C      C
   5      4
```

FORMULA   C10H6O2S   MOL WT   190.22
SOLVENT   CDCL3
ORIG ST   C4H12SI   TEMP   AMB

134.30   124.75   111.90   142.34   145.24
 1/2   2/2   7/2   8/1   10/1

A.J.JONES,P.J.GARRATT,K.P.C.VOLLHARDT
ANGEW CHEM   85, 260 (1973)

1465   Q 000220 BENZOCYCLOBUTA(1,2-C)THIOPHENE

```
   7      1
   C      C
  ╱ ╲  ╱ ╲
 6╱  8C-C10 C2
S    |  ‖   |
  ╲  9C-C11 C3
   ╲ ╱  ╲ ╱
   C      C
   5      4
```

FORMULA   C10H6S   MOL WT   158.22
SOLVENT   CDCL3
ORIG ST   C4H12SI   TEMP   AMB

118.13   128.44   112.51   146.68   145.56
 1/2   2/2   5/2   8/1   10/1

A.J.JONES,P.J.GARRATT,K.P.C.VOLLHARDT
ANGEW CHEM   85, 260 (1973)

1466   Q 000220 BIPHENYLENE

```
   8      1
   C      C
  ╱ ╲  ╱ ╲
 7C  9C-C11 C2
 ‖   |  |   ‖
 6C 10C-C12 C3
  ╲ ╱  ╲ ╱
   C      C
   5      4
```

FORMULA   C12H8   MOL WT   152.20
SOLVENT   CDCL3
ORIG ST   C4H12SI   TEMP   AMB

117.83   128.47   151.70
 1/2   2/2   9/1

A.J.JONES,P.J.GARRATT,K.P.C.VOLLHARDT
ANGEW CHEM   85, 260 (1973)

1467　Q 000300　1,2,3,3,4,5,5-HEPTACHLOROCYCLOPENTENE

```
    CL  CL
     \  /
  H   C5  CL
   \4/  \ /
  CL-C    C1
    |     ||
  CL-C---C
    3|    2\
    CL     CL
```

FORMULA	C5HCL7	MOL WT 309.23
SOLVENT	CHCL3	
ORIG ST	C4H12SI	TEMP AMB

135.90	86.90	77.60
1/S	3/S	4/D

G.E.HAWKES,R.A.SMITH,J.D.ROBERTS
J ORG CHEM　　　　　　　　　39, 1276 (1974)

1468　Q 000300　1,2,3,4-TETRACHLOROCYCLOPENTA-1,3-DIENE

```
      5
  CL  C  CL
   \4/  \1/
    C    C
    ||   ||
    C---C
   /3    2\
  CL      CL
```

FORMULA	C5H2CL4	MOL WT 203.88
SOLVENT	CHCL3	
ORIG ST	C4H12SI	TEMP AMB

125.20	128.20	47.10
1/S	2/S	5/T

G.E.HAWKES,R.A.SMITH,J.D.ROBERTS
J ORG CHEM　　　　　　　　　39, 1276 (1974)

1469　Q 000300　CYCLOPENTADIENE

```
  3   2
  C---C
  ||  ||
 4C   C1
   \ /
    C
    5
```

FORMULA	C5H6	MOL WT 66.10
SOLVENT	C5H6	
ORIG ST	C4H12SI	TEMP AMB

133.00	133.40	42.20
1/2	2/2	5/3

Y.K.GRISHIN,N.M.SERGEYEV,Y.A.USTYNYUK
ORG MAGN RESON　　　　　　　4,　377 (1972)

1470　Q 000300　CYCLOPENTANONE

```
  3   2
  C---C
  |   |
 C4   C=O
   \5/1
    C
```

FORMULA	C5H8O	MOL WT 84.12
SOLVENT	CDCL3	
ORIG ST	C4H12SI	TEMP 308

219.60	37.90	22.70	22.70	37.90
1/1	2/3	3/3	4/3	5/3

G.E.HAWKES,K.HERWIG,J.D.ROBERTS
J ORG CHEM　　　　　　　　　39, 1017 (1974)

1471　Q 000300　CYCLOPENTANONE

```
     O
     ||
     C1
    / \
  5C   C2
   |   |
  4C---C3
```

FORMULA	C5H8O	MOL WT 84.12
SOLVENT	C4H8O2	
ORIG ST	C4H12SI	TEMP AMB

218.20	38.00	23.50	23.50	38.00
1/1	2/3	3/3	4/3	5/3

L.F.JOHNSON,W.C.JANKOWSKI
CARBON-13 NMR SPECTRA,JOHN
WILEY AND SONS,NEW YORK　　　113 (1972)

1472   Q 000300 CYCLOPENTANONE

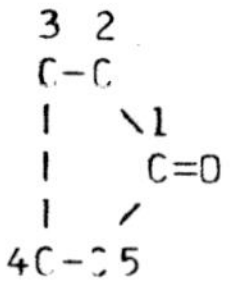

```
  3 2
  C-C
  |  \1
  |   C=0
  |  /
4C-C5
```

FORMULA	C5H8O
SOLVENT	CDCL3
ORIG ST	C4H12SI

MOL WT   84.12

TEMP   AMB

219.40	38.10	23.20	23.20	38.10
1/1	2/3	3/3	4/3	5/3

J.B. STOTHERS, C.T. TAN
CAN J CHEM   52, 308 (1974)

1473   Q 000300 CYCLOPENTANONE OXIME

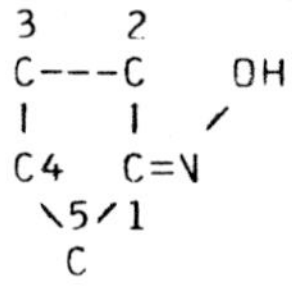

```
 3    2
 C---C   OH
 |   |  /
 C4  C=N
  \5/1
   C
```

FORMULA	C5H9NO
SOLVENT	CDCL3
ORIG ST	C4H12SI

MOL WT   99.13

TEMP   308

167.10	27.10	25.10	24.40	30.60
1/1	2/3	3/3	4/3	5/3

G.E.HAWKES, K.HERWIG, J.D.ROBERTS
J ORG CHEM   39, 1017 (1974)

1474   Q 000300 CYCLOPENTANOL

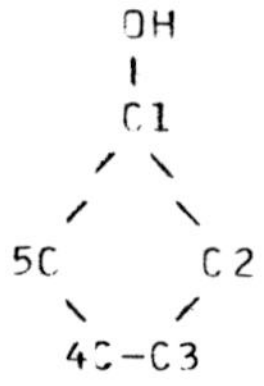

```
   OH
   |
   C1
  / \
5C    C2
  \  /
  4C-C3
```

FORMULA	C5H10O
SOLVENT	H2O
ORIG ST	C4H12SI

MOL WT   86.13

TEMP   AMB

75.10	36.00	24.10
1/2	2/3	3/3

R.G.S.RITCHIE, N.CYR, B.KORSCH, H.J.KOCH,
A.S.PERLIN
CAN J CHEM   53, 1424 (1975)

1475   Q 000300 CIS-CYCLOPENTANE-1,2-DIOL

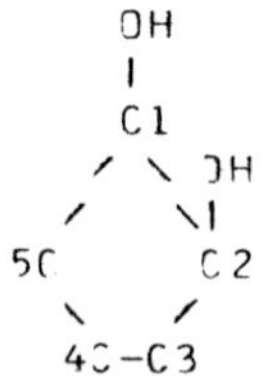

```
   OH
   |
   C1
  / \ OH
 /   \|
5C    C2
 \   /
 4C-C3
```

FORMULA	C5H10O2
SOLVENT	H2O
ORIG ST	C4H12SI

MOL WT   102.13

TEMP   AMB

74.80	31.20	20.30
1/2	3/3	4/3

R.G.S.RITCHIE, N.CYR, B.KORSCH, H.J.KOCH,
A.S.PERLIN
CAN J CHEM   53, 1424 (1975)

1476   Q 000300 TRANS-CYCLOPENTANE-1,2-DIOL

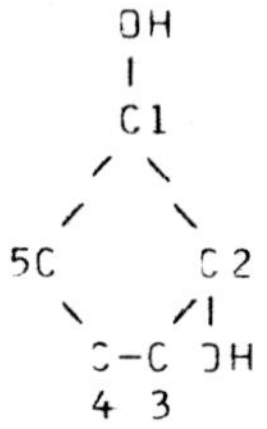

```
   OH
   |
   C1
  / \
5C    C2
 \   /|
  C-C OH
  4  3
```

FORMULA	C5H10O2
SOLVENT	H2O
ORIG ST	C4H12SI

MOL WT   102.13

TEMP   AMB

97.70	32.40	21.30
1/2	3/3	4/3

R.G.S.RITCHIE, N.CYR, B.KORSCH, H.J.KOCH,
A.S.PERLIN
CAN J CHEM   53, 1424 (1975)

1477   Q 000300   CIS-CYCLOPENTANE-1,3-DIOL

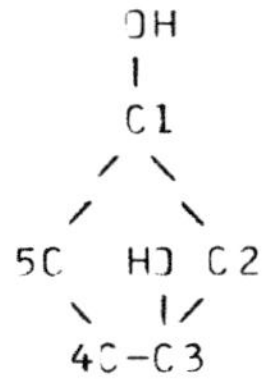

```
     OH
      |
     C1
    / \
   /   \
 5C  HO C2
   \   |/
    4C-C3
```

FORMULA	C5H10O2		MOL WT	102.13
SOLVENT	H2O			
ORIG ST	C4H12SI		TEMP	AMB

73.40	44.60	34.00
1/2	2/3	4/3

R.G.S.RITCHIE,N.CYR,B.KORSCH,H.J.KOCH,
A.S.PERLIN
CAN J CHEM   53, 1424 (1975)

1478   Q 000300   TRANS-CYCLOPENTANE-1,3-DIOL

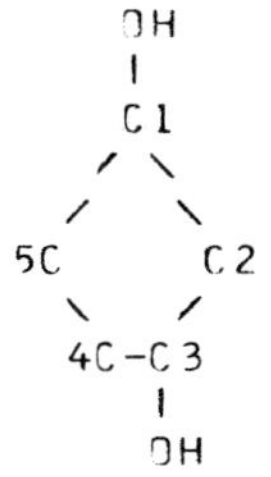

```
     OH
      |
     C1
    / \
   /   \
 5C     C2
   \   /
    4C-C3
      |
     OH
```

FORMULA	C5H10O2		MOL WT	102.13
SOLVENT	H2O			
ORIG ST	C4H12SI		TEMP	AMB

73.20	44.90	33.70
1/2	2/3	4/3

R.G.S.RITCHIE,N.CYR,B.KORSCH,H.J.KOCH,
A.S.PERLIN
CAN J CHEM   53, 1424 (1975)

1479   Q 000300   CIS-,CIS-CYCLOPENTANE-1,2,3-TRIOL

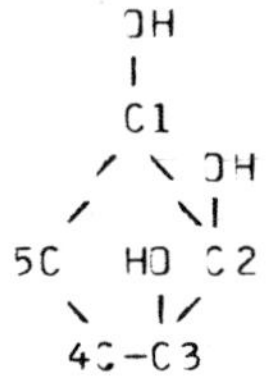

```
     OH
      |
     C1
    / \ OH
   /   \|
 5C  HO C2
   \   |/
    4C-C3
```

FORMULA	C5H10O3		MOL WT	118.13
SOLVENT	H2O			
ORIG ST	C4H12SI		TEMP	AMB

72.80	74.80	29.90
1/2	2/2	4/3

R.G.S.RITCHIE,N.CYR,B.KORSCH,H.J.KOCH,
A.S.PERLIN
CAN J CHEM   53, 1424 (1975)

1480   Q 000300   CIS-,TRANS-CYCLOPENTANE-1,3,2-TRIOL

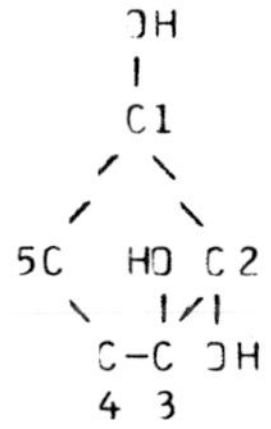

```
     OH
      |
     C1
    / \
   /   \
 5C  HO C2
   \   |/|
   C-C OH
   4 3
```

FORMULA	C5H10O3		MOL WT	118.13
SOLVENT	H2O			
ORIG ST	C4H12SI		TEMP	AMB

76.60	85.10	29.10
1/2	2/2	4/3

R.G.S.RITCHIE,N.CYR,B.KORSCH,H.J.KOCH,
A.S.PERLIN
CAN J CHEM   53, 1424 (1975)

1481   Q 000300   CIS-,TRANS-CYCLOPENTANE-1,2,3-TRIOL

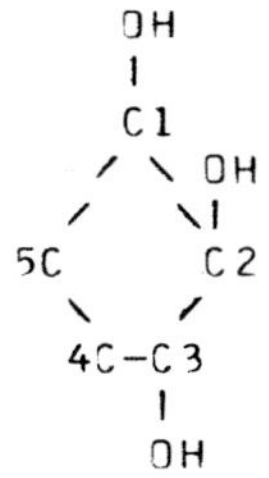

```
     OH
      |
     C1
    / \ OH
   /   \|
 5C     C2
   \   /
    4C-C3
      |
     OH
```

FORMULA	C5H10O3		MOL WT	118.13
SOLVENT	H2O			
ORIG ST	C4H12SI		TEMP	AMB

72.50	79.90	76.80	29.00	29.00
1/2	2/2	3/2	4/3	5/3

R.G.S.RITCHIE,N.CYR,B.KORSCH,H.J.KOCH,
A.S.PERLIN
CAN J CHEM   53, 1424 (1975)

1482 Q 000300 CIS-,TRANS-CYCLOPENTANE-1,4,2-TRIOL

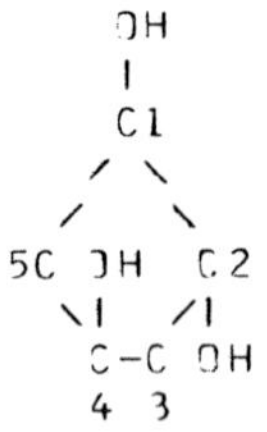

```
      OH
      |
      C1
     / \
    /   \
  5C OH  C2
   \|  /|
    C-C OH
    4  3
```

FORMULA   C5H10O3   MOL WT 118.13
SOLVENT   H2O
ORIG ST   C4H12SI   TEMP   AMB

78.40   78.20   41.80   70.10   41.90
1/2   2/2   3/3   4/2   5/3

R.G.S.RITCHIE,N.CYR,B.KORSCH,H.J.KOCH,
A.S.PERLIN
CAN J CHEM   53, 1424 (1975)

1483 Q 000300 CIS-,CIS-,TRANS-CYCLOPENTANETETROL

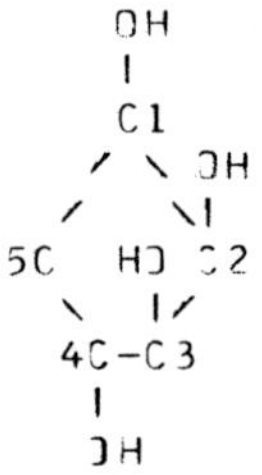

```
      OH
      |
      C1
     / \ OH
    /   \|
  5C HO  C2
   \   |/
    4C-C3
      |
      OH
```

FORMULA   C5H10O4   MOL WT 134.13
SOLVENT   H2O
ORIG ST   C4H12SI   TEMP   AMB

70.70   74.20   79.30   76.40   39.10
1/2   2/2   3/2   4/2   5/3

R.G.S.RITCHIE,N.CYR,B.KORSCH,H.J.KOCH,
A.S.PERLIN
CAN J CHEM   53, 1424 (1975)

1484 Q 000300 CIS-,TRANS-,CIS-CYCLOPENTANETETROL

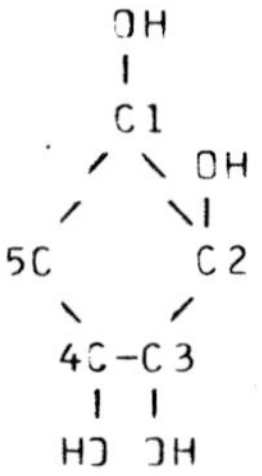

```
      OH
      |
      C1
     / \ OH
    /   \|
  5C     C2
   \    /
    4C-C3
    | |
   HO OH
```

FORMULA   C5H10O4   MOL WT 134.13
SOLVENT   H2O
ORIG ST   C4H12SI   TEMP   AMB

69.50   77.10   39.20
1/2   2/2   5/3

R.G.S.RITCHIE,N.CYR,B.KORSCH,H.J.KOCH,
A.S.PERLIN
CAN J CHEM   53, 1424 (1975)

1485 Q 000300 CIS-,TRANS-,TRANS-CYCLOPENTANETETROL

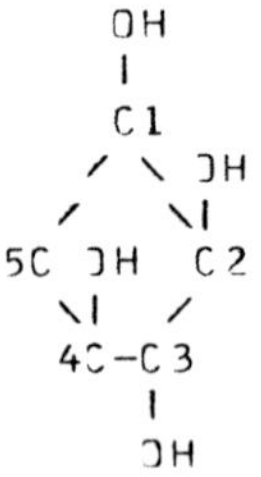

```
      OH
      |
      C1
     / \ OH
    /   \|
  5C OH  C2
   \|  /
    4C-C3
      |
      OH
```

FORMULA   C5H10O4   MOL WT 134.13
SOLVENT   H2O
ORIG ST   C4H12SI   TEMP   AMB

69.90   76.90   82.50   74.00   38.40
1/2   2/2   3/2   4/2   5/3

R.G.S.RITCHIE,N.CYR,B.KORSCH,H.J.KOCH,
A.S.PERLIN
CAN J CHEM   53, 1424 (1975)

1486 Q 000300 TRANS-,CIS-,TRANS-CYCLOPENTANETETROL

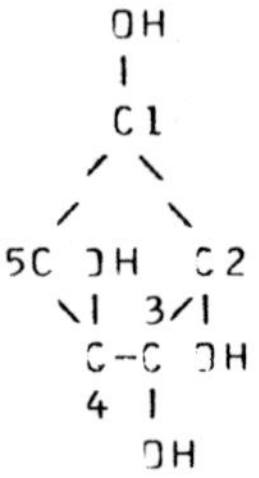

```
      OH
      |
      C1
     / \
    /   \
  5C OH  C2
   \| 3/|
    C-C OH
    4  |
      OH
```

FORMULA   C5H10O4   MOL WT 134.13
SOLVENT   H2O
ORIG ST   C4H12SI   TEMP   AMB

75.80   78.60   38.80
1/2   2/2   5/3

R.G.S.RITCHIE,N.CYR,B.KORSCH,H.J.KOCH,
A.S.PERLIN
CAN J CHEM   53, 1424 (1975)

1487   Q 000300 TRANS-,TRANS-,TRANS-CYCLOPENTANETETROL

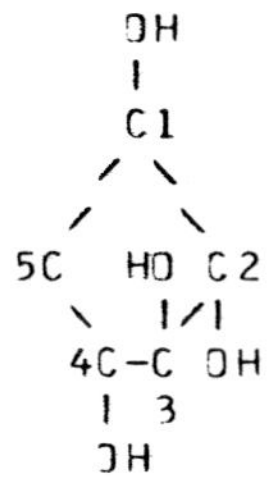

```
    OH
    |
    C1
   / \
  /   \
 5C  HO C2
  \  |/|
  4C-C OH
  | 3
  OH
```

FORMULA C5H10O4   MOL WT 134.13
SOLVENT H2O
ORIG ST C4H12SI   TEMP   AMB

    74.10   83.00   38.40
     1/2   2/2   5/3

R.G.S.RITCHIE,N.CYR,B.KORSCH,H.J.KOCH,
A.S.PERLIN
CAN J CHEM   53, 1424 (1975)

1488   Q 000300 CIS-,CIS-,CIS-,TRANS-CYCLOPENTANEPENTOL

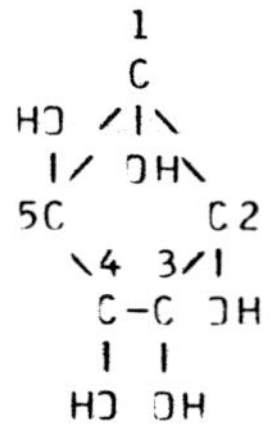

```
    1
    C
 HO /|\
  |/ OH\
  5C    C2
   \4 3/|
    C-C OH
    | |
   HO OH
```

FORMULA C5H10O5   MOL WT 150.13
SOLVENT H2O
ORIG ST C4H12SI   TEMP   AMB

    76.50   72.00   84.10
     1/2   2/2   5/2

R.G.S.RITCHIE,N.CYR,B.KORSCH,H.J.KOCH,
A.S.PERLIN
CAN J CHEM   53, 1424 (1975)

1489   Q 000300 HEXABROMOCYCLOPENTADIENE

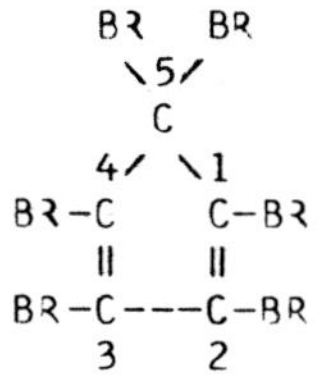

```
  BR   BR
   \5/
    C
   4/ \1
 BR-C   C-BR
   ||  ||
 BR-C---C-BR
   3    2
```

FORMULA C5BR6   MOL WT 539.48
SOLVENT CHCL3
ORIG ST C4H12SI   TEMP   AMB

   130.00   122.80   57.10
     1/S   2/S   5/S

G.E.HAWKES,R.A.SMITH,J.D.ROBERTS
J ORG CHEM   39, 1276 (1974)

1490   Q 000300 1,2-DICHLORO-3,3,4,4,5,5-HEXAFLUOROPENTENE

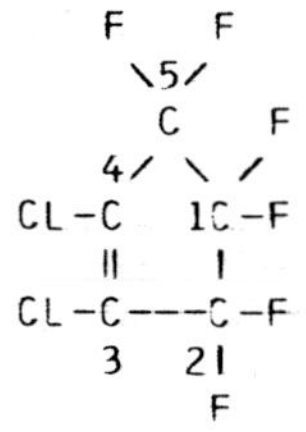

```
   F   F
    \5/
     C   F
   4/ \ /
 CL-C  1C-F
   ||   |
 CL-C---C-F
   3   21
       F
```

FORMULA C5CL2F6   MOL WT 244.95
SOLVENT C6H12
ORIG ST C4H12SI   TEMP   AMB

   102.70   105.20   127.80
     1/T   2/T   3/S

G.E.HAWKES,R.A.SMITH,J.D.ROBERTS
J ORG CHEM   39, 1276 (1974)

1491   Q 000300 TETRACHLOROCYCLOPENT-2-ENE-1,4-DIONE

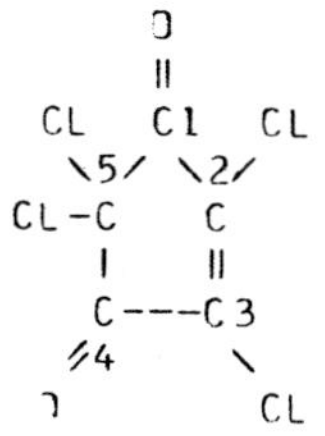

```
      O
      ||
 CL  Cl  CL
  \5/ \2/
 CL-C   C
   |   ||
    C---C3
   /4   \
  O      CL
```

FORMULA C5CL4O2   MOL WT 233.87
SOLVENT CHCL3
ORIG ST C4H12SI   TEMP   AMB

   179.50   148.70   70.30
     1/S   2/S   5/S

G.E.HAWKES,R.A.SMITH,J.D.ROBERTS
J ORG CHEM   39, 1276 (1974)

1492   Q 000300 TETRACHLOROPENT-2-EN-1-ONE-4,5-SULFATE

```
        O
        ‖
    CL  C1  CL
     \5/  \ /
  O  O---C   C2
    \|   |   ‖
     S   C---C3
    ⁄ \ ⁄|4   \
   O   O CL    CL
```

FORMULA   C5CL4O5S   MOL WT 313.93
SOLVENT   CHCL3
ORIG ST   C4H12SI   TEMP   AMB

175.80   136.60   154.70   102.60   97.30
1/S   2/S   3/S   4/S   5/S

G.E.HAWKES,R.A.SMITH,J.D.ROBERTS
J ORG CHEM   39, 1276 (1974)

1493   Q 000300 HEXACHLOROCYCLOPENTADIENE

```
      CL  CL
       \5/
        C
      4⁄ \1
   CL-C    C-CL
      ‖    ‖
   CL-C---C-CL
      3    2
```

FORMULA   C5CL6   MOL WT 272.77
SOLVENT   CHCL3
ORIG ST   C4H12SI   TEMP   AMB

132.00   131.00   82.20
1/S   2/S   5/S

G.E.HAWKES,R.A.SMITH,J.D.ROBERTS
J ORG CHEM   39, 1276 (1974)

1494   Q 000300 HEXACHLOROCYCLOPENTADIENE

```
      CL  CL
       \ /
    CL  C3  CL
     \ /  \ /
     4C    C2
     ‖      ‖
    5C---C1
    ⁄        \
   CL         CL
```

FORMULA   C5CL6   MOL WT 272.77
SOLVENT   C4H8O2
ORIG ST   C4H12SI   TEMP   AMB

132.80   128.50   81.50   128.50   132.80
1/1   2/1   3/1   4/1   5/1

L.F.JOHNSON,W.C.JANKOWSKI
CARBON-13 NMR SPECTRA,JOHN
WILEY AND SONS,NEW YORK   102 (1972)

1495   Q 000300 HEXACHLOROCYCLOPENT-2-EN-1-ONE

```
        O
        ‖
    CL  C1  CL
     \5/  \2/
   CL-C    C
      |    ‖
   CL-C---C
      |     \
     CL     CL
```

FORMULA   C5CL6O   MOL WT 288.77
SOLVENT   CHCL3
ORIG ST   C4H12SI   TEMP   AMB

178.10   131.00   157.10   90.40   87.40
1/S   2/S   3/S   4/S   5/S

G.E.HAWKES,R.A.SMITH,J.D.ROBERTS
J ORG CHEM   39, 1276 (1974)

1496   Q 000300 HEXACHLOROCYCLOPENT-3-EN-1-ONE

```
      CL  CL
       \ /
    CL  C5  O
     \4/ \ ⁄
     C    C1
     ‖    |
    3C---C-CL
    ⁄      |2
   CL     CL
```

FORMULA   C5CL6O   MOL WT 288.77
SOLVENT   CHCL3
ORIG ST   C4H12SI   TEMP   AMB

182.80   76.40   136.80
1/S   2/S   3/S

G.E.HAWKES,R.A.SMITH,J.D.ROBERTS
J ORG CHEM   39, 1276 (1974)

1497   Q 000300 HEXACHLOROCYCLOPENTANE-1,3-DIONE

```
        O                FORMULA  C5CL6O2              MOL WT  304.77
        ||               SOLVENT  CHCL3
    CL  C1  CL           ORIG ST  C4H12SI             TEMP      AMB
      \5/ \2/
    CL-C    C-CL            182.20   87.40   67.10
      |    |                 1/S     2/S     5/S
      C---C-CL
     /4   |3              G.E.HAWKES,R.A.SMITH,J.D.ROBERTS
     O     CL             J ORG CHEM                   39, 1276 (1974)
```

1498   Q 000300 OCTACHLOROCYCLOPENTENE

```
      CL  CL             FORMULA  C5CL8               MOL WT  343.68
        \3/              SOLVENT  CHCL3
         C  CL           ORIG ST  C4H12SI             TEMP      AMB
      2/ \4/
    CL-C   C-CL            135.20   92.90   99.40
      ||   |                1/S     3/S     4/S
    CL-C---C-CL
      1   |5             G.E.HAWKES,R.A.SMITH,J.D.ROBERTS
          CL             J ORG CHEM                   39, 1276 (1974)
```

1499   Q 000300 1,2,3,3,4,5,5-HEPTACHLORO-4-METHYLCYCLOPENTENE

```
      CL  CL             FORMULA  C6H3CL7             MOL WT  323.26
     6  \ /              SOLVENT  CHCL3
    H3C   C1  CL         ORIG ST  C4H12SI             TEMP      AMB
      \5/ \2/
    CL-C   C               134.70  127.40   69.90   24.10
      |    ||               1/S     2/S     5/S     6/Q
    CL-C---C 3
      |4   \            G.E.HAWKES,R.A.SMITH,J.D.ROBERTS
      CL    CL           J ORG CHEM                   39, 1276 (1974)
```

1500   Q 000300 FULVENE

```
                         FORMULA  C6H6                MOL WT   78.11
        1                SOLVENT  CDCL3
    2C==C                ORIG ST  C4H12SI             TEMP       273
     |   \
     |    C=CH2            124.90  134.30  134.30  124.90  152.60  123.40
     |   /5 6               1/2     2/2     3/2     4/2     5/1     6/3
    3C==C
        4                R.HOLLENSTEIN,W.V.PHILIPSBORN,R.VOEGELI,
                         M.NEUENSCHWANDER
                         HELV CHIM ACTA               56,  847 (1973)
```

1501   Q 000300 1-METHYLCYCLOPENTADIENE

```
     3  2
     C---C               FORMULA  C6H8                MOL WT   80.13
     ||  ||              SOLVENT  C6H8
    4C  C1               ORIG ST  C4H12SI             TEMP      AMB
      \ / \
       C   CH3             144.60  128.40  134.10  130.70   45.00   15.40
       5   6               1/1     2/2     3/2     4/2     5/3     6/4

                         Y.K.GRISHIN,N.M.SERGEYEV,Y.A.USTYNYUK
                         ORG MAGN RESON               4,  377 (1972)
```

```
1502              Q  000300  2-METHYLCYCLOPENTADIENE

              6
            CH3                FORMULA  C6H8                    MOL WT   80.13
       3    2/               SOLVENT  C6H8
       C---C                 ORIG ST  C4H12SI                    TEMP     AMB
       ||   ||
      4C    C1                 127.30   142.60   136.50   133.40   40.70   14.80
        \  /                    1/2      2/1      3/2      4/2      5/3     6/4
         C
         5                   Y.K.GRISHIN,N.M.SERGEYEV,Y.A.USTYNYUK
                             ORG MAGN RESON                   4,   377 (1972)
```

```
1503              Q  000300  1-METHYLDICHLOROSILYLCYCLOPENTADIENE

       3    2
       C---C                 FORMULA  C6H8CL2SI                MOL WT  179.12
       ||   ||               SOLVENT  C6H8CL2SI
      4C    C1  CL           ORIG ST  C4H12SI                    TEMP     AMB
        \  \ / 6
         C   SI-CH3            141.00   148.30   142.00   133.80   45.60    7.10
         5    \                 1/1      2/2      3/2      4/2      5/3     6/4
              CL
                             Y.K.GRISHIN,N.M.SERGEYEV,Y.A.USTYNYUK
                             ORG MAGN RESON                   4,   377 (1972)
```

```
1504              Q  000300  5-METHYLDICHLOROSILYLCYCLOPENTADIENE

      3    2
      C---C                  FORMULA  C6H8CL2SI                MOL WT  179.12
      ||   ||                SOLVENT  C6H8CL2SI
     4C    C1                ORIG ST  C4H12SI                    TEMP     263
       \5/
        C                      131.90   135.60   53.30    2.40
        |                       1/2      2/2      5/2     6/4
   CL-SI-CL
        |                    Y.K.GRISHIN,N.M.SERGEYEV,Y.A.USTYNYUK
      6CH3                   ORG MAGN RESON                   4,   377 (1972)
```

```
1505              Q  000300  2-METHYLCYCLOPENTANONE

            6
          CH3                FORMULA  C6H10O                   MOL WT   98.15
       3  2/                 SOLVENT  CDCL3
       C-C                   ORIG ST  C4H12SI                    TEMP     AMB
       |  \1
       |   C=O                 220.90   43.90    31.90    20.70   37.50   14.20
       |  /                     1/1      2/2      3/3      4/3     5/3     6/4
      4C-C5
                             J.B.STOTHERS,C.T.TAN
                             CAN J CHEM                       52,  308 (1974)
```

```
1506              Q  000300  3-METHYLCYCLOPENTANONE

      6  3  2
    H3C-C-C                  FORMULA  C6H10O                   MOL WT   98.15
       |   \1                SOLVENT  CDCL3
       |    C=O              ORIG ST  C4H12SI                    TEMP     AMB
       |   /
      4C-C5                    218.70   46.70    31.80    31.40   38.50   20.30
                               1/1      2/3      3/2      4/3     5/3     6/4

                             J.B.STOTHERS,C.T.TAN
                             CAN J CHEM                       52,  308 (1974)
```

1507   Q 000300   3-METHYLCYCLOPENTANONE

```
         O
         ‖
         C1
       ╱   ╲
    5C      C2
     |       |
     C---C-CH3
     4    3  6
```

FORMULA	C6H10O			MOL WT	98.15
SOLVENT	CDCL3				
ORIG ST	C4H12SI			TEMP	AMB

| 219.00 | 46.70 | 31.70 | 31.30 | 38.40 | 20.30 |
| 1/1 | 2/3 | 3/2 | 4/3 | 5/3 | 6/4 |

L.F.JOHNSON,W.C.JANKOWSKI
CARBON-13 NMR SPECTRA,JOHN
WILEY AND SONS,NEW YORK   180 (1972)

1508   Q 000300   METHYLCYCLOPENTANE

```
      6
     CH3
      |
     C1
   ╱   ╲
 5C     C2
  |      |
  4C---C3
```

FORMULA	C6H12			MOL WT	84.16
SOLVENT	CDCL3				
ORIG ST	C4H12SI			TEMP	AMB

| 34.80 | 34.90 | 25.50 | 25.50 | 34.90 | 20.70 |
| 1/2 | 2/3 | 3/3 | 4/3 | 5/3 | 6/4 |

L.F.JOHNSON,W.C.JANKOWSKI
CARBON-13 NMR SPECTRA,JOHN
WILEY AND SONS,NEW YORK   186 (1972)

1509   Q 000300   CIS-2-METHYLCYCLOPENTANOL

```
       OH
        |
       C1  6
      ╱ ╲ CH3
     ╱   ╲|
   5C      C2
     ╲   ╱
      4C-C3
```

FORMULA	C6H12O			MOL WT	100.16
SOLVENT	H2O				
ORIG ST	C4H12SI			TEMP	AMB

| 76.30 | 40.90 | 35.60 | 23.20 | 32.10 | 15.80 |
| 1/2 | 2/2 | 3/3 | 4/3 | 5/3 | 6/4 |

R.G.S.RITCHIE,N.CYR,B.KORSCH,H.J.KOCH,
A.S.PERLIN
CAN J CHEM   53, 1424 (1975)

1510   Q 000300   TRANS-2-METHYLCYCLOPENTANOL

```
       OH
        |
       C1
      ╱ ╲
     ╱   ╲
   5C      C2
     ╲   ╱|
      C-C CH3
      4  3 6
```

FORMULA	C6H12O			MOL WT	100.16
SOLVENT	H2O				
ORIG ST	C4H12SI			TEMP	AMB

| 80.90 | 43.30 | 35.00 | 22.60 | 32.90 | 19.40 |
| 1/2 | 2/2 | 3/3 | 4/3 | 5/3 | 6/4 |

R.G.S.RITCHIE,N.CYR,B.KORSCH,H.J.KOCH,
A.S.PERLIN
CAN J CHEM   53, 1424 (1975)

1511   Q 000300   CIS-3-METHYLCYCLOPENTANOL

```
       OH
        |
       C1
      ╱ ╲
     ╱  6╲
   5C H3C C2
     ╲   ╱
      4C-C3
```

FORMULA	C6H12O			MOL WT	100.16
SOLVENT	H2O				
ORIG ST	C4H12SI			TEMP	AMB

| 74.30 | 45.20 | 34.10 | 33.50 | 36.50 | 22.20 |
| 1/2 | 2/3 | 3/2 | 4/3 | 5/3 | 6/4 |

R.G.S.RITCHIE,N.CYR,B.KORSCH,H.J.KOCH,
A.S.PERLIN
CAN J CHEM   53, 1424 (1975)

1512   Q 000300   TRANS-3-METHYLCYCLOPENTANOL

```
     OH
     |
     C1
    / \
  5C    C2
   \    /
   4C—C3
     |
    6CH3
```

FORMULA   C6H12O   MOL WT   100.16
SOLVENT   H2O
ORIG ST   C4H12SI   TEMP   AMB

74.30	45.40	33.00	33.80	36.30	21.80
1/2	2/3	3/2	4/3	5/3	6/4

R.G.S.RITCHIE,N.CYR,B.KORSCH,H.J.KOCH,
A.S.PERLIN
CAN J CHEM   53, 1424 (1975)

1513   Q 000300   HEXACHLOROFULVENE

```
  CL  CL
   \  /
    C6
    ||
    C5
   4/ \1
CL—C    C—CL
   ||   ||
CL—C———C—CL
   3    2
```

FORMULA   C6CL6   MOL WT   284.78
SOLVENT   CHCL3
ORIG ST   C4H12SI   TEMP   AMB

131.40	117.80	137.20	130.80
1/S	2/S	5/S	6/S

G.E.HAWKES,R.A.SMITH,J.D.ROBERTS
J ORG CHEM   39, 1276 (1974)

1514   Q 000300   1-DICHLOROMETHYLENEHEXACHLOROCYCLOPENT-3-ENE

```
CL  CL  CL
 |  \2/
 C6   C
/ \1/ \
CL  C  3C—CL
 |  ||
CL—C———C—CL
 5|    4
 CL
```

FORMULA   C6CL8   MOL WT   355.69
SOLVENT   CHCL3
ORIG ST   C4H12SI   TEMP   AMB

137.60	82.10	135.70	136.80
1/S	2/S	3/S	6/S

G.E.HAWKES,R.A.SMITH,J.D.ROBERTS
J ORG CHEM   39, 1276 (1974)

1515   Q 000300   PENTACHLORO-5-TRICHLOROMETHYLCYCLOPENTADIENE

```
  CL  CL
   \  /
    C6  CL
   / \5/
  CL   C
  4/  \1
CL—C    C—CL
  3||   ||2
CL—C———C—CL
```

FORMULA   C6CL8   MOL WT   355.69
SOLVENT   CHCL3
ORIG ST   C4H12SI   TEMP   AMB

133.00	130.70	77.60	97.20
1/S	2/S	5/S	6/S

G.E.HAWKES,R.A.SMITH,J.D.ROBERTS
J ORG CHEM   39, 1276 (1974)

1516   Q 000300   HEPTACHLORO-1-TRICHLOROMETHYLCYCLOPENTENE

```
   CL  CL
    \  |
CL—C—C—CL  CL
  4|  5\  6/
   |    C—C—CL
  3|   /1  \
CL—C—C2    CL
   |   \
  CL   CL
```

FORMULA   C6CL10   MOL WT   426.60
SOLVENT   CHCL3
ORIG ST   C4H12SI   TEMP   AMB

142.00	136.90	93.20	100.60	91.70	88.10
1/S	2/S	3/S	4/S	5/S	6/S

G.E.HAWKES,R.A.SMITH,J.D.ROBERTS
J ORG CHEM   39, 1276 (1974)

```
1517            Q 000300  1,1-DIMETHOXYTETRACHLOROCYCLOPENTA-2,4-DIENE
        7        6
    H3C-O    O-CH3        FORMULA   C7H6CL4O2              MOL WT  263.94
        \5/                SOLVENT   CHCL3
     CL  C  CL             ORIG ST   C4H12SI               TEMP      AMB
      \ / \ /
      4C   C1              129.00   128.50   104.50    51.60
      ||   ||               1/S      2/S      5/S      6/Q
      3C---C2
      /      \             G.E.HAWKES,R.A.SMITH,J.D.ROBERTS
    CL        CL           J ORG CHEM                   39, 1276 (1974)
```

```
1518            Q 000300  6-ACETOXYFULVENE

                          FORMULA   C7H8O2                MOL WT  124.14
             O            SOLVENT   CDCL3
      1      ||           ORIG ST   C4H12SI               TEMP      273
  2C==C    O-C-CH3
    |   \   /  7 8        117.50   133.10   131.40   124.10   130.00   138.80
    |    C=C               1/2      2/2      3/2      4/2      5/1      6/2
    |   /5 6\             167.30    20.70
  3C==C      H             7/1      8/4
    4
                          R.HOLLENSTEIN,W.V.PHILIPSBORN,R.VOEGELI,
                          M.NEUENSCHWANDER
                          HELV CHIM ACTA               56,  847 (1973)
```

```
1519            Q 000300  1-CHLORODIMETHYLSILYLCYCLOPENTADIENE
      3    2
      C---C               FORMULA   C7H11CLSI             MOL WT  158.70
      ||   ||  6          SOLVENT   C7H11CLSI
     4C   C1  CH3         ORIG ST   C4H12SI               TEMP      AMB
      \  / \ /
       C    SI-CL         144.30   145.70   140.50   133.90    45.70    3.30
       5     \             1/1      2/2      3/2      4/2      5/3      6/4
             CH3
              8           Y.K.GRISHIN,N.M.SERGEYEV,Y.A.USTYNYUK
                          ORG MAGN RESON               4,  377 (1972)
```

```
1520            Q 000300  5-CHLORODIMETHYLSILYLCYCLOPENTADIENE
       3    2
       C---C              FORMULA   C7H11CLSI             MOL WT  158.70
       ||   ||            SOLVENT   C7H11CLSI
      4C   C1             ORIG ST   C4H12SI               TEMP      263
       \5/
        C                 133.00   133.70    54.00    1.00
      7 | 6                1/2      2/2      5/2      6/4
    H3C-SI-CH3
        |                 Y.K.GRISHIN,N.M.SERGEYEV,Y.A.USTYNYUK
        CL                ORG MAGN RESON               4,  377 (1972)
```

```
1521            Q 000300  CIS-2,3-DIMETHYLCYCLOPENTANONE
        7 6
     H3C CH3              FORMULA   C7H12O                MOL WT  112.17
       | |                SOLVENT   CDCL3
     3C-C2                ORIG ST   C4H12SI               TEMP      AMB
      |  \1
      |    C=O            220.90    48.00    34.40    28.10    35.20    9.50
      |   /                1/1      2/2      3/2      4/3      5/3      6/4
     4C-C5                 14.80
                           7/4

                          J.B.STOTHERS,C.T.TAN
                          CAN J CHEM                   52,  308 (1974)
```

1522   Q 000300 TRANS-2,3-DIMETHYLCYCLOPENTANONE

```
        6
        CH3              FORMULA   C7H12O              MOL WT   112.17
     3  /               SOLVENT   CDCL3
     C-C2               ORIG ST   C4H12SI              TEMP       AMB
    7/|   \1
   H3C |    C=O         220.30   51.80   39.80   29.60   37.50   12.00
       |   /             1/1      2/2     3/2     4/3     5/3     6/4
      4C-C5             19.10
                         7/4
```

J.B. STOTHERS, C.T. TAN
CAN J CHEM   52,   308 (1974)

1523   Q 000300 CIS-2,4-DIMETHYLCYCLOPENTANONE

```
        6
        CH3              FORMULA   C7H12O              MOL WT   112.17
     3  /               SOLVENT   CDCL3
     C-C2               ORIG ST   C4H12SI              TEMP       AMB
     |   \1
     |    C=O           220.00   45.60   40.90   29.60   46.10   14.00
     |R  /               1/1      2/2     3/3     4/2     5/3     6/4
      4C-C5             20.30
                7        7/4
          R  -CH3
```

J.B. STOTHERS, C.T. TAN
CAN J CHEM   52,   308 (1974)

1524   Q 000300 TRANS-2,4-DIMETHYLCYCLOPENTANONE

```
        6                FORMULA   C7H12O              MOL WT   112.17
        CH3             SOLVENT   CDCL3
        |               ORIG ST   C4H12SI              TEMP       AMB
     3C-C2
     |   \1             221.00   41.50   39.10   28.00   45.90   15.30
     |    C=O            1/1      2/2     3/3     4/2     5/3     6/4
     |   /              20.80
     4C-C5               7/4
     |
    H3C                 J.B. STOTHERS, C.T. TAN
     7                  CAN J CHEM              52,   308 (1974)
```

1525   Q 000300 CIS-3,4-DIMETHYLCYCLOPENTANONE

```
                         FORMULA   C7H12O              MOL WT   112.17
     6 3 2             SOLVENT   CDCL3
   H3C-C-C             ORIG ST   C4H12SI              TEMP       AMB
     |   \1
     |    C=O           218.60   45.50   34.10   34.10   45.50   15.00
    7 |  /              1/1      2/3     3/2     4/2     5/3     6/4
   H3C-C-C5            15.00
     4                  7/4
```

J.B. STOTHERS, C.T. TAN
CAN J CHEM   52,   308 (1974)

1526   Q 000300 TRANS-3,4-DIMETHYLCYCLOPENTANONE

```
        6
        CH3             FORMULA   C7H12O              MOL WT   112.17
        /               SOLVENT   CDCL3
     3C-C2             ORIG ST   C4H12SI              TEMP       AMB
     |   \1
     |    C=O           217.30   47.30   39.20   39.20   47.30   18.10
     |   /              1/1      2/3     3/2     4/2     5/3     6/4
     C-C5              18.10
    /4                  7/4
   H3C
     7                  J.B. STOTHERS, C.T. TAN
                        CAN J CHEM              52,   308 (1974)
```

1527 — Q 000300 CIS-2,5-DIMETHYLCYCLOPENTANONE

```
        6
       CH3
        |
      3C-C2
      |  \1
      |   C=O
      | 3/
      4C-C5
        7
      R -CH3
```

FORMULA	C7H12O			MOL WT	112.17
SOLVENT	CDCL3				
ORIG ST	C4H12SI			TEMP	AMB

222.70	42.50	28.90	28.90	42.50	15.30
1/1	2/2	3/3	4/3	5/2	6/4
15.30					
7/4					

J.B.STOTHERS,C.T.TAN
CAN J CHEM   52, 308 (1974)

1528 — Q 000300 TRANS-2,5-DIMETHYLCYCLOPENTANONE

```
        6
       CH3
     3 |
     C-C
     | 2\1
     |   C=O
     |  /
    4C-C5
      |
     CH3
      7
```

FORMULA	C7H12O			MOL WT	112.17
SOLVENT	CDCL3				
ORIG ST	C4H12SI			TEMP	AMB

222.30	43.60	30.10	30.10	43.60	14.70
1/1	2/2	3/3	4/3	5/2	6/4
14.70					
7/4					

J.B.STOTHERS,C.T.TAN
CAN J CHEM   52, 308 (1974)

1529 — Q 000300 DECACHLORO-1-VINYLCYCLOPENTENE

```
    CL  CL
     \ /
  CL  C5  CL  CL
   \4/ \1 | /
  CL-C   C-C=C7
   |   || 6 \
  CL-C---C   CL
   31   2\
   CL    CL
```

FORMULA	C7CL10			MOL WT	438.61
SOLVENT	CHCL3				
ORIG ST	C4H12SI			TEMP	AMB

140.40	134.80	92.90	100.40	91.80	128.50
1/S	2/S	3/S	4/S	5/S	6/S
117.80					
7/S					

G.E.HAWKES,R.A.SMITH,J.D.ROBERTS
J ORG CHEM   39, 1276 (1974)

1530 — Q 000300 6-DIMETHYLAMINOFULVENE

```
          7
         CH3
    1     /
  2C==C  N
   |  \ / \
   |   C=C  CH3
   |  /5 6\ 8
  3C==C    H
    4
```

FORMULA	C8H11N			MOL WT	121.18
SOLVENT	CDCL3				
ORIG ST	C4H12SI			TEMP	243

114.00	125.10	119.40	124.40	116.60	149.10
1/2	2/2	3/2	4/2	5/1	6/2
47.10	40.30				
7/4	8/4				

R.HOLLENSTEIN,W.V.PHILIPSBORN,R.VOEGELI,
M.NEUENSCHWANDER
HELV CHIM ACTA   56, 847 (1973)

1531 — Q 000300 2,2,4-TRIMETHYLCYCLOPENTANONE

```
      7
     CH3
     | 6
    3C-C-CH3
    | 2\1
    |   C=O
    |  /
   4C-C5
    |
   CH3
    8
```

FORMULA	C8H14O			MOL WT	126.20
SOLVENT	CDCL3				
ORIG ST	C4H12SI			TEMP	AMB

222.00	46.50	47.50	27.50	45.80	20.50
1/1	2/1	3/3	4/2	5/3	8/4
24.30	24.40				
6/4	7/4				

J.B.STOTHERS,C.T.TAN
CAN J CHEM   52, 308 (1974)

1532   Q 000300   2,2,5-TRIMETHYLCYCLOPENTANONE

```
        7
       CH3
    3  |  6
    C-C-CH3
    |  2\1
    |    C=O
    |   /
   4C-C5
    |
    8CH3
```

FORMULA	C8H14O			MOL WT	126.20
SOLVENT	CDCL3				
ORIG ST	C4H12SI			TEMP	AMB

224.10	44.70	36.60	28.00	43.10	15.20
1/1	2/1	3/3	4/3	5/2	8/4
24.10	24.80				
6/4	7/4				

J.B.STOTHERS,C.T.TAN
CAN J CHEM   52, 308 (1974)

1533   Q 000300   2,4,4-TRIMETHYLCYCLOPENTANONE

```
      6
     CH3
   3  |
   C-C2
   |   \1
   |    C=O
 7 |4 /
H3C-C-C5
   |
   8CH3
```

FORMULA	C8H14O			MOL WT	126.20
SOLVENT	CDCL3				
ORIG ST	C4H12SI			TEMP	AMB

220.20	42.70	46.10	33.80	52.50	14.90
1/1	2/2	3/3	4/1	5/3	6/4
27.90	29.80				
7/4	8/4				

J.B.STOTHERS,C.T.TAN
CAN J CHEM   52, 308 (1974)

1534   Q 000300   1-TRIMETHYLSILYLCYCLOPENTADIENE

```
  3     2
  C---C
  ||   ||  6
 4C   C1  CH3
    \ / \ /
     C   SI-CH3
     5    \
          CH3
          8
```

FORMULA	C8H14SI			MOL WT	138.29
SOLVENT	C8H14SI				
ORIG ST	C4H12SI			TEMP	AMB

147.90	142.70	138.40	140.00	45.70	0.20
1/1	2/2	3/2	4/2	5/3	6/4

Y.K.GRISHIN,N.M.SERGEYEV,Y.A.USTYNYUK
ORG MAGN RESON   4, 377 (1972)

1535   Q 000300   5-TRIMETHYLSILYLCYCLOPENTADIENE

```
  3     2
  C---C
  ||   ||
 4C   C1
   \5/
    C
  8 | 6
 H3C-SI-CH3
    |
    7CH3
```

FORMULA	C8H14SI			MOL WT	138.29
SOLVENT	C8H14SI				
ORIG ST	C4H12SI			TEMP	253

134.00	131.30	52.50	1.20
1/2	2/2	5/2	6/4

Y.K.GRISHIN,N.M.SERGEYEV,Y.A.USTYNYUK
ORG MAGN RESON   4, 377 (1972)

1536   Q 000300   N-PROPYLCYCLOPENTANE

```
     2C
    / \
  3C   1C-CH2-CH2-CH3
   |  16  7  8
  C---C
  4    5
```

FORMULA	C8H16			MOL WT	112.22
SOLVENT	CDCL3				
ORIG ST	C4H12SI			TEMP	AMB

40.20	32.80	25.30	25.00	32.80	38.80
1/2	2/3	3/3	4/3	5/3	6/3
22.00	14.40				
7/3	8/4				

L.F.JOHNSON,W.C.JANKOWSKI
CARBON-13 NMR SPECTRA,JOHN
WILEY AND SONS,NEW YORK   312 (1972)

1537 — Q 000300 6-PROPYLFULVENE

```
    1      7   8   9
  2C==C    CH2-CH2-CH3
  |    \  /
  |     C=C
  |    /5  6\
  3C==C      H
      4
```

FORMULA   C9H12
SOLVENT   CDCL3
ORIG ST   C4H12SI

MOL WT   120.20
TEMP   273

119.30	133.00	130.90	125.70	146.00	143.70
1/2	2/2	3/2	4/2	5/1	6/2
33.10	22.70	14.00			
7/3	8/3	9/4			

R.HOLLENSTEIN,W.V.PHILIPSBORN,R.VOEGELI,
M.NEUENSCHWANDER
HELV CHIM ACTA   56, 847 (1973)

1538 — Q 000300 1-ACETOXY-1-ETHYNYLCYCLOPENTANE

```
      O
      \8  9
       C-CH3
      /
      O
  1 2 |3
  HC≡C-C---C4
      |   |
     7C   C5
       \ /
        C6
```

FORMULA   C9H12O2
SOLVENT   CDCL3
ORIG ST   C4H12SI

MOL WT   152.19
TEMP   AMB

73.20	84.30	80.10	40.40	23.30	169.20
1/2	2/1	3/1	4/3	5/3	8/1
21.60					
9/4					

M.T.W.HEARN
TETRAHEDRON   32, 115 (1976)

1539 — Q 000300 2,2,4,4-TETRAMETHYLCYCLOPENTANONE

```
      7
      CH3
    | 6
  3C-C-CH3
  | 2\1
  |    C=O
  8 |4 /
  H3C-C-C5
  |
  9CH3
```

FORMULA   C9H16O
SOLVENT   CDCL3
ORIG ST   C4H12SI

MOL WT   140.23
TEMP   AMB

222.50	45.20	52.30	33.10	52.60	27.40
1/1	2/1	3/3	4/1	5/3	6/4
27.40	30.10	30.10			
7/4	8/4	9/4			

J.B.STOTHERS,C.T.TAN
CAN J CHEM   52, 308 (1974)

1540 — Q 000300 2,2,5,5-TETRAMETHYLCYCLOPENTANONE

```
      7
      CH3
  3 | 6
  C-C-CH3
  | 2\1
  |    C=O
  | 5/9
  4C-C-CH3
  |
  8CH3
```

FORMULA   C9H16O
SOLVENT   CDCL3
ORIG ST   C4H12SI

MOL WT   140.23
TEMP   AMB

226.40	45.20	34.90	34.90	45.20	24.90
1/1	2/1	3/3	4/3	5/1	6/4
24.90	24.90	24.90			
7/4	8/4	9/4			

J.B.STOTHERS,C.T.TAN
CAN J CHEM   52, 308 (1974)

1541 — Q 000300 6,6-TETRAMETHYLENEFULVENE

FORMULA   C10H12
SOLVENT   CDCL3
ORIG ST   C4H12SI

MOL WT   911.07
TEMP   273

```
    1      7   8
  2C==C    C--C
  |    \  /   |
  |     C=C   |
  |    /5  6\ |
  3C==C    C--C
    4      10  9
```

121.10	129.30	129.30	121.10	138.00	162.50
1/2	2/2	3/2	4/2	5/1	6/1
33.00	26.00	26.00	33.00		
7/3	8/3	9/3	10/3		

R.HOLLENSTEIN,W.V.PHILIPSBORN,R.VOEGELI,
M.NEUENSCHWANDER
HELV CHIM ACTA   56, 847 (1973)

1542   Q 000300 6-TERT-BUTYLFULVENE

```
              8
             CH3                FORMULA   C10H14              MOL WT   134.22
      1      /9                 SOLVENT   CDCL3
   2C==C    7C-CH3             ORIG ST   C4H12SI             TEMP       273
   |     \  / \
   |      C=C   CH3             119.60  133.80  128.50  128.30  141.80  153.90
   |     /5 6\  10               1/2     2/2     3/2     4/2     5/1     6/2
   3C==C     H                  35.60   30.80   30.80   30.80
       4                         7/1     8/4     9/4    10/4
```

R.HOLLENSTEIN,W.V.PHILIPSBORN,R.VOEGELI,
M.NEUENSCHWANDER
HELV CHIM ACTA   56,   847 (1973)

1543   Q 000300 DECACHLORO-1,1'-DICYCLOPENTADIENYL

```
       CL       CL
       |        |              FORMULA   C10CL10             MOL WT   474.64
   CL  C6      C1  CL          SOLVENT   CHCL3
     \ #    CL/ \ /           ORIG ST   C4H12SI             TEMP       AMB
    7C   \  1/   C2
     |  10C--C5  |             132.00  131.00   73.10
    8C  /|   \   C3             1/S     2/S      5/S
    / \ /  CL  \ # \
   CL 9C       C4 CL
      |        |               G.E.HAWKES,R.A.SMITH,J.D.ROBERTS
      CL       CL              J ORG CHEM                 39, 1276 (1974)
```

1544   Q 000300 DECACHLORO-2,3'-DICYCLOPENTADIENYL

```
       CL      CL CL
       |        \ /            FORMULA   C10CL10             MOL WT   474.64
   CL  C6      C5  CL          SOLVENT   CHCL3
     \ / \    / \ /           ORIG ST   C4H12SI             TEMP       AMB
   CL-C10 \  /   C4
     |  7C--C1  ||            142.90  137.90  136.90  133.40   83.10  130.90
    9C  /     \  C3            1/S     2/S     3/S     40S      5/S     6/S
    / \ /     \ / \           129.90  128.30  126.30   82.40
   CL  C8      C2  CL          7/S     8/S     9/S     10/S
      |        |
      CL       CL             G.E.HAWKES,R.A.SMITH,J.D.ROBERTS
                              J ORG CHEM                 39, 1276 (1974)
```

1545   Q 000300 DECACHLORO-2,2'-DICYCLOPENTADIENYL

```
       CL       CL
       |        |              FORMULA   C10CL10             MOL WT   474.64
   CL 10C-CL  CL-C5  CL        SOLVENT   CHCL3
     \ /      / \ /           ORIG ST   C4H12SI             TEMP       AMB
    9C   \   /   C4
     ||  6C--C1  ||           138.50  138.30  132.40  128.10   83.40
    8C  #    \   C3            1/S     2/S     3/S     4/S      5/S
    / \ #    \ / \
   CL 7C      C2  CL
      |        |              G.E.HAWKES,R.A.SMITH,J.D.ROBERTS
      CL       CL             J ORG CHEM                 39, 1276 (1974)
```

1546   Q 000300 DECACHLORO-3,3'-DICYCLOPENTADIENYL

```
       CL       CL
       |        |              FORMULA   C10CL10             MOL WT   474.64
   CL 8C       C3  CL          SOLVENT   CHCL3
     \ # \    / \ /           ORIG ST   C4H12SI             TEMP       AMB
    9C   \   /   C4
     |  7C--C2  ||            141.00  131.70  128.50  125.80   82.30
   CL-C10 #  \  5C-CL          1/S     2/S     3/S     4/S      5/1
    / \ #    \ / \
   CL 6C      C1  CL
      |        |              G.E.HAWKES,R.A.SMITH,J.D.ROBERTS
      CL       CL             J ORG CHEM                 39, 1276 (1974)
```

1547   Q 000300   PERCHLORO-1,1'-DICYCLOPENTENYL

```
     CL        CL  CL
      |         \  /
  CL  7C         C5  CL
   \  / \\     //  \ /
  CL-C8   \\   /    4C-CL
    |  6C--C1   |
  CL-C9  /    \\  3C-CL
    / \ /      \\ /
  CL   C10       C2
    / \          |
  CL   CL        CL
```

FORMULA	C10CL14			MOL WT	616.45
SOLVENT	CHCL3				
ORIG ST	C4H12SI			TEMP	AMB

144.50	131.00	93.10	101.20	92.10
1/S	2/S	3/S	4/S	5/S

G.E.HAWKES,R.A.SMITH,J.D.ROBERTS
J ORG CHEM   39, 1276 (1974)

1548   Q 000300   6,6-PENTAMETHYLENEFULVENE

```
   1       7 8
 2C==C    C-C
   |   \   /  \
   |    C=C    C9
   |   /5 6\  /
 3C==C    C-C
   4      11 10
```

FORMULA	C11H14			MOL WT	146.23
SOLVENT	CDCL3				
ORIG ST	C4H12SI			TEMP	273

119.70	130.60	130.60	119.70	139.70	158.00
1/2	2/2	3/2	4/2	5/1	6/1
33.70	28.60	26.50	28.60	33.70	
7/3	8/3	9/3	10/3	11/3	

R.HOLLENSTEIN,W.V.PHILIPSBORN,R.VOEGELI,
M.NEUENSCHWANDER
HELV CHIM ACTA   56, 847 (1973)

1549   Q 000300   6,6-HEXAMETHYLENEFULVENE

```
   1       7 8 9
 2C==C    C-C-C
   |   \   /    |
   |    C=C     |
   |   /5 6\    |
 3C==C   12C-C-C
   4        11 10
```

FORMULA	C12H16			MOL WT	160.26
SOLVENT	CDCL3				
ORIG ST	C4H12SI			TEMP	273

120.00	130.00	130.00	120.00	141.70	160.20
1/2	2/2	3/2	4/2	5/1	6/1
34.30	29.00	27.90	27.90	29.00	34.30
7/3	8/3	9/3	10/3	11/3	12/3

R.HOLLENSTEIN,W.V.PHILIPSBORN,R.VOEGELI,
M.NEUENSCHWANDER
HELV CHIM ACTA   56, 847 (1973)

1550   Q 000400   1,4-CYCLOHEXADIENE

```
    1 6
    C=C
   2/   \5
   C     C
    \   /
    3C=C4
```

FORMULA	C6H8			MOL WT	80.13
SOLVENT	CDCL3				
ORIG ST	C4H12SI			TEMP	AMB

124.50	26.00	124.50	124.50	26.00	124.50
1/2	2/3	3/2	4/2	5/3	6/2

L.F.JOHNSON,W.C.JANKOWSKI
CARBON-13 NMR SPECTRA,JOHN
WILEY AND SONS,NEW YORK   166 (1972)

1551   Q 000400   1,4-CYCLOHEXADIENE

```
    5
    C
  4 // \ 5
  C     C
  |     |
  C     C
  3 \ // 1
    C
    2
```

FORMULA	C6H8			MOL WT	80.13
SOLVENT	CCL4/CDCL3				
ORIG ST	C4H12SI			TEMP	AMB

124.10	25.70
1/2	3/3

H.GUENTHER,G.JIKELI
CHEM BER   106, 1863 (1973)

1552   Q 000400 CYCLOHEXENE

```
      6 1
      C=C
    /     \
  5C       C2
    \     /
      C-C
      4 3
```

FORMULA	C6H10			MOL WT	82.15
SOLVENT	CDCL3				
ORIG ST	C4H12SI			TEMP	AMB

127.20	25.30	22.90	22.90	25.30	127.20
1/2	2/3	3/3	4/3	5/3	6/2

L.F.JOHNSON,W.C.JANKOWSKI
CARBON-13 NMR SPECTRA,JOHN
WILEY AND SONS,NEW YORK   176 (1972)

1553   Q 000400 CYCLOHEXANONE

```
      3 2
      C-C
    /    \1
  4C      C=O
    \    /
      C-C
      5 6
```

FORMULA	C6H10O			MOL WT	98.15
SOLVENT	CDCL3				
ORIG ST	C4H12SI			TEMP	AMB

211.30	41.90	27.10	25.10	27.10	41.90
1/1	2/3	3/3	4/3	5/3	6/3

L.F.JOHNSON,W.C.JANKOWSKI
CARBON-13 NMR SPECTRA,JOHN
WILEY AND SONS,NEW YORK   179 (1972)

1554   Q 000400 CYCLOHEXANONE

```
      3 2
      C-C
    /    \1
  4C      C=O
    \    /
      C-C
      5 6
```

FORMULA	C6H100			MOL WT	98.15
SOLVENT	CDCL3				
ORIG ST	C4H12SI			TEMP	AMB

211.50	41.90	27.10	25.00	27.10	41.90
1/1	2/3	3/3	4/3	5/3	6/3

J.B.STOTHERS,C.T.TAN
CAN J CHEM   52, 308 (1974)

1555   Q 000400 CYCLOHEXANONE

```
      3 2
      C-C
    4/   \1
    C      C=O
    \5 6/
      C-C
```

FORMULA	C6H100			MOL WT	98.15
SOLVENT	CDCL3				
ORIG ST	C4H12SI			TEMP	308

209.70	41.50	26.60	24.60	26.60	41.50
1/1	2/3	3/3	4/3	5/3	6/3

G.E.HAWKES,K.HERWIG,J.D.ROBERTS
J ORG CHEM   39, 1017 (1974)

1556   Q 000400 CYCLOHEXENE OXIDE

```
         O
        / \
     6C---C1
     /       \
   5C         C2
     \       /
     4C---C3
```

FORMULA	C6H100			MOL WT	98.15
SOLVENT	CDCL3				
ORIG ST	C4H12SI			TEMP	AMB

52.20	24.70	19.60	19.60	24.70	52.20
1/2	2/3	3/3	4/3	5/3	6/2

N.R.EASTON,JR.,F.A.L.ANET,P.A.BURNS,C.S.FOOTE
J AM CHEM SOC   96, 3945 (1974)

1557   Q 000400 1,1-DIFLUOROCYCLOHEXANE

```
    3  2
    C—C    F
   /    \1/
 4C      C
   \    / \
    C—C    F
    5  6
```

FORMULA   C6H10F2   MOL WT   120.14
SOLVENT   C4H8O2
ORIG ST   C4H12SI   TEMP   AMB

123.70	34.60	23.50	24.90	23.50	34.60
1/1	2/3	3/3	4/3	5/3	6/3

L.F.JOHNSON,W.C.JANKOWSKI
CARBON-13 NMR SPECTRA,JOHN
WILEY AND SONS,NEW YORK   177 (1972)

1558   Q 000400 BROMOCYCLOHEXANE

```
    5  6
    C—C
   /    \
  C      C—BR
 4\     /1
    C—C
    3  2
```

FORMULA   C6H11BR   MOL WT   163.06
SOLVENT   CDCL3
ORIG ST   C4H12SI   TEMP   AMB

53.00	37.50	25.80	25.20	25.80	37.50
1/2	2/3	3/3	4/3	5/3	6/3

L.F.JOHNSON,W.C.JANKOWSKI
CARBON-13 NMR SPECTRA,JOHN
WILEY AND SONS,NEW YORK   184 (1972)

1559   Q 000400 BROMOCYCLOHEXANE

```
    5  6
    C—C
   /    \1
 4C      C—BR
   \    /
    C—C
    3  2
```

FORMULA   C6H11BR   MOL WT   163.06
SOLVENT   C4H10O
ORIG ST   C4H12SI   TEMP   AMB

52.60	37.90	26.10	25.60
1/2	2/3	3/3	4/3

T.PEHK,E.LIPPMAA
ORG MAGN RESON   3, 679 (1971)

1560   Q 000400 CHLOROCYCLOHEXANE

```
    5  6
    C—C
   /    \1
 4C      C—CL
   \    /
    C—C
    3  2
```

FORMULA   C6H11CL   MOL WT   118.61
SOLVENT   C4H10O
ORIG ST   C4H12SI   TEMP   AMB

59.80	37.20	25.20	25.60
1/2	2/3	3/3	4/3

T.PEHK,E.LIPPMAA
ORG MAGN RESON   3, 679 (1971)

1561   O 000400 TRICHLOROSILYLCYCLOHEXANE

```
    5  6
    C—C    CL
   /    \1 /
 4C      C—SI—CL
   \    / 
    C—C    CL
    3  2
```

FORMULA   C6H11CL3SI   MOL WT   217.60
SOLVENT   C4H10O
ORIG ST   C4H12SI   TEMP   AMB

33.90	27.10	25.90	26.40
1/2	2/3	3/3	4/3

T.PEHK,E.LIPPMAA
ORG MAGN RESON   3, 679 (1971)

1562   Q 000400   FLUOROCYCLOHEXANE

```
    5 6
    C-C
   /    \1
 4C      C-F
   \    /
    C-C
    3 2
```

FORMULA	C6H11F			MOL WT	102.15
SOLVENT	C4H10O				
ORIG ST	C4H12SI			TEMP	AMB

90.50	33.10	23.50	26.00
1/3	2/3	3/3	4/3

T.PEHK,E.LIPPMAA
ORG MAGN RESON   3, 679 (1971)

1563   Q 000400   IODOCYCLOHEXANE

```
    5 6
    C-C
   /    \1
 4C      C-J
   \    /
    C-C
    3 2
```

FORMULA	C6H11I			MOL WT	210.06
SOLVENT	C4H10O				
ORIG ST	C4H12SI			TEMP	AMB

31.80	39.80	27.40	25.50
1/2	2/3	3/3	4/3

T.PEHK,E.LIPPMAA
ORG MAGN RESON   3, 679 (1971)

1564   Q 000400   CYCLOHEXANONE OXIME

```
    3 2
    C-C      OH
   4/   \1 /
   C      C=N
    \5 6/
    C-C
```

FORMULA	C6H11NO				MOL WT	113.16
SOLVENT	CDCL3					
ORIG ST	C4H12SI				TEMP	308

160.40	25.70	25.40	24.40	26.70	31.90
1/1	2/3	3/3	4/3	5/3	6/3

G.E.HAWKES,K.HERWIG,J.D.ROBERTS
J ORG CHEM   39, 1017 (1974)

1565   Q 000400   NITROCYCLOHEXANE

```
    5 6
    C-C
   /    \1
 4C      C-NO2
   \    /
    C-C
    3 2
```

FORMULA	C6H11NO2			MOL WT	129.16
SOLVENT	C4H10O				
ORIG ST	C4H12SI			TEMP	AMB

84.60	31.40	24.70	25.50
1/2	2/3	3/3	4/3

T.PEHK,E.LIPPMAA
ORG MAGN RESON   3, 679 (1971)

1566   Q 000400   CYCLOHEXANE

```
    5 6
    C-C
   /    \
 4C      C 1
   \    /
    C-C
    3 2
```

FORMULA	C6H12	MOL WT	84.16
SOLVENT	C4H10O		
ORIG ST	C4H12SI	TEMP	AMB

27.60
1/3

T.PEHK,E.LIPPMAA
ORG MAGN RESON   3, 679 (1971)

1567   Q 000400   CYCLOHEXANOL

```
   5 6
   C-C
  /    \1
4C      C-OH
  \    /
   C-C
   3 2
```

FORMULA C6H12O				MOL WT 100.16

SOLVENT C4H10O
ORIG ST C4H12SI   TEMP   AMB

70.00	36.00	25.00	26.40
1/2	2/3	3/3	4/3

T.PEHK,E.LIPPMAA
ORG MAGN RESON   3, 679 (1971)

1568   Q 000400   CYCLOHEXANOL

```
   3 2
   C-C
  /    \1
4C      C-OH
  \    /
   C-C
   5 6
```

FORMULA C6H12O					MOL WT 100.16

SOLVENT CDCL3
ORIG ST C4H12SI   TEMP   AMB

70.00	35.50	24.30	25.70	24.30	35.50
1/2	2/3	3/3	4/3	5/3	6/3

L.F.JOHNSON,W.C.JANKOWSKI
CARBON-13 NMR SPECTRA,JOHN
WILEY AND SONS,NEW YORK   191 (1972)

1569   Q 000400   MYO-INOSITOL

```
     OH
     |
   2C---C3
   /    I\
1C OH HO C4
 I\I   5/I
 HO C---C OH
   6    I
       OH
```

FORMULA C6H12O6					MOL WT 180.16

SOLVENT H2O
ORIG ST C4H12SI   TEMP   AMB

75.20	72.00	73.30	73.10	73.30	72.00
1/2	2/2	3/2	4/2	5/2	6/2

L.F.JOHNSON,W.C.JANKOWSKI
CARBON-13 NMR SPECTRA,JOHN
WILEY AND SONS,NEW YORK   196 (1972)

1570   Q 000400   CYCLOHEXANETHIOL

```
   5 6
   C-C
  /    \1
4C      C-SH
  \    /
   C-C
   3 2
```

FORMULA C6H12S				MOL WT 116.23

SOLVENT C4H10O
ORIG ST C4H12SI   TEMP   AMB

38.50	38.50	26.80	25.90
1/2	2/3	3/3	4/3

T.PEHK,E.LIPPMAA
ORG MAGN RESON   3, 679 (1971)

1571   Q 000400   AMINOCYCLOHEXANE

```
   5 6
   C-C
  /    \1
4C      C-NH2
  \    /
   C-C
   3 2
```

FORMULA C6H13N				MOL WT 99.18

SOLVENT C4H10O
ORIG ST C4H12SI   TEMP   AMB

51.10	37.70	25.80	26.50
1/2	2/3	3/3	4/3

T.PEHK,E.LIPPMAA
ORG MAGN RESON   3, 679 (1971)

1572 Q 000400 CYCLOHEXYLAMINE

```
    5 6
    C-C
  /     \
 C       C-NH2
 4\     /1
   C-C
   3 2
```

FORMULA	C6H13N			MOL WT	99.18
SOLVENT	CDCL3				
ORIG ST	C4H12SI			TEMP	AMB

50.40	36.70	25.10	25.70	25.10	36.70
1/2	2/3	3/3	4/3	5/3	6/3

L.F.JOHNSON,W.C.JANKOWSKI
CARBON-13 NMR SPECTRA,JOHN
WILEY AND SONS,NEW YORK 200 (1972)

1573 Q 000400 AMINOCYCLOHEXANE HYDROCHLORIDE

```
    5 6
    C-C
  /    \1
 4C      C-NH2.HCL
  \     /
   C-C
   3 2
```

FORMULA	C6H13N			MOL WT	99.18
SOLVENT	CH4O/C4H10O				
ORIG ST	C4H12SI			TEMP	AMB

51.50	33.40	25.60	26.00
1/2	2/3	3/3	4/3

T.PEHK,E.LIPPMAA
ORG MAGN RESON 3, 679 (1971)

1574 Q 000400 1,2,3,4-TETRACHLORO-5,5,6,6-TETRAFLUORO-1,3-CYCLOHEXADIENE

```
   F   CL
   1|  2/
F  F-C-C
  \  /   \3
  6C      C-CL
  / \    /
 F  5C=C4
   CL   CL
```

FORMULA	C6CL4F4			MOL WT	289.87
SOLVENT	CHCL3				
ORIG ST	C4H12SI			TEMP	AMB

110.40	126.20	132.40
1/T	2/S	3/S

G.E.HAWKES,R.A.SMITH,J.D.ROBERTS
J ORG CHEM 39, 1276 (1974)

1575 Q 000400 HEXACHLOROCYCLOHEXA-2,4-DIEN-1-ONE

```
      ]
      ||
  CL  C1  CL
   \6/ \ /
 CL-C   C2
   |    ||
   5C   C3
   / \4/ \
  CL  C  CL
```

FORMULA	C6CL6O			MOL WT	300.78
SOLVENT	C4H8O2				
ORIG ST	C4H12SI			TEMP	AMB

175.60	143.70	135.10	128.50	126.40	76.80
1/S	2/S	3/S	4/S	5/S	6/S

G.E.HAWKES,R.A.SMITH,J.D.ROBERTS
J ORG CHEM 39, 1276 (1974)

1576 Q 000400 1,2,3,4-TETRACHLORO-5-METHYLCYCLOHEXA-1,3-DIENE

```
        7
  CL   CH3
   \6 1/
    C-C
  //    \2
 CL-C5   C
   \    /
    4C=C3
   /    \
  CL     CL
```

FORMULA	C7H6CL4			MOL WT	231.94
SOLVENT	C6H12				
ORIG ST	C4H12SI			TEMP	AMB

37.30	39.30	127.30	123.80	123.80	134.20
1/D	2/T	3/S	4/S	5/S	6/S
16.70					
7/Q					

G.E.HAWKES,R.A.SMITH,J.D.ROBERTS
J ORG CHEM 39, 1276 (1974)

1577   Q 000400 CYCLOHEXANECARBOXYLIC ACID CHLORIDE

```
    5 6
    C-C     O
   /    \1 7/
 4C      C-C
   \    /    \
    C-C      CL
    3 2
```

FORMULA   C7H11CLO   MOL WT 146.62
SOLVENT   C4H10O
ORIG ST   C4H12SI   TEMP   AMB

55.40	29.70	25.50	25.90	176.30
1/2	2/3	3/3	4/3	7/1

T.PEHK,E.LIPPMAA
ORG MAGN RESON   3, 679 (1971)

1578   Q 000400 CYANOCYCLOHEXANE

```
    5 6
    C-C
   /    \1 7
 4C      C-C≡N
   \    /
    C-C
    3 2
```

FORMULA   C7H11N   MOL WT 109.17
SOLVENT   C4H10O
ORIG ST   C4H12SI   TEMP   AMB

28.30	30.10	24.60	25.80	122.40
1/2	2/3	3/3	4/3	7/1

T.PEHK,E.LIPPMAA
ORG MAGN RESON   3, 679 (1971)

1579   Q 000400 CYCLOHEXANE CARBOXYLATE

```
    5 6
    C-C     O
   /    \1 7/
 4C      C-C
   \    /    \
    C-C      -
    3 2
```

FORMULA   C7H11O2   MOL WT 127.16
SOLVENT   C4H10O
ORIG ST   C4H12SI   TEMP   AMB

47.20	30.90	26.90	26.90	185.40
1/2	2/3	3/3	4/3	7/1

T.PEHK,E.LIPPMAA
ORG MAGN RESON   3, 679 (1971)

1580   Q 000400 CYCLOHEXANECARBALDEHYDE

```
    4 3
    C-C      O
  5/   \2 1/
 C      C-C
  \6 7/    \
   C-C      H
```

FORMULA   C7H12O   MOL WT 112.17
SOLVENT   CDCL3
ORIG ST   C4H12SI   TEMP   308

204.70	50.10	26.10	25.20	25.20	25.20
1/2	2/2	3/3	4/3	5/3	6/3
26.10					
7/3					

G.E.HAWKES,K.HERWIG,J.D.ROBERTS
J ORG CHEM   39, 1017 (1974)

1581   Q 000400 2-METHYLCYCLOHEXANONE

```
      7
     CH3
    3 2/
    C-C
   /    \1
 4C      C=O
   \    /
    C-C
    5 6
```

FORMULA   C7H12O   MOL WT 112.17
SOLVENT   CDCL3
ORIG ST   C4H12SI   TEMP   AMB

212.90	45.30	36.20	25.20	28.00	41.80
1/1	2/2	3/3	4/3	5/3	6/3
14.70					
7/4					

J.B.STOTHERS,C.T.TAN
CAN J CHEM   52, 308 (1974)

1582   Q 000400 2-METHYLCYCLOHEXANONE

```
        ]
   7    ||
  H3C   C1
    \  / \
    C2   C6
    |    |
    C3   C5
    \   /
     .C4
```

FORMULA	C7H12O			MOL WT	112.17
SOLVENT	CDCL3				
ORIG ST	C4H12SI			TEMP	308

212.90	45.50	36.40	25.30	28.00	42.00
1/1	2/2	3/3	4/3	5/3	6/3
14.80					
7/4					

G.E.HAWKES,K.HERWIG,J.D.ROBERTS
J ORG CHEM   39, 1017 (1974)

1583   Q 000400 3-METHYLCYCLOHEXANONE

```
     7
    H3C
     \3 2
      C-C
    4/   \1
   C      C=O
    \    /
     C-C
     5 6
```

FORMULA	C7H12O			MOL WT	112.17
SOLVENT	CDCL3				
ORIG ST	C4H12SI			TEMP	AMB

211.00	49.90	34.10	33.30	25.30	41.00
1/1	2/3	3/2	4/3	5/3	6/3
22.00					
7/4					

J.B.STOTHERS,C.T.TAN
CAN J CHEM   52, 308 (1974)

1584   Q 000400 3-METHYLCYCLOHEXANONE

```
     ]
     ||
     C1
    / \
  C2   C6
  |    |
  C3   C5
 7/ \ /
H3C   C4
```

FORMULA	C7H12O			MOL WT	112.17
SOLVENT	CDCL3				
ORIG ST	C4H12SI			TEMP	308

211.60	50.00	34.20	33.40	25.30	41.10
1/1	2/3	3/2	4/3	5/3	6/3
22.10					
7/4					

G.E.HAWKES,K.HERWIG,J.D.ROBERTS
J ORG CHEM   39, 1017 (1974)

1585   Q 000400 4-METHYLCYCLOHEXANONE

```
      3 2
      C-C
  7 4/   \1
 H3C-C     C=O
    \     /
     C-C
     5 6
```

FORMULA	C7H12O			MOL WT	112.17
SOLVENT	CDCL3				
ORIG ST	C4H12SI			TEMP	AMB

211.40	40.60	34.70	31.10	34.70	40.60
1/1	2/3	3/3	4/2	5/3	6/3
20.90					
7/4					

J.B.STOTHERS,C.T.TAN
CAN J CHEM   52, 308 (1974)

1586   Q 000400 CYCLOHEXANECARBOXYLIC ACID

```
    5 6
    C-C      ]
   /   \1 7/
  4C     C-C
   \    /  \
    C-C    OH
    3 2
```

FORMULA	C7H12O2			MOL WT	128.17
SOLVENT	C4H10O				
ORIG ST	C4H12SI			TEMP	AMB

43.70	29.60	26.20	26.60	182.10
1/2	2/3	3/3	4/3	7/1

T.PEHK,E.LIPPMAA
ORG MAGN RESON   3, 679 (1971)

1587   Q 000400   CHLOROMETHYLCYCLOHEXANE

```
      5 6
      C-C
    /     \ 1 7
  4C       C-CH2-CL
    \     /
      C-C
      3 2
```

FORMULA   C7H13CL   MOL WT 132.63
SOLVENT   C4H10O
ORIG ST   C4H12SI   TEMP   AMB

40.90	31.20	26.40	26.80	50.70
1/2	2/3	3/3	4/3	7/3

T.PEHK,E.LIPPMAA
ORG MAGN RESON   3, 679 (1971)

1588   Q 000400   ANTI-CYCLOHEXANECARBALDOXIME

```
              OH
    4  3     /
    C-C    N
  5/    \2 ||
  C      C-CH
   \6 7/   1
    C-C
```

FORMULA   C7H13NO   MOL WT 127.19
SOLVENT   CDCL3
ORIG ST   C4H12SI   TEMP   308

155.80	38.50	30.30	25.50	25.60	25.50
1/2	2/2	3/3	4/3	5/3	6/3
30.30					
7/3					

G.E.HAWKES,K.HERWIG,J.D.ROBERTS
J ORG CHEM   39, 1017 (1974)

1589   Q 000400   SYN-CYCLOHEXANECARBALDOXIME

```
              HO
    4  3     \
    C-C    N
  5/    \2 ||
  C      C-CH
   \6 7/   1
    C-C
```

FORMULA   C7H13NO   MOL WT 127.19
SOLVENT   CDCL3
ORIG ST   C4H12SI   TEMP   308

156.30	33.90	29.50	25.60	25.90	25.60
1/2	2/2	3/3	4/3	5/3	6/3
29.50					
7/3					

G.E.HAWKES,K.HERWIG,J.D.ROBERTS
J ORG CHEM   39, 1017 (1974)

1590   Q 000400   ANTI-2-METHYLCYCLOHEXANONE OXIME

```
      OH
     /
    N
    ||
  7
H3C   C1
  \2/  \
    C    C6
    |    |
    C3   C5
     \  /
      C4
```

FORMULA   C7H13NO   MOL WT 127.19
SOLVENT   CDCL3
ORIG ST   C4H12SI   TEMP   308

163.00	37.20	35.70	23.90	24.80	26.10
1/1	2/2	3/3	4/3	5/3	6/3
16.90					
7/4					

G.E.HAWKES,K.HERWIG,J.D.ROBERTS
J ORG CHEM   39, 1017 (1974)

1591   Q 000400   ANTI-3-METHYLCYCLOHEXANONE OXIME

```
      OH
     /
    N
    ||
    C1
   /  \
  C2   C6
  |    |
  C3   C5
 7/ \  /
H3C   C4
```

FORMULA   C7H13NO   MOL WT 127.19
SOLVENT   CDCL3
ORIG ST   C4H12SI   TEMP   308

160.50	40.10	33.40	34.30	23.90	25.30
1/1	2/3	3/2	4/3	5/3	6/3
21.90					
7/4					

G.E.HAWKES,K.HERWIG,J.D.ROBERTS
J ORG CHEM   39, 1017 (1974)

```
1592                    Q 000400  SYN-2-METHYLCYCLOHEXANONE OXIME

        HO
          \             FORMULA   C7H13NO              MOL WT   127.19
           N            SOLVENT   CDCL3
      7    ||           ORIG ST   C4H12SI              TEMP        308
     H3C   C1
        \2/ \             163.60    26.80   26.70   20.50   28.40   31.70
         C   C6             1/1      2/2     3/3     4/3     5/3     6/3
         |   |            16.20
         C3  C5             7/4
          \ /
           C4             G.E.HAWKES,K.HERWIG,J.D.ROBERTS
                          J ORG CHEM                    39, 1017 (1974)
```

```
1593                    Q 000400  SYN-3-METHYLCYCLOHEXANONE OXIME

        HO
          \             FORMULA   C7H13NO              MOL WT   127.19
           N            SOLVENT   CDCL3
           ||           ORIG ST   C4H12SI              TEMP        308
           C1
         / \              160.40    34.20   32.50   32.30   24.50   31.70
       C2   C6             1/1      2/3     3/2     4/3     5/3     6/3
       |    |            22.10
       C3   C5             7/4
      7/ \ /
     H3C   C4             G.E.HAWKES,K.HERWIG,J.D.ROBERTS
                          J ORG CHEM                    39, 1017 (1974)
```

```
1594                    Q 000400  METHYLCYCLOHEXANE

                        FORMULA   C7H14                MOL WT    98.19
      3 2               SOLVENT   CDCL3
      C-C               ORIG ST   C4H12SI              TEMP        AMB
    /    \1 7
   4C      C-CH3          33.00    35.60   26.60   26.60   26.60   35.60
    \    /                 1/2      2/3     3/3     4/3     5/3     6/3
      C-C               22.90
      5 6                 7/4

                        L.F.JOHNSON,W.C.JANKOWSKI
                        CARBON-13 NMR SPECTRA,JOHN
                        WILEY AND SONS,NEW YORK           265 (1972)
```

```
1595                    Q 000400  METHYLCYCLOHEXANE

      5 6               FORMULA   C7H14                MOL WT    98.19
      C-C               SOLVENT   C4H10O
    /    \1 7           ORIG ST   C4H12SI              TEMP        AMB
   4C      C-CH3
    \    /                34.40    36.00   27.10   27.00   23.20
      C-C                  1/2      2/3     3/3     4/3     7/4
      3 2
                        T.PEHK,E.LIPPMAA
                        ORG MAGN RESON                  3,  679 (1971)
```

```
1596                    Q 000400  HYDROXYMETHYLCYCLOHEXANE

      5 6               FORMULA   C7H14O               MOL WT   114.19
      C-C               SOLVENT   C4H10O
    /    \1 7           ORIG ST   C4H12SI              TEMP        AMB
   4C      C-CH2-OH
    \    /                41.30    30.50   26.70   27.40   68.10
      C-C                  1/2      2/3     3/3     4/3     7/3
      3 2
                        T.PEHK,E.LIPPMAA
                        ORG MAGN RESON                  3,  679 (1971)
```

1597   Q 000400 METHOXYCYCLOHEXANE

```
  5 6
  C-C
 /   \1  7
4C    C-O-CH3
 \   /
  C-C
  3 2
```

FORMULA C7H14O   MOL WT 114.19
SOLVENT C4H10O
ORIG ST C4H12SI   TEMP   AMB

78.60	32.30	24.30	26.70	55.10
1/2	2/3	3/3	4/3	7/4

T.PEHK,E.LIPPMAA
ORG MAGN RESON   3, 679 (1971)

1598   Q 000400 1-METHYLCYCLOHEXANOL

```
     OH
     | 7
    1C-CH3
    6/|
 5C---C |
  | C---C
  1/3   2
 4C
```

FORMULA C7H14O   MOL WT 114.19
SOLVENT CDCL3
ORIG ST C4H12SI   TEMP   318

69.77	39.67	22.87	25.96	29.54
1/1	2/3	3/3	4/3	7/4

Y.SENDA,J.ISHIYAMA,S.IMAIZUMI
TETRAHEDRON   31, 1601 (1975)

1599   Q 000400 CIS-2-METHYLCYCLOHEXANOL

```
  5 6
  C-C
 /   \
4C    C1
 \ 2/|
 3C-C OH
    |
   7CH3
```

FORMULA C7H14O   MOL WT 114.19
SOLVENT CDCL3
ORIG ST C4H12SI   TEMP   AMB

70.90	35.90	29.00	24.40	21.00	32.40
1/2	2/2	3/3	4/3	5/3	6/3
16.70					
7/4					

L.F.JOHNSON,W.C.JANKOWSKI
CARBON-13 NMR SPECTRA,JOHN
WILEY AND SONS,NEW YORK   272 (1972)

1600   Q 000400 TRANS-2-METHYLCYCLOHEXANOL

```
    6
  5C-C OH
 /    \|
4C     C1
 \    /
  3C-C2
     |
    CH3
     7
```

FORMULA C7H14O   MOL WT 114.19
SOLVENT CDCL3
ORIG ST C4H12SI   TEMP   AMB

76.20	40.20	33.80	25.80	25.30	35.50
1/2	2/2	3/3	4/3	5/3	6/3
18.70					
7/4					

L.F.JOHNSON,W.C.JANKOWSKI
CARBON-13 NMR SPECTRA,JOHN
WILEY AND SONS,NEW YORK   272 (1972)

1601   Q 000400 CIS-3-METHYLCYCLOHEXANOL

```
   7
  H3C
   \3 2
    C-C  OH
   /   \ /
  4C    C1
   \   /
    C-C
    5 6
```

FORMULA C7H14O   MOL WT 114.19
SOLVENT CDCL3
ORIG ST C4H12SI   TEMP   AMB

70.40	44.60	31.60	35.30	24.30	34.30
1/2	2/3	3/2	4/3	5/3	6/3
22.40					
7/4					

L.F.JOHNSON,W.C.JANKOWSKI
CARBON-13 NMR SPECTRA,JOHN
WILEY AND SONS,NEW YORK   267 (1972)

1602 — Q 000400 TRANS-3-METHYLCYCLOHEXANOL

```
    7
  H3C
     \3  2
      C-C
     /    \
   4C      C1
     \    / \
      C-C   OH
      5  6
```

FORMULA	C7H14O			MOL WT	114.19
SOLVENT	CDCL3				
ORIG ST	C4H12SI			TEMP	AMB

66.40	41.50	26.50	34.30	20.10	33.10
1/2	2/3	3/2	4/3	5/3	6/3
21.90					
7/4					

L.F.JOHNSON,W.C.JANKOWSKI
CARBON-13 NMR SPECTRA,JOHN
WILEY AND SONS,NEW YORK   267 (1972)

1603 — Q 000400 CIS-4-METHYLCYCLOHEXANOL

```
     5  6
      C-C
    4/   \
    C      C1
   / \    / \
 H3C  C-C   OH
  7   3 2
```

FORMULA	C7H14O			MOL WT	114.19
SOLVENT	CDCL3				
ORIG ST	C4H12SI			TEMP	AMB

66.70	32.10	29.10	31.10	29.10	32.10
1/2	2/3	3/3	4/2	5/3	6/3
21.50					
7/4					

L.F.JOHNSON,W.C.JANKOWSKI
CARBON-13 NMR SPECTRA,JOHN
WILEY AND SONS,NEW YORK   271 (1972)

1604 — Q 000400 TRANS-4-METHYLCYCLOHEXANOL

```
     5  6
      C-C   OH
    4/   \ /
    C      C1
   / \    /
 H3C  C-C
  7   3 2
```

FORMULA	C7H14O			MOL WT	114.19
SOLVENT	CDCL3				
ORIG ST	C4H12SI			TEMP	AMB

70.50	33.50	35.50	31.90	35.50	33.50
1/2	2/3	3/3	4/2	5/3	6/3
21.90					
7/4					

L.F.JOHNSON,W.C.JANKOWSKI
CARBON-13 NMR SPECTRA,JOHN
WILEY AND SONS,NEW YORK   271 (1972)

1605 — Q 000400 N-METHYLCYCLOHEXYLAMINE

```
   3  2    7
    C-C    CH3
   /   \1 /
 4C      C-N
   \   / \
    C-C   H
    5  6
```

FORMULA	C7H15N			MOL WT	113.20
SOLVENT	CDCL3				
ORIG ST	C4H12SI			TEMP	AMB

58.60	33.30	25.10	26.30	25.10	33.30
1/2	2/3	3/3	4/3	5/3	6/3
33.50					
7/4					

L.F.JOHNSON,W.C.JANKOWSKI
CARBON-13 NMR SPECTRA,JOHN
WILEY AND SONS,NEW YORK   275 (1972)

1606 — Q 000400 4-DICHLOROMETHYL-4-METHYLCYCLOHEXA-2,5-DIENONE

```
     O
     ||
     C1
    / \
  6C    C2
   ||    ||
  5C    C3
    \  /
     C4
    8/ \7
  H3C  HC-CL
        |
        CL
```

FORMULA	C8H8CL2O			MOL WT	191.06
SOLVENT	CDCL3				
ORIG ST	C4H12SI			TEMP	305

| 184.40 | 130.60 | 148.40 | 48.10 | 76.60 | 22.70 |
| 1/1 | 2/2 | 3/2 | 4/1 | 7/2 | 8/4 |

R.HOLLENSTEIN,W.VON PHILIPSBORN
HELV CHIM ACTA   55, 2030 (1972)

1607   Q 000400   6-DICHLOROMETHYL-6-METHYLCYCLOHEXA-2,4-DIENONE

```
   CL   O
   |    ||
CL-CH  C1
  7\  / \
      C6   C2
   8/ |    ||
  H3C C5   C3
      \   /
       C4
```

FORMULA	C8H8CL2O			MOL WT	191.06
SOLVENT	CDCL3				
ORIG ST	C4H12SI			TEMP	305

201.40	125.30	141.50	123.30	140.00	57.90
1/1	2/2	3/2	4/2	5/2	6/1
76.80	24.00				
7/2	8/4				

R.HOLLENSTEIN,W.VON PHILIPSBORN
HELV CHIM ACTA   55, 2030 (1972)

1608   Q 000400   4-VINYLCYCLOHEXENE-1

```
      3
    C-C2
   //   \
 4C      C-CH=CH2
   \    /1 7 8
    C-C
    5 6
```

FORMULA	C8H12			MOL WT	108.18
SOLVENT	CDCL3				
ORIG ST	C4H12SI			TEMP	AMB

37.70	126.00	126.80	143.70	112.30	24.90
1/2	3/2	4/2	7/2	8/3	
28.50	31.10				

L.F.JOHNSON,W.C.JANKOWSKI
CARBON-13 NMR SPECTRA,JOHN
WILEY AND SONS,NEW YORK   306 (1972)

1609   Q 000400   1-ETHYNYLCYCLOHEXANOL

```
      4 5
      C-C
  1 2 3/  \
HC≡C-C-OH  C6
        \  /
        C-C
        8 7
```

FORMULA	C8H12O			MOL WT	124.18
SOLVENT	CDCL3				
ORIG ST	C4H12SI			TEMP	AMB

72.20	87.80	68.60	39.70	25.10	25.10
1/2	2/1	3/1	4/3	5/3	6/3

M.T.W.HEARN
TETRAHEDRON   32, 115 (1976)

1610   Q 000400   ACETYLCYCLOHEXANE

```
   5 6
   C-C      O
  /   \1 7//
4C     C-C
  \   /  \8
   C-C    CH3
   3 2
```

FORMULA	C8H14O			MOL WT	126.20
SOLVENT	C4H10O				
ORIG ST	C4H12SI			TEMP	AMB

51.50	29.00	26.60	26.30	209.40	27.60
1/2	2/3	3/3	4/3	7/1	8/4

T.PEHK,E.LIPPMAA
ORG MAGN RESON   3, 679 (1971)

1611   Q 000400   2,2-DIMETHYLCYCLOHEXANONE

```
    8    7
   H3C  CH3
     \  /
     3C-C2
    /    \1
  4C      C=O
    \    /
    C-C
    5 6
```

FORMULA	C8H14O			MOL WT	126.20
SOLVENT	CDCL3				
ORIG ST	C4H12SI			TEMP	AMB

215.00	45.10	41.10	21.50	27.60	38.20
1/1	2/1	3/3	4/3	5/3	6/3
25.20	25.20				
7/4	8/4				

J.B.STOTHERS,C.T.TAN
CAN J CHEM   52, 308 (1974)

1612   Q 000400 CIS-2,5-DIMETHYLCYCLOHEXANONE

```
            7
           CH3                FORMULA   C8H14O              MOL WT  126.20
            I                 SOLVENT   CDCL3
          3C-C2               ORIG ST   C4H12SI             TEMP      AMB
          /8  \1
        4C CH3 C=O            213.00    44.50   31.30   30.10   32.80   47.50
          \I  /               1/1       2/2     3/3     4/3     5/2     6/3
          C-C                 15.30     19.80
          5  6                7/4       8/4

                              J.B.STOTHERS,C.T.TAN
                              CAN J CHEM                    52,   308 (1974)
```

1613   Q 000400 TRANS-2,5-DIMETHYLCYCLOHEXANONE

```
            7
           CH3                FORMULA   C8H14O              MOL WT  126.20
         3  I                 SOLVENT   CDCL3
          C-C                 ORIG ST   C4H12SI             TEMP      AMB
          /  2\1
        4C      C=O           211.80    44.30   35.20   34.20   35.60   50.20
          \5  /               1/1       2/2     3/3     4/3     5/2     6/3
          C-C                 14.40     22.40
          I  6                7/4       8/4
          CH3
          8                   J.B.STOTHERS,C.T.TAN
                              CAN J CHEM                    52,   308 (1974)
```

1514   Q 000400 CIS-2,6-DIMETHYLCYCLOHEXANONE

```
            7
           CH3                FORMULA   C8H14O              MOL WT  126.20
            I                 SOLVENT   CDCL3
          3C-C                ORIG ST   C4H12SI             TEMP      AMB
          /  2\1
        4C H3C 8C=O           213.60    45.30   37.30   25.60   37.30   45.30
          \   I/               1/1       2/2     3/3     4/3     5/3     6/2
          5C-C6               14.60     14.60
                              7/4       8/4

                              J.B.STOTHERS,C.T.TAN
                              CAN J CHEM                    52,   308 (1974)
```

1615   Q 000400 TRANS-2,6-DIMETHYLCYCLOHEXANONE

```
            7
           CH3                FORMULA   C8H14O              MOL WT  126.20
         3  I                 SOLVENT   CDCL3
          C-C                 ORIG ST   C4H12SI             TEMP      AMB
          /  2\1
        4C      C=O           216.20    42.70   34.80   20.30   34.80   42.70
          \   /               1/1       2/2     3/3     4/3     5/3     6/2
          5C-C6               16.00     16.00
            I                 7/4       8/4
          8CH3
                              J.B.STOTHERS,C.T.TAN
                              CAN J CHEM                    52,   308 (1974)
```

1616   Q 000400 CIS-3,5-DIMETHYLCYCLOHEXANONE

```
            7
          H3C                 FORMULA   C8H14O              MOL WT  126.20
           I  2               SOLVENT   CDCL3
         3C-C                 ORIG ST   C4H12SI             TEMP      AMB
         /8  \1
       4C CH3 C=O             210.40    49.30   33.20   42.70   33.20   49.30
         \I  /                 1/1       2/3     3/2     4/3     5/2     6/3
         C-C                  22.30     22.30
         5  6                 7/4       8/4

                              J.B.STOTHERS,C.T.TAN
                              CAN J CHEM                    52,   308 (1974)
```

1617 — Q 000400 TRANS-3,5-DIMETHYLCYCLOHEXANONE

```
      7
     CH3
     | 2
   3C-C
   /    \1
 4C      C=O
   \5  /
    C-C
    | 6
   CH3
    8
```

FORMULA	C8H14O			MOL WT	126.20
SOLVENT	CDCL3				
ORIG ST	C4H12SI			TEMP	AMB

211.00	48.70	29.60	39.60	29.60	48.70
1/1	2/3	3/2	4/3	5/2	6/3
20.80	20.80				
7/4	8/4				

J.B.STOTHERS,C.T.TAN
CAN J CHEM   52, 308 (1974)

1618 — Q 000400 ACETOXYCYCLOHEXANE

```
   5  6
   C-C        O
  /    \1    || 8
 4C      C-O-C-CH3
  \    /     7
   C-C
   3 2
```

FORMULA	C8H14O2			MOL WT	142.20
SOLVENT	C4H10O				
ORIG ST	C4H12SI			TEMP	AMB

72.30	32.20	24.40	26.10	169.20	21.00
1/2	2/3	3/3	4/3	7/1	8/4

T.PEHK,E.LIPPMAA
ORG MAGN RESON   3, 679 (1971)

1619 — Q 000400 CYCLOHEXYLACETATE

```
   3 2
   C-C
  /    \1  7 8
 4C      C-O-C-CH3
  \    /     ||
   C-C        O
   5 6
```

FORMULA	C8H14O2			MOL WT	142.20
SOLVENT	CDCL3				
ORIG ST	C4H12SI			TEMP	AMB

72.50	31.70	23.90	25.50	23.90	31.70
1/2	2/3	3/3	4/3	5/3	6/3
170.10	21.90				
7/1	8/4				

L.F.JOHNSON,W.C.JANKOWSKI
CARBON-13 NMR SPECTRA,JOHN
WILEY AND SONS,NEW YORK   311 (1972)

1620 — Q 000400 METHOXYCARBONYLCYCLOHEXANE

```
   5 6
   C-C        O
  /    \1 7//
 4C      C-C
  \    /    \ 8
   C-C       O-CH3
   3 2
```

FORMULA	C8H14O2			MOL WT	142.20
SOLVENT	C4H10O				
ORIG ST	C4H12SI			TEMP	AMB

43.40	29.60	26.00	26.40	175.30	51.00
1/2	2/3	3/3	4/3	7/1	8/4

T.PEHK,E.LIPPMAA
ORG MAGN RESON   3, 679 (1971)

1621 — Q 000400 ETHYLCYCLOHEXANE

```
   5 6
   C-C
  /    \1 7  8
 4C      C-CH2-CH3
  \    /
   C-C
   3 2
```

FORMULA	C8H16			MOL WT	112.22
SOLVENT	C4H10O				
ORIG ST	C4H12SI			TEMP	AMB

40.20	33.70	27.10	27.40	30.70	11.50
1/2	2/3	3/3	4/3	7/3	8/4

T.PEHK,E.LIPPMAA
ORG MAGN RESON   3, 679 (1971)

1622　　　Q 000400 CIS-1,2-DIMETHYLCYCLOHEXANOL

```
        7CH3        FORMULA   C8H16O                MOL WT   128.22
         |          SOLVENT   CDCL3
        1C-OH       ORIG ST   C4H12SI               TEMP        318
         6/|
      5C---C | 8      72.98    42.28    32.21   25.48    24.26    41.49
       | C---C-CH3     1/1      2/2      3/3     4/3      5/3      6/3
       1/3    2       20.80    15.47
      4C                7/4      8/4
```

Y. SENDA, J. ISHIYAMA, S. IMAIZUMI
TETRAHEDRON　　　　　　　　　31, 1601 (1975)

1623　　　Q 000400 TRANS-1,2-DIMETHYLCYCLOHEXANOL

```
                    FORMULA   C8H16O                MOL WT   128.22
        OH          SOLVENT   CDCL3
         | 7        ORIG ST   C4H12SI               TEMP        318
        1C-CH3
         6/|          70.98    40.46    30.69   26.08    22.14    40.10
      5C---C | 8       1/1      2/2      3/3     4/3      5/3      6/3
       | C---C-CH3    28.75    15.28
       1/3    2        7/4      8/4
      4C
```

Y. SENDA, J. ISHIYAMA, S. IMAIZUMI
TETRAHEDRON　　　　　　　　　31, 1601 (1975)

1624　　　Q 000400 CIS-1,3-DIMETHYLCYCLOHEXANOL

```
        OH          FORMULA   C8H16O                MOL WT   128.22
         | 7        SOLVENT   CDCL3
        1C-CH3      ORIG ST   C4H12SI               TEMP        318
         6/|
      5C---C |        69.77    47.68    27.78   34.58    21.72    38.40
       |3C---C2        1/1      2/3      3/2     4/3      5/3      6/3
       1/|            31.85    22.56
      4C CH3           7/4      8/4
         8
```

Y. SENDA, J. ISHIYAMA, S. IMAIZUMI
TETRAHEDRON　　　　　　　　　31, 1601 (1975)

1625　　　Q 000400 TRANS-1,3-DIMETHYLCYCLOHEXANOL

```
         7
        CH3         FORMULA   C8H16O                MOL WT   128.22
         |          SOLVENT   CDCL3
        1C-OH       ORIG ST   C4H12SI               TEMP        318
         6/|
      5C---C |        71.16    49.50    30.63   34.76    23.84    40.16
       |3C---C2        1/1      2/3      3/2     4/3      5/3      6/3
       1/|            26.02    22.69
      4C CH3           7/4      8/4
         8
```

Y. SENDA, J. ISHIYAMA, S. IMAIZUMI
TETRAHEDRON　　　　　　　　　31, 1601 (1975)

1626　　　Q 000400 CIS-1,4-DIMETHYLCYCLOHEXANOL

```
        7CH3        FORMULA   C8H16O                MOL WT   128.22
         |          SOLVENT   CDCL3
        1C-OH       ORIG ST   C4H12SI               TEMP        318
         6/|
      5C---C |        70.56    39.80    32.39   31.91    32.39    39.80
       | C---C2        1/1      2/3      3/3     4/2      5/3      6/3
       8 1/3         25.90    21.65
      H3C-C            7/4      8/4
         4
```

Y. SENDA, J. ISHIYAMA, S. IMAIZUMI
TETRAHEDRON　　　　　　　　　31, 1601 (1975)

1627 Q 000400 TRANS-1,4-DIMETHYLCYCLOHEXANOL

```
        OH
        | 7
      1C-CH3
       6/|
   5C---C |
    | C---C2
   8 |/3
  H3C-C
     4
```

FORMULA	C8H16O		MOL WT	128.22
SOLVENT	CDCL3			
ORIG ST	C4H12SI		TEMP	318

68.80	38.64	30.45	31.36	30.45	38.64
1/1	2/3	3/3	4/2	5/3	6/3
31.97	22.26				
7/4	8/4				

Y.SENDA,J.ISHIYAMA,S.IMAIZUMI
TETRAHEDRON   31, 1601 (1975)

1628 Q 000400 ISOPHORONE

```
    8     9
  H3C   CH3
    \   /
     5C-C4
    /     \
  6C      C-CH3
    \    //3 7
     C-C
    //1 2
    O
```

FORMULA	C9H14O		MOL WT	138.21
SOLVENT	CDCL3			
ORIG ST	C4H12SI		TEMP	AMB

199.00	125.30	159.70	45.10	33.30	50.70
1/1	2/2	3/1	4/3	5/1	6/3
24.30	28.20	28.20			
7/4	8/4	9/4			

L.F.JOHNSON,W.C.JANKOWSKI
CARBON-13 NMR SPECTRA,JOHN
WILEY AND SONS,NEW YORK   357 (1972)

1629 Q 000400 2,2,6-TRIMETHYLCYCLOHEXANONE

```
    8     7
  H3C   CH3
    \   /
     3C-C2
    /     \1
  4C      C=O
    \    /
     5C-C6
      |
     9CH3
```

FORMULA	C9H16O		MOL WT	140.23
SOLVENT	CDCL3			
ORIG ST	C4H12SI		TEMP	AMB

216.40	45.00	41.80	21.60	36.70	40.60
1/1	2/1	3/3	4/3	5/3	6/2
25.20	25.60	15.00			
7/4	8/4	9/4			

J.B.STOTHERS,C.T.TAN
CAN J CHEM   52, 308 (1974)

1630 Q 000400 3,3,5-TRIMETHYLCYCLOHEXANONE

```
    47
    CH3
  8 | 2
  H3C-C-C
   /3  \1
  4C      C=O
   \    /
    5C-C
    | 6
    CH3
    9
```

FORMULA	C9H16O		MOL WT	140.23
SOLVENT	CDCL3			
ORIG ST	C4H12SI		TEMP	AMB

210.50	53.90	35.10	47.20	29.50	49.00
1/1	2/3	3/1	4/3	5/2	6/3
25.70	32.00	22.40			
7/4	8/4	9/4			

J.B.STOTHERS,C.T.TAN
CAN J CHEM   52, 308 (1974)

1631 Q 000400 3,3,5-TRIMETHYLCYCLOHEXANOL

```
    3  4
    C-C
   /3  \5
 2C8CH3 C
   \1  /|
   1C-C OH
    | 6
   7CH3     9
         8 -CH3
```

FORMULA	C9H18O		MOL WT	142.24
SOLVENT	CDCL3			
ORIG ST	C4H12SI		TEMP	AMB

32.20	48.10	27.20	44.60	67.50	97.70
1/1	2/3	3/2	4/3	5/2	6/3
33.10	25.70	22.30			
7/4	8/4	9/4			

L.F.JOHNSON,W.C.JANKOWSKI
CARBON-13 NMR SPECTRA,JOHN
WILEY AND SONS,NEW YORK   360 (1972)

1632   Q 000400   4-ALLYL-2-FLUORO-4-METHYLCYCLOHEXA-2,5-DIENONE

```
        O
        ||
       C1   F
      / \ /
    6C    C2
    ||    ||
    5C    C3
      \ /   8
       C4  CH
     10/ \7/ \9
    H3C  CH2 CH2
```

FORMULA	C10H11FO			MOL WT	166.20
SOLVENT	CDCL3				
ORIG ST	C4H12SI			TEMP	305

178.30	152.10	129.80	42.90	155.60	127.80
1/1	2/2	3/2	4/1	5/2	6/2
44.60	131.70	119.20	25.10		
7/3	8/2	9/3	10/4		

R.HOLLENSTEIN,W.VON PHILIPSBORN
HELV CHIM ACTA   55, 2030 (1972)

1633   Q 000400   4-ALLYL-3-FLUORO-4-METHYLCYCLOHEXA-2,5-DIENONE

```
        O
        ||
       C1
      / \
    6C    C2
    ||    ||
    5C    3C-F
      \ /
       C4  8CH
     10/ \7/ \9
    H3C  CH2 CH2
```

FORMULA	C10H11FO			MOL WT	166.20
SOLVENT	CDCL3				
ORIG ST	C4H12SI			TEMP	305

187.90	109.70	180.00	43.30	150.00	128.00
1/1	2/2	3/2	4/1	5/2	6/2
41.70	131.20	119.20	22.90		
7/3	8/2	9/3	10/4		

R.HOLLENSTEIN,W.VON PHILIPSBORN
HELV CHIM ACTA   55, 2030 (1972)

1634   Q 000400   4-ALLYL-4-METHYLCYCLOHEXA-2,5-DIENONE

```
        O
        ||
       C1
      / \
    C6    C2
    ||    ||
    5C    C3
      \ /  8
       C4  CH
     10/ \7/ \9
    H3C  CH2 CH2
```

FORMULA	C10H12O			MOL WT	148.21
SOLVENT	CDCL3				
ORIG ST	C4H12SI			TEMP	305

185.70	128.40	155.00	41.50	48.40	132.30
1/1	2/2	3/2	4/1	7/3	8/2
118.50	25.00				
9/3	10/4				

R.HOLLENSTEIN,W.VON PHILIPSBORN
HELV CHIM ACTA   55, 2030 (1972)

1635   Q 000400   6-ALLYL-6-METHYLCYCLOHEXA-2,4-DIENONE

```
            O
     10     ||
     H3C    C1
        \ / \
    8CH  C6   C2
    / \ /|    ||
  H2C H2C C5  C3
   9   7  \ /
           C4
```

FORMULA	C10H12O			MOL WT	148.21
SOLVENT	CDCL3				
ORIG ST	C4H12SI			TEMP	305

204.60	125.70	141.20	120.20	147.00	50.70
1/1	2/2	3/2	4/2	5/2	6/1
44.20	132.80	117.60	24.00		
7/3	8/2	9/3	10/4		

R.HOLLENSTEIN,W.VON PHILIPSBORN
HELV CHIM ACTA   55, 2030 (1972)

1636   Q 000400   3,4,6,6-TETRAMETHYLCYCLOHEXA-2,4-DIENONE

```
            O
     10     ||
     H3C    C1
        \ / \
         C6   C2
     9/|   ||
   H3C C5   C3
      \ / \7
       C4   CH3
        |
       8CH3
```

FORMULA	C10H14O			MOL WT	150.22
SOLVENT	CDCL3				
ORIG ST	C4H12SI			TEMP	305

205.20	123.20	155.00	127.80	143.70	46.00
1/1	2/2	3/1	4/1	5/2	6/1
21.00	18.80	25.40			
7/4	8/4	9/4			

R.HOLLENSTEIN,W.VON PHILIPSBORN
HELV CHIM ACTA   55, 2030 (1972)

1637   Q 000400 1-ACETOXY-1-ETHYNYLCYCLOHEXANE

```
               O
               ‖
         O-C
   1 2 3/   9\10
   HC≡C-C-C4    CH3
       ∕   \
     8C      C5
       \   ∕
        C-C
        7 6
```

FORMULA	C10H14O2			MOL WT	166.22
SOLVENT	CDCL3				
ORIG ST	C4H12SI			TEMP	AMB
74.30	83.60	75.00	36.90	22.40	25.10
1/2	2/1	3/1	4/3	5/3	6/3
168.90	21.70				
9/1	10/4				

M.T.W.HEARN
TETRAHEDRON   32, 115 (1976)

1638   Q 000400 4-TERT-BUTYLCYCLOHEXANONE

```
    9    3 2
   H3C   C-C
   8 |7 ∕   \1
   H3C-C-C4     C=O
     |   \   ∕
    H3C    C-C
    10    5 6
```

FORMULA	C10H18O			MOL WT	154.25
SOLVENT	CDCL3				
ORIG ST	C4H12SI			TEMP	AMB
211.50	41.20	27.60	46.70	27.60	41.20
1/1	2/3	3/3	4/2	5/3	6/3
32.40	27.60	27.60	27.60		
7/1	8/4	9/4	10/4		

J.B.STOTHERS,C.T.TAN
CAN J CHEM   52, 308 (1974)

1639.   Q 000400 4-TERT-BUTYLCYCLOHEXANONE

```
   10    3 2
   H3C   C-C
    | 4∕   \1
  H3C-C-C      C=O
   8 |7 \   ∕
   H3C    C-C
    9    5 6
```

FORMULA	C10H18O			MOL WT	154.25
SOLVENT	CDCL3				
ORIG ST	C4H12SI			TEMP	AMB
211.50	41.10	27.50	46.60	27.50	41.10
1/1	2/3	3/3	4/2	5/3	6/3
32.30	27.50	27.50	27.50		
7/1	8/4	9/4	10/4		

L.F.JOHNSON,W.C.JANKOWSKI
CARBON-13 NMR SPECTRA,JOHN
WILEY AND SONS,NEW YORK   405 (1972)

1640   Q 000400 ISOMENTHONE

```
      H3C10
       |
       C5
      ∕ \
    C4   C6
     |   |
    C3   C1
     \ ∕ \
      C2   O
       |
      HC7
     8∕ \9
   H3C    CH3
```

FORMULA	C10H18O			MOL WT	154.25
SOLVENT	CDCL3				
ORIG ST	C4H12SI			TEMP	308
214.10	57.20	27.00	29.60	34.40	48.20
1/1	2/2	3/3	4/3	5/2	6/3
26.80	21.00	19.90	21.40		
7/2	8/4	9/4	10/4		

G.E.HAWKES,K.HERWIG,J.D.ROBERTS
J ORG CHEM   39, 1017 (1974)

1641   Q 000400 MENTHONE

```
      H3C10
       |
       C5
      ∕ \
    C4   C6
     |   |
    C3   C1
     \ ∕ \
      C2   O
       |
      HC7
     8∕ \9
   H3C    CH3
```

FORMULA	C10H18O			MOL WT	154.25
SOLVENT	CDCL3				
ORIG ST	C4H12SI			TEMP	308
212.00	56.10	28.40	34.10	35.60	51.00
1/1	2/2	3/3	4/3	5/2	6/3
26.00	21.30	18.80	22.30		
7/2	8/4	9/4	10/4		

G.E.HAWKES,K.HERWIG,J.D.ROBERTS
J ORG CHEM   39, 1017 (1974)

```
1642                    Q 000400  2,2,6,6-TETRAMETHYLCYCLOHEXANONE
              8
            CH3                FORMULA   C10H18O                    MOL WT   154.25
          3 | 7               SOLVENT   CDCL3
          C-C-CH3             ORIG ST   C4H12SI                    TEMP      AMB
          /  2\1
        4C     C=O             220.00    44.40    40.20    18.10    40.20    44.40
          \  6/10              1/1       2/1      3/3      4/3      5/3      6/1
          5C-C-CH3             27.60     27.60    27.60    27.60
            |                  7/4       8/4      9/4      10/4
            9CH3
                               J.B.STOTHERS,C.T.TAN
                               CAN J CHEM                          52,   308 (1974)
```

```
1643                    Q 000400  3,3,5,5-TETRAMETHYLCYCLOHEXANONE
              7
            CH3                FORMULA   C10H18O                    MOL WT   154.25
         8 |3               SOLVENT   CDCL3
        H3C-C-C2            ORIG ST   C4H12SI                    TEMP      AMB
           /   \1
        4C       C=O          212.00    53.80    36.00    51.50    36.00    53.80
          \5 6/               1/1       2/3      3/1      4/3      5/1      6/3
        H3C-C-C               31.30     31.30    31.30    31.30
         9 |                  7/4       8/4      9/4      10/4
            CH3
           10                  J.B.STOTHERS,C.T.TAN
                               CAN J CHEM                          52,   308 (1974)
```

```
1544                    Q 000400  ANTI-ISOMENTHONE OXIME
          H3C10               FORMULA   C10H19NO                   MOL WT   169.27
            |                SOLVENT   CDCL3
           C5               ORIG ST   C4H12SI                    TEMP      308
          / \
        C4   C6              163.00    47.50    26.70    28.30    34.50    29.70
        |    |               1/1       2/2      3/3      4/3      5/2      6/3
        C3   C1   OH         26.70     21.20    20.60    22.50
         \  / \ /            7/2       8/4      9/4      10/4
          C2   N
           |
          HC7                G.E.HAWKES,K.HERWIG,J.D.ROBERTS
         8/  \9              J ORG CHEM                          39,  1017 (1974)
        H3C   CH3
```

```
1645                    Q 000400  SYN-ISOMENTHONE OXIME
          H3C10               FORMULA   C10H19NO                   MOL WT   169.27
            |                SOLVENT   CDCL3
           C5               ORIG ST   C4H12SI                    TEMP      308
          / \
        C4   C6              162.60    39.30    27.30    37.50    33.20    29.50
        |    |               1/1       2/2      3/3      4/3      5/2      6/3
        C3   C1              29.70     20.70    20.40    22.40
         \  / \              7/2       8/4      9/4      10/4
          C2   N
           |
          HC7      OH         G.E.HAWKES,K.HERWIG,J.D.ROBERTS
         8/  \9              J ORG CHEM                          39,  1017 (1974)
        H3C   CH3
```

```
1646                    Q 000400  ANTI-MENTHONE OXIME
          H3C10               FORMULA   C10H19NO                   MOL WT   169.27
            |                SOLVENT   CDCL3
           C5               ORIG ST   C4H12SI                    TEMP      308
          / \
        C4   C6              161.00    48.80    26.90    32.80    32.40    31.90
        |    |               1/1       2/2      3/3      4/3      5/2      6/3
        C3   C1   OH         26.40     21.80    19.10    21.40
         \  / \ /            7/2       8/4      9/4      10/4
          C2   N
           |
          HC7                G.E.HAWKES,K.HERWIG,J.D.ROBERTS
         8/  \9              J ORG CHEM                          39,  1017 (1974)
        H3C   CH3
```

1647 — Q 000400 SYN-MENTHONE OXIME

```
H3C10
  |
  C5
 / \
C4  C6
|   |
C3  C1
 \ / \\
  C2  N
  |    \
  HC7   OH
  8/  \9
 H3C   CH3
```

FORMULA	C10H19NO			MOL WT	169.27
SOLVENT	CDCL3				
ORIG ST	C4H12SI			TEMP	308

161.20	40.00	22.00	26.70	29.60	35.20
1/1	2/2	3/3	4/3	5/2	6/3
26.60	20.40	18.10	20.80		
7/2	8/4	9/4	10/4		

G.E.HAWKES,K.HERWIG,J.D.ROBERTS
J ORG CHEM   39, 1017 (1974)

1648 — Q 000400 N-BUTYLCYCLOHEXANE

```
 3 2
 C-C
/   \1
4C   C-CH2-CH2-CH2-CH3
\   / 7  8   9   10
 C-C
 5 6
```

FORMULA	C10H20			MOL WT	140.27
SOLVENT	CDCL3				
ORIG ST	C4H12SI			TEMP	AMB

37.90	33.60	26.60	26.90	26.60	33.60
1/2	2/3	3/3	4/3	5/3	6/3
37.40	29.30	23.20	14.10		
7/3	8/3	9/3	10/4		

L.F.JOHNSON,W.C.JANKOWSKI
CARBON-13 NMR SPECTRA,JOHN
WILEY AND SONS,NEW YORK   409 (1972)

1549 — Q 000400 6-ALLYL-4,6-DIMETHYLCYCLOHEXA-2,4-DIENONE

```
         O
  11     ||
 H3C     C1
   \   / \
  9CH  C6  C2
 10/ \ /|   ||
 H2C H2C C5  C3
   8    \ /
        C4
        |
       7CH3
```

FORMULA	C11H14O			MOL WT	162.23
SOLVENT	CDCL3				
ORIG ST	C4H12SI			TEMP	305

205.20	125.40	145.60	127.70	141.20	50.30
1/1	2/2	3/2	4/1	5/2	6/1
20.80	44.60	133.10	117.60	24.50	
7/4	8/3	9/2	10/3	11/4	

R.HOLLENSTEIN,W.VON PHILIPSBORN
HELV CHIM ACTA   55, 2030 (1972)

1650 — Q 000400 6-ALLYL-5,6-DIMETHYLCYCLOHEXA-2,4-DIENONE

```
         O
  11     ||
 H3C     C1
   \   / \
  9CH  C6  C2
 10/ \8/|   ||
 H2C H2C C5  C3
     / \ /
   H3C7  C4
```

FORMULA	C11H14O			MOL WT	162.23
SOLVENT	CDCL3				
ORIG ST	C4H12SI			TEMP	305

205.80	123.60	141.80	119.40	155.60	54.70
1/1	2/2	3/2	4/2	5/1	6/1
19.30	43.20	133.10	116.90	24.80	
7/4	8/3	9/2	10/3	11/4	

R.HOLLENSTEIN,W.VON PHILIPSBORN
HELV CHIM ACTA   55, 2030 (1972)

1651 — Q 000400 SPIRO(5.5)UNDECANE

```
    10
     C
    / \
  11C   C9
   |    |
  6C   C8
 4  5/|\  /
 C---C | C
  | C---C 7
  1/2   1
  3C
```

FORMULA	C11H20			MOL WT	152.28
SOLVENT	C11H20				
ORIG ST	C4H12SI			TEMP	AMB

37.61	22.02	27.50	32.71
1/3	2/3	3/3	6/1

D.K.DALLING,D.M.GRANT,E.G.PAUL
J AM CHEM SOC   95, 3718 (1973)

1652   Q 000400 2,4-DIMETHYL-6-(2-PROPYNYL)-CYCLOHEXA-2,4
                                      DIENONE

```
            O
     11     ‖
    H3C     C1          FORMULA   C11H12O                   MOL WT   160.22
 10  9     \ / \        SOLVENT   CDCL3
HC≡C-H2C-C6    C2       ORIG ST   C4H12SI                   TEMP        305
      8 |      ‖
        C5    C3           204.40   123.30   142.00   119.60   154.40   54.10
     7/ \ /                  1/1      2/2      3/2      4/2      5/1      6/1
    H3C   C4                19.30    26.70    80.00    69.50    24.80
                             7/4      8/3      9/1     10/2     11/4
```

R.HOLLENSTEIN,W.VON PHILIPSBORN
HELV CHIM ACTA   55, 2030 (1972)

1653   Q 000400 CIS-2-TERT-BUTYL-1-METHYLCYCLOHEXANOL

```
     7CH3              FORMULA   C11H22O                   MOL WT   170.30
       |              SOLVENT   CDCL3
     1C-OH             ORIG ST   C4H12SI                   TEMP        318
    6/ |  9CH3
  5C---C | 8/10            75.47    56.36    27.24    27.24    24.75    46.89
   | C---C-C-CH3            1/1      2/2      3/3      4/3      5/3      6/3
   1/3    2  \             22.56    34.40    30.39
  4C          CH3           7/4      8/1      9/4
             11
```

Y.SENDA,J.ISHIYAMA,S.IMAIZUMI
TETRAHEDRON   31, 1601 (1975)

1554   Q 000400 TRANS-2-TERT-BUTYL-1-METHYLCYCLOHEXANOL

```
     OH               FORMULA   C11H22O                   MOL WT   170.30
      | 7             SOLVENT   CDCL3
     1C-CH3            ORIG ST   C4H12SI                   TEMP        318
    6/ |
  5C---C | 8 9            73.83    54.84    24.45    27.11    22.14    44.59
   | C---C-C-CH3           1/1      2/2      3/3      4/3      5/3      6/3
   1/3   2 |\             32.88    34.76    31.18
  4C      H3C CH3          7/4      8/1      9/4
          11 10
```

Y.SENDA,J.ISHIYAMA,S.IMAIZUMI
TETRAHEDRON   31, 1601 (1975)

1655   Q 000400 CIS-3-TERT-BUTYL-1-METHYLCYCLOHEXANOL

```
     OH               FORMULA   C11H22O                   MOL WT   170.30
      | 7             SOLVENT   CDCL3
     1C-CH3            ORIG ST   C4H12SI                   TEMP        318
    6/ |
  5C---C |                70.07    39.98    42.77    26.57    22.14    38.64
   | C---C2                 1/1      2/3      3/2      4/3      5/3      6/3
   1/ \ 9                  32.15    32.15    27.48
  4C    C-CH3              7/4      8/1      9/4
        |\
     H3C CH3           Y.SENDA,J.ISHIYAMA,S.IMAIZUMI
     11 10             TETRAHEDRON                       31, 1601 (1975)
```

1656   Q 000400 TRANS-3-TERT-BUTYL-1-METHYLCYCLOHEXANOL

```
     7
     CH3              FORMULA   C11H22O                   MOL WT   170.30
      |              SOLVENT   CDCL3
     1C-OH             ORIG ST   C4H12SI                   TEMP        318
    6/ |
  5C---C |                71.59    41.98    45.80    26.87    24.02    40.04
   | C---C2                 1/1      2/3      3/2      4/3      5/3      6/3
   1/3\8 9                 26.08    32.09    26.27
  4C    C-CH3              7/4      8/1      9/4
        |\
     H3C CH3           Y.SENDA,J.ISHIYAMA,S.IMAIZUMI
     11 10             TETRAHEDRON                       31, 1601 (1975)
```

1657   Q 000400 CIS-4-TERT-BUTYL-1-METHYLCYCLOHEXANOL

```
        7
       CH3              FORMULA   C11H22O              MOL WT  170.30
        |               SOLVENT   CDCL3
       1C-OH            ORIG ST   C4H12SI              TEMP       318
       6/|
  9   5C---C |           71.23   40.96   25.13   47.94   25.13   40.96
 H3C   | C---C2           1/1     2/3     3/3     4/2     5/3     6/3
  10\8 1/3               25.42   32.47   27.86
 H3C-C-C                  7/4     8/1     9/4
  11/  4
 H3C                     Y.SENDA,J.ISHIYAMA,S.IMAIZUMI
                         TETRAHEDRON                   31, 1601 (1975)
```

1658   Q 000400 TRANS-4-TERT-BUTYL-1-METHYLCYCLOHEXANOL

```
        OH
        | 7             FORMULA   C11H22O              MOL WT  170.30
       1C-CH3           SOLVENT   CDCL3
       6/|              ORIG ST   C4H12SI              TEMP       318
  9   5C---C |
 H3C   | C---C2          68.86   39.37   22.69   47.74   22.69   39.37
  10\8 1/3               1/1     2/3     3/3     4/2     5/3     6/3
 H3C-C-C                 31.42   32.39   27.76
  11/  4                 7/4     8/1     9/4
 H3C
                         Y.SENDA,J.ISHIYAMA,S.IMAIZUMI
                         TETRAHEDRON                   31, 1601 (1975)
```

1659   Q 000400 6-SEC-BUTYNYL-2,6-DIMETHYLCYCLOHEXA-2,4-DIENONE THREO

```
        O
  12    ||    7          FORMULA   C12H14O             MOL WT  174.24
 H3C    C1   CH3         SOLVENT   CDCL3
     \ / \ /             ORIG ST   C4H12SI             TEMP       305
      C6  C2
  10 9 8/|    ||         204.40  133.00  138.00  121.70  142.10   53.20
 HC≡C-HC C5   C3          1/1     2/1     3/2     4/2     5/2     6/1
      |  \  /            15.30   34.80   85.40   71.30   17.00   24.80
     H3C   C4             7/4     8/2     9/1    10/2    11/4    12/4
  11
                         R.HOLLENSTEIN,W.VON PHILIPSBORN
                         HELV CHIM ACTA                55, 2030 (1972)
```

1660   Q 000400 6-SEC-BUTYNYL-3,6-DIMETHYLCYCLOHEXA-2,4-DIENONE ERYTHRO

```
        O
  12    ||              FORMULA   C12H14O             MOL WT  174.24
 H3C    C1              SOLVENT   CDCL3
     \ / \              ORIG ST   C4H12SI             TEMP       305
      C6  C2
  10 9 8/|    ||        203.40  123.60  152.80  125.50  143.10   51.50
 HC≡C-HC C5   C3         1/1     2/2     3/1     4/2     5/2     6/1
      |  \ / \7         22.60   33.40   84.90   69.40   15.00   21.40
     H3C   C4  CH3       7/4     8/2     9/1    10/2    11/4    12/4
  11
                        R.HOLLENSTEIN,W.VON PHILIPSBORN
                        HELV CHIM ACTA                55, 2030 (1972)
```

1661   Q 000400 6-SEC-BUTYNYL-3,6-DIMETHYLCYCLOHEXA-2,4-DIENONE THREO

```
        O
  12    ||              FORMULA   C12H14O             MOL WT  174.24
 H3C    C1              SOLVENT   CDCL3
     \ / \              ORIG ST   C4H12SI             TEMP       305
      C6  C2
  10 9 8/|    ||        203.40  123.50  153.70  125.60  143.80   52.20
 HC≡C-HC C5   C3         1/1     2/2     3/1     4/2     5/2     6/1
      |  \ / \7         22.70   34.40   85.30   71.50   17.10   24.80
     H3C   C4  CH3       7/4     8/2     9/1    10/2    11/4    12/4
  11
                        R.HOLLENSTEIN,W.VON PHILIPSBORN
                        HELV CHIM ACTA                55, 2030 (1972)
```

1662 — Q 000400 6-(2-PROPYNYL)-2,4,6-TRIMETHYLCYCLOHEXA-2,4-DIENONE

```
            O
   12      II      7
  H3C      C1     CH3
 11 10      \ / \ /
HC≡C-H2C-C6    C2
    9  I        II
     5C        C3
        \   /
         C4
          I
       8CH3
```

FORMULA	C12H14O			MOL WT	174.24
SOLVENT	CDCL3				
ORIG ST	C4H12SI			TEMP	305

203.90	132.20	142.30	127.90	137.40	48.10
1/1	2/1	3/2	4/1	5/2	6/1
15.20	21.10	28.40	80.40	70.30	24.20
7/4	8/4	9/3	10/1	11/2	12/4

R.HOLLENSTEIN,W.VON PHILIPSBORN
HELV CHIM ACTA　　　　　55, 2030 (1972)

1663 — Q 000400 6-ALLYL-2,4,6-TRIMETHYLCYCLOHEXA-2,4-DIENONE

```
            O
   12      II      7
  H3C      C1     CH3
            \ / \ /
   10CH    C6    C2
   ⁄ \   ⁄ I     II
 H2C H2C C5    C3
  11   9    \ /
           C4
            I
         8CH3
```

FORMULA	C12H16O			MOL WT	176.26
SOLVENT	CDCL3				
ORIG ST	C4H12SI			TEMP	305

205.20	132.50	142.20	127.50	138.80	49.80
1/1	2/1	3/2	4/1	5/2	6/1
15.20	21.10	44.90	133.50	117.20	24.70
7/4	8/4	9/3	10/2	11/3	12/4

R.HOLLENSTEIN,W.VON PHILIPSBORN
HELV CHIM ACTA　　　　　55, 2030 (1972)

1664 — Q 000400 2,3,4,5,6,6-HEXAMETHYLCYCLOHEXA-2,4-DIENONE

```
            O
   12      II      7
  H3C      C1     CH3
            \ / \ /
   H3C-C6    C2
   11  I     II
      C5    C3
   10⁄  \ ⁄ \8
  H3C    C4    CH3
          I
       9CH3
```

FORMULA	C12H18O			MOL WT	178.28
SOLVENT	CDCL3				
ORIG ST	C4H12SI			TEMP	305

204.60	126.30	149.90	124.20	145.20	48.30
1/1	2/1	3/1	4/1	5/1	6/1
11.50	18.10	15.70	15.70	25.00	
7/4	8/4	9/4	10/4	11/4	

R.HOLLENSTEIN,W.VON PHILIPSBORN
HELV CHIM ACTA　　　　　55, 2030 (1972)

1665 — Q 000400 HEXAMETHYLCYCLOHEXA-2,5-DIENONE

```
            O
   12      II      7
  H3C      C1     CH3
            \ / \ /
      6C    C2
      II    II
      5C    C3
   11⁄  \  ⁄ \8
  H3C    C4    CH3
      10⁄  \9
     H3C    CH3
```

FORMULA	C12H18O			MOL WT	178.28
SOLVENT	CDCL3				
ORIG ST	C4H12SI			TEMP	305

184.70	129.60	157.10	42.30	11.60	16.50
1/1	2/1	3/1	4/1	7/4	8/4
24.30					
9/4					

R.HOLLENSTEIN,W.VON PHILIPSBORN
HELV CHIM ACTA　　　　　55, 2030 (1972)

1666 — Q 000400 2-CYCLOHEXYLCYCLOHEXANONE

```
          O
   9 8    \2 3
   C-C     C-C
  ⁄   \  1⁄   \
 10C    C-C    C4
   \  ⁄7 \   ⁄
   C-C     C-C
  11 12   6 5
```

FORMULA	C12H20O			MOL WT	180.29
SOLVENT	CDCL3				
ORIG ST	C4H12SI			TEMP	AMB

56.40	212.50	41.90	27.90	24.20	31.50
1/2	2/1	3/3	4/3	5/3	6/3
36.10	29.40	26.50	26.50	26.50	29.40
7/2	8/3	9/3	10/3	11/3	12/3

L.F.JOHNSON,W.C.JANKOWSKI
CARBON-13 NMR SPECTRA,JOHN
WILEY AND SONS,NEW YORK　　　441 (1972)

```
1667              Q 000400  6-ALLYL-2,3,5,6-TETRAMETHYLCYCLOHEXA-2,4-
                                    DIENONE
            O
     13    ||    7        FORMULA  C13H18O              MOL WT  190.29
    H3C   C1   CH3        SOLVENT  CDCL3
  12    \ / \ /          ORIG ST  C4H12SI              TEMP       305
  H2C  H2C-C6  C2
     \ /10|    ||         204.50   127.00   147.70   124.40   150.10    52.30
   11CH  C5   C3           1/1      2/1      3/1      4/2      5/1       6/1
      / \ / \8             10.60    20.60    18.90    43.50   133.70   116.50
    H3C9   C   CH3          7/4      8/4      9/4     10/3     11/2      12/3
        4                  25.20
                           13/4
                         R.HOLLENSTEIN,W.VON PHILIPSBORN
                         HELV CHIM ACTA                  55, 2030 (1972)
```

```
1668              Q 000400  BETA-DAMASCENON
     12  13
   H3C H3C  O            FORMULA  C13H18O              MOL WT  190.29
     \ /  ||  2          SOLVENT  CDCL3
   10C   C4  CH          ORIG ST  C4H12SI              TEMP       305
    / \ / \ / \1
   9C   C5  CH  CH3       19.40   145.80   134.40   200.60   127.90   139.20
    |   ||  3              1/4      2/2      3/2      4/1      5/1       6/1
   8C   C6               127.10   127.90    39.40    33.80    18.30    26.30
    \ / \11               7/2      8/2      9/3     10/1     11/4      12/4
    C7   CH3
                         R.HOLLENSTEIN,W.VON PHILIPSBORN
                         HELV CHIM ACTA                  55, 2030 (1972)
```

```
1669              Q 000400  BETA-DAMASCON
     12  13
   H3C H3C  O            FORMULA  C13H20O              MOL WT  192.30
     \ /  ||  2          SOLVENT  CDCL3
   10C   C4  CH          ORIG ST  C4H12SI              TEMP       305
    / \ / \ / \1
   9C   C5  CH  CH3       21.30   145.50   134.60   201.90   130.30   140.20
    |   ||  3              1/4      2/2      3/2      4/1      5/1       6/1
   8C   C6                31.20    19.00    38.00    33.40    18.30    28.80
    \ / \11               7/3      8/3      9/3     10/1     11/4      12/4
    C7   CH3
                         R.HOLLENSTEIN,W.VON PHILIPSBORN
                         HELV CHIM ACTA                  55, 2030 (1972)
```

```
1670              Q 000400  BETA-IONONE
    12    13    1
   H3C   CH3   CH3        FORMULA  C13H20O              MOL WT  192.30
     \ /     |           SOLVENT  CDCL3
   10C   4CH   C2         ORIG ST  C4H12SI              TEMP       305
    / \ / \ / \
   9C   C5   CH   O        27.00   197.90   131.30   142.70   135.60   135.60
    |   ||   3              1/4      2/1      3/2      4/2      5/1       6/1
   8C   C6                 33.30    18.70    39.60    33.90    21.50    28.60
    \ / \11                7/3      8/3      9/3     10/1     11/4      12/4
    C7   CH3
                         R.HOLLENSTEIN,W.VON PHILIPSBORN
                         HELV CHIM ACTA                  55, 2030 (1972)
```

```
1671              Q 000400  2,2,6-TRIMETHYL-4-TERT-BUTYLCYCLOHEXANONE
      E8   A7
    H3C    CH3           FORMULA  C13H24O              MOL WT  196.34
  10     \ /             SOLVENT  CDCL3
  H3C    3C-C2           ORIG ST  C4H12SI              TEMP       AMB
  11\9 4/    \1
  H3C-C-C       C=O       216.80    44.30    43.00    42.40    37.80    40.00
  12/  \      /            1/1      2/1      3/3      4/2      5/3       6/2
  H3C    5C-C6            26.00    27.50    32.00    27.50    27.50    27.50
         |                 7/4      8/4      9/1     10/4     11/4      12/4
       13CH3              15.00
                          13/4
                         J.B.STOTHERS,C.T.TAN
                         CAN J CHEM                      52,  308 (1974)
```

1672   Q 000400 METHANAL DICYCLOHEXYL ACETAL

```
   12 13                  2 3      FORMULA   C13H24O2              MOL WT   212.34
    C-C                    C-C     SOLVENT   C4H10O
  11/   \8  7         1/     \     ORIG ST   C4H12SI               TEMP         AMB
   C        C-O-CH2-O-C        C4
    \   /              \   /           74.40    33.20    24.60    26.50    91.10
     C-C                C-C            1/2      2/3      3/3      4/3      7/3
    10 9               6 5
                                 T.PEHK,E.LIPPMAA
                                 ORG MAGN RESON              3,  679 (1971)
```

1673   Q 000400 DICYLOHEXYL PERCARBONATE

```
   13 14                  2 3      FORMULA   C14H22O6              MOL WT   286.33
    C-C      O    O        C-C     SOLVENT   CCL4/C6H12
   /    \    ||   ||    1/    \4   ORIG ST   C4H12SI               TEMP         AMB
 C12      C-O-C-O-O-C-O-C        C
   \    /9   8    7    \    /          80.30    31.90    23.90    25.80   152.00
    C-C                 C-C            1/2      2/3      3/3      4/3      7/1
   11 10               6 5
                                 T.PEHK,E.LIPPMAA
                                 ORG MAGN RESON              3,  679 (1971)
```

1574   Q 000400 2,2,6,6-TETRAMETHYL-4-TERT-BUTYLCYCLOHEXANONE

```
             ⋀8
             CH3                 FORMULA   C14H26O               MOL WT   210.36
   10     3 |  7E               SOLVENT   CDCL3
  H3C      C-C-CH3             ORIG ST   C4H12SI               TEMP         AMB
  11\9 4/   2\1
  H3C-C-C        C=O               220.10    44.10    41.60    38.80    41.60    44.10
     /    \  6/14E                 1/1      2/1      3/3      4/2      5/3      6/1
  H3C      5C-C-CH3                27.90    28.40    31.80    27.40    27.40    27.40
  12        |                      7/4      8/4      9/1     10/4     11/4     12/4
           CH3                     28.40    27.90
           ⋀13                     13/4     14/4

                                 J.B.STOTHERS,C.T.TAN
                                 CAN J CHEM                 52,  308 (1974)
```

1675   Q 000400 OVALICIN

```
       14
        C                        FORMULA   C16H24O5              MOL WT   296.37
      /|    15        12        SOLVENT   CDCL3
   O |    CH3       CH3         ORIG ST   C4H12SI               TEMP         AMB
    \| OH|   9    |
    6C | C7  CH2 C11              78.20    85.90   206.20    36.40    30.00    60.20
    /  \|/ \|/ \  / \ /  \        1/S      2/D      3/S      4/T      5/T      6/S
  5C  1C O-C   C    CH3           60.00    56.50    26.80   117.80   135.00    17.70
   |    |   8   10  13            7/S      8/D      9/T     10/D     11/S     12/Q
  4C    C2                        25.40    51.00    14.10    58.90
   \  / \                         13/Q     14/T     15/Q     16/Q
    C3   O-CH3
    ||    16                    D.E.CANE,R.H.LEVIN
    O                           J AM CHEM SOC              97, 1282 (1975)
```

1676   C22   Q 000400 TETRACYCLOHEXYL TIN

```
          / \
        C21 C23
         |   |               FORMULA   C24H44SN              MOL WT   411.58
        C20 C24              SOLVENT   C4H10O
    17    \ /    2 3         ORIG ST   C4H12SI               TEMP         AMB
    C-C18 C19 C-C
  16/    \ | 1/    \             27.10    33.30    30.00    27.90
   C    13C-SN-C        C4        1/2      2/3      3/3      4/3
    \    /  |   \    /
    C-C14  C7   C-C
   15   / \     6 5
      C12 C8
       |   |
      C11 C9                 T.PEHK,E.LIPPMAA
       \  /                  ORG MAGN RESON              3,  679 (1971)
        C10
```

1677   Q 000500 2,2-DIMETHYLCYCLOHEPTANONE

```
            O
      7   //
      C--C1
     6/   | 8
  5C-C   2C-CH3
    \   /|
     C--C  CH3
     4  3  9
```

FORMULA	C9H16O			MOL WT	140.23
SOLVENT	C4H8O2				
ORIG ST	C4H12SI			TEMP	313

214.20	46.40	38.50	24.30	30.00	25.80
1/1	2/1	3/3	4/3	5/3	6/3
39.00	24.90				
7/3	8/4				

M.CHRISTL,J.D.ROBERTS
J ORG CHEM   37, 3443 (1972)

1678   Q 000500 TRANS-1,2-DIMETHYLCYCLOHEPTANOL

```
        OH
     7  11
     C--C
    6/  |\8
 5C-C   2C CH3
   \   /|
    C--C CH3
    4  3 9
```

FORMULA	C9H18O			MOL WT	142.24
SOLVENT	C4H8O2				
ORIG ST	C4H12SI			TEMP	313

72.40	43.20	30.50	27.50	28.30	21.40
1/1	2/2	3/3	4/3	5/3	6/3
42.60	28.70	16.60			
7/3	8/4	9/4			

M.CHRISTL,J.D.ROBERTS
J ORG CHEM   37, 3443 (1972)

1679   Q 000500 TRANS-2-METHYLCYCLOHEPTANOL

```
        OH
     7  |
     C--C1
    6/  |
 5C-C   C2
   \   /\8
    C--C   CH3
    4  3
```

FORMULA	C8H16O			MOL WT	128.22
SOLVENT	C4H8O2				
ORIG ST	C4H12SI			TEMP	313

77.30	41.70	31.70	25.70	27.80	21.80
1/2	2/2	3/3	4/3	5/3	6/3
36.00	20.20				
7/3	8/4				

M.CHRISTL,J.D.ROBERTS
J ORG CHEM   37, 3443 (1972)

1680   Q 000500 TROPOLONE

```
    3 2
    C-C
   //   \
  4C     \1
   |      C-OH
  5C     /
   \    /
    C-C
    6 7\
         O
```

FORMULA	C7H6O2			MOL WT	122.12
SOLVENT	CDCL3				
ORIG ST	C6H6			TEMP	AMB

172.60	123.50	137.10	128.10
1/1	2/2	3/2	4/2

L.WEILER
CAN J CHEM   50, 1975 (1972)

1681   Q 000500 1,4-CYCLOHEPTADIENE

```
    1   2
     C=C
  7 /    \
   C      \
   |       C3
   C      /
  6 \    /
     C=C
    5   4
```

FORMULA	C7H10			MOL WT	94.16
SOLVENT	CCL4/CDCL3				
ORIG ST	C4H12SI			TEMP	AMB

130.90	128.20	27.10	28.80
1/2	2/2	3/3	6/3

H.GUENTHER,G.JIKELI
CHEM BER   106, 1863 (1973)

1682   Q 000500 CYCLOHEPTANONE

```
    3 2
    C-C
  4/    \1
  C       \1
  |        C=O
  C5      /
   \6 7/
    C-C
```

	FORMULA	C7H12O		MOL WT	112.17
	SOLVENT	CDCL3			
	ORIG ST	C4H12SI		TEMP	308

215.00	43.90	24.40	30.60	30.60	24.40
1/1	2/3	3/3	4/3	5/3	6/3
43.90					
7/3					

G.E.HAWKES,K.HERWIG,J.D.ROBERTS
J ORG CHEM   39, 1017 (1974)

1683   Q 000500 CYCLOHEPTANONE OXIME

```
    3 2
    C-C
  4/    \    OH
  C      \1  /
  |        C=N
  C5      /
   \6 7/
    C-C
```

	FORMULA	C7H13NO		MOL WT	127.19
	SOLVENT	CDCL3			
	ORIG ST	C4H12SI		TEMP	308

164.30	30.49	27.50	30.44	28.90	24.70
1/1	2/3	3/3	4/3	5/3	6/3
33.90					
7/3					

G.E.HAWKES,K.HERWIG,J.D.ROBERTS
J ORG CHEM   39, 1017 (1974)

1684   Q 000500 OCTACHLORO-1,3,5-CYCLOHEPTATRIENE

```
      5 6
   CL-C=C-CL
    /    \  CL
  CL-C4    \ /
    ||      7C
  CL-C3    / \
    \   /  CL
   CL-C=C-CL
      2 1
```

	FORMULA	C7CL8		MOL WT	367.70
	SOLVENT	CHCL3			
	ORIG ST	C4H12SI		TEMP	AMB

135.10	134.70	129.80	85.20
1/S	2/S	3/S	7/S

G.E.HAWKES,R.A.SMITH,J.D.ROBERTS
J ORG CHEM   39, 1276 (1974)

1685   Q 000500 CYCLOOCTANONE

```
    3 2
    C-C   O
  4/   \1/
  C      C
  |      |
  C5     C8
   \6 7/
    C-C
```

	FORMULA	C8H14O		MOL WT	126.20
	SOLVENT	CDCL3			
	ORIG ST	C4H12SI		TEMP	308

218.00	42.00	27.40	25.80	24.90	25.80
1/1	2/3	3/3	4/3	5/3	6/3
27.40	42.00				
7/3	8/3				

G.E.HAWKES,K.HERWIG,J.D.ROBERTS
J ORG CHEM   39, 1017 (1974)

1686   Q 000500 CYCLOOCTENE OXIDE

```
        O
       / \
    8C---C1
    /      \
  7C        C2
  |         |
  6C        C3
   \       /
    5C---C4
```

	FORMULA	C8H14O		MOL WT	126.20
	SOLVENT	CDCL3			
	ORIG ST	C4H12SI		TEMP	AMB

55.90	26.80	26.60	25.80	25.80	26.60
1/2	2/3	3/3	4/3	5/3	6/3
26.80	55.90				
7/3	8/2				

N.R.EASTON,JR.,F.A.L.ANET,P.A.BURNS,C.S.FOOTE
J AM CHEM SOC   96, 3945 (1974)

1687 — Q 000500 2-METHYLCYCLOHEPTANONE

```
        O
    7  ⁄⁄
    C--C1
  6⁄   |
5C-C    C2
  \   ⁄ \8
   C--C   CH3
   4  3
```

FORMULA	C8H14O			MOL WT	126.20
SOLVENT	C4H8O2				
ORIG ST	C4H12SI			TEMP	313

212.40	45.30	32.60	28.00	29.20	23.70
1/1	2/2	3/3	4/3	5/3	6/3
41.60	16.40				
7/3	8/4				

M.CHRISTL,J.D.ROBERTS
J ORG CHEM   37, 3443 (1972)

1688 — Q 000500 3-METHYLCYCLOHEPTANONE

```
        O
    7  ⁄⁄
    C--C1
  6⁄   |
5C-C    C2
  \    ⁄
   4C--C3
      \8
      CH3
```

FORMULA	C8H14O			MOL WT	126.20
SOLVENT	C4H8O2				
ORIG ST	C4H12SI			TEMP	313

210.20	50.80	30.50	38.60	28.00	23.60
1/1	2/3	3/2	4/3	5/3	6/3
43.10	22.60				
7/3	8/4				

M.CHRISTL,J.D.ROBERTS
J ORG CHEM   37, 3443 (1972)

1689 — Q 000500 4-METHYLCYCLOHEPTANONE

```
        O
    7  ⁄⁄
    C--C1
  6⁄   |
5C-C    C2
  \    ⁄
   4C--C3
  8⁄
  H3C
```

FORMULA	C8H14O			MOL WT	126.20
SOLVENT	C4H8O2				
ORIG ST	C4H12SI			TEMP	313

211.40	41.40	31.70	36.00	38.20	22.50
1/1	2/3	3/3	4/2	5/3	6/3
42.90	22.50				
7/3	8/4				

M.CHRISTL,J.D.ROBERTS
J ORG CHEM   37, 3443 (1972)

1690 — Q 000500 CYCLOOCTANONE OXIME

```
         OH
   3  2  |
   C-C   N
  4⁄  \1⁄⁄
  C    C
  |    |
  C5   C8
   \6 7⁄
    C-C
```

FORMULA	C8H15NO			MOL WT	141.21
SOLVENT	CDCL3				
ORIG ST	C4H12SI			TEMP	308

163.70	27.30	26.90	26.80	24.90	25.50
1/1	2/3	3/3	4/3	5/3	6/3
24.40	33.10				
7/3	8/3				

G.E.HAWKES,K.HERWIG,J.D.ROBERTS
J ORG CHEM   39, 1017 (1974)

1691 — Q 000500 METHYLCYCLOHEPTANE

```
      8
      CH3
   7 ⁄
   C--C1
  6⁄   |
5C-C    C2
  \   ⁄
   C--C
   4  3
```

FORMULA	C8H16			MOL WT	112.22
SOLVENT	CS2/C4H12SI				
ORIG ST	C4H12SI			TEMP	313

| 34.90 | 37.50 | 26.90 | 28.90 | 24.40 |
| 1/2 | 2/3 | 3/3 | 4/3 | 8/4 |

M.CHRISTL,J.D.ROBERTS
J ORG CHEM   37, 3443 (1972)

1692   Q 000500 1-METHYLCYCLOHEPTANOL

```
        OH
    7  11  8
    C--C-CH3
   6/    |
  5C-C    C2
    \    /
     C--C
     4  3
```

FORMULA	C8H16O			MOL WT	128.22
SOLVENT	C4H8O2				
ORIG ST	C4H12SI			TEMP	313

72.20	42.40	22.00	29.20	30.50
1/1	2/3	3/3	4/3	8/4

M.CHRISTL,J.D.ROBERTS
J ORG CHEM   37, 3443 (1972)

1693   Q 000500 CIS-2-METHYLCYCLOHEPTANOL

```
        OH
    7  11  8
    C--C CH3
   6/   |/
  5C-C    C2
    \    /
     C--C
     4  3
```

FORMULA	C8H16O			MOL WT	128.22
SOLVENT	C4H8O2				
ORIG ST	C4H12SI			TEMP	313

73.20	38.40	29.40	25.50	27.70	21.90
1/2	2/2	3/3	4/3	5/3	6/3
34.30	17.30				
7/3	8/4				

M.CHRISTL,J.D.ROBERTS
J ORG CHEM   37, 3443 (1972)

1594   Q 000500 CIS-3-METHYLCYCLOHEPTANOL

```
        OH
    7   |
    C--C1
   6/    |
  5C-C    C2
    \    /
     C--C-CH3
     4  3 8
```

FORMULA	C8H16O			MOL WT	128.22
SOLVENT	C4H8O2				
ORIG ST	C4H12SI			TEMP	313

70.60	46.50	29.60	36.20	25.60	22.10
1/2	2/3	3/2	4/3	5/3	6/3
37.20	23.60				
7/3	8/4				

M.CHRISTL,J.D.ROBERTS
J ORG CHEM   37, 3443 (1972)

1695   Q 000500 TRANS-3-METHYLCYCLOHEPTANOL

```
        OH
    7   |
    C--C1
   6/    |
  5C-C    C2
    \    /
    4C--C3
        |
      8CH3
```

FORMULA	C8H16O			MOL WT	128.22
SOLVENT	C4H8O2				
ORIG ST	C4H12SI			TEMP	313

68.20	44.90	27.40	36.80	27.40	23.40
1/2	2/3	3/2	4/3	5/3	6/3
36.80	23.40				
7/3	8/4				

M.CHRISTL,J.D.ROBERTS
J ORG CHEM   37, 3443 (1972)

1696   Q 000500 CIS-4-METHYLCYCLOHEPTANOL

```
        OH
    7   |
    C--C1
   6/    |
  5C-C    C2
    \    /
 H3C-C--C
  8  4   3
```

FORMULA	C8H16O			MOL WT	128.22
SOLVENT	C4H8O2				
ORIG ST	C4H12SI			TEMP	313

70.60	33.80	29.40	33.50	37.10	22.10
1/2	2/3	3/3	4/2	5/3	6/3
37.40	23.20				
7/3	8/4				

M.CHRISTL,J.D.ROBERTS
J ORG CHEM   37, 3443 (1972)

1697   Q 000500 TRANS-4-METHYLCYCLOHEPTANOL

```
        OH
     7   |
      C--C1
    6/   |
   5C-C     C2
     \    /
      4C--C3
       |
      H3C8
```

FORMULA   C8H16O   MOL WT   128.22
SOLVENT   C4H8O2
ORIG ST   C4H12SI   TEMP   313

71.50   35.40   31.40   33.90   36.10   20.20
1/2   2/3   3/3   4/2   5/3   6/3
36.80   23.20
7/3   8/4

M.CHRISTL,J.D.ROBERTS
J ORG CHEM   37, 3443 (1972)

1698   Q 000500 CIS-3,5-DIMETHYLCYCLOHEPTANONE

```
        O
     7   //
      C--C1
    6/   |
   5C-C     C2
   |  \    /
  H3C  C--C3
   9  4  |
       8CH3
```

FORMULA   C9H16O   MOL WT   140.23
SOLVENT   C4H8O2
ORIG ST   C4H12SI   TEMP   313

210.50   51.30   30.30   47.90   35.50   32.10
1/1   2/3   3/2   4/3   5/2   6/3
42.10   23.60   23.60
7/3   8/4   9/4

M.CHRISTL,J.D.ROBERTS
J ORG CHEM   37, 3443 (1972)

1699   Q 000500 3,3-DIMETHYLCYCLOHEPTANONE

```
        O
     7   //
      C--C1
    6/   |
   5C-C     C2
     \   3/
      4C--C-CH3
       |  8
      9CH3
```

FORMULA   C9H16O   MOL WT   140.23
SOLVENT   C4H8O2
ORIG ST   C4H12SI   TEMP   313

209.40   55.00   31.70   44.10   23.60   24.40
1/1   2/3   3/1   4/3   5/3   6/3
43.00   28.60
7/3   8/4

M.CHRISTL,J.D.ROBERTS
J ORG CHEM   37, 3443 (1972)

1700   Q 000500 4,4-DIMETHYLCYCLOHEPTANONE

```
        O
     7   //
      C--C1
    6/   |
   5C-C     C2
     \4   /
   H3C-C--C3
    9  |
      8CH3
```

FORMULA   C9H16O   MOL WT   140.23
SOLVENT   C4H8O2
ORIG ST   C4H12SI   TEMP   313

211.10   38.40   35.60   32.60   42.70   19.20
1/1   2/3   3/3   4/1   5/3   6/3
42.70   27.90
7/3   8/4

M.CHRISTL,J.D.ROBERTS
J ORG CHEM   37, 3443 (1972)

1701   Q 000500 CIS-1,3-DIMETHYLCYCLOHEPTANE

```
     7  1
      C--C
    6/   |\8
   5C-C   2C CH3
     \   /
      4C--C3
       |
      9CH3
```

FORMULA   C9H18   MOL WT   126.24
SOLVENT   CS2/C4H12SI
ORIG ST   C4H12SI   TEMP   313

34.20   46.90   37.40   26.50   24.90
1/2   2/3   4/3   5/3   8/4

M.CHRISTL,J.D.ROBERTS
J ORG CHEM   37, 3443 (1972)

1702   Q 000500 CIS-1,4-DIMETHYLCYCLOHEPTANE

```
        7
       C--C1              FORMULA   C9H18                    MOL WT   126.24
      6/   I\8            SOLVENT   CS2/C4H12SI
     5C-C    2C CH3       ORIG ST   C4H12SI                  TEMP        313
      \    /
       4C--C3                34.20   33.50   38.40   27.00   24.30
        I                     1/2     2/3     5/3     6/3     8/4
       9CH3
                          M.CHRISTL,J.D.ROBERTS
                          J ORG CHEM                       37, 3443 (1972)
```

1703   Q 000500 1,1-DIMETHYLCYCLOHEPTANE

```
        8
       CH3                FORMULA   C9H18                    MOL WT   126.24
     7 I 9                SOLVENT   CS2/C4H12SI
    C--C-CH3              ORIG ST   C4H12SI                  TEMP        313
   6/  II
  5C-C    C2                33.30   42.60   23.80   30.80   30.80
   \    /                    1/1     2/3     3/3     4/3     8/4
    C--C
    4   3                 M.CHRISTL,J.D.ROBERTS
                          J ORG CHEM                       37, 3443 (1972)
```

1704   Q 000500 CIS-1,2-DIMETHYLCYCLOHEPTANE

```
                          FORMULA   C9H18                    MOL WT   126.24
        7                 SOLVENT   CS2/C4H12SI
       C--C1              ORIG ST   C4H12SI                  TEMP        313
      6/  I\8
     5C-C    2C CH3         37.50   34.00   26.60   29.20   17.90
      \    / \               1/2     3/3     4/3     5/3     8/4
       4C--C3 CH3
             9            M.CHRISTL,J.D.ROBERTS
                          J ORG CHEM                       37, 3443 (1972)
```

1705   Q 000500 TRANS-1,2-DIMETHYLCYCLOHEPTANE

```
        8
       CH3                FORMULA   C9H18                    MOL WT   126.24
        I                 SOLVENT   CS2/C4H12SI
     7C--C1               ORIG ST   C4H12SI                  TEMP        313
    6/   I
  5C-C    C2                41.30   35.80   26.70   29.70   22.60
   \    / \9                 1/2     3/3     4/3     5/3     8/4
    C--C   CH3
    4   3                 M.CHRISTL,J.D.ROBERTS
                          J ORG CHEM                       37, 3443 (1972)
```

1706   Q 000500 TRANS-1,3-DIMETHYLCYCLOHEPTANE

```
        8
       CH3                FORMULA   C9H18                    MOL WT   126.24
     7  I                 SOLVENT   CS2/C4H12SI
      C--C1               ORIG ST   C4H12SI                  TEMP        313
     6/  I
   5C-C    C2               31.10   44.70   37.50   29.10   24.30
    \    /                   1/2     2/3     4/3     5/3     8/4
     4C--C3
       I                  M.CHRISTL,J.D.ROBERTS
      9CH3                J ORG CHEM                       37, 3443 (1972)
```

1707   Q 000500   TRANS-1,4-DIMETHYLCYCLOHEPTANE

```
        8
       CH3                FORMULA   C9H18                MOL WT  126.24
    7   |                 SOLVENT   CS2/C4H12SI
    C--C1                 ORIG ST   C4H12SI              TEMP      313
   6/   |
  5C-C    C2               35.20    36.60    36.80    24.10    24.30
    \   /                  1/2      2/3      5/3      6/3      8/4
    4C--C3
      |                   M.CHRISTL,J.D.ROBERTS
     9CH3                 J ORG CHEM                     37, 3443 (1972)
```

1708   Q 000500   CIS-1,3-DIMETHYLCYCLCHEPTANOL

```
        OH                FORMULA   C9H18O               MOL WT  142.24
    7   |                 SOLVENT   C4H8O2
    C--C1                 ORIG ST   C4H12SI              TEMP      313
   6/   |\8
  5C-C    2C CH3          71.40    51.40    29.50    37.90    28.20    22.00
    \   /                 1/1      2/3      3/2      4/3      5/3      6/3
    C--C-CH3              42.30    31.80    24.40
    4  3 9                7/3      8/4      9/4

                         M.CHRISTL,J.D.ROBERTS
                         J ORG CHEM                     37, 3443 (1972)
```

1709   Q 000500   TRANS-1,3-DIMETHYLCYCLOHEPTANOL

```
        OH                FORMULA   C9H18O               MOL WT  142.24
    7   |                 SOLVENT   C4H8O2
    C--C1                 ORIG ST   C4H12SI              TEMP      313
   6/   |\8
  5C-C    2C CH3          71.80    51.40    27.10    38.50    28.90    22.00
    \   /                 1/1      2/3      3/2      4/3      5/3      6/3
    4C--C3                42.70    29.50    24.50
      |                   7/3      8/4      9/4
     9CH3
                         M.CHRISTL,J.D.ROBERTS
                         J ORG CHEM                     37, 3443 (1972)
```

1710   Q 000500   CIS-1,4-DIMETHYLCYCLOHEPTANOL

```
        OH                FORMULA   C9H18O               MOL WT  142.24
    7   |                 SOLVENT   C4H8O2
    C--C1                 ORIG ST   C4H12SI              TEMP      313
   6/   |\8
  5C-C    2C CH3          72.10    39.80    30.70    35.30    38.50    20.90
    \   /                 1/1      2/3      3/3      4/2      5/3      6/3
  H3C-C--C                42.80    30.30    23.00
   9  4  3                7/3      8/4      9/4

                         M.CHRISTL,J.D.ROBERTS
                         J ORG CHEM                     37, 3443 (1972)
```

1711   Q 000500   TRANS-1,4-DIMETHYLCYCLOHEPTANOL

```
        OH                FORMULA   C9H18O               MOL WT  142.24
    7   |                 SOLVENT   C4H8O2
    C--C1                 ORIG ST   C4H12SI              TEMP      313
   6/   |\8
  5C-C    2C CH3          71.90    40.50    29.80    35.10    38.10    21.10
    \   /                 1/1      2/3      3/3      4/2      5/3      6/3
    4C--C                 42.40    30.90    23.20
    |   3                 7/3      8/4      9/4
   H3C
    9                    M.CHRISTL,J.D.ROBERTS
                         J ORG CHEM                     37, 3443 (1972)
```

1712   Q 000500 2,2-DIMETHYLCYCLOHEPTANOL

```
        OH
     7  11  8
      C--C CH3
    6/     1/
   5C-C    C2
    \     /1
      C--C CH3
     4  3  9
```

FORMULA	C9H18O
SOLVENT	C4H8O2
ORIG ST	C4H12SI

MOL WT   142.24

TEMP   313

78.70	36.90	38.90	21.50	27.90	23.90
1/2	2/1	3/3	4/3	5/3	6/3
32.60	22.00	27.90			
7/3	8/4	9/4			

M.CHRISTL,J.D.ROBERTS
J ORG CHEM   37, 3443 (1972)

1713   Q 000500 3,3-DIMETHYLCYCLOHEPTANOL

```
        OH
         |
      7C--C1
     6/    |
    5C-C   C2
     \  3/
      C--C-CH3
     4  | 8
       CH3
        9
```

FORMULA	C9H18O
SOLVENT	C4H8O2
ORIG ST	C4H12SI

MOL WT   142.24

TEMP   313

67.60	51.30	30.50	41.80	22.40	25.70
1/2	2/3	3/1	4/3	5/3	6/3
39.00	31.50	28.80			
7/3	8/4	9/4			

M.CHRISTL,J.D.ROBERTS
J ORG CHEM   37, 3443 (1972)

1714   Q 000500 CIS,CIS-3,5-DIMETHYLCYCLOHEPTANOL

```
          OH
   9   7   |
  H3C    C--C1
   |  6/    |
   5C-C    C2
    \    /
      C--C-CH3
     4  3  8
```

FORMULA	C9H18O
SOLVENT	C4H8O2
ORIG ST	C4H12SI

MOL WT   142.24

TEMP   313

69.90	46.80	29.90	46.80	32.70	29.90
1/2	2/3	3/2	4/3	5/2	6/3
34.00	24.00	23.40			
7/3	8/4	9/4			

M.CHRISTL,J.D.ROBERTS
J ORG CHEM   37, 3443 (1972)

1715   Q 000500 TRANS,TRANS-3,5-DIMETHYLCYCLOHEPTANOL

```
         OH
      7   |
       C--C1
     6/    |
    5C-C   C2
    |\     /
   H3C C--C3
    9  4  |
        CH3
         8
```

FORMULA	C9H18O
SOLVENT	C4H8O2
ORIG ST	C4H12SI

MOL WT   142.24

TEMP   313

68.80	45.30	27.00	46.80	34.90	32.40
1/2	2/3	3/2	4/3	5/2	6/3
34.90	23.40	24.00			
7/3	8/4	9/4			

M.CHRISTL,J.D.ROBERTS
J ORG CHEM   37, 3443 (1972)

1716   Q 000500 4,4-DIMETHYLCYCLOHEPTANOL

```
         OH
      7   |
       C--C1
     6/    |
    5C-C   C2
     \4    /
   H3C-C--C3
    9  |
     8CH3
```

FORMULA	C9H18O
SOLVENT	C4H8O2
ORIG ST	C4H12SI

MOL WT   142.24

TEMP   313

71.90	31.30	39.10	32.20	41.70	18.50
1/2	2/3	3/3	4/1	5/3	6/3
34.90	30.10	30.10			
7/3	8/4	9/4			

M.CHRISTL,J.D.ROBERTS
J ORG CHEM   37, 3443 (1972)

1717   Q 000500   1,1,2–TRIMETHYLCYCLOHEPTANE

```
        8
        CH3              FORMULA   C10H20              MOL WT   140.27
     7   |              SOLVENT   CS2/C4H12SI
      C--C1             ORIG ST   C4H12SI              TEMP        313
     6/   |\9
   5C-C    2C CH3           35.60    43.20    32.40    29.70    30.40    22.80
     \    / \               1/1      2/2      3/3      4/3      5/3      6/3
      C--C   CH3            43.90    29.70    23.70    18.80
      4  3   10             7/3      8/4      9/4     10/4
```

M.CHRISTL,J.D.ROBERTS
J ORG CHEM   37, 3443 (1972)

1718   Q 000500   1,1,3–TRIMETHYLCYCLOHEPTANE

```
        8CH3              FORMULA   C10H20             MOL WT   140.27
     7   |               SOLVENT   CS2/C4H12SI
      C--C1              ORIG ST   C4H12SI             TEMP        313
     6/   |\9
   5C-C    2C CH3            32.40    51.40    29.50    39.40    30.10    23.50
     \    /                  1/1      2/3      3/2      4/3      5/3      6/3
    4C--C3                   42.50    29.50    32.40    25.70
      |                      7/3      8/4      9/4     10/4
      CH3
      10
```

M.CHRISTL,J.D.ROBERTS
J ORG CHEM   37, 3443 (1972)

1719   Q 000500   1,1,4–TRIMETHYLCYCLOHEPTANE

```
        8
        CH3              FORMULA   C10H20             MOL WT   140.27
     7   |              SOLVENT   CS2/C4H12SI
      C--C1             ORIG ST   C4H12SI             TEMP        313
     6/   |\9
   5C-C    2C CH3           33.30    40.60    32.00    36.90    40.00    22.50
     \    /                 1/1      2/3      3/3      4/2      5/3      6/3
    4C--C3                  42.60    31.20    30.60    24.10
      |                     7/3      8/4      9/4     10/4
      CH3
      10
```

M.CHRISTL,J.D.ROBERTS
J ORG CHEM   37, 3443 (1972)

1720   Q 000500   1,2,2–TRIMETHYLCYCLOHEPTANOL

```
        OH               FORMULA   C10H20O            MOL WT   156.27
     7  11 8             SOLVENT   C4H8O2
      C--C-CH3           ORIG ST   C4H12SI            TEMP        313
     6/   |
   5C-C    2C-CH3            74.60    39.70    37.00    21.30    25.90    20.50
     \    /| 9               1/1      2/1      3/3      4/3      5/3      6/3
      C--C  CH3             38.90    25.20    24.30    24.50
      4  3  10              7/3      8/4      9/4     10/4
```

M.CHRISTL,J.D.ROBERTS
J ORG CHEM   37, 3443 (1972)

1721   Q 000500   1,3,3–TRIMETHYLCYCLOHEPTANOL

```
        OH               FORMULA   C10H20O            MOL WT   156.27
     7   |              SOLVENT   C4H8O2
      C--C1             ORIG ST   C4H12SI             TEMP        313
     6/   |\8
   5C-C    2C CH3           72.10    53.90    32.60    42.30    24.60    24.20
     \   3/                 1/1      2/3      3/1      4/3      5/3      6/3
      C--C-CH3            42.90    32.40    31.50    30.20
      4  |  9              7/3      8/4      9/4     10/4
      CH3
      10
```

M.CHRISTL,J.D.ROBERTS
J ORG CHEM   37, 3443 (1972)

```
1722                    Q 000500  1,4,4-TRIMETHYLCYCLOHEPTANOL

           OH           FORMULA   C10H20O              MOL WT   156.27
        7   |           SOLVENT   C4H8O2
        C--C1           ORIG ST   C4H12SI              TEMP     313
       6/   |\8
      5C-C    2C CH3       71.00    35.60    34.10    32.30    42.80    18.60
        \ 4  /             1/1      2/3      3/3      4/1      5/3      6/3
     H3C-C--C3            44.50    31.10    29.90    29.90
      10 |                 7/3      8/4      9/4      10/4
        CH3
         9               M.CHRISTL,J.D.ROBERTS
                         J ORG CHEM                      37,  3443 (1972)
```

```
1723                    Q 000500  1,6-DIMETHOXYCARBONYL-1,3,5-CYCLOHEPTATRIENE
        8C-]-CH3
      2  / ＼ 10         FORMULA   C11H12O4             MOL WT   208.22
      C=C1   ]           SOLVENT   CCL4/CDCL3
    3 /   \              ORIG ST   C4H12SI              TEMP     AMB
    C      \
    ||      C7            125.10   132.60   133.00    25.40   165.00    51.80
    C      /              1/1      2/2      3/2       7/3      8/1      10/4
   4 \    /
     C=C6  ]
    5  \ ⁄ 11           H.GUENTHER,G.JIKELI
       9C-]-CH3          CHEM BER                      106,  1863 (1973)
```

```
1724                    Q 000600  1,3,4,5,6,6,7-HEPTACHLOROBICYCLO(3.2.0)HEPT-
                                  3-EN-2-ONE
         O
         ||             FORMULA   C7HCL7O              MOL WT   349.26
      CL C2  CL         SOLVENT   CHCL3
     7 |/ \ /           ORIG ST   C4H12SI              TEMP     AMB
   CL-C-C1  C3
    6|  |   ||            71.90   185.00   135.60   158.10    78.10    87.80
   CL-C-C---C4            1/S      2/S      3/S      4/S      5/S      6/S
     | |5    \           67.10
    CL CL    CL           7/D

                         G.E.HAWKES,R.A.SMITH,J.D.ROBERTS
                         J ORG CHEM                      39,  1276 (1974)
```

```
1725                    Q 000600  1,2,3,4,4,5,6,7,7-NONACHLOROBICYCLO(3.2.0)-
                                  HEPT-2-ENE
        CL
        |               FORMULA   C7HCL9               MOL WT   404.16
    CL CL C2            SOLVENT   CHCL3
     \7 |/ ＼3          ORIG ST   C4H12SI              TEMP     AMB
   CL-C-C1  C-CL
    |  |    |             79.70   138.10   133.60    91.90    81.10    70.30
   CL-C-C---C-CL          1/S      2/S      3/S      4/S      5/S      6/D
    6 |5   |4            87.40
      CL   CL             7/S

                         G.E.HAWKES,R.A.SMITH,J.D.ROBERTS
                         J ORG CHEM                      39,  1276 (1974)
```

```
1726                    Q 000600  1,2,3,4,5,6,6,7,7-NONACHLORO-2-NORBORNENE

                        FORMULA   C7HCL9               MOL WT   404.16
     CL      CL         SOLVENT   CHCL3
      \1   2/           ORIG ST   C4H12SI              TEMP     AMB
   CL  C---C
    \ / \7  ＼3           87.70   135.90   133.20    81.90    73.20    91.50
    6C  RCR  C-CL         1/S      2/S      3/S      4/S      5/D      6/S
   / \  \ /              98.50
  CL 5C---C4    R -CL     7/S
    /       \
  CL        CL          G.E.HAWKES,R.A.SMITH,J.D.ROBERTS
                         J ORG CHEM                      39,  1276 (1974)
```

1727

Q 000600   6,7-DIBROMO-1,2,3,4,4,5-HEXACHLOROBICYCLO-
(3.2.0)HEPT-2-ENE

```
        CL
        |
   CL   C2
   7 1/  \3
  BR-C-C1  C-CL
   |  |    |
  BR-C-C---C-CL
   6 |5   |4
     CL   CL
```

FORMULA	C7H2BR2CL6			MOL WT	458.62
SOLVENT	CHCL3				
ORIG ST	C4H12SI			TEMP	AMB

72.70	137.20	133.20	91.50	82.00	53.20
1/S	2/S	3/S	4/S	5/S	6/D
51.20					
7/D					

G.E.HAWKES,R.A.SMITH,J.D.ROBERTS
J ORG CHEM   39, 1276 (1974)

1728

Q 000600   1,2,4,5,6,7-HEXACHLOROBICYCLO(3.2.0)HEPTA-2,6-
DIENE

```
        CL
        |
   CL   C2
   7\  / \
  CL-C-C1  C3
   ||  |   |
  CL-C-C---C4
   6 |5   |
     CL   CL
```

FORMULA	C7H2CL6			MOL WT	298.81
SOLVENT	CHCL3				
ORIG ST	C4H12SI			TEMP	AMB

81.20	135.50	128.00	64.00	77.70	134.70
1/S	2/S	3/D	4/D	5/S	6/S
131.60					
7/S					

G.E.HAWKES,R.A.SMITH,J.D.ROBERTS
J ORG CHEM   39, 1276 (1974)

1729

Q 000600   1,2,3,4,7,7-HEXACHLORO-2,5-NORBORNADIENE

```
   CL      CL
    \1    /
     C---C2
    / \7  \3
  6C  RCR  C-CL
    \   \ /
    5C---C4    R -CL
        \
        CL
```

FORMULA	C7H2CL6			MOL WT	298.81
SOLVENT	CHCL3				
ORIG ST	C4H12SI			TEMP	AMB

83.70	137.80	141.10	116.30
1/S	2/S	5/D	7/S

G.E.HAWKES,R.A.SMITH,J.D.ROBERTS
J ORG CHEM   39, 1276 (1974)

1730

Q 000600   ENDO-(Z)-6-FLUORO-1,2,3,4,5,7,7-HEPTACHLORO-
2-NORBORNENE

```
   CL      CL
    \1    2/
     C---C
   6/ \7  \3
   C  RCR  C-CL
  / \   \ /
 F  5C---C4    R -CL
    |     \
    CL    CL
```

FORMULA	C7H2CL7F			MOL WT	353.26
SOLVENT	C6H12				
ORIG ST	C4H12SI			TEMP	AMB

80.80	132.50	130.90	82.10	63.10	91.90
1/S	2/S	3/S	4/S	5/D	6/T
98.40					
7/S					

G.E.HAWKES,R.A.SMITH,J.D.ROBERTS
J ORG CHEM   39, 1276 (1974)

1731

Q 000600   ANTI-2,3,4,5,5,6,6,7-OCTACHLORO-2-NORBORNENE

```
         CL
     1  2/
  CL  C---C
   \6/ \7  \3
   C  RC   C-CL
  / \   \ /
 CL 5C---C4    R -CL
   / \    \
  CL  CL  CL
```

FORMULA	C7H2CL8			MOL WT	369.72
SOLVENT	C6H12				
ORIG ST	C4H12SI			TEMP	AMB

63.40	134.30	131.40	81.20	94.80	91.10
1/S	2/S	3/S	4/S	5/S	6/S
64.20					
7/D					

G.E.HAWKES,R.A.SMITH,J.D.ROBERTS
J ORG CHEM   39, 1276 (1974)

1732   Q 000600   ENDO-(Z)-1,2,3,4,5,6,7,7-OCTACHLORO-
2-NORBORNENE

```
    CL    CL
     \1   2/
      C---C
     6/ \7  \3
      C  RCR  C-CL
     / \    \ /
   CL 5C---C4   R -CL
      |     \
      CL    CL
```

FORMULA   C7H2CL8   MOL WT   369.72
SOLVENT   CHCL3
ORIG ST   C4H12SI   TEMP   AMB

82.50	132.50	64.70	99.00
1/S	2/S	5/D	7/S

G.E.HAWKES,R.A.SMITH,J.D.ROBERTS
J ORG CHEM   39, 1276 (1974)

1733   Q 000600   ENDO-EXO-1,2,3,4,5,6,7,7-OCTACHLORO-
2-NORBORNENE

```
    CL      CL
     \1    2/
  CL  C---C
  \ / \7  \3
   6C  RCR  C-CL
    \    \ /
    5C---C4   R -CL
    |     \
    CL    CL
```

FORMULA   C7H2CL8   MOL WT   369.72
SOLVENT   CHCL3
ORIG ST   C4H12SI   TEMP   AMB

81.20	134.20	133.30	82.00	69.00	66.70
1/S	2/S	3/S	4/S	5/D	6/D
99.60					
7/D					

G.E.HAWKES,R.A.SMITH,J.D.ROBERTS
J ORG CHEM   39, 1276 (1974)

1734   Q 000600   EXO-(Z)-1,2,3,4,5,6,7,7-OCTACHLORO-2-NORBORNENE

```
    CL      CL
     \1    2/
  CL  C---C
  \6/ \7  \3
    C  RCR  C-CL
     \    \ /
   CL-C---C4   R -CL
      5     \
            CL
```

FORMULA   C7H2CL8   MOL WT   369.72
SOLVENT   CHCL3
ORIG ST   C4H12SI   TEMP   AMB

81.50	134.80	61.60	99.00
1/S	2/S	5/D	7/S

G.E.HAWKES,R.A.SMITH,J.D.ROBERTS
J ORG CHEM   39, 1276 (1974)

1735   Q 000600   1,2,3,4,4,5,6,7-OCTACHLOROBICYCLO(3.2.0)HEPT-
2-ENE

```
        CL
        |
   CL   C2
    7\ / \3
  CL-C-C1  C-CL
   | | |   |
  CL-C-C---C-CL
   6 |5  |4
     CL  CL
```

FORMULA   C7H2CL8   MOL WT   369.72
SOLVENT   CHCL3
ORIG ST   C4H12SI   TEMP   AMB

73.00	137.80	132.50	91.60	81.30	63.90
1/S	2/S	3/S	4/S	5/S	6/D
83.90					
7/D					

G.E.HAWKES,R.A.SMITH,J.D.ROBERTS
J ORG CHEM   39, 1276 (1974)

1736   Q 000600   1,2,3,4,6,6,7,7-OCTACHLORO-2-NORBORNENE

```
    CL      CL
     \1    2/
  CL  C---C
  \ / \7  \3
   6C  RCR  C-CL
  / \    \ /
 CL 5C---C4   R -CL
        \
         CL
```

FORMULA   C7H2CL8   MOL WT   369.72
SOLVENT   C6H12
ORIG ST   C4H12SI   TEMP   AMB

87.90	134.20	133.40	77.10	54.70	87.50
1/S	2/S	3/S	4/S	5/T	6/S
100.30					
7/S					

G.E.HAWKES,R.A.SMITH,J.D.ROBERTS
J ORG CHEM   39, 1276 (1974)

1737 Q 000600 SYN-2,3,4,5,5,6,6,7-OCTACHLORO-2-NORBORNENE

```
          CL
    1   2/
CL  C---C
   \ / \7 \3
   6C  CR  C-CL
   / \   \ /
CL 5C---C4    R -CL
  / \   \
CL  CL  CL
```

FORMULA	C7H2CL8			MOL WT	369.72
SOLVENT	C6H12				
ORIG ST	C4H12SI			TEMP	AMB

69.10	131.80	133.00	87.60	97.00	92.40
1/D	2/S	3/S	4/X	5/S	6/S
70.40					
7/D					

G.E.HAWKES,R.A.SMITH,J.D.ROBERTS
J ORG CHEM   39, 1276 (1974)

1738 Q 000600 1,2,3,4,5,6,7,7-OCTACHLOROTRICYCLO(2.2.1.0)-HEPTANE

```
      C.
      |
    3C---C2
   /  CL\ \
  4/  7| | \ 1
CL-C----C-X--C-CL
  \  | | /
   \ CL| /
   5C---C6
   /      \
  CL      CL
```

FORMULA	C7H2CL8			MOL WT	369.72
SOLVENT	CHCL3				
ORIG ST	C4H12SI			TEMP	AMB

59.60	39.60	64.00	78.20	93.90	57.70
1/S	2/D	3/D	4/S	5/S	6/S
91.70					
7/S					

G.E.HAWKES,R.A.SMITH,J.D.ROBERTS
J ORG CHEM   39, 1276 (1974)

1739 Q 000600 ENDO-6-FLUORO-1,2,3,4,7,7-HEXACHLORO-2-NORBORNENE

```
  CL      CL
   \1   2/
    C---C
  6/ \   \3
  C  RCR  C-CL
 / \   \ /
F  5C---C4    R -CL
      \
      CL
```

FORMULA	C7H3CL6F			MOL WT	318.82
SOLVENT	C6H12				
ORIG ST	C4H12SI			TEMP	AMB

81.70	133.00	129.90	78.20	42.90	95.40
1/S	2/S	3/S	4/S	5/T	6/T
100.90					
7/S					

G.E.HAWKES,R.A.SMITH,J.D.ROBERTS
J ORG CHEM   39, 1276 (1974)

1740 Q 000600 ENDO-1,2,3,4,6,7,7-HEPTACHLORO-2-NORBORNENE

```
   CL      CL
    \1   2/
     C---C
   6/ \7 \3
   C  RCR  C-CL
  / \   \ /
CL 5C---C4    R -CL
      \
      CL
```

FORMULA	C7H3CL7			MOL WT	335.27
SOLVENT	C6H12				
ORIG ST	C4H12SI			TEMP	AMB

83.10	131.20	132.50	77.80	45.20	61.30
1/S	2/S	3/S	4/S	5/T	6/D
101.00					
7/S					

G.E.HAWKES,R.A.SMITH,J.D.ROBERTS
J ORG CHEM   39, 1276 (1974)

1741 Q 000600 3-NORTRICYCLYL CATION

```
   3   7
   C---C
  / \ / \
 4C  2C  C6
  \  |+ /
   \ 1C /
    \ |/
     C
     5
```

FORMULA	C7H9			MOL WT	93.15
SOLVENT	CLFO2S/F5SB				
ORIG ST	CS2			TEMP	195

257.30	85.10	110.40	45.40	41.20	45.40
1/2	2/2	3/2	4/3	5/2	6/3
110.40					
7/2					

G.A.OLAH,G.LIANG
J AM CHEM SOC   95, 3792 (1973)

1742   Q 000600 3—NORTRICYCLANONE   PROTONATED

```
    3    7
    C———C
   / \ / \
 4C  2C   C6
   \ |+ /
    \1CR/
     \|/
      C
      5
```

FORMULA   C7H9O   MOL WT   109.15
SOLVENT   HFO3S/F5SB
ORIG ST   CS2   TEMP   195

222.60   40.10   37.00   29.20   42.60   29.20
1/1   2/2   3/2   4/3   5/2   6/3
37.00
7/2

R —OH

G.A.OLAH,G.LIANG
J AM CHEM SOC   95, 3792 (1973)

1743   Q 000600 3—NORTRICYCLANONE   PROTONATED

```
    3    7
    C———C
   / \ / \
 4C  2C   C6
   \ |+ /
    \1CR/
     \|/
      C
      5
```

FORMULA   C7H9O   MOL WT   109.15
SOLVENT   HFO3S/F5SB
ORIG ST   CS2   TEMP   195

220.30   40.10   37.00   29.20   42.60   29.20
1/1   2/2   3/2   4/3   5/2   6/3
37.00
7/2

R —OH

G.A.OLAH,G.LIANG
J AM CHEM SOC   95, 3792 (1973)

1744   Q 000600 BICYCLO(2.2.1)HEPTAN—2—ONE

```
 5C———C6
 / 7 \
4C———C———C1
 \       /
 3C———C2
      \\
       O
```

FORMULA   C7H10O   MOL WT   110.16
SOLVENT   CDCL3
ORIG ST   C4H12SI   TEMP   308

49.30   216.80   44.70   34.80   26.70   23.70
1/2   2/1   3/3   4/2   5/3   6/3
37.10
7/3

G.E.HAWKES,K.HERWIG,J.D.ROBERTS
J ORG CHEM   39, 1017 (1974)

1745   Q 000600 ANTI—BICYCLO(2.2.1)HEPTAN—2—ONE OXIME

```
 5C———C6
 / 7 \
4C———C———C1
 \       /
 3C———C2
      \\
       N
      /
    HO
```

FORMULA   C7H11NO   MOL WT   125.17
SOLVENT   CDCL3
ORIG ST   C4H12SI   TEMP   308

38.50   166.30   37.20   35.60   26.00   27.40
1/2   2/1   3/3   4/2   5/3   6/3
38.30
7/3

G.E.HAWKES,K.HERWIG,J.D.ROBERTS
J ORG CHEM   39, 1017 (1974)

1746   Q 000600 SYN—BICYCLO(2.2.1)HEPTAN—2—ONE OXIME

```
 5C———C6
 / 7 \
4C———C———C1
 \       /
 3C———C2  OH
      \\ /
       N
```

FORMULA   C7H11NO   MOL WT   125.17
SOLVENT   CDCL3
ORIG ST   C4H12SI   TEMP   308

42.00   167.40   34.90   35.50   27.10   27.80
1/2   2/1   3/3   4/2   5/3   6/3
39.10
7/3

G.E.HAWKES,K.HERWIG,J.D.ROBERTS
J ORG CHEM   39, 1017 (1974)
```
```

1747   Q 000600 HEXACHLOROBICYCLO(3.2.0)HEPTA-3,6-DIEN-2-ONE

```
          O
          ‖
      CL  C2  CL
     7\  /  \3/
  CL-C-C1   C
     ‖ |    ‖
  CL-C-C---C4
   6 |5    \
      CL    CL
```

FORMULA	C7CL6O			MOL WT	312.80
SOLVENT	CHCL3				
ORIG ST	C4H12SI			TEMP	AMB

73.80	182.00	132.00	156.80	75.80	136.10
1/S	2/S	3/S	4/S	5/S	6/S
131.50					
7/S					

G.E.HAWKES,R.A.SMITH,J.D.ROBERTS
J ORG CHEM   39, 1276 (1974)

1748   Q 000600 OCTACHLOROBICYCLO(3.2.0)HEPTA-2,6-DIENE

```
          CL
          |
    CL CL C2
     \  |/ \3
    7C-C1   C-CL
     ‖ |    |
    6C-C---C-CL
     / |5   |4
   CL  CL   CL
```

FORMULA	C7CL8			MOL WT	367.70
SOLVENT	CHCL3				
ORIG ST	C4H12SI			TEMP	AMB

78.90	137.50	133.40	87.90	82.00	133.40
1/S	2/S	3/S	4/S	5/S	6/S
60.40					
7/S					

G.E.HAWKES,R.A.SMITH,J.D.ROBERTS
J ORG CHEM   39, 1276 (1974)

1749   Q 000600 OCTACHLORO-2,5-NORBORNADIENE

```
    CL      CL
     \1   2/
      C---C
    6/ \7  \3
  CL-C  RCR  C-CL
     \   \ /
     5C---C4     R -CL
     /      \
    CL      CL
```

FORMULA	C7CL8			MOL WT	367.70
SOLVENT	CHCL3				
ORIG ST	C4H12SI			TEMP	AMB

86.00	137.40	113.10
1/S	2/S	7/S

G.E.HAWKES,R.A.SMITH,J.D.ROBERTS
J ORG CHEM   39, 1276 (1974)

1750   Q 000600 DECACHLOROBICYCLO(3.2.0)HEPT-2-ENE

```
          CL
          |
    CL CL C2
     \7 |/ \3
    CL-C-C1  C-CL
     | |    |
    CL-C-C---C-CL
     /6 |5  |4
    CL CL   CL
```

FORMULA	C7CL10			MOL WT	438.61
SOLVENT	CHCL3				
ORIG ST	C4H12SI			TEMP	AMB

79.40	139.80	132.20	90.80	86.40	95.20
1/S	2/S	3/S	4/S	5/S	6/S
95.20					
7/S					

G.E.HAWKES,R.A.SMITH,J.D.ROBERTS
J ORG CHEM   39, 1276 (1974)

1751   Q 000600 DECACHLORO-2-NORBORNENE

```
    CL      CL
     \1   2/
    CL  C---C
     \ / \7  \3
     6C  RCR  C-CL
     / \   \ /     R -CL
    CL 5C---C4
     / \    \
    CL CL   CL
```

FORMULA	C7CL10			MOL WT	438.61
SOLVENT	CHCL3				
ORIG ST	C4H12SI			TEMP	AMB

87.80	137.10	97.40	97.40
1/1	2/1	5/1	7/1

G.E.HAWKES,R.A.SMITH,J.D.ROBERTS
J ORG CHEM   39, 1276 (1974)

1752   Q 000600 DECACHLORO-2-NORBORNENE

```
    CL        CL
     \1    2/
  CL   C---C
   \  /  \7  \3
   6C   RCR   C-CL
   / \      \ /
  CL  5C---C4     R -CL
    / \       \
  CL  CL  CL
```

FORMULA   C7CL10   MOL WT   438.61
SOLVENT   C2CL4/C4H8O2
ORIG ST   C4H12SI   TEMP   AMB

88.70	138.30	98.60	98.60
1/1	2/1	4/1	7/1

G.E.HAWKES,R.A.SMITH,J.D.ROBERTS
J ORG CHEM   39, 1276 (1974)

1753   Q 000600 1,2,3,4,5,6,7,8,8-NONACHLOROBICYCLO(3.2.1)-OCTA-2,6-DIENE

```
    CL        CL
     \5    /
  CL   C----C4
   \  / \    \
   6C   \8    \3
   II CL-C-CL C-CL
   7C   /      //
   / \ /      //
  CL   C----C
    /1      2\
   CL        CL
```

FORMULA   C8HCL9   MOL WT   416.17
SOLVENT   CHCL3
ORIG ST   C4H12SI   TEMP   AMB

82.00	129.10	130.90	65.70	81.20	133.10
1/S	2/S	3/S	4/D	5/S	6/S
136.60	99.00				
7/S	8/S				

G.E.HAWKES,R.A.SMITH,J.D.ROBERTS
J ORG CHEM   39, 1276 (1974)

1754   Q 000600 1,2,3,3,4,5,6,7,8,8-DECACHLOROBICYCLO(3.2.1)-OCT-6-ENE

```
    CL        CL
     \5    4/
  CL   C----C
   \  / \    \   CL
   6C   \8    \3/
   II CL-C-CL  C
   7C   /     / \
   / \ /     /   CL
  CL   C----C
    /1      2\
   CL        CL
```

FORMULA   C8H2CL10   MOL WT   452.64
SOLVENT   CHCL3
ORIG ST   C4H12SI   TEMP   AMB

80.50	70.50	90.70	132.00	100.20
1/S	2/D	3/S	6/S	8/S

G.E.HAWKES,R.A.SMITH,J.D.ROBERTS
J ORG CHEM   39, 1276 (1974)

1755   Q 000600 1,2,3,4,4,5,6,7,7,8-DECACHLOROBICYCLO(3.3.0)-OCT-2-ENE

```
     CL   CL
      I   I
   CL  C8   C2
    \7/ \1/ \3    R -CL
   CL-C    CR   C-CL
    I    I    I
   CL-C---C---C-CL
    6   I5   I4
      CL   CL
```

FORMULA   C8H2CL10   MOL WT   452.64
SOLVENT   CHCL3
ORIG ST   C4H12SI   TEMP   AMB

81.20	134.80	134.80	94.00	85.50	74.10
1/1	2/1	3/1	4/1	5/1	6/2
99.90	119.70				
7/1	8/2				

G.E.HAWKES,R.A.SMITH,J.D.ROBERTS
J ORG CHEM   39, 1276 (1974)

1756   Q 000600 4-METHOXYPENTACHLOROBICYCLO(3.2.0)HEPTA-3,6-DIEN-2-ONE

```
      O
      II
   CL  CL  C2  CL
    \   I/ \3/
    7C-C1   C
    II  I   II
   6C-C---C4
   /  I5   \   8
  CL  CL    O-CH3
```

FORMULA   C8H3CL5O2   MOL WT   309.38
SOLVENT   CHCL3
ORIG ST   C4H12SI   TEMP   AMB

72.90	183.40	107.20	171.40	73.50	134.80
1/S	2/S	3/S	4/S	5/S	6/S
133.80	61.40				
7/S	8/Q				

G.E.HAWKES,R.A.SMITH,J.D.ROBERTS
J ORG CHEM   39, 1276 (1974)

1757

```
   C.        CL
    \3        /
   CL-C---C2
    /    |\     CL
  4/   7 | \1 8/
 CL-C----C-X--C-CH
    \    | /    \
     \   |/      CL
     CL-C---C
     /5   6\
    CL      CL
```

Q 000600 1-DICHLOROMETHYL-2,3,3,4,5,5,6-HEPTACHLOROTRI-CYCLO(2.2.1.0)HEPTANE

FORMULA   C8H3CL9   MOL WT 418.19
SOLVENT   C6H6
ORIG ST   C4H12SI   TEMP   AMB

43.50	60.90	93.20	76.50	39.50	66.30
1/S	2/S	3/S	4/D	7/T	8/D

G.E.HAWKES,R.A.SMITH,J.D.ROBERTS
J ORG CHEM   39, 1276 (1974)

1758

```
   CL        CL
    \3        /
    C---C2
    /    |\     CL
  4/   7 | \1 /
 CL-C----C-X--C-CH
    \    | /    \
     \   |/      CL
     C---C
     /5   6\
    CL      CL
```

Q 000600 1-DICHLOROMETHYL-2,3,3,4,5,5,6-HEPTACHLOROTRI-CYCLO(2.2.1.0)HEPTANE

FORMULA   C8H3CL9   MOL WT 418.19
SOLVENT   C6H12
ORIG ST   C4H12SI   TEMP   AMB

43.20	61.30	92.90	76.20	39.50	65.90
1/S	2/S	3/S	4/S	7/T	8/D

G.E.HAWKES,R.A.SMITH,J.D.ROBERTS
J ORG CHEM   39, 1276 (1974)

1759

```
   CL        CL
    \1    2/
     C---C
  8 6/ \7 \3
 H2C=C  RCR  C-CL
    \    \ /
    5C---C4    R -CL
         \
          CL
```

Q 000600 1,2,3,4,7,7-HEXACHLORO-6-METHYLENE-2-NORBORNENE

FORMULA   C8H4CL6   MOL WT 312.84
SOLVENT   C6H12
ORIG ST   C4H12SI   TEMP   AMB

84.10	130.70	132.20	78.10	40.70	140.30
1/S	2/S	3/S	4/S	5/T	6/S
101.60	111.40				
7/S	8/T				

G.E.HAWKES,R.A.SMITH,J.D.ROBERTS
J ORG CHEM   39, 1276 (1974)

1760

```
   CL        CL
    \3        /
   CL-C---C2
    /    |\
  4/   7 | \1 8
 CL-C----C-X--C-CH2-CL
    \    | /
     \   |/
     CL-C---C
     /5   6\
    CL      CL
```

Q 000600 2,3,3,4,5,5,6-HEPTACHLORO-1-MONOCHLOROMETHYL-TRICYCLO(2.2.1.0)HEPTANE

FORMULA   C8H4CL8   MOL WT 383.75
SOLVENT   C6H6
ORIG ST   C4H12SI   TEMP   AMB

38.20	61.40	93.60	76.60	41.90	38.20
1/S	2/S	3/S	4/S	7/T	8/T

G.E.HAWKES,R.A.SMITH,J.D.ROBERTS
J ORG CHEM   39, 1276 (1974)

1761

```
   CL        CL
    \3        /
    CL-C---C2
    /    |\
  4/   7 | \1 8
 CL-C----C-X--C-CH2-CL
    \    | /
     \   |/
     CL-C---C
     /5   6\
    CL      CL
```

Q 000600 2,3,3,4,5,5,6-HEPTACHLORO-1-MONOCHLOROMETHYL-TRICYCLO(2.2.1.0)HEPTANE

FORMULA   C8H4CL8   MOL WT 383.75
SOLVENT   C6H12
ORIG ST   C4H12SI   TEMP   AMB

37.90	61.30	93.10	76.30	41.90	37.90
1/S	2/S	3/S	4/S	7/T	8/T

G.E.HAWKES,R.A.SMITH,J.D.ROBERTS
J ORG CHEM   39, 1276 (1974)

1762   Q 000600   1-METHYL-3-NORTRICYCLYL CATION

```
      3     7
      C---C
     / \ / \
   4C   2C   C6
    \  |+ /
     \1CR/
      \|/
       C
       5         8
          R —CH3
```

FORMULA	C8H11			MOL WT	107.18
SOLVENT	CLFO2S/F5SB				
ORIG ST	CS2			TEMP	195

292.00	66.30	82.50	42.50	45.80	42.50
1/1	2/2	3/2	4/3	5/2	6/3
82.50	32.50				
7/2	8/4				

G.A.OLAH,G.LIANG
J AM CHEM SOC   95, 3792 (1973)

1763   Q 000600   HEXACHLOROBICYCLO(3.3.0)HEPTA-1,3,7-TRIEN-2-ONE

```
    CL   O
    |    ||
 CL C8  C2 CL
  \7/ \1/ \ /
   C   C   C3
   |   ||  ||
 CL-C---C---C4
  61   5    \
   CL        CL
```

FORMULA	C8CL6O			MOL WT	324.81
SOLVENT	CHCL3				
ORIG ST	C4H12SI			TEMP	AMB

144.30	170.00	140.70	132.10	126.70	86.70
1/S	2/S	3/S	4/S	5/S	6/S
125.40	123.50				
7/S	8/S				

G.E.HAWKES,R.A.SMITH,J.D.ROBERTS
J ORG CHEM   39, 1276 (1974)

1764   Q 000600   OCTACHLOROBICYCLO(3.3.0)OCTA-1,4,6-TRIENE

```
       CL
       |
      C7  CL
     / \8/
  CL-C6  C-CL
     |   |
    5C---C1
     ||  ||
  CL-C4 2C-CL
      \ /
    CL-C-CL
       3
```

FORMULA	C8CL8			MOL WT	379.71
SOLVENT	CHCL3				
ORIG ST	C4H12SI			TEMP	AMB

140.80	139.80	82.40	135.10	133.20	127.70
1/S	2/S	3/S	4/S	5/S	6/S
127.50	86.60				
7/S	8/S				

G.E.HAWKES,R.A.SMITH,J.D.ROBERTS
J ORG CHEM   39, 1276 (1974)

1765   Q 000600   OCTACHLOROBICYCLO(3.3.0)OCTA-2,4,6-TRIENE

```
       CL
       |
      C7  CL
     / \8/
  CL-C6  C-CL
     |   ||
    5C---C-CL
     ||  |
  CL-C4  C-CL
      \ //2
    CL-C3
```

FORMULA	C8CL8			MOL WT	379.71
SOLVENT	CHCL3				
ORIG ST	C4H12SI			TEMP	AMB

77.10	146.20	135.30	132.30	132.10	126.90
1/S	2/S	3/S	4/S	5/S	6/S
125.00	92.40				
7/S	8/S				

G.E.HAWKES,R.A.SMITH,J.D.ROBERTS
J ORG CHEM   39, 1276 (1974)

1766   Q 000600   DECACHLOROBICYCLO(3.3.0)OCTA-1,6-DIENE

```
   CL-C7  CL
     / \8/
  CL-C6  C-CL
    |5  |
  CL-C---C1
     |   ||
  CL-C4 2C-CL
      \ /
       C3
      / \
    CL   CL
```

FORMULA	C8CL10			MOL WT	450.62
SOLVENT	CHCL3				
ORIG ST	C4H12SI			TEMP	AMB

138.00	136.80	96.00	97.90	78.90	136.40
1/S	2/S	3/S	4/S	5/S	6/S
134.50	81.20				
7/S	8/S				

G.E.HAWKES,R.A.SMITH,J.D.ROBERTS
J ORG CHEM   39, 1276 (1974)

© 1979 HEYDEN & SON LIMITED

1767   Q 000600 DECACHLOROBICYCLO(3.3.0)OCTA-2,6-DIENE

```
    CL-C7  CL
     6/ \8/
   CL-C   C-CL
     |5  |1
   CL-C---C-CL
     |    |
   CL-C  2C-CL
    /4\  //
   CL 3C-CL
```

FORMULA   C8CL10   MOL WT 450.62
SOLVENT   CHCL3
ORIG ST   C4H12SI   TEMP   AMB

89.00   136.50   133.50   92.40
1/S   2/S   3/S   4/S

G.E.HAWKES,R.A.SMITH,J.D.ROBERTS
J ORG CHEM   39, 1276 (1974)

1768   Q 000600 5,6-DIMETHYLENE-1,2,3,4,7,7-HEXACHLORO-
                            2-NORBORNENE

```
        7
   CL     CH2
    \     //
    4C---C5
   3/ \   \6 8
 CL-C  RCR C=CH2
    \  \ /
    2C---C1      R -CL
   /      \
  CL      CL
```

FORMULA   C9H4CL6   MOL WT 324.85
SOLVENT   C4H8O2
ORIG ST   C4H12SI   TEMP   AMB

82.90   131.90   139.30   101.90   108.60
1/S   2/S   5/S   7/T   8/T

G.E.HAWKES,R.A.SMITH,J.D.ROBERTS
J ORG CHEM   39, 1276 (1974)

1769   Q 000600 ENDO-5-NORBORNENE-2,3-DICARBOXYLIC ANHYDRIDE

```
   2    O
   C    ||
  /|\   C1
 / | \5/ \
3C | C   \
 || C4 |    O
6C | C   /
  \ | /8\ /
   \|/  C9
    C   ||
    7    O
```

FORMULA   C9H8O3   MOL WT 164.16
SOLVENT   C2D6OS
ORIG ST   C4H12SI   TEMP   AMB

172.20   45.30   135.30   52.30   47.00   135.30
1/1   2/2   3/2   4/3   5/2   6/2
45.30   47.00   172.20
7/2   8/2   9/1

L.F.JOHNSON,W.C.JANKOWSKI
CARBON-13 NMR SPECTRA,JOHN
WILEY AND SONS,NEW YORK   339 (1972)

1770   Q 000600 1-ETHYL-3-NORTRICYCLYL CATION

```
   3    7
   C---C
  / \ / \
4C  2C  C6
  \  |+ /
   \1CR/
    \|/
     C
     5      8   9
     R -CH2-CH3
```

FORMULA   C9H13   MOL WT 121.20
SOLVENT   CLFO2S/F5SB
ORIG ST   CS2   TEMP   195

294.00   63.20   81.40   42.30   45.00   42.30
1/1   2/2   3/2   4/3   5/2   6/3
81.40   41.50   8.80
7/2   8/3   9/4

G.A.OLAH,G.LIANG
J AM CHEM SOC   95, 3792 (1973)

1771   Q 000600 ENDO-2-CHLOROBICYCLO(3.3.1)NONAN-9-ONE

```
   1    2
   C---C-CL
  / \   \
 8C 9C=O C3
 |   |   |
 7C  C---C4
  \ /5
   6C
```

FORMULA   C9H13CLO   MOL WT 172.66
SOLVENT   CDCL3
ORIG ST   C4H12SI   TEMP   AMB

53.90   61.70   31.50   29.80   44.90   34.00
1/2   2/2   3/3   4/3   5/2   6/3
20.40   28.80   215.80
7/3   8/3   9/1

A.HEUMANN,H.KOLSHORN
TETRAHEDRON   31, 1571 (1975)

1772   Q 000600 EXO-2-CHLOROBICYCLO(3.3.1)NONAN-9-ONE

```
              CL
        1     |
        C---C2
       / \   \
     8C   9C=0 C3
      |    |   |
     7C    C---C4
       \  /5
        6C
```

FORMULA	C9H13CLO			MOL WT	172.66
SOLVENT	CDCL3				
ORIG ST	C4H12SI			TEMP	AMB

54.60	65.50	28.40	29.90	45.50	34.70
1/2	2/2	3/3	4/3	5/2	6/3
19.60	32.70	216.10			
7/3	8/3	9/1			

A.HEUMANN,H.KOLSHORN
TETRAHEDRON   31, 1571 (1975)

1773   Q 000600 BICYCLO(3.3.1)NONAN-9-ONE

```
        1    2
        C---C
       / \   \
     8C   9C=0 C3
      |    |   |
     7C    C---C4
       \  /5
        6C
```

FORMULA	C9H14O			MOL WT	138.21
SOLVENT	CDCL3				
ORIG ST	C4H12SI			TEMP	AMB

| 46.50 | 34.20 | 20.50 | 221.70 |
| 1/2 | 2/3 | 3/3 | 9/1 |

A.HEUMANN,H.KOLSHORN
TETRAHEDRON   31, 1571 (1975)

1774   Q 000600 ENDO-2-HYDROXYBICYCLO(3.3.1)NONAN-9-ONE

```
        1    2
        C---C-OH
       / \   \
     8C   9C=0 C3
      |    |   |
     7C    C---C4
       \  /5
        6C
```

FORMULA	C9H14O2			MOL WT	154.21
SOLVENT	CDCL3				
ORIG ST	C4H12SI			TEMP	AMB

54.40	73.40	29.50	27.70	45.20	34.00
1/2	2/2	3/3	4/3	5/2	6/3
20.80	27.20	219.30			
7/3	8/3	9/1			

A.HEUMANN,H.KOLSHORN
TETRAHEDRON   31, 1571 (1975)

1775   Q 000600 EXO-2-HYDROXYBICYCLO(3.3.1)NONAN-9-ONE

```
              OH
        1     |
        C---C2
       / \   \
     8C   9C=0 C3
      |    |   |
     7C    C---C4
       \  /5
        6C
```

FORMULA	C9H14O2			MOL WT	154.21
SOLVENT	CDCL3				
ORIG ST	C4H12SI			TEMP	AMB

54.70	76.80	28.50	29.10	46.20	34.50
1/2	2/2	3/3	4/3	5/2	6/3
19.60	30.60	220.10			
7/3	8/3	9/1			

A.HEUMANN,H.KOLSHORN
TETRAHEDRON   31, 1571 (1975)

1776   Q 000600 BICYCLO(3.3.1)NONANE

```
        1    2
        C---C
       / \   \
     8C   C9  C3
      |    |   |
     7C    C---C4
       \  /5
        6C
```

FORMULA	C9H16			MOL WT	124.23
SOLVENT	CDCL3				
ORIG ST	C4H12SI			TEMP	AMB

| 27.90 | 31.60 | 22.50 | 35.10 |
| 1/2 | 2/3 | 3/3 | 9/3 |

A.HEUMANN,H.KOLSHORN
TETRAHEDRON   31, 1571 (1975)

1777   Q 000600   ENDO-(Z)-1,2,3,4,5,6,7,8,9,10-DECACHLORO-TRICYCLO(5.2.1.0)DECA-4,8-DIENE

```
    CL    CL    CL
     \4   I  10/
      C---C---C
     3/ \  5\6  \9
  CL-C  7CR  CR  C-CL
     \    \ / \ //
      2C---C1  C8  R -CL
     /      \     \
    CL      CL   CL
```

FORMULA	C10H2CL10			MOL WT	476.66
SOLVENT	CHCL3				
ORIG ST	C4H12SI			TEMP	AMB
82.50	131.90	129.80	81.90	82.30	86.40
1/S	2/S	3/S	4/S	5/S	6/S
79.00	134.60	134.90	65.00		
7/D	8/S	9/S	10/D		

G.E.HAWKES,R.A.SMITH,J.D.ROBERTS
J ORG CHEM   39, 1276 (1974)

1778   Q 000600   ENDO-1,2,4,5,6,7,8,9-OCTACHLOROTRICYCLO-(5.2.1.0)DECA-4,8-DIENE

```
    CL    CL
     \4   I  10
      C---C---C
     3/ \  5\   \9
  CL-C  7C  RC6  C-CL
     \    \ / \ //
      2C---C1  8C   R -CL
     /      \     \
    CL      CL   CL
```

FORMULA	C10H4CL8			MOL WT	407.77
SOLVENT	CHCL3				
ORIG ST	C4H12SI			TEMP	AMB
77.00	132.70	132.20	76.20	80.60	89.60
1/S	2/S	3/S	4/S	5/S	6/S
59.70	136.90	132.70	47.70		
7/T	8/S	9/S	10/T		

G.E.HAWKES,R.A.SMITH,J.D.ROBERTS
J ORG CHEM   39, 1276 (1974)

1779   Q 000600   DICYCLOPENTADIENE

```
       3 2
       C=C
      /10 \
   4C--C--C1
     \     /
     5C-C9
     /    \
   6C      C8
     \    //
      \  /
       C
       7
```

FORMULA	C10H12			MOL WT	132.21
SOLVENT	C4H8O2				
ORIG ST	C4H12SI			TEMP	AMB
132.10	132.10	34.80	136.00	131.90	50.40
2/2	3/2	6/3	7/2	8/2	10/3
41.50	45.50	46.40	55.10		

L.F.JOHNSON,W.C.JANKOWSKI
CARBON-13 NMR SPECTRA,JOHN
WILEY AND SONS,NEW YORK   372 (1972)

1780   Q 000600   1,6-METHANOBICYCLO(4.3.0)NONA-2,4-DIENE

```
    2C    C9
   3 / \1/ \
    C   C   \
    I  I\    \
    I  I  C10 C8
    I  I/   /
    C   C  /
   4 \ /6\ /
     5C    C7
```

FORMULA	C10H12			MOL WT	132.21
SOLVENT	CCL4/CDCL3				
ORIG ST	C4H12SI			TEMP	AMB
37.70	129.00	119.20	32.30	19.70	15.70
1/1	2/2	3/2	7/3	8/3	10/3

H.GUENTHER,G.JIKELI
CHEM BER   106, 1863 (1973)

1781   Q 000600   ADAMANTAN-2-ONE

```
            O
     1     2//
     C-----C
    / \   / \
   /   C10  \
  /    \5 4 \
 9C        C--C--C3
   \     6/    /
    \   C    /
     \ / \  /
      7C-----C8
```

FORMULA	C10H14O			MOL WT	150.22
SOLVENT	CDCL3				
ORIG ST	C4H12SI			TEMP	308
47.10	217.90	39.40	27.60	36.40	
1/2	2/1	4/3	5/2	6/3	

G.E.HAWKES,K.HERWIG,J.D.ROBERTS
J ORG CHEM   39, 1017 (1974)

1782　　　Q 000600　1-CHLOROCAMPHENE

```
      5  6
      C-C
     4/ 7 \1
     C--C--C-CL
    10\   /
    H3C-C-C
      /3 2\
    H3C     CH2
     9       8
```

FORMULA　C10H15CL　　　MOL WT　170.68
SOLVENT　CDCL3
ORIG ST　C4H12SI　　　TEMP　　308

73.40	163.00	42.70	44.90	25.70	37.90
1/1	2/1	3/1	4/2	5/3	6/3
46.30	101.30	29.80	26.30		
7/3	8/3	9/4	10/4		

D.G.MORRIS,A.M.MURRAY
J CHEM SOC PERKIN TRANS II　　　539 (1975)

1783　　　Q 000600　ADAMANTAN-2-ONE OXIME

```
          H]
           |
           N
      1     2/
      C-----C
     / \     \
    /  C10    \
   /    \5  4  \
  C9     C--C--C3
   \     6/    /
    \    C    /
     \  / \  /
      7C-----C8
```

FORMULA　C10H15NO　　　MOL WT　165.24
SOLVENT　CDCL3
ORIG ST　C4H12SI　　　TEMP　　308

29.10	167.40	36.20	37.70	27.90	36.60
1/2	2/1	3/2	4/3	5/2	6/3
39.00					
9/3					

G.E.HAWKES,K.HERWIG,J.D.ROBERTS
J ORG CHEM　　　39, 1017 (1974)

1784　　　Q 000600　1-NITROCAMPHENE

```
      5  6
      C-C
     4/ 7 \1
     C--C--C-NJ2
    10\   /
    H3C-C-C
      /3 2\
    H3C     CH2
     9       8
```

FORMULA　C10H15NO2　　　MOL WT　181.24
SOLVENT　CDCL3
ORIG ST　C4H12SI　　　TEMP　　308

96.20	158.80	43.00	44.70	24.70	32.20
1/1	2/1	3/1	4/2	5/3	6/3
41.10	101.00	29.60	26.10		
7/3	8/3	9/4	10/4		

D.G.MORRIS,A.M.MURRAY
J CHEM SOC PERKIN TRANS II　　　539 (1975)

1785　　　Q 000600　CAMPHENE

```
      5  6
      C-C
     / 7 \
    4C--C--C1
   10\   /
   H3C-C-C
     /3 2\
   H3C     CH2
    9       8
```

FORMULA　C10H16　　　MOL WT　136.24
SOLVENT　CDCL3
ORIG ST　C4H12SI　　　TEMP　　308

48.20	165.90	41.70	47.00	23.80	28.90
1/2	2/1	3/1	4/2	5/3	6/3
37.40	99.10	29.40	25.80		
7/3	8/3	9/4	10/4		

D.G.MORRIS,A.M.MURRAY
J CHEM SOC PERKIN TRANS II　　　539 (1975)

1786　　　Q 000600　1-ACETYL-BICYCLO(2.2.2)OCTANE

```
        2  3
    ]    C--C
  10 || 1/6 5\
  H3C-C-C-C--C-C4
    9  \8 7/
        C--C
```

FORMULA　C10H16O　　　MOL WT　152.24
SOLVENT　CDCL3
ORIG ST　C4H12SI　　　TEMP　　308

44.50	27.50	25.50	24.10	25.50	27.50
1/1	2/3	3/3	4/2	7/3	8/3
214.20	25.50				
9/1	10/4				

G.E.HAWKES,K.HERWIG,J.D.ROBERTS
J ORG CHEM　　　39, 1017 (1974)

1787   Q 000600 1—HYDROXYCAMPHENE

                              FORMULA   C10H16O   MOL WT   152.24
        5 6   SOLVENT   CDCL3
        C-C   ORIG ST   C4H12SI   TEMP   308
       4/ 7 \1
       C--C--C-]H   84.70   165.90   42.00   44.10   24.90   33.60
      10\   1/1   2/1   3/1   4/2   5/3   6/3
      H3C-C-C   43.40   97.70   29.40   26.30
        /3 2\   7/3   8/3   9/4   10/4
      H3C   CH2
       9   8   D.G.MORRIS,A.M.MURRAY
                              J CHEM SOC PERKIN TRANS II   539 (1975)

1788   Q 000600 1,7,7—TRIMETHYL—BICYCLO(2.2.1)HEPTAN—2—ONE

                              FORMULA   C10H16O   MOL WT   152.24
      5C---C6   SOLVENT   CDCL3
       /   \   ORIG ST   C4H12SI   TEMP   308
      / ?7 \1 10
    4C-----C-----C-CH3   57.70   219.10   43.30   43.30   27.20   30.00
     \   Z   /   1/1   2/1   3/3   4/2   5/3   6/3
      \   /   46.80   9.30   19.20   19.80
     3 C---C2   7/1   8/4   9/4   10/4
        \\
    R -CH3   O   G.E.HAWKES,K.HERWIG,J.D.ROBERTS
    8 Z -CH3   J ORG CHEM   39, 1017 (1974)
        9

1789   Q 000600 1—AMINOCAMPHENE

                              FORMULA   C10H17N   MOL WT   151.25
        5 6   SOLVENT   CDCL3
        C-C   ORIG ST   C4H12SI   TEMP   308
       4/ 7 \1
       C--C--C-NH2   66.50   168.20   42.40   45.50   25.30   35.30
      10\   /   1/1   2/1   3/1   4/2   5/3   6/3
      H3C-C-C   45.50   97.20   29.60   26.30
        /3 2\   7/3   8/3   9/4   10/4
      H3C   CH2
       9   8   D.G.MORRIS,A.M.MURRAY
                              J CHEM SOC PERKIN TRANS II   539 (1975)

1790   Q 000600 1—ACETYL—BICYCLO(2.2.2)OCTANE OXIME

      H]   FORMULA   C10H17NO   MOL WT   167.25
       \   2 3   SOLVENT   CDCL3
        N   C--C   ORIG ST   C4H12SI   TEMP   308
     10 || 1/6 5\
     H3C-C-C-C--C-C4   37.10   28.30   25.70   24.00   25.70   28.30
       9 \8 7/   1/1   2/3   3/3   4/2   7/3   8/3
          C--C   163.60   10.10
                               9/1   10/4

                              G.E.HAWKES,K.HERWIG,J.D.ROBERTS
                              J ORG CHEM   39, 1017 (1974)

1791   Q 000600 1,7,7—TRIMETHYL—BICYCLO(2.2.1)HEPTAN—2—ONE
                         OXIME

      5C---C6   FORMULA   C10H17NO   MOL WT   167.25
       /   \   SOLVENT   CDCL3
      / ?7 \1 10   ORIG ST   C4H12SI   TEMP   308
    4C-----C-----C-CH3
     \   Z   /   51.70   169.50   33.10   43.80   27.30   32.70
      \   /   1/1   2/1   3/3   4/2   5/3   6/3
     3C---C2]H   48.20   11.10   18.50   19.50
        \\   7/1   8/4   9/4   10/4
    R -CH3   N
    8 Z -CH3   G.E.HAWKES,K.HERWIG,J.D.ROBERTS
        9   J ORG CHEM   39, 1017 (1974)

```
1792                    Q 000600 OCTACHLOROTRICYCLO(4.2.1.1)DECA-3,7-DIEN-
       CL            CL          9,10-DIONE
        |            |
    CL 9C-----------C4  CL   FORMULA  C10CL8O2          MOL WT  435.73
      \ / \      / \ /        SOLVENT  CHCL3
     8C   \    /    C3        ORIG ST  C4H12SI          TEMP      AMB
      ||  10C=O O=C5  ||
     7C   /       \   C2         71.10  136.80  178.60
     / \ /         \ / \         1/S    2/S     5/S
    CL 6C-----------C1  CL
        |            |
       CL            CL     G.E.HAWKES,R.A.SMITH,J.D.ROBERTS
                           J ORG CHEM                   39, 1276 (1974)
```

```
1793                    Q 000600 OCTACHLOROTRICYCLO(5.3.0.0)DECA-4,8-DIEN-
      O         O              3,10-DIONE
      ||        ||
    10C CL  CL C3         FORMULA  C10CL8O2          MOL WT  435.73
     / \|    |/ \          SOLVENT  C2CL4/C4H8O2
    /  1C---C2   \         ORIG ST  C4H12SI          TEMP      AMB
   CL-C  |    |  4C-CL
   9\  7C---C6  /             70.70  181.50  138.10  155.60  78.50
    \ /|    |\ /              1/S    2/S     3/S     4/S     5/S
    8C CL  CL C5
     |         |
     CL        CL        G.E.HAWKES,R.A.SMITH,J.D.ROBERTS
                         J ORG CHEM                   39, 1276 (1974)
```

```
1794                    Q 000600 OCTACHLOROTRICYCLO(5.3.0.0)DECA-4,8-DIEN-
      O         O              3,10-DIONE
      ||        ||
    10C CL  CL C3         FORMULA  C10CL8O2          MOL WT  435.73
     / \|    |/ \          SOLVENT  CHCL3
    /  1C---C2   \         ORIG ST  C4H12SI          TEMP      AMB
   CL-C  |    |  4C-CL
   9\  7C---C6  /             70.50  181.90  137.20  158.70  75.90
    \ /|    |\ /              1/S    2/S     3/S     4/S     5/S
    8C CL  CL C5
     |         |
     CL        CL        G.E.HAWKES,R.A.SMITH,J.D.ROBERTS
                         J ORG CHEM                   39, 1276 (1974)
```

```
1795                    Q 000600 ENDO-PERCHLOROTRICYCLO(3.2.1.0)DECA-4,8-DIENE
     CL   CL  CL         FORMULA  C10CL12          MOL WT  545.55
       \4   |  10|        SOLVENT  CHCL3
       C---C---C-CL       ORIG ST  C4H12SI          TEMP      AMB
     3/ \7  5\   \9
    CL-C  RCR  C6  C-CL      85.30  140.10  138.50  84.20  87.60  86.40
     \   \/|\  /             1/S    2/S     3/S     4/S    5/S    6/S
      C---C R C8   R -CL    105.40  136.40  134.10  95.30
      /2   1\    \           7/S    8/S     9/S     10/S
     CL      CL  CL

                         G.E.HAWKES,R.A.SMITH,J.D.ROBERTS
                         J ORG CHEM                   39, 1276 (1974)
```

```
1796                    Q 000600 EXO-PERCHLOROTRICYCLO(3.2.1.0)DECA-4,8-DIENE
        CL   CL          FORMULA  C10CL12          MOL WT  545.55
          \ /9            SOLVENT  CHCL3
     CL CL 10C-C-CL       ORIG ST  C4H12SI          TEMP      AMB
       \4   \ /  ||
       C---C5  C8            93.80  141.30  135.40  83.30  88.80  86.70
      3/ \7  \ / \           1/S    2/S     3/S     4/S    5/S    6/S
    CL-C  RCR  C6  CL        97.90  138.70  131.90  93.30
     \   \/ \     R -CL      7/S    8/S     9/S     10/S
      2C---C1  CL
      /        \          G.E.HAWKES,R.A.SMITH,J.D.ROBERTS
     CL        CL        J ORG CHEM                   39, 1276 (1974)
```

1797   Q 000600 CAMPHENE—1—CARBOXYLIC ACID

```
                        FORMULA   C11H16O2              MOL WT  180.25
        5  6            SOLVENT   CDCL3
        C-C       O     ORIG ST   C4H12SI              TEMP       308
      4/ 7 \1  //
      C--C--C-C11            59.50   162.40    42.90    47.40    24.50    31.90
     10\   /   \             1/1      2/1      3/1      4/2      5/3      6/3
    H3C-C-C        OH        40.70   101.00    29.60    26.00   180.70
       /3 2\                 7/3      8/3      9/4     10/4     11/1
     H3C     CH2
      9       8          D.G.MORRIS,A.M.MURRAY
                        J CHEM SOC PERKIN TRANS II              539 (1975)
```

1798   Q 000600 ENDO—2—ACETOXYBICYCLO(3.3.1)NONAN—9—ONE

```
            O
            ||              FORMULA   C11H16O3              MOL WT  196.25
           C 10             SOLVENT   CDCL3
      1   2  / \            ORIG ST   C4H12SI              TEMP       AMB
      C---C-O   CH3
     / \    \    11             51.20    74.70    28.80    27.70    45.50    34.50
    8C  9C=O C3                 1/2      2/2      3/3      4/3      5/2      6/3
    |   |    |                  20.90    26.70   216.60   169.90    21.20
    7C    C---C4                7/3      8/3      9/1     10/1     11/4
     \  /5
      6C                   A.HEUMANN,H.KOLSHORN
                          TETRAHEDRON                        31, 1571 (1975)
```

1799   Q 000600 EXO—2—ACETOXYBICYCLO(3.3.1)NONAN—9—ONE

```
         O
         ||  11             FORMULA   C11H16O3              MOL WT  196.25
       10C-CH3              SOLVENT   CDCL3
         /                 ORIG ST   C4H12SI              TEMP       AMB
        O
     1   |                     51.10    78.20    26.20    28.80    45.80    34.70
     C---C2                    1/2      2/2      3/3      4/3      5/2      6/3
    / \    \                   19.50    30.80   217.80   169.90    21.00
   8C  9C=O C3                 7/3      8/3      9/1     10/1     11/4
   |   |    |
   7C    C---C4
    \  /5                  A.HEUMANN,H.KOLSHORN
     6C                    TETRAHEDRON                        31, 1571 (1975)
```

1800   Q 000600 CAMPHENE—1—CARBOXAMIDE

```
                        FORMULA   C11H17NO              MOL WT  179.26
        5  6            SOLVENT   CDCL3
        C-C       O     ORIG ST   C4H12SI              TEMP       308
      4/ 7 \1  //
      C--C--C-C11            60.70   163.90    43.00    47.70    24.60    31.40
     10\   /   \             1/1      2/1      3/1      4/2      5/3      6/3
    H3C-C-C        NH2       41.60   101.00    29.60    26.00
       /3 2\                 7/3      8/3      9/4     10/4     11/1
     H3C     CH2
      9       8          D.G.MORRIS,A.M.MURRAY
                        J CHEM SOC PERKIN TRANS II              539 (1975)
```

1801   Q 000600 1—METHYLCAMPHENE

```
        5  6            FORMULA   C11H18                MOL WT  150.27
        C-C             SOLVENT   CDCL3
      4/ 7 \1 11        ORIG ST   C4H12SI              TEMP       308
      C--C--C-CH3
     10\   /                49.90   169.20    42.80    47.40    25.50    35.60
    H3C-C-C                 1/1      2/1      3/1      4/2      5/3      6/3
       /3 2\                44.30    96.80    29.60    26.10    18.30
     H3C     CH2            7/3      8/3      9/4     10/4     11/4
      9       8
                        D.G.MORRIS,A.M.MURRAY
                        J CHEM SOC PERKIN TRANS II              539 (1975)
```

1802   Q 000600   1-HYDROXYMETHYLCAMPHENE

```
                          FORMULA   C11H18O           MOL WT   166.27
        5 6               SOLVENT   CDCL3
        C-C               ORIG ST   C4H12SI           TEMP        308
       4/ 7 \1 11
       C--C--C-CH2-OH      56.20   165.20    43.20    47.00    24.60    31.00
      10\   /              1/1      2/1      3/1      4/2      5/3      6/3
      H3C-C-C              39.60    98.30    29.20    25.90    63.80
       /3 2\               7/3      8/3      9/4     10/4     11/3
      H3C     CH2
       9       8          D.G.MORRIS,A.M.MURRAY
                          J CHEM SOC PERKIN TRANS II            539 (1975)
```

1803   Q 000600   CAMPHENE-1-CARBOXYLIC ACID METHYL ESTER

```
                          FORMULA   C12H18O2          MOL WT   194.28
        5 6               SOLVENT   CDCL3
        C-C     O         ORIG ST   C4H12SI           TEMP        308
       4/ 7 \1 //
       C--C--C-C11         59.70   163.00    42.80    47.30    24.50    31.90
      10\   /   \ 12       1/1      2/1      3/1      4/2      5/3      6/3
      H3C-C-C    O-CH3     40.60   100.60    29.60    26.00   174.10    47.30
       /3 2\               7/3      8/3      9/4     10/4     11/1     12/4
      H3C     CH2
       9       8          D.G.MORRIS,A.M.MURRAY
                          J CHEM SOC PERKIN TRANS II            539 (1975)
```

1804   Q 000600   CEDROL

```
    12   10
   H3C   C OH             FORMULA   C15H26O           MOL WT   222.37
     | 9/ \|11 13         SOLVENT   CDCL3
    6C C C C---CH3        ORIG ST   C4H12SI           TEMP        AMB
    / \|/4\|
   7C   C5  C3             56.50    43.30    61.00    54.00    41.40    74.80
    |   |   | 14           1/2      2/1      3/2      5/1      6/2     11/1
   8C---C---C-CH3         15.50    25.30    27.60    28.90    30.30    31.60
     1   2|               12/4
        CH3               35.20    37.00    41.90
         15
                          L.F.JOHNSON,W.C.JANKOWSKI
                          CARBON-13 NMR SPECTRA,JOHN
                          WILEY AND SONS,NEW YORK              466 (1972)
```

1805   Q 000600   ENDO-2-TOSYLOXYBICYCLO(3.3.1)NONAN-9-ONE

```
                          FORMULA   C16H20O4S         MOL WT   308.40
      11 12               SOLVENT   CDCL3
       O   C=C            ORIG ST   C4H12SI           TEMP        AMB
   1   2  ||  /  \ 16
   C---C-O-S-C10 13C-CH3   51.90    81.70    28.20    27.50    44.80    34.00
   / \  \3 ||  \  //       1/2      2/2      3/3      4/3      5/2      6/3
  8C  9C=O C O   C-C      20.30    26.70   215.00   144.90   133.80   129.90
   |   |    15 14          7/3      8/3      9/1     10/1     11/2     12/2
  7C   C---C4             127.50    21.50
   \  /5                  13/1     16/4
    6C
                          A.HEUMANN,H.KOLSHORN
                          TETRAHEDRON                    31, 1571 (1975)
```

1806   Q 000600   EXO-2-TOSYLOXYBICYCLO(3.3.1)NONAN-9-ONE

```
      14 15
       C=C   O            FORMULA   C16H20O4S         MOL WT   308.40
   16 /   \10||           SOLVENT   CDCL3
  H3C-C13   C-S=O         ORIG ST   C4H12SI           TEMP        AMB
      \\ // |
      C-C11 O              51.70    86.40    27.10    27.80    45.50    34.80
      12     |             1/2      2/2      3/3      4/3      5/2      6/3
       1C---C2            19.30    30.90   215.50   143.90   134.20   129.90
       / \   \             7/3      8/3      9/1     10/1     11/2     12/2
      8C  9C=O C3         127.50    21.60
       |   |   |          13/1     16/4
      7C   C---C4
       \  /5              A.HEUMANN,H.KOLSHORN
        6C               TETRAHEDRON                   31, 1571 (1975)
```

1807 Q 000800 TOLUENE

```
    3 2
    C=C
   /   \1 7
 4C      C-CH3
   \   //
    C-C
    5 6
```

FORMULA	C7H8			MOL WT	92.14
SOLVENT	C4H8O2				
ORIG ST	C4H12SI			TEMP	AMB

137.70	129.10	128.30	125.50	128.30	129.10
1/1	2/2	3/2	4/2	5/2	6/2
21.20					
7/4					

L.F.JOHNSON,W.C.JANKOWSKI
CARBON-13 NMR SPECTRA,JOHN
WILEY AND SONS,NEW YORK   243 (1972)

1808 Q 000800 TOLUENE

```
    5 6
    C-C
   //   \1 7
 4C      C-CH3
   \   /
    C=C
    3 2
```

FORMULA	C7H8			MOL WT	92.14
SOLVENT	CDCL3				
ORIG ST	C4H12SI			TEMP	AMB

137.80	129.30	128.50	125.60	21.30
1/1	2/2	3/2	4/2	7/4

G.W.BUCHANAN,G.MONTAUDO,P.FINOCCHIARO
CAN J CHEM   52, 3196 (1974)

1809 Q 000800 ETHYLBENZENE

```
    5 6
    C-C
   //   \
 C      C-CH2-CH3
 4\   /1 7   8
    C=C
    3 2
```

FORMULA	C8H10			MOL WT	106.17
SOLVENT	C4H8O2				
ORIG ST	C4H12SI			TEMP	AMB

144.20	128.00	128.40	125.80	128.40	128.00
1/1	2/2	3/2	4/2	5/2	6/2
29.10	15.70				
7/3	8/4				

L.F.JOHNSON,W.C.JANKOWSKI
CARBON-13 NMR SPECTRA,JOHN
WILEY AND SONS,NEW YORK   298 (1972)

1810 Q 000800 META—XYLENE

```
      8
     CH3
      |
  6C=C1
   \   \2
 5C      C-CH3
   \   // 7
    C-C
    4 3
```

FORMULA	C8H10			MOL WT	106.17
SOLVENT	C4H8O2				
ORIG ST	C4H12SI			TEMP	AMB

130.00	137.60	126.20	128.30	126.20	137.60
1/2	2/1	3/2	4/2	5/2	6/1
21.20	21.20				
7/4	8/4				

L.F.JOHNSON,W.C.JANKOWSKI
CARBON-13 NMR SPECTRA,JOHN
WILEY AND SONS,NEW YORK   297 (1972)

1811 Q 000800 ORTHO—XYLENE

```
    5 6
    C-C
   //   \1 7
 4C      C-CH3
   \   /
    C=C2
    3  \8
       CH3
```

FORMULA	C8H10			MOL WT	106.17
SOLVENT	CDCL3				
ORIG ST	C4H12SI			TEMP	AMB

136.40	129.90	126.10	19.60
1/1	3/2	4/2	7/4

G.W.BUCHANAN,G.MONTAUDO,P.FINOCCHIARO
CAN J CHEM   52, 3196 (1974)

1812   Q 000800 PARA-XYLENE

```
     5 6
     C-C
  8 4⁄   ⧵1 7
 H3C-C      C-CH3
     ⧵   ⁄
      C=C
     3 2
```

FORMULA	C8H10		MOL WT	106.17
SOLVENT	CDCL3			
ORIG ST	C4H12SI		TEMP	AMB

134.50	129.10	20.90
1/1	2/2	7/4

G.W.BUCHANAN,G.MONTAUDO,P.FINOCCHIARO
CAN J CHEM   52, 3196 (1974)

1813   Q 000800 DIMETHYLPHENYLCARBENIUM ION

```
    8 9
    C=C
   ⁄   ⧵4 1+
 7C    C-C-CH3
   ⧵  I 2
    C-C   CH3
    6 5   3
```

FORMULA	C9H11		MOL WT	119.19
SOLVENT	CLFO2S/F5SB			
ORIG ST	CS2		TEMP	198

253.10	33.70	33.70	138.80	141.20	132.10
1/1	2/4	3/4	4/1	5/2	6/2
154.70	132.10	141.20			
7/2	8/2	9/2			

G.A.OLAH,P.W.WESTERMAN
J AM CHEM SOC   95, 7530 (1973)

1814   Q 000800 CUMENE

```
   5 6   8
   C-C   CH3
  ⫽  ⧵  I 7
 C⧵   ⁄1 I
 4⧵  ⁄1   CH3
   C=C   CH3
   3 2   9
```

FORMULA	C9H12		MOL WT	120.20
SOLVENT	CDCL3			
ORIG ST	C4H12SI		TEMP	AMB

148.70	126.30	128.20	125.70	128.20	126.30
1/1	2/2	3/2	4/2	5/2	6/2
34.20	23.90	23.90			
7/2	8/4	9/4			

L.F.JOHNSON,W.C.JANKOWSKI
CARBON-13 NMR SPECTRA,JOHN
WILEY AND SONS,NEW YORK   352 (1972)

1815   Q 000800 MESITYLENE

```
      7
   H3C
     ⧵1 2
     C-C
    ⫽  ⧵3 8
  6C    C-CH3
    ⧵  ⁄
     C=C
    ⁄5 4
  H3C
    9
```

FORMULA	C9H12		MOL WT	120.20
SOLVENT	C4H8O2			
ORIG ST	C4H12SI		TEMP	AMB

137.40	127.10	137.40	127.10	137.40	127.10
1/1	2/2	3/1	4/2	5/1	6/2
21.00	21.00	21.00			
7/4	8/4	9/4			

L.F.JOHNSON,W.C.JANKOWSKI
CARBON-13 NMR SPECTRA,JOHN
WILEY AND SONS,NEW YORK   351 (1972)

1816   Q 000800 1,2,4-TRIMETHYLBENZENE

```
     5 6
     C-C
  9 4⁄  ⧵1 7
 H3C-C      C-CH3
     ⧵   ⁄
     3C=C2
       ⧵8
        CH3
```

FORMULA	C9H12		MOL WT	120.20
SOLVENT	CDCL3			
ORIG ST	C4H12SI		TEMP	AMB

133.40	136.30	130.50	135.20	126.70	129.80
1/1	2/1	3/2	4/1	5/2	6/2
19.10	19.50	20.90			
7/4	8/4	9/4			

G.W.BUCHANAN,G.MONTAUDO,P.FINOCCHIARO
CAN J CHEM   52, 3196 (1974)

1817   Q 000800   1,2,4-TRIMETHYLBENZENE

```
                         FORMULA   C9H12              MOL WT  120.20
     5  6                SOLVENT   CDCL3
     C-C                 ORIG ST   C4H12SI            TEMP       AMB
  9 4/    \ 7
 H3C-C      C-CH3        19.10    19.50    20.80   126.40  129.50  130.40
    \    /1              7/4      8/4      9/4
     C=C 2              133.10   135.00   136.10
    3   \8
        CH3
                        L.F.JOHNSON,W.C.JANKOWSKI
                        CARBON-13 NMR SPECTRA,JOHN
                        WILEY AND SONS,NEW YORK               350 (1972)
```

1818   Q 000800   1,2,3,5-TETRAMETHYLBENZENE

```
    10
    H3C                 FORMULA   C10H14             MOL WT  134.22
      \5  6             SOLVENT   CDCL3
       C-C              ORIG ST   C4H12SI            TEMP       AMB
      //   \1 7
     4C       C-CH3     136.00   131.60   136.00   128.90  134.30   20.30
       \    /           1/1      2/1      3/1      4/2     5/1      7/4
        3C=C2           14.60    20.30    20.90
       9/   \8          8/4      9/4      10/4
      H3C    CH3
                        G.W.BUCHANAN,G.MONTAUDO,P.FINOCCHIARO
                        CAN J CHEM                   52, 3196 (1974)
```

1819   Q 000800   PARA-TERT-BUTYLTOLUENE

```
    3  2    9           FORMULA   C11H16             MOL WT  148.25
    C-C    CH3          SOLVENT   CDCL3
 11 4/    \1 17         ORIG ST   C4H12SI            TEMP       AMB
 H3C-C      C-C-CH3
     \    / I 8         148.00   125.00   128.70   134.50  128.70  125.00
      C=C    CH3        1/1      2/2      3/2      4/1     5/2     6/2
      5  6    10        34.20    31.40    31.40    31.40   20.70
                        7/1      8/4      9/4      10/4    11/4

                        L.F.JOHNSON,W.C.JANKOWSKI
                        CARBON-13 NMR STECTRA,JOHN
                        WILEY AND SONS,NEW YORK               422 (1972)
```

1820   Q 000800   PENTAMETHYLBENZENE

```
    11
    H3C                 FORMULA   C11H16             MOL WT  148.25
      \5  6             SOLVENT   CDCL3
       C-C              ORIG ST   C4H12SI            TEMP       AMB
  10 4/    \1 7
 H3C-C      C-CH3       133.00   132.10   134.50   131.50   20.70   15.80
     \    /             1/1      2/1      3/1      6/2      7/4     8/4
      3C=C2             16.10
     9/   \8            9/4
    H3C    CH3
                        G.W.BUCHANAN,G.MONTAUDO,P.FINOCCHIARO
                        CAN J CHEM                   52, 3196 (1974)
```

1821   Q 000900   1-BROMO-4-NITROBENZENE

```
                        FORMULA   C6H4BRNO2          MOL WT  202.01
    2  3                SOLVENT   CDCL3
    C-C                 ORIG ST   C4H12SI            TEMP       AMB
   1/   \4
 BR-C     C-NO2         129.80   132.50   124.90   147.00  132.50  124.90
    \    /              1/1      2/2      3/2      4/1     5/2     6/2
     C=C
     6  5               L.F.JOHNSON,W.C.JANKOWSKI
                        CARBON-13 NMR SPECTRA,JOHN
                        WILEY AND SONS,NEW YORK               147 (1972)
```

1822   Q 000900 1,3,5-TRIBROMOBENZENE

```
  BR   C2  BR
   \  /  \  /
    3C    C1
    |     ||
    4C    C6
     \   /
      5C
      |
      BR
```

FORMULA C6H3BR3   MOL WT 314.80
SOLVENT CDCL3
ORIG ST C4H12SI   TEMP   AMB

123.20   132.90   123.20   132.90   123.20   132.90
 1/1   2/2   3/1   4/2   5/1   6/2

L.F.JOHNSON,W.C.JANKOWSKI
CARBON-13 NMR SPECTRA,JOHN
WILEY AND SONS,NEW YORK   145 (1972)

1823   Q 000900 1,2,3-TRICHLOROBENZENE

```
        CL
     5  6/
      C-C
     //   \\
    4C     C-CL
     \    /1
      C=C
     3  2\
        CL
```

FORMULA C6H3CL3   MOL WT 181.45
SOLVENT CDCL3
ORIG ST C4H12SI   TEMP   AMB

131.60   134.30   128.60   127.50   128.60   134.30
 1/1   2/1   3/2   4/2   5/2   6/1

L.F.JOHNSON,W.C.JANKOWSKI
CARBON-13 NMR SPECTRA,JOHN
WILEY AND SONS,NEW YORK   146 (1972)

1824   Q 000900 META-DICHLOROBENZENE

```
     C_
     |
   2C=C1
   /    \
  3C     C-CL
   \    //6
    C-C
    4  5
```

FORMULA C6H4CL2   MOL WT 147.00
SOLVENT C4H8O2
ORIG ST C4H12SI   TEMP   AMB

128.90   135.10   127.00   130.60   127.00   135.10
 1/2   2/1   3/2   4/2   5/2   6/1

L.F.JOHNSON,W.C.JANKOWSKI
CARBON-13 NMR SPECTRA,JOHN
WILEY AND SONS,NEW YORK   149 (1972)

1825   Q 000900 ORTHO-DICHLOROBENZENE

```
        CL
     2  1/
      C-C
     /    \
    3C     C-CL
     \    /6
      C=C
      4  5
```

FORMULA C6H4CL2   MOL WT 147.00
SOLVENT CDCL3
ORIG ST C4H12SI   TEMP   AMB

132.60   130.50   127.70   127.70   130.50   132.60
 1/1   2/2   3/2   4/2   5/2   6/1

L.F.JOHNSON,W.C.JANKOWSKI
CARBON-13 NMR SPECTRA,JOHN
WILEY AND SONS,NEW YORK   148 (1972)

1826   Q 000900 PARA-DIFLUOROBENZENE

```
     5  6
      C-C
     //   \\
  F-C       C-F
    4\    /1
      C=C
      3  2
```

FORMULA C6H4F2   MOL WT 114.10
SOLVENT CDCL3
ORIG ST C4H12SI   TEMP   AMB

159.10   116.50   116.50   159.10   116.50   116.50
 1/1   2/2   3/2   4/1   5/2   6/2

L.F.JOHNSON,W.C.JANKOWSKI
CARBON-13 NMR SPECTRA,JOHN
WILEY AND SONS,NEW YORK   150 (1972)

1827   Q 000900 BROMOBENZENE

```
      5 6
      C-C
    //    \\
   C      C-BR
  4\     /1
    C=C
    3 2
```

FORMULA	C6H5BR			MOL WT	157.01
SOLVENT	CDCL3				
ORIG ST	C4H12SI			TEMP	AMB

122.40	131.40	129.80	126.70	129.80	131.40
1/1	2/2	3/2	4/2	5/2	6/2

L.F.JOHNSON,W.C.JANKOWSKI
CARBON-13 NMR SPECTRA,JOHN
WILEY AND SONS,NEW YORK   152 (1972)

1828   Q 000900 CHLOROBENZENE

```
      3 2
      C-C
    4//   \\1
   C      C-CL
    \     /
    C=C
    5 6
```

FORMULA	C6H5CL			MOL WT	112.56
SOLVENT	C4H8O2				
ORIG ST	C4H12SI			TEMP	AMB

134.30	128.60	129.80	126.50	129.80	128.60
1/1	2/2	3/2	4/2	5/2	6/2

L.F.JOHNSON,W.C.JANKOWSKI
CARBON-13 NMR SPECTRA,JOHN
WILEY AND SONS,NEW YORK   153 (1972)

1829   Q 000900 IODOBENZENE

```
      5 6
      C-C
    //   \\
   C      C-J
  4\     /1
    C=C
    3 2
```

FORMULA	C6H5I			MOL WT	204.01
SOLVENT	C4H8O2				
ORIG ST	C4H12SI			TEMP	AMB

94.40	137.20	129.90	127.10	129.90	137.20
1/1	2/2	3/2	4/2	5/2	6/2

L.F.JOHNSON,W.C.JANKOWSKI
CARBON-13 NMR SPECTRA,JOHN
WILEY AND SONS,NEW YORK   156 (1972)

1830   Q 000900 2,6-DICHLOROBENZALDEHYDE

```
          CL
      6 7/
      C-C      O
    //   \\   //
  4C     C-C
    \    /1 5\
    C=C      H
    3 2\
        CL
```

FORMULA	C7H4CL2O			MOL WT	175.02
SOLVENT	CDCL3				
ORIG ST	C4H12SI			TEMP	AMB

130.30	136.60	129.60	133.50	188.30	129.60
1/1	2/1	3/2	4/2	5/1	6/2
136.60					
7/1					

L.F.JOHNSON,W.C.JANKOWSKI
CARBON-13 NMR SPECTRA,JOHN
WILEY AND SONS,NEW YORK   223 (1972)

1831   Q 000900 ALPHA,ALPHA,ALPHA-TRIFLUOROTOLUENE

```
      6 7
      C-C      F
    //   \\   /
  4C     C-C-F
    \    /1 5\
    C=C      F
    3 2
```

FORMULA	C7H5F3			MOL WT	146.11
SOLVENT	CDCL3				
ORIG ST	C4H12SI			TEMP	AMB

131.10	125.40	128.90	131.90	124.60	128.90
1/1	2/2	3/2	4/2	5/1	6/2
125.40					
7/2					

L.F.JOHNSON,W.C.JANKOWSKI
CARBON-13 NMR SPECTRA,JOHN
WILEY AND SONS,NEW YORK   225 (1972)

1832   Q 000900 META–CHLOROTOLUENE

```
      CL
      | 2
    3C=C
    /    \1 7
  4C      C-CH3
    \\   //
     C-C
     5 6
```

FORMULA	C7H7CL			MOL WT	126.59
SOLVENT	CDCL3				
ORIG ST	C4H12SI			TEMP	AMB

139.70	129.10	134.00	125.50	129.30	127.10
1/1	2/2	3/1	4/2	5/2	6/2
21.00					
7/4					

L.F.JOHNSON,W.C.JANKOWSKI
CARBON–13 NMR SPECTRA,JOHN
WILEY AND SONS,NEW YORK   233 (1972)

1833   Q 000900 ORTHO–CHLOROTOLUENE

```
     5 6
     C-C
    //   \\
   C      C-CH3
  4\    /1 7
    C=C
   3 2\
       CL
```

FORMULA	C7H7CL			MOL WT	126.59
SOLVENT	CDCL3				
ORIG ST	C4H12SI			TEMP	AMB

135.90	134.40	129.00	126.40	127.00	130.90
1/1	2/1	3/2	4/2	5/2	6/2
19.90					
7/4					

L.F.JOHNSON,W.C.JANKOWSKI
CARBON–13 NMR STECTRA,JOHN
WILEY AND SONS,NEW YORK   232 (1972)

1834   Q 000900 PARA–CHLOROTOLUENE

```
       5 6
       C-C
      //   \\
  CL-C      C-CH3
    4\    /1 7
      C=C
      3 2
```

FORMULA	C7H7CL			MOL WT	126.59
SOLVENT	CDCL3				
ORIG ST	C4H12SI			TEMP	AMB

136.20	130.40	128.30	131.20	128.30	130.40
1/1	2/2	3/2	4/1	5/2	6/2
20.70					
7/4					

L.F.JOHNSON,W.C.JANKOWSKI
CARBON–13 NMR STECTRA,JOHN
WILEY AND SONS,NEW YORK   234 (1972)

1835   Q 000900 PARA–CHLOROANISOLE

```
       3 2
       C=C
      4/   \1 7
  CL-C      C-O-CH3
      \\   //
       C-C
       5 6
```

FORMULA	C7H7CLO			MOL WT	142.59
SOLVENT	CDCL3				
ORIG ST	C4H12SI			TEMP	AMB

158.20	115.20	129.20	125.40	129.20	115.20
1/1	2/2	3/2	4/1	5/2	6/2
55.30					
7/4					

L.F.JOHNSON,W.C.JANKOWSKI
CARBON–13 NMR SPECTRA,JOHN
WILEY AND SONS,NEW YORK   235 (1972)

1836   Q 000900 ALPHA–BETA–DIBROMOETHYLBENZENE

```
     5 6
     C-C   BR
    //   \ |
   C      C-CH-CH2-BR
  4\    /1 7  8
    C=C
    3 2
```

FORMULA	C8H8BR2			MOL WT	263.96
SOLVENT	CDCL3				
ORIG ST	C4H12SI			TEMP	AMB

138.40	127.50	128.60	128.90	128.60	127.50
1/1	2/2	3/2	4/2	5/2	6/2
50.80	34.90				
7/2	8/3				

L.F.JOHNSON,W.C.JANKOWSKI
CARBON–13 NMR SPECTRA,JOHN
WILEY AND SONS,NEW YORK   286 (1972)

1837 Q 000900 1-CHLORO-2,4-DIMETHOXY-5-NITROBENZENE

FORMULA C8H8CLNO4   MOL WT 217.61
SOLVENT CDCL3
ORIG ST C4H12SI   TEMP AMB

```
        6
 O2N    C    CL
   \  / \  /
   5C     C1
   ||     |
   4C     C2
   /  \  / \
 H3C-O   C   O-CH3
  8      3      7
```

113.80	159.90	97.20	154.50	131.90	127.70
1/1	2/1	3/2	4/1	5/1	6/2
56.90	56.80
7/4	8/4

L.F.JOHNSON,W.C.JANKOWSKI
CARBON-13 NMR SPECTRA,JOHN
WILEY AND SONS,NEW YORK   287 (1972)

1838 Q 000900 1-BROMO-3-PHENYLPROPANE

FORMULA C9H11BR   MOL WT 199.09
SOLVENT C4H8O2
ORIG ST C4H12SI   TEMP AMB

```
    3 2
    C-C
   //   \1
 4C      C-CH2-CH2-CH2-BR
   \   /  7   8    9
    C=C
    5 6
```

140.60	128.40	128.40	126.00	128.40	128.40
1/1	2/2	3/2	4/2	5/2	6/2
34.30	34.00	32.90
7/3	8/3	9/3

L.F.JOHNSON,W.C.JANKOWSKI
CARBON-13 NMR SPECTRA,JOHN
WILEY AND SONS,NEW YORK   347 (1972)

1839 Q 001000 NITROBENZENE

FORMULA C6H5NO2   MOL WT 123.11
SOLVENT CDCL3
ORIG ST C4H12SI   TEMP AMB

```
    3 2
    C-C
   //   \1
 4C      C-NO2
   \   /
    C=C
    5 6
```

| 148.20 | 123.40 | 129.40 | 134.60 | 129.40 | 123.40 |
| 1/1 | 2/2 | 3/2 | 4/2 | 5/2 | 6/2 |

L.F.JOHNSON,W.C.JANKOWSKI
CARBON-13 NMR SPECTRA,JOHN
WILEY AND SONS,NEW YORK   157 (1972)

1840 Q 001200 N-METHYLANILINE

FORMULA C7H9N   MOL WT 107.16
SOLVENT C4H8O2
ORIG ST C4H12SI   TEMP AMB

```
    5 6
    C-C    H
   //   \  |
  C      C-N-CH3
 4\    /1   7
   C=C
   3 2
```

| 150.20 | 112.30 | 129.20 | 116.70 | 129.20 | 112.30 |
| 1/1 | 2/2 | 3/2 | 4/2 | 5/2 | 6/2 |
| 30.20 |
| 7/4 |

L.F.JOHNSON,W.C.JANKOWSKI
CARBON-13 NMR SPECTRA,JOHN
WILEY AND SONS,NEW YORK   250 (1972)

1841 Q 001200 META-FLUOROANILINE

FORMULA C6H6FN   MOL WT 111.12
SOLVENT CDCL3
ORIG ST C4H12SI   TEMP AMB

```
    F
    |
    C-C
  //3 2\\
 C      C-NH2
 4\    /1
   C=C
   5 6
```

| 148.60 | 102.00 | 163.90 | 104.80 | 130.50 | 110.80 |
| 1/1 | 2/2 | 3/1 | 4/2 | 5/2 | 6/2 |

L.F.JOHNSON,W.C.JANKOWSKI
CARBON-13 NMR SPECTRA,JOHN
WILEY AND SONS,NEW YORK   158 (1972)

1842   Q 001200 N-METHYL-N-NITROSOANILINE

```
    5  6    7          FORMULA  C7H8N2O              MOL WT  136.15
    C-C     CH3        SOLVENT  CDCL3
   ⫽  ⟍  |            ORIG ST  C4H12SI              TEMP      AMB
   C      C-N-NO
  4⟍  ⟋1              142.20  119.00  129.30  127.10  129.30  119.00
    C=C                 1/1     2/2     3/2     4/2     5/2     6/2
    3  2              31.10
                        7/4
```

L.F.JOHNSON,W.C.JANKOWSKI
CARBON-13 NMR SPECTRA,JOHN
WILEY AND SONS,NEW YORK   244 (1972)

1843   Q 001200 BENZYLAMINE

```
    3  2              FORMULA  C7H9N                MOL WT  107.16
    C=C                SOLVENT  CDCL3
   4⟋  ⟍1 7           ORIG ST  C4H12SI              TEMP      AMB
   C      C-CH2-NH2
   ⟍  ⟋              143.40  126.90  128.30  126.50  128.30  126.90
    C-C                1/1     2/2     3/2     4/2     5/2     6/2
    5  6              46.30
                        7/3
```

L.F.JOHNSON,W.C.JANKOWSKI
CARBON-13 NMR SPECTRA,JOHN
WILEY AND SONS,NEW YORK   251 (1972)

1844   Q 001200 ACETANILIDE

```
    3  2              FORMULA  C8H9NO               MOL WT  135.17
    C-C     H         SOLVENT  CDCL3
   ⫽  ⟍| | 7 8       ORIG ST  C4H12SI              TEMP      AMB
  4C      C-N-C-CH3
   ⟍  ⟋    ‖         138.20  120.40  128.70  124.10  128.70  120.40
    C=C      O          1/1     2/2     3/2     4/2     5/2     6/2
    5  6              169.50  24.10
                        7/1     8/4
```

L.F.JOHNSON,W.C.JANKOWSKI
CARBON-13 NMR SPECTRA,JOHN
WILEY AND SONS,NEW YORK   295 (1972)

1845   Q 001200 5-CHLORO-2,4-DIMETHOXYANILINE

```
         6            FORMULA  C8H10CLNO2           MOL WT  187.63
   CL    C    NH2     SOLVENT  CDCL3
    ⟍  ⫽ ⟍ ⟋         ORIG ST  C4H12SI              TEMP      AMB
    5C    C1
    |     ‖          130.90  146.50   98.90  147.50  113.80  116.00
    4C    C2           1/1     2/1     3/2     4/1     5/1     6/2
   ⟋ ⟍ ⟋ ⟍          55.70   57.30
  H3C-O   C   O-CH3     7/4     8/4
   8     3     7
```

L.F.JOHNSON,W.C.JANKOWSKI
CARBON-13 NMR SPECTRA,JOHN
WILEY AND SONS,NEW YORK   299 (1972)

1846   Q 001200 N-BENZYLMETHYLAMINE

```
    5  6       8      FORMULA  C8H11N               MOL WT  121.18
    C-C       CH3     SOLVENT  C4H8O2
   ⫽  ⟍       |      ORIG ST  C4H12SI              TEMP      AMB
   C      C-CH2-N
  4⟍  ⟋1 7    |      141.10  128.20  128.20  126.70  128.20  128.20
    C=C       H         1/1     2/2     3/2     4/2     5/2     6/2
    3  2              56.10   35.90
                        7/3     8/4
```

L.F.JOHNSON,W.C.JANKOWSKI
CARBON-13 NMR SPECTRA,JOHN
WILEY AND SONS,NEW YORK   304 (1972)

1847   Q 001200 2,3-DIMETHYLANILINE

```
        NH2          FORMULA   C8H11N              MOL WT  121.18
        |   7        SOLVENT   CDCL3
        C1  CH3      ORIG ST   C4H12SI             TEMP      AMB
       / \ /
      6C   C2         144.50  120.60  136.80  120.40  125.90  113.10
      |    ||          1/1     2/1     3/1     4/2     5/2     6/2
      5C   C3         12.40   20.30
       \ / \           7/4     8/4
        C   CH3
        4   8        L.F.JOHNSON,W.C.JANKOWSKI
                     CARBON-13 NMR SPECTRA,JOHN
                     WILEY AND SONS,NEW YORK           303 (1972)
```

1848   Q 001200 2-ETHYLANILINE

```
                     FORMULA   C8H11N              MOL WT  121.18
      7    8         SOLVENT   CDCL3
      CH2-CH3        ORIG ST   C4H12SI             TEMP      AMB
        |
     3C=C2            144.10  127.80  128.20  118.50  126.60  115.20
     /   \1            1/1     2/1     3/2     4/2     5/2     6/2
    4C     C-NH2      23.90   12.90
     \    //           7/3     8/4
      C-C
      5 6            L.F.JOHNSON,W.C.JANKOWSKI
                     CARBON-13 NMR SPECTRA,JOHN
                     WILEY AND SONS,NEW YORK           305 (1972)
```

1849   Q 001200 N-SEC-BUTYLANILINE

```
          H          FORMULA   C10H15N             MOL WT  149.24
    3 2   |          SOLVENT   CDCL3
    C-C   N          ORIG ST   C4H12SI             TEMP      AMB
   //  \1/ \7 8 9
  4C     C  CH-CH2-CH3   147.70  113.10  129.10  116.60  129.10  113.10
   \   /   |             1/1     2/2     3/2     4/2     5/2     6/2
    C=C   CH3           49.70   29.60   10.20   20.10
    5 6   10             7/2     8/3     9/4     10/4

                     L.F.JOHNSON,W.C.JANKOWSKI
                     CARBON-13 NMR SPECTRA,JOHN
                     WILEY AND SONS,NEW YORK           389 (1972)
```

1850   Q 001200 2,6-DIETHYLANILINE

```
      7    8
      CH2-CH3        FORMULA   C10H15N             MOL WT  149.24
    3 2/             SOLVENT   CDCL3
    C-C              ORIG ST   C4H12SI             TEMP      AMB
   //  \1
  4C     C-NH2        141.50  127.40  125.90  118.10  125.90  127.40
   \    /              1/1     2/1     3/2     4/2     5/2     6/1
    C=C               24.20   12.90   24.20   12.90
    5 6\               7/3     8/4     9/3     10/4
    CH2-CH3
    9    10          L.F.JOHNSON,W.C.JANKOWSKI
                     CARBON-13 NMR SPECTRA,JOHN
                     WILEY AND SONS,NEW YORK           391 (1972)
```

1851   Q 001200 N,N-DIETHYLANILINE

```
                     FORMULA   C10H15N             MOL WT  149.24
    3 2   9   10     SOLVENT   CDCL3
    C-C   CH2-CH3    ORIG ST   C4H12SI             TEMP      AMB
   //  \1 /
  C.     C-N          147.80  112.00  129.10  115.50  129.10  112.00
   \   /   \           1/1     2/2     3/2     4/2     5/2     6/2
    C=C   CH2-CH3     44.20   12.50   44.20   12.50
    5 6   7   8        7/3     8/4     9/3     10/4

                     L.F.JOHNSON,W.C.JANKOWSKI
                     CARBON-13 NMR SPECTRA,JOHN
                     WILEY AND SONS,NEW YORK           390 (1972)
```

1852   Q 001200   BETA-(3,4-DIMETHOXYPHENYL)-ETHYLAMINE

```
        10
      H3C-O
           \2  3
           C=C
    9   1/    \4  7   8
  H3C-O-C        C-CH2-CH2-NH2
         \      //
          6C-C5
```

FORMULA	C10H15NO2			MOL WT	181.24
SOLVENT	CDCL3				
ORIG ST	C4H12SI			TEMP	AMB
147.60	149.10	111.70	132.60	120.70	112.40
1/1	2/1	3/2	4/1	5/2	6/2
39.80	43.70	55.90	55.80		
7/3	8/3	9/4	10/4		

L.F.JOHNSON,W.C.JANKOWSKI
CARBON-13 NMR SPECTRA,JOHN
WILEY AND SONS,NEW YORK   392 (1972)

1853   Q 001200   DIBENZYL METHYL AMINE

```
          15
          CH3
  10  9   14  |      2  3
  C-C     CH2-N-CH2  C-C
 //  \ /        7\  /    \\
C     C8      1C      C4
11\   /          \   /
  C=C             C=C
 12  13           6  5
```

FORMULA	C15H17N			MOL WT	211.31
SOLVENT	CDCL3				
ORIG ST	C4H12SI			TEMP	AMB
139.20	128.70	128.00	126.70	128.00	128.70
1/1	2/2	3/2	4/2	5/2	6/2
61.80	139.20	128.70	128.00	126.70	128.00
7/3	8/1	9/2	10/2	11/2	12/2
128.70	61.80	42.10			
13/2	14/3	15/4			

L.F.JOHNSON,W.C.JANKOWSKI
CARBON-13 NMR SPECTRA,JOHN
WILEY AND SONS,NEW YORK   465 (1972)

1854   Q 001500   PHENYL HYDRAZINE

```
    5   6
    C-C
   //   \\
  C      C-NH-NH2
  4\    /1
    C=C
    3   2
```

FORMULA	C6H8N2			MOL WT	108.14
SOLVENT	CDCL3				
ORIG ST	C4H12SI			TEMP	AMB
151.30	112.00	129.00	118.90	129.00	112.00
1/1	2/2	3/2	4/2	5/2	6/2

L.F.JOHNSON,W.C.JANKOWSKI
CARBON-13 NMR SPECTRA,JOHN
WILEY AND SONS,NEW YORK   168 (1972)

1855   Q 001600   PARA-DIMETHYLAMINOBENZENE DIAZONIUM
                                         TETRAFLUOROBORATE

```
    7       5  6
  H3C      C-C
    \     //   \\1 +            -
     N-C4      C-N≡N   3F4
  8/     \    /
  H3C     C=C
          3  2
```

FORMULA	C8H10BF4N3			MOL WT	234.99
SOLVENT	C2D6OS				
ORIG ST	C4H12SI			TEMP	AMB
88.80	133.70	113.70	156.00	40.30	
1/S	2/D	3/D	4/S	7/Q	

R.RADEGLIA,K.JACKOWSKI,H.-J.JESSEN
Z CHEM   15,   311 (1975)

1856   Q 001600   3-METHOXY-PARA-DIMETHYLAMINOBENZENE
                                         DIAZONIUM TETRAFLUOROBORATE

```
    9       5  6
  H3C      C-C
    \     //   \\1 +            -
     N-C4      C-N≡N   3F4
  8/     \    /
  H3C     C=C
      7  /3  2
    H3C-O
```

FORMULA	C9H12BF4N3O			MOL WT	265.02
SOLVENT	C2D6OS				
ORIG ST	C4H12SI			TEMP	AMB
90.10	111.20	147.40	151.10	115.90	150.40
1/S	2/D	3/S	4/S	5/D	6/D
56.50	43.40				
7/Q	8/Q				

R.RADEGLIA,K.JACKOWSKI,H.-J.JESSEN
Z CHEM   15,   311 (1975)

1857   Q 001600 2-METHOXY-PARA-DIETHYLAMINOBENZENE
DIAZONIUM TETRAFLUOROBORATE

```
                          FORMULA  C11H16BF4N3O           MOL WT  293.07
  11 10      5 6         SOLVENT  C2D6OS
H3C-CH2    C-C           ORIG ST  C4H12SI                TEMP       AMB
      \    ∥   ∖1 +    -
       N-C4    C-N≡N   3F4     79.30  162.80   93.90  156.80  109.90  131.90
   9 8/    \   /                1/S     2/S     3/D     4/S     5/D     6/D
H3C-CH2    C=C                 57.80   45.70   12.40
      3 2\  7                   7/Q     8/T     9/Q
          O-CH3
```

R.RADEGLIA,K.JACKOWSKI,H.-J.JESSEN
Z CHEM   15, 311 (1975)

1858   Q 001600 AZOBENZENE

```
    9 8        2 3       FORMULA  C12H10N2              MOL WT  182.23
    C-C   N=N   C-C      SOLVENT  CDCL3
   ∥   ∖  ∕ ∖ ∕   ∥  ∖   ORIG ST  C4H12SI              TEMP       AMB
 10C     C   1C    C4
    \   ∕7     \   ∕     152.50  122.70  128.80  130.70  128.80  122.70
    C=C         C=C       1/1     2/2     3/2     4/2     5/2     6/2
   11 12        6 5      152.50  122.70  128.80  130.70  128.80  122.70
                          7/1     8/2     9/2    10/2    11/2    12/2
```

L.F.JOHNSON,W.C.JANKOWSKI
CARBON-13 NMR SPECTRA,JOHN
WILEY AND SONS,NEW YORK   431 (1972)

1859   Q 001600 AZOXYBENZENE

```
    3 2        8 9       FORMULA  C12H10N2O             MOL WT  198.23
    C-C   O    C-C       SOLVENT  CDCL3
   ∥   ∖1 |   7∥   ∖     ORIG ST  C4H12SI              TEMP       AMB
  4C    C-N=N-C    C10
    \   ∕     \   ∕       148.20  125.40  128.50  129.40  128.50  125.40
    C=C        C=C         1/1     2/2     3/2     4/2     5/2     6/2
    5 6        12 11      144.00  122.20  128.50  131.30  128.50  122.20
                           7/1     8/2     9/2    10/2    11/2    12/2
```

L.F.JOHNSON,W.C.JANKOWSKI
CARBON-13 NMR SPECTRA,JOHN
WILEY AND SONS,NEW YORK   432 (1972)

1860   Q 001600 PARA-(PARA-ETHOXYPHENYLAZO)PHENYL HEXANOATE

```
   12 11  10  9    8
  H3C-CH2-CH2-CH2-CH2    FORMULA  C20H24N2O3            MOL WT  340.43
                    \    SOLVENT  CDCL3
                    C=O  ORIG ST  C4H12SI              TEMP       AMB
                   ∕7
  17 18      6 5 0       150.20  124.70  122.00  152.20  122.00  124.70
  C-C        C-C |        1/1     2/2     3/2     4/1     5/2     6/2
 ∥   ∖      ∥   ∖|       171.70   34.30   24.50   31.20   22.20   14.70
16C    C-N=N-C    C4      7/1     8/3     9/3    10/3    11/3    12/4
 ∕ ∖  ∕13    1∖  ∕       146.70  123.50  114.60  161.50  114.60  123.50
O   C=C        C=C       13/1    14/2    15/2    16/1    17/2    18/2
|  15 14       2 3        63.70   13.80
|                        19/3    20/4
CH2-CH3
19  20
```

L.F.JOHNSON,W.C.JANKOWSKI
CARBON-13 NMR SPECTRA,JOHN
WILEY AND SONS,NEW YORK   487 (1972)

1861   Q 001600 4,4-BIS(HEXYLOXY)AZOXYBENZENE

```
  12 11  10  9   8   7
H3C-CH2-CH2-CH2-CH2-CH2
                      |
  15 14       2  3  0
   C-C     0   C-C |
  ⁄⁄  \13 ‖ 1⁄  \|
16C     C-N=N-C    C4
  ⁄ \   ⁄      \  ⁄
 0   C=C       C=C
 |  17 18      6 5
 |
CH2-CH2-CH2-CH2-CH2-CH3
 19  20  21  22  23  24
```

FORMULA	C24H34N2O3		MOL WT	398.55
SOLVENT	CDCL3			
ORIG ST	C4H12SI		TEMP	AMB

141.40	127.70	114.10	159.70	114.10	127.70
1/1	2/2	3/2	4/1	5/2	6/2
68.40	13.90	137.90	123.50	113.90	161.30
7/3	12/4	13/1	14/2	15/2	16/1
113.90	123.50	68.10	29.10	25.60	31.50
17/2	18/2	19/3	20/3	21/3	22/3
22.50	13.90				
23/3	24/4				

L.F.JOHNSON,W.C.JANKOWSKI
CARBON-13 NMR SPECTRA,JOHN
WILEY AND SONS,NEW YORK   493 (1972)

1862   Q 001800 PHENOL

```
    5  6
    C-C
   ⁄⁄   \
  C      C-OH
  4\    ⁄1
    C=C
    3  2
```

FORMULA	C6H6O		MOL WT	94.11
SOLVENT	CDCL3			
ORIG ST	C4H12SI		TEMP	AMB

154.90	115.40	129.70	121.00	129.70	115.40
1/1	2/2	3/2	4/2	5/2	6/2

L.F.JOHNSON,W.C.JANKOWSKI
CARBON-13 NMR SPECTRA,JOHN
WILEY AND SONS,NEW YORK   160 (1972)

1863   Q 001800 META-CRESOL

```
   OH
   | 2
  3C=C
  ⁄     \1 7
 4C      C-CH3
  \     ⁄⁄
   C-C
   5  6
```

FORMULA	C7H8O		MOL WT	108.14
SOLVENT	CDCL3			
ORIG ST	C4H12SI		TEMP	AMB

139.80	116.20	155.00	112.50	129.40	121.80
1/1	2/2	3/1	4/2	5/2	6/2
21.10					
7/4					

L.F.JOHNSON,W.C.JANKOWSKI
CARBON-13 NMR SPECTRA,JOHN
WILEY AND SONS,NEW YORK   247 (1972)

1864   Q 001800 2-ISOPROPYLPHENOL

```
     OH
   6  |     8
   C-C1    CH3
  ⁄⁄  \2 7⁄
 5C     C-CH
  \   ⁄   \
   C=C    CH3
   4  3    9
```

FORMULA	C9H12O		MOL WT	136.20
SOLVENT	CDCL3			
ORIG ST	C4H12SI		TEMP	AMB

152.50	134.70	126.60	121.00	126.40	115.40
1/1	2/1	3/2	4/2	5/2	6/2
26.80	22.50	22.50			
7/2	8/4	9/4			

L.F.JOHNSON,W.C.JANKOWSKI
CARBON-13 NMR SPECTRA,JOHN
WILEY AND SONS,NEW YORK   353 (1972)

1865   Q 001800   2-TERT-BUTYLPHENOL

```
      OH
   3 |    10
   C-C2   CH3
  //   \1 | 9
 4C      C-C-CH3
  \    / 7|
   C=C    CH3
   5 6    8
```

FORMULA	C10H14O			MOL WT	150.22
SOLVENT	CDCL3				
ORIG ST	C4H12SI			TEMP	AMB
136.10	154.00	116.60	126.90	120.50	126.90
1/1	2/1	3/2	4/2	5/2	6/2
34.40	29.50	29.50	29.50		
7/1	8/4	9/4	10/4		

L.F.JOHNSON,W.C.JANKOWSKI
CARBON-13 NMR SPECTRA,JOHN
WILEY AND SONS,NEW YORK   385 (1972)

1866   Q 001800   2-METHYL-6-TERT-BUTYLPHENOL

```
    7
  H3C    OH
    \2 1/   11
    C-C    CH3
   //  \   /10
  3C    C-C-CH3
   \   /6 8\
    C=C    CH3
    4 5    9
```

FORMULA	C11H16O			MOL WT	164.25
SOLVENT	CDCL3				
ORIG ST	C4H12SI			TEMP	AMB
152.60	122.90	128.40	119.90	124.90	135.60
1/1	2/1	3/2	4/2	5/2	6/1
15.60	34.40	29.70	29.70	29.70	
7/4	8/1	9/4	10/4	11/4	

L.F.JOHNSON,W.C.JANKOWSKI
CARBON-13 NMR SPECTRA,JOHN
WILEY AND SONS,NEW YORK   425 (1972)

1867   Q 001800   1,6-DIISOPROPYLPHENOL

```
    12CH3
      |
    10CH
   5 6/ \11
    C-C   CH3
   //  \1
  4C     C-OH
   \   / 9
  3C=C2  CH3
     \ /
     7CH
      |
     8CH3
```

FORMULA	C12H18O			MOL WT	178.28
SOLVENT	CDCL3				
ORIG ST	C4H12SI			TEMP	AMB
149.90	133.70	123.40	120.60	123.40	133.70
1/1	2/1	3/2	4/2	5/2	6/1
27.10	22.70	22.70	27.10	22.70	22.70
7/2	8/4	9/4	10/2	11/4	12/4

L.F.JOHNSON,W.C.JANKOWSKI
CARBON-13 NMR SPECTRA,JOHN
WILEY AND SONS,NEW YORK   438 (1972)

1868   Q 001800   PARA-(1-INDANYL)-PHENOL

```
   C8     5 6
   / \    C-C
  9/  \  // \\
 C    C-C    C-OH
  \  /7 4\  /1
  10C-C15  C=C
  /  \    3 2
 11C   C14
  \\  //
  12C-C13
```

FORMULA	C15H14O			MOL WT	210.28
SOLVENT	CDCL3				
ORIG ST	C4H12SI			TEMP	AMB
153.40	115.30	129.10	137.70	129.10	115.30
1/1	2/2	3/2	4/1	5/2	6/2
50.70	36.50	31.60	147.00	144.00	124.20
7/2	8/3	9/3	10/2	15/2	
124.70	126.20	126.30			

L.F.JOHNSON,W.C.JANKOWSKI
CARBON-13 NMR SPECTRA,JOHN
WILEY AND SONS,NEW YORK   464 (1972)

1869   Q 001900   CATECHOL

```
      OH
    2 1/
    C-C
   //  \\
  3C    C-OH
   \   /6
    C=C
    4 5
```

FORMULA	C6H6O2			MOL WT	110.11
SOLVENT	H2O				
ORIG ST	C4H12SI			TEMP	AMB
145.00	117.30	122.10	122.10	117.30	145.00
1/1	2/2	3/2	4/2	5/2	6/1

L.F.JOHNSON,W.C.JANKOWSKI
CARBON-13 NMR SPECTRA,JOHN
WILEY AND SONS,NEW YORK   161 (1972)

```
1870              Q 002000 2,4,5-TRICHLOROPHENOL
       CL
        \5 6              FORMULA   C6H3CL3O            MOL WT   197.45
         C=C             SOLVENT   C2D6OS
        /    \           ORIG ST   C4H12SI              TEMP      300
     CL-C    C-OH
       4\    /1           153.00   130.55   119.85   121.05   121.05   117.50
         C-C              1/1      2/1      3/2      4/1      5/1      6/2
         3 2\
            CL           W.VOELTER,S.FUCHS,R.H.SEUFFER,K.ZECH
                         MONATSH CHEM                   105, 1110 (1974)
```

```
1871              Q 002100 PARA-NITROPHENOL
                          FORMULA   C6H5NO3             MOL WT   139.11
        3 2             SOLVENT   C2D6OS
        C=C             ORIG ST   C4H12SI              TEMP      300
       4/   \1
    O2N-C      C-OH       164.10   115.90   126.35   139.85   126.35   115.90
       \\    //           1/1      2/2      3/2      4/1      5/2      6/2
         C-C
        5 6             W.VOELTER,S.FUCHS,R.H.SEUFFER,K.ZECH
                         MONATSH CHEM                   105, 1110 (1974)
```

```
1872              Q 002300 BENZYL ALCOHOL
        5 6             FORMULA   C7H8O               MOL WT   108.14
        C-C             SOLVENT   CDCL3
       //   \\          ORIG ST   C4H12SI              TEMP      AMB
      C      C-CH2-OH
      4\    /1 7          140.80   126.80   128.20   127.20   128.20   126.80
         C=C              1/1      2/2      3/2      4/2      5/2      6/2
         3 2              64.50
                          7/3

                         L.F.JOHNSON,W.C.JANKOWSKI
                         CARBON-13 NMR SPECTRA,JOHN
                         WILEY AND SONS,NEW YORK          246 (1972)
```

```
1873              Q 002300 ALPHA-METHYLBENZYL ALCOHOL
        5 6    8        FORMULA   C8H10O              MOL WT   122.17
        C-C   CH3       SOLVENT   CDCL3
       //   \\  |       ORIG ST   C4H12SI              TEMP      AMB
      C      C-CH-OH
      4\    /1 7          145.90   125.30   128.20   127.00   128.20   125.30
         C=C              1/1      2/2      3/2      4/2      5/2      6/2
         3 2              69.90    25.00
                          7/2      8/4

                         L.F.JOHNSON,W.C.JANKOWSKI
                         CARBON-13 NMR SPECTRA,JOHN
                         WILEY AND SONS,NEW YORK          300 (1972)
```

```
1874              Q 002300 ALPHA,ALPHA-DIMETHYLBENZYL ALCOHOL
                          FORMULA   C9H12O              MOL WT   136.20
        5 6             SOLVENT   CDCL3
        C-C   OH        ORIG ST   C4H12SI              TEMP      AMB
       //   \1 17
      4C      C-C-CH3     149.10   124.40   128.00   126.40   128.00   124.40
        \    /1 | 8       1/1      2/2      3/2      4/2      5/2      6/2
         C=C   CH3        72.20    31.50    31.50
         3 2   9          7/1      8/4      9/4

                         L.F.JOHNSON,W.C.JANKOWSKI
                         CARBON-13 NMR SPECTRA,JOHN
                         WILEY AND SONS,NEW YORK          354 (1972)
```

1875   Q 002400 PHENYLPROPANOLAMINE HYDROCHLORIDE

```
    3 2
    C-C   OH
   ⁄⁄  ＼1 |   8  9
 4C      C-CH-CH-CH3
   ＼  ⁄   7 |
    C=C      NH2
    5 6      •
         HCL
```

FORMULA   C9H13NO   MOL WT 151.21
SOLVENT   H2O
ORIG ST   C4H12SI   TEMP   AMB

139.40	127.10	129.70	129.40	129.70	127.10
1/1	2/2	3/2	4/2	5/2	6/2
73.70	53.30	13.30			
7/2	8/2	9/4			

L.F.JOHNSON,W.C.JANKOWSKI
CARBON-13 NMR SPECTRA,JOHN
WILEY AND SONS,NEW YORK   355 (1972)

1876   Q 002700 DIPHENYLETHER

```
   9 8      2 3
   C-C  O  C-C
  ⁄⁄  ＼ ⁄ ＼ ⁄⁄  ＼
10C    C7 1C    C4
  ＼  ⁄      ＼  ⁄
   C=C      C=C
  11 12      6 5
```

FORMULA   C12H10O   MOL WT 170.21
SOLVENT   C4H8O2
ORIG ST   C4H12SI   TEMP   AMB

157.60	119.00	129.80	123.20	129.80	119.00
1/1	2/2	3/2	4/2	5/2	6/2
157.60	119.00	129.80	123.20	129.80	119.00
7/1	8/2	9/2	10/2	11/2	12/2

L.F.JOHNSON,W.C.JANKOWSKI
CARBON-13 NMR SPECTRA,JOHN
WILEY AND SONS,NEW YORK   433 (1972)

1877   Q 002700 2,4-DINITROANISOLE

```
      7
   C6  O-CH3
  ⁄ ＼ ⁄
 5C    C1
 ‖    |
 4C    C2
  ⁄ ＼ ⁄⁄ ＼
O2N   C3  NO2
```

FORMULA   C7H6N2O5   MOL WT 198.14
SOLVENT   CDCL3
ORIG ST   C4H12SI   TEMP   AMB

157.00	139.30	122.20	140.70	129.80	114.70
1/1	2/1	3/2	4/1	5/2	6/2
58.80					
7/4					

G.W.BUCHANAN,G.MONTAUDO,P.FINOCCHIARO
CAN J CHEM   52, 767 (1974)

1878   Q 002700 2-NITROANISOLE

```
      7
   C6  O-CH3
  ⁄ ＼ ⁄
 5C    C1
 ‖    |
 4C    C2
  ＼ ⁄⁄
   C3  NO2
```

FORMULA   C7H7NO3   MOL WT 153.14
SOLVENT   CDCL3
ORIG ST   C4H12SI   TEMP   AMB

153.00	139.90	125.80	120.90	134.50	114.30
1/1	2/1	3/2	4/2	5/2	6/2
57.40					
7/4					

G.W.BUCHANAN,G.MONTAUDO,P.FINOCCHIARO
CAN J CHEM   52, 767 (1974)

1879   Q 002700 4-NITROANISOLE

```
      7
   C6  O-CH3
  ⁄ ＼ ⁄
 5C    C1
 ‖    |
 4C    C2
  ⁄ ＼ ⁄⁄
O2N   C3
```

FORMULA   C7H7NO3   MOL WT 153.14
SOLVENT   CDCL3
ORIG ST   C4H12SI   TEMP   AMB

163.80	113.60	125.10	141.00	125.10	113.60
1/1	2/2	3/2	4/1	5/2	6/2
55.90					
7/4					

G.W.BUCHANAN,G.MONTAUDO,P.FINOCCHIARO
CAN J CHEM   52, 767 (1974)

1880   Q 002700 ANISOLE

```
   5  6
   C-C
  //    \
 C      C-O-CH3
  4\   /1   7
    C=C
    3  2
```

FORMULA	C7H8O			MOL WT	108.14
SOLVENT	C4H8O2				
ORIG ST	C4H12SI			TEMP	AMB

160.20	114.10	129.50	120.70	129.50	114.10
1/1	2/2	3/2	4/2	5/2	6/2
54.70					
7/4					

L.F.JOHNSON,W.C.JANKOWSKI
CARBON-13 NMR SPECTRA,JOHN
WILEY AND SONS,NEW YORK   248 (1972)

1881   Q 002700 ANISOLE

```
         7
   C6  O-CH3
  / \ /
 5C    C1
 ||    |
 4C    C2
  \  //
   C3
```

FORMULA	C7H8O			MOL WT	108.14
SOLVENT	CDCL3				
ORIG ST	C4H12SI			TEMP	AMB

159.20	113.80	128.90	120.10	128.90	113.80
1/1	2/2	3/2	4/2	5/2	6/2
55.10					
7/4					

G.W.BUCHANAN,G.MONTAUDO,P.FINOCCHIARO
CAN J CHEM   52, 767 (1974)

1882   Q 002700 2-METHYLANISOLE

```
         7
   C6  O-CH3
  / \ /
 5C    C1
 ||    |
 4C    C2
  \  // \8
   C3   CH3
```

FORMULA	C8H10O			MOL WT	122.17
SOLVENT	CDCL3				
ORIG ST	C4H12SI			TEMP	AMB

157.60	130.50	126.50	120.30	126.50	110.20
1/1	2/1	3/2	4/2	5/2	6/2
55.30	16.00				
7/4	8/4				

G.W.BUCHANAN,G.MONTAUDO,P.FINOCCHIARO
CAN J CHEM   52, 767 (1974)

1883   Q 002700 2,6-DIMETHYLANISOLE

```
   9
  CH3
   |     7
  C6  O-CH3
  / \ /
 5C    C1
 ||    |
 4C    C2
  \  // \8
   C3   CH3
```

FORMULA	C9H12O			MOL WT	136.20
SOLVENT	CS2				
ORIG ST	C4H12SI			TEMP	AMB

156.20	129.30	127.60	122.60	127.60	129.30
1/1	2/1	3/2	4/2	5/2	6/1
57.90	15.00				
7/4	8/4				

G.W.BUCHANAN,G.MONTAUDO,P.FINOCCHIARO
CAN J CHEM   52, 767 (1974)

1884   Q 002700 2-ISOPROPYLANISOLE

```
         7
   C6  O-CH3
  / \ /
 5C    C1
 ||    |   9
 4C    C2  CH3
  \  // \8/
   C3   CH
        |
       CH3
       10
```

FORMULA	C10H14O			MOL WT	150.22
SOLVENT	CDCL3				
ORIG ST	C4H12SI			TEMP	AMB

156.70	137.00	126.30	120.50	125.90	110.60
1/1	2/1	3/2	4/2	5/2	6/2
55.60	27.30	23.00			
7/4	8/2	9/4			

G.W.BUCHANAN,G.MONTAUDO,P.FINOCCHIARO
CAN J CHEM   52, 767 (1974)

1885   Q 002700 2□,4'-DINITRODIPHENYL ETHER

```
         NO2
          /
      9C=C8      2C=C3
     10/   \7  1/   \
   □2N-C      C-□-C      □4
      \   //   \   //
       C-C       6C-C5
      11 12
```

FORMULA	C12H8N2O5			MOL WT	260.21
SOLVENT	CDCL3				
ORIG ST	C4H12SI			TEMP	AMB
152.90	119.60	130.10	125.90	155.30	138.90
1/1	2/2	3/2	4/2	7/1	8/1
121.00	140.70	128.10	118.00		
9/2	10/1	11/2	12/2		

G.W.BUCHANAN,G.MONTAUDO,P.FINOCCHIARO
CAN J CHEM   52, 767 (1974)

1886   Q 002700 2-NITRODIPHENYL ETHER

```
         NO2
      9 8/
      C=C      2C=C3
    10/   \7  1/   \
    C       C-□-C      C4
     \   //   \   //
      C-C       6C-C5
     11 12
```

FORMULA	C12H9NO3			MOL WT	215.21
SOLVENT	CDCL3				
ORIG ST	C4H12SI			TEMP	AMB
154.90	118.60	129.30	122.70	129.30	118.60
1/1	2/2	3/2	4/2	5/2	6/2
149.70	140.80	124.90	123.90	133.40	120.10
7/1	8/1	9/2	10/2	11/2	12/2

G.W.BUCHANAN,G.MONTAUDO,P.FINOCCHIARO
CAN J CHEM   52, 767 (1974)

1887   Q 002700 4-NITRODIPHENYL ETHER

```
      9C=C8      2C=C3
    10/   \7  1/   \
  □2N-C      C-O-C      □4
     \   //   \   //
      C-C       6C-C5
     11 12
```

FORMULA	C12H9NO3			MOL WT	215.21
SOLVENT	CDCL3				
ORIG ST	C4H12SI			TEMP	AMB
153.70	119.90	129.50	124.70	129.50	119.90
1/1	2/2	3/2	4/2	5/2	6/2
162.00	116.80	125.10	141.80	125.10	116.80
7/1	8/2	9/2	10/1	11/2	12/2

G.W.BUCHANAN,G.MONTAUDO,P.FINOCCHIARO
CAN J CHEM   52, 767 (1974)

1888   Q 002700 DIPHENYL ETHER

```
      9C=C8      2C=C3
    10/   \7  1/   \
    C       C-□-C      C4
     \   //   \   //
      C-C       6C-C5
     11 12
```

FORMULA	C12H10O			MOL WT	170.21
SOLVENT	CDCL3				
ORIG ST	C4H12SI			TEMP	AMB
156.10	117.30	128.20	121.60		
1/1	2/2	3/2	4/2		

G.W.BUCHANAN,G.MONTAUDO,P.FINOCCHIARO
CAN J CHEM   52, 767 (1974)

1889   Q 002700 2',4'-DINITRO-2-METHYLDIPHENYL ETHER

```
           13
       NO2 CH3
        /    \
      9C=C8      2C=C3
    10/   \7  1/   \
  □2N-C      C-□-C      □4
     \   //   \   //
      C-C       6C-C5
     11 12
```

FORMULA	C13H10N2O5			MOL WT	274.24
SOLVENT	CDCL3				
ORIG ST	C4H12SI			TEMP	AMB
150.50	129.70	127.50	126.30	131.70	120.10
1/1	2/1	3/2	4/2	5/2	6/2
154.90	138.20	121.30	140.40	128.30	116.80
7/1	8/1	9/2	10/1	11/2	12/2
15.80					
13/4					

G.W.BUCHANAN,G.MONTAUDO,P.FINOCCHIARO
CAN J CHEM   52, 767 (1974)

1890 — Q 002700 2-METHYL-4'-NITRODIPHENYL ETHER

```
              13
            H3C
               \
      9C=C8        2C=C3
    10/    \7   1/      \
  O2N-C       C-O-C        C4
       \      //    \      //
        C-C          6C-C5
       11 12
```

FORMULA	C13H11NO3			MOL WT	229.24
SOLVENT	CDCL3				
ORIG ST	C4H12SI			TEMP	AMB
151.70	130.10	127.30	124.50	131.50	120.70
1/1	2/1	3/2	4/2	5/2	6/2
162.40	115.80	125.50	141.80	125.50	115.80
7/1	8/2	9/2	10/1	11/2	12/2
15.80					
13/4					

G.W.BUCHANAN,G.MONTAUDO,P.FINOCCHIARO
CAN J CHEM   52, 767 (1974)

1891 — Q 002700 2-METHYLDIPHENYL ETHER

```
              13
            H3C
     9  8      \2
     C=C        C=C3
   10/   \7  1/    \
   C       C-O-C      C4
     \     //    \    //
      C-C          6C-C5
     11 12
```

FORMULA	C13H12O			MOL WT	184.24
SOLVENT	CDCL3				
ORIG ST	C4H12SI			TEMP	AMB
154.50	131.10	126.80	121.80	129.70	119.60
1/1	2/1	3/2	4/2	5/2	6/2
157.80	117.40	129.30	122.10	129.30	117.40
7/1	8/2	9/2	10/2	11/2	12/2
16.00					
13/4					

G.W.BUCHANAN,G.MONTAUDO,P.FINOCCHIARO
CAN J CHEM   52, 767 (1974)

1892 — Q 002700 2,6-DIISOPROPYLANISOLE

```
    H3C    CH3
    13\   /12
     11CH
       |      7
      C6   O-CH3
      / \ /
     5C   C1
     ||   |   9
     4C   C2  CH3
       \  // \8/
        C3  CH
         |
        10CH3
```

FORMULA	C13H20O			MOL WT	192.30
SOLVENT	CDCL3				
ORIG ST	C4H12SI			TEMP	AMB
154.70	141.40	123.90	123.30	123.90	141.40
1/1	2/1	3/2	4/2	5/2	6/1
61.60	26.80	23.80			
7/4	8/2	9/4			

G.W.BUCHANAN,G.MONTAUDO,P.FINOCCHIARO
CAN J CHEM   52, 767 (1974)

1893 — Q 002700 2,6-DIMETHYL-2'-NITRODIPHENYL ETHER

```
              13
         NO2 CH3
         /    \
      9C=C8       2C=C3
    10/   \7   1/     \
    C       C-O-C        C4
     \     //    \      //
      C-C          6C-C5
     11 12       /
              H3C
              14
```

FORMULA	C14H13NO3			MOL WT	243.26
SOLVENT	CDCL3				
ORIG ST	C4H12SI			TEMP	AMB
150.10	130.50	128.90	125.50	149.10	138.40
1/1	2/1	3/2	4/2	7/1	8/1
125.70	120.70	133.40	114.80	16.20	
9/2	10/2	11/2	12/2	13/4	

G.W.BUCHANAN,G.MONTAUDO,P.FINOCCHIARO
CAN J CHEM   52, 767 (1974)

1894 — Q 002700 2,6-DIMETHYL-4'-NITRODIPHENYL ETHER

```
              13
            H3C
               \
      9C=C8        2C=C3
    10/    \7   1/      \
  O2N-C       C-O-C        C4
       \      //    \      //
        C-C          6C-C5
       11 12       /
              H3C
              14
```

FORMULA	C14H13NO3			MOL WT	243.26
SOLVENT	CDCL3				
ORIG ST	C4H12SI			TEMP	AMB
150.30	130.30	128.90	126.00	162.60	114.80
1/1	2/1	3/2	4/2	7/1	8/2
126.00	142.80	16.00			
9/2	10/1	13/4			

G.W.BUCHANAN,G.MONTAUDO,P.FINOCCHIARO
CAN J CHEM   52, 767 (1974)

1895   Q 002700   2,2'-DIMETHYLDIPHENYL ETHER

```
      14      13
     H3C     CH3
      /       \
   9C=C8     2C=C3
  10/    \7    1/    \
   C      C-O-C       C4
    \\   //    \\    //
    C-C        6C-C5
   11 12
```

FORMULA	C14H14O			MOL WT	198.27
SOLVENT	CDCL3				
ORIG ST	C4H12SI			TEMP	AMB

154.60	130.70	126.30	122.40	128.10	117.10
1/1	2/1	3/2	4/2	5/2	6/2
15.50					
13/4					

G.W.BUCHANAN,G.MONTAUDO,P.FINOCCHIARO
CAN J CHEM   52, 767 (1974)

1896   Q 002700   2',4'-DINITRO-2-ISOPROPYLDIPHENYL ETHER

```
        14      15
       H3C     CH3
        \      /
        NO2   CH
        /   13\
     9C=C8     2C=C3
    10/    \7    1/    \
  O2N-C      C-O-C       C4
      \\   //    \\    //
      C-C        6C-C5
     11 12
```

FORMULA	C15H14N2O5			MOL WT	302.29
SOLVENT	CDCL3				
ORIG ST	C4H12SI			TEMP	AMB

150.50	140.60	126.70	126.40	127.30	111.90
1/1	2/1	3/2	4/2	5/2	6/2
155.90	139.00	121.60	141.00	128.10	117.20
7/1	8/1	9/2	10/1	11/2	12/2
27.90	22.90				
13/2	14/4				

G.W.BUCHANAN,G.MONTAUDO,P.FINOCCHIARO
CAN J CHEM   52, 767 (1974)

1897   Q 002700   2-ISOPROPYL-4'-NITRODIPHENYL ETHER

```
           15
          CH3
       14  |
      H3C-CH
         13\
     9C=C8     2C=C3
    10/    \7    1/    \
  O2N-C      C-O-C       C4
      \\   //    \\    //
      C-C        6C-C5
     11 12
```

FORMULA	C15H15NO3			MOL WT	257.29
SOLVENT	CDCL3				
ORIG ST	C4H12SI			TEMP	AMB

151.30	140.30	126.90	124.80	127.30	109.80
1/1	2/1	3/2	4/2	5/2	6/2
163.20	116.20	125.50	142.40	125.50	116.20
7/1	8/2	9/2	10/1	11/2	12/2
27.50	23.10				
13/2	14/4				

G.W.BUCHANAN,G.MONTAUDO,P.FINOCCHIARO
CAN J CHEM   52, 767 (1974)

1898   Q 002700   2,6-DIISOPROPYL-2',4'DINITRODIPHENYL ETHER

```
       14   15
      H3C    CH3
        \   /
        NO2 CH
        /  13\
     9C=C8     2C=C3
    10/    \7    1/    \
  O2N-C      C-O-C       C4
      \\   //    \\    //
      C-C        6C-C5
     11 12   16/
          H3C-CH
          17 |
          18CH3
```

FORMULA	C18H20N2O5			MOL WT	344.37
SOLVENT	CDCL3				
ORIG ST	C4H12SI			TEMP	AMB

147.00	140.80	125.70	128.20	155.80	138.40
1/1	2/1	3/2	4/2	7/1	8/1
122.50	141.40	130.20	116.50	27.50	22.40
9/2	10/1	11/2	12/2	13/2	14/4
24.10					
15/4					

G.W.BUCHANAN,G.MONTAUDO,P.FINOCCHIARO
CAN J CHEM   52, 767 (1974)

1899   Q 002700   2,6-DIISOPROPYL-4'-NITRODIPHENYL ETHER

```
         15CH3
       14  |
      H3C-CH
         13\
     9C=C8     2C=C3
    10/    \7    1/    \
  O2N-C      C-O-C       C4
      \\   //    \\    //
      C-C        6C-C5
     11 12   16/
          H3C-CH
          17 |
          18CH3
```

FORMULA	C18H21NO3			MOL WT	299.37
SOLVENT	CDCL3				
ORIG ST	C4H12SI			TEMP	AMB

148.10	141.80	125.30	127.10	164.40	115.60
1/1	2/1	3/2	4/2	7/1	8/2
126.30	142.80	28.10	23.90		
9/2	10/1	13/2	14/4		

G.W.BUCHANAN,G.MONTAUDO,P.FINOCCHIARO
CAN J CHEM   52, 767 (1974)

1900   Q 002900 PARA-FLUOROANISOLE

```
        3  2
        C=C
      4/    \1   7
   F-C        C-O-CH3
      \      //
        C-C
       5  6
```

FORMULA	C7H7FO			MOL WT	126.13
SOLVENT	CDCL3				
ORIG ST	C4H12SI			TEMP	AMB

156.00	114.90	115.80	157.40	115.80	114.90
1/1	2/2	3/2	4/1	5/2	6/2
55.50					
7/4					

L.F.JOHNSON,W.C.JANKOWSKI
CARBON-13 NMR SPECTRA,JOHN
WILEY AND SONS,NEW YORK   237 (1972)

1901   Q 002900 PARA-NITROANISOLE

```
        3  2
        C=C
      4/    \1   7
  O2N-C        C-O-CH3
      \      //
        C-C
       5  6
```

FORMULA	C7H7NO3			MOL WT	153.14
SOLVENT	CDCL3				
ORIG ST	C4H12SI			TEMP	AMB

164.50	125.70	114.00	164.70	114.00	125.70
1/1	2/2	3/2	4/1	5/2	6/2
55.90					
7/4					

L.F.JOHNSON,W.C.JANKOWSKI
CARBON-13 NMR SPECTRA,JOHN
WILEY AND SONS,NEW YORK   241 (1972)

1902   Q 002900 PARA-IODOANISOLE

```
        3  2
        C=C
      4/    \1   7
   J-C        C-O-CH3
      \      //
        C-C
       5  6
```

FORMULA	C7H7IO			MOL WT	234.04
SOLVENT	CDCL3				
ORIG ST	C4H12SI			TEMP	AMB

159.30	116.20	138.00	82.60	138.00	116.20
1/1	2/2	3/2	4/1	5/2	6/2
55.10					
7/4					

L.F.JOHNSON,W.C.JANKOWSKI
CARBON-13 NMR SPECTRA,JOHN
WILEY AND SONS,NEW YORK   239 (1972)

1903   Q 002900 2-CHLORO-4-METHOXYANISOLE

```
          CL
          |
       3C=C2
      /     \
 H3C-O-C        C-O-CH3
   8   4\    //1   7
        C-C
       5  6
```

FORMULA	C8H9CLO2			MOL WT	172.61
SOLVENT	CDCL3				
ORIG ST	C4H12SI			TEMP	AMB

149.40	123.00	153.90	56.60	55.70	112.70
1/1	2/1	4/1	7/4	8/4	
113.30	116.20				

L.F.JOHNSON,W.C.JANKOWSKI
CARBON-13 NMR STECTRA,JOHN
WILEY AND SONS,NEW YORK   294 (1972)

1904   Q 002900 ANISE ALCOHOL

```
        3  2
        C=C
    8  4/    \1   7
 HO-CH2-C        C-O-CH3
      \      //
        C-C
       5  6
```

FORMULA	C8H10O2			MOL WT	138.17
SOLVENT	CDCL3				
ORIG ST	C4H12SI			TEMP	AMB

158.80	113.70	128.40	133.20	128.40	113.70
1/1	2/2	3/2	4/1	5/2	6/2
55.00	64.20				
7/4	8/3				

L.F.JOHNSON,W.C.JANKOWSKI
CARBON-13 NMR SPECTRA,JOHN
WILEY AND SONS,NEW YORK   301 (1972)

1905   Q 002900   METHOXYCHLOR

```
              CL
             | 16
      10  9CL-C-CL   3  2
      C-C    |      C-C
  14  11/   \ |  4/    \1
 H3C-O-C      C-CH-C      C-O
      \    /8 7  \   /   |
      12C=C13        C=C   CH3
         5  6        15
```

FORMULA C16H15CL3O2				MOL WT	345.66
SOLVENT CDCL3					
ORIG ST C4H12SI			TEMP		AMB
159.00	113.50	131.00	130.50	131.00	113.50
1/1	2/2	3/2	4/1	5/2	6/2
69.70	130.50	131.00	113.50	159.00	113.50
7/2	8/1	9/2	10/2	11/1	12/2
131.00	55.00	55.00	102.50		
13/2	14/4	15/4	16/1		

L.F.JOHNSON,W.C.JANKOWSKI
CARBON-13 NMR SPECTRA,JOHN
WILEY AND SONS,NEW YORK   470 (1972)

1906   Q 003200   BENZALDEHYDE

```
        3  2
        C=C        O
       /   \1  7/
     4C      C-C
       \\    //   \
        C-C        H
        5  6
```

FORMULA C7H6O				MOL WT	106.13
SOLVENT CDCL3					
ORIG ST C4H12SI			TEMP		AMB
136.40	129.50	128.90	134.20	128.90	129.50
1/1	2/2	3/2	4/2	5/2	6/2
192.00					
7/1					

L.F.JOHNSON,W.C.JANKOWSKI
CARBON-13 NMR SPECTRA,JOHN
WILEY AND SONS,NEW YORK   229 (1972)

1907   Q 003200   SALICYLALDEHYDE

```
         OH
        3 |
        C=C2       O
       /   \1  7/
     4C      C-C
       \\    //   \
        C-C        H
        5  6
```

FORMULA C7H6O2				MOL WT	122.12
SOLVENT C4H8O2					
ORIG ST C4H12SI			TEMP		AMB
121.00	161.40	117.40	136.60	119.60	133.60
1/1	2/1	3/2	4/2	5/2	6/2
196.70					
7/1					

L.F.JOHNSON,W.C.JANKOWSKI
CARBON-13 NMR SPECTRA,JOHN
WILEY AND SONS,NEW YORK   231 (1972)

1908   Q 003200   PHENYLACETALDEHYDE

```
    5  4
    C-C           O
  6/   \3  2   1/
  C      C-CH2-C
   \7  8/        \
    C=C           H
```

FORMULA C8H8O				MOL WT	120.15
SOLVENT CDCL3					
ORIG ST C4H12SI			TEMP		308
199.30	50.60	132.00	129.70	129.10	127.50
1/2	2/3	3/1	4/2	5/2	6/2
129.10	129.70				
7/2	8/2				

G.E.HAWKES,K.HERWIG,J.D.ROBERTS
J ORG CHEM   39, 1017 (1974)

1909   Q 003200   VANILLIN

```
    H3C-O
    8 |
      2C=C3    O
    /    \  ||
  HO-C      C-CH
    1\   /4  7
      C-C
      6  5
```

FORMULA C8H8O3				MOL WT	152.15
SOLVENT CDCL3					
ORIG ST C4H12SI			TEMP		AMB
152.30	147.50	109.40	129.50	127.40	114.80
1/1	2/1	3/2	4/1	5/2	6/2
191.30	56.00				
7/2	8/4				

L.F.JOHNSON,W.C.JANKOWSKI
CARBON-13 NMR SPECTRA,JOHN
WILEY AND SONS,NEW YORK   292 (1972)

1910   Q 003200   TRANS-CINNAMALDEHYDE

```
         H
          \
      ]=C      H
     9\   / 6 5
      C=C   C-C
     /8 7\  /   \
    H   1C     C4
         \    /
         2C=C3
```

FORMULA   C9H8O   MOL WT   132.16
SOLVENT   CDCL3
ORIG ST   C4H12SI   TEMP   AMB

133.90	128.40	128.90	128.40	128.90	128.40
1/1	2/2	3/2	4/2	5/2	6/2
152.30	131.00	193.20			
7/2	8/2	9/2			

L.F.JOHNSON,W.C.JANKOWSKI
CARBON-13 NMR SPECTRA,JOHN
WILEY AND SONS,NEW YORK   336 (1972)

1911   Q 003200   HYDROCINNAMALDEHYDE

```
     3 2
      C-C             ]
     /   \1          /
   4C      C-CH2-CH2-C
     \    /  7   8   9\
      C=C               H
      5 6
```

FORMULA   C9H10O   MOL WT   134.18
SOLVENT   CDCL3
ORIG ST   C4H12SI   TEMP   AMB

140.40	128.20	128.40	126.10	128.40	128.20
1/1	2/2	3/2	4/2	5/2	6/2
28.00	45.00	201.10			
7/3	8/3	9/2			

L.F.JOHNSON,W.C.JANKOWSKI
CARBON-13 NMR SPECTRA,JOHN
WILEY AND SONS,NEW YORK   343 (1972)

1912   Q 003200   3,5-DIMETHOXYBENZALDEHYDE

```
           8
        ]-CH3
         |
        C-C    ]
       /3 2\   ||
     4C      C-CH
       \    /1 7
       5C=C6
        | 9
        ]-CH3
```

FORMULA   C9H10O3   MOL WT   166.18
SOLVENT   CDCL3
ORIG ST   C4H12SI   TEMP   AMB

138.40	107.00	161.20	107.00	161.20	107.00
1/1	2/2	3/1	4/2	5/1	6/2
191.60	55.40	55.40			
7/2	8/4	9/4			

L.F.JOHNSON,W.C.JANKOWSKI
CARBON-13 NMR SPECTRA,JOHN
WILEY AND SONS,NEW YORK   346 (1972)

1913   Q 003200   PARA-DIMETHYLAMINOBENZALDEHYDE

```
   9        3 2
  H3C      C-C      O
     \    /4  \1 7/
      N-C      C-C
     /    \    /   \
   H3C      C=C      H
   8        5 6
```

FORMULA   C9H11NO   MOL WT   149.19
SOLVENT   CDCL3
ORIG ST   C4H12SI   TEMP   AMB

154.10	110.80	131.60	124.90	131.60	110.80
1/1	2/2	3/2	4/1	5/2	6/2
189.70	39.70	39.70			
7/2	8/4	9/4			

L.F.JOHNSON,W.C.JANKOWSKI
CARBON-13 NMR SPECTRA,JOHN
WILEY AND SONS,NEW YORK   349 (1972)

1914   Q 003400   ANTI-PHENYLACETALDOXIME

```
              ]H
    5 4        /
     C-C      N
    6/  \3 2 /
   C      C-CH2-CH
    \7 8/        1
     C=C
```

FORMULA   C8H9NO   MOL WT   135.17
SOLVENT   CDCL3
ORIG ST   C4H12SI   TEMP   308

150.90	36.00	136.90	128.90	128.90	126.90
1/2	2/3	3/1	4/2	5/2	6/2
128.90	128.90				
7/2	8/2				

G.E.HAWKES,K.HERWIG,J.D.ROBERTS
J ORG CHEM   39, 1017 (1974)

1915   Q 003400 SYN-PHENYLACETALDOXIME

```
        HO
  5 4      \
  C-C       N
 6//  \3 2  ||
 C      C-CH2-CH
  \7 8/       1
   C=C
```

FORMULA	C8H9NO			MOL WT	135.17
SOLVENT	CDCL3				
ORIG ST	C4H12SI			TEMP	308

150.90	31.90	136.90	128.90	128.90	126.90
1/2	2/3	3/1	4/2	5/2	6/2
128.90	128.90				
7/2	8/2				

G.E.HAWKES,K.HERWIG,J.D.ROBERTS
J ORG CHEM   39, 1017 (1974)

1916   Q 003500 ACETOPHENONE

```
   5 6
   C-C    O
  //  \\  ||
 C      C-C-CH3
 4\  /1 7 8
   C=C
   3 2
```

FORMULA	C8H8O			MOL WT	120.15
SOLVENT	CDCL3				
ORIG ST	C4H12SI			TEMP	AMB

137.10	128.20	128.40	132.90	128.40	128.20
1/1	2/2	3/2	4/2	5/2	6/2
197.60	26.30				
7/1	8/4				

L.F.JOHNSON,W.C.JANKOWSKI
CARBON-13 NMR SPECTRA,JOHN
WILEY AND SONS,NEW YORK   288 (1972)

1917   Q 003500 ORTHO-HYDROXYACETOPHENONE

```
    3C=C4
   /    \
  2C     C5
   \\    //
   1C-C6
   /    \
 HO      C-CH3
         //7 8
          O
```

FORMULA	C8H8O2			MOL WT	136.15
SOLVENT	CDCL3				
ORIG ST	C4H12SI			TEMP	AMB

162.30	118.80	136.30	118.20	130.70	119.70
1/1	2/2	3/2	4/2	5/2	6/1
204.40	26.30				
7/1	8/4				

L.F.JOHNSON,W.C.JANKOWSKI
CARBON-13 NMR STECTRA,JOHN
WILEY AND SONS,NEW YORK   290 (1972)

1918   Q 003500 3-FLUORO-4-METHOXYACETOPHENONE

```
     6 5
     C-C    O
    //  \\  ||
 H3C-O-C    C-C-CH3
  7  1\  /4 8 9
     2C=C3
       |
       F
```

FORMULA	C9H9FO2			MOL WT	168.17
SOLVENT	CDCL3				
ORIG ST	C4H12SI			TEMP	AMB

151.90	152.00	115.60	130.60	112.40	125.80
1/1	2/1	3/2	4/1	5/2	6/2
56.20	195.50	26.10			
7/4	8/1	9/4			

L.F.JOHNSON,W.C.JANKOWSKI
CARBON-13 NMR SPECTRA,JOHN
WILEY AND SONS,NEW YORK   340 (1972)

1919   Q 003500 2-PROPIONYLBENZENE

```
         5 6
   O      C-C
 1 || 3  4//   \7
 H3C-C-CH2-C    C
   2      \9 8/
           C=C
```

FORMULA	C9H10O			MOL WT	134.18
SOLVENT	CDCL3				
ORIG ST	C4H12SI			TEMP	308

29.00	206.00	50.70	134.30	129.30	128.60
1/4	2/1	3/3	4/1	5/2	6/2
126.90	128.60	129.30			
7/2	8/2	9/2			

G.E.HAWKES,K.HERWIG,J.D.ROBERTS
J ORG CHEM   39, 1017 (1974)

1920   Q 003500   BUTYROPHENONE

```
          O
      2   ||
      C   C7
    ⁄ \1⁄ \
  3C    C   CH2-CH2-CH3
   |    ||   8   9   10
  4C   C6
    \ ⁄
     C
     5
```

FORMULA	C10H12O			MOL WT	148.21
SOLVENT	CDCL3				
ORIG ST	C4H12SI			TEMP	AMB

137.10	127.90	128.40	132.70	128.40	127.90
1/1	2/2	3/2	4/2	5/2	6/2
199.80	40.40	17.70	13.80		
7/1	8/3	9/3	10/4		

L.F.JOHNSON,W.C.JANKOWSKI
CARBON-13 NMR SPECTRA,JOHN
WILEY AND SONS,NEW YORK   373 (1972)

1921   Q 003500   BENZIL

```
           7  6
      O    C-C
  11 10 ||  ⁄    \
  C-C   C8  C     C5
 ⁄   \9⁄ \ ⁄2\   ⁄
12C    C  1C   C=C
   \    ||   3 4
    C=C   O
   13 14
```

FORMULA	C14H10O2			MOL WT	210.23
SOLVENT	CDCL3				
ORIG ST	C4H12SI			TEMP	AMB

194.30	133.00	129.70	128.90	134.70	128.90
1/1	2/1	3/2	4/2	5/2	6/2
129.70	194.30	133.00	129.70	128.90	134.70
7/2	8/1	9/1	10/2	11/2	12/2
128.90	129.70				
13/2	14/2				

L.F.JOHNSON,W.C.JANKOWSKI
CARBON-13 NMR SPECTRA,JOHN
WILEY AND SONS,NEW YORK   457 (1972)

1922   Q 003500   DIBENZOYLMETHANE

```
       O    O
  12 11 ||  ||    2 3
  C-C   C9 7C    C-C
 ⁄   \ ⁄ \ ⁄ \  ⁄    \
C     C   CH2 C     C4
13\  ⁄10  8   1\   ⁄
  C=C            C=C
 14 15          6 5
```

FORMULA	C15H12O2			MOL WT	224.26
SOLVENT	CDCL3				
ORIG ST	C4H12SI			TEMP	AMB

135.40	127.00	128.50	132.30	128.50	127.00
1/1	2/2	3/2	4/2	5/2	6/2
185.50	93.00	185.50	135.40	127.00	128.50
7/1	8/3	9/1	10/1	11/2	12/2
132.30	128.50	127.00			
13/2	14/2	15/2			

L.F.JOHNSON,W.C.JANKOWSKI
CARBON-13 NMR SPECTRA,JOHN
WILEY AND SONS,NEW YORK   463 (1972)

1923   Q 003700   ANTI-2-PROPIONYLBENZENE OXIME

```
  HO
    \        5  6
     N      C-C
  1 || 3  4⁄    \7
 H3C-C-CH2-C     C
   2        \9 8⁄
          C=C
```

FORMULA	C9H11NO			MOL WT	149.19
SOLVENT	CDCL3				
ORIG ST	C4H12SI			TEMP	308

13.40	157.60	42.00	136.90	129.10	128.60
1/4	2/1	3/3	4/1	5/2	6/2
126.40	128.60	129.10			
7/2	8/2	9/2			

G.E.HAWKES,K.HERWIG,J.D.ROBERTS
J ORG CHEM   39, 1017 (1974)

1924   Q 003700   SYN-2-PROPIONYLBENZENE OXIME

```
     OH
    ⁄        5  6
   N        C-C
  1 || 3  4⁄    \7
 H3C-C-CH2-C     C
   2        \9 8⁄
          C=C
```

FORMULA	C9H11NO			MOL WT	149.19
SOLVENT	CDCL3				
ORIG ST	C4H12SI			TEMP	308

19.50	156.90	34.90	136.70	129.10	128.60
1/4	2/1	3/3	4/1	5/2	6/2
126.40	128.60	129.10			
7/2	8/2	9/2			

G.E.HAWKES,K.HERWIG,J.D.ROBERTS
J ORG CHEM   39, 1017 (1974)

1925   Q 003900 PARA-QUINONE

```
        3  2
        C=C
     4/    \1
   O=C     C=O
     \    /
        C=C
       5  6
```

FORMULA   C6H4O2   MOL WT 108.10
SOLVENT   CDCL3
ORIG ST   C4H12SI   TEMP   AMB

187.10   136.40   136.40   187.10   136.40   136.40
1/1   2/2   3/2   4/1   5/2   6/2

L.F.JOHNSON,W.C.JANKOWSKI
CARBON-13 NMR SPECTRA,JOHN
WILEY AND SONS,NEW YORK   151 (1972)

1926   Q 003900 1,4-NAPHTHOQUINONE

```
      6C-C5
     //    \
   7C       C4
     \     /
      8C=C3
     /     \
   O=C     C=O
    9\    /2
    10C=C1
```

FORMULA   C10H6O2   MOL WT 158.16
SOLVENT   CDCL3
ORIG ST   C4H12SI   TEMP   AMB

138.50   184.70   131.80   126.20   133.70   133.70
1/2   2/1   3/1   4/2   5/2   6/2
126.20   131.80   184.70   138.50
7/2   8/1   9/1   10/2

L.F.JOHNSON,W.C.JANKOWSKI
CARBON-13 NMR SPECTRA,JOHN
WILEY AND SONS,NEW YORK   366 (1972)

1927   Q 003900 2,5-DIAZIRIDINO-PARA-BENZOQUINONE

```
        O     C7
        ||   /|
        C1  N |
       / \ / \|
      6C   C2  C8
      ||   ||
    C  5C   C3
    |\ / \ /
    | N   C4
    |/    ||
    C     O
```

FORMULA   C10H10N2O2   MOL WT 190.20
SOLVENT   CDCL3
ORIG ST   C4H12SI   TEMP   298

183.10   156.70   115.30   27.50
1/1   2/1   3/2   7/3

R.RADEGLIA,S.DAEHNE
Z CHEM   13, 474 (1973)

1928   Q 003900 4,5-DI-N,N-DIMETHYLAMINO-ORTHO-BENZOQUINONE

```
    CH3
     |    6
     N   C   O
    / \ / \ /
  H3C  5C   C1
     |    |
  H3C  4C   C2
    8\ / \ / \
     N   C   O
     |   3
    7CH3
```

FORMULA   C10H14N2O2   MOL WT 194.24
SOLVENT   CDCL3
ORIG ST   C4H12SI   TEMP   AMB

178.40   103.80   157.10   41.70
1/1   3/2   4/1   7/4

R.RADEGLIA,S.DAEHNE
Z CHEM   13, 474 (1973)

1929   Q 003900 2,5-DI-N,N-DIMETHYLAMINO-PARA-BENZOQUINONE

```
    O   7CH3
    ||   |
    C1   N
   / \ / 
  6C   C2  CH3
  ||   ||  8
 H3C  5C   C3
   \ / \ /
    N   C4
    |   ||
   H3C  O
```

FORMULA   C10H14N2O2   MOL WT 194.24
SOLVENT   CDCL3
ORIG ST   C4H12SI   TEMP   298

181.30   152.00   101.80   42.70
1/1   2/1   3/2   7/4

R.RADEGLIA,S.DAEHNE
Z CHEM   13, 474 (1973)

1930   Q 003900   3,6-DIMETHOXYTHYMOQUINONE

```
       8        7
     H3C      O-CH3
        \3 2/
         C=C
        4/   \1
     O=C       C=O
        \   /  11
        5C=C6  CH3
       9 /   \ /
     H3C-O   10CH
              \12
              CH3
```

FORMULA	C12H16O4			MOL WT	224.26
SOLVENT	CDCL3				
ORIG ST	C4H12SI			TEMP	AMB

183.80	155.70	126.00	184.50	155.50	135.30
1/1	2/1	3/1	4/1	5/1	6/1
60.80	8.20	61.00	24.60	20.50	20.50
7/4	8/4	9/4	10/2	11/4	12/4

L.F.JOHNSON,W.C.JANKOWSKI
CARBON-13 NMR SPECTRA,JOHN
WILEY AND SONS,NEW YORK   436 (1972)

1931   Q 003900   2,5-DI-N-PYRROLIDINO-PARA-BENZOQUINONE

```
      O   7C---C 8
      ||   |    |
      C1   N    C9
     / \  / \  /
    6C   C2   C10
     ||   ||
     C   5C    C3
    / \ / \ /
   C    N    C4
   |    |    ||
   C---C     O
```

FORMULA	C14H18N2O2			MOL WT	246.31
SOLVENT	CDCL3				
ORIG ST	C4H12SI			TEMP	298

180.30	149.60	99.80	51.20	50.70	26.70
1/1	2/1	3/2	7/3	7/3	8/3
23.80					
8/3					

R.RADEGLIA,S.DAEHNE
Z CHEM   13, 474 (1973)

1932   Q 003900   2,5-DI-N-PIPERIDINO-PARA-BENZOQUINONE

```
              C8
             / \
      O   7C    C9
      ||   |    |
      C1   N    C10
     / \  / \  /
    6C    C2   C11
     ||    ||
     C   5C    C3
    / \ / \ /
   C    N    C4
   |    |    ||
   C    C    O
    \  /
     C
```

FORMULA	C16H22N2O2			MOL WT	274.37
SOLVENT	CDCL3				
ORIG ST	C4H12SI			TEMP	298

182.50	153.20	105.50	50.40	26.00	24.50
1/1	2/1	3/2	7/3	8/3	9/3

R.RADEGLIA,S.DAEHNE
Z CHEM   13, 474 (1973)

1933   Q 003900   2,5-DI-N-METHYLANILIDO-PARA-BENZOQUINONE

```
      O   7CH3
      ||   |    9
    C     C1   N    C
   / \   / \  / \  / \
  C   C  6C   C2  8C   C10
  ||  |  ||   ||  |    ||
  C   C  5C   C3  C13 C11
   \ / \ / \  / \     / \ /
    C    N    C4      C
    |    ||        12
   H3C   O
```

FORMULA	C20H18N2O2			MOL WT	318.38
SOLVENT	CDCL3				
ORIG ST	C4H12SI			TEMP	298

181.70	151.40	107.10	43.20	148.00	125.40
1/1	2/1	3/2	7/4	8/1	9/2
129.50	126.30				
10/2	11/2				

R.RADEGLIA,S.DAEHNE
Z CHEM   13, 474 (1973)

1934   Q 004200   BENZOIC ACID

```
     5  6
     C-C   O
    /   \  ||
   C     C-C-OH
   4\   /1 7
     C=C
     3  2
```

FORMULA	C7H6O2			MOL WT	122.12
SOLVENT	CDCL3				
ORIG ST	C4H12SI			TEMP	AMB

129.40	130.20	128.40	133.70	128.40	130.20
1/1	2/2	3/2	4/2	5/2	6/2
172.60					
7/1					

L.F.JOHNSON,W.C.JANKOWSKI
CARBON-13 NMR SPECTRA,JOHN
WILEY AND SONS,NEW YORK   230 (1972)

1935   Q 004200   2,4-DICHLOROPHENOXYACETIC ACID SODIUM SALT

```
      7 8
      C=C              O
    6/   \3   2   1/
 CL-C      C-O-CH2-C
    \5 4/            \ -  +
     C-C             O NA
      \
       CL
```

FORMULA	C8H5CL2NAO3			MOL WT	243.02
SOLVENT	C2D6OS/CCL4				
ORIG ST	C4H12SI			TEMP	AMB
172.50	69.80	154.50	123.00	130.00	125.00
1/1	2/3	3/1	4/1	5/2	6/1
128.40	118.10				
7/2	8/2				

P.JUNGHANS,H.SPRINZ
Z CHEM   15, 153 (1975)

1936   Q 004200   2,4,5-TRICHLOROPHENOXYACETIC ACID

```
      CL
      \7 8
      C=C              O
    6/   \3   2   1/
 CL-C      C-O-CH2-C
    \5 4/            \
     C-C             OH
      \
       CL
```

FORMULA	C8H5CL3O3			MOL WT	255.49
SOLVENT	C2D6OS/CCL4				
ORIG ST	C4H12SI			TEMP	AMB
169.80	67.00	153.60	122.30	131.40	124.10
1/1	2/3	3/1	4/1	5/2	6/1
131.40	116.70				
7/1	8/2				

P.JUNGHANS,H.SPRINZ
Z CHEM   15, 153 (1975)

1937   Q 004200   2,4,6-TRICHLOROPHENOXYACETIC ACID

```
         CL
      7 8/
      C86C             O
    6/   \3   2   1/
 CL-C      C-O-CH2-C
    \5 4/            \
     C-C             OH
      \
       CL
```

FORMULA	C8H5CL3O3			MOL WT	255.49
SOLVENT	C2D6OS/CCL4				
ORIG ST	C4H12SI			TEMP	AMB
169.80	70.50	150.30	129.90	130.30	129.90
1/1	2/3	3/1	4/1	5/2	6/1
130.30	129.90				
7/2	8/1				

P.JUNGHANS,H.SPRINZ
Z CHEM   15, 153 (1975)

1938   Q 004200   2-CHLOROPHENOXYACETIC ACID SODIUM SALT

```
      7 8
      C=C              O
    6/   \3   2   1/
   C      C-O-CH2-C
    \5 4/            \ -  +
     C-C             O NA
      \
       CL
```

FORMULA	C8H6CLNAO3			MOL WT	208.58
SOLVENT	C2D6OS/CCL4				
ORIG ST	C4H12SI			TEMP	AMB
172.60	69.50	155.20	122.20	130.40	121.50
1/1	2/3	3/1	4/1	5/2	6/2
128.60	115.20				
7/2	8/2				

P.JUNGHANS,H.SPRINZ
Z CHEM   15, 153 (1975)

1939   Q 004200   4-CHLOROPHENOXYACETIC ACID SODIUM SALT

```
      7 8
      C=C              O
    6/   \3   2   1/
 CL-C      C-O-CH2-C
    \5 4/            \ -  +
     C-C             O VA
```

FORMULA	C8H6CLNAO3			MOL WT	208.58
SOLVENT	C2D6OS/CCL4				
ORIG ST	C4H12SI			TEMP	AMB
173.40	69.00	155.20	117.30	129.80	124.50
1/1	2/3	3/1	4/2	5/2	6/1
129.80	117.30				
7/2	8/2				

P.JUNGHANS,H.SPRINZ
Z CHEM   15, 153 (1975)

1940   Q 004200 2-BROMOPHENOXYACETIC ACID

```
  7 8
  C=C              ]
6/   \3   2   1/
C         C-]-CH2-C
 \5 4/            \
  C-C              ]H
     \
      B]
```

FORMULA	C8H7BRO3			MOL WT	231.05
SOLVENT	C2D6OS/CCL4				
ORIG ST	C4H12SI			TEMP	AMB

170.40	66.90	155.30	112.40	134.00	123.50
1/1	2/3	3/1	4/1	5/2	6/2
128.80	114.90				
7/2	8/2				

P.JUNGHANS,H.SPRINZ
Z CHEM   15, 153 (1975)

1941   Q 004200 3-BROMOPHENOXYACETIC ACID

```
  7 8
  C=C              ]
6/   \3   2   1/
C         C-]-CH2-C
 \5 4/            \
  C-C              ]H
 /
BR
```

FORMULA	C8H7BRO3			MOL WT	231.05
SOLVENT	C2D6OS/CCL4				
ORIG ST	C4H12SI			TEMP	AMB

170.40	66.40	159.80	119.10	123.10	125.10
1/1	2/3	3/1	4/2	5/1	6/2
131.90	115.00				
7/2	8/2				

P.JUNGHANS,H.SPRINZ
Z CHEM   15, 153 (1975)

1942   Q 004200 4-BROMOPHENOXYACETIC ACID

```
   7 8
   C=C                ]
  6/   \3   2   1/
BR-C         C-]-CH2-C
  \5 4/              \
   C-C                ]H
```

FORMULA	C8H7BRO3			MOL WT	231.05
SOLVENT	C2D6OS/CCL4				
ORIG ST	C4H12SI			TEMP	AMB

170.70	66.40	158.30	118.10	133.30	113.70
1/1	2/3	3/1	4/2	5/2	6/1
133.30	118.10				
7/2	8/2				

P.JUNGHANS,H.SPRINZ
Z CHEM   15, 153 (1975)

1943   Q 004200 PARA-TOLUIC ACID

```
      5 6
      C-C      ]
  8 4/   \1 7/
H3C-C         C-C
      \   /     \
       C=C       ]H
       3 2
```

FORMULA	C8H8O2			MOL WT	136.15
SOLVENT	C2D6OS				
ORIG ST	C4H12SI			TEMP	300

128.50	129.70	129.40	143.30	167.80	21.40
1/1	2/2	3/2	4/1	7/1	8/4

U.HOLLSTEIN,E.BREITMAIER,G.JUNG
J AM CHEM SOC   96, 8036 (1974)

1944   Q 004200 PHENOXYACETIC ACID

```
  7 8
  C=C              ]
6/   \3   2   1/
C         C-]-CH2-C
 \5 4/            \
  C-C              ]H
```

FORMULA	C8H8O3			MOL WT	152.15
SOLVENT	C2D6OS/CCL4				
ORIG ST	C4H12SI			TEMP	AMB

173.10	68.50	155.80	115.30	129.70	120.50
1/1	2/3	3/1	4/2	5/2	6/2
129.70	115.30				
7/2	8/2				

P.JUNGHANS,H.SPRINZ
Z CHEM   15, 153 (1975)

1945   Q 004200 4-CHLORO-2-METHYLPHENOXYACETIC ACID SODIJM
                                  SALT

```
    7 8
    C=C                    D        FORMULA  C9H8CLNAO3          MOL WT  222.61
  6/   \3    2   1/                 SOLVENT  C2D6OS/CCL4
CL-C      C-O-CH2-C                 ORIG ST  C4H12SI            TEMP      AMB
    \5 4/          \
    C-C            ONA              173.30   69.40   157.10  126.70  129.90  124.40
      \9                            1/1      2/3     3/1     4/1     5/2     6/1
      CH3                           129.20  114.20
                                    7/2      8/2
```

                                  P.JUNGHANS,H.SPRINZ
                                  Z CHEM   15, 153 (1975)

1946   Q 004200 2-METHYLPHENOXYACETIC ACID SODIUM SALT

```
    7 8
    C=C                    D        FORMULA  C9H9NAO3            MOL WT  188.16
  6/   \3    2   1/                 SOLVENT  C2D6OS/CCL4
  C      C-D-CH2-C                  ORIG ST  C4H12SI            TEMP      AMB
    \5 4/          \  -  +
    C-C            O NA             173.90   69.60   158.20  127.00  131.00  120.60
      \9                            1/1      2/3     3/1     4/1     5/2     6/2
      CH3                           127.70  113.10
                                    7/2      8/2     9/4
```

                                  P.JUNGHANS,H.SPRINZ
                                  Z CHEM   15, 153 (1975)

1947   Q 004200 3-METHYLPHENOXYACETIC ACID SODIUM SALT

```
    7 8
    C=C                    D        FORMULA  C9H9NAO3            MOL WT  188.16
  6/   \3    2   1/                 SOLVENT  C2D6OS/CCL4
  C      C-D-CH2-C                  ORIG ST  C4H12SI            TEMP      AMB
    \5 4/          \  -  +
    C-C            D NA             170.70   65.60   158.70  116.40  140.00  122.80
  9/                                1/1      2/3     3/1     4/2     5/1     6/2
  CH3                               127.50  112.50
                                    7/2      8/2     9/4
```

                                  P.JUNGHANS,H.SPRINZ
                                  Z CHEM   15, 153 (1975)

1948   Q 004200 HYDROCINNAMIC ACID

```
    5 6
    C-C                    O        FORMULA  C9H10O2             MOL WT  150.18
   //  \                   II       SOLVENT  CDCL3
  C      C-CH2-CH2-C-OH             ORIG ST  C4H12SI            TEMP      AMB
  4\   /1 7    8   9
    C=C                             140.10  128.20   128.50  126.30  128.50  128.20
    3 2                             1/1      2/2     3/2     4/2     5/2     6/2
                                    30.50   35.50   179.50
                                    7/3      8/3     9/1
```

                                  L.F.JOHNSON,W.C.JANKOWSKI
                                  CARBON-13 NMR SPECTRA,JOHN
                                  WILEY AND SONS,NEW YORK   344 (1972)

1949   Q 004300 DIETHYLAMINOETHYL PARA-CHLOROBENZOATE
                                  HYDROCHLORIDE

```
      2
  CL  C                             FORMULA  C13H18CLNO2         MOL WT  255.75
   \ // \                           SOLVENT  H2O
   1C    C3                         ORIG ST  C4H12SI            TEMP      AMB
    I    II       9  10  11
   6C    C4  D   CH2 CH2-CH3        138.80  127.70   129.90  126.20  164.90  49.50
    \ / \ / \ / \ /                 1/1      2/2     3/2     4/1     7/1     8/3
      C   C7  CH2 N.HCL             58.70   47.60   7.30
      5   II  8   I                 9/3      10/3    11/4
      D       12CH2
              I                     O.A.GANSOW,W.M.BECKENBAUGH,R.L.SASS
              13CH3                 TETRAHEDRON                    28, 2691 (1972)
```

```
1950                    Q 004300 DIETHYLAMINOETHYL PARA-FLUOROBENZOATE
                                 HYDROCHLORIDE

      2
   F   C              FORMULA   C13H18FNO2              MOL WT  239.29
    \ // \            SOLVENT   H2O
    1C   C3           ORIG ST   C4H12SI                TEMP      AMB
     |   ||       9  10  11
    6C   C4  O  CH2 CH2-CH3   159.40  114.20  131.20  124.10  165.20   49.50
     \ / \ / \ / \ /           1/4     2/2     3/2     4/1     7/1     8/3
      C   C7  CH2 N.HCL        58.70   47.70    7.50
      5   ||   8   |           9/3    10/3    11/4
          O      12CH2
                   |           O.A.GANSOW,W.M.BECKENBAUGH,R.L.SASS
                 13CH3         TETRAHEDRON                    28, 2691 (1972)
```

```
1951                    Q 004350 DIETHYLAMINOETHYL PARA-NITROBENZOATE
                                 HYDROCHLORIDE

      2
  O2N   C             FORMULA   C13H18N2O4             MOL WT  266.30
    \ // \            SOLVENT   H2O
    1C   C3           ORIG ST   C4H12SI                TEMP      AMB
     |   ||       9  10  11
    6C   C4  O  CH2 CH2-CH3   149.40  122.40  129.90  133.30  164.30   49.40
     \ / \ / \ / \ /           1/1     2/2     3/2     4/1     7/1     8/3
      C   C7  CH2 N.HCL        59.40   47.65    7.30
      5   ||   8   |           9/3    10/3    11/4
          O      12CH2
                   |           O.A.GANSOW,W.M.BECKENBAUGH,R.L.SASS
                 13CH3         TETRAHEDRON                    28, 2691 (1972)
```

```
1952                    Q 004400 PARA-AMINOBENZOIC ACID

           O
           ||
        C6   C7          FORMULA   C7H7NO2              MOL WT  137.14
       // \ / \          SOLVENT   C2H6OS
      5C   C1  OH        ORIG ST   C4H12SI              TEMP       300
       |   ||
      4C   C2            118.70  132.40  113.70  154.10  168.70
       \ / \ /            1/1     2/2     3/2     4/1     7/1
      H2N   C3
                         U.EWERS,H.GUENTHER,L.JAENICKE
                         CHEM BER                    106, 3951 (1973)
```

```
1953                    Q 004400 DIETHYLAMINOETHYL PARA-AMINOBENZOATE
                                 HYDROCHLORIDE

      2
  H2N   C             FORMULA   C13H20N2O2             MOL WT  236.32
    \ // \            SOLVENT   H2O
    1C   C3           ORIG ST   C4H12SI                TEMP      AMB
     |   ||       9  10  11
    6C   C4  O  CH2 CH2-CH3   152.20  113.20  130.90  115.80  166.40   49.40
     \ / \ / \ / \ /           1/1     2/2     3/2     4/1     7/1     8/3
      C   C7  CH2 N.HCL        57.00   47.55    7.50
      5   ||   8   |           9/3    10/3    11/4
          O      12CH2
                   |           O.A.GANSOW,W.M.BECKENBAUGH,R.L.SASS
                 13CH3         TETRAHEDRON                    28, 2691 (1972)
```

```
1954                    Q 004400 DIETHYLAMINOETHYL PARA-AMINOBENZAMIDE
                                 HYDROCHLORIDE

      2
  H2N   C             FORMULA   C13H21N3O              MOL WT  235.33
    \ // \            SOLVENT   H2O
    1C   C3  H        ORIG ST   C4H12SI                TEMP      AMB
     |   ||       9  10  11
    6C   C4  N  CH2 CH2-CH3   154.50  113.90  128.20  120.60  169.70   34.30
     \ / \ / \ / \ /           1/1     2/2     3/2     4/1     7/1     8/3
      C   C7  CH2 N.HCL        50.80   47.30    7.40
      5   ||   8   |           9/3    10/3    11/4
          O      12CH2
                   |           O.A.GANSOW,W.M.BECKENBAUGH,R.L.SASS
                 13CH3         TETRAHEDRON                    28, 2691 (1972)
```

1955   Q 004700 PHTHALIC ANHYDRIDE

```
    2     7
    C    C=O
  / \1/ \
 3C    C   \
 ||    |    )
 4C    C   /
  \   /8\ /
    C    C=O
    5     6
```

FORMULA	C8H4O3			MOL WT	148.12
SOLVENT	C2D6OS				
ORIG ST	C4H12SI			TEMP	AMB
131.10	125.30	136.10	136.10	125.30	163.10
1/1	2/2	3/2	4/2	5/2	6/1
163.10	131.10				
7/1	8/1				

L.F.JOHNSON,W.C.JANKOWSKI
CARBON-13 NMR SPECTRA,JOHN
WILEY AND SONS,NEW YORK   281 (1972)

1956   Q 004800 BENZOYL CHLORIDE

```
    5 6
    C-C    O
   //   \  ||
  C      C-C-CL
  4\    /1 7
    C=C
    3 2
```

FORMULA	C7H5CLO			MOL WT	140.57
SOLVENT	CDCL3				
ORIG ST	C4H12SI			TEMP	AMB
133.10	131.30	128.90	135.30	128.90	131.30
1/1	2/2	3/2	4/2	5/2	6/2
168.00					
7/1					

L.F.JOHNSON,W.C.JANKOWSKI
CARBON-13 NMR SPECTRA,JOHN
WILEY AND SONS,NEW YORK   224 (1972)

1957   Q 004900 N-(2-BROMOETHYL)PHTHALIMIDE

```
      O
      ||
  3C  1C
  // \ / \
 4C   2C   \
 |    ||    N-CH2-CH2-BR
 5C   C   / 9   10
  \  /7\ /
    C   C8
    6   ||
        O
```

FORMULA	C10H8BRNO2			MOL WT	254.08
SOLVENT	CDCL3				
ORIG ST	C4H12SI			TEMP	AMB
167.50	131.70	123.30	134.00	134.00	123.30
1/1	2/1	3/2	4/2	5/2	6/2
131.70	167.50	39.20	28.20		
7/1	8/1	9/3	10/3		

L.F.JOHNSON,W.C.JANKOWSKI
CARBON-13 NMR SPECTRA,JOHN
WILEY AND SONS,NEW YORK   368 (1972)

1958   Q 005100 N-(3-BROMOPROPYL)PHTHALIMIDE

```
      O
    2 ||
    C  C7
  // \1/ \
 3C   C   \  9   10   11
 |    ||    N-CH2-CH2-CH2
 4C   C   /               |
  \  /6\ /               BR
    C  C8
    5  ||
       O
```

FORMULA	C11H10BRNO2			MOL WT	268.11
SOLVENT	CDCL3				
ORIG ST	C4H12SI			TEMP	AMB
131.90	123.10	133.90	133.90	123.10	131.90
1/1	2/2	3/2	4/2	5/2	6/1
167.90	167.90	36.60	31.60	29.80	
7/1	8/1	9/3	10/3	11/3	

L.F.JOHNSON,W.C.JANKOWSKI
CARBON-13 NMR SPECTRA,JOHN
WILEY AND SONS,NEW YORK   417 (1972)

1959   Q 005100 PARA-TOLUOYLTHREONINE METHYL ESTER

```
             13
       O   O-CH3
   5 6    \ /
   C-C   O  C9
 7 4//  \1 ||  |10
H3C-C      C-C-NH-CH
    \   / 8     |
     C=C      11CH-OH
     3 2        |
             12CH3
```

FORMULA	C13H17NO4			MOL WT	251.28
SOLVENT	CDCL3				
ORIG ST	C4H12SI			TEMP	300
130.30	128.70	126.90	141.90	20.90	168.10
1/1	2/2	3/2	4/1	7/4	8/1
171.00	57.85	67.35	19.65	52.00	
9/1	10/2	11/2	12/4	13/4	

U.HOLLSTEIN,E.BREITMAIER,G.JUNG
J AM CHEM SOC   96, 8036 (1974)

1960   Q 005200 4-BROMOBENZONITRILE

```
      6  5
      C=C
     /    \
 BR-C      C-CN
   1 \   / 4  7
      C-C
      2  3
```

FORMULA	C7H4BRN			MOL WT	182.02
SOLVENT	CDCL3				
ORIG ST	C4H12SI			TEMP	AMB
127.80	133.30	132.50	111.20	132.50	133.30
1/1	2/2	3/2	4/1	5/2	6/2
117.80					
7/1					

L.F.JOHNSON,W.C.JANKOWSKI
CARBON-13 NMR STECTRA,JOHN
WILEY AND SONS,NEW YORK   222 (1972)

1961   Q 005200 BENZONITRILE

```
      5  6
      C-C
     //   \\
    C      C-C≡N
   4 \   / 1  7
      C=C
      3  2
```

FORMULA	C7H5N			MOL WT	103.12
SOLVENT	CDCL3				
ORIG ST	C4H12SI			TEMP	AMB
112.30	132.00	129.10	132.70	129.10	132.00
1/1	2/2	3/2	4/2	5/2	6/2
118.70					
7/1					

L.F.JOHNSON,W.C.JANKOWSKI
CARBON-13 NMR SPECTRA,JOHN
WILEY AND SONS,NEW YORK   226 (1972)

1962   Q 005200 PHENYLACETONITRILE

```
      5  6
      C-C
     //   \\
    C      C-CH2-CN
   4 \   / 1  7   8
      C=C
      3  2
```

FORMULA	C8H7N			MOL WT	117.15
SOLVENT	CDCL3				
ORIG ST	C4H12SI			TEMP	AMB
130.10	127.80	129.00	127.80	129.00	127.80
1/1	2/2	3/2	4/2	5/2	6/2
23.20	118.00				
7/3	8/1				

L.F.JOHNSON,W.C.JANKOWSKI
CARBON-13 NMR SPECTRA,JOHN
WILEY AND SONS,NEW YORK   284 (1972)

1963   Q 005300 METHYL BENZOATE

```
      3  2
      C-C       O
     /   \1    //
   4C      C-C
     \   / 7 \
      C=C     O-CH3
      5  6     8
```

FORMULA	C8H8O2			MOL WT	136.15
SOLVENT	CDCL3				
ORIG ST	C4H12SI			TEMP	AMB
130.30	129.50	128.30	132.80	128.30	129.50
1/1	2/2	3/2	4/2	5/2	6/2
166.80	51.80				
7/1	8/4				

L.F.JOHNSON,W.C.JANKOWSKI
CARBON-13 NMR SPECTRA,JOHN
WILEY AND SONS,NEW YORK   291 (1972)

1964   Q 005300 METHYL SALICYLATE

```
          OH
          |
     3C-C2   O
     //   \1 ||
   4C      C-C-O-CH3
     \   / 7   8
      C=C
      5  6
```

FORMULA	C8H8O3			MOL WT	152.15
SOLVENT	CDCL3				
ORIG ST	C4H12SI			TEMP	AMB
112.40	161.70	117.50	135.50	119.00	129.90
1/1	2/1	3/2	4/2	5/2	6/2
170.50	52.10				
7/1	8/4				

L.F.JOHNSON,W.C.JANKOWSKI
CARBON-13 NMR SPECTRA,JOHN
WILEY AND SONS,NEW YORK   293 (1972)

1965   Q 005300   METHYL-ANTHRANILATE

```
      NH2
       I
    3C-C2  ]
   //   \  ||
  4C      C-C-]-CH3
   \   /1  7  8
    C=C
    5 6
```

FORMULA	C8H9NO2			MOL WT	151.17
SOLVENT	CDCL3				
ORIG ST	C4H12SI			TEMP	AMB
110.00	150.60	116.60	134.00	116.00	131.10
1/1	2/1	3/2	4/2	5/2	6/2
168.50	51.30				
7/1	8/4				

L.F.JOHNSON,W.C.JANKOWSKI
CARBON-13 NMR STECTRA,JOHN
WILEY AND SONS,NEW YORK   296 (1972)

1966   Q 005300   BENZYLACETATE

```
            ]
   3  2  7  ||
   C-C  CH2 C8
  //  \ / \ / \
 4C    ]1  ]   CH3
  \   /        9
   C=C
   5 6
```

FORMULA	C9H10O2			MOL WT	150.18
SOLVENT	CDCL3				
ORIG ST	C4H12SI			TEMP	AMB
136.20	128.10	128.40	128.10	128.40	128.10
1/1	2/2	3/2	4/2	5/2	6/2
66.10	170.50	20.70			
7/3	8/1	9/4			

L.F.JOHNSON,W.C.JANKOWSKI
CARBON-13 NMR SPECTRA,JOHN
WILEY AND SONS,NEW YORK   345 (1972)

1967   Q 005300   BUTYL BENZOATE

```
      ]
 3 2  ||      9   11
 C-C  C7  CH2 CH3
 //  \]1/ \  / \ /
4C    C  ]-CH2 CH2
 \   /   8   10
  C=C
  5 6
```

FORMULA	C11H14O2			MOL WT	178.23
SOLVENT	CDCL3				
ORIG ST	C4H12SI			TEMP	AMB
130.70	129.50	128.20	132.60	128.20	129.20
1/1	2/2	3/2	4/2	5/2	6/2
166.40	64.70	30.90	19.30	13.70	
7/1	8/3	9/3	10/3	11/4	

L.F.JOHNSON,W.C.JANKOWSKI
CARBON-13 NMR SPECTRA,JOHN
WILEY AND SONS,NEW YORK   421 (1972)

1968   Q 005300   DIETHYLAMINOETHYL BENZOATE HYDROCHLORIDE

```
    2
    C
   // \
  1C    C3
  |    ||        9  10  11
  6C    C4  ]  CH2 CH2-CH3
   \  / \ / \ / \ /
    C   C7  CH2 N.HCL
    5  ||  8  |
       ]     12CH2
             |
           13CH3
```

FORMULA	C13H19NO2			MOL WT	221.30
SOLVENT	H2O				
ORIG ST	C4H12SI			TEMP	AMB
133.00	113.00	128.40	125.70	166.20	49.50
1/2	2/2	3/2	4/1	7/1	8/3
58.60	48.50	7.40			
9/3	10/3	11/4			

C.A.GANSOW,W.M.BECKENBAUGH,R.L.SASS
TETRAHEDRON   28, 2691 (1972)

1969   Q 005300   BIS-(2-METHOXYETHYL)-PHTHALATE

```
       ]
   2   ||
   C   C7
  // \ /1\ \  9   10    11
 3C   C   ]-CH2-CH2-0-CH3
 |    ||
 4C   C   ]-CH2-CH2-0-CH3
  \  /6\ /  12  13    14
   C   C8
   5   ||
       ]
```

FORMULA	C14H18O6			MOL WT	282.30
SOLVENT	CDCL3				
ORIG ST	C4H12SI			TEMP	AMB
132.00	131.00	128.90	128.90	131.00	132.00
1/1	2/2	3/2	4/2	5/2	6/1
167.30	167.30	64.60	70.20	58.70	64.60
7/1	8/1	9/3	10/3	11/4	12/3
70.20	58.70				
13/3	14/4				

L.F.JOHNSON,W.C.JANKOWSKI
CARBON-13 NMR SPECTRA,JOHN
WILEY AND SONS,NEW YORK   461 (1972)

1970   Q 005500 BENZENESULFONYL CHLORIDE

```
                              FORMULA   C6H5CLO2S              MOL WT  176.62
          3  2                SOLVENT   CDCL3
          C-C    )            ORIG ST   C4H12SI                TEMP       AMB
        4//    \\1 ||
        C        C-S-CL       144.10  126.80  129.70  135.30  129.70  126.80
          \    /   ||           1/1     2/2     3/2     4/2     5/2     6/2
          C=C      O
          5  6                 L.F.JOHNSON,W.C.JANKOWSKI
                               CARBON-13 NMR SPECTRA,JOHN
                               WILEY AND SONS,NEW YORK             155 (1972)
```

1971   Q 005500 BENZENESULFONIC ACID

```
                              FORMULA   C6H6O3S                MOL WT  158.18
          3  2                SOLVENT   H2O
          C-C    )            ORIG ST   C4H12SI                TEMP       AMB
        //    \\1 ||
       4C        C-S-OH       143.50  126.30  129.80  132.30  129.80  126.30
          \    /   ||           1/1     2/2     3/2     4/2     5/2     6/2
          C=C      O
          5  6                 L.F.JOHNSON,W.C.JANKOWSKI
                               CARBON-13 NMR SPECTRA,JOHN
                               WILEY AND SONS,NEW YORK             163 (1972)
```

1972   Q 005500 TOSYLCHLORIDE

```
                              FORMULA   C7H7CLO2S              MOL WT  190.65
          5  6                SOLVENT   C2D6OS
          C=C    )            ORIG ST   C4H12SI                TEMP       300
        /    \   ||
   H3C-C        C-S-CL        144.45  128.80  126.00  139.30  126.00  128.80
    7  4\    //1 ||             1/1     2/2     3/2     4/2     5/2     6/2
          C-C    )
          3  2                 21.25
                               7/4

                               W.VOELTER,S.FUCHS,R.H.SEUFFER,K.ZECH
                               MONATSH CHEM            105, 1110 (1974)
```

1973   Q 005750 BENZYL DISULFIDE

```
                                              FORMULA   C14H14S2          MOL WT  246.40
   11 10                     3  4             SOLVENT   CDCL3
   C-C                       C-C              ORIG ST   C4H12SI           TEMP       AMB
  12//   \\9 8       1    2//    \\
   C       C-CH2-S-S-CH2-C        C5  43.10  137.20  129.20  128.20  127.20  128.20
    \     /              \    /         1/3     2/1     3/2     4/2     5/2     6/2
     C=C                  C=C     7 6  129.20   43.10  137.20  129.20  128.20  127.20
    13 14                 7  6         7/2     8/3     9/1    10/2    11/2    12/2
                                      128.20  129.20
                                      13/2    14/2
                                      L.F.JOHNSON,W.C.JANKOWSKI
                                      CARBON-13 NMR SPECTRA,JOHN
                                      WILEY AND SONS,NEW YORK             459 (1972)
```

1974   Q 005800 THIOPHENOL

```
          5  6                FORMULA   C6H6S                  MOL WT  110.18
          C-C                 SOLVENT   CDCL3
        //    \\              ORIG ST   C4H12SI                TEMP       AMB
        C        C-SH
       4\    //1              130.70  129.20  128.90  125.40  128.90  129.20
          C=C                   1/1     2/2     3/2     4/2     5/2     6/2
          3  2
                               L.F.JOHNSON,W.C.JANKOWSKI
                               CARBON-13 NMR SPECTRA,JOHN
                               WILEY AND SONS,NEW YORK             164 (1972)
```

1975   Q 005800 4-BROMOTHIOANISOLE

```
                    FORMULA    C7H7BRS              MOL WT   203.10
       5 6          SOLVENT    CDCL3
       C-C          ORIG ST    C4H12SI              TEMP        AMB
      4/   \1
   BR-C      C-S      137.90   128.60   131.80   118.90   131.80   128.60
      \3 2/    \7      1/1      2/2      3/2      4/1      5/2      6/2
       C=C       CH3   16.10
                       7/4
```

G.W.BUCHANAN,C.REYES-ZAMORA,D.E.CLARKE
CAN J CHEM   52, 3895 (1974)

1976   Q 005800 4-CHLOROTHIOANISOLE

```
                    FORMULA    C7H7CLS              MOL WT   158.65
       5 6          SOLVENT    CDCL3
       C-C          ORIG ST    C4H12SI              TEMP        AMB
      4/   \1
   CL-C      C-S      137.30   128.20   128.90   131.10   128.90   128.20
      \3 2/    \7      1/1      2/2      3/2      4/1      5/2      6/2
       C=C       CH3   16.10
                       7/4
```

G.W.BUCHANAN,C.REYES-ZAMORA,D.E.CLARKE
CAN J CHEM   52, 3895 (1974)

1977   Q 005800 4-FLUOROTHIOANISOLE

```
                    FORMULA    C7H7FS               MOL WT   142.20
       5 6          SOLVENT    CDCL3
       C-C          ORIG ST    C4H12SI              TEMP        AMB
      4/   \1
   F-C       C-S      133.80   129.80   116.00   161.50   116.00   129.80
      \3 2/    \7      1/1      2/2      3/2      4/1      5/2      6/2
       C=C       CH3   17.10
                       7/4
```

G.W.BUCHANAN,C.REYES-ZAMORA,D.E.CLARKE
CAN J CHEM   52, 3895 (1974)

1978   Q 005800 4-NITROTHIOANISOLE

```
                    FORMULA    C7H7NO2S             MOL WT   169.20
       5 6          SOLVENT    CDCL3
       C-C          ORIG ST    C4H12SI              TEMP        AMB
      4/   \1
   O2N-C     C-S      145.30   125.50   123.90   148.90   123.90   125.50
      \3 2/    \7      1/1      2/2      3/2      4/1      5/2      6/2
       C=C       CH3   15.00
                       7/4
```

G.W.BUCHANAN,C.REYES-ZAMORA,D.E.CLARKE
CAN J CHEM   52, 3895 (1974)

1979   Q 005800 THIOANISOLE

```
                    FORMULA    C7H8S                MOL WT   124.21
       3 2          SOLVENT    CDCL3
       C=C          ORIG ST    C4H12SI              TEMP        AMB
      4/   \1  7
   C        C-S-CH3   138.40   126.50   128.60   124.80   128.60   126.50
     \    //          1/1      2/2      3/2      4/2      5/2      6/2
       C-C            15.60
       5 6            7/4
```

L.F.JOHNSON,W.C.JANKOWSKI
CARBON-13 NMR SPECTRA,JOHN
WILEY AND SONS,NEW YORK   249 (1972)

1980   Q 005800 THIOANISOLE

```
   5 6
   C-C
 4/   \1
 C      C-S
  \3 2/   \7
  C=C    CH3
```

FORMULA	C7H8S				MOL WT	124.21
SOLVENT	CDCL3					
ORIG ST	C4H12SI				TEMP	AMB

138.60	126.80	128.80	125.00	128.80	126.80
1/1	2/2	3/2	4/2	5/2	6/2
15.90					
7/4					

G.W.BUCHANAN,C.REYES-ZAMORA,D.E.CLARKE
CAN J CHEM   52, 3895 (1974)

1981   Q 005800 4-METHOXYTHIOANISOLE

```
     5 6
     C-C
  8   4/   \1
H3C-O-C      C-S
    \3 2/   \7
    C=C    CH3
```

FORMULA	C8H10OS				MOL WT	154.23
SOLVENT	CDCL3					
ORIG ST	C4H12SI				TEMP	AMB

129.20	130.30	114.80	158.50	114.80	130.30
1/1	2/2	3/2	4/1	5/2	6/2
17.90	55.30				
7/4	8/4				

G.W.BUCHANAN,C.REYES-ZAMORA,D.E.CLARKE
CAN J CHEM   52, 3895 (1974)

1982   Q 005800 2-METHYLTHIOANISOLE

```
   5 6
   C-C
 4/   \1
 C      C-S
  \3 2/   \7
  C=C    CH3
     \8
     CH3
```

FORMULA	C8H10S				MOL WT	138.23
SOLVENT	CDCL3					
ORIG ST	C4H12SI				TEMP	AMB

137.80	136.00	129.90	124.80	125.40	126.60
1/1	2/1	3/2	4/2	5/2	6/2
16.00	19.90				
7/4	8/4				

G.W.BUCHANAN,C.REYES-ZAMORA,D.E.CLARKE
CAN J CHEM   52, 3895 (1974)

1983   Q 005800 3-METHYLTHIOANISOLE

```
   5 6
   C-C
 4/   \1
 C      C-S
  \3 2/   \7
  C=C    CH3
 8/
H3C
```

FORMULA	C8H10S				MOL WT	138.23
SOLVENT	CDCL3					
ORIG ST	C4H12SI				TEMP	AMB

138.60	127.90	138.50	126.10	128.80	124.30
1/1	2/2	3/1	4/2	5/2	6/2
16.10	21.20				
7/4	8/4				

G.W.BUCHANAN,C.REYES-ZAMORA,D.E.CLARKE
CAN J CHEM   52, 3895 (1974)

1984   Q 005800 4-METHYLTHIOANISOLE

```
     5 6
     C-C
  8  4/   \1
H3C-C      C-S
    \3 2/   \7
    C=C    CH3
```

FORMULA	C8H10S				MOL WT	138.23
SOLVENT	CDCL3					
ORIG ST	C4H12SI				TEMP	AMB

135.10	127.80	129.60	134.90	129.60	127.80
1/1	2/2	3/2	4/1	5/2	6/2
16.50	20.80				
7/4	8/4				

G.W.BUCHANAN,C.REYES-ZAMORA,D.E.CLARKE
CAN J CHEM   52, 3895 (1974)

1985 — Q 005800 2,4,6-TRIMETHYLTHIOANISOLE

```
        CH3
     5 6/10
     C-C
   9 4/   \1
  H3C-C      C-S
    \3 2/    \7
     C=C      CH3
      \8
      CH3
```

FORMULA	C10H14S			MOL WT	166.29
SOLVENT	CDCL3				
ORIG ST	C4H12SI			TEMP	AMB
137.90	142.50	128.90	132.10	128.90	142.50
1/1	2/1	3/2	4/1	5/2	6/1
18.30	21.50	20.90	21.50		
7/4	8/4	9/4	10/4		

G.W.BUCHANAN,C.REYES-ZAMORA,D.E.CLARKE
CAN J CHEM   52, 3895 (1974)

1986 — Q 005800 2,4,6-TRIISOPROPYLTHIOANISOLE

```
     H3C
      15\
        CH-CH3
   14    5 6/10 16
  H3C      C-C
    \9 4/   \1
    CH-C      C-S
   13/  \3 2/   \7
  H3C   C\C    CH3
         \8
        CH-CH3
       12/   11
      H3C
```

FORMULA	C16H26S			MOL WT	250.45
SOLVENT	CDCL3				
ORIG ST	C4H12SI			TEMP	AMB
130.50	152.80	121.80	149.40	121.80	152.50
1/1	2/1	3/2	4/1	5/2	6/1
21.10	31.50	34.40	31.50	24.50	24.50
7/4	8/2	9/2	10/2	11/4	12/4
23.90	23.90	24.50	24.50		
13/4	14/4	15/4	16/4		

G.W.BUCHANAN,C.REYES-ZAMORA,D.E.CLARKE
CAN J CHEM   52, 3895 (1974)

1987 — Q 006000 4-BROMOMETHYL PHENYL SULFOXIDE

```
     5 6
     C-C      ]
   4/   \1  /
  BR-C      C-S
    \3 2/    \7
     C=C      CH3
```

FORMULA	C7H7BROS			MOL WT	219.10
SOLVENT	CDCL3				
ORIG ST	C4H12SI			TEMP	AMB
145.20	125.20	132.60	125.30	132.60	125.20
1/1	2/2	3/2	4/1	5/2	6/2
44.00					
7/4					

G.W.BUCHANAN,C.REYES-ZAMORA,D.E.CLARKE
CAN J CHEM   52, 3895 (1974)

1988 — Q 006000 4-BROMOMETHYL PHENYL SULFONE

```
     5 6
     C-C
   4/   \1   7
  BR-C      C-SO2-CH3
    \3 2/
     C=C
```

FORMULA	C7H7BRO2S			MOL WT	235.10
SOLVENT	CDCL3				
ORIG ST	C4H12SI			TEMP	AMB
140.20	128.90	132.70	128.70	132.70	128.90
1/1	2/2	3/2	4/1	5/2	6/2
44.40					
7/4					

G.W.BUCHANAN,C.REYES-ZAMORA,D.E.CLARKE
CAN J CHEM   52, 3895 (1974)

1989 — Q 006000 4-CHLOROMETHYL PHENYL SULFOXIDE

```
     5 6
     C-C      ]
   4/   \1  /
  CL-C      C-S
    \3 2/    \7
     C=C      CH3
```

FORMULA	C7H7CLOS			MOL WT	174.65
SOLVENT	CDCL3				
ORIG ST	C4H12SI			TEMP	AMB
144.90	125.00	129.50	137.00	129.50	125.00
1/1	2/2	3/2	4/1	5/2	6/2
44.00					
7/4					

G.W.BUCHANAN,C.REYES-ZAMORA,D.E.CLARKE
CAN J CHEM   52, 3895 (1974)

1990   Q 006000   4-CHLOROMETHYL PHENYL SULFONE

```
                              FORMULA    C7H7CLO2S            MOL WT   190.65
         5  6                 SOLVENT    CDCL3
         C-C                  ORIG ST    C4H12SI              TEMP        AMB
       4/    \1     7
     CL-C      C-SO2-CH3       139.60   128.90   129.60   140.10   129.60   128.90
        \3  2/                  1/1      2/2      3/2      4/1      5/2      6/2
         C=C                   44.30
                               7/4
```

G.W.BUCHANAN,C.REYES-ZAMORA,D.E.CLARKE
CAN J CHEM   52, 3895 (1974)

1991   Q 006000   4-FLUOROMETHYL PHENYL SULFOXIDE

```
                              FORMULA    C7H7FOS              MOL WT   158.20
         5  6                 SOLVENT    CDCL3
         C-C      O           ORIG ST    C4H12SI              TEMP        AMB
       4/    \1   //
     F-C      C-S              142.40   126.30   116.90   164.70   116.90   126.30
        \3  2/    \7            1/1      2/2      3/2      4/1      5/2      6/2
         C=C      CH3          44.30
                               7/4
```

G.W.BUCHANAN,C.REYES-ZAMORA,D.E.CLARKE
CAN J CHEM   52, 3895 (1974)

1992   Q 006000   4-FLUOROMETHYL PHENYL SULFONE

```
                              FORMULA    C7H7FO2S             MOL WT   174.20
         5  6                 SOLVENT    CDCL3
         C-C                  ORIG ST    C4H12SI              TEMP        AMB
       4/    \1     7
     F-C      C-SO2-CH3        137.50   130.30   116.70   165.80   116.70   130.30
        \3  2/                  1/1      2/2      3/2      4/1      5/2      6/2
         C=C                   44.50
                               7/4
```

G.W.BUCHANAN,C.REYES-ZAMORA,D.E.CLARKE
CAN J CHEM   52, 3895 (1974)

1993   Q 006000   METHYL-4-NITROPHENYL SULFOXIDE

```
                              FORMULA    C7H7NO3S             MOL WT   185.20
         5  6                 SOLVENT    CDCL3
         C-C      O           ORIG ST    C4H12SI              TEMP        AMB
       4/    \1   //
   O2N-C      C-S              154.00   124.90   124.60   150.10   124.60   124.90
        \3  2/    \7            1/1      2/2      3/2      4/1      5/2      6/2
         C=C      CH3          44.20
                               7/4
```

G.W.BUCHANAN,C.REYES-ZAMORA,D.E.CLARKE
CAN J CHEM   52, 3895 (1974)

1994   Q 006000   METHYL-4-NITROPHENYL SULFONE

```
                              FORMULA    C7H7NO4S             MOL WT   201.20
         5  6                 SOLVENT    CDCL3
         C-C                  ORIG ST    C4H12SI              TEMP        AMB
       4/    \1     7
   O2N-C      C-SO2-CH3        146.90   128.90   124.60   151.30   124.60   128.90
        \3  2/                  1/1      2/2      3/2      4/1      5/2      6/2
         C=C                   44.30
                               7/4
```

G.W.BUCHANAN,C.REYES-ZAMORA,D.E.CLARKE
CAN J CHEM   52, 3895 (1974)

1995   Q 006000 METHYLPHENYL SULFOXIDE

```
                           FORMULA   C7H8OS              MOL WT   140.21
       5 6                 SOLVENT   CDCL3
       C-C      ]          ORIG ST   C4H12SI             TEMP       AMB
     4⁄   ∖1   ⁄
     C       C-S             146.30  123.60  129.40  131.00  129.40  123.60
       ∖3 2⁄   ∖7              1/1     2/2     3/2     4/2     5/2     6/2
       C=C      CH3            44.00
                              7/4
```

G.W.BUCHANAN,C.REYES-ZAMORA,D.E.CLARKE
CAN J CHEM   52, 3895 (1974)

1996   Q 006000 METHYL PHENYL SULFONE

```
                           FORMULA   C7H8O2S             MOL WT   156.20
       5 6                 SOLVENT   CDCL3
       C-C                 ORIG ST   C4H12SI             TEMP       AMB
     4⁄   ∖1   7
     C       C-SO2-CH3       141.00  127.20  129.30  133.50  129.30  127.20
       ∖3 2⁄                   1/1     2/2     3/2     4/2     5/2     6/2
       C=C                     44.30
                              7/4
```

G.W.BUCHANAN,C.REYES-ZAMORA,D.E.CLARKE
CAN J CHEM   52, 3895 (1974)

1997   Q 006000 METHYL-2-METHYLPHENYL SULFOXIDE

```
                           FORMULA   C8H10OS             MOL WT   154.23
       5 6                 SOLVENT   CDCL3
       C-C      ]          ORIG ST   C4H12SI             TEMP       AMB
     4⁄   ∖1   ⁄
     C       C-S             144.20  134.00  130.70  130.60  127.50  123.10
       ∖3 2⁄   ∖7              1/1     2/1     3/2     4/2     5/2     6/2
       C=C      CH3            42.20   18.40
         ∖8                   7/4     8/4
          CH3
```

G.W.BUCHANAN,C.REYES-ZAMORA,D.E.CLARKE
CAN J CHEM   52, 3895 (1974)

1998   Q 006000 METHYL-3-METHYLPHENYL SULFOXIDE

```
                           FORMULA   C8H10OS             MOL WT   154.23
       5 6                 SOLVENT   CDCL3
       C-C      ]          ORIG ST   C4H12SI             TEMP       AMB
     4⁄   ∖1   ⁄
     C       C-S             145.90  123.70  139.40  131.60  129.20  120.60
       ∖3 2⁄   ∖7              1/1     2/2     3/1     4/2     5/2     6/2
       C=C      CH3            43.90   21.30
      8⁄                      7/4     8/4
     H3C
```

G.W.BUCHANAN,C.REYES-ZAMORA,D.E.CLARKE
CAN J CHEM   52, 3895 (1974)

1999   Q 006000 METHYL-4-METHYLPHENYL SULFOXIDE

```
                           FORMULA   C8H10OS             MOL WT   154.23
        5 6                SOLVENT   CDCL3
        C-C      ]         ORIG ST   C4H12SI             TEMP       AMB
     8 4⁄   ∖1   ⁄
   H3C-C       C-S           143.40  123.70  130.10  141.50  130.10  123.70
       ∖3 2⁄   ∖7              1/1     2/2     3/2     4/1     5/2     6/2
       C=C      CH3            44.10   21.30
                              7/4     8/4
```

G.W.BUCHANAN,C.REYES-ZAMORA,D.E.CLARKE
CAN J CHEM   52, 3895 (1974)

```
2000              Q  006000  METHYL-2-METHYLPHENYL SULFONE

                              FORMULA   C8H10O2S              MOL WT   170.23
          5  6                SOLVENT   CDCL3
          C-C                 ORIG ST   C4H12SI               TEMP       AMB
        4∕   ∖1      7
        C        C-SO2-CH3      139.60   137.80   132.40   133.60   126.90   129.30
          ∖3 2∕                  1/1      2/1      3/2      4/2      5/2      6/2
          C=C                   43.70    20.00
            ∖8                    7/4      8/4
            CH3
                              G.W.BUCHANAN,C.REYES-ZAMORA,D.E.CLARKE
                              CAN J CHEM                    52, 3895 (1974)
```

```
2001              Q  006000  METHYL-3-METHYLPHENYL SULFONE

                              FORMULA   C8H10O2S              MOL WT   170.23
          5  6                SOLVENT   CDCL3
          C-C                 ORIG ST   C4H12SI               TEMP       AMB
        4∕   ∖1      7
        C        C-SO2-CH3      141.10   127.70   139.70   134.40   129.30   124.50
          ∖3 2∕                  1/1      2/2      3/1      4/2      5/2      6/2
          C=C                   44.50    21.20
         8∕                      7/4      8/4
        H3C
                              G.W.BUCHANAN,C.REYES-ZAMORA,D.E.CLARKE
                              CAN J CHEM                    52, 3895 (1974)
```

```
2002              Q  006000  METHYL-4-METHYLPHENYL SULFONE

                              FORMULA   C8H10O2S              MOL WT   170.23
          5  6                SOLVENT   CDCL3
          C-C                 ORIG ST   C4H12SI               TEMP       AMB
      8 4∕   ∖1      7
    H3C-C        C-SO2-CH3      138.50   127.50   130.00   144.60   130.00   127.50
          ∖3 2∕                  1/1      2/2      3/2      4/1      5/2      6/2
          C=C                   44.60    21.50
                                 7/4      8/4

                              G.W.BUCHANAN,C.REYES-ZAMORA,D.E.CLARKE
                              CAN J CHEM                    52, 3895 (1974)
```

```
2003              Q  006000  METHYL-4-METHOXYPHENYL SULFOXIDE

                              FORMULA   C8H10O2S              MOL WT   170.23
          5  6                SOLVENT   CDCL3
          C-C      O           ORIG ST   C4H12SI               TEMP       AMB
      8  4∕  ∖1     ∕
    H3C-O-C     C-S            137.40   125.50   115.00   162.20   115.00   125.50
          ∖3 2∕   ∖7             1/1      2/2      3/2      4/1      5/2      6/2
          C=C      CH3          44.00    55.60
                                 7/4      8/4

                              G.W.BUCHANAN,C.REYES-ZAMORA,D.E.CLARKE
                              CAN J CHEM                    52, 3895 (1974)
```

```
2004              Q  006000  METHYL-4-METHOXYPHENYL SULFONE

                              FORMULA   C8H10O3S              MOL WT   186.23
          5  6                SOLVENT   CDCL3
          C-C                 ORIG ST   C4H12SI               TEMP       AMB
      8  4∕  ∖1     7
    H3C-O-C     C-SO2-CH3      132.50   129.50   114.60   163.80   114.60   129.50
          ∖3 2∕                  1/1      2/2      3/2      4/1      5/2      6/2
          C=C                   44.80    55.70
                                 7/4      8/4

                              G.W.BUCHANAN,C.REYES-ZAMORA,D.E.CLARKE
                              CAN J CHEM                    52, 3895 (1974)
```

2005   Q 006000 METHYL-2,4,6-TRIMETHYLPHENYL SULFOXIDE

```
          CH3          FORMULA  C10H14OS              MOL WT  182.29
       5 6/10          SOLVENT  CDCL3
        C-C     J      ORIG ST  C4H12SI               TEMP       AMB
     9 4/   \1 /
    H3C-C     C-S       143.20   139.60   132.30   134.90   132.30   139.60
      \3 2/   \7          1/1      2/1      3/2      4/1      5/2      6/1
       C=C     CH3       44.40             20/90
        \8                 7/4      8/4      9/4     10/4
        CH3
```

G.W.BUCHANAN,C.REYES-ZAMORA,D.E.CLARKE
CAN J CHEM   52, 3895 (1974)

2006   Q 006000 METHYL-2,4,6-TRIMETHYLPHENYL SULFONE

```
          CH3          FORMULA  C10H14O2S             MOL WT  198.29
       5 6/10          SOLVENT  CDCL3
        C-C            ORIG ST  C4H12SI               TEMP       AMB
     9 4/   \1    7
    H3C-C     C-SO2-CH3  142.40   138.80   131.50   134.50   131.50   138.80
      \3 2/              1/1      2/1      3/2      4/1      5/2      6/1
       C=C               44.40    22.40    20.80    22.40
        \8                 7/4      8/4      9/4     10/4
        CH3
```

G.W.BUCHANAN,C.REYES-ZAMORA,D.E.CLARKE
CAN J CHEM   52, 3895 (1974)

2007   Q 006000 METHYL-2,4,6-TRIISOPROPYLPHENYL SULFOXIDE

```
       H3C             FORMULA  C16H26OS              MOL WT  266.45
      15\10 16         SOLVENT  CDCL3
        CH-CH3         ORIG ST  C4H12SI               TEMP       AMB
   14      5 6/
  H3C       C-C     J    136.00   149.80   123.20   152.30   123.20   149.80
    \9 4/   \1 /          1/1      2/1      3/2      4/1      5/2      6/1
    CH-C     C-S          41.10    28.10    34.30    28.10    24.70    24.20
  13/  \3 2/   \7          7/4      8/2      9/2     10/2     11/4     12/4
  H3C   C=C     CH3       23.80    23.80    24.70    24.40
        \8                13/4     14/4     15/4     16/4
        CH-CH3
       12/   11    G.W.BUCHANAN,C.REYES-ZAMORA,D.E.CLARKE
       H3C          CAN J CHEM                     52, 3895 (1974)
```

2008   Q 006000 METHYL-2,4,6-TRIISOPROPYLPHENYL SULFONE

```
       H3C             FORMULA  C16H26O2S             MOL WT  282.45
      15\10 16         SOLVENT  CDCL3
        CH-CH3         ORIG ST  C4H12SI               TEMP       AMB
   14      5 6/
  H3C       C-C          134.20   150.90   124.10   153.70   124.10   150.90
    \9 4/   \1    7        1/1      2/1      3/2      4/1      5/2      6/1
    CH-C     C-SO2-CH3    46.60    29.60    34.30    29.60    25.00    25.00
  13/  \3 2/               7/4      8/2      9/2     10/2     11/4     12/4
  H3C   C=C               23.60    23.60    25.00    25.00
        \8  11            13/4     14/4     15/4     16/4
        CH-CH3
       12/         G.W.BUCHANAN,C.REYES-ZAMORA,D.E.CLARKE
       H3C          CAN J CHEM                     52, 3895 (1974)
```

2009   Q 006700 PHENYLISOCYANATE

```
                       FORMULA  C7H5NO                MOL WT  119.12
       5 6             SOLVENT  CDCL3
        C-C            ORIG ST  C4H12SI               TEMP       AMB
       //   \\
      C      C-N=C=O    133.60   124.70   129.50   125.70   129.50   124.70
     4\  /1    7          1/1      2/2      3/2      4/2      5/2      6/2
       C=C               129.50
       3 2                 7/1
```

L.F.JOHNSON,W.C.JANKOWSKI
CARBON-13 NMR SPECTRA,JOHN
WILEY AND SONS,NEW YORK   228 (1972)

2010   Q 006700   TOLYLENE-2,4-DIISOCYANATE

```
          NCO
         / 9
     5C-C4
     ∥    ∖
   6C      C3
     ∖    /
     1C=C2
     /    ∖
   CH3    NCO
    7      8
```

FORMULA	C9H6N2O2			MOL WT	174.16
SOLVENT	CDCL3				
ORIG ST	C4H12SI			TEMP	AMB

136.60	133.30	120.90	132.20	122.20	131.40
1/1	2/1	3/2	4/1	5/2	6/2
17.70	125.00	125.00			
7/4	8/1	9/1			

L.F.JOHNSON,W.C.JANKOWSKI
CARBON-13 NMR SPECTRA,JOHN
WILEY AND SONS,NEW YORK   332 (1972)

2011   Q 007000   2,3,4,5,6,2'-HEXAXHLOROBIPHENYL

```
    CL   CL  CL
    9 |   | 3/
    C-C8  2C-C
  10/  ∖7 1/  ∖4
   C      C-C    C-CL
    ∖    /    ∖   /
    C=C      C=C
   11 12   /6  5∖
       CL       CL
```

FORMULA	C12H4CL6			MOL WT	360.88
SOLVENT	C2D4BR2				
ORIG ST	C4H12SI			TEMP	313

137.28	132.36	135.26	132.36	131.28	133.00
1/1	2/1	3/1	4/1	7/1	8/1
129.84	129.20	126.70	129.72		
9/2	10/2	11/2	12/2		

N.K.WILSON
J AM CHEM SOC   97, 3573 (1975)

2012   Q 007000   2,3,4,5,6,3'-HEXACHLOROBIPHENYL

```
   CL      CL  CL
    ∖9 8    ∖2 3/
    C-C      C-C
  10/  ∖7 1/  ∖4
   C      C-C    C-CL
    ∖    /    ∖   /
    C=C      C=C
   11 12   /6  5∖
       CL       CL
```

FORMULA	C12H4CL6			MOL WT	360.88
SOLVENT	C2D4BR2				
ORIG ST	C4H12SI			TEMP	313

138.25	133.79	137.77	132.85	131.42	128.48
1/1	2/1	3/1	4/1	7/1	8/2
132.10	128.48	129.52	126.66		
9/1	10/2	11/2	12/2		

N.K.WILSON
J AM CHEM SOC   97, 3573 (1975)

2013   Q 007000   2,3,4,5,6,4'-HEXACHLOROBIPHENYL

```
      CL   CL
    9 8    ∖2 3/
    C-C      C-C
  10/  ∖7 1/  ∖4
  CL-C     C-C    C-CL
    ∖    /    ∖   /
    C=C      C=C
   11 12   /6  5∖
       CL       CL
```

FORMULA	C12H4CL6			MOL WT	360.88
SOLVENT	C2D4BR2				
ORIG ST	C4H12SI			TEMP	313

138.57	134.09	134.65	132.74	132.18	134.09
1/1	2/1	3/1	4/1	5/1	6/1
131.42	128.49	129.87	132.18	128.49	129.87
7/1	8/2	9/2	10/1	11/2	12/2

N.K.WILSON
J AM CHEM SOC   97, 3573 (1975)

2014   Q 007000   2,3,4,2',3',4'-HEXACHLOROBIPHENYL

```
   CL  CL    CL  CL
    ∖9 |     | 3/
    C-C8    2C-C
  10/  ∖7 1/  ∖4
  CL-C     C-C    C-CL
    ∖    /    ∖   /
    C=C      C=C
   11 12    6  5
```

FORMULA	C12H4CL6			MOL WT	360.88
SOLVENT	C2D4BR2				
ORIG ST	C4H12SI			TEMP	313

136.99	131.71	133.02	133.62	127.74	128.34
1/1	2/1	3/1	4/1	5/2	6/2

N.K.WILSON
J AM CHEM SOC   97, 3573 (1975)

2015   Q 007000 2,3,6,2',3',6'-HEXACHLOROBIPHENYL

```
   CL  CL    CL  CL
    \9 |    | 3/          FORMULA   C12H4CL6           MOL WT   360.88
     C-C8  2C-C           SOLVENT   C2D4BR2
   10/   \7 1/   \        ORIG ST   C4H12SI            TEMP        313
    C     C-C      C4
     \   /   \   /        136.00   132.66   132.30   130.56   127.98   131.35
      C=C12  6C=C          1/1      2/1      3/1      4/2      5/2      6/1
    11 |     | 5
      CL     CL           N.K.WILSON
                          J AM CHEM SOC                 97, 3573 (1975)
```

2016   Q 007000 2,4,5,2',4',5'-HEXACHLOROBIPHENYL

```
    CL      CL
    9 |     | 3           FORMULA   C12H4CL6           MOL WT   360.88
    C-C8  2C-C            SOLVENT   C2D4BR2
   10/   \7 1/   \4       ORIG ST   C4H12SI            TEMP        313
  CL-C    C-C     C-CL
    \    /   \   /        135.30   130.70   131.70   133.90   131.90   130.60
    11C=C    C=C           1/1      2/1      3/2      4/1      5/1      6/2
    / 12    6 5\
   CL        CL           N.K.WILSON
                          J AM CHEM SOC                 97, 3573 (1975)
```

2017   Q 007000 2,4,6,2',4',6'-HEXACHLOROBIPHENYL

```
    CL      CL
    9 |     | 3           FORMULA   C12H4CL6           MOL WT   360.88
    C-C8  2C-C            SOLVENT   C2D4BR2
   10/   \7 1/   \4       ORIG ST   C4H12SI            TEMP        313
  CL-C    C-C     C-CL
    \    /   \   /        132.20   135.00   127.80   134.90
     C=C12  6C=C           1/1      2/1      3/2      4/1
    11 |     | 5
      CL     CL           N.K.WILSON
                          J AM CHEM SOC                 97, 3573 (1975)
```

2018   Q 007000 3,4,5,3',4',5'-HEXACHLOROBIPHENYL

```
   CL         CL
    \9 8   2 3/           FORMULA   C12H4CL6           MOL WT   360.88
     C-C    C-C           SOLVENT   C2D4BR2
   10/   \7 1/   \4       ORIG ST   C4H12SI            TEMP        313
  CL-C    C-C     C-CL
    \    /   \   /        136.60   126.40   134.20   133.60
    11C=C    C=C           1/1      2/2      3/1      4/1
    / 12    6 5\
   CL        CL           N.K.WILSON
                          J AM CHEM SOC                 97, 3573 (1975)
```

2019   Q 007000 2,3,4,5,6-PENTACHLOROBIPHENYL

```
        CL   CL           FORMULA   C12H5CL5           MOL WT   326.44
    9 8   \2 3/           SOLVENT   C2D4BR2
    C-C    C-C            ORIG ST   C4H12SI            TEMP        313
  10/   \7 1/   \4
    C     C-C     C-CL    139.80   132.36   136.33   132.26   131.23   128.10
     \   /   \   /         1/1      2/1      3/1      4/1      7/1      8/2
      C=C    C=C          128.25   128.15
    11 12   /6 5\          9/2      10/2
      CL    CL
                          N.K.WILSON
                          J AM CHEM SOC                 97, 3573 (1975)
```

2020 — Q 007000 2,4,5,2',5'-PENTACHLOROBIPHENYL

```
        CL    CL
      9 |    | 3
      C-C8  2C-C
    10/    \7 1/   \4
     C     C-C    C-CL
      \   /   \   /
    11C=C     C=C
      /  12   6 5\
    CL            CL
```

FORMULA	C12H5CL5			MOL WT	326.44
SOLVENT	C2D4BR2				
ORIG ST	C4H12SI			TEMP	313

136.50	131.50	130.56	133.00	132.22	129.61
1/1	2/1	3/2	4/1	5/1	6/2
137.19	130.72	131.73	130.37	132.00	130.46
7/1	8/1	9/2	10/2	11/1	12/2

N.K.WILSON
J AM CHEM SOC　　　　　　　97, 3573 (1975)

2021 — Q 007000 2,3,4,5-TETRACHLOROBIPHENYL

```
        CL    CL
      9 8  \2 3/
      C-C   C-C
    10/   \7 1/  \4
     C    C-C   C-CL
      \  /   \  /
      C=C    C=C
    11 12   6 5\
              CL
```

FORMULA	C12H6CL4			MOL WT	291.99
SOLVENT	C2D4BR2				
ORIG ST	C4H12SI			TEMP	313

140.34	130.60	132.98	130.97	131.18	129.37
1/1	2/1	3/1	4/1	5/1	6/2
136.74	127.74	128.43	127.95		
7/1	8/2	9/2	10/2		

N.K.WILSON
J AM CHEM SOC　　　　　　　97, 3573 (1975)

2022 — Q 007000 2,3,5,6-TETRACHLOROBIPHENYL

```
        CL    CL
      9 8  \2 3/
      C-C   C-C
    10/   \7 1/  \
     C    C-C   C4
      \  /   \  /
      C=C    C=C
    11 12   /6 5\
       CL        CL
```

FORMULA	C12H6CL4			MOL WT	291.99
SOLVENT	C2D4BR2				
ORIG ST	C4H12SI			TEMP	313

141.80	131.35	131.15	129.36	136.42	127.93
1/1	2/1	3/1	4/2	7/1	8/2
128.19	127.93				
9/2	10/2				

N.K.WILSON
J AM CHEM SOC　　　　　　　97, 3573 (1975)

2023 — Q 007000 2,4,2',4'-TETRACHLOROBIPHENYL

```
        CL    CL
      9 |    | 3
      C-C8  2C-C
    10/    \7 1/   \4
   CL-C     C-C   C-CL
      \   /   \   /
      C=C     C=C
    11 12     6 5
```

FORMULA	C12H6CL4			MOL WT	291.99
SOLVENT	C2D4BR2				
ORIG ST	C4H12SI			TEMP	313

135.20	133.80	131.50	134.20	126.60	129.00
1/1	2/1	3/2	4/1	5/2	6/2

N.K.WILSON
J AM CHEM SOC　　　　　　　97, 3573 (1975)

2024 — Q 007000 2,5,2',5'-TETRACHLOROBIPHENYL

```
        CL    CL
      9 |    | 3
      C-C8  2C-C
    10/    \7 1/   \
     C     C-C    C4
      \   /   \   /
    11C=C     C=C
      /  12   6 5\
    CL            CL
```

FORMULA	C12H6CL4			MOL WT	291.99
SOLVENT	C2D4BR2				
ORIG ST	C4H12SI			TEMP	313

137.70	131.14	130.29	129.17	131.64	130.09
1/1	2/1	3/2	4/2	5/1	6/2

N.K.WILSON
J AM CHEM SOC　　　　　　　97, 3573 (1975)

2025   Q 007000 2,6,2',6'—TETRACHLOROBIPHENYL

```
      CL    CL
   9 |    | 3
    C-C8  2C-C
 10/   \7 1/   \
  C    C-C    C4
   \  /   \  /
   C=C12  6C=C
  11 |      | 5
    CL      CL
```

FORMULA   C12H6CL4   MOL WT   291.99
SOLVENT   C2D4BR2
ORIG ST   C4H12SI   TEMP   313

134.40	134.30	127.40	129.80
1/1	2/1	3/2	4/2

N.K.WILSON
J AM CHEM SOC   97, 3573 (1975)

2026   Q 007000 3,4,3',4'—TETRACHLOROBIPHENYL

```
   CL          CL
    \9 8    2 3/
     C-C    C-C
  10/   \7 1/   \4
 CL-C    C-C    C-CL
     \  /   \  /
     C=C    C=C
    11 12   6 5
```

FORMULA   C12H6CL4   MOL WT   291.99
SOLVENT   C2D4BR2
ORIG ST   C4H12SI   TEMP   313

137.80	128.30	132.50	131.80	130.40	125.70
1/1	2/2	3/1	4/1	5/2	6/2

N.K.WILSON
J AM CHEM SOC   97, 3573 (1975)

2027   Q 007000 3,5,3',5'—TETRACHLOROBIPHENYL

```
   CL          CL
    \9 8    2 3/
     C-C    C-C
  10/   \7 1/   \4
  C    C-C    C
   \  /   \  /
 11C=C    C=C
  / 12    6 5\
 CL          CL
```

FORMULA   C12H6CL4   MOL WT   291.99
SOLVENT   C2D4BR2
ORIG ST   C4H12SI   TEMP   313

140.40	125.10	134.90	127.90
1/1	2/2	3/1	4/2

N.K.WILSON
J AM CHEM SOC   97, 3573 (1975)

2028   Q 007000 2,6—DICHLOROBIPHENYL

```
       CL
   9 8    \2 3
    C-C    C-C
 10/   \7 1/   \C4
  C    C-C    
   \  /   \  /
   C=C    C=C
  11 12   /6 5
       CL
```

FORMULA   C12H8CL2   MOL WT   223.10
SOLVENT   C2D4BR2
ORIG ST   C4H12SI   TEMP   313

136.00	134.15	127.55	128.42	138.72	127.43
1/1	2/1	3/2	4/2	7/1	8/2
128.87	127.43				
9/2	10/2				

N.K.WILSON
J AM CHEM SOC   97, 3573 (1975)

2029   Q 007000 3,4—DICHLOROBIPHENYL

```
          CL
   9 8    2 3/
    C-C    C-C
 10/   \7 1/   \4
  C    C-C    C-CL
   \  /   \  /
   C=C    C-C
  11 12   6 5
```

FORMULA   C12H8CL2   MOL WT   223.10
SOLVENT   C2D4BR2
ORIG ST   C4H12SI   TEMP   313

140.29	128.35	132.07	130.60	130.04	125.77
1/1	2/2	3/1	4/1	5/2	6/2
137.69	126.28	128.35	127.54		
7/1	8/2	9/2	10/2		

N.K.WILSON
J AM CHEM SOC   97, 3573 (1975)

2030   Q 007000 2,2'—DICHLOROBIPHENYL

```
    CL      CL
   9 18   21 3
    C-C     C-C
  10/    \7 1/   \
  C       C-C     C 4
   \     /   \   /
    C=C       C=C
   11 12     6 5
```

FORMULA	C12H8CL2			MOL WT	223.10
SOLVENT	C2D4BR2				
ORIG ST	C4H12SI			TEMP	313

137.60	132.80	130.70	128.70	126.00	128.90
1/1	2/1	3/2	4/2	5/2	6/2

N.K.WILSON
J AM CHEM SOC   97, 3573 (1975)

2031   Q 007000 3,3'—DICHLOROBIPHENYL

```
  CL            CL
  \9 8     2 3/
   C-C     C-C
  10/   \7 1/   \
  C       C-C     C 4
   \     /   \   /
    C=C       C=C
   11 12     6 5
```

FORMULA	C12H8CL2			MOL WT	223.10
SOLVENT	C2D4BR2				
ORIG ST	C4H12SI			TEMP	313

140.70	127.40	134.20	126.80	129.60	124.80
1/1	2/2	3/1	4/2	5/2	6/2

N.K.WILSON
J AM CHEM SOC   97, 3573 (1975)

2032   Q 007000 4,4'—DICHLOROBIPHENYL

```
    9 8     2 3
    C-C     C-C
  10/   \7 1/   \4
  CL-C     C-C     C-CL
   \     /   \   /
    C=C       C=C
   11 12     6 5
```

FORMULA	C12H8CL2		MOL WT	223.10
SOLVENT	C2D4BR2			
ORIG ST	C4H12SI		TEMP	313

137.50	127.70	128.50	132.90
1/1	2/2	3/2	4/1

N.K.WILSON
J AM CHEM SOC   97, 3573 (1975)

2033   Q 007000 2-CHLOROBIPHENYL

```
          CL
    9 8    \2 3
    C-C     C-C
  10/   \7 1/   \
  C       C-C     C 4
   \     /   \   /
    C=C       C=C
   11 12     6 5
```

FORMULA	C12H9CL			MOL WT	188.66
SOLVENT	C2D4BR2				
ORIG ST	C4H12SI			TEMP	313

138.50	131.75	130.81	127.95	126.28	129.34
1/1	2/1	3/2	4/2	5/2	6/2
139.75	127.44	128.85	126.98		
7/1	8/2	9/2	10/2		

N.K.WILSON
J AM CHEM SOC   97, 3573 (1975)

2034   Q 007000 3-CHLOROBIPHENYL

```
           CL
    9 8     2 3/
    C-C     C-C
  10/   \7 1/   \
  C       C-C     C 4
   \     /   \   /
    C=C       C=C
   11 12     6 5
```

FORMULA	C12H9CL			MOL WT	188.66
SOLVENT	C2D4BR2				
ORIG ST	C4H12SI			TEMP	313

142.15	127.31	133.86	126.68	129.43	124.71
1/1	2/2	3/1	4/2	5/2	6/2
138.77	126.50	128.29	126.68		
7/1	8/2	9/2	10/2		

N.K.WILSON
J AM CHEM SOC   97, 3573 (1975)

2035 Q 007000 4-CHLOROBIPHENYL

```
        9 8       2 3
        C-C       C-C
     10╱   ╲7 1╱   ╲4
       C     C-C     C-CL
       ╲   ╱   ╲   ╱
        C=C       C=C
       11 12      6 5
```

FORMULA	C12H9CL			MOL WT	188.66
SOLVENT	C2D4BR2				
ORIG ST	C4H12SI			TEMP	313

138.93	127.86	128.41	132.73	139.13	126.43
1/1	2/2	3/2	4/1	7/1	8/2
128.41	127.11				
9/2	10/2				

N.K.WILSON
J AM CHEM SOC   97, 3573 (1975)

2036 Q 007000 4-IODOBIPHENYL

```
        9 8       3 2
        C-C       C-C
      ╱    ╲7 4╱   ╲1
    10C     C-C     C-J
       ╲   ╱   ╲   ╱
        C=C       C=C
       11 12      5 6
```

FORMULA	C12H9I			MOL WT	280.11
SOLVENT	COCL3				
ORIG ST	C4H12SI			TEMP	AMB

92.90	137.60	128.70	139.80	128.70	137.60
1/1	2/2	3/2	4/1	5/2	6/2
140.40	126.60	128.70	127.40	128.70	126.60
7/1	8/2	9/2	10/2	11/2	12/2

L.F.JOHNSON,W.C.JANKOWSKI
CARBON-13 NMR SPECTRA,JOHN
WILEY AND SONS,NEW YORK   430 (1972)

2037 Q 007000 BIPHENYL

```
        9 8       2 3
        C-C       C-C
     10╱   ╲7 1╱   ╲
       C     C-C     C4
       ╲   ╱   ╲   ╱
        C=C       C=C
       11 12      6 5
```

FORMULA	C12H10		MOL WT	154.21
SOLVENT	C2D4BR2			
ORIG ST	C4H12SI		TEMP	313

140.60	126.75	128.37	126.87
1/1	2/2	3/2	4/2

N.K.WILSON
J AM CHEM SOC   97, 3573 (1975)

2038 Q 007100 DIPHENYLCARBENIUM ION

```
      4 3         9 10
      C=C         C=C
    ╱   ╲2 1+╱      ╲
  5C     C-C-C       C11
    ╲   ╱ | 8╲      ╱
      C-C   H   C-C
      6 7       13 12
```

FORMULA	C13H11			MOL WT	167.23
SOLVENT	CLFO2S/F5SB				
ORIG ST	C S2			TEMP	198

199.40	137.10	147.90	132.50	149.70	132.50
1/2	2/1	3/2	4/2	5/2	6/2
142.20	137.10	147.90	132.50	149.70	132.50
7/2	8/1	9/2	10/2	11/2	12/2
242.80					
13/2					

G.A.OLAH,P.W.WESTERMAN
J AM CHEM SOC   95, 7530 (1973)

2039 Q 007100 HYDROXYDIPHENYLCARBENIUM ION

```
      4 3
      C=C
    ╱   ╲
  5C     C2
    ╲   ╱ ╲ +
    C-C  1C-OH
    6 7   |
         8C
        ╱ ╲
      9C   C13
      ||   |
     10C   C12
        ╲ ╱
        11C
```

FORMULA	C13H11O			MOL WT	183.23
SOLVENT	CLFO2S/F5SB				
ORIG ST	C S2			TEMP	198

207.90	130.20	138.90	131.20	144.20	131.20
1/1	2/1	3/2	4/2	5/2	6/2
138.90	129.30	135.40	131.20	142.10	131.20
7/2	8/1	9/2	10/2	11/2	12/2
135.40					
13/2					

G.A.OLAH,P.W.WESTERMAN
J AM CHEM SOC   95, 7530 (1973)

2040   Q 007100 DIPHENYLMETHANE

```
    9 8         2 3
    C-C         C-C
  10╱   ╲7 13 1╱   ╲
   C      C-CH2-C     C4
    ╲   ╱       ╲   ╱
    C=C         C=C
    11 12       6 5
```

FORMULA	C13H12			MOL WT	168.24
SOLVENT	CDCL3				
ORIG ST	C4H12SI			TEMP	AMB

140.80	128.70	128.30	125.70	42.00
1/1	2/2	3/2	4/2	13/3

G.W.BUCHANAN,G.MONTAUDO,P.FINOCCHIARO
CAN J CHEM   52, 3196 (1974)

2041   Q 007100 DIPHENYLMETHANE

```
   10 9        3 4
   C-C         C-C
  ╱   ╲8 1   2╱   ╲
 11C      C-CH2-C    C5
   ╲   ╱       ╲   ╱
    C=C         C=C
    12 13       7 6
```

FORMULA	C13H12				MOL WT	168.24
SOLVENT	CCL4					
ORIG ST	C4H12SI				TEMP	AMB

41.70	140.40	128.40	127.90	125.60	127.90
1/3	2/1	3/2	4/2	5/2	6/2
128.40	140.40	128.40	127.90	125.60	127.90
7/2	8/1	9/2	10/2	11/2	12/2
128.40					
13/2					

D.H.HUNTER,J.B.STOTHERS
CAN J CHEM   51, 2884 (1973)

2042   Q 007100 BENZOHYDROL

```
   12 13        7 6
   C-C   OH  C-C
  ╱   ╲ | | ╱   ╲
 11C     C-C-C     C5
   ╲  ╱8 | 2╲   ╱
    C=C   H   C=C
    10 9       3 4
```

FORMULA	C13H12O				MOL WT	184.24
SOLVENT	CDCL3					
ORIG ST	C4H12SI				TEMP	AMB

75.80	143.70	126.40	128.20	127.20	128.20
1/2	2/1	3/2	4/2	5/2	6/2
126.40	143.70	126.40	128.20	127.20	128.20
7/2	8/1	9/2	10/2	11/2	12/2
126.40					
13/2					

L.F.JOHNSON,W.C.JANKOWSKI
CARBON-13 NMR SPECTRA,JOHN
WILEY AND SONS,NEW YORK   453 (1972)

2043   Q 007100 DIPHENYLMETHYLCARBENIUM ION

```
       13 14
       C-C
      ╱   ╲
   12C      C9
     ╲   ╱ ╲ +
      C=C  1C-CH3
     11 10  |  2
          3C
         ╱ ╲
       4C    C8
        ||   |
       5C    C7
         ╲  ╱
          6C
```

FORMULA	C14H13				MOL WT	181.26
SOLVENT	CLFO2S/F5SB					
ORIG ST	CS2				TEMP	198

228.00	30.00	140.30	140.00	130.30	146.90
1/1	2/4	3/1	4/2	5/2	6/2
130.30	140.00	140.30	140.00	130.30	146.90
7/2	8/2	9/1	10/2	11/2	12/2
130.30	140.00				
13/2	14/2				

G.A.OLAH,P.W.WESTERMAN
J AM CHEM SOC   95, 7530 (1973)

2044   Q 007100 2-METHYLDIPHENYLMETHANE

```
        14
       H3C
    9 8      ╲2 3
    C-C       C-C
  10╱   ╲7 13 1╱   ╲
   C      C-CH2-C     C4
    ╲   ╱       ╲   ╱
    C=C         C=C
    11 12       6 5
```

FORMULA	C14H14				MOL WT	182.27
SOLVENT	CDCL3					
ORIG ST	C4H12SI				TEMP	AMB

138.50	136.10	129.60	126.10	127.00	130.00
1/1	2/1	3/2	4/2	5/2	6/2
140.00	128.40	128.10	125.60	39.50	19.60
7/1	8/2	9/2	10/2	13/3	14/4

G.W.BUCHANAN,G.MONTAUDO,P.FINOCCHIARO
CAN J CHEM   52, 3196 (1974)

2045   Q 007100   4-METHYLDIPHENYLMETHANE

```
      9 8           2 3
      C-C           C-C
   10⁄   ⁊7 13 1⁄    ⁊4 14
    C      C-CH2-C      C-CH3
     \   /       \   /
      C=C         C=C
     11 12        6 5
```

FORMULA C14H14				MOL WT	182.27
SOLVENT CDCL3					
ORIG ST C4H12SI				TEMP	AMB
137.80	128.80	128.60	134.50	141.20	128.40
1/1	2/2	3/2	4/1	7/1	8/2
128.10	125.60	41.60	21.00		
9/2	10/2	13/3	14/4		

G.W.BUCHANAN,G.MONTAUDO,P.FINOCCHIARO
CAN J CHEM   52, 3196 (1974)

2046   Q 007100   2,5-DIMETHYLDIPHENYLMETHANE

```
              14
             H3C
      9 8       \2 3
      C-C        C-C
   10⁄   ⁊7 13 1⁄   \
    C      C-CH2-C      C4
     \   /       \   /
      C=C         C=C
     11 12        6 5\15
                    CH3
```

FORMULA C15H16				MOL WT	196.29
SOLVENT CDCL3					
ORIG ST C4H12SI				TEMP	AMB
138.30	133.10	129.90	126.80	134.90	130.50
1/1	2/1	3/2	4/2	5/1	6/2
140.30	128.40	128.00	125.50	39.50	19.20
7/1	8/2	9/2	10/2	13/3	14/4
21.00					
15/4					

G.W.BUCHANAN,G.MONTAUDO,P.FINOCCHIARO
CAN J CHEM   52, 3196 (1974)

2047   Q 007100   4,4-DIMETHOXYBENZHYDROL

```
      6 7          14 13
      C=C   OH     C=C
    5⁄   \   I   /    \12
    O-C      C-CH-C      C-O
    ⁄   \   ⁄2 1 9\   ⁄    \
  CH3  C-C       C-C   H3C
   8    4 3      10 11   15
```

FORMULA C15H16O3				MOL WT	244.29
SOLVENT C2D6OS					
ORIG ST C4H12SI				TEMP	300
73.80	127.75	138.45	113.60	158.40	113.60
1/2	2/1	3/2	4/2	5/1	6/2
138.45	55.15	127.75	138.45	113.60	158.40
7/2	8/4	9/1	10/2	11/2	12/1
113.60	138.45	55.15			
13/2	14/2	15/4			

W.VOELTER,S.FUCHS,R.H.SEUFFER,K.ZECH
MONATSH CHEM   105, 1110 (1974)

2048   Q 007100   2,3,5,6-TETRAMETHYLDIPHENYLMETHANE

```
        14        15
        H3C       CH3
      9 8   \2 3⁄
      C-C      C-C
   10⁄   ⁊7 13 1⁄    \
    C      C-CH2-C      C4
     \   /       \   /
      C=C         C=C
     11 12    17⁄6 5\16
            H3C       CH3
```

FORMULA C17H20				MOL WT	224.35
SOLVENT CDCL3					
ORIG ST C4H12SI				TEMP	AMB
136.20	132.70	133.30	129.80	140.00	128.00
1/1	2/1	3/2	4/2	7/1	8/2
127.60	125.30	35.60	15.80	20.60	
9/2	10/2	13/3	14/4	15/4	

G.W.BUCHANAN,G.MONTAUDO,P.FINOCCHIARO
CAN J CHEM   52, 3196 (1974)

2049   Q 007100   2,5,2',5'-TETRAMETHYLDIPHENYLMETHANE

```
        16    14
        CH3 H3C
      9 8⁄       \2 3
      C-C        C-C
   10⁄   ⁊7 13 1⁄    \
    C      C-CH2-C      C4
     \   /       \   /
     11C=C         C=C
    17⁄   12      6 5\15
    H3C                CH3
```

FORMULA C17H20				MOL WT	224.35
SOLVENT CDCL3					
ORIG ST C4H12SI				TEMP	AMB
138.00	133.00	129.80	126.60	135.00	129.70
1/1	2/1	3/2	4/2	5/1	6/2
36.60	19.20	21.20			
13/3	14/4	15/4			

G.W.BUCHANAN,G.MONTAUDO,P.FINOCCHIARO
CAN J CHEM   52, 3196 (1974)

2050　Q 007100　2,3,5,6,2'-PENTAMETHYLDIPHENYLMETHANE

```
    18    14       15
   CH3  H3C      CH3
  9 8/       \2 3/
  C-C          C-C
 10/    \7 13 1/    \
  C     C-CH2-C     C4
   \   /      \    /
   C=C         C=C
  11 12      17/6 5\16
    H3C          CH3
```

FORMULA	C18H22			MOL WT	238.38
SOLVENT	CDCL3				
ORIG ST	C4H12SI			TEMP	AMB
136.10	132.90	133.30	129.70	137.80	135.60
1/1	2/1	3/1	4/2	7/1	8/1
129.40	125.70	126.40	125.40	33.30	15.70
9/2	10/2	11/2	12/2	13/3	14/4
20.50	19.20				
15/4	18/4				

G.W.BUCHANAN,G.MONTAUDO,P.FINOCCHIARO
CAN J CHEM　　　　　　　　52, 3196 (1974)

2051　Q 007100　2,3,5,6,4'-PENTAMETHYLDIPHENYLMETHANE

```
        14      15
      H3C      CH3
  9 8      \2 3/
  C-C        C-C
 18 /   \7 13 1/   \
H3C-C10   C-CH2-C    C4
   \   /      \    /
   C=C         C=C
  11 12      17/6 5\16
    H3C          CH3
```

FORMULA	C18H22			MOL WT	238.38
SOLVENT	CDCL3				
ORIG ST	C4H12SI			TEMP	AMB
136.50	132.70	133.30	129.70	137.00	128.70
1/1	2/1	3/1	4/2	7/1	8/2
127.50	134.60	35.30	15.80	20.60	20.90
9/2	10/1	13/3	14/4	15/4	18/4

G.W.BUCHANAN,G.MONTAUDO,P.FINOCCHIARO
CAN J CHEM　　　　　　　　52, 3196 (1974)

2052　Q 007100　2,4,6,2',5'-PENTAMETHYLDIPHENYLMETHANE

```
    17    14
   CH3  H3C
  9 8/       \2 3
  C-C          C-C
 10/    \7 13 1/    \415
  C     C-CH2-C     C-CH3
   \   /      \    /
 11C=C         C=C
  /  12       /6 5
 H3C        H3C
 18          16
```

FORMULA	C18H22			MOL WT	238.38
SOLVENT	CDCL3				
ORIG ST	C4H12SI			TEMP	AMB
135.00	136.70	128.60	134.90	137.40	133.40
1/1	2/1	3/2	4/1	7/1	8/1
129.20	126.50	133.60	126.00	32.10	20.20
9/2	10/2	11/1	12/2	13/3	14/4
21.10	19.30	21.20			
15/4	17/4	18/4			

G.W.BUCHANAN,G.MONTAUDO,P.FINOCCHIARO
CAN J CHEM　　　　　　　　52, 3196 (1974)

2053　Q 007100　2,4,6,2',4',6'-HEXAMETHYLDIPHENYLMETHANE

```
    CH3 H3C
  9 8/17   14\2 3
  C-C          C-C
 18 /   \7 13 1/   \4 15
H3C-C10   C-CH2-C    C-CH3
   \   /      \    /
   C=C12        C=C
  11  \19   16/6 5
    CH3 H3C
```

FORMULA	C19H24			MOL WT	252.40
SOLVENT	CDCL3				
ORIG ST	C4H12SI			TEMP	AMB
134.60	136.30	129.00	134.50	31.20	20.90
1/1	2/1	3/2	4/1	13/3	14/4
20.90					
15/4					

G.W.BUCHANAN,G.MONTAUDO,P.FINOCCHIARO
CAN J CHEM　　　　　　　　52, 3196 (1974)

2054　Q 007300　TRIPHENYLCARBENIUM ION

```
  12 13        7 6
   C-C        C-C
  //   \ 1+ /    \
 11C    C-C-C     C5
   \   /8 | 2\   /
   C=C  C14 C=C
  10 9  / \  3 4
    19C   C15
     ||   |
    18C   C16
     \  //
       C
       17
```

FORMULA	C19H15			MOL WT	243.33
SOLVENT	CLFO2S/F5SB				
ORIG ST	CS2			TEMP	198
210.60	139.60	143.00	130.00	142.80	130.00
1/1	2/1	3/2	4/2	5/2	6/2
143.00	139.60	143.00	130.00	142.80	130.00
7/2	8/1	9/2	10/2	11/2	12/2
143.00	139.60	143.00	130.00	142.80	130.00
13/2	14/1	15/2	16/2	17/2	18/2
143.00					
19/2					

G.A.OLAH,P.W.WESTERMAN
J AM CHEM SOC　　　　　　　95, 7530 (1973)

2055 — Q 007600 PHENYLCYCLOPROPYLCARBENIUM ION

```
        10 9
3C           C=C
  I\2 1+ /      \
  I C-C-C        C8
  I/  I 5\       /
4C   H   C-C
         6 7
```

FORMULA	C10H11			MOL WT	131.20
SOLVENT	CLFO2S/F5SB				
ORIG ST	CS2			TEMP	198

225.10	43.90	43.90	43.90	136.40	144.00
1/2	2/2	3/3	4/3	5/1	6/2
131.00	147.80	131.00	135.20		
7/2	8/2	9/2	10/2		

G.A.OLAH,P.W.WESTERMAN
J AM CHEM SOC   95, 7530 (1973)

2056 — Q 007600 PHENYLCYCLOPROPYLMETHYLCARBENIUM ION

```
    C5
   /I
  / I  1+
C--C-C-CH3
4  3 I 2
     C6
    / \
 11C    C7
  II    I
 10C    C8
   \  //
     C9
```

FORMULA	C11H13			MOL WT	145.23
SOLVENT	CLFO2S/F5SB				
ORIG ST	CS2			TEMP	198

245.00	22.20	44.60	43.80	43.80	138.60
1/1	2/4	3/2	4/3	5/3	6/1
133.60	130.00	144.30	130.00	133.60	
7/2	8/2	9/2	10/2	11/2	

G.A.OLAH,P.W.WESTERMAN
J AM CHEM SOC   95, 7530 (1973)

2057 — Q 007600 DICYCLOPROPYLPHENYLCARBENIUM ION

```
  C7    4C
 /I     I\
/ I  1+ I \
C--C--C--C--C
6  5 I 2 3
    8C
   / \
 9C    C13
  II    I
 10C    C12
   \   //
   11C
```

FORMULA	C13H15			MOL WT	171.26
SOLVENT	CLFO2S/F5SB				
ORIG ST	CS2			TEMP	198

259.80	41.50	35.80	35.80	41.50	35.80
1/1	2/2	3/3	4/3	5/2	6/3
35.80	134.10	128.40	126.80	135.30	126.80
7/3	8/1	9/2	10/2	11/2	12/2
128.40					
13/2					

G.A.OLAH,P.W.WESTERMAN
J AM CHEM SOC   95, 7530 (1973)

2058 — Q 007600 DIPHENYLCYCLOPROPYLCARBENIUM ION

```
 13 12
  C=C
 /   \11
14C    C
  \  / \1+
   C-C   C-C--C4
  15 16  I 2\ I
       5C   \I
        / \   C3
      6C    C10
       II    I
      7C    C9
        \  //
        8C
```

FORMULA	C16H15			MOL WT	207.30
SOLVENT	CLFO2S/F5SB				
ORIG ST	CS2			TEMP	198

233.80	39.70	34.70	34.70	140.70	137.80
1/1	2/2	3/3	4/3	5/1	6/2
130.20	149.00	130.20	137.80	140.70	137.80
7/2	8/2	9/2	10/2	11/1	12/2
130.20	149.00	130.20	137.80		
13/2	14/2	15/2	16/2		

G.A.OLAH,P.W.WESTERMAN
J AM CHEM SOC   95, 7530 (1973)

2059 — Q 007800 CIS-DECALIN

```
     C7
 9  8/I
 C---C I
 /I C---C6
/ I/5
1C  C10
 \  \
  \  \
  2C  C4
  I /
  I/
  C3
```

FORMULA	C10H18		MOL WT	138.25
SOLVENT	C10H18			
ORIG ST	C4H12SI		TEMP	AMB

| 29.80 | 24.60 | 36.95 |
| 1/3 | 2/3 | 9/2 |

D.K.DALLING,D.M.GRANT
J AM CHEM SOC   96, 1827 (1974)

2060 Q 007800 TRANS-DECALIN

```
            C7
      9    8/1
      C---C 1
 2   1/1 C---C6
 C---C 1/5
 I C---C10
 1/4
 C3
```

FORMULA	C10H18		MOL WT	138.25
SOLVENT	C10H18			
ORIG ST	C4H12SI		TEMP	AMB

| 34.75 | 27.15 | 44.20 |
| 1/3 | 2/3 | 9/2 |

D.K.DALLING,D.M.GRANT
J AM CHEM SOC 96, 1827 (1974)

2061 Q 007800 1-METHYLNAPHTHALENE

```
      1CH3
    9   I
    C   C2
  ⁄ \ / \
 8C   C10 C3
 I   II   I
 7C   C11 C4
  \ ⁄ \ ⁄
   C   C
   6   5
```

FORMULA	C11H10		MOL WT	142.20
SOLVENT	CDCL3			
ORIG ST	C4H12SI		TEMP	AMB

19.10	123.90	125.30	125.50	126.30	126.40
1/4					
128.40	132.60	133.50	134.00		

L.F.JOHNSON,W.C.JANKOWSKI
CARBON-13 NMR SPECTRA,JOHN
WILEY AND SONS,NEW YORK 415 (1972)

2062 Q 007800 2-METHYLNAPHTALENE

```
   7C-C8
  ⁄／  \
 6C      C9
  \    ⁄
   5C=C10
  ⁄    \
 4C      C1
  \    ⁄
   3C-C2
       \
        CH3
        11
```

FORMULA	C11H10		MOL WT	142.20
SOLVENT	CDCL3			
ORIG ST	C4H12SI		TEMP	AMB

135.10	131.70	133.60	21.40	124.80	125.70
2/1	5/1	10/1	11/4		
126.70	127.10	127.50	127.60	127.90	

L.F.JOHNSON,W.C.JANKOWSKI
CARBON-13 NMR SPECTRA,JOHN
WILEY AND SONS,NEW YORK 416 (1972)

2063 Q 007800 ACENAPHTHENE

```
    1    2
    C---C
    I    I
   9C    C10
  ⁄ \ ⁄ \
 8C   C11 C3
 I   II   I
 7C   C12 C4
  \ ⁄ \ ⁄
   C    C
   6    5
```

FORMULA	C12H10		MOL WT	154.21
SOLVENT	CCL4			
ORIG ST	C4H12SI		TEMP	AMB

30.10	30.10	118.60	127.10	121.90	121.90
1/3	2/3	3/2	4/2	5/2	6/2
127.10	118.60	145.00	145.00	139.10	131.40
7/2	8/2	9/1	10/1	11/1	12/1

D.H.HUNTER,J.B.STOTHERS
CAN J CHEM 51, 2884 (1973)

2064 Q 007800 1,8-DIMETHYLNAPHTHALENE

```
   11   12
   CH3 CH3
    I   I
   8C   C1
  ⁄ \ ⁄ \
 7C   C9 C2
 I   II   I
 6C   C10 C3
  \ ⁄ \ ⁄
   C   C
   5   4
```

FORMULA	C12H12		MOL WT	156.23
SOLVENT	CCL4			
ORIG ST	C4H12SI		TEMP	AMB

134.40	128.80	124.30	127.50	127.50	124.30
1/1	2/2	3/2	4/2	5/2	6/2
128.80	134.40	132.80	135.20	25.60	25.60
7/2	8/1	9/1	10/1	11/4	12/4

D.H.HUNTER,J.B.STOTHERS
CAN J CHEM 51, 2884 (1973)

2065   Q 007800 1,1,2,2-TETRAMETHYLACENAPHTHENE

```
  13  15
  CH3 CH3                FORMULA   C16H18              MOL WT   210.32
 14 |   | 16            SOLVENT   CCL4
H3C-C---C-CH3           ORIG ST   C4H12SI              TEMP      AMB
  |1  2|
  9C   C10              49.40    49.40   116.90   122.50   127.50   127.50
  ⁄  ╲  ⁄  ╲            1/1      2/1     3/2      4/2      5/2      6/2
 8C   C11 C3            122.50   116.90  152.60   152.60   134.60   131.20
  |    ‖   |            7/2      8/2     9/1      10/2     11/1     12/1
 7C   C12 C4            25.80    25.80   25.80    25.80
  ╲  ⁄  ╲  ⁄            13/4     14/4    15/4     16/4
    C    C
    6    5              D.H.HUNTER,J.B.STOTHERS
                        CAN J CHEM                          51, 2884 (1973)
```

2066   Q 008500 1-NAPHTHOL

```
     OH
  8   |                 FORMULA   C10H8O              MOL WT   144.17
  C   C1                SOLVENT   CDCL3
 ⁄ ╲ ⁄ ╲                ORIG ST   C4H12SI              TEMP      309
7C   C9  C2
 |   ‖   |              151.23   108.79  125.84   120.79   127.71   126.45
6C   C10 C3            1/1      2/2     3/2      4/2      5/2      6/2
 ╲ ⁄ ╲ ⁄                125.31   121.49  124.40   134.78
   C   C                7/2      8/2     9/1      10/1
   5   4
                        L.ERNST
                        CHEM BER                           108, 2030 (1975)
```

2067   Q 008500 1-NAPHTHOL

```
     OH
  8   |                 FORMULA   C10H8O              MOL WT   144.17
  C   C1                SOLVENT   C2D6O
 ⁄ ╲ ⁄ ╲                ORIG ST   C4H12SI              TEMP      309
7C   C9  C2
 |   ‖   |              153.88   108.81  126.93   119.82   128.17   126.78
6C   C10 C3            1/1      2/2     3/2      4/2      5/2      6/2
 ╲ ⁄ ╲ ⁄                125.27   122.87  125.81   135.75
   C   C                7/2      8/2     9/1      10/1
   5   4
                        L.ERNST
                        CHEM BER                           108, 2030 (1975)
```

2068   Q 008500 2-NAPHTHOL

```
  8   1                 FORMULA   C10H8O              MOL WT   144.17
  C   C   OH            SOLVENT   CDCL3
 ⁄ ╲ ⁄ ╲ ⁄              ORIG ST   C4H12SI              TEMP      309
7C   C9  C2
 |   ‖   |              109.61   153.35  117.78   129.91   127.82   123.70
6C   C10 C3            1/2      2/1     3/2      4/2      5/2      6/2
 ╲ ⁄ ╲ ⁄                126.58   126.43  134.67   129.07
   C   C                7/2      8/2     9/1      10/1
   5   4
                        L.ERNST
                        CHEM BER                           108, 2030 (1975)
```

2069   Q 008500 2-NAPHTHOL

```
  8   1                 FORMULA   C10H8O              MOL WT   144.17
  C   C   OH            SOLVENT   C2D6O
 ⁄ ╲ ⁄ ╲ ⁄              ORIG ST   C4H12SI              TEMP      309
7C   C9  C2
 |   ‖   |              109.76   156.04  119.12   130.20   128.39   123.60
6C   C10 C3            1/2      2/1     3/2      4/2      5/2      6/2
 ╲ ⁄ ╲ ⁄                126.90   126.90  135.89   129.32
   C   C                7/2      8/2     9/1      10/1
   5   4
                        L.ERNST
                        CHEM BER                           108, 2030 (1975)
```

2070 — Q 008500 1,5-NAPHTHALENEDIOL

```
        OH
     8   |
      C   C1
    ╱ ╲ ╱ ╲
   7C   C9  C2
    |   ||   |
   6C   C10 C3
    ╲ ╱ ╲ ╱
     C5   C
     |    4
     OH
```

FORMULA	C10H8O2			MOL WT	160.17
SOLVENT	C2D6O				
ORIG ST	C4H12SI			TEMP	309

153.64	109.29	125.61	114.21	127.20
1/1	2/2	3/2	4/2	9/1

L.ERNST
CHEM BER   108, 2030 (1975)

2071 — Q 008500 2,6-NAPHTHALENEDIOL

```
     8   1
      C   C   OH
    ╱ ╲ ╱ ╲ ╱
   7C   C9  C2
    |   ||   |
   6C   C10 C3
    ╱ ╲ ╱ ╲ ╲
  HO  C   C
      5   4
```

FORMULA	C10H8O2			MOL WT	160.17
SOLVENT	C2D6O				
ORIG ST	C4H12SI			TEMP	309

109.95	153.76	119.38	128.37	130.47
1/2	2/1	3/2	4/2	9/1

L.ERNST
CHEM BER   108, 2030 (1975)

2072 — Q 008500 2,7-NAPHTHALENEDIOL

```
     8   1
  HO  C   C   OH
    ╲ ╱ ╲ ╱ ╲ ╱
   7C   C9  C2
    |   ||   |
   6C   C10 C3
    ╲ ╱ ╲ ╱
     C   C
     5   4
```

FORMULA	C10H8O2			MOL WT	160.17
SOLVENT	C2D6O				
ORIG ST	C4H12SI			TEMP	309

108.46	156.42	116.09	130.10	137.54	124.39
1/2	2/1	3/2	4/2	9/1	10/1

L.ERNST
CHEM BER   108, 2030 (1975)

2073 — Q 008900 1-METHOXYNAPHTHALENE

```
     CH3
      ╱11
     O
     8   |
      C   C1
    ╱ ╲ ╱ ╲
   7C   C9  C2
    |   ||   |
   6C   C10 C3
    ╲ ╱ ╲ ╱
     C   C
     5   4
```

FORMULA	C11H10O			MOL WT	158.20
SOLVENT	CDCL3				
ORIG ST	C4H12SI			TEMP	309

155.47	103.80	125.86	120.24	127.45	126.38
1/1	2/2	3/2	4/2	5/2	6/2
125.14	122.03	125.69	134.55	55.33	
7/2	8/2	9/1	10/1	11/4	

L.ERNST
CHEM BER   108, 2030 (1975)

2074 — Q 008900 2-METHOXYNAPHTHALENE

```
     8   1    11
      C   C   O-CH3
    ╱ ╲ ╱ ╲ ╱
   7C   C9  C2
    |   ||   |
   6C   C10 C3
    ╲ ╱ ╲ ╲
     C   C
     5   4
```

FORMULA	C11H10O			MOL WT	158.20
SOLVENT	CDCL3				
ORIG ST	C4H12SI			TEMP	309

105.82	157.66	118.71	129.37	127.65	123.57
1/2	2/1	3/2	4/2	5/2	6/2
126.33	126.76	134.60	129.04	55.11	
7/2	8/2	9/1	10/1	11/4	

L.ERNST
CHEM BER   108, 2030 (1975)

2075   Q 009300   1-NAPHTHALDEHYDE

```
    7C-C8
    //   \\
  6C      C9
    \    /
    5C=C10   O
    /   \1  //
  4C      C-C11
    \    //  \
     C-C      H
     3 2
```

FORMULA	C11H8O			MOL WT	156.19
SOLVENT	COCL3				
ORIG ST	C4H12SI			TEMP	AMB

131.10	136.20	134.90	133.40	130.20	193.00
1/1	2/2	4/2	5/1	10/1	11/2
128.70	124.60	126.60	128.20		

L.F.JOHNSON,W.C.JANKOWSKI
CARBON-13 NMR SPECTRA,JOHN
WILEY AND SONS,NEW YORK   414 (1972)

2076   Q 009300   2-ACETONAPHTHONE

```
     C-C
    //  \\
   C      C
    \    /
    C=C
    /    \
   C      C
    \\   //
     C-C    O
        \  //
         1C
          |
         2CH3
```

FORMULA	C12H10O			MOL WT	170.21
SOLVENT	COCL3				
ORIG ST	C4H12SI			TEMP	AMB

197.50	26.30	123.70	126.50	127.60	128.20
1/1	2/4				
129.40	129.90	132.30	134.30	135.30	

L.F.JOHNSON,W.C.JANKOWSKI
CARBON-13 NMR SPECTRA,JOHN
WILEY AND SONS,NEW YORK   434 (1972)

2077   Q 010100   TETRALIN

```
     9   7
     C   C
    //\8/ \
  10C  C   C6
    |  ||  |
   1C  C   C5
    \ /3\ /
     C   C
     2   4
```

FORMULA	C10H12			MOL WT	132.21
SOLVENT	C4H8O2				
ORIG ST	C4H12SI			TEMP	AMB

129.00	125.50	136.80	29.50	23.60	23.60
1/2	2/2	3/1	4/3	5/3	6/3
29.50	136.80	125.50	129.00		
7/3	8/1	9/2	10/2		

L.F.JOHNSON,W.C.JANKOWSKI
CARBON-13 NMR SPECTRA,JOHN
WILEY AND SONS,NEW YORK   371 (1972)

2078   Q 010100   CIS-DECALIN

```
    8   1
    C   C
   / \9/ \
  7C  C   C2
  |   |   |
  6C  C10 C3
   \ / \ /
    C   C
    5   4
```

FORMULA	C10H18			MOL WT	138.25
SOLVENT	C4H8O2				
ORIG ST	C4H12SI			TEMP	AMB

36.80	29.70	24.50
9/2	1/3	2/3

L.F.JOHNSON,W.C.JANKOWSKI
CARBON-13 NMR SPECTRA,JOHN
WILEY AND SONS,NEW YORK   401 (1972)

2079   Q 010100   TRANS-DECALIN

```
    8   1
    C   C
   / \9/ \
  7C  C   C2
  |   |   |
  6C  C10 C3
   \ / \ /
    C   C
    5   4
```

FORMULA	C10H18			MOL WT	138.25
SOLVENT	C4H8O2				
ORIG ST	C4H12SI			TEMP	AMB

34.60	27.10	44.00
1/3	2/3	9/2

L.F.JOHNSON,W.C.JANKOWSKI
CARBON-13 NMR SPECTRA,JOHN
WILEY AND SONS,NEW YORK   401 (1972)

2080   Q 010100 CIS—SYN—1—METHYLDECALIN

```
        C6
       /|
      / |
    5C  C7
      \  \
       \  \
      10C  C8
      4/|  /
   3C---C 1/
    | C---C9
    1/11
   2C  CH3
      11
```

FORMULA	C11H20			MOL WT	152.28
SOLVENT	C11H20				
ORIG ST	C4H12SI			TEMP	AMB

37.18	29.54	27.42	25.77	33.55	21.89
1/2	2/3	3/3	4/3	5/3	6/3
27.42	20.04	42.98	38.70	19.78	
7/3	8/3	9/2	10/2	11/4	

D.K.DALLING,D.M.GRANT,E.G.PAUL
J AM CHEM SOC   95, 3718 (1973)

2081   Q 010100 CIS—SYN—2—METHYLDECALIN

```
        C6
       /|
      / |
    5C  C7
      \  \
       \  \
      10C  C8
      4/|  /
   3C---C 1/
    | C---C9
  11  1/1
  H3C--C2
```

FORMULA	C11H20			MOL WT	152.28
SOLVENT	C11H20				
ORIG ST	C4H12SI			TEMP	AMB

35.63	34.22	30.33	33.22	26.37	27.63
1/3	2/2	3/3	4/3	5/3	6/3
21.43	33.22	36.98	36.47	23.16	
7/3	8/3	9/2	10/2	11/4	

D.K.DALLING,D.M.GRANT,E.G.PAUL
J AM CHEM SOC   95, 3718 (1973)

2082   Q 010100 CIS—9—METHYLDECALIN

```
        C6
       /|
      / |
    5C  C7
      \  \
       \  \
      10C  C8
      4/|  /
   3C---C 1/
    | C---C9
    1/1   |
   2C    CH3
        11
```

FORMULA	C11H20			MOL WT	152.28
SOLVENT	C11H20				
ORIG ST	C4H12SI			TEMP	AMB

36.39	22.81	24.48	28.10	32.93	41.76
1/3	2/3	3/3	4/3	9/1	10/2
28.22					
11/4					

D.K.DALLING,D.M.GRANT,E.G.PAUL
J AM CHEM SOC   95, 3718 (1973)

2083   Q 010100 TRANS—ANTI—1—METHYLDECALIN

```
         .C6
    10    5/1
    C---C  |
    4/| C---C 7
 3C---C 1/8
  | C---C9
  1/|11
 2C |
   CH3
   11
```

FORMULA	C11H20			MOL WT	152.28
SOLVENT	C11H20				
ORIG ST	C4H12SI			TEMP	AMB

38.42	37.12	27.45	35.39	35.29	27.16
1/2	2/3	3/3	4/3	5/3	6/3
26.82	31.02	50.60	44.05	19.74	
7/3	8/3	9/2	10/2	11/4	

D.K.DALLING,D.M.GRANT,E.G.PAUL
J AM CHEM SOC   95, 3718 (1973)

2084   Q 010100 TRANS—SYN—2—METHYLDECALIN

```
          C6
    10    5/1
    C---C  |
    4/| C---C 7
 3C---C 1/8
  | C---C9
 11  1/1
 H3C--C
    2
```

FORMULA	C11H20			MOL WT	152.28
SOLVENT	C11H20				
ORIG ST	C4H12SI			TEMP	AMB

43.31	33.06	35.66	34.30	34.30	27.02
1/3	2/2	3/3	4/3	5/3	6/3
27.02	34.15	43.49	43.49	22.85	
7/3	8/3	9/2	10/2	11/4	

D.K.DALLING,D.M.GRANT,E.G.PAUL
J AM CHEM SOC   95, 3718 (1973)

```
2085                    Q 010100  TRANS-9-METHYLDECALIN

              C6           FORMULA   C11H20                MOL WT   152.28
        10    5/1          SOLVENT   C11H20
        C---C I            ORIG ST   C4H12SI               TEMP       AMB
       4/I C--C7
     3C---C I/8               42.44    22.24    27.42    29.36    34.78    46.17
      I C---C9                 1/3      2/3      3/3      4/3      9/1     10/2
     1/1    I                 15.75
    2C      I                 11/4
         CH3
         11                D.K.DALLING,D.M.GRANT,E.G.PAUL
                           J AM CHEM SOC                   95, 3718 (1973)
```

```
2086                    Q 010100  TRANS-ANTI-1-SYN-2-DIMETHYLDECALIN

              C6           FORMULA   C12H22                MOL WT   166.31
        10    5/1          SOLVENT   C12H22
        C---C I            ORIG ST   C4H12SI               TEMP       AMB
       4/I C---C7
     3C---C I/8               44.34    39.27    35.82    34.80    34.51    27.14
      I C---C                  1/2      2/2      3/3      4/3      5/3      6/3
    12  I/II 9                26.58    30.96    49.36    43.30    16.10    20.94
   H3C--C21                    7/3      8/3      9/2     10/2     11/4     12/4
         CH3
         11                D.K.DALLING,D.M.GRANT,E.G.PAUL
                           J AM CHEM SOC                   95, 3718 (1973)
```

```
2087                    Q 010100  TRANS-ANTI-1-SYN-3-DIMETHYLDECALIN

              C6           FORMULA   C12H22                MOL WT   166.31
        10    5/1          SOLVENT   C12H22
        C---C I            ORIG ST   C4H12SI               TEMP       AMB
       4/I C---C7
     3C---C I/8               37.71    45.42    32.44    42.99    34.63    27.16
    12/I C---C                 1/2      2/3      3/2      4/3      5/3      6/3
   H3C I/II 9                 26.91    30.63    49.44    43.67    19.74    22.84
     C21                       7/3      8/3      9/2     10/2     11/4     12/4
        CH3
        11                 D.K.DALLING,D.M.GRANT,E.G.PAUL
                           J AM CHEM SOC                   95, 3718 (1973)
```

```
2088                    Q 010100  TRANS-ANTI-1-SYN-8-DIMETHYLDECALIN

              C6           FORMULA   C12H22                MOL WT   166.31
    11  10    5/1          SOLVENT   C12H22
   H3C    C---C I          ORIG ST   C4H12SI               TEMP       AMB
    I 4/I C---C7
   3C-X-C I/I8               34.04    35.60    28.54    35.48    35.37    20.95
    I C---C I                 1/2      2/3      3/3      4/3      5/3      6/3
    I/1    9 CH3             34.59    26.60    52.42    36.84    19.19    12.50
    C2     12                 7/3      8/2      9/2     10/2     11/4     12/4

                           D.K.DALLING,D.M.GRANT,E.G.PAUL
                           J AM CHEM SOC                   95, 3718 (1973)
```

```
2089                    Q 010100  TRANS-SYN-2-SYN-3-DIMETHYLDECALIN

              C6           FORMULA   C12H22                MOL WT   166.31
        10    5/1          SOLVENT   C12H22
        C---C I            ORIG ST   C4H12SI               TEMP       AMB
       4/I C---C7
     3C---C I/8               44.18    39.78    34.32    27.14    44.18    20.30
    12/I C---C                 1/3      2/2      5/3      6/3      9/2     11/4
   H3C I/1    9
   H3C--C2                 D.K.DALLING,D.M.GRANT,E.G.PAUL
    11                     J AM CHEM SOC                   95, 3718 (1973)
```

2090   Q 010100   TRANS-SYN-2-SYN-7-DIMETHYLDECALIN

```
          C6         FORMULA   C12H22                    MOL WT   166.31
     10    5/| CH3   SOLVENT   C12H22
      C---C 1/12     ORIG ST   C4H12SI                   TEMP       AMB
    4/| C---C7
  3C---C 1/8             43.05   32.89   35.60   34.05   42.84   42.60
   | C---C9               1/3     2/2     3/3     4/3     9/2    10/2
   |/1                   23.00
 H3C--C                  11/4
  11
```

D.K.DALLING,D.M.GRANT,E.G.PAUL
J AM CHEM SOC   95, 3718 (1973)

2091   Q 010200   ACENAPHTHENE

```
    11  10
     C-C          FORMULA   C12H10                       MOL WT   154.21
    //   \9       SOLVENT   CDCL3
  12C    C--C8    ORIG ST   C4H12SI                       TEMP      AMB
    \  /  |
    2C=C1  |         139.10  131.50  122.00  127.60  118.90  145.70
    /   \ 6 |7        1/1     2/1     3/2     4/2     5/2     6/1
  3C       C--C       30.10   30.10  145.70  118.90  127.60  122.00
    \\    //          7/3     8/3     9/1    10/2    11/2    12/2
    4C-C5
```

L.F.JOHNSON,W.C.JANKOWSKI
CARBON-13 NMR SPECTRA,JOHN
WILEY AND SONS,NEW YORK   440 (1972)

2092   Q 010400   BIPHENYLENE

```
   8           1
   C           C        FORMULA   C12H8                  MOL WT   152.20
  //  \      /  \       SOLVENT   CS2
 7C 12C---C9  C2        ORIG ST   C6H6                    TEMP      AMB
  |   ||   ||  |
 6C 11C---C10 C3          117.80  128.49  151.70
  \\ /      \ //           1/2     2/2     9/1
   C           C
   5           4
```

A.J.JONES,D.M.GRANT
CHEM COMMUN   1670 (1968)

2093   Q 010400   2,3-DIHYDROPHENALENE

```
     2
     C            FORMULA   C13H12                       MOL WT   168.24
    / \           SOLVENT   CCL4
  1C   C3         ORIG ST   C4H12SI                       TEMP      AMB
   |   |
  10C   C11          31.10   22.90   31.10  123.30  124.80  125.60
   // \ / \\         1/3     2/3     3/3     4/2     5/2     6/2
  9C   C12 C4       125.60  124.80  123.30  135.30  135.30  129.80
   |   ||  |         7/2     8/2     9/2    10/1    11/1    12/1
  8C   C13 C5       133.50
   \\ / \ //        13/1
    C   C
    7   6
```

D.H.HUNTER,J.B.STOTHERS
CAN J CHEM   51, 2884 (1973)

2094   Q 010500   INDENE

```
                  FORMULA   C9H8                         MOL WT   116.16
    4             SOLVENT   CDCL3
    C             ORIG ST   C4H12SI                       TEMP      AMB
   // \8
  5C   C---C3        38.85  133.65  131.90  123.50  124.40  126.05
   |   ||  ||         1/3     2/2     3/2     4/2     5/2     6/2
  6C   C   C2       120.80  144.60  143.35
   \ /9\ /           7/2     8/1     9/1
    C   C
    7   1
```

N.PLATZER,J.-J.BASSELIER,P.DEMERSEMAN
BULL SOC CHIM FRANCE   5, 905 (1974)

2095   Q 010500 INDANE OXIDE

```
         4
         C
       ⁄  \8
     5C    C---C3
     |     ||   |
     6C    C    C2
       \  ⁄9\ ⁄ \
         C    C---O
         7        1
```

58.70	57.30	34.40	140.80	143.40	124.90
1/2	2/2	3/3	8/1	9/1	
125.80	125.90	128.20			

FORMULA C9H8O   MOL WT 132.16
SOLVENT CDCL3
ORIG ST C4H12SI   TEMP   AMB

L.F.JOHNSON,W.C.JANKOWSKI
CARBON-13 NMR SPECTRA,JOHN
WILEY AND SONS,NEW YORK   337 (1972)

2096   Q 010500 1-INDANONE

```
         5
         C
       ⁄ \ 4
     6C   C---C3
     |    ||   |
     7C   C    C2
       \ ⁄9\  ⁄
         C   C1
         8   ||
             O
```

206.20	36.00	25.60	154.90	134.30	134.30
1/1	2/3	3/3	4/1	6/2	7/2
136.90	123.30	126.60	127.00		
9/1					

FORMULA C9H8O   MOL WT 132.16
SOLVENT CDCL3
ORIG ST C4H12SI   TEMP   AMB

L.F.JOHNSON,W.C.JANKOWSKI
CARBON-13 NMR SPECTRA,JOHN
WILEY AND SONS,NEW YORK   338 (1972)

2097   Q 010500 INDANE

```
      5C-C6
     ⁄⁄   \\
   4C      C 7
     \     ⁄
     3C=C8
     ⁄    \
   2C      C 9
     \     ⁄
      \   ⁄
       1C
```

25.30	32.80	143.90	125.90	124.20	124.20
1/3	2/3	3/1	4/2	5/2	6/2
125.90	143.90	32.80			
7/2	8/1	9/3			

FORMULA C9H10   MOL WT 118.18
SOLVENT CDCL3
ORIG ST C4H12SI   TEMP   AMB

L.F.JOHNSON,W.C.JANKOWSKI
CARBON-13 NMR SPECTRA,JOHN
WILEY AND SONS,NEW YORK   342 (1972)

2098   Q 010500 1-AMINOINDANE

```
       C-C
      ⁄⁄   \\
    6C      C9
      \     ⁄
      5C=C4
      ⁄    \
  H2N-C      C3
    1 \     ⁄
       \   ⁄
        C
        2
```

57.10	37.30	29.90	147.30	142.70	123.20
1/2	2/3	3/3	4/1	5/1	
124.40	126.30	126.90			

FORMULA C9H11N   MOL WT 133.19
SOLVENT CDCL3
ORIG ST C4H12SI   TEMP   AMB

L.F.JOHNSON,W.C.JANKOWSKI
CARBON-13 NMR SPECTRA,JOHN
WILEY AND SONS,NEW YORK   348 (1972)

2099   Q 010500 5-ACETYLINDANE

```
      O   1C   C8
   11 ||  2⁄ \9⁄ \
   H3C-C-C   C    \
   10 |    ||     C
      3C   C    ⁄7
       \  ⁄5\  ⁄
        C    C
        4    6
```

135.70	149.90	32.90	25.30	32.40	144.40
2/1	5/1	6/3	7/3	8/3	9/1
197.40	26.40	124.10	126.70		
10/1	11/4				

FORMULA C11H12O   MOL WT 160.22
SOLVENT CDCL3
ORIG ST C4H12SI   TEMP   AMB

L.F.JOHNSON,W.C.JANKOWSKI
CARBON-13 NMR SPECTRA,JOHN
WILEY AND SONS,NEW YORK   420 (1972)

```
2100                    Q 010500 2,2,3,3-TETRAMETHYLINDANONE-1
     5      12
     C      CH3          FORMULA    C13H16O              MOL WT   188.27
   ⁄ \6  7| 13          SOLVENT    CDCL3
  4C    C---C-CH3        ORIG ST    C4H12SI              TEMP        AMB
   |    ||   | 10
  3C    C   C-CH3         133.20   161.50   44.60   53.30   190.00   26.20
   \ ⁄1\ ⁄3\11            1/1      6/1      7/1     8/1     9/1      10/4
    C    C9  CH3          26.20    21.60    21.60   123.10  123.70   127.20
   2     ||               11/4     12/4     13/4
         J                134.40
```

L.F.JOHNSON,W.C.JANKOWSKI
CARBON-13 NMR SPECTRA,JOHN
WILEY AND SONS,NEW YORK   454 (1972)

```
2101                    Q 010600 FLUORENE
    6    1    13
    C    C    C          FORMULA    C13H10               MOL WT   166.22
  ⁄ \5⁄ \2⁄ \            SOLVENT    CDCL3
 7C   C   C   C12        ORIG ST    C4H12SI              TEMP        AMB
  |   ||  ||  |
 8C   C---C   C11         36.80    143.10   141.60  141.60  143.10   119.70
  \ ⁄4   3\ ⁄             1/3      2/1      3/1     4/1     5/1
    C       C             124.80   126.50
    9       10
```

L.F.JOHNSON,W.C.JANKOWSKI
CARBON-13 NMR SPECTRA,JOHN
WILEY AND SONS,NEW YORK   452 (1972)

```
2102                    Q 010800 ANTHRACENE
   8    9    1
   C    C    C           FORMULA    C14H10               MOL WT   178.24
  ⁄ \ ⁄ \ ⁄ \            SOLVENT    C6H6
 7C 14C  C11 C2          ORIG ST    C6H6                 TEMP        AMB
  |  ||   |   |
 6C 13C  C12 C3           130.10   125.52   132.56  132.24
  \ ⁄ \ ⁄ \ ⁄             1/2      2/2      9/2     11/1
    C    C    C
    5    10   4
```

R.J.PUGMIRE,D.M.GRANT,M.J.ROBINS,R.K.ROBINS
J AM CHEM SOC   91, 6381 (1969)

```
2103                    Q 010800 ANTHRACENE
                        FORMULA    C14H10               MOL WT   178.24
   C8  C9  C1           SOLVENT    CDCL3
  ⁄ \ ⁄ \ ⁄ \           ORIG ST    C4H12SI              TEMP        313
 C7  C14 C11 C2
  |   |   ||  |          127.90   125.10   125.10  127.90  127.90   125.10
 C6  C13 C12 C3          1/2      2/2      3/2     4/2     5/2      6/2
  \ ⁄ \ ⁄ \ ⁄            125.10   127.90   126.00  126.00  131.50   131.50
   C5   C   C4           7/2      8/2      9/2     10/2    11/1     12/1
        10               131.50   131.50
                         13/1     14/1
```

M.L.CASPAR,J.B.STOTHERS,N.K.WILSON
CAN J CHEM   53, 1958 (1975)

```
2104                    Q 010800 CIS-ANTI-CIS-PERHYDROANTHRACENE
     C7
    ⁄ \                 FORMULA    C14H24               MOL WT   192.35
   6C   C8             SOLVENT    CS2
   |    |              ORIG ST    C4H12SI              TEMP        AMB
   5C   C12
    \ ⁄|                 29.70    24.30    32.40   33.90
  9C----C |              1/3      2/3      9/3     11/2
   | C----C 10
   |⁄ \
  14C   C1
   |    |
  4C    C2
   \ ⁄
    C3
```

D.K.DALLING,D.M.GRANT
J AM CHEM SOC   96, 1827 (1974)

2105   Q 010800   CIS-SYN-CIS-PERHYDROANTHRACENE

```
        11C
   14    9/1 \
    C----C  I  \
   /I C----C  C5
  / I/10  /12/
 4C  C13  / /
   \  \ 8C  C6
    \  \  \  I
    3C  C1 \I
     I /    C7
     I/
     C2
```

FORMULA	C14H24			MOL WT	192.35
SOLVENT	CS2				
ORIG ST	C4H12SI			TEMP	AMB

32.28	25.30	33.26	37.59
1/3	2/3	9/3	11/2

D.K.DALLING,D.M.GRANT
J AM CHEM SOC   96, 1827 (1974)

2106   Q 010800   CIS-TRANS-PERHYDROANTHRACENE

```
       C6
      /I
     / I
   5C  C7
    \  \
     \  \
    11C  C8
   14  10/I /
    C---C I/
 3  4/I C---C
 C---C I/9  12
 I C---C
 I/1   13
 2C
```

FORMULA	C14H24			MOL WT	192.35
SOLVENT	CS2				
ORIG ST	C4H12SI			TEMP	AMB

34.43	27.16	27.16	34.35	27.07	27.57
1/3	2/3	3/3	4/3	5/3	6/3
21.37	32.62	33.88	40.78	37.09	37.50
7/3	8/3	9/3	10/3	11/2	12/2
44.18	36.60				
13/2	14/2				

D.K.DALLING,D.M.GRANT
J AM CHEM SOC   96, 1827 (1974)

2107   Q 010800   TRANS-ANTI-TRANS-PERHYDROANTHRACENE

```
             7
             C
 3         5  6/1
 C      \4 14   C---C I
 I\4 14      I C---C8
 I C---C     I C---C8
 C---C I\9 111/I12
 2  1\I C---C I
    C-------C
    13      10
```

FORMULA	C14H24			MOL WT	192.35
SOLVENT	CS2				
ORIG ST	C4H12SI			TEMP	AMB

35.47	26.81	38.45	36.44
1/3	2/3	9/3	11/2

D.K.DALLING,D.M.GRANT
J AM CHEM SOC   96, 1827 (1974)

2108   Q 010800   TRANS-SYN-TRANS-PERHYDROANTHRACENE

```
           C6
   11    5/1
    C---C I
  14  10/I C---C7
   C---C I/8
 3   4/I C---C
 C---C I/9  12
 I C---C
 I/1  13
 2C
```

FORMULA	C14H24			MOL WT	192.35
SOLVENT	CS2				
ORIG ST	C4H12SI			TEMP	AMB

34.38	27.10	42.14	43.82
1/3	2/3	9/3	11/2

D.K.DALLING,D.M.GRANT
J AM CHEM SOC   96, 1827 (1974)

2109   Q 010800   1-METHYLANTHRACENE

```
       15
       CH3
        I
  C8  C9  C1
 / \ / \ / \
 C7  C14 C11 C2
 I   I   II  I
 C6  C13 C12 C3
 \ / \ / \ /
 C5  C   C4
     10
```

FORMULA	C15H12			MOL WT	192.26
SOLVENT	CDCL3				
ORIG ST	C4H12SI			TEMP	313

133.90	125.40	124.90	126.50	127.70	125.10
1/1	2/2	3/2	4/2	5/2	6/2
125.00	128.30	122.50	126.60	131.20	131.70
7/2	8/2	9/2	10/2	11/1	12/1
131.40	131.20	19.70			
13/1	14/1	15/4			

M.L.CASPAR,J.B.STOTHERS,N.K.WILSON
CAN J CHEM   53, 1958 (1975)

2110 Q 010800 2-METHYLANTHRACENE

```
                              FORMULA   C15H12            MOL WT   192.26
                              SOLVENT   CDCL3
                    15        ORIG ST   C4H12SI           TEMP       313
      C8  C9  C1  CH3
     ∥  ＼ ∥ ＼ ∕ ＼ ∕         126.30   134.90   128.20   128.00   128.10   124.90
     C7  C14 C11 C2            1/2      2/1      3/2      4/2      5/2      6/2
     |   |   ‖   |            125.20   128.20   125.20   125.90   132.10   130.40
     C6  C13 C12 C3            7/2      8/2      9/2      10/2     11/1     12/1
     ＼ ∕ ＼ ∕ ＼ ∥           131.30   131.90   21.90
      C5   C    C4             13/1     14/1     15/4
           10
```

M.L.CASPAR,J.B.STOTHERS,N.K.WILSON
CAN J CHEM   53, 1958 (1975)

2111 Q 010800 9-METHYLANTHRACENE

```
           15                 FORMULA   C15H12            MOL WT   192.26
           CH3                SOLVENT   CDCL3
           |                  ORIG ST   C4H12SI           TEMP       313
      C8  C9  C1
     ∥  ＼ ∥ ＼ ∕ ＼           124.30   124.80   124.40   128.70   128.70   124.40
     C7  C14 C11 C2            1/2      2/2      3/2      4/2      5/2      6/2
     |   |   ‖   |            124.80   124.30   129.70   124.30   129.80   131.10
     C6  C13 C12 C3            7/2      8/2      9/1      10/2     11/1     12/1
     ＼ ∕ ＼ ∕ ＼ ∥           131.10   129.80   13.70
      C5   C    C4             13/1     14/1     15/4
           10
```

M.L.CASPAR,J.B.STOTHERS,N.K.WILSON
CAN J CHEM   53, 1958 (1975)

2112 Q 010800 1,4-DIMETHYLANTHRACENE

```
           15                 FORMULA   C16H14            MOL WT   206.29
           CH3                SOLVENT   CDCL3
           |                  ORIG ST   C4H12SI           TEMP       313
      C8  C9  C1
     ∥  ＼ ∥ ＼ ∕ ＼           132.10   125.20   125.20   132.10   128.10   125.10
     C7  C14 C11 C2            1/1      2/2      3/2      4/1      5/2      6/2
     |   |   ‖   |            125.10   128.10   123.00   123.00   131.00   131.00
     C6  C13 C12 C3            7/2      8/2      9/2      10/2     11/1     12/1
     ＼ ∕ ＼ ∕ ＼ ∥           131.30   131.30   19.70    19.70
      C5   C    C4             13/1     14/1     15/4     16/4
           10   |
                CH3           M.L.CASPAR,J.B.STOTHERS,N.K.WILSON
                16            CAN J CHEM               53, 1958 (1975)
```

2113 Q 010800 1,8-DIMETHYLANTHRACENE

```
      16        15            FORMULA   C16H14            MOL WT   206.29
      CH3       CH3           SOLVENT   CDCL3
      |         |             ORIG ST   C4H12SI           TEMP       313
      C8  C9  C1
     ∥  ＼ ∥ ＼ ∕ ＼           134.10   125.40   124.90   126.30   126.30   124.90
     C7  C14 C11 C2            1/1      2/2      3/2      4/2      5/2      6/2
     |   |   ‖   |            125.40   134.10   118.70   127.10   130.90   131.40
     C6  C13 C12 C3            7/2      8/1      9/2      10/2     11/1     12/1
     ＼ ∕ ＼ ∕ ＼ ∥           131.40   130.90   19.70    19.70
      C5   C    C4             13/1     14/1     15/4     16/4
           10
```

M.L.CASPAR,J.B.STOTHERS,N.K.WILSON
CAN J CHEM   53, 1958 (1975)

2114 Q 010800 9,10-DIMETHYLANTHRACENE

```
           15                 FORMULA   C16H14            MOL WT   206.29
           CH3                SOLVENT   CDCL3
           |                  ORIG ST   C4H12SI           TEMP       313
      C8  C9  C1
     ∥  ＼ ∥ ＼ ∕ ＼           125.00   124.40   124.40   125.00   125.00   124.40
     C7  C14 C11 C2            1/2      2/2      3/2      4/2      5/2      6/2
     |   |   ‖   |            124.40   125.00   128.10   128.10   129.60   129.60
     C6  C13 C12 C3            7/2      8/2      9/1      10/1     11/1     12/1
     ＼ ∕ ＼ ∕ ＼ ∥           129.60   129.60   14.10    14.10
      C5  C10  C4              13/1     14/1     15/4     16/4
           |
           CH3                G.W.BUCHANAN,R.S.OZUBKO,N.K.WILSON
           16                 CAN J CHEM               53, 1958 (1975)
```

2115   Q 010800   1,4,9-TRIMETHYLANTHRACENE

```
        17   15        FORMULA   C17H16                 MOL WT   220.32
        CH3  CH3       SOLVENT   CDCL3
         |    |        ORIG ST   C4H12SI                TEMP       313
       C8   C9   C1
      ⁄ \  ⁄ \  ⁄ \     133.00   128.50   125.10   132.50   128.70   124.70
     C7   C14  C11  C2    1/1      2/2      3/2      4/1      5/2      6/2
      |    |    ||   |   124.30   124.50   131.50   122.00   131.70   132.10
     C6   C13  C12  C3    7/2      8/2      9/1     10/2     11/1     12/1
      \  ⁄ \  ⁄ \  ⁄    130.40   131.20    27.10    19.60    20.30
       C5    C    C4     13/1     14/1     15/4     16/4     17/4
            10    |
                 CH3
                  16     M.L.CASPAR,J.B.STOTHERS,N.K.WILSON
                         CAN J CHEM                    53, 1958 (1975)
```

2116   Q 010800   2,7,9-TRIMETHYLANTHRACENE

```
         17            FORMULA   C17H16                 MOL WT   220.32
         CH3           SOLVENT   CDCL3
    16    |     15     ORIG ST   C4H12SI                TEMP       313
   H3C  C8   C9   C1  CH3
    \ ⁄  \ ⁄  \ ⁄  \ ⁄   122.70   134.30   126.90   128.70   128.70   126.90
     C7   C14  C11  C2    1/2      2/1      3/2      4/2      5/2      6/2
      |    |    ||   |   134.30   122.70   127.50   124.50   130.30   129.40
     C6   C13  C12  C3    7/1      8/2      9/1     10/2     11/1     12/1
      \ ⁄  \ ⁄  \ ⁄     129.40   130.30    22.40    22.40    13.70
       C5    C    C4     13/1     14/1     15/4     16/4     17/4
            10

                         M.L.CASPAR,J.B.STOTHERS,N.K.WILSON
                         CAN J CHEM                    53, 1958 (1975)
```

2117   Q 010800   1,4,5,8-TETRAMETHYLANTHRACENE

```
    18        15        FORMULA   C18H18                 MOL WT   234.34
    CH3       CH3       SOLVENT   CDCL3
     |         |        ORIG ST   C4H12SI                TEMP       313
    C8   C9   C1
   ⁄ \  ⁄ \  ⁄ \         132.20   125.20   125.20   132.20   132.20   125.20
  C7   C14  C11  C2       1/1      2/2      3/2      4/1      5/1      6/2
   |    |    ||   |      125.20   132.20   119.60   119.60   130.60   130.60
  C6   C13  C12  C3       7/2      8/1      9/2     10/2     11/1     12/1
   \ ⁄  \ ⁄  \ ⁄        130.60   130.60    19.60    19.60    19.60    19.60
    C5    C    C4         13/1     14/1     15/4     16/4     17/4     18/4
    |    10    |
    CH3       CH3
    17        16          M.L.CASPAR,J.B.STOTHERS,N.K.WILSON
                          CAN J CHEM                    53, 1958 (1975)
```

2118   Q 010800   1,4,5,9-TETRAMETHYLANTHRACENE

```
    18   15             FORMULA   C18H18                 MOL WT   234.34
    CH3  CH3            SOLVENT   CDCL3
     |    |             ORIG ST   C4H12SI                TEMP       313
    C8   C9   C1
   ⁄ \  ⁄ \  ⁄ \         133.10   128.80   125.00   133.00   134.60   125.30
  C7   C14  C11  C2       1/1      2/2      3/2      4/1      5/1      6/2
   |    |    ||   |      125.00   123.00   131.72   118.30   131.68   132.20
  C6   C13  C12  C3       7/2      8/2      9/1     10/2     11/1     12/1
   \ ⁄  \ ⁄  \ ⁄        130.20   132.00    27.00    19.90    20.00    20.30
    C5    C    C4         13/1     14/1     15/4     16/4     17/4     18/4
    |    10    |
    CH3       CH3
    17        16          M.L.CASPAR,J.B.STOTHERS,N.K.WILSON
                          CAN J CHEM                    53, 1958 (1975)
```

2119   Q 010800   1,4,5,8,9-PENTAMETHYLANTHRACENE

```
    18   19   15        FORMULA   C19H20                 MOL WT   248.37
    CH3  CH3  CH3       SOLVENT   CDCL3
     |    |    |        ORIG ST   C4H12SI                TEMP       313
    C8   C9   C1
   ⁄ \  ⁄ \  ⁄ \         134.30   128.50   124.80   132.20   132.20   124.80
  C7   C14  C11  C2       1/1      2/2      3/2      4/1      5/1      6/2
   |    |    ||   |      128.50   134.30   133.40   118.00   132.30   131.00
  C6   C13  C12  C3       7/2      8/1      9/1     10/2     11/1     12/1
   \ ⁄  \ ⁄  \ ⁄        131.00   132.30    25.90    20.00    20.00    25.90
    C5    C    C4         13/1     14/1     15/4     16/4     17/4     18/4
    |    10    |          26.80
    CH3       CH3        19/4
    17        16          M.L.CASPAR,J.B.STOTHERS,N.K.WILSON
                          CAN J CHEM                    53, 1958 (1975)
```

2120 Q 010900 PHENANTHRENE

```
    8  7
     C=C
 10 9/    \6 5
  C-C        C-C
  //  \    //  \
11C   14C-C      C4
  \   /  1\   /
   C=C      C=C
  12  13    2  3
```

FORMULA	C14H10		MOL WT	178.24
SOLVENT	CDCL3			
ORIG ST	C4H12SI		TEMP	AMB

131.90	122.40	126.30	126.30	128.30	130.10
1/1	2/2	3/2	4/2	5/2	6/1
126.60	126.60	130.10	128.30	126.30	126.30
7/2	8/2	9/1	10/2	11/2	12/2
122.40	131.90				
13/2	14/1				

L.F.JOHNSON, W.C.JANKOWSKI
CARBON-13 NMR SPECTRA, JOHN
WILEY AND SONS, NEW YORK   456 (1972)

2121 Q 010900 CIS-SYN-CIS-PERHYDROPHENANTHRENE

```
       C10
      / \
    /12  \9
  11C--C  C--C14
  I / \  / \
  IC1  \ / \
  II  13C    C5
  4CI   I    I
  / I   I    I
3C--C2 8C    C6
      \  /
       \ /
       C7
```

FORMULA	C14H24		MOL WT	192.35
SOLVENT	C14H24			
ORIG ST	C4H12SI		TEMP	AM2

32.42	24.61	26.22	29.45	28.38	38.06
1/3	2/3	3/3	4/3	9/3	11/2
41.68					
12/2					

D.K.DALLING, D.M.GRANT
J AM CHEM SOC   96, 1827 (1974)

2122 Q 010900 TRANS-ANTI-CIS-PERHYDROPHENANTHRENE

```
  C3        6C--C7
 / \        I /
/   \4 12   IC8
2C--C  C--C   II
 1\  / \  5CI
  \ / \ / I
   C--C  C--C14
  11 10\13 /
       \ /
        C
        9
```

FORMULA	C14H24		MOL WT	192.35
SOLVENT	CS2			
ORIG ST	C4H12SI		TEMP	AMB

34.99	27.21	28.18	30.44	27.21	20.74
1/3	2/3	3/3	4/3	5/3	6/3
27.61	26.98	32.86	29.84	44.20	38.47
7/3	8/3	9/3	10/3	11/2	12/2
42.27	37.96				
13/2	14/2				

D.K.DALLING, D.M.GRANT
J AM CHEM SOC   96, 1827 (1974)

2123 Q 010900 TRANS-ANTI-TRANS-PERHYDROPHENANTHRENE

```
     C3
    / \
   /   \4 12     5
  2C--C  C--C    C
   1\  / \  / \
    \ / \ /14 \6
     C--C  C--C  C--C7
    11 10\13 / \  /
         \ /     \ /
          C        C8
          9
```

FORMULA	C14H24		MOL WT	192.35
SOLVENT	C14H24			
ORIG ST	C4H12SI		TEMP	AMB

34.59	26.84	27.14	30.13	35.02	43.61
1/3	2/3	3/3	4/3	9/3	10/2
48.21					
12/2					

D.K.DALLING, D.M.GRANT
J AM CHEM SOC   96, 1827 (1974)

2124 Q 010900 TRANS-SYN-CIS-PERHYDROPHENANTHRENE

```
    C3
   / \
  /   \4 12
 2C--C  C--C
  1\  / \
   \ / \13 14
    C--C  C--C
   11 10\/ / \
    /\ /    \
  8C  C9   C5
    \      /
     \    /
      7C--C6
```

FORMULA	C14H24		MOL WT	192.35
SOLVENT	C14H24			
ORIG ST	C4H12SI		TEMP	AMB

35.33	27.48	27.48	31.03	21.81	27.08
1/3	2/3	3/3	4/3	5/3	6/3
21.81	33.05	26.08	35.40	36.47	47.70
7/3	8/3	9/3	10/3	11/2	12/2
42.35	38.38				
13/2	14/2				

D.K.DALLING, D.M.GRANT
J AM CHEM SOC   96, 1827 (1974)

2125   Q 011200 PYRENE

```
        10
      C    C1
    9/  \ / \
   C    C14 C2
   |    ||  |
   C13 C15 C3
   / \ / \ //
  C8  C16 C11
  ||  |   |
  C7  C12 C4
   \  / \ /
    C6   C5
```

FORMULA   C16H10   MOL WT 209.19
SOLVENT   CDCL3
ORIG ST   C4H12SI   TEMP   305

124.60	125.50	124.60	127.00	127.00	124.60
1/2	2/2	3/2	4/2	5/2	6/2
125.50	124.60	127.00	127.00	130.90	130.90
7/2	8/2	9/2	10/2	11/1	12/1
130.90	130.90	124.60	124.60		
13/1	14/1	15/1	16/1		

G.W.BUCHANAN,R.S.OZUBKO
CAN J CHEM   53, 1829 (1975)

2126   Q 011200 PYRENE

```
      4  5
      C=C
   11/    \12
  3C-C      C-C6
   //  \   //  \
 2C  15C-C16   C7
   \   /  \   /
  1C=C      C=C8
   14\    //13
       C=C
      10  9
```

FORMULA   C16H10   MOL WT 202.26
SOLVENT   CDCL3
ORIG ST   C4H12SI   TEMP   AMB

124.60	125.50	124.60	127.00	127.00	124.60
1/2	2/2	3/2	4/2	5/2	6/2
125.50	124.60	127.00	127.00	130.90	130.90
7/2	8/2	9/2	10/2	11/1	12/1
130.90	130.90	124.60	124.60		
13/1	14/1	15/1	16/1		

L.F.JOHNSON,W.C.JANKOWSKI
CARBON-13 NMR SPECTRA,JOHN
WILEY AND SONS,NEW YORK   469 (1972)

2127   Q 011200 1,3,6-TRIDEUTERO-BENZO(A)PYRENE

```
        D
  12    |
   C    C1
  11/ \ / \
  C    C18 C2
10 |   ||  |
  C    C17 C19 C3
  / \ / \ / \ //
 C9  C16 C20 C13 D
 |   ||  |   |
 C8  C15 C14 C4
  \ / \ / \ //
   C7  C6  C5
        |
        D
```

FORMULA   C20H9D3   MOL WT 255.34
SOLVENT   CDCL3
ORIG ST   C4H12SI   TEMP   305

124.30	125.67	124.37	127.28	127.61	125.06
1/3	2/2	3/3	4/2	5/2	6/3
128.71	125.88	125.91	122.87	122.00	127.95
7/2	8/2	9/2	10/2	11/2	12/2
131.14	129.62	131.39	127.21	128.10	131.20
13/1	14/1	15/1	16/1	17/1	18/1

G.W.BUCHANAN,R.S.OZUBKO
CAN J CHEM   53, 1829 (1975)

2128   Q 011200 BENZO(A)PYRENE

```
       12
       C    C1
     11/ \ / \
     C    C18 C2
  10 |   ||  |
     C    C17 C19 C3
     / \ / \ / \ //
    C9  C16 C20 C13
    |   ||  |   |
    C8  C15 C14 C4
     \ / \ / \ //
      C7  C6  C5
```

FORMULA   C20H12   MOL WT 923.08
SOLVENT   CDCL3
ORIG ST   C4H12SI   TEMP   305

124.61	125.78	124.68	127.33	127.66	125.38
1/2	2/2	3/2	4/2	5/2	6/2
128.77	125.88	125.91	122.87	122.00	128.00
7/2	8/2	9/2	10/2	11/2	12/2
131.21	129.71	131.46	127.22	128.10	131.27
13/1	14/1	15/1	16/1	17/1	18/1
125.27	123.58				
19/1	20/1				

G.W.BUCHANAN,R.S.OZUBKO
CAN J CHEM   53, 1829 (1975)

```
2129        C11           Q 011300  TRIPHENYLENE
         10C    C12
          ||     |     1          FORMULA   C18H12              MOL WT   228.30
         9C    C18  C               SOLVENT   CS2
          \  // \  /               ORIG ST   C6H6               TEMP      AMB
         17C    C13 C2
          |     ||   |              123.55   127.37   130.02
         16C    C14 C3                1/2      2/2      13/1
          \  /  \  //
         8C    C15  C
          ||     |    4
         7C    C5              A.J.JONES,D.M.GRANT
          \  //                CHEM COMMUN                      1670 (1968)
            C6
```

```
2130
                         R 000300  OXIRANE
         3      2
          C---C                     FORMULA   C2H4O              MOL WT   44.05
           \  /                     SOLVENT   C3H6O3
            O                       ORIG ST   C6H6               TEMP      AMB
            1
                                    39.50
                                     2/3

                                 G.E.MACIEL,G.B.SAVITSKY
                                 J PHYS CHEM                     69, 3925 (1965)
```

```
2131                     R 000300  EPIBROMOHYDRIN

                                    FORMULA   C3H5BRO            MOL WT   136.98
         1    2    3                SOLVENT   CDCL3
        CH2-CH-CH2-BR              ORIG ST   C4H12SI            TEMP      AMB
           \  /
            O                       48.70    52.00    33.0
                                     1/3      2/2      3/3

                                 N.R.EASTON,JR.,F.A.L.ANET,P.A.BURNS,C.S.FOOTE
                                 J AM CHEM SOC                   96, 3945 (1974)
```

```
2132                     R 000300  EPICHLOROHYDRIN

                                    FORMULA   C3H5CLO            MOL WT   92.53
         1    2    3                SOLVENT   CDCL3
        CH2-CH-CH2-CL              ORIG ST   C4H12SI            TEMP      AMB
           \  /
            O                       47.00    51.60    45.50
                                     1/3      2/2      3/3

                                 N.R.EASTON,JR.,F.A.L.ANET,P.A.BURNS,C.S.FOOTE
                                 J AM CHEM SOC                   96, 3945 (1974)
```

```
2133                     R 000300  PROPYLENE OXIDE

                                    FORMULA   C3H6O              MOL WT   58.08
              O                     SOLVENT   C4H8O2
           /  \                     ORIG ST   C4H12SI            TEMP      AMB
        C---C-CH3
        1    2   3                  47.30    47.60    18.10
                                     1/3      2/2      3/4

                                 L.F.JOHNSON,W.C.JANKOWSKI
                                 CARBON-13 NMR SPECTRA,JOHN
                                 WILEY AND SONS,NEW YORK        29 (1972)
```

2134   R 000300   TRANS-1,2-DIVINYLOXIRANE

```
    3  4
   CH=CH2
    /
   C1
  /|
 ) |
  \|
   C2
    \5  6
    CH=CH2
```

FORMULA	C6H8O		MOL WT	96.13
SOLVENT	CCL4/CDCL3			
ORIG ST	C4H12SI		TEMP	AMB

59.70	135.10	118.20
1/2	3/2	4/3

H.GUENTHER,G.JIKELI
CHEM BER   106, 1863 (1973)

2135   R 000300   1,2-EPOXYETHYLBENZENE

```
   3 2
   C-C      )
  // \1 / \
4C     C-C---C
  \  / 7 8
   C=C
   5 6
```

FORMULA	C8H8O			MOL WT	120.15
SOLVENT	CDCL3				
ORIG ST	C4H12SI			TEMP	AMB

137.70	125.40	128.40	128.00	128.40	125.40
1/1	2/2	3/2	4/2	5/2	6/2
52.10	50.80
7/2	8/3

L.F.JOHNSON,W.C.JANKOWSKI
CARBON-13 NMR SPECTRA,JOHN
WILEY AND SONS,NEW YORK   289 (1972)

2136   R 000300   STYRENE OXIDE

```
     6 5
     C-C
  7 8 1/  \
 CH2-CH-C     C4
   \ /  \  /
    )    2C=C3
```

FORMULA	C8H8O			MOL WT	120.15
SOLVENT	CDCL3				
ORIG ST	C4H12SI			TEMP	AMB

138.50	126.00	129.00	128.70	129.00	126.00
1/1	2/2	3/2	4/2	5/2	6/2
51.00	52.40
7/3	8/2

N.R.EASTON,JR.,F.A.L.ANET,P.A.BURNS,C.S.FOOTE
J AM CHEM SOC   96, 3945 (1974)

2137   R 000300   1,6-EPOXYBICYCLO(4,3,0)NONA-2,4-DIENE

```
   2C    C9
  3 / \1/ \
  C   \    \
  |   |\    \
  |   | )  C8
  |   |/   /
  C   C  /
 4 \ /6\ /
   5C    C7
```

FORMULA	C9H10O			MOL WT	134.18
SOLVENT	CCL4/CDCL3				
ORIG ST	C4H12SI			TEMP	AMB

69.90	128.30	126.30	29.50	18.20
1/1	2/2	3/2	7/3	8/3

H.GUENTHER,G.JIKELI
CHEM BER   106, 1863 (1973)

2138   R 000300   ENDRIN

```
         CL
    9     |
    C    C1
  8/|\  /|\   CL
  C | \ / | \2/
 /| 110C  111C
 ) | C12I RCR ||
 \| | C  | C3
  C | /5\ | / \
  7\|/  \|/   CL
    C    C4
    6    |   R -CL
         CL
```

FORMULA	C12H8CL6O			MOL WT	380.91
SOLVENT	CDCL3				
ORIG ST	C4H12SI			TEMP	AMB

79.40	132.30	132.30	79.40	39.20	54.50
1/1	2/1	3/1	4/1	5/2	6/2
47.00	47.00	54.50	39.20	108.70	29.70
7/2	8/2	9/2	10/2	11/1	12/3

L.F.JOHNSON,W.C.JANKOWSKI
CARBON-13 NMR SPECTRA,JOHN
WILEY AND SONS,NEW YORK   429 (1972)

2139   R 000300 TRANS—STILBENE OXIDE

```
        H   O   H
     14  \ / \ /   2 3
   13C-C  8C---C7  C-C
    //  \ /    \1/   \\
  12C    C9    C    C4
    \   /      \   /
   11C=C10      C=C
      6 5
```

FORMULA	C14H12O			MOL WT	196.25
SOLVENT	CDCL3				
ORIG ST	C4H12SI			TEMP	AMB

137.10	125.60	128.80	128.50	128.80	125.60
1/1	2/2	3/2	4/2	5/2	6/2
63.00	63.00	137.10	125.60	128.80	128.50
7/2	8/2	9/1	10/2	11/2	12/2
128.80	125.60				
13/2	14/2				

N.R.EASTON,JR.,F.A.L.ANET,P.A.BURNS,C.S.FOOTE
J AM CHEM SOC   96, 3945 (1974)

2140   R 000400 N-METHYLAZIRIDINE

```
   3     2
   C---C
    \ /
     N1
     |
     CH3
     4
```

FORMULA	C3H7N	MOL WT	57.10
SOLVENT	C3H7N		
ORIG ST	C6H6	TEMP	AMB

| 28.50 | 48.60 |
| 2/3 | 4/4 |

G.E.MACIEL,G.B.SAVITSKY
J PHYS CHEM   69, 3925 (1965)

2141   R 000500 THIIRANE

```
   3     2
   C---C
    \ /
     S
     1
```

FORMULA	C2H4S	MOL WT	60.12
SOLVENT	C2H4S		
ORIG ST	C6H6	TEMP	AMB

| 18.70 |
| 2/3 |

G.E.MACIEL,G.B.SAVITSKY
J PHYS CHEM   69, 3925 (1965)

2142   R 000550 THIETE SULFONE   THIACYCLOBUTENE-1,1-DIOXIDE

```
  3C=C2
  | |
  4C-SO2
     1
```

FORMULA	C3H4O2S	MOL WT	104.13
SOLVENT	C3H4O2S		
ORIG ST	C4H12SI	TEMP	AMB

| 148.50 | 138.60 | 72.80 |
| 2/2 | 3/2 | 4/3 |

G.C.LEVY,D.C.DITTMER
ORG MAGN RESON   4, 107 (1972)

2143   R 000550 OXETANE

```
   3     2
   C---C
   |   |
   C---O
   4     1
```

FORMULA	C3H6O	MOL WT	58.08
SOLVENT	C3H6O		
ORIG ST	C6H6	TEMP	AMB

| 72.60 | 22.90 |
| 2/3 | 3/3 |

G.E.MACIEL,G.B.SAVITSKY
J PHYS CHEM   69, 3925 (1965)

2144   R 000550 THIETANE

```
  3     2
  C---C
  |     |
  C---S
  4     1
```

FORMULA	C3H6S	MOL WT	74.15
SOLVENT	C3H6S		
ORIG ST	C6H6	TEMP	AMB

```
27.50   29.70
 2/3     3/3
```

G.E.MACIEL,G.B.SAVITSKY
J PHYS CHEM   69, 3925 (1965)

2145   R 000550 N—METHYLAZETIDINE

```
  3     2
  C---C
  |     |
  C---N1
  4     \
          CH3
          5
```

FORMULA	C4H9N	MOL WT	71.12
SOLVENT	C4H9N		
ORIG ST	C6H6	TEMP	AMB

```
57.70   17.50   46.40
 2/3     3/3     5/4
```

G.E.MACIEL,G.B.SAVITSKY
J PHYS CHEM   69, 3925 (1965)

2146   R 000600 FURAN

```
  4     3
  C---C
  ||    ||
 5C    C2
   \  /
    O
    1
```

FORMULA	C4H4O	MOL WT	68.08
SOLVENT	C4H4O		
ORIG ST	CS2	TEMP	AMB

```
142.70  109.60
 2/2     3/2
```

T.F.PAGE,JR.,T.ALGER,D.M.GRANT
J AM CHEM SOC   87, 5333 (1965)

2147   R 000600 FURAN

```
 4C---C3
  ||    ||
 5C    C2
   \  /
    O
    1
```

FORMULA	C4H4O	MOL WT	68.08
SOLVENT	CDCL3		
ORIG ST	C4H12SI	TEMP	AMB

```
142.60  109.60  109.60  142.60
 2/2     3/2     4/2     5/2
```

N.PLATZER,J.—J.BASSELIER,P.DEMERSEMAN
BULL SOC CHIM FRANCE   5, 905 (1974)

2148   R 000600 2,5—DIHYDROFURAN

```
 3C===C2
  |     |
 4C    C1
   \  /
    O
```

FORMULA	C4H6O	MOL WT	70.09
SOLVENT	CDCL3		
ORIG ST	C4H12SI	TEMP	AMB

```
75.30  126.30  126.30  75.30
 1/3    2/2     3/2     4/3
```

L.F.JOHNSON,W.C.JANKOWSKI
CARBON-13 NMR SPECTRA,JOHN
WILEY AND SONS,NEW YORK   59 (1972)

2149   R 000600   TETRAHYDROFURAN

```
        O
       / \
     1C   C4
     |    |
     2C---C3
```

FORMULA	C4H8O			MOL WT	72.11
SOLVENT	COCL3				
ORIG ST	C4H12SI			TEMP	AMB

68.60	26.70	26.70	68.60
1/3	2/3	3/3	4/3

N.R.EASTON,JR.,F.A.L.ANET,P.A.BURNS,C.S.FOOTE
J AM CHEM SOC   96, 3945 (1974)

2150   R 000600   TETRAHYDROFURAN

```
        O1
       / \
     5C   C2
     |    |
     4C---C3
```

FORMULA	C4H8O	MOL WT	72.11
SOLVENT	D2O		
ORIG ST	C4H12SI	TEMP	AMB

68.60	26.20
2/3	3/3

R.G.S.RITCHIE,N.CYR,B.KORSCH,H.J.KOCH,
A.S.PERLIN
CAN J CHEM   53, 1424 (1975)

2151   R 000600   TETRAHYDROFURAN

```
  4      3
  C---C
  |    |
  5C    C2
   \  /
    O
    1
```

FORMULA	C4H8O	MOL WT	72.11
SOLVENT	C4H8O		
ORIG ST	C6H6	TEMP	AMB

68.40	26.50
2/3	3/3

G.F.MACIEL,G.B.SAVITSKY
J PHYS CHEM   69, 3925 (1965)

2152   R 000600   TETRAHYDROFURAN

```
  3C---C2
  |    |
  4C    C1
   \  /
    O
```

FORMULA	C4H8O			MOL WT	72.11
SOLVENT	COCL3				
ORIG ST	C4H12SI			TEMP	AMB

67.90	25.80	25.80	67.90
1/3	2/3	3/3	4/3

L.F.JOHNSON,W.C.JANKOWSKI
CARBON-13 NMR SPECTRA,JOHN
WILEY AND SONS,NEW YORK   73 (1972)

2153   R 000600   FURAN

```
  3    2
  C---C
  ||   ||
  4C    C
   \  /1
    O
```

FORMULA	C4H4O			MOL WT	68.08
SOLVENT	C4H8O2				
ORIG ST	C4H12SI			TEMP	AMB

143.00	109.70	109.70	143.00
1/2	2/2	3/2	4/2

L.F.JOHNSON,W.C.JANKOWSKI
CARBON-13 NMR SPECTRA,JOHN
WILEY AND SONS,NEW YORK   48 (1972)

2154   R 000600 1,4-ANHYDROERYTHRITOL

```
         O1
        / \
       /   \
     5C     C2
       \   /
       4C-C3
        |  |
       HO OH
```

FORMULA	C4H8O3			MOL WT	104.11
SOLVENT	D2O				
ORIG ST	C4H12SI			TEMP	AMB

72.70	72.00
2/3	3/2

R.G.S.RITCHIE,N.CYR,B.KORSCH,H.J.KOCH,
A.S.PERLIN
CAN J CHEM   53, 1424 (1975)

2155   R 000600 1,4-ANHYDROTHREITOL

```
         O1
        / \
       /   \
     5C  HO C2
       \  | /
       4C-C3
        |
       HO
```

FORMULA	C4H8O3			MOL WT	104.11
SOLVENT	D2O				
ORIG ST	C4H12SI			TEMP	AMB

73.80	77.40
2/3	3/2

R.G.S.RITCHIE,N.CYR,B.KORSCH,H.J.KOCH,
A.S.PERLIN
CAN J CHEM   53, 1424 (1975)

2156   R 000600 FURFURAL

```
   3     2
   C---C
   ||    || 5
  4C    C-CH
    \  /1 ||
     O    O
```

FORMULA	C5H4O2			MOL WT	96.09
SOLVENT	C4H8O2				
ORIG ST	C4H12SI			TEMP	AMB

153.80	121.60	112.90	148.70	178.20
1/1	2/2	3/2	4/2	5/2

L.F.JOHNSON,W.C.JANKOWSKI
CARBON-13 NMR SPECTRA,JOHN
WILEY AND SONS,NEW YORK   104 (1972)

2157   R 000600 2-METHYLFURAN

```
   4   3
   C---C
   ||   ||
  5C    C2
    \  / \
     O   CH3
     1    6
```

FORMULA	C5H6O			MOL WT	82.10
SOLVENT	C5H6O				
ORIG ST	CS2			TEMP	AMB

152.20	106.20	110.90	141.20	13.40
2/1	3/2	4/2	5/2	6/4

T.F.PAGE,JR.,T.ALGER,D.M.GRANT
J AM CHEM SOC   87, 5333 (1965)

2158   R 000600 ALPHA-METHYLTETRAHYDROFURAN

```
   3C---C2
    |    |
   4C    C1
    \   / \
     O   CH3
          5
```

FORMULA	C5H10O			MOL WT	86.13
SOLVENT	C4H8O2				
ORIG ST	C4H12SI			TEMP	AMB

75.00	33.50	26.20	67.20	21.00
1/2	2/3	3/3	4/3	5/4

L.F.JOHNSON,W.C.JANKOWSKI
CARBON-13 NMR SPECTRA,JOHN
WILEY AND SONS,NEW YORK   123 (1972)

2159 R 000600 BETA—METHYL TETRAHYDROFURAN

```
        5CH3
          /
    3C---C2
    I    I
    4C   C1
      \ /
       J
```

FORMULA	C5H10O			MOL WT	86.13
SOLVENT	C4H8O2				
ORIG ST	C4H12SI			TEMP	AMB

74.70	34.00	34.70	67.60	17.90
1/3	2/2	3/3	4/3	5/4

L.F.JOHNSON,W.C.JANKOWSKI
CARBON-13 NMR SPECTRA,JOHN
WILEY AND SONS,NEW YORK   125 (1972)

2160 R 000600 METHYL FUROATE

```
   3    2
   C---C
   II   II 5  6
   C    C-C-O-CH3
   4\  /1 II
    O   O
```

FORMULA	C6H6O3			MOL WT	126.11
SOLVENT	CDCL3				
ORIG ST	C4H12SI			TEMP	AMB

144.80	117.90	111.90	146.40	159.00	51.70
1/1	2/2	3/2	4/2	5/1	6/4

L.F.JOHNSON,W.C.JANKOWSKI
CARBON-13 NMR SPECTRA,JOHN
WILEY AND SONS,NEW YORK   162 (1972)

2161 R 000600 2,5-DIMETHYLFURAN

```
   4    3
   C---C
   II   II
   5C   C2
  / \  / \
H3C  O  CH3
 7   1   6
```

FORMULA	C6H8O			MOL WT	96.13
SOLVENT	C6H8O				
ORIG ST	CS2			TEMP	AMB

149.90	106.20	12.90
2/1	3/2	6/4

T.F.PAGE,JR.,T.ALGER,D.M.GRANT
J AM CHEM SOC   87, 5333 (1965)

2162 R 000600 3,3,5,5-TETRAMETHYL-2-ISOPROPYLTETRAHYDROFJRAN

```
        8
        CH3
  10 3  2/
  H3C C---C-CH3
    \I   I 9 6
     4C   C1  CH3
  11/ \ / \ /
  H3C   O   5CH
         \
          CH3
          7
```

FORMULA	C11H22O			MOL WT	170.30
SOLVENT	CDCL3				
ORIG ST	C4H12SI			TEMP	AMB

108.30	44.30	50.90	81.30	67.80	21.30
1/2	2/1	3/3	4/1	5/2	
23.20	23.40	28.00	30.50	32.50	

L.F.JOHNSON,W.C.JANKOWSKI
CARBON-13 NMR SPECTRA,JOHN
WILEY AND SONS,NEW YORK   427 (1972)

2163 R 000600 5-HEXA-2,4-DIYNYLIDENE-2,5-DIHYDROFURANESPIRO-5O-(4'-HYDROXY-4',5'-DIHYDROFURANE)

```
    8   9
    C===C O
 1 2 3 4 5 6 I  I/ \13
H3C-C≡C-C≡C-CH=C7  C10 C
       \  /I   II
        O C---C
       /11 12
       HO
```

FORMULA	C13H10O3			MOL WT	214.22
SOLVENT					
ORIG ST	C4H12SI			TEMP	305

4.50	80.60	64.80	77.50	70.20	82.60
1/Q	2/S	3/S	4/S	5/S	6/D
168.50	133.90	126.70	117.90	76.00	105.80
7/S	8/D	9/D	10/S	11/D	12/D
146.30					
13/D					

R.ZEISBERG,F.BOHLMANN
CHEM BER   107, 3800 (1974)

2164   R 000600   5-HEXA-2,4-DIYNYLIDENE-3,4-EPOXY-2,5-DIHYDRO-
FURANESPIRO-5'-(4'-HYDROXY-4',5'-DIHYDROFURANE)

FORMULA   C13H10O4   MOL WT   230.22
SOLVENT
ORIG ST   C4H12SI   TEMP   AMB

```
         O
        8/ \9
        C===C O
 1 2 3 4 5 6 |   1/ \13
H3C-C≡C-C≡C-CH=C7  C10 C
        \ /I   II
         O C---C
        /11  12
       HO
```

4.75	80.50	64.60	77.40	68.50	87.40
1/Q	2/S	3/S	4/S	5/S	6/D
162.60	57.70	52.90	111.10	75.50	105.30
7/S	8/S	9/S	10/S	11/D	12/D
147.40
13/D

R.ZEISBERG,F.BOHLMANN
CHEM BER   107, 3800 (1974)

2165   R 000600   5-HEXA-2,4-DIYNYLIDENE-2,5-DIHYDROFURANESPIRO-
2'-TETRAHYDROFURANE   (E)-ISOMER

FORMULA   C13H12O2   MOL WT   200.24
SOLVENT
ORIG ST   C4H12SI   TEMP   AMB

```
        8   9
        C===C O
 1 2 3 4 5 6 |   1/ \13
H3C-C≡C-C≡C-CH=C7  C10 C
        \ /I   I
        'O C---C
        11  12
```

4.60	80.70	65.40	79.00	71.00	78.50
1/Q	2/S	3/S	4/S	5/S	6/D
167.40	135.60	127.30	121.20	35.70	24.60
7/S	8/D	9/D	10/S	11/T	12/T
69.50
13/T

R.ZEISBERG,F.BOHLMANN
CHEM BER   107, 3800 (1974)

2166   R 000600   5-HEXA-2,4-DIYNYLIDENE-2,5-DIHYDROFURANESPIRO-
2'-TETRAHYDROFURANE   (Z)-ISOMER

FORMULA   C13H12O2   MOL WT   200.24
SOLVENT
ORIG ST   C4H12SI   TEMP   AMB

```
        8   9
        C===C O
 1 2 3 4 5 6 |   1/ \13
H3C-C≡C-C≡C-CH=C7  C10 C
        \ /I   I
         O C---C
        11  12
```

4.40	79.40	65.10	76.40	71.40	79.40
1/Q	2/S	3/S	4/S	5/S	6/D
168.80	136.10	125.50	120.80	35.40	24.40
7/S	8/D	9/D	10/S	11/T	12/T
69.70
13/T

R.ZEISBERG,F.BOHLMANN
CHEM BER   107, 3800 (1974)

2167   R 000700   BENZOFURAN

FORMULA   C8H6O   MOL WT   118.14
SOLVENT   CDCL3
ORIG ST   C4H12SI   TEMP   AMB

```
     4
     C
   // \8
  5C   C---C3
  I   II   II
  6C   C   C2
   \ /9\ /
    C  O
    7  1
```

144.80	106.50	121.15	122.70	124.25	111.45
2/2	3/2	4/2	5/2	6/2	7/2
127.45	155.00
8/1	9/1

N.PLATZER,J.-J.BASSELIER,P.DEMERSEMAN
BULL SOC CHIM FRANCE   5, 905 (1974)

2168   R 000700   2-METHYLBENZOFURAN

FORMULA   C9H8O   MOL WT   132.16
SOLVENT   CDCL3
ORIG ST   C4H12SI   TEMP   AMB

```
     4
     C
   // \8
  5C   C---C3
  I   II   II
  6C   C   C-CH3
   \ /9\ /2 10
    C  O
    7  1
```

155.20	102.60	120.10	122.40	123.15	110.65
2/1	3/2	4/2	5/2	6/2	7/2
129.30	154.90	13.70
8/1	9/1	10/4

N.PLATZER,J.-J.BASSELIER,P.DEMERSEMAN
BULL SOC CHIM FRANCE   5, 905 (1974)

2169   R 000700 3-METHYLBENZOFURAN

```
                          FORMULA   C9H80                  MOL WT   132.16
      4        10         SOLVENT   CDCL3
      C        CH3        ORIG ST   C4H12SI                TEMP       AMB
     ⁄ \8
   5C    C---C3           141.30  115.50  119.40  122.20  124.05  111.35
    |    ||   ||           2/2     3/1     4/2     5/2     6/2     7/2
   6C    C    C2          129.10  155.40    7.65
    \  ⁄9\ ⁄               8/1     9/1    10/4
      C    O
      7    1              N.PLATZER,J.—J.BASSELIER,P.DEMERSEMAN
                          BULL SOC CHIM FRANCE              5,   905 (1974)
```

2170   R 000700 4-METHYLBENZOFURAN

```
      10
      CH3                 FORMULA   C9H80                  MOL WT   132.16
       |                  SOLVENT   CDCL3
      4C                  ORIG ST   C4H12SI                TEMP       AMB
     ⁄ \8
   5C    C---C3           144.15  105.05  130.95  123.00  124.15  108.80
    |    ||   ||           2/2     3/2     4/1     5/2     6/2     7/2
   6C    C    C2          127.25  154.85   18.45
    \  ⁄9\ ⁄               8/1     9/1    10/4
      C    O
      7    1              N.PLATZER,J.—J.BASSELIER,P.DEMERSEMAN
                          BULL SOC CHIM FRANCE              5,   905 (1974)
```

2171   R 000700 5-METHYLBENZOFURAN

```
                          FORMULA   C9H80                  MOL WT   132.16
   10    4                SOLVENT   CDCL3
  H3C    C                ORIG ST   C4H12SI                TEMP       AMB
    \  ⁄ \8
   5C    C---C3           144.85  106.30  121.05  131.90  125.50  110.85
    |    ||   ||           2/2     3/2     4/2     5/1     6/2     7/2
   6C    C    C2          127.65  153.55   21.15
    \  ⁄9\ ⁄               8/1     9/1    10/4
      C    O
      7    1              N.PLATZER,J.—J.BASSELIER,P.DEMERSEMAN
                          BULL SOC CHIM FRANCE              5,   905 (1974)
```

2172   R 000700 6-METHYLBENZOFURAN

```
                          FORMULA   C9H80                  MOL WT   132.16
      4                   SOLVENT   CDCL3
      C                   ORIG ST   C4H12SI                TEMP       AMB
     ⁄ \8
   5C    C---C3           144.25  106.45  120.65  124.25  134.35  111.75
    |    ||   ||           2/2     3/2     4/2     5/2     6/1     7/2
   6C    C    C2          125.05  155.60   21.45
    ⁄ \  ⁄9\ ⁄             8/1     9/1    10/4
  H3C    C    O
   10    7    1           N.PLATZER,J.—J.BASSELIER,P.DEMERSEMAN
                          BULL SOC CHIM FRANCE              5,   905 (1974)
```

2173   R 000700 7-METHYLBENZOFURAN

```
      4                   FORMULA   C9H80                  MOL WT   132.16
      C                   SOLVENT   CDCL3
     ⁄ \8                 ORIG ST   C4H12SI                TEMP       AMB
   5C    C---C3
    |    ||   ||          144.45  106.70  118.60  122.80  125.15  121.55
   6C    C    C2           2/2     3/2     4/2     5/2     6/2     7/1
    \  ⁄9\ ⁄              126.95  154.15   14.95
   7C    O                 8/1     9/1    10/4
    |    1
    CH3                   N.PLATZER,J.—J.BASSELIER,P.DEMERSEMAN
    10                    BULL SOC CHIM FRANCE              5,   905 (1974)
```

© 1979 HEYDEN & SON LIMITED

2174   R 000700 2,3-DIMETHYLBENZOFURAN

```
    4         11
    C        CH3        FORMULA  C10H10O              MOL WT  146.19
   ⁄ \8    3⁄          SOLVENT  CDCL3
 5C   C---C             ORIG ST  C4H12SI              TEMP       AMB
  |   ||   || 10
 6C   C   C-CH3         150.35  109.60  118.55  122.00  123.10  110.45
   \ ⁄9\ ⁄2              2/1     3/1     4/2     5/2     6/2     7/2
    C   O               130.55  154.05   11.55    7.65
    7   1                8/1     9/1     10/4    11/4
```

N.PLATZER,J.—J.BASSELIER,P.DEMERSEMAN
BULL SOC CHIM FRANCE   5,   905 (1974)

2175   R 000700 2,4-DIMETHYLBENZOFURAN

```
    11
    CH3                 FORMULA  C10H10O              MOL WT  146.19
    |                   SOLVENT  CDCL3
    4C                  ORIG ST  C4H12SI              TEMP       AMB
   ⁄ \8     3
 5C   C---C             154.60  101.25  129.80  122.80  123.10  108.10
  |   ||   || 10         2/1     3/2     4/1     5/2     6/2     7/2
 6C   C   C-CH3         129.00  154.75   13.90   18.45
   \ ⁄9\ ⁄2              8/1     9/1     10/4    11/4
    C   O
    7    1
```

N.PLATZER,J.—J.BASSELIER,P.DEMERSEMAN
BULL SOC CHIM FRANCE   5,   905 (1974)

2176   R 000700 2,5-DIMETHYLBENZOFURAN

```
   11    4
  H3C    C              FORMULA  C10H10O              MOL WT  146.19
   \ ⁄ \8    3          SOLVENT  CDCL3
 5C   C---C             ORIG ST  C4H12SI              TEMP       AMB
  |   ||   || 10
 6C   C   C-CH3         155.40  102.50  120.20  131.65  124.35  110.20
   \ ⁄9\ ⁄2              2/1     3/2     4/2     5/1     6/2     7/2
    C   O               129.50  153.45   13.90   21.25
    7    1               8/1     9/1     10/4    11/4
```

N.PLATZER,J.—J.BASSELIER,P.DEMERSEMAN
BULL SOC CHIM FRANCE   5,   905 (1974)

2177   R 000700 2,6-DIMETHYLBENZOFURAN

```
    4
    C                   FORMULA  C10H10O              MOL WT  146.19
   ⁄ \8    3            SOLVENT  CDCL3
 5C    C---C            ORIG ST  C4H12SI              TEMP       AMB
  |   ||   || 10
 6C    C   C-CH3        154.65  102.40  119.60  123.85  133.00  111.05
  ⁄ \ ⁄9\ ⁄2             2/1     3/2     4/2     5/2     6/1     7/2
 H3C   C   O            126.85  155.40   13.80   21.55
  11   7   1             8/1     9/1     10/4    11/4
```

N.PLATZER,J.—J.BASSELIER,P.DEMERSEMAN
BULL SOC CHIM FRANCE   5,   905 (1974)

2178   R 000700 2,3,4-TRIMETHYLBENZOFURAN

```
    12
    CH3                 FORMULA  C11H12O              MOL WT  160.22
    |        11         SOLVENT  CDCL3
    4C       CH3        ORIG ST  C4H12SI              TEMP       AMB
   ⁄ \8    3⁄
 5C    C---C            149.85  110.25  130.70  122.90  123.55  108.35
  |   ||   || 10         2/1     3/1     4/1     5/2     6/2     7/2
 6C    C   C-CH3        128.40  154.00   11.50   10.55   19.05
   \ ⁄9\ ⁄2              8/1     9/1     10/4    11/4    12/4
    C   O
    7   1
```

N.PLATZER,J.—J.BASSELIER,P.DEMERSEMAN
BULL SOC CHIM FRANCE   5,   905 (1974)

2179　　　　　　R 000700　2,3,5-TRIMETHYLBENZOFURAN

```
                        FORMULA   C11H12O              MOL WT   160.22
    12    4        11   SOLVENT   CDCL3
    H3C   C        CH3  ORIG ST   C4H12SI              TEMP       AMB
       \ // \8  3/
       5C    C---C        150.45  109.40  118.55  131.15  124.15  109.90
       |    ||   || 10      2/1     3/1     4/2     5/1     6/2     7/2
       6C    C   C-CH3    130.65  152.40   11.65    7.75   21.35
        \ /9\ /2            8/1     9/1    10/4    11/4    12/4
          C   O
          7   1          N.PLATZER,J.-J.BASSELIER,P.DEMERSEMAN
                         BULL SOC CHIM FRANCE          5,  905 (1974)
```

2180　　　　　　R 000700　2,3,6-TRIMETHYLBENZOFURAN

```
                        FORMULA   C11H12O              MOL WT   160.22
        4        11      SOLVENT   CDCL3
        C        CH3     ORIG ST   C4H12SI              TEMP       AMB
       // \8  3/
       5C    C---C        149.60  109.40  117.95  123.30  132.90  110.75
       |    ||   || 10      2/1     3/1     4/2     5/2     6/1     7/2
       6C    C   C-CH3    128.1   154.3    11.55    7.75   21.44
      / \ /9\ /2           8/1     9/1    10/4    11/4    12/4
    H3C   C   O
    12    7   1          N.PLATZER,J.-J.BASSELIER,P.DEMERSEMAN
                         BULL SOC CHIM FRANCE          5,  905 (1974)
```

2181　　　　　　R 000700　2,3,7-TRIMETHYLBENZOFURAN

```
        4        11      FORMULA   C11H12O              MOL WT   160.22
        C        CH3     SOLVENT   CDCL3
       // \8  3/         ORIG ST   C4H12SI              TEMP       AMB
       5C    C---C
       |    ||   || 10    150.00  109.90  116.10  122.10  124.15  120.45
       6C    C   C-CH3      2/1     3/1     4/2     5/2     6/2     7/1
        \ /9\ /2          130.10  153.00   11.65    7.85   14.95
       7C    O             8/1     9/1    10/4    11/4    12/4
       |     1
       CH3                N.PLATZER,J.-J.BASSELIER,P.DEMERSEMAN
       12                 BULL SOC CHIM FRANCE          5,  905 (1974)
```

2182　　　　　　R 000700　3,4,7-TRIMETHYLBENZOFURAN

```
    11CH3
       |         10       FORMULA   C11H12O              MOL WT   160.22
      4C         CH3      SOLVENT   CDCL3
       // \8  3/          ORIG ST   C4H12SI              TEMP       AMB
      5C    C---C
      |    ||   ||         141.00  116.55  129.05  123.60  124.90  119.05
      6C    C   C2           2/2     3/1     4/1     5/2     6/2     7/1
       \ /9\ /             126.60  154.80   10.40   18.70   14.65
      7C    O               8/1     9/1    10/4    11/4    12/4
      |     1
      CH3                  N.PLATZER,J.-J.BASSELIER,P.DEMERSEMAN
      12                   BULL SOC CHIM FRANCE          5,  905 (1974)
```

2183　　　　　　R 000700　2,3,4,6-TETRAMETHYLBENZOFURAN

```
    12CH3
      |          11       FORMULA   C12H14O              MOL WT   174.24
      4C         CH3      SOLVENT   CDCL3
       // \8  3/          ORIG ST   C4H12SI              TEMP       AMB
      5C    C---C
      |    ||   || 10      148.90  110.00  129.90  124.95  132.50  108.70
      6C    C   C-CH3        2/1     3/1     4/1     5/2     6/1     7/2
     / \ /9\ /2            126.10  154.65   11.15   10.20   18.85   21.25
    H3C   C   O             8/1     9/1    10/4    11/4    12/4    13/4
    13    7   1
                          N.PLATZER,J.-J.BASSELIER,P.DEMERSEMAN
                          BULL SOC CHIM FRANCE          5,  905 (1974)
```

2184   R 000700 2,3,4,7-TETRAMETHYLBENZOFURAN

```
  12CH3
    |        11        FORMULA  C12H14O              MOL WT  174.24
   4C        CH3       SOLVENT  CDCL3
   // \8   3/          ORIG ST  C4H12SI              TEMP      AMB
  5C   C---C
   |   ||  || 10       149.40  110.55  127.95  123.55  123.85  118.15
  6C   C   C-CH3        2/1     3/1     4/1     5/2     6/2     7/1
   \\ /9\ /2            127.95  153.00  11.45   10.40   18.85   14.75
   7C    O              8/1     9/1    10/4    11/4    12/4    13/4
    |    1
   CH3                 N.PLATZER,J.-J.BASSELIER,P.DEMERSEMAN
   13                  BULL SOC CHIM FRANCE            5,  905 (1974)
```

2185   R 000700 2,3,5,6-TETRAMETHYLBENZOFURAN

```
                       FORMULA  C12H14O              MOL WT  174.24
   12    4     11       SOLVENT  CDCL3
  H3C    C     CH3      ORIG ST  C4H12SI              TEMP      AMB
    \ // \8  3/
    5C    C---C         149.40  109.10  118.80  130.05  131.55  111.05
     |   ||  || 10       2/1     3/1     4/2     5/1     6/1     7/2
    6C   C   C-CH3      128.40  152.90  11.55   7.55    19.80   20.20
    / \ /9\ /2           8/1     9/1    10/4    11/4    12/4    13/4
  H3C   C   O
   13   7   1          N.PLATZER,J.-J.BASSELIER,P.DEMERSEMAN
                       BULL SOC CHIM FRANCE            5,  905 (1974)
```

2186   R 000700 2,3,5,7-TETRAMETHYLBENZOFURAN

```
   12    4     11       FORMULA  C12H14O              MOL WT  174.24
  H3C    C     CH3      SOLVENT  CDCL3
    \ // \8  3/         ORIG ST  C4H12SI              TEMP      AMB
    5C    C---C
     |   6   6 10       150.05  109.60  116.00  131.25  125.40  119.90
    6C   C   C-CH3       2/1     3/1     4/2     5/1     6/2     7/1
     \ /9\ /2           130.05  151.35  11.65   7.85    21.35   14.85
    7C    O              8/1     9/1    10/4    11/4    12/4    13/4
     |    1
    CH3                N.PLATZER,J.-J.BASSELIER,P.DEMERSEMAN
    13                 BULL SOC CHIM FRANCE            5,  905 (1974)
```

2187   R 000700 2,3,6,7-TETRAMETHYLBENZOFURAN

```
     4      11          FORMULA  C12H14O              MOL WT  174.24
     C      CH3         SOLVENT  CDCL3
   // \8  3/            ORIG ST  C4H12SI              TEMP      AMB
  5C   C---C
   |   ||  || 10        149.45  109.80  115.10  124.00  131.15  118.95
  6C   C   C-CH3         2/1     3/1     4/2     5/2     6/1     7/1
   / \ /9\ /2           127.90  153.30  11.65   7.85    19.00   11.50
  H3C  7C   O            8/1     9/1    10/4    11/4    12/4    13/4
  12   |    1
      CH3               N.PLATZER,J.-J.BASSELIER,P.DEMERSEMAN
      13                BULL SOC CHIM FRANCE            5,  905 (1974)
```

2188   R 000900 2,5-DIBROMOTHIOPHENE

```
    3    2              FORMULA  C4H2BR2S             MOL WT  241.93
    C---C               SOLVENT  CDCL3
    ||  ||              ORIG ST  C4H12SI              TEMP      AMB
   4C   C1
   / \ / \              111.40  130.10  130.10  111.40
  BR   S   BR            1/1     2/2     3/2     4/1

                       L.F.JOHNSON,W.C.JANKOWSKI
                       CARBON-13 NMR SPECTRA,JOHN
                       WILEY AND SONS,NEW YORK          44 (1972)
```

2189

```
    4   3
    C---C
    ||  ||
   5C   C2
     \ / \
      S   BR
      1
```

R 000900 2-BROMOTHIOPHENE

FORMULA   C4H3BRS   MOL WT 163.04
SOLVENT   C4H3BRS
ORIG ST   CS2   TEMP   AMB

112.60	130.00	127.70	127.10
2/1	3/2	4/2	5/2

K.TAKAHASHI,T.SONE,K.FUJIEDA
J PHYS CHEM   74, 2765 (1970)

2190

```
         BR
    4   3\
    C---C
    ||  ||
   5C   C2
     \ /
      S
      1
```

R 000900 3-BROMOTHIOPHENE

FORMULA   C4H3BRS   MOL WT 163.04
SOLVENT   C4H3BRS
ORIG ST   CS2   TEMP   AMB

126.90	110.40	130.10	123.00
2/2	3/1	4/2	5/2

K.TAKAHASHI,T.SONE,K.FUJIEDA
J PHYS CHEM   74, 2765 (1970)

2191

R 000900 2-CHLOROTHIOPHENE

```
    3   2
    C---C
    ||   \
    C    C-CL
   4\   /1
      S
```

FORMULA   C4H3CLS   MOL WT 118.59
SOLVENT   CDCL3
ORIG ST   C4H12SI   TEMP   AMB

129.90	126.40	125.90	123.90
1/1			

L.F.JOHNSON,W.C.JANKOWSKI
CARBON-13 NMR SPECTRA,JOHN
WILEY AND SONS,NEW YORK   46 (1972)

2192

R 000900 2-IODOTHIOPHENE

```
    4   3
    C---C
    ||  ||
    C   C-J
   5\  /2
     S1
```

FORMULA   C4H3IS   MOL WT 210.04
SOLVENT   CDCL3
ORIG ST   C4H12SI   TEMP   AMB

75.10	138.00	130.00	132.60
2/S	3/D	4/D	5/D

R.ZEISBERG,F.BOHLMANN
CHEM BER   108, 1040 (1975)

2193

```
    4   3
    C---C
    ||  ||
   5C   C2
     \ / \
      S   J
      1
```

R 000900 2-IODOTHIOPHENE

FORMULA   C4H3IS   MOL WT 210.04
SOLVENT   C4H3IS
ORIG ST   CS2   TEMP   AMB

74.80	137.70	129.70	132.30
2/1	3/2	4/2	5/2

K.TAKAHASHI,T.SONE,K.FUJIEDA
J PHYS CHEM   74, 2765 (1970)

2194   R 000900 3-IODOTHIOPHENE

```
        J
   4   3/
   C---C
   ||  ||
  5C   C2
    \  /
     S
     1
```

FORMULA C4H3IS
SOLVENT C4H3IS
ORIG ST CS2

MOL WT 210.04

TEMP   AMB

128.10	78.70	135.20	129.20
2/2	3/1	4/2	5/2

K.TAKAHASHI,T.SONE,K.FUJIEDA
J PHYS CHEM   74, 2765 (1970)

2195   R 000900 THIOPHENE

```
   3    2
   C---C
   ||  ||
   C   C
  4\  /1
    S
```

FORMULA C4H4S
SOLVENT CDCL3
ORIG ST C4H12SI

MOL WT 84.14

TEMP   AMB

124.90	126.70	126.70	124.90
1/2	2/2	3/2	4/2

L.F.JOHNSON,W.C.JANKOWSKI
CARBON-13 NMR SPECTRA,JOHN
WILEY AND SONS,NEW YORK   50 (1972)

2196   R 000900 THIOPHENE

```
   4   3
   C---C
   ||  ||
  5C   C2
    \  /
     S
     1
```

FORMULA C4H4S
SOLVENT C4H4S
ORIG ST CS2

MOL WT 84.14

TEMP   AMB

124.40	126.20
2/2	3/2

K.TAKAHASHI,T.SONE,K.FUJIEDA
J PHYS CHEM   74, 2765 (1970)

2197   R 000900 THIOPHENE

```
  4C---C 3
   ||  ||
  5C   C2
    \  /
     S
     1
```

FORMULA C4H4S
SOLVENT C)CL3
ORIG ST C4H12SI

MOL WT 84.14

TEMP   AMB

125.40	127.20	127.20	125.40
2/2	3/2	4/2	5/2

N.PLATZER,J.-J.BASSELIER,P.DEMERSEMAN
BULL SOC CHIM FRANCE   5, 905 (1974)

2198   R 000900 TETRAHYDROTHIOPHENE

```
   4    3
   C---C
   |   |
  5C   C2
    \  /
     S
     1
```

FORMULA C4H8S
SOLVENT C4H8O
ORIG ST C6H6

MOL WT 88.17

TEMP   AMB

32.00	32.00
2/3	3/3

G.E.MACIEL,G.B.SAVITSKY
J PHYS CHEM   69, 3925 (1965)

2199   R 000900 TETRAMETHYLENE SULFONE

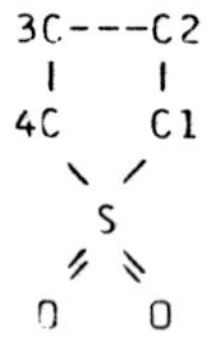

```
3C---C2
 |    |
4C    C1
  \  /
   S
  // \\
 O    O
```

FORMULA   C4H8O2S   MOL WT   120.17
SOLVENT   C4H8O2
ORIG ST   C4H12SI   TEMP   AMB

        51.50   22.80   22.80   51.50
        1/3   2/3   3/3   4/3

L.F.JOHNSON,W.C.JANKOWSKI
CARBON-13 NMR SPECTRA,JOHN
WILEY AND SONS,NEW YORK   77 (1972)

2200   R 000900 TETRAHYDROTHIOPHENE

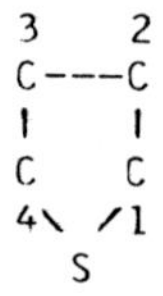

```
3     2
C---C
|    |
C    C
4\  /1
   S
```

FORMULA   C4H8S   MOL WT   88.17
SOLVENT   C4H8O2
ORIG ST   C4H12SI   TEMP   AMB

        31.20   31.40   31.40   31.20
        1/3   2/3   3/3   4/3

L.F.JOHNSON,W.C.JANKOWSKI
CARBON-13 NMR SPECTRA,JOHN
WILEY AND SONS,NEW YORK   78 (1972)

2201   R 000900 THIOPHENE-2-ALDEHYDE

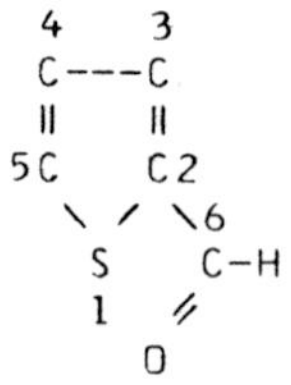

```
4   3
C---C
||  ||
5C   C2
  \ / \6
   S   C-H
   1   //
       O
```

FORMULA   C5H4OS   MOL WT   112.15
SOLVENT   C5H4OS
ORIG ST   CS2   TEMP   AMB

    143.30   136.40   128.10   134.60   182.80
     2/1   3/2   4/2   5/2   6/2

K.TAKAHASHI,T.SONE,K.FUJIEDA
J PHYS CHEM   74, 2765 (1970)

2202   R 000900 THIOPHENE-3-ALDEHYDE

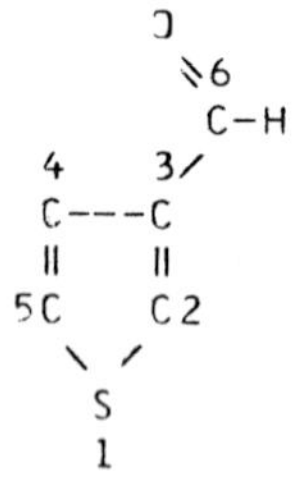

```
    O
     \6
      C-H
4   3/
C---C
||  ||
5C   C2
  \ /
   S
   1
```

FORMULA   C5H4OS   MOL WT   112.15
SOLVENT   C5H4OS
ORIG ST   CS2   TEMP   AMB

    137.10   142.60   124.90   127.30   184.70
     2/2   3/1   4/2   5/2   6/2

K.TAKAHASHI,T.SONE,K.FUJIEDA
J PHYS CHEM   74, 2765 (1970)

2203   R 000900 2-METHOXYTHIOPHENE

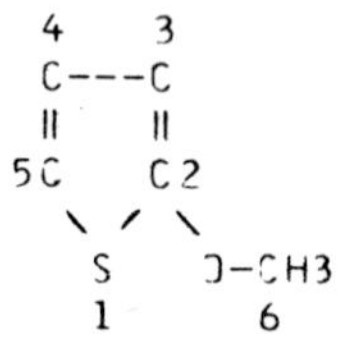

```
4   3
C---C
||  ||
5C   C2
  \ /
   S    O-CH3
   1    6
```

FORMULA   C5H6OS   MOL WT   114.17
SOLVENT   C5H6OS
ORIG ST   CS2   TEMP   AMB

    166.70   103.60   124.50   111.40
     2/1   3/2   4/2   5/2   6/4

K.TAKAHASHI,T.SONE,K.FUJIEDA
J PHYS CHEM   74, 2765 (1970)

2204   R 000900   3-METHOXYTHIOPHENE

```
        O-CH3
  4   3/  6
  C---C
  ||   ||
  5C   C2
   \  /
    S
    1
```

FORMULA	C_5H_6OS		MOL WT	114.17
SOLVENT	C_5H_6OS			
ORIG ST	CS2		TEMP	AMB

96.70	158.90	119.20	124.50	
2/2	3/1	4/2	5/2	6/4

K.TAKAHASHI,T.SONE,K.FUJIEDA
J PHYS CHEM   74, 2765 (1970)

2205   R 000900   2-METHYLTHIOPHENE

```
  4   3
  C---C
  ||   ||
  5C   C2
   \  / \
    S    CH3
    1    6
```

FORMULA	C_5H_6S		MOL WT	98.17
SOLVENT	C_5H_6S			
ORIG ST	CS2		TEMP	AMB

139.00	124.70	126.40	122.60	14.40
2/1	3/2	4/2	5/2	6/4

K.TAKAHASHI,T.SONE,K.FUJIEDA
J PHYS CHEM   74, 2765 (1970)

2206   R 000900   3-METHYLTHIOPHENE

```
        CH3
  4   3/6
  C---C
  ||   ||
  5C   C2
   \  /
    S
    1
```

FORMULA	C_5H_6S		MOL WT	98.17
SOLVENT	C_5H_6S			
ORIG ST	CS2		TEMP	AMB

120.10	136.80	128.80	124.70	
2/2	3/1	4/2	5/2	6/4

K.TAKAHASHI,T.SONE,K.FUJIEDA
J PHYS CHEM   74, 2765 (1970)

2207   R 000900   2-METHOXYCARBONYLTHIOPHENE

```
  4   3
  C---C
  ||   ||
  5C   C2
   \  / \6
    S    C-O
    1  /   \7
      O     CH3
```

FORMULA	$C_6H_6O_2S$		MOL WT	142.18
SOLVENT	$C_6H_6O_2S$			
ORIG ST	CS2		TEMP	AMB

133.20	132.80	127.10	131.90	161.80	
2/1	3/2	4/2	5/2	6/1	7/4

K.TAKAHASHI,T.SONE,K.FUJIEDA
J PHYS CHEM   74, 2765 (1970)

2208   R 000900   3-METHOXYCARBONYLTHIOPHENE

```
      O   7CH3
       \6 /
        C-O
  4   3/
  C---C
  ||   ||
  5C   C2
   \  /
    S
    1
```

FORMULA	$C_6H_6O_2S$		MOL WT	142.18
SOLVENT	$C_6H_6O_2S$			
ORIG ST	CS2		TEMP	AMB

131.90	132.90	127.10	125.40	161.90	
2/2	3/1	4/2	5/2	6/1	7/4

K.TAKAHASHI,T.SONE,K.FUJIEDA
J PHYS CHEM   74, 2765 (1970)

2209 — R 000900 2,6-DIMETHYLTHIOPHENE

```
    4    3
    C---C
    ||   ||
   5C    C2
   / \  / \
 H3C  S  CH3
  7   1   6
```

FORMULA	C6H8S		MOL WT	112.19
SOLVENT	C6H8S			
ORIG ST	CS2		TEMP	AMB

137.10	125.00	14.90
2/1	3/2	6/4

T.F.PAGE,JR.,T.ALGER,D.M.GRANT
J AM CHEM SOC　　　　　87, 5333 (1965)

2210 — R 000900 2,9-DIIODO-DI-ALPHA-THIOPHENE

```
  3   4 7   8
  C---C C---C
  ||  || ||  ||
J-C   C-C   C-J
 2\  /5 6\  /9
   S1      S10
```

FORMULA	C8H4I2S2		MOL WT	418.06
SOLVENT	CDCL3			
ORIG ST	C4H12SI		TEMP	AMB

72.80	137.90	125.30	142.30	142.30	125.30
2/S	3/D	4/D	5/S	6/S	7/D
137.90	72.80				
8/D	9/S				

R.ZEISBERG,F.BOHLMANN
CHEM BER　　　　　108, 1040 (1975)

2211 — R 000900 BENZOTHIOPHENE

```
      4
      C
    // \8
  5C    C---C3
  |    ||   ||
  6C    C    C2
   \  /9\  /
      C  S
      7  1
```

FORMULA	C8H6S		MOL WT	134.20
SOLVENT	CDCL3			
ORIG ST	C4H12SI		TEMP	AMB

126.10	123.65	123.50	124.00	124.05	122.30
2/2	3/2	4/2	5/2	6/2	7/2
139.45	139.60				
8/1	9/1				

N.PLATZER,J.-J.BASSELIER,P.DEMERSEMAN
BULL SOC CHIM FRANCE　　　5, 905 (1974)

2212 — R 000900 2-METHYLBENZOTHIOPHENE

```
      4
      C
    / \8   3
  5C    C---C
  |    ||   || 10
  6C    C    C-CH3
   \  /9\  /2
      C  S
      7  1
```

FORMULA	C9H8S		MOL WT	148.23
SOLVENT	CDCL3			
ORIG ST	C4H12SI		TEMP	AMB

140.65	121.65	122.50	123.95	123.30	121.90
2/1	3/2	4/2	5/2	6/2	7/2
139.80	140.65	15.90			
8/1	9/1	10/4			

N.PLATZER,J.-J.BASSELIER,P.DEMERSEMAN
BULL SOC CHIM FRANCE　　　5, 905 (1974)

2213 — R 000900 3-METHYLBENZOTHIOPHENE

```
      4       10
      C       CH3
    // \8    /
  5C    C---C3
  |    ||   ||
  6C    C    C2
   \  /9\  /
      C  S
      7  1
```

FORMULA	C9H8S		MOL WT	148.23
SOLVENT	CDCL3			
ORIG ST	C4H12SI		TEMP	AMB

121.55	131.80	122.60	123.90	123.70	121.55
2/2	3/1	4/2	5/2	6/2	7/2
140.30	139.60	13.70			
8/1	9/1	10/4			

N.PLATZER,J.-J.BASSELIER,P.DEMERSEMAN
BULL SOC CHIM FRANCE　　　5, 905 (1974)

2214   R 000900 3-THIENYLPROPINAL DIMETHYL ACETAL

```
4    3         9
C---C        O-CH3
||   || 6 7 8/
C    C-C≡C-CH
5\  /2         \
  S1         O-CH3
               10
```

FORMULA	C9H10O2S			MOL WT	182.24
SOLVENT	CDCL3				
ORIG ST	C4H12SI			TEMP	AMB
121.60	132.90	127.00	127.80	79.00	87.80
2/S	3/D	4/D	5/D	6/S	7/S
93.60	52.40	52.40			
8/D	9/Q	10/Q			

R.ZEISBERG,F.BOHLMANN
CHEM BER   108, 1040 (1975)

2215   R 000900 2-ACETYL-DI-ALPHA-THIOPHENE

```
8    7 4    3
C---C C---C    O
||   || || || //
C    C-C  C-C11
9\  /6 5\ /2  \12
  S10   S1   CH3
```

FORMULA	C10H8OS2			MOL WT	208.30
SOLVENT	CDCL3				
ORIG ST	C4H12SI			TEMP	AMB
142.40	133.40	124.10	145.60	136.30	125.60
2/S	3/D	4/D	5/S	6/S	7/D
128.20	126.50	190.10	26.30		
8/D	9/D	11/S	12/Q		

R.ZEISBERG,F.BOHLMANN
CHEM BER   108, 1040 (1975)

2216   R 000900 5-(5-METHYLTHIENYL)-THIOPHENE-2-CARBALDEHYDE

```
9    8 4    3
C---C C---C    O
12 ||   || || || 6//
H3C-C10 C-C   C-C
    \ /7 5\ /2  \
     S11   S1   H
```

FORMULA	C10H8OS2			MOL WT	208.30
SOLVENT	CDCL3				
ORIG ST	C4H12SI			TEMP	AMB
141.10	137.40	123.40	147.70	182.30	133.70
2/S	3/D	4/D	5/S	6/D	7/S
126.20	126.20	142.40	15.40		
8/D	9/D	10/S	12/Q		

R.ZEISBERG,F.BOHLMANN
CHEM BER   108, 1040 (1975)

2217   R 000900 2,3-DIMETHYLBENZOTHIOPHENE

```
   4        11
   C        CH3
  / \8  3/
5C   C---C
 |   ||  || 10
6C   C   C-CH3
  \ /9\ /2
   C   S
   7   1
```

FORMULA	C10H10S			MOL WT	162.26
SOLVENT	CDCL3				
ORIG ST	C4H12SI			TEMP	AMB
133.35	126.75	120.95	123.50	123.25	121.75
2/1	3/1	4/2	5/2	6/2	7/2
140.90	138.10	10.95	13.40		
8/1	9/1	10/4	11/4		

N.PLATZER,J.-J.BASSELIER,P.DEMERSEMAN
BULL SOC CHIM FRANCE   5, 905 (1974)

2218   R 000900 3,5-DIMETHYLBENZOTHIOPHENE

```
 11    4        10
 H3C   C        CH3
   \ / \8    /
    5C   C---C3
    |   ||  ||
    6C   C   C2
     \ /9\ /
      C   S
      7   1
```

FORMULA	C10H10S			MOL WT	162.26
SOLVENT	CDCL3				
ORIG ST	C4H12SI			TEMP	AMB
121.55	131.45	121.55	133.20	125.60	122.90
2/2	3/1	4/2	5/1	6/2	7/2
139.80	137.45	13.60	21.25		
8/1	9/1	10/4	11/4		

N.PLATZER,J.-J.BASSELIER,P.DEMERSEMAN
BULL SOC CHIM FRANCE   5, 905 (1974)

2219 R 000900 2,3,5-TRIMETHYLBENZOTHIOPHENE

```
 12     4        11
H3C    C         CH3
  \  / \8   3/
   5C    C---C
   |    ||   || 10
   6C    C   C-CH3
    \  /9\  /2
     C    S
     7    1
```

FORMULA	C11H12S			MOL WT	176.28
SOLVENT	CDCL3				
ORIG ST	C4H12SI			TEMP	AMB

133.55	126.50	121.15	133.05	124.95	121.45
2/1	3/1	4/2	5/1	6/2	7/2
141.20	135.30	11.15	13.60	21.45	
8/1	9/1	10/4	11/4	12/4	

N.PLATZER,J.-J.BASSELIER,P.DEMERSEMAN
BULL SOC CHIM FRANCE   5, 905 (1974)

2220 R 000900 2,3,7-TRIMETHYLBENZOTHIOPHENE

```
    4        11
    C         CH3
  / \8   3/
 5C    C---C
 |    ||   || 10
 6C    C   C-CH3
  \  /9\  /2
  7C    S
  |    1
  CH3
  12
```

FORMULA	C11H12S			MOL WT	176.28
SOLVENT	CDCL3				
ORIG ST	C4H12SI			TEMP	AMB

132.90	127.45	118.70	124.15	123.75	131.05
2/1	3/1	4/2	5/2	6/2	7/1
140.85	138.15	11.25	13.50	20.00	
8/1	9/1	10/4	11/4	12/4	

N.PLATZER,J.-J.BASSELIER,P.DEMERSEMAN
BULL SOC CHIM FRANCE   5, 905 (1974)

2221 R 000900 2-(3-BUTEN-1-YNYL)-DI-ALPHA-THIOPHENE

```
8   7 4    3
C---C C---C
||  || ||   || 11  13 14
C   C-C    C-C≡C-CH=CH2
9\ /6 5\ /2    12
 S10    S1
```

FORMULA	C12H8S2			MOL WT	216.33
SOLVENT	CDCL3				
ORIG ST	C4H12SI			TEMP	AMB

121.60	132.80	123.30	138.90	136.40	124.10
2/S	3/D	4/D	5/S	6/S	7/D
127.80	124.80	83.60	93.20	116.70	126.70
8/D	9/D	11/S	12/S	13/D	14/T

R.ZEISBERG,F.BOHLMANN
CHEM BER   108, 1040 (1975)

2222 R 000900 2-(1-PROPYNYL)-DI-ALPHA-THIENYLETHYNE

```
   3    4        9   10
   C---C        C---C
15  13 ||    || 6 7 ||    ||
H3C-C≡C-C    C-C≡C-C    C
 14    2\ /5      8\ /11
        S1          S12
```

FORMULA	C13H8S2			MOL WT	228.34
SOLVENT	CDCL3				
ORIG ST	C4H12SI			TEMP	AMB

126.00	130.70	131.50	122.60	86.10	86.70
2/S	3/D	4/D	5/S	6/S	7/S
122.50	132.10	129.00	127.60	72.90	91.60
8/S	9/D	10/D	11/D	13/S	14/S
4.40					
15/Q					

R.ZEISBERG,F.BOHLMANN
CHEM BER   108, 1040 (1975)

2223 R 001200 PYRROLE

```
 3   2
 C---C
 ||   ||
 C   C
 4\ /1
   N
```

FORMULA	C4H5N		MOL WT	67.09
SOLVENT	CDCL3			
ORIG ST	C4H12SI		TEMP	AMB

| 117.90 | 107.90 | 107.90 | 117.90 |
| 1/2 | 2/2 | 3/2 | 4/2 |

L.F.JOHNSON,W.C.JANKOWSKI
CARBON-13 NMR SPECTRA,JOHN
WILEY AND SONS,NEW YORK   54 (1972)

2224   R 001200 PYRROLE

```
4C---C3
 ||  ||
5C    C2
  \  /
   N
   1
```

FORMULA C4H5N   MOL WT 67.09
SOLVENT CDCL3
ORIG ST C4H12SI   TEMP   AMB

118.50	108.20	108.20	118.50
2/2	3/2	4/2	5/2

N.PLATZER,J.—J.BASSELIER,P.DEMERSEMAN
BULL SOC CHIM FRANCE   5, 905 (1974)

2225   R 001200 2-PYRROLIDONE

```
3C---C2
 |    |
4C    C1
  \  / \\
   N    O
```

FORMULA C4H7NO   MOL WT 85.11
SOLVENT CDCL3
ORIG ST C4H12SI   TEMP   AMB

179.40	30.30	20.80	42.40
1/1	2/3	3/3	4/3

L.F.JOHNSON,W.C.JANKOWSKI
CARBON-13 NMR SPECTRA,JOHN
WILEY AND SONS,NEW YORK   67 (1972)

2226   R 001200 PYRROLIDINE

```
 3   2
 C---C
 |   |
 C   C
4\  /1
  N
```

FORMULA C4H9N   MOL WT 71.12
SOLVENT CDCL3
ORIG ST C4H12SI   TEMP   AMB

47.10	25.70	25.70	47.10
1/3	2/3	3/3	4/3

L.F.JOHNSON,W.C.JANKOWSKI
CARBON-13 NMR SPECTRA,JOHN
WILEY AND SONS,NEW YORK   84 (1972)

2227   R 001200 PYRROLIDINE

```
 4     3
 C---C
 |   |
5C    C2
  \1/
   N
```

FORMULA C4H9N   MOL WT 71.12
SOLVENT C4H9N
ORIG ST C4H12SI   TEMP   300

47.05	25.75
2/3	3/3

E.BREITMAIER,W.VOELTER
UNPUBLISHED   (1974)

2228   R 001200 2-METHYLPYRROLE

```
 4   3
 C---C
 ||  ||
5C    C2
  \1/ \
   N   CH3
       6
```

FORMULA C5H7N   MOL WT 81.12
SOLVENT C5H7N
ORIG ST CS2   TEMP   AMB

127.50	105.80	108.00	116.90	11.50
2/1	3/2	4/2	5/2	6/4

T.F.PAGE,JR.,T.ALGER,D.M.GRANT
J AM CHEM SOC   87, 5333 (1965)

```
2229          R  001200  N-METHYLPYRROLIDINE
    4     3
    C---C              FORMULA   C5H11N              MOL WT   85.15
    |   |              SOLVENT   C5H11N
   5C   C2             ORIG ST   C6H6                TEMP       AMB
     \1/
      N                  56.70    24.40    42.70
      |                   2/3      3/3      6/4
     CH3
      6                G.E.MACIEL,G.B.SAVITSKY
                       J PHYS CHEM                    69, 3925 (1965)
```

```
2230          R  001200  2,5-DIMETHYLPYRROLE
    4     3
    C---C              FORMULA   C6H9N               MOL WT   95.15
    ||  ||             SOLVENT   C6H9N
   5C   C2             ORIG ST   CS2                 TEMP       AMB
  / \1/ \
H3C   N   CH3           125.90   106.30   12.90
  7       6              2/1      3/2      6/4

                       T.F.PAGE,JR.,T.ALGER,D.M.GRANT
                       J AM CHEM SOC                  87, 5333 (1965)
```

```
2231          R  001200  N-VINYLPYRROLIDONE

                       FORMULA   C6H9NO              MOL WT   111.14
                       SOLVENT   CDCL3
     O                 ORIG ST   C4H12SI             TEMP       AMB
     ||   5   6
    4C   CH=CH2          44.40   17.20    31.10   172.90   129.20   93.80
   / \  /                1/3     2/3      3/3      4/1      5/1      6/3
  3C   N
   |   |               L.F.JOHNSON,W.C.JANKOWSKI
   C---C               CARBON-13 NMR SPECTRA,JOHN
   2   1               WILEY AND SONS,NEW YORK             173 (1972)
```

```
2232          R  001200  PYRROLNITRIN
       C10
      // \              FORMULA   C10H6CL2N2O2        MOL WT   257.08
   9C    C11            SOLVENT   CHCL3
   |     ||             ORIG ST   C4H12SI             TEMP       AMB
  8C    C6       CL
 / \ / \    /            117.50   111.70   115.30   116.70   127.80   148.30
CL  C7 4C---C3           2/2      3/1      4/1      5/2      6/1      7/1
    |   ||   ||         124.80   130.20   130.20   128.50
   NO2 C5   C2           8/1      9/2      10/2     11/2
        \1/
         N             L.L.MARTIN,C.-J.CHANG,H.G.FLOSS,J.A.MABE,
                       E.W.HAGAMAN,E.WENKERT
                       J AM CHEM SOC                  94, 8942 (1972)
```

```
2233          R  001200  3-(3,4-DICHLOROPHENYL)-PYRROLE
       C10
      // \              FORMULA   C10H7CL2N           MOL WT   212.08
  CL-C9  C11            SOLVENT   CHCL3
    |    ||             ORIG ST   C4H12SI             TEMP       AMB
   8C    C6
  / \ / \               118.10   106.10   122.30   114.90   135.90   126.40
CL  C7 4C---C3           2/2      3/2      4/1      5/2      6/1      7/1
    ||   ||   ||        132.20   128.40   130.20   123.10
   5C    C2              8/1      9/2      10/2     11/2
        \1/
         N             L.L.MARTIN,C.-J.CHANG,H.G.FLOSS,J.A.MABE,
                       E.W.HAGAMAN,E.WENKERT
                       J AM CHEM SOC                  94, 8942 (1972)
```

2234   R 001200 AMINOPYRROLNITRIN

```
        C10
      //  \
    9C     C11
    |      ||
    8C     C6        CL
   /  \  /  \      /
  CL   C7 4C---C3
       |   ||   ||
       NH2 C5   C2
              \1/
               N
```

FORMULA	C10H8CL2N2			MOL WT	227.09
SOLVENT	CHCL3				
ORIG ST	C4H12SI			TEMP	AMB

117.80	111.50	118.70	117.20	120.20	141.40
2/2	3/1	4/1	5/2	6/1	7/1
119.30	130.20	116.00	128.20		
8/1	9/2	10/2	11/2		

L.L.MARTIN,C.-J.CHANG,H.G.FLOSS,J.A.MABE,
E.W.HAGAMAN,E.WENKERT
J AM CHEM SOC   94, 8942 (1972)

2235   R 001200 3-PHENYLPYRROLE

```
        C10
      //  \
    9C     C11
    |      ||
    8C     C6
     \  /  \
      C    4C---C3
      7    ||   ||
           5C   C2
             \1/
              N
```

FORMULA	C10H9N			MOL WT	143.19
SOLVENT	CHCL3				
ORIG ST	C4H12SI			TEMP	AMB

118.80	105.60	123.80	114.50	135.50	124.90
2/2	3/2	4/1	5/2	6/1	7/2
128.40	124.50				
8/2	9/2				

L.L.MARTIN,C.-J.CHANG,H.G.FLOSS,J.A.MABE,
E.W.HAGAMAN,E.WENKERT
J AM CHEM SOC   94, 8942 (1972)

2236   R 001400 INDOLE

```
      3
      C
    /  \7
  4C    C---C2
  ||    |   ||
  5C    C   C1
    \  /8\ /
      C   N
      6   |
          H
```

FORMULA	C8H7N			MOL WT	117.15
SOLVENT	CDCL3				
ORIG ST	C4H12SI			TEMP	AMB

124.10	102.10	120.50	121.70	119.60	111.00
1/2	2/2	3/2	4/2	5/2	6/2
127.60	135.50				
7/1	8/1				

L.F.JOHNSON,W.C.JANKOWSKI
CARBON-13 NMR SPECTRA,JOHN
WILEY AND SONS,NEW YORK   283 (1972)

2237   R 001400 INDOLE

```
      4
      C
    /  \8
  5C    C---C3
  |     ||  ||
  6C    C   C2
    \  /9\ /
      C   N
      7   1
```

FORMULA	C8H7N			MOL WT	117.15
SOLVENT	CDCL3				
ORIG ST	C4H12SI			TEMP	AMB

125.15	102.65	121.35	122.30	120.25	111.85
2/2	3/2	4/2	5/2	6/2	7/2
128.75	136.15				
8/1	9/1				

N.PLATZER,J.-J.BASSELIER,P.DEMERSEMAN
BULL SOC CHIM FRANCE   5, 905 (1974)

2238   R 001400 3-METHYLINDOLE

```
      3      9
      C      CH3
    /  \7   /
  4C    C---C2
  ||    ||  ||
  5C    C   C1
    \  /8\ /
      C   N
      6
```

FORMULA	C9H9N			MOL WT	131.18
SOLVENT	CDCL3				
ORIG ST	C4H12SI			TEMP	AMB

121.60	110.90	118.60	121.60	118.90	110.90
1/2	2/1	3/2	4/2	5/2	6/2
128.00	136.00	9.40			
7/1	8/1	9/4			

L.F.JOHNSON,W.C.JANKOWSKI
CARBON-13 NMR SPECTRA,JOHN
WILEY AND SONS,NEW YORK   341 (1972)

2239　　　　　　　　　　R 001800　PYRAZOLE

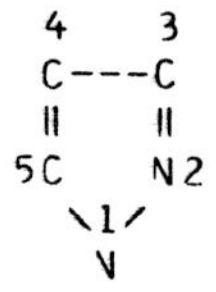

```
    4   3
    C---C
    ||  ||
   5C   N2
    \1/
     N
```

FORMULA　C3H4N2　　　　　MOL WT　68.08
SOLVENT　C3H6O
ORIG ST　C6H6　　　　　　TEMP　　AMB

134.30　105.20
3/2　　　4/2

R.G.REES,M.J.GREEN
J CHEM SOC B　　　　　　　　　387 (1968)

2240　　　　　　　　　　R 001800　1—METHYLPYRAZOLE

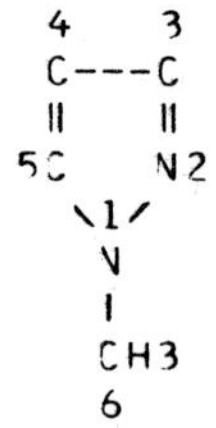

```
    4   3
    C---C
    ||  ||
   5C   N2
    \1/
     N
     |
    CH3
     6
```

FORMULA　C4H6N2　　　　　MOL WT　82.11
SOLVENT　C4H6N2
ORIG ST　C6H6　　　　　　TEMP　　AMB

139.20　105.70　130.30　38.10
3/2　　　4/2　　　5/2　　　6/4

R.G.REES,M.J.GREEN
J CHEM SOC B　　　　　　　　　387 (1968)

2241　　　　　　　　　　R 001800　1—PHENYLPYRAZOLE

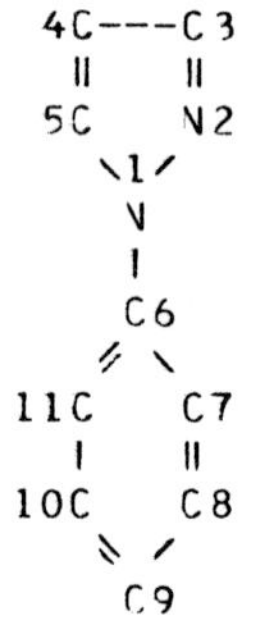

```
  4C---C 3
   ||  ||
  5C   N2
   \1/
    N
    |
    C6
   // \
 11C   C7
   |   ||
 10C   C8
   \  /
    C9
```

FORMULA　C9H8N2　　　　　MOL WT　144.18
SOLVENT　C9H8N2
ORIG ST　C6H6　　　　　　TEMP　　AMB

141.30　107.60　127.00　140.30　118.80　129.40
3/2　　　4/2　　　5/2　　　6/1　　　7/2　　　8/2
126.10
9/2

R.G.REES,M.J.GREEN
J CHEM SOC B　　　　　　　　　387 (1968)

2242　　　12CH3　　　R 001800　1—PHENYL—3—METHYLPYRAZOLE

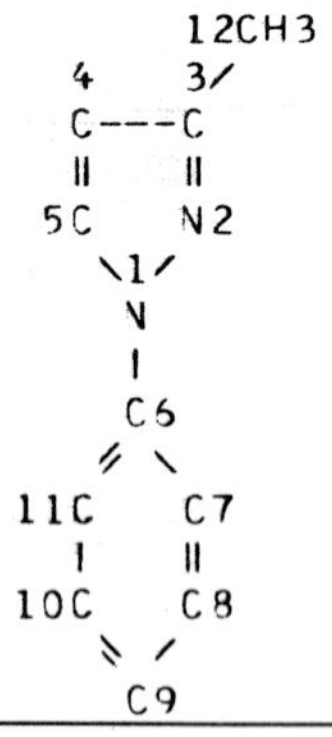

```
   4    3/
   C---C
   ||  ||
  5C   N2
   \1/
    N
    |
    C6
   // \
 11C   C7
   |   ||
 10C   C8
   \  /
    C9
```

FORMULA　C10H10N2　　　　MOL WT　158.20
SOLVENT　CHCL3
ORIG ST　C6H6　　　　　　TEMP　　AMB

150.50　107.70　127.30　140.50　118.80　129.50
3/1　　　4/2　　　5/2　　　6/1　　　7/2　　　8/2
131.00　13.70
9/2　　　12/4

R.G.REES,M.J.GREEN
J CHEM SOC B　　　　　　　　　387 (1968)

2243　　　　　　　　　　R 001900　ISOPILOCARPINE HYDROCHLORIDE

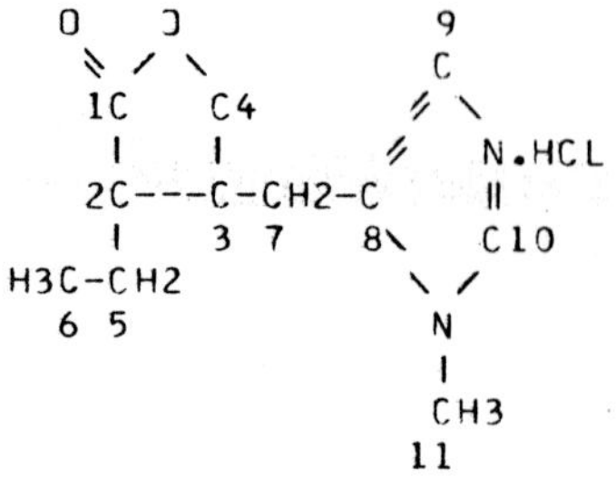

```
   O   O        9
   \ / \        C
   1C   C4     // \
   |    |    //   N.HCL
  2C---C-CH2-C   ||
   |   3 7  8\   C10
 H3C-CH2       \ /
   6 5          N
                |
               CH3
               11
```

FORMULA　C11H16N2O2　　　MOL WT　208.26
SOLVENT　H2O
ORIG ST　C4H12SI　　　　TEMP　　AMB

182.70　46.90　38.20　72.60　22.40　11.00
1/1　　　2/2　　3/2　　4/3　　5/3　　6/4
26.40　132.80　117.70　136.10　34.10
7/3　　　8/1　　　9/2　　　10/2　　11/4

L.F.JOHNSON,W.C.JANKOWSKI
CARBON—13 NMR SPECTRA,JOHN
WILEY AND SONS,NEW YORK
　　　　　　　　　　　　　　　　423 (1972)

2244   R 001900 PILOCARPINE HYDROCHLORIDE

```
    O   O
    \1/ \4 7  10
H3C-H2C  C    C  CH2  C
  6   5\|   I/ \ \ /
     2C---C3 8C    N.HCL
          I    II
          N---C
          I    9
        H3C
         11
```

FORMULA	C11H16N2O2			MOL WT	208.26
SOLVENT	H2O				
ORIG ST	C4H12SI			TEMP	AMB
182.30	45.00	36.80	71.80	18.60	12.20
1/1	2/1	3/2	4/3	5/3	6/4
21.50	133.20	136.10	117.60	34.10	
7/3	8/1	9/2	10/2	11/4	

L.F.JOHNSON,W.C.JANKOWSKI
CARBON-13 NMR SPECTRA,JOHN
WILEY AND SONS,NEW YORK   424 (1972)

2245   R 002100 MONOMETHYLOL DIMETHYLHYDANTOIN

```
    6CH2-OH
  5   |
  H3C   N
  4\ / \
H3C-C3 1C=O
    |   |
  O=C---N
     2
```

FORMULA	C6H10N2O3			MOL WT	158.16
SOLVENT	H2O				
ORIG ST	C4H12SI			TEMP	AMB
157.10	180.40	59.90	24.30	24.30	61.90
1/1	2/1	3/1	4/4	5/4	6/3

L.F.JOHNSON,W.C.JANKOWSKI
CARBON-13 NMR SPECTRA,JOHN
WILEY AND SONS,NEW YORK   178 (1972)

2246   R 002200 ISOXAZOLE

```
     O1
    / \
  5C   N2
  ||   ||
  4C---C3
```

FORMULA	C3H3NO		MOL WT	69.06
SOLVENT	C3D6O			60.0
ORIG ST	C4H12SI		TEMP	301
150.00	104.50	158.90		
3/D	4/D	5/D		

R.FAURE,J.-R.LLINAS,E.-J.VINCENT,M.RAJZMANN
CAN J CHEM   53, 1677 (1975)

2247   R 002300 BENZOXAZOLE

```
     5
     C   N
   / \6/ \
  4C   C   \
  |   ||   C1
  3C   C   /
   \ /7\ /
     C   O
     2
```

FORMULA	C7H5NO			MOL WT	119.12
SOLVENT	CDCL3				
ORIG ST	C4H12SI			TEMP	AMB
152.60	110.80	124.40	125.40	120.50	140.10
1/2	2/2	3/2	4/2	5/2	6/1
150.00					
7/1					

L.F.JOHNSON,W.C.JANKOWSKI
CARBON-13 NMR SPECTRA,JOHN
WILEY AND SONS,NEW YORK   227 (1972)

2248   R 002300 2,2'-BISBENZOXAZOLINYL

```
  13   12      3
   C  NH    NH  C4
 14/ \ / \  / \ /
  C   C18 \  / C9  C5
  |   ||  11C-C2  ||  |
  C   C17 /   \  C8  C6
 15\ / \ /     \ / \ /
   C   O      O   C
   16  10     1   7
```

FORMULA	C14H12N2O2			MOL WT	240.26
SOLVENT	C2D6OS				
ORIG ST	C4H12SI			TEMP	300
75.60	116.50	121.70	119.10	114.60	141.00
2/2	4/2	5/2	6/2	7/2	8/1
130.30					
9/1					

U.HOLLSTEIN,E.BREITMAIER,G.JUNG
J AM CHEM SOC   96, 8036 (1974)

2249   R 002600 ISOTHIAZOLE

```
        S1
       ╱ ╲
     5C   N2
     ‖    ‖
     4C───C3
```

FORMULA C3H3NS   MOL WT   85.13
SOLVENT C3D6O   60.0
ORIG ST C4H12SI   TEMP   301

   157.70 124.10 148.70
    3/D   4/D   5/D

R.FAURE,J.-R.LLINAS,E.-J.VINCENT,M.RAJZMANN
CAN J CHEM   53, 1677 (1975)

2250   R 002600 THIAZOLE

```
   2C───N
   ‖    ‖
   3C   C1
    ╲  ╱
     S
```

FORMULA C3H3NS   MOL WT   85.13
SOLVENT CDCL3
ORIG ST C4H12SI   TEMP   AMB

   152.70 143.20 118.60
    1/2   2/2   3/2

L.F.JOHNSON,W.C.JANKOWSKI
CARBON-13 NMR SPECTRA,JOHN
WILEY AND SONS,NEW YORK   17 (1972)

2251   R 002600 THIAZOLE

```
   4    3
   C───N
   ‖    ‖
  5C    C2
    ╲  ╱
     S
     1
```

FORMULA C3H3NS   MOL WT   85.13
SOLVENT C3H3NS
ORIG ST CS2   TEMP   AMB

   152.20 142.50 118.50
    2/2   4/2   5/2

R.GARNIER,R.FAURE,A.BABADJAMIAN,E.-J.VINCENT
BULL SOC CHIM FRANCE   1040 (1972)

2252   R 002600 THIAZOLIDINE

```
      1
      S
     ╱ ╲
   5C    C2
   │     │
    C───N
    4    3
```

FORMULA C3H7NS   MOL WT   89.16
SOLVENT C6D6
ORIG ST C4H12SI   TEMP   AMB

   55.60   53.10   34.00
    2/3   3/3   4/3

R.FAURE,J.-R.LLINAS,E.-J.VINCENT,J.-L.LARICE
COMPT REND C   279, 717 (1974)

2253   R 002600 4-CYANOISOTHIAZOLE

```
        S1
       ╱ ╲
     5C   N2
   6 ‖    ‖
  N≡C─C───C3
     4
```

FORMULA C4H2N2S   MOL WT 110.14
SOLVENT C3D6O
ORIG ST C4H12SI   TEMP   301

   159.20 110.10 158.80 113.40
    3/D   4/S   5/D   6/S

R.FAURE,J.-R.LLINAS,E.-J.VINCENT,M.RAJZMANN
CAN J CHEM   53, 1677 (1975)

2254   R 002600 5-FORMYLISOTHIAZOLE

```
    O
    ‖
  HC6   S1
    \  / \
     5C    N2
     ‖     ‖
     4C---C3
```

FORMULA   C4H3NOS
SOLVENT   C3D6O
ORIG ST   C4H12SI

MOL WT   113.14
             60.0
TEMP   301

159.60   129.40   165.30   183.00
 3/D   4/D   5/S   6/D

R.FAURE,J.-R.LLINAS,E.-J.VINCENT,M.RAJZMANN
CAN J CHEM   53, 1677 (1975)

2255   R 002600 2-METHYLTHIAZOLE

```
  4    3
  C---N
  ‖    ‖
 5C    C2
  \  / \
   5    CH3
   1    6
```

FORMULA   C4H5NS
SOLVENT   C4H5NS
ORIG ST   CS2

MOL WT   99.16

TEMP   AMB

164.00   141.10   118.10   16.90
 2/1   4/2   5/2   6/4

R.GARNIER,R.FAURE,A.BABADJAMIAN,E.-J.VINCENT
BULL SOC CHIM FRANCE   1040 (1972)

2256   R 002600 3-METHYLISOTHIAZOLE

```
      S1
     / \
   5C    N2
   ‖     ‖
   4C---C3
       \6
        CH3
```

FORMULA   C4H5NS
SOLVENT   C3D6O
ORIG ST   C4H12SI

MOL WT   99.16
             60.0
TEMP   301

167.10   124.50   149.10   18.50
 3/S   4/D   5/D   6/Q

R.FAURE,J.-R.LLINAS,E.-J.VINCENT,M.RAJZMANN
CAN J CHEM   53, 1677 (1975)

2257   R 002600 4-METHYLISOTHIAZOLE

```
      S1
     / \
   5C    N2
   ‖     ‖
   4C---C3
  6/
 H3C
```

FORMULA   C4H5NS
SOLVENT   C3D6O
ORIG ST   C4H12SI

MOL WT   99.16
             60.0
TEMP   301

159.30   134.80   144.30   11.80
 3/D   4/S   5/D   6/Q

R.FAURE,J.-R.LLINAS,E.-J.VINCENT,M.RAJZMANN
CAN J CHEM   53, 1677 (1975)

2258   R 002600 4-METHYLTHIAZOLE

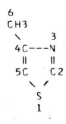

```
   6
  CH3
    \    3
    4C---N
    ‖    ‖
   5C    C2
    \   /
     S
     1
```

FORMULA   C4H5NS
SOLVENT   C4H5NS
ORIG ST   CS2

MOL WT   99.16

TEMP   AMB

151.90   153.00   113.20   16.10
 2/2   4/1   5/2   6/4

R.GARNIER,R.FAURE,A.BABADJAMIAN,E.-J.VINCENT
BULL SOC CHIM FRANCE   1040 (1972)

2259   R 002600   5-METHYLISOTHIAZOLE

```
H3C6   S1
   \  / \
    5C    N2
    ||   ||
    4C---C3
```

FORMULA   C4H5NS   MOL WT   99.16
SOLVENT   C3D6O   60.0
ORIG ST   C4H12SI   TEMP   301

158.30	124.20	163.70	12.40
3/D	4/D	5/S	6/Q

R.FAURE,J.-R.LLINAS,E.-J.VINCENT,M.RAJZMANN
CAN J CHEM   53, 1677 (1975)

2260   R 002600   2-METHYLTHIAZOLIDINE

```
       1
       S
     / \2 6
    5C   C-CH3
    |    |
    4C---N3
```

FORMULA   C4H9NS   MOL WT   103.19
SOLVENT   C6D6
ORIG ST   C4H12SI   TEMP   AMB

66.90	52.80	36.60	22.20
2/2	4/3	5/3	6/4

R.FAURE,J.-R.LLINAS,E.-J.VINCENT,J.-L.LARICE
COMPT REND C   279, 717 (1974)

2261   R 002600   4-METHYLTHIAZOLIDINE

```
       1
       S
      / \
    5C   C2
    |    |
    4C---N3
    /
  H3C6
```

FORMULA   C4H9NS   MOL WT   103.19
SOLVENT   C6D6
ORIG ST   C4H12SI   TEMP   AMB

54.60	61.10	40.70	18.50
2/3	4/2	5/3	6/4

R.FAURE,J.-R.LLINAS,E.-J.VINCENT,J.-L.LARICE
COMPT REND C   279, 717 (1974)

2262   R 002600   5-METHYLTHIAZOLIDINE

```
        1
        S
   6 5/ \
  H3C-C   C2
     |    |
    4C---N3
```

FORMULA   C4H9NS   MOL WT   103.19
SOLVENT   C6D6
ORIG ST   C4H12SI   TEMP   AMB

56.30	60.50	45.80	21.80
2/3	4/3	5/2	6/4

R.FAURE,J.-R.LLINAS,E.-J.VINCENT,J.-L.LARICE
COMPT REND C   279, 717 (1974)

2263   R 002600   2,4-DIMETHYLTHIAZOLE

```
   7
  CH3
    \   3
    4C---N
    ||   ||
    5C    C2
    \   / \
     S    CH3
     1    6
```

FORMULA   C5H7NS   MOL WT   113.18
SOLVENT   C5H7NS
ORIG ST   CS2   TEMP   AMB

163.30	151.50	111.90	17.45	15.65
2/1	4/1	5/2	6/4	7/4

R.GARNIER,R.FAURE,A.BABADJAMIAN,E.-J.VINCENT
BULL SOC CHIM FRANCE   1040 (1972)

```
2264                    R 002600  4,5-DIMETHYLTHIAZOLE
    6
   CH3                        FORMULA   C5H7NS                    MOL WT   113.18
    \       3                 SOLVENT   C5H7NS
    4C---N                    ORIG ST   CS2                       TEMP        AMB
    ||   ||
    5C    C2                    148.10  149.00  125.35   13.30     9.60
    / \  /                       2/2     4/1     5/1     6/4       7/4
  CH3 S
    7   1                      R.GARNIER,R.FAURE,A.BABADJAMIAN,E.-J.VINCENT
                              BULL SOC CHIM FRANCE               1040 (1972)
```

```
2265                    R 002600  2-ETHYLTHIAZOLE
   4   3
   C---N                       FORMULA   C5H7NS                    MOL WT   113.18
   ||   ||                     SOLVENT   C5H7NS
   5C    C2                    ORIG ST   CS2                       TEMP        AMB
    \   / \
     S   CH2-CH3                170.70  141.40  117.25   25.80    13.20
     1   6   7                    2/1     4/2     5/2     6/3      7/4

                              R.GARNIER,R.FAURE,A.BABADJAMIAN,E.-J.VINCENT
                              BULL SOC CHIM FRANCE               1040 (1972)
```

```
2266                    R 002600  4-ETHYLTHIAZOLE
   7   6
  CH3-CH2                      FORMULA   C5H7NS                    MOL WT   113.18
    \       3                 SOLVENT   C5H7NS
    4C---N                    ORIG ST   CS2                       TEMP        AMB
    ||   ||
    5C    C2                    151.90  159.10  111.75   24.30    13.10
    \   /                        2/2     4/1     5/2     6/3      7/4
      S
      1                        R.GARNIER,R.FAURE,A.BABADJAMIAN,E.-J.VINCENT
                              BULL SOC CHIM FRANCE               1040 (1972)
```

```
2267                    R 002600  2-ISOPROPYLTHIAZOLE
    4   3
    C---N                        FORMULA   C6H9NS                   MOL WT   127.21
    ||   ||   7                  SOLVENT   C6H9NS
    5C    C2  CH3                ORIG ST   CS2                      TEMP        AMB
     \  / \  /
      S   6CH                     175.80  141.40  116.65   32.30    22.30
      1    \                        2/1     4/2     5/2     6/2      7/4
           CH3
            8                    R.GARNIER,R.FAURE,A.BABADJAMIAN,E.-J.VINCENT
                                BULL SOC CHIM FRANCE               1040 (1972)
```

```
2268                    R 002600  2,2,4-TRIMETHYLTHIAZOLIDINE

                             FORMULA   C6H13NS                    MOL WT   131.24
     1 6                     SOLVENT   C6D6
     S CH3                   ORIG ST   C4H12SI                     TEMP        AMB
    / \| 7
   5C  2C-CH3                  75.60   59.50   44.60   33.40    31.60    19.10
   8 | |                        2/1     4/2     5/3     6/4      7/4      8/4
  H3C-C---N3
      4                       R.FAURE,J.-R.LLINAS,E.-J.VINCENT,J.-L.LARICE
                             COMPT REND C                      279,  717 (1974)
```

2269   R 002600 2,5,5-TRIMETHYLTHIAZOLIDINE

```
    7    1
  H3C   S
    8\5/ \2 6
  H3C-C   C-CH3
      |   |
      4C---N3
```

FORMULA	C6H13NS			MOL WT	131.24
SOLVENT	C6D6				
ORIG ST	C4H12SI			TEMP	AMB
67.80	66.20	58.30	22.70	31.80	29.60
2/2	4/3	5/1	6/4	7/4	8/4

R.FAURE,J.—R.LLINAS,E.—J.VINCENT,J.—L.LARICE
COMPT REND C   279, 717 (1974)

2270   R 002600 N-(2-THIAZOLYL)-ACETOACETAMIDE

```
    J    OH  N---C2
    ||   ||  ||
   C6   C4 1C   C3
  / \ / \ / \
 H3C   C5  N   S
  7    |   |
       H   H
```

FORMULA	C7H8N2O2S			MOL WT	184.22
SOLVENT	H2O				
ORIG ST	C4H12SI			TEMP	AMB
161.40	137.20	113.30	169.50	88.80	188.10
1/1	2/2	3/2	4/1	5/2	6/1
27.60					
7/4					

L.F.JOHNSON,W.C.JANKOWSKI
CARBON-13 NMR SPECTRA,JOHN
WILEY AND SONS,NEW YORK   245 (1972)

2271   R 002600 2,5-DIMETHYL-4-ETHYLTHIAZOLE

```
  8   7
 CH3-CH2
     \    3
      4C---N
      ||  ||
      5C   C2
     / \ / \
   9CH3 S   CH3
     1   6
```

FORMULA	C7H11NS			MOL WT	141.24
SOLVENT	C7H11NS				
ORIG ST	CS2			TEMP	AMB
159.45	152.60	123.30	17.70	21.50	9.70
2/1	4/1	5/1	6/4	7/3	8/4
13.00					
9/4					

R.GARNIER,R.FAURE,A.BABADJAMIAN,E.—J.VINCENT
BULL SOC CHIM FRANCE   1040 (1972)

2272   R 002600 2-TERT-BUTYLTHIAZOLE

```
    4   3
    C---N
    ||  ||   7
   5C   C2  CH3
    \ / \ /8
     S   6C-CH3
     1    \
          CH3
           9
```

FORMULA	C7H11NS			MOL WT	141.24
SOLVENT	C7H11NS				
ORIG ST	CS2			TEMP	AMB
179.25	141.30	116.60	36.90	30.30	
2/1	4/2	5/2	6/1	7/4	

R.GARNIER,R.FAURE,A.BABADJAMIAN,E.—J.VINCENT
BULL SOC CHIM FRANCE   1040 (1972)

2273   R 002600 2,2,4,4-TETRAMETHYLTHIAZOLIDINE

```
     1   6
     S   CH3
    / \2/7
   5C   C-CH3
  8 |4  |
 H3C-C---N3
     |
    9CH3
```

FORMULA	C7H15NS			MOL WT	145.27
SOLVENT	C6D6				
ORIG ST	C4H12SI			TEMP	AMB
75.00	66.70	49.20	33.80	33.80	29.20
2/1	4/1	5/3	6/4	7/4	8/4
29.20					
9/4					

R.FAURE,J.—R.LLINAS,E.—J.VINCENT,J.—L.LARICE
COMPT REND C   279, 717 (1974)

2274   R 002600 2,5-DIMETHYL-4-ISOPROPYLTHIAZOLE

```
    8CH3
      \
      7CH
     / \4    3
  9CH3 C---N
      ||   ||
      5C   C2
     / \ / \
  10CH3 S   CH3
       1    6
```

FORMULA	C8H13NS			MOL WT	155.26
SOLVENT	C8H13NS				
ORIG ST	CS2			TEMP	AMB

158.85	155.75	121.80	17.70	26.80	21.30
2/1	4/1	5/1	6/4	7/2	8/4
9.30					
10/4					

R.GARNIER,R.FAURE,A.BABADJAMIAN,E.-J.VINCENT
BULL SOC CHIM FRANCE   1040 (1972)

2275   R 002600 2-METHYL-4-TERT-BUTYLTHIAZOLE

```
  9CH3 CH3
    \ /8
    C7
   / \4    3
 10CH3 C---N
     ||   ||
     5C   C2
      \ / \
       S   CH3
       1    6
```

FORMULA	C8H13NS			MOL WT	155.26
SOLVENT	C8H13NS				
ORIG ST	CS2			TEMP	AMB

162.90	165.20	108.60	18.30	34.10	29.40
2/1	4/1	5/2	6/4	7/1	8/4

R.GARNIER,R.FAURE,A.BABADJAMIAN,E.-J.VINCENT
BULL SOC CHIM FRANCE   1040 (1972)

2276   R 002600 2-TERT-BUTYL-4-METHYLTHIAZOLE

```
  10CH3
    \     3
    4C---N
    ||   ||  7
    5C   C2 CH3
     \ / \ /
      S   C6
      1  / \
       CH3 CH3
        8   9
```

FORMULA	C8H13NS			MOL WT	155.26
SOLVENT	C8H13NS				
ORIG ST	CS2			TEMP	AMB

178.60	151.20	111.10	36.70	30.30	16.40
2/1	4/1	5/2	6/1	7/4	10/4

R.GARNIER,R.FAURE,A.BABADJAMIAN,E.-J.VINCENT
BULL SOC CHIM FRANCE   1040 (1972)

2277   R 002600 2,4-DIISOPROPYLTHIAZOLE

```
  10CH3
    \
    9CH
   / \4    3
 11CH3 C---N
     ||   ||
     5C   C2 7CH3
      \ / \ /
       S   6CH
       1     \
             8CH3
```

FORMULA	C9H15NS			MOL WT	169.29
SOLVENT	C9H15NS				
ORIG ST	CS2			TEMP	AMB

174.80	161.90	108.20	32.60	22.30	30.45
2/1	4/1	5/2	6/2	7/4	9/2
21.60					
10/4					

R.GARNIER,R.FAURE,A.BABADJAMIAN,E.-J.VINCENT
BULL SOC CHIM FRANCE   1040 (1972)

2278   R 002600 2-ETHYL-4-TERT-BUTYLTHIAZOLE

```
  10CH3 CH3
    \ /9
    C8
   / \4    3
 11CH3 C---N
     ||   ||
     5C   C2
      \ / \
       S   CH2-CH3
       1   6   7
```

FORMULA	C9H15NS			MOL WT	169.29
SOLVENT	C9H15NS				
ORIG ST	CS2			TEMP	AMB

169.80	165.20	107.80	26.30	13.20	33.90
2/1	4/1	5/2	6/3	7/4	8/1
29.30					
9/4					

R.GARNIER,R.FAURE,A.BABADJAMIAN,E.-J.VINCENT
BULL SOC CHIM FRANCE   1040 (1972)

```
2279              R 002600 2-ISOPROPYL-4-TERT-BUTYLTHIAZOLE
  11CH3 CH3
     \ /10              FORMULA  C10H17NS           MOL WT   183.32
      9C               SOLVENT  C10H17NS
     / \4   3          ORIG ST  CS2                TEMP     AMB
  12CH3 C---N
      ||   ||           175.10  165.20  107.55  32.60  22.40  34.20
     5C   C2 7CH3        2/1     4/1     5/2     6/2    7/4    9/1
      \ / \ /            29.60
       S  6CH           10/4
       |    \
           8CH3         R.GARNIER,R.FAURE,A.BABADJAMIAN,E.-J.VINCENT
                        BULL SOC CHIM FRANCE              1040 (1972)
```

```
2280              R 002600 2,4-DI-TERT-BUTYLTHIAZOLE
  12CH3 CH3
     \ /11              FORMULA  C11H19NS           MOL WT   197.34
      C10               SOLVENT  C11H19NS
     / \4   3           ORIG ST  CS2                TEMP     AMB
  13CH3 C---N
      ||   ||  7         177.90  164.70  107.40  36.70  30.10  33.90
     5C   C2  CH3         2/1     4/1     5/2     6/1    7/4    10/1
      \ / \ /             29.50
       S   C6            11/4
       |   / \
       CH3 CH3           R.GARNIER,R.FAURE,A.BABADJAMIAN,E.-J.VINCENT
        8   9            BULL SOC CHIM FRANCE              1040 (1972)
```

```
2281              R 002700 2-METHYLBENZOTHIAZOLE
                         FORMULA  C8H7NS             MOL WT   149.22
        2                SOLVENT  CDCL3
        C   S            ORIG ST  C4H12SI            TEMP     AMB
       / \6/ \
     3C   C   \1 8        166.40  135.50  153.30  19.80  121.10  122.20
      ||   |   C-CH3       1/1     6/1     7/1     8/4
     4C   C   //          124.40  125.70
      \  /7\ \
        C   N
        5                L.F.JOHNSON,W.C.JANKOWSKI
                         CARBON-13 NMR SPECTRA,JOHN
                         WILEY AND SONS,NEW YORK           285 (1972)
```

```
2282              R 002800 1,2,3-TRIAZOLE
     4    3
     C---N               FORMULA  C2H3N3             MOL WT   69.07
     ||   ||             SOLVENT  C3H6O
     5C   N2             ORIG ST  CS2                TEMP     AMB
      \1/
       N                  130.10
                          4/2

                         F.J.WEIGERT,J.D.ROBERTS
                         J AM CHEM SOC               90, 3543 (1968)
```

```
2283              R 002800 1,2,4-TRIAZOLE
     4    3
     N---C               FORMULA  C2H3N3             MOL WT   69.07
     ||   ||             SOLVENT  C3H6O
     5C   N2             ORIG ST  CS2                TEMP     AMB
      \1/
       N                  147.30
                          3/2

                         F.J.WEIGERT,J.D.ROBERTS
                         J AM CHEM SOC               90, 3543 (1968)
```

2284   R 003100 TETRAZOLE

```
  4    3
  N---N
  ||   ||
 5C    N2
   \1/
    N
```

FORMULA	CH2N4	MOL WT	70.05
SOLVENT	C2H6OS		
ORIG ST	C S2	TEMP	AMB

143.70
5/2

F.J.WEIGERT,J.D.ROBERTS
J AM CHEM SOC   90, 3543 (1968)

2285   R 003400 2-PYRONE

```
     C4
    / \
  C3   C5
  |    ||
  C2   C6
    / \1/
   O   O
```

FORMULA	C5H4O2	MOL WT	96.09
SOLVENT	C5H4O2/C3D6O		
ORIG ST	C4H12SI	TEMP	AMB

| 162.00 | 116.70 | 144.30 | 106.80 | 153.30 |
| 2/S | 3/D | 4/D | 5/D | 6/D |

W.V.TURNER,W.H.PIRKLE
J ORG CHEM   39, 1935 (1974)

2286   R 003400 3-HYDROXY-2-PYRONE

```
 HO    C4
   \ / \
   C3   C5
   |    ||
   C2   C6
    / \1/
   O   O
```

FORMULA	C5H4O3	MOL WT	112.09
SOLVENT	CDCL3		
ORIG ST	C4H12SI	TEMP	AMB

| 161.80 | 142.70 | 115.20 | 107.30 | 142.30 |
| 2/S | 3/S | 4/D | 5/D | 6/D |

W.V.TURNER,W.H.PIRKLE
J ORG CHEM   39, 1935 (1974)

2287   R 003400 PYRAN-2-THIONE

```
     C4
    / \
  C3   C5
  |    ||
  C2   C6
    / \1/
   S   O
```

FORMULA	C5H4OS	MOL WT	112.15
SOLVENT	CDCL3/CCL4		
ORIG ST	C4H12SI	TEMP	AMB

| 196.90 | 131.80 | 134.40 | 109.60 | 155.60 |
| 2/1 | 3/2 | 4/2 | 5/2 | 6/2 |

W.V.TURNER,W.H.PIRKLE
J ORG CHEM   39, 1935 (1974)

2288   R 003400 5,6-DIHYDROPYRAN

```
    4
    C
   / \
  5C   C3
  |    ||
  6C   C2
   \ /
    O
    1
```

FORMULA	C5H8O	MOL WT	84.12
SOLVENT	CCL4		
ORIG ST	C4H12SI	TEMP	AMB

| 144.10 | 99.40 | 19.40 | 22.60 | 64.80 |
| 2/2 | 3/2 | 4/3 | 5/3 | 6/3 |

E.WENKERT,D.W.COCHRAN,E.W.HAGAMAN,F.M.SCHELL,
N.NEUSS,A.S.KATNER,P.POTIER,C.KAN,M.PLAT,
M.KOCH,H.MEHRI,J.POISSON,N.KUNESCH,Y.ROLLAND
J AM CHEM SOC   95, 4990 (1973)

2289

R 003400 TETRAHYDROPYRAN

```
    2 1
    C-C
   /    \
 3C      O
   \    /
    C-C
    4 5
```

FORMULA	C5H10O
SOLVENT	C4H8O2
ORIG ST	C4H12SI

MOL WT 86.13

TEMP AMB

| 68.60 | 27.20 | 24.20 | 27.20 | 68.60 |
| 1/3 | 2/3 | 3/3 | 4/3 | 5/3 |

L.F.JOHNSON,W.C.JANKOWSKI
CARBON—13 NMR SPECTRA,JOHN
WILEY AND SONS,NEW YORK   124 (1972)

2290

R 003400 TETRAHYDROPYRAN

```
    2 1
    C-C
   /    \
 3C      O
   \    /
    C-C
    4 5
```

FORMULA	C5H10O
SOLVENT	CDCL3
ORIG ST	C4H12SI

MOL WT 86.13

TEMP AMB

| 69.70 | 27.90 | 25.10 | 27.90 | 69.70 |
| 1/3 | 2/3 | 3/3 | 4/3 | 5/3 |

N.R.EASTON,JR.,F.A.L.ANET,P.A.BURNS,C.S.FOOTE
J AM CHEM SOC   96, 3945 (1974)

2291

R 003400 TETRAHYDROPYRAN

```
    4
    C
   / \
 5C   C3
  |   |
 6C   C2
   \ /
    O
    1
```

FORMULA	C5H10O
SOLVENT	CCL4
ORIG ST	C4H12SI

MOL WT 86.13

TEMP AMB

| 67.90 | 26.60 | 23.60 |
| 2/3 | 3/3 | 4/3 |

E.WENKERT,D.W.COCHRAN,E.W.HAGAMAN,F.M.SCHELL,
N.NEUSS,A.S.KATNER,P.POTIER,C.KAN,M.PLAT,
M.KOCH,H.MEHRI,J.POISSON,N.KUNESCH,Y.ROLLAND
J AM CHEM SOC   95, 4990 (1973)

2292

R 003400 TETRAHYDROPYRAN

```
    4
    C
   / \
 5C   C3
  |   |
 6C   C2
   \ /
    O
    1
```

FORMULA	C5H10O
SOLVENT	C5H10O
ORIG ST	C6H6

MOL WT 86.13

TEMP AMB

| 69.50 | 27.70 | 24.90 |
| 2/3 | 3/3 | 4/3 |

G.E.MACIEL,G.B.SAVITSKY
J PHYS CHEM   69, 3925 (1965)

2293

R 003400 COUMALYL CHLORIDE

```
    1 5
   O-C   O
  /   \5 7/
O=C2    C-C
  \3 4/    \
   C=C      O-CL
```

FORMULA	C6H3CLO4
SOLVENT	CDCL3
ORIG ST	C4H12SI

MOL WT 174.54

TEMP AMB

| 158.20 | 115.30 | 140.30 | 116.60 | 163.00 |
| 2/S | 3/D | 4/D | 5/S | 6/D |

W.V.TURNER,W.H.PIRKLE
J ORG CHEM   39, 1935 (1974)

2294 R 003400 4—CHLORO—6—METHYL—2—PYRONE

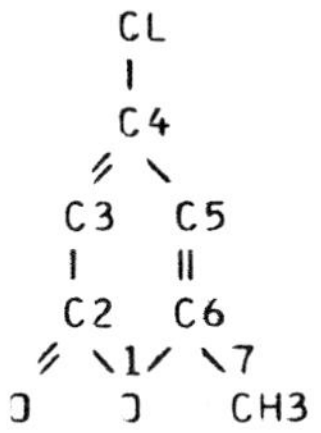

```
    CL
    |
    C4
   ∕ ╲
 C3   C5
 |    ‖
 C2   C6
 ∕ ╲1∕ ╲7
O    O   CH3
```

FORMULA	C6H5CLO2			MOL WT	144.56
SOLVENT	CDCL3				
ORIG ST	C4H12SI			TEMP	AMB

160.90	110.80	151.80	106.10	162.90	19.80
2/S	3/D	4/S	5/D	6/S	7/Q

W.V.TURNER,W.H.PIRKLE
J ORG CHEM   39, 1935 (1974)

2295 R 003400 4—METHYL—6—CHLORO—2—PYRONE

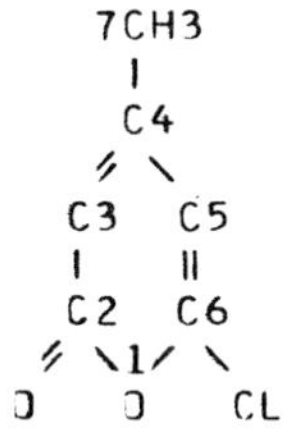

```
   7CH3
    |
    C4
   ∕ ╲
 C3   C5
 |    ‖
 C2   C6
 ∕ ╲1∕ ╲
O    O   CL
```

FORMULA	C6H5CLO2			MOL WT	144.56
SOLVENT	CDCL3				
ORIG ST	C4H12SI			TEMP	AMB

160.60	110.40	157.50	107.10	148.70	21.40
2/S	3/D	4/S	5/D	6/S	7/Q

W.V.TURNER,W.H.PIRKLE
J ORG CHEM   39, 1935 (1974)

2296 R 003400 4—METHYL—2—PYRONE

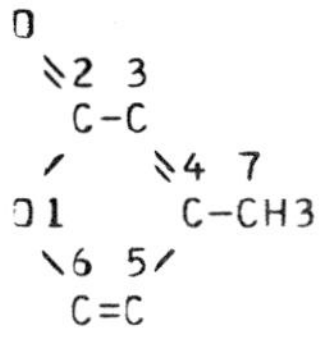

```
 O
  ╲2 3
  C—C
 ∕  ╲4 7
O1   C—CH3
  ╲6 5∕
   C=C
```

FORMULA	C6H6O2			MOL WT	110.11
SOLVENT	CDCL3				
ORIG ST	C4H12SI			TEMP	AMB

161.80	113.70	156.10	109.30	151.10	21.10
2/S	3/D	4/S	5/D	6/D	7/Q

W.V.TURNER,W.H.PIRKLE
J ORG CHEM   39, 1935 (1974)

2297 R 003400 5—METHYL—2—PYRONE

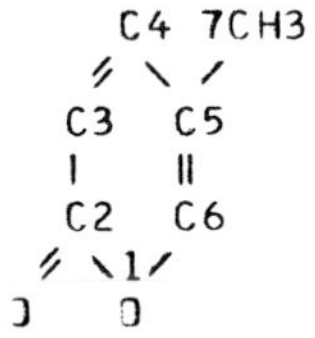

```
   C4 7CH3
   ∕ ╲ ∕
 C3   C5
 |    ‖
 C2   C6
 ∕ ╲1∕
O    O
```

FORMULA	C6H6O2			MOL WT	110.11
SOLVENT	CDCL3				
ORIG ST	C4H12SI			TEMP	AMB

161.20	115.70	146.50	114.70	148.00	14.40
2/S	3/D	4/D	5/S	6/D	7/Q

W.V.TURNER,W.H.PIRKLE
J ORG CHEM   39, 1935 (1974)

2298 R 003400 6—METHYL—2—PYRONE

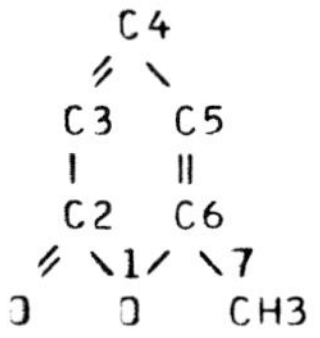

```
    C4
   ∕ ╲
 C3   C5
 |    ‖
 C2   C6
 ∕ ╲1∕ ╲7
O    O   CH3
```

FORMULA	C6H6O2			MOL WT	110.11
SOLVENT	CDCL3				
ORIG ST	C4H12SI			TEMP	AMB

162.00	112.60	144.10	103.40	162.90	19.80
2/S	3/D	4/D	5/D	6/S	7/Q

W.V.TURNER,W.H.PIRKLE
J ORG CHEM   39, 1935 (1974)

2299   R 003400   3-METHOXY-2-PYRONE

```
      7
  H3C-O    C4
     \  /  \
       C3   C5
       |    ||
       C2   C6
      /  \1/
     O    O
```

FORMULA	C6H6O3			MOL WT	126.11
SOLVENT	CDCL3				
ORIG ST	C4H12SI			TEMP	AMB

158.60	145.90	112.90	106.00	142.90	56.00
2/1	3/1	4/2	5/2	6/2	7/4

W.V.TURNER,W.H.PIRKLE
J ORG CHEM   39, 1935 (1974)

2300   R 003400   4-METHYLPYRAN-2-THIONE

```
     S
    \2 3
     C-C
    /   \4 7
   O1    C-CH3
    \6 5/
     C=C
```

FORMULA	C6H6OS			MOL WT	126.18
SOLVENT	CDCL3				
ORIG ST	C4H12SI			TEMP	AMB

197.00	130.20	149.20	113.20	155.00	21.10
2/S	3/D	4/S	5/D	6/D	7/Q

W.V.TURNER,W.H.PIRKLE
J ORG CHEM   39, 1935 (1974)

2301   R 003400   3-ACETOXY-2-PYRONE

```
    8  7
  H3C-C-O    C4
      ||  \  /  \
      O    C3   C5
           |    ||
           C2   C6
          /  \1/
         O    O
```

FORMULA	C7H6O4			MOL WT	154.12
SOLVENT	CDCL3				
ORIG ST	C4H12SI			TEMP	AMB

157.60	137.30	131.20	105.40	149.10	20.20
2/1	3/1	4/2	5/2	6/2	8/4

W.V.TURNER,W.H.PIRKLE
J ORG CHEM   39, 1935 (1974)

2302   R 003400   4-METHOXY-6-METHYL-2-PYRONE

```
      O-CH3
      |  7
      C4
     /  \
    C3   C5
    |    ||
    C2   C6
   /  \1/  \8
  O    O    CH3
```

FORMULA	C7H8O3			MOL WT	140.14
SOLVENT	CDCL3				
ORIG ST	C4H12SI			TEMP	AMB

162.10	87.30	171.40	100.30	164.60	55.90
2/S	3/D	4/S	5/D	6/S	7/Q
19.70					
8/Q					

W.V.TURNER,W.H.PIRKLE
J ORG CHEM   39, 1935 (1974)

2303   R 034000   5-ETHYL-5,6-DIHYDROPYRAN

```
    4    7
    C    CH2
   / \  / \
  5C   C3  CH3
  |    ||  8
  6C   C2
   \  /
    O
    1
```

FORMULA	C7H12O			MOL WT	112.17
SOLVENT	CCL4				
ORIG ST	C4H12SI			TEMP	AMB

138.20	112.70	25.80	22.60	64.20	22.20
2/2	3/1	4/3	5/3	6/3	7/3
12.40					
8/4					

E.WENKERT,D.W.COCHRAN,E.W.HAGAMAN,F.M.SCHELL,
N.NEUSS,A.S.KATNER,P.POTIER,C.KAN,M.PLAT,
M.KOCH,H.MEHRI,J.POISSON,N.KUNESCH,Y.ROLLAND
J AM CHEM SOC   95, 4990 (1973)

2304   R 003400 4,6-DIMETHYL-5-CARBETHOXYPYRAN-2-THIONE

```
  10CH3    O
    |      ||
    C4  O-C-CH2-CH3
   ⁄ \ ⁄  7  8  9
  C3   C5
  |    ||
  C2   C6
 ⁄ \1⁄ \11
 S   O   CH3
```

FORMULA	C10H12O3S			MOL WT	212.27
SOLVENT	CDCL3				
ORIG ST	C4H12SI			TEMP	AMB

195.70	128.60	146.80	116.60	168.40	164.60
2/S	3/D	4/S	5/S	6/S	7/S
		20.50	19.90		
8/3	9/4	10/Q	11/Q		

W.V.TURNER,W.H.PIRKLE
J ORG CHEM   39, 1935 (1974)

2305   R 003400 4,6-DIMETHYL-5-CARBETHOXY-2-PYRONE

```
  10CH3    O
    |      ||
    C4  O-C-CH2-CH3
   ⁄ \ ⁄  7  8  9
  C3   C5
  |    ||
  C2   C6
 ⁄ \1⁄ \11
 O   O   CH3
```

FORMULA	C10H12O4			MOL WT	196.20
SOLVENT	C10H12O4/C3D6O				
ORIG ST	C4H12SI			TEMP	AMB

159.60	111.90	154.40	112.90	165.10	165.20
2/S	3/D	4/S	5/S	6/S	7/S
		21.00	19.60		
8/3	9/4	10/Q	11/Q		

W.V.TURNER,W.H.PIRKLE
J ORG CHEM   39, 1935 (1974)

2306   R 003400 4-METHOXY-6-(4-METHOXYSTYRYL)-2-PYRONE

```
  O-CH3
  | 15
  C4
 ⁄ \
 C3   C5         C10
 |    || 7  8  9⁄ \
 C2 6C-CH=CH-C    C11
⁄ \1⁄        |    |
O   O       14C    C12
        \ ⁄ \ 16
        C13 O-CH3
```

FORMULA	C15H14O4			MOL WT	258.28
SOLVENT	CDCL3				
ORIG ST	C4H12SI			TEMP	AMB

160.90	88.40	171.30	100.40	163.90	116.60
2/1	3/2	4/1	5/2	6/1	7/2
135.40	128.20	129.00	114.50	159.20	114.50
8/2	9/1	10/2	11/2	12/1	13/2
129.00	55.80	55.40			
14/2	15/4	16/4			

W.V.TURNER,W.H.PIRKLE
J ORG CHEM   39, 1935 (1974)

2307   R 003500 CHROMONE

```
       O
       ||
   C5   C4
  ⁄ \ ⁄ \
 6C   C9   C3
 |    ||   ||
 7C   C10  C2
  \ ⁄ \ ⁄
   C8   O1
```

FORMULA	C9H6O2			MOL WT	146.15
SOLVENT	CDCL3				
ORIG ST	C4H12SI			TEMP	305

145.50	113.00	177.60	125.30	125.80	133.80
2/D	3/D	4/S	5/D	6/D	7/D
118.30	125.00	156.60			
8/D	9/S	10/S			

M.S.CHAUHAN,I.W.J.STILL
CAN J CHEM   53, 2880 (1975)

2308   R 003500 COUMARIN

```
   C5   C4
  ⁄ \ ⁄ \
 C6   C9   C3
 |    ||   |
 C7   C10  C2
  \8⁄ \1⁄ \
   C    O    O
```

FORMULA	C9H6O2			MOL WT	146.15
SOLVENT	CDCL3				
ORIG ST	C4H12SI			TEMP	AMB

160.50	116.40	143.50	116.50	124.40	128.00
2/S	3/D	4/D	5/D	6/D	7/D
131.70	118.80	153.90			
8/D	9/S	10/S			

W.V.TURNER,W.H.PIRKLE
J ORG CHEM   39, 1935 (1974)

2309   R 003500   BENZOPYRAN-2-THIONE

```
      C5  C4
     ⁄  \ ⁄ \
    C6  C9  C3
    |   ||  |
    C7  C10 C2
    \8⁄ \1⁄ \
      C   O   S
```

FORMULA	C9H6OS			MOL WT	162.21
SOLVENT	C3H6O				
ORIG ST	C4H12SI			TEMP	AMB

198.50	130.20	133.00	116.90	126.30	129.00
2/S	3/D	4/D	5/D	6/D	7/D
135.60	121.30	157.20			
8/D	9/S	10/S			

W.V. TURNER, W.H. PIRKLE
J ORG CHEM   39, 1935 (1974)

2310   R 003500   4H-CHROMENE-4-THIONE

```
        S
        ||
      C5  C4
     ⁄ \ ⁄ \
   6C  C9  C3
   |   ||  ||
   7C  C10 C2
    \ ⁄ \ ⁄
      C8  O1
```

FORMULA	C9H6OS			MOL WT	162.21
SOLVENT	CDCL3				
ORIG ST	C4H12SI			TEMP	305

146.00	126.20	202.80	128.60	124.80	134.20
2/D	3/D	4/S	5/D	6/D	7/D
118.60	130.90	151.40			
8/D	9/S	10/S			

M.S. CHAUHAN, I.W.J. STILL
CAN J CHEM   53, 2880 (1975)

2311   R 003500   CHROMAN-4-ONE

```
        O
        ||
      C5  C4
     ⁄ \ ⁄ \
   6C  C9  C3
   |   ||  |
   7C  C10 C2
    \ ⁄ \ ⁄
      C8  O1
```

FORMULA	C9H8O2			MOL WT	148.16
SOLVENT	CDCL3				
ORIG ST	C4H12SI			TEMP	305

66.90	37.70	191.40	127.00	121.20	135.80
2/T	3/T	4/S	5/D	6/D	7/D
117.80	121.40	161.80			
8/D	9/S	10/S			

M.S. CHAUHAN, I.W.J. STILL
CAN J CHEM   53, 2880 (1975)

2312   R 003500   2,2-DIMETHYLCHROMANOL

```
    8        11
    C   O1  CH3
   ⁄ \ ⁄ \ ⁄
  7C  C10 C2
  |   ||  |\12
  6C  C9  C CH3
   ⁄ \ ⁄ \ ⁄3
  HO  C   C
      5   4
```

FORMULA	C11H14O2			MOL WT	178.23
SOLVENT	CDCL3				
ORIG ST	C4H12SI			TEMP	AMB

74.00	32.80	22.60	114.50	148.40	115.50
2/1	3/3	4/3	5/2	6/1	7/2
117.60	121.70	147.50	26.70		
8/2	9/1	10/1	11/4		

M. MATSUO, S. URANO
TETRAHEDRON   32, 229 (1976)

2313   R 003500   2,2,5,7-TETRAMETHYLCHROMANOL

```
   14         11
  H3C  C8  O1  CH3
    \ ⁄ \ ⁄ \ ⁄
    7C  C10 C2
    |   ||  |\12
    6C  C9  C CH3
     ⁄ \ ⁄ \ ⁄3
   HO  C5  C
       |   4
      13CH3
```

FORMULA	C13H18O2			MOL WT	206.29
SOLVENT	CDCL3				
ORIG ST	C4H12SI			TEMP	AMB

72.60	33.00	20.80	122.00	145.00	122.40
2/1	3/3	4/3	5/1	6/1	7/1
116.10	117.80	147.20	26.50	11.40	16.00
8/2	9/1	10/1	11/4	13/4	14/4

M. MATSUO, S. URANO
TETRAHEDRON   32, 229 (1976)

2314   R 003500   2,2,5,8-TETRAMETHYLCHROMANOL

```
    14CH3
     |        11
    C8   01  CH3
    ∥ \ / \ /
   7C   C10 C2
    |   ∥   |\12
   6C   C9  C CH3
   / \ / \ /3
 HO   C5  C
     |    4
    13CH3
```

FORMULA	C13H18O2			MOL WT	206.29
SOLVENT	CDCL3				
ORIG ST	C4H12SI			TEMP	AMB
72.40	33.00	21.10	119.40	145.60	115.50
2/1	3/3	4/3	5/1	6/1	7/2
123.80	120.00	145.90	26.60	11.00	15.30
8/1	9/1	10/1	11/4	13/4	14/4

M.MATSUO,S.URANO
TETRAHEDRON   32, 229 (1976)

2315   R 003500   2,2,7,8-TETRAMETHYLCHROMANOL

```
      14
      CH3
  13   |      11
  H3C  C8  01  CH3
   \ ∥ \ / \ /
    7C   C10 C2
    |   ∥   |\12
    6C   C9  C CH3
   / \ / \ /3
  HO   C   C
       5   4
```

FORMULA	C13H18O2			MOL WT	206.29
SOLVENT	CDCL3				
ORIG ST	C4H12SI			TEMP	AMB
73.30	33.00	22.60	112.30	146.10	121.80
2/1	3/3	4/3	5/2	6/1	7/1
125.60	117.90	145.70	26.90	11.90	11.90
8/1	9/1	10/1	11/4	13/4	14/4

M.MATSUO,S.URANO
TETRAHEDRON   32, 229 (1976)

2316   R 003500   2,2,5,7,8-PENTAMETHYLCHROMANOL

```
      15
      CH3
  14   |      11
  H3C  C8  01  CH3
   \ ∥ \ / \ /
    7C   C10 C2
    |   ∥   |\12
    6C   C9 3C CH3
   / \ / \ /
  HO   C5  C
       |   4
       CH3
       13
```

FORMULA	C14H20O2			MOL WT	220.31
SOLVENT	CDCL3				
ORIG ST	C4H12SI			TEMP	AMB
72.30	33.10	21.10	118.60	144.40	121.10
2/1	3/3	4/3	5/1	6/1	7/1
122.40	116.90	145.50	26.70	11.30	12.10
8/1	9/1	10/1	11/4	13/4	14/4
11.80					
15/4					

M.MATSUO,S.URANO
TETRAHEDRON   32, 229 (1976)

2317   R 003700   THIAPYRAN-2-ONE

```
      C4
     ∥ \
    C3  C5
    |   ∥
    C2  C6
    ∥ \1/
    0   S
```

FORMULA	C5H4OS			MOL WT	112.15
SOLVENT	CDCL3/CCL4				
ORIG ST	C4H12SI			TEMP	AMB
183.70	124.90	140.90	118.70	137.40	
2/S	3/D	4/D	5/D	6/D	

W.V.TURNER,W.H.PIRKLE
J ORG CHEM   39, 1935 (1974)

2318   R 003700   THIAPYRAN-2-THIONE

```
      C4
     ∥ \
    C3  C5
    |   ∥
    C2  C6
    ∥ \1/
    S   S
```

FORMULA	C5H4S2			MOL WT	128.22
SOLVENT	CDCL3				
ORIG ST	C4H12SI			TEMP	AMB
205.20	139.30	132.60	122.70	141.80	
2/S	3/D	4/D	5/D	6/D	

W.V.TURNER,W.H.PIRKLE
J ORG CHEM   39, 1935 (1974)

2319   R 003700 PENTAMETHYLENE SULFIDE

```
      4
      C
    / \
  5C   C3
  |    |
  6C   C2
   \  /
    S
    1
```

FORMULA	C5H10S	MOL WT 102.20
SOLVENT	C5H10S	
ORIG ST	C6H6	TEMP AMB

29.00	29.00	29.00
2/3	3/3	4/3

G.E.MACIEL,G.B.SAVITSKY
J PHYS CHEM   69, 3925 (1965)

2320   R 003700 4-METHYLTHIAPYRAN-2-ONE

```
    O
    ‖2 3
    C-C
  /    ‖4 7
 S1    C-CH3
  \6 5/
    C=C
```

FORMULA	C6H6OS	MOL WT 126.18
SOLVENT	CDCL3	
ORIG ST	C4H12SI	TEMP AMB

184.30	123.80	153.40	122.20	136.00	24.10
2/S	3/D	4/S	5/D	6/D	7/Q

W.V.TURNER,W.H.PIRKLE
J ORG CHEM   39, 1935 (1974)

2321   R 003700 THIOCHROMONE

```
      O
      ‖
    C5  C4
   ∥ \ / \
  6C   C9  C3
  |    ‖   ‖
  7C   C10 C2
   \ / \ /
    C8  S1
```

FORMULA	C9H6OS	MOL WT 162.21
SOLVENT	CDCL3	
ORIG ST	C4H12SI	TEMP 305

137.90	126.60	179.40	128.50	125.70	131.40
2/D	3/D	4/S	5/D	6/D	7/D
127.70	132.20	137.50			
8/D	9/S	10/S			

M.S.CHAUHAN,I.W.J.STILL
CAN J CHEM   53, 2880 (1975)

2322   R 003700 THIOCHROMONE-1,1-DIOXIDE

```
      O
      ‖
    C5  C4
   ∥ \ / \
  6C   C9  C3
  |    ‖   ‖
  7C   C10 C2
   \ / \ /
    C8  S1
      ∥ ∥
      O  O
```

FORMULA	C9H6O3S	MOL WT 194.21
SOLVENT	CDCL3	
ORIG ST	C4H12SI	TEMP 305

140.60	132.20	177.80	128.50	135.00	133.00
2/D	3/D	4/S	5/D	6/D	7/D
123.10	128.20	141.10			
8/D	9/S	10/S			

M.S.CHAUHAN,I.W.J.STILL
CAN J CHEM   53, 2880 (1975)

2323   R 003700 3-BROMOTHIOCHROMAN-4-ONE 1-OXIDE

```
      O
      ‖
    C5  C4  BR
   ∥ \ / \ /
  6C   C9  C3
  |    ‖   |
  7C   C10 C2
   \ / \ /
    C8  S1
      ‖
      O
```

FORMULA	C9H7BRO2S	MOL WT 259.12
SOLVENT	CDCL3	
ORIG ST	C4H12SI	TEMP 305

54.90	42.10	186.10	130.50	132.80	135.30
2/T	3/D	4/S	5/D	6/D	7/D
129.50	130.30	144.20			
8/D	9/S	10/S			

M.S.CHAUHAN,I.W.J.STILL
CAN J CHEM   53, 2880 (1975)

2324   R 003700 THIOCHROMAN-4-ONE

```
        O
        ||
   C5   C4
  ∥ \  / \
 6C   C9  C3
 |    ||   |
 7C   C10 C2
  \  / \ /
   C8   S1
```

FORMULA C9H8OS				MOL WT	164.23
SOLVENT CDCL3					
ORIG ST C4H12SI				TEMP	305

26.60	39.50	193.80	129.10	124.90	133.10
2/T	3/T	4/S	5/D	6/D	7/D
127.50	130.90	142.10			
8/D	9/S	10/S			

M.S.CHAUHAN,I.W.J.STILL
CAN J CHEM   53, 2880 (1975)

2325   R 003700 THIOCHROMAN-4-ONE 1-OXIDE

```
        O
        ||
   C5   C4
  ∥ \  / \
 6C   C9  C3
 |    ||   |
 7C   C10 C2
  \  / \ /
   C8   S1
        ||
        O
```

FORMULA C9H8O2S				MOL WT	180.23
SOLVENT CDCL3					
ORIG ST C4H12SI				TEMP	305

47.10	30.70	191.90	129.70	132.00	134.40
2/T	3/T	4/S	5/D	6/D	7/D
128.30	129.10	145.60			
8/D	9/S	10/S			

M.S.CHAUHAN,I.W.J.STILL
CAN J CHEM   53, 2880 (1975)

2326   R 003700 THIOCHROMAN-4-ONE 1,1-DIOXIDE

```
        O
        ||
   C5   C4
  ∥ \  / \
 6C   C9  C3
 |    ||   |
 7C   C10 C2
  \  / \ /
   C8   S1
        ∥ \
        O   O
```

FORMULA C9H8O3S				MOL WT	196.23
SOLVENT CDCL3					
ORIG ST C4H12SI				TEMP	305

49.30	36.80	190.10	128.80	134.90	133.30
2/T	3/T	4/S	5/D	6/D	7/D
123.70	130.30	141.50			
8/D	9/S	10/S			

M.S.CHAUHAN,I.W.J.STILL
CAN J CHEM   53, 2880 (1975)

2327   R 003700 2-METHYLTHIOCHROMONE

```
        O
        ||
   C5   C4
  ∥ \  / \
 6C   C9  C3
 |    ||   ||
 7C   C10 C2
  \  / \ / \11
   C8   S1  CH3
```

FORMULA C10H8OS				MOL WT	176.24
SOLVENT CDCL3					
ORIG ST C4H12SI				TEMP	305

151.00	124.60	179.90	128.20	125.90	131.20
2/S	3/D	4/S	5/D	6/D	7/D
127.30	130.50	137.40	23.00		
8/D	9/S	10/S	11/Q		

M.S.CHAUHAN,I.W.J.STILL
CAN J CHEM   53, 2880 (1975)

2328   R 003700 3-METHYLTHIOCHROMONE

```
        O    11
        ||
   C5   C4  CH3
  ∥ \  / \ /
 6C   C9  C3
 |    ||   ||
 7C   C10 C2
  \  / \ /
   C8   S1
```

FORMULA C10H8OS				MOL WT	176.24
SOLVENT CDCL3					
ORIG ST C4H12SI				TEMP	305

132.70	133.00	179.30	128.50	126.30	131.00
2/D	3/S	4/S	5/D	6/D	7/D
127.00	130.60	137.40	18.90		
8/D	9/S	10/S	11/Q		

M.S.CHAUHAN,I.W.J.STILL
CAN J CHEM   53, 2880 (1975)

2329　　　R 003700　2—METHYLTHIOCHROMAN—4—ONE

```
          O
          ‖
       C5   C4
      ⁄ \  ⁄ \
    6C    C9  C3
    |    |
    7C   C10 C2
     \ ⁄ \ ⁄ \11
      C8  S1  CH3
```

FORMULA	C10H100S			MOL WT	178.25
SOLVENT	CDCL3				
ORIG ST	C4H12SI			TEMP	305

36.20	47.60	193.90	128.80	124.60	133.10
2/D	3/T	4/S	5/D	6/D	7/D
127.30	130.20	141.60	20.30		
8/D	9/S	10/S	11/Q		

M.S.CHAUHAN,I.W.J.STILL
CAN J CHEM　　　　　　　　　　53, 2880 (1975)

2330　　　R 003700　3—METHYLTHIOCHROMAN—4—ONE

```
          O
          ‖   11
       C5   C4  CH3
      ⁄ \  ⁄ \ ⁄
    6C    C9  C3
    |    ‖   |
    7C   C10 C2
     \ ⁄ \ ⁄
      C8  S1
```

FORMULA	C10H100S			MOL WT	178.25
SOLVENT	CDCL3				
ORIG ST	C4H12SI			TEMP	305

33.00	42.10	196.20	129.40	124.70	133.30
2/T	3/D	4/S	5/D	6/D	7/D
127.20	130.40	141.80	14.90		
8/D	9/S	10/S	11/Q		

M.S.CHAUHAN,I.W.J.STILL
CAN J CHEM　　　　　　　　　　53, 2880 (1975)

2331　　　R 003700　6—METHYLTHIOCHROMAN—4—ONE

```
               O
      11       ‖
     H3C   C5   C4
       \ ⁄ \  ⁄ \
      6C    C9  C3
      |    ‖   |
      7C   C10 C2
       \ ⁄ \ ⁄
        C8  S1
```

FORMULA	C10H100S			MOL WT	178.25
SOLVENT	CDCL3				
ORIG ST	C4H12SI			TEMP	305

26.70	39.80	194.30	129.30	134.90	134.40
2/T	3/T	4/S	5/D	6/S	7/D
127.50	130.80	138.80	20.80		
8/D	9/S	10/S	11/Q		

M.S.CHAUHAN,I.W.J.STILL
CAN J CHEM　　　　　　　　　　53, 2880 (1975)

2332　　　R 003700　8—METHYLTHIOCHROMAN—4—ONE

```
          O
          ‖
       C5   C4
      ⁄ \  ⁄ \
    6C    C9  C3
    |    ‖   |
    7C   C10 C2
     \ ⁄ \ ⁄
      C8  S1
      |
     11CH3
```

FORMULA	C10H100S			MOL WT	178.25
SOLVENT	CDCL3				
ORIG ST	C4H12SI			TEMP	305

26.00	39.10	194.30	126.90	124.10	134.30
2/T	3/T	4/S	5/D	6/D	7/D
135.40	131.10	141.70	20.00		
8/S	9/S	10/S	11/Q		

M.S.CHAUHAN,I.W.J.STILL
CAN J CHEM　　　　　　　　　　53, 2880 (1975)

2333　　　R 003700　6—METHOXYTHIOCHROMAN—4—ONE

```
               O
      11        ‖
     H3C-O   C5   C4
        \ ⁄ \  ⁄ \
       6C    C9  C3
       |    ‖   |
       7C   C10 C2
        \ ⁄ \ ⁄
         C8  S1
```

FORMULA	C10H1002S			MOL WT	194.25
SOLVENT	CDCL3				
ORIG ST	C4H12SI			TEMP	305

26.80	39.60	193.60	111.30	157.30	121.90
2/T	3/T	4/S	5/D	6/S	7/D
128.70	131.50	133.40	55.40		
8/D	9/S	10/S	11/Q		

M.S.CHAUHAN,I.W.J.STILL
CAN J CHEM　　　　　　　　　　53, 2880 (1975)

2334 — R 003700 2-METHYLTHIOCHROMAN-4-ONE 1-OXIDE ISOMERS

```
        O
        ‖
   C5   C4
  ╱ ╲ ╱ ╲
6C    C9   C3
 |    ‖    |
7C    C10  C2
  ╲ ╱ ╲ ╱ ╲11
   C8   S1   CH3
        ‖
        O
```

FORMULA	C10H10O2S			MOL WT	194.25
SOLVENT	CDCL3				
ORIG ST	C4H12SI			TEMP	305

55.80	50.50	40.30	37.10	192.50	191.60
2/D	2/D	3/T	3/T	4/S	4/S
131.10	129.90	132.30	134.60	128.40	127.00
5/D	5/D	6/D	7/D	8/D	8/D
129.30	129.10	146.90	143.30	15.40	13.50
9/S	9/S	10/S	10/S	11/Q	11/Q

M.S.CHAUHAN,I.W.J.STILL
CAN J CHEM   53, 2880 (1975)

2335 — R 003700 3-METHYLTHIOCHROMAN-4-ONE 1-OXIDE

```
        O
        ‖   11
   C5   C4  CH3
  ╱ ╲ ╱ ╲ ╱
6C    C9   C3
 |    ‖    |
7C    C10  C2
  ╲ ╱ ╲ ╱
   C8   S1
        ‖
        O
```

FORMULA	C10H10O2S			MOL WT	194.25
SOLVENT	CDCL3				
ORIG ST	C4H12SI			TEMP	305

51.80	32.10	195.30	130.40	132.80	134.40
2/T	3/D	4/S	5/D	6/D	7/D
129.40	129.40	142.50	14.90		
8/D	9/S	10/S	11/Q		

M.S.CHAUHAN,I.W.J.STILL
CAN J CHEM   53, 2880 (1975)

2336 — R 003700 6-METHYLTHIOCHROMAN-4-ONE 1-OXIDE

```
           O
  11       ‖
 H3C   C5   C4
  ╲ ╱ ╲ ╱ ╲
6C    C9   C3
 |    ‖    |
7C    C10  C2
  ╲ ╱ ╲ ╱
   C8   S1
        ‖
        O
```

FORMULA	C10H10O2S			MOL WT	194.25
SOLVENT	CDCL3				
ORIG ST	C4H12SI			TEMP	305

46.60	30.10	192.40	129.40	143.10	135.30
2/T	3/T	4/S	5/D	6/S	7/D
128.90	129.10	142.40	21.40		
8/D	9/S	10/S	11/Q		

M.S.CHAUHAN,I.W.J.STILL
CAN J CHEM   53, 2880 (1975)

2337 — R 003700 8-METHYLTHIOCHROMAN-4-ONE 1-OXIDE

```
        O
        ‖
   C5   C4
  ╱ ╲ ╱ ╲
6C    C9   C3
 |    ‖    |
7C    C10  C2
  ╲ ╱ ╲ ╱
   C8   S1
   |    ‖
  H3C11 O
```

FORMULA	C10H10O2S			MOL WT	194.25
SOLVENT	CDCL3				
ORIG ST	C4H12SI			TEMP	305

43.50	27.00	193.10	126.70	132.30	136.50
2/T	3/T	4/S	5/D	6/D	7/D
139.30	130.20	141.00	18.90		
8/S	9/S	10/S	11/Q		

M.S.CHAUHAN,I.W.J.STILL
CAN J CHEM   53, 2880 (1975)

2338 — R 003700 6-METHOXYTHIOCHROMAN-4-ONE 1-OXIDE

```
            O
  11        ‖
 H3C-O   C5   C4
   ╲ ╱ ╲ ╱ ╲
6C    C9   C3
 |    ‖    |
7C    C10  C2
  ╲ ╱ ╲ ╱
   C8   S1
        ‖
        O
```

FORMULA	C10H10O3S			MOL WT	210.25
SOLVENT	CDCL3				
ORIG ST	C4H12SI			TEMP	305

46.20	29.30	192.40	112.60	162.70	120.90
2/T	3/T	4/S	5/D	6/S	7/D
131.20	131.10	136.60	55.80		
8/D	9/S	10/S	11/Q		

M.S.CHAUHAN,I.W.J.STILL
CAN J CHEM   53, 2880 (1975)

2339   R 003700 2-METHYLTHIOCHROMAN-4-ONE 1,1-DIOXIDE

```
        O
        ‖
   C5   C4
    ╲  ╱ ╲
  6C   C9   C3
   |   ‖   |
  7C   C10  C2
    ╲ ╱ ╲ ╱ ╲11
    C8   S1   CH3
       ╱╱ ╲
      O   O
```

FORMULA   C10H10O3S   MOL WT   210.25
SOLVENT   CDCL3
ORIG ST   C4H12SI   TEMP   305

54.50	44.20	190.40	128.50	134.90	133.20
2/D	3/T	4/S	5/D	6/D	7/D
124.20	130.60	136.80	11.60		
8/D	9/S	10/S	11/Q		

M.S.CHAUHAN,I.W.J.STILL
CAN J CHEM   53, 2880 (1975)

2340   R 003700 3-METHYLTHIOCHROMAN-4-ONE 1,1-DIOXIDE

```
        O
        ‖   11
   C5   C4   CH3
   ╱ ╲ ╱ ╲ ╱
  6C   C9   C3
   |   ‖   |
  7C   C10  C2
    ╲ ╱ ╲ ╱
    C8   S1
       ╱╱ ╲
      O   O
```

FORMULA   C10H10O3S   MOL WT   210.25
SOLVENT   CDCL3
ORIG ST   C4H12SI   TEMP   305

55.60	41.40	193.20	128.70	134.70	133.20
2/T	3/D	4/S	5/D	6/D	7/D
123.40	129.30	141.90	15.20		
8/D	9/S	10/S	11/Q		

M.S.CHAUHAN,I.W.J.STILL
CAN J CHEM   53, 2880 (1975)

2341   R 003700 5-METHYLTHIOCHROMAN-4-ONE 1,1-DIOXIDE

```
  H3C11  O
   |     ‖
   C5    C4
   ╱ ╲  ╱ ╲
  6C   C9   C3
   |   ‖   |
  7C   C10  C2
    ╲ ╱ ╲ ╱
    C8   S1
       ╱╱ ╲
      O   O
```

FORMULA   C10H10O3S   MOL WT   210.25
SOLVENT   CDCL3
ORIG ST   C4H12SI   TEMP   305

48.40	37.40	192.00	142.60	137.30	133.30
2/T	3/T	4/S	5/S	6/D	7/D
121.90	129.20	142.40	22.70		
8/D	9/S	10/S	11/Q		

M.S.CHAUHAN,I.W.J.STILL
CAN J CHEM   53, 2880 (1975)

2342   R 003700 6-METHYLTHIOCHROMAN-4-ONE 1,1-DIOXIDE

```
         O
         ‖
   11    C4
  H3C   C5
    ╲ ╱ ╲ ╱ ╲
  6C   C9   C3
   |   ‖   |
  7C   C10  C2
    ╲ ╱ ╲ ╱
    C8   S1
       ╱╱ ╲
      O   O
```

FORMULA   C10H10O3S   MOL WT   210.25
SOLVENT   CDCL3
ORIG ST   C4H12SI   TEMP   305

49.40	36.80	190.50	129.10	144.50	135.50
2/T	3/T	4/S	5/D	6/S	7/D
123.80	130.20	138.80	21.60		
8/D	9/S	10/S	11/Q		

M.S.CHAUHAN,I.W.J.STILL
CAN J CHEM   53, 2880 (1975)

2343   R 003700 7-METHYLTHIOCHROMAN-4-ONE 1,1-DIOXIDE

```
        O
        ‖
   C5   C4
   ╱ ╲ ╱ ╲
  6C   C9   C3
   |   ‖   |
  7C   C10  C2
  11╱ ╲ ╱ ╲ ╱
  H3C   C8   S1
          ╱╱ ╲
         O   O
```

FORMULA   C10H10O3S   MOL WT   210.25
SOLVENT   CDCL3
ORIG ST   C4H12SI   TEMP   305

49.40	36.70	189.90	128.90	134.10	146.70
2/T	3/T	4/S	5/D	6/D	7/S
121.70	128.00	141.30	21.90		
8/D	9/S	10/S	11/Q		

M.S.CHAUHAN,I.W.J.STILL
CAN J CHEM   53, 2880 (1975)

2344　　R 003700 2,3-DIMETHYLTHIOCHROMONE

```
        O
        ‖    12
    C5  C4  CH3
   ∕ ∖ ∕ ∖ ∕
 6C   C9  C3
  |   ‖   ‖
 7C   C10 C2
  ∖ ∕ ∖ ∕ ∖11
    C8  S1  CH3
```

FORMULA　C11H10OS
SOLVENT　CDCL3
ORIG ST　C4H12SI

MOL WT　190.27
TEMP　305

144.30	129.80	179.30	128.90	125.30	130.50
2/S	3/S	4/S	5/D	6/D	7/D
126.90	130.70	136.80	21.60	13.00	
8/D	9/S	10/S	11/Q	12/Q	

M.S.CHAUHAN,I.W.J.STILL
CAN J CHEM　　　　53, 2880 (1975)

2345　　R 003700 2-METHOXYCARBONYLTHIOCHROMAN-4-ONE 1,1-DIOXIDE

```
        O
        ‖
    C5  C4
   ∕ ∖ ∕ ∖
 6C   C9  C3
  |   ‖   |
 7C   C10 C2   O
  ∖ ∕ ∖ ∕ ∖ ∕ ∖12
    C8  S1  C11 CH3
        ∕ ∖  ‖
       O  O  O
```

FORMULA　C11H10O5S
SOLVENT　CDCL3
ORIG ST　C4H12SI

MOL WT　254.26
TEMP　305

64.10	39.80	188.10	128.40	134.70	133.70
2/D	3/T	4/S	5/D	6/D	7/D
124.10	130.90	139.30	164.90	53.90	
8/D	9/S	10/S	11/S	12/Q	

M.S.CHAUHAN,I.W.J.STILL
CAN J CHEM　　　　53, 2880 (1975)

2346　　R 003700 3-METHOXYCARBONYLTHIOCHROMAN-4-ONE 1,1-DIOXIDE

```
     HO   O
      |   ‖
    C5  C4  C11
   ∕ ∖ ∕ ∖ ∖  12
 6C   C9  C3  O-CH3
  |   ‖   |
 7C   C10 C2
  ∖ ∕ ∖ ∕
    C8  S1
        ∕ ∖
       O  O
```

FORMULA　C11H10O5S
SOLVENT　CDCL3
ORIG ST　C4H12SI

MOL WT　254.26
TEMP　305

47.50	93.20	171.00	127.20	133.30	132.30
2/T	3/S	4/S	5/D	6/D	7/D
123.20	129.80	138.10	162.90	52.80	
8/D	9/S	10/S	11/S	12/Q	

M.S.CHAUHAN,I.W.J.STILL
CAN J CHEM　　　　53, 2880 (1975)

2347　　R 003700 2,2-DIMETHYLTHIOCHROMAN-4-ONE

```
        O
        ‖
    C5  C4
   ∕ ∖ ∕ ∖
 6C   C9  C3
  |   ‖   |  11
 7C   C10 C-CH3
  ∖ ∕ ∖ ∕2∖12
    C8  S1  CH3
```

FORMULA　C11H12OS
SOLVENT　CDCL3
ORIG ST　C4H12SI

MOL WT　192.28
TEMP　305

44.50	53.80	194.50	128.60	124.60	133.50
2/S	3/T	4/S	5/D	6/D	7/D
127.50	129.70	141.20	28.50		
8/D	9/S	10/S	11/Q		

M.S.CHAUHAN,I.W.J.STILL
CAN J CHEM　　　　53, 2880 (1975)

2348　　R 003700 2,3-DIMETHYLTHIOCHROMAN-4-ONE　CIS-TRANS ISOMER

```
        O
        ‖
    C5  C4
   ∕ ∖ ∕ ∖3 12
 6C   C9  C-CH3
  |   ‖   |  11
 7C   C10 C-CH3
  ∖ ∕ ∖ ∕2
    C8  S1
```

FORMULA　C11H12OS
SOLVENT　CDCL3
ORIG ST　C4H12SI

MOL WT　192.28
TEMP　305

41.70	40.60	48.70	46.90	196.70	196.30
2/D	2/D	3/D	3/D	4/S	4/S
129.30	124.60	133.00	127.30	129.90	129.70
5/D	6/D	7/D	8/D	9/S	9/S
140.80	140.40	19.90	15.90	13.40	11.10
10/S	10/S	11/Q	11/Q	12/Q	12/Q

M.S.CHAUHAN,I.W.J.STILL
CAN J CHEM　　　　53, 2880 (1975)

2349   R 003700 3,3-DIMETHYLTHIOCHROMAN-4-ONE

```
        O
        ‖    11
   C5   C4   CH3
   ⁄ \  ⁄  \3⁄12
 6C    C9  C-CH3
 |     ‖    |
 7C   C10  C2
   \  ⁄  \ ⁄
    C8   S1
```

FORMULA   C11H12OS   MOL WT 192.28
SOLVENT   CDCL3
ORIG ST   C4H12SI   TEMP   305

39.10	40.90	198.30	130.00	124.80	132.70
2/T	3/S	4/S	5/D	6/D	7/D
127.10	129.50	141.40	23.50		
8/D	9/S	10/S	11/Q		

M.S.CHAUHAN,I.W.J.STILL
CAN J CHEM   53, 2880 (1975)

2350   R 003700 2,2-DIMETHYLTHIOCHROMAN-4-ONE 1-OXIDE

```
        O
        ‖
   C5   C4
   ⁄ \  ⁄ \
 6C    C9  C3
 |     ‖    | 11
 7C   C10  C-CH3
   \  ⁄  \ ⁄2\12
    C8   S1  CH3
         ‖
         O
```

FORMULA   C11H12O2S   MOL WT 208.23
SOLVENT   CDCL3
ORIG ST   C4H12SI   TEMP   305

56.50	45.70	192.10	129.10	131.40	134.90
2/S	3/T	4/S	5/D	6/D	7/D
128.20	129.10	144.00	19.30	24.30	
8/D	9/S	10/S	11/Q	12/Q	

M.S.CHAUHAN,I.W.J.STILL
CAN J CHEM   53, 2880 (1975)

2351   R 003700 3,3-DIMETHYLTHIOCHROMAN-4-ONE 1-OXIDE

```
        O
        ‖    11
   C5   C4   CH3
   ⁄ \  ⁄  \3⁄12
 6C    C9  C-CH3
 |     ‖    |
 7C   C10  C2
   \  ⁄  \ ⁄
    C8   S1
         ‖
         O
```

FORMULA   C11H12O2S   MOL WT 208.28
SOLVENT   CDCL3
ORIG ST   C4H12SI   TEMP   305

60.70	43.00	196.70	129.60	131.10	134.20
2/T	3/S	4/S	5/D	6/D	7/D
126.70	127.90	147.20	26.90	24.50	
8/D	9/S	10/S	11/Q	12/Q	

M.S.CHAUHAN,I.W.J.STILL
CAN J CHEM   53, 2880 (1975)

2352   R 003700 2,2-DIMETHYLTHIOCHROMAN-4-ONE 1,1-DIOXIDE

```
        O
        ‖
   C5   C4
   ⁄ \  ⁄ \3
 6C    C9  C
 |     ‖    | 11
 7C   C10  C-CH3
   \  ⁄  \ ⁄2\12
    C8   S1  CH3
         ⁄⁄ \
        O    O
```

FORMULA   C11H12O3S   MOL WT 224.28
SOLVENT   CDCL3
ORIG ST   C4H12SI   TEMP   305

58.60	50.50	190.50	128.10	135.20	133.10
2/S	3/T	4/S	5/D	6/D	7/D
124.80	130.50	139.10	21.10		
8/D	9/S	10/S	11/Q		

M.S.CHAUHAN,I.W.J.STILL   •
CAN J CHEM   53, 2880 (1975)

2353   R 003700 3,3-DIMETHYLTHIOCHROMAN-4-ONE 1,1-DIOXIDE

```
        O
        ‖    11
   C5   C4   CH3
   ⁄ \  ⁄  \3⁄12
 6C    C9  C-CH3
 |     ‖    |
 7C   C10  C2
   \  ⁄  \ ⁄
    C8   S1
         ⁄⁄ \
        O    O
```

FORMULA   C11H12O3S   MOL WT 224.28
SOLVENT   CDCL3
ORIG ST   C4H12SI   TEMP   305

59.70	45.10	196.10	129.30	134.60	133.20
2/T	3/S	4/S	5/D	6/D	7/D
123.20	129.30	141.90	26.40		
8/D	9/S	10/S	11/Q		

M.S.CHAUHAN,I.W.J.STILL
CAN J CHEM   53, 2880 (1975)

2354

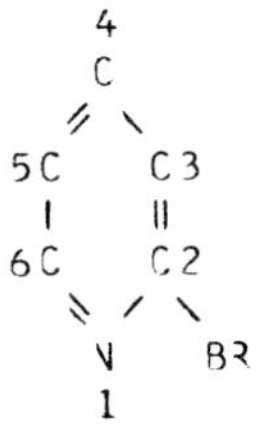

```
      4
      C
    // \
  5C    C3
  |     ||
  6C    C2
    \ // \
      N    BR
      1
```

R 003800 2-BROMOPYRIDINE

FORMULA	C5H4BRN			MOL WT	158.00
SOLVENT	C5H4BRN				
ORIG ST	C6H6			TEMP	AMB

142.90	129.00	139.50	123.70	151.00
2/1	3/2	4/2	5/2	6/2

G.MIYAJIMA,Y.SASAKI,M.SUZUKI
CHEM PHARM BULL   20, 429 (1972)

2355

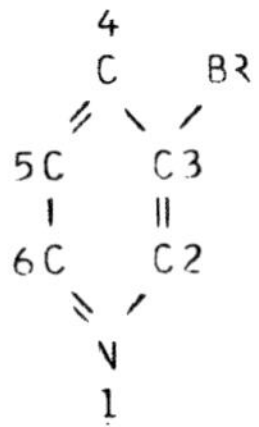

```
      4
      C    BR
    // \ /
  5C    C3
  |     ||
  6C    C2
    \ /
      N
      1
```

R 003800 3-BROMOPYRIDINE

FORMULA	C5H4BRN			MOL WT	158.00
SOLVENT	C5H4BRN				
ORIG ST	C6H6			TEMP	AMB

151.70	121.60	139.10	125.40	148.70
2/2	3/1	4/2	5/2	6/2

G.MIYAJIMA,Y.SASAKI,M.SUZUKI
CHEM PHARM BULL   20, 429 (1972)

2356

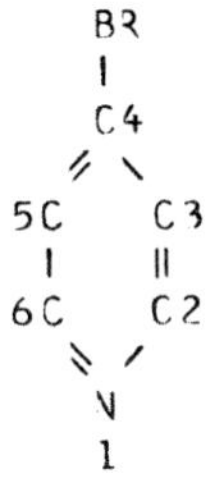

```
     BR
     |
     C4
   // \
 5C    C3
 |     ||
 6C    C2
   \ /
     N
     1
```

R 003800 4-BROMOPYRIDINE

FORMULA	C5H4BRN			MOL WT	158.00
SOLVENT	C5H4BRN				
ORIG ST	C6H6			TEMP	AMB

152.60	127.60	133.20
2/2	3/2	4/1

G.MIYAJIMA,Y.SASAKI,M.SUZUKI
CHEM PHARM BULL   20, 429 (1972)

2357

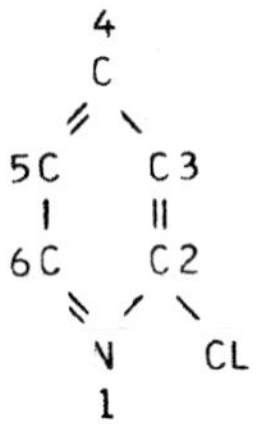

```
      4
      C
    // \
  5C    C3
  |     ||
  6C    C2
    \ / \
      N    CL
      1
```

R 003800 2-CHLOROPYRIDINE

FORMULA	C5H4CLN			MOL WT	113.55
SOLVENT	C5H4CLN				
ORIG ST	C6H6			TEMP	AMB

151.90	124.90	139.50	123.00	150.20
2/1	3/2	4/2	5/2	6/2

G.MIYAJIMA,Y.SASAKI,M.SUZUKI
CHEM PHARM BULL   20, 429 (1972)

2358

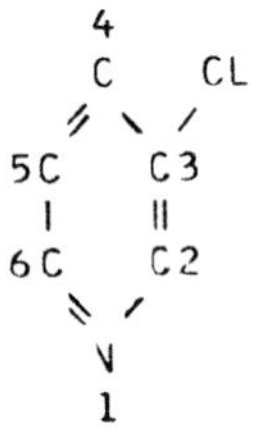

```
      4
      C    CL
    // \ /
  5C    C3
  |     ||
  6C    C2
    \ /
      N
      1
```

R 003800 3-CHLOROPYRIDINE

FORMULA	C5H4CLN			MOL WT	113.55
SOLVENT	C5H4CLN				
ORIG ST	C6H6			TEMP	AMB

149.30	132.40	136.00	124.90	148.20
2/2	3/1	4/2	5/2	6/2

G.MIYAJIMA,Y.SASAKI,M.SUZUKI
CHEM PHARM BULL   20, 429 (1972)

2359   R 003800 2-FLUOROPYRIDINE

```
      4
      C
    ⁄  ╲
  5C    C3
  |     ‖
  6C    C2
    ╲ ⁄ ╲
     N    F
     1
```

FORMULA	C5H4FN				MOL WT	97.09
SOLVENT	C5H4FN					
ORIG ST	C6H6				TEMP	AMB

164.00	109.50	141.30	121.50	147.90
2/1	3/2	4/2	5/2	6/2

G.MIYAJIMA,Y.SASAKI,M.SUZUKI
CHEM PHARM BULL   20, 429 (1972)

2360   R 003800 3-IODOPYRIDINE

```
      4
      C    J
    ⁄ ╲  ⁄
  5C    C3
  |     ‖
  6C    C2
    ╲ ⁄
     N
     1
```

FORMULA	C5H4IN				MOL WT	205.00
SOLVENT	C5H4IN					
ORIG ST	C6H6				TEMP	AMB

156.70	95.80	145.30	126.60	149.30
2/2	3/1	4/2	5/2	6/2

G.MIYAJIMA,Y.SASAKI,M.SUZUKI
CHEM PHARM BULL   20, 429 (1972)

2361   R 003800 2-NITROPYRIDINE

```
      4
      C
    ⁄  ╲
  5C    C3
  |     ‖
  6C    C2
    ╲ ⁄ ╲
     N    NO2
     1
```

FORMULA	C5H4N2O2				MOL WT	124.10
SOLVENT	C5H4N2O2					
ORIG ST	C6H6				TEMP	AMB

157.60	119.10	141.70	130.80	150.00
2/1	3/2	4/2	5/2	6/2

G.MIYAJIMA,Y.SASAKI,M.SUZUKI
CHEM PHARM BULL   20, 429 (1972)

2362   R 003800 PYRIDINE

```
    4 5
    C=C
   3⁄   ╲
   C     N
    ╲   ⁄
    C—C
    2 1
```

FORMULA	C5H5N				MOL WT	79.10
SOLVENT	CDCL3					
ORIG ST	C4H12SI				TEMP	AMB

149.80	123.60	135.70	123.60	149.80
1/2	2/2	3/2	4/2	5/2

L.F.JOHNSON,W.C.JANKOWSKI
CARBON-13 NMR SPECTRA,JOHN
WILEY AND SONS,NEW YORK   106 (1972)

2363   R 003800 PYRIDINE

```
      4
      C
    ⁄  ╲
  5C    C3
  |     ‖
  6C    C2
    ╲ ⁄
     N
     1
```

FORMULA	C5H5N				MOL WT	79.10
SOLVENT	CDCL3					
ORIG ST	C4H12SI				TEMP	AMB

149.59	123.46	135.58	123.46	146.59
2/2	3/2	4/2	5/2	6/2

R.J.CUSHLEY,D.NAUGLER,C.ORTIZ
CAN J CHEM   53, 3419 (1975)

2364   R 003800 2-HYDROXYPYRIDINE

```
      4
      C
    ⁄ ╲
  5C    C3
  |     ‖
  6C    C2
    ╲ ⁄ ╲
     N   OH
     1
```

FORMULA	C5H5NO			MOL WT	95.10
SOLVENT	C2H6OS				
ORIG ST	C6H6			TEMP	AMB

| 164.50 | 107.00 | 136.60 | 121.10 | 142.80 |
| 2/1 | 3/2 | 4/2 | 5/2 | 6/2 |

G.MIYAJIMA,Y.SASAKI,M.SUZUKI
CHEM PHARM BULL   20, 429 (1972)

2365   R 003800 3-HYDROXYPYRIDINE

```
      4
      C   OH
    ⁄ ╲ ⁄
  5C    C3
  |     ‖
  6C    C2
    ╲ ⁄
     N
     1
```

FORMULA	C5H5NO			MOL WT	95.10
SOLVENT	C2H6OS				
ORIG ST	C6H6			TEMP	AMB

| 138.90 | 155.60 | 124.00 | 125.50 | 141.00 |
| 2/2 | 3/1 | 4/2 | 5/2 | 6/2 |

G.MIYAJIMA,Y.SASAKI,M.SUZUKI
CHEM PHARM BULL   20, 429 (1972)

2366   R 003800 PYRIDINE N-OXIDE

```
      C4
    ⁄ ╲
  5C    C3
  |     ‖
  6C    C2
    ╲+⁄
     N1
     |
     O-
```

FORMULA	C5H5NO			MOL WT	95.10
SOLVENT	CDCL3				
ORIG ST	C4H12SI			TEMP	AMB

| 139.09 | 125.65 | 126.10 | 125.65 | 139.09 |
| 2/2 | 3/2 | 4/2 | 5/2 | 6/2 |

R.J.CUSHLEY,D.NAUGLER,C.ORTIZ
CAN J CHEM   53, 3419 (1975)

2367   R 003800 2-PYRIDONE

```
      C3
    ⁄ ╲
  4C    C2
  ‖     |
  5C    C1
    ╲ ⁄ ╲
     N   O
```

FORMULA	C5H5NO			MOL WT	95.10
SOLVENT	CDCL3				
ORIG ST	C4H12SI			TEMP	AMB

| 165.30 | 120.10 | 134.80 | 106.70 | 141.60 |
| 1/1 | 2/2 | 3/2 | 4/2 | 5/2 |

L.F.JOHNSON,W.C.JANKOWSKI
CARBON-13 NMR SPECTRA,JOHN
WILEY AND SONS,NEW YORK   107 (1972)

2368   R 003800 2-AMINOPYRIDINE

```
      4
      C
    ⁄ ╲
  5C    C3
  |     ‖
  6C    C2
    ╲ ⁄ ╲
     N   NH2
     1
```

FORMULA	C5H6N2			MOL WT	94.12
SOLVENT	C2H6OS				
ORIG ST	C6H6			TEMP	AMB

| 160.90 | 109.50 | 138.50 | 113.60 | 148.70 |
| 2/1 | 3/2 | 4/2 | 5/2 | 6/2 |

G.MIYAJIMA,Y.SASAKI,M.SUZUKI
CHEM PHARM BULL   20, 429 (1972)

2369   R 003800 2-AMINOPYRIDINE

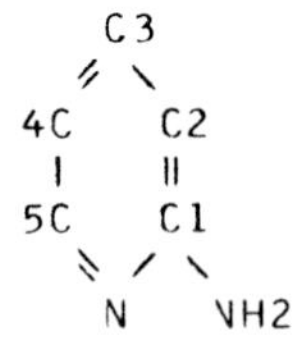

```
        C3
      ⁄ ╲
   4C    C2
    |    ||
   5C    C1
     ╲ ⁄ ╲
      N   NH2
```

FORMULA   C5H6N2   MOL WT   94.12
SOLVENT   CDCL3
ORIG ST   C4H12SI   TEMP   AMB

158.90   108.50   137.50   113.30   147.70
 1/1   2/2   3/2   4/2   5/2

L.F.JOHNSON,W.C.JANKOWSKI
CARBON-13 NMR SPECTRA,JOHN
WILEY AND SONS,NEW YORK   109 (1972)

2370   R 003800 3-AMINOPYRIDINE

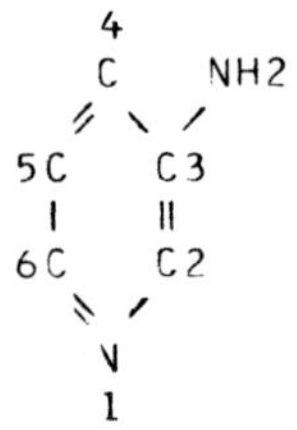

```
       4
      C   NH2
     ⁄ ╲ ⁄
   5C    C3
    |    ||
   6C    C2
     ╲ ⁄
      N
      1
```

FORMULA   C5H6N2   MOL WT   94.12
SOLVENT   C2H6OS
ORIG ST   C6H6   TEMP   AMB

137.70   145.70   122.00   125.10   138.80
 2/2   3/1   4/2   5/2   6/2

G.MIYAJIMA,Y.SASAKI,M.SUZUKI
CHEM PHARM BULL   20, 429 (1972)

2371   R 003800 4-AMINOPYRIDINE

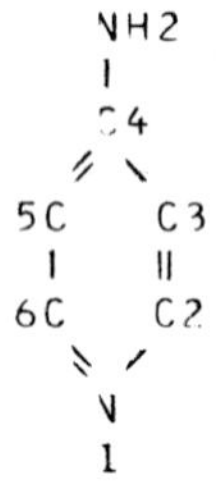

```
     NH2
      |
     C4
    ⁄ ╲
  5C    C3
   |    ||
  6C    C2
    ╲ ⁄
     N
     1
```

FORMULA   C5H6N2   MOL WT   94.12
SOLVENT   C2H6OS
ORIG ST   C6H6   TEMP   AMB

150.50   110.40   155.80
 2/2   3/2   4/1

G.MIYAJIMA,Y.SASAKI,M.SUZUKI
CHEM PHARM BULL   20, 429 (1972)

2372   R 003800 PIPERIDINE

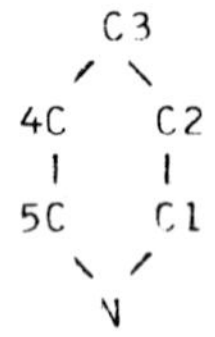

```
        C3
      ⁄ ╲
   4C    C2
    |    |
   5C    C1
     ╲ ⁄
      N
```

FORMULA   C5H11N   MOL WT   85.15
SOLVENT   C4H8O2
ORIG ST   C4H12SI   TEMP   AMB

 47.90   27.80   25.90   27.80   47.90
 1/3   2/3   3/3   4/3   5/3

L.F.JOHNSON,W.C.JANKOWSKI
CARBON-13 NMR SPECTRA,JOHN
WILEY AND SONS,NEW YORK   133 (1972)

2373   R 003800 PIPERIDINE

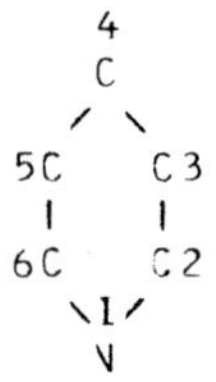

```
      4
      C
    ⁄ ╲
  5C    C3
   |    |
  6C    C2
    ╲ 1 ⁄
     N
```

FORMULA   C5H11N   MOL WT   85.15
SOLVENT   C5H11N
ORIG ST   C6H12   TEMP   AMB

 46.95   26.95   24.95
 2/3   3/3

G.ELLIS,R.G.JONES
J CHEM SOC PERKIN TRANS II   437 (1972)

2374   R 003800 2-CYANOPYRIDINE

```
      4
      C
    ⁄   \
 5C      C3
  |      ‖
 6C      C2
   \  ⁄  \7
    N      C≡N
    1
```

FORMULA   C6H4N2   MOL WT   104.11
SOLVENT   C2D6OS
ORIG ST   C4H12SI   TEMP   AMB

133.60	129.30	137.90	127.90	151.50	117.10
2/1	3/2	4/2	5/2	6/2	7/1

Y.TAKEUCHI,N.DENNIS
J AM CHEM SOC   96, 3657 (1974)

2375   R 003800 3-CYANOPYRIDINE

```
      4    7
      C    C≡N
    ⁄  \  ⁄
 5C      C3
  |      ‖
 6C      C2
   \   ⁄
    N
    1
```

FORMULA   C6H4N2   MOL WT   104.11
SOLVENT   C2D6OS
ORIG ST   C4H12SI   TEMP   AMB

153.30	110.00	140.80	124.90	154.10	117.80
2/2	3/1	4/2	5/2	6/2	7/1

Y.TAKEUCHI,N.DENNIS
J AM CHEM SOC   96, 3657 (1974)

2376   R 003800 4-CYANOPYRIDINE

```
   7C≡N
    |
    C4
  ⁄   \
 5C      C3
  |      ‖
 6C      C2
   \   ⁄
    N
    1
```

FORMULA   C6H4N2   MOL WT   104.11
SOLVENT   C2D6OS
ORIG ST   C4H12SI   TEMP   AMB

151.40	126.70	120.90	117.40
2/2	3/2	4/1	7/1

Y.TAKEUCHI,N.DENNIS
J AM CHEM SOC   96, 3657 (1974)

2377   R 003800 2-CHLORO-6-METHOXY-3-NITROPYRIDINE

```
 O2N      CL
   \    ⁄
    C—C
   ⁄2 1\
  C      N
  3\    ⁄
    C=C
   4 5\  6
      O—CH3
```

FORMULA   C6H5CLNO3   MOL WT   174.56
SOLVENT   CDCL3
ORIG ST   C4H12SI   TEMP   AMB

143.10	138.30	137.40	110.20	164.60	55.30
1/1	2/1	3/2	4/2	5/1	6/4

L.F.JOHNSON,W.C.JANKOWSKI
CARBON-13 NMR SPECTRA,JOHN
WILEY AND SONS,NEW YORK   154 (1972)

2378   R 003800 PYRIDINE-2-ALDEHYDE

```
      4
      C
    ⁄   \
 5C      C3
  |      ‖
 6C      C2  H
   \  ⁄  \7⁄
    N     C
    1     ‖
          O
```

FORMULA   C6H5NO   MOL WT   107.11
SOLVENT   C6H5NO
ORIG ST   C6H6   TEMP   AMB

153.10	121.60	137.50	128.30	150.30
2/1	3/2	4/2	5/2	6/2

G.MIYAJIMA,Y.SASAKI,M.SUZUKI
CHEM PHARM BULL   20, 429 (1972)

2379   R 003800 PYRIDINE-3-ALDEHYDE

```
         O
    4    ||
    C    C7
   ∥ \ ∕ \
  5C   C3  H
  |    |
  6C   C2
   \ ∕
    N
    1
```

FORMULA	C6H5NO			MOL WT	107.11
SOLVENT	C6H5NO				
ORIG ST	CS2			TEMP	AMB

151.00	131.10	135.20	123.80	154.00	191.00
2/2	3/1	4/2	5/2	6/2	7/2

H.L.RETCOFSKY,R.A.FRIEDEL
J PHYS CHEM   72, 290 (1968)

2380   R 003800 PYRIDINE-4-ALDEHYDE

```
  H   O
   \ ∕
    C7
    |
    C4
   ∕ \
  5C   C3
  |    ||
  6C   C2
   \1∕
    N
```

FORMULA	C6H5NO			MOL WT	107.11
SOLVENT	C6H5NO				
ORIG ST	CS2			TEMP	AMB

150.30	121.60	140.70	191.50
2/2	3/2	4/1	7/2

H.L.RETCOFSKY,R.A.FRIEDEL
J PHYS CHEM   71, 3592 (1967)

2381   R 003800 NICOTINAMIDE

```
    4  3
    C—C    O
   ∕   \2 ||
  5C     C—C—NH2
   \   ∕  6
    N=C
     1
```

FORMULA	C6H6N2O			MOL WT	122.13
SOLVENT	H2O				
ORIG ST	C4H12SI			TEMP	AMB

152.50	129.60	137.00	124.90	148.30	170.60
1/2	2/1	3/2	4/2	5/2	6/1

L.F.JOHNSON,W.C.JANKOWSKI
CARBON-13 NMR SPECTRA,JOHN
WILEY AND SONS,NEW YORK   159 (1972)

2382   R 003800 ALPHA-PICOLINE

```
    3
    C
   ∕ \
  4C   C2
  |    ||
  5C   C—CH3
   \ ∕1 6
    N
```

FORMULA	C6H7N			MOL WT	93.13
SOLVENT	C4H8O2				
ORIG ST	C4H12SI			TEMP	AMB

158.60	123.00	135.90	120.60	149.40	24.30
1/1	2/2	3/2	4/2	5/2	6/4

L.F.JOHNSON,W.C.JANKOWSKI
CARBON-13 NMR SPECTRA,JOHN
WILEY AND SONS,NEW YORK   165 (1972)

2383   R 003800 2-METHYLPYRIDINE

```
    4
    C
   ∕ \
  5C   C3
  |    ||
  6C   C2
   \ ∕ \7
    N   CH3
    1
```

FORMULA	C6H7N			MOL WT	93.13
SOLVENT	CDCL3				
ORIG ST	C4H12SI			TEMP	AMB

158.10	123.01	135.95	120.46	148.88	24.36
2/1	3/2	4/2	5/2	6/2	7/4

R.J.CUSHLEY,D.NAUGLER,C.ORTIZ
CAN J CHEM   53, 3419 (1975)

2384   R 003800 2-METHYLPYRIDINE

```
      4
      C
    ⁄⁄  ＼
  5C    C3
  |     ||
  6C    C2
   ＼ ⁄  ＼7
     N    CH3
     1
```

FORMULA C6H7N				MOL WT	93.13
SOLVENT C6H7N					
ORIG ST C6H6				TEMP	AMB
158.70	123.20	136.10	120.80	149.50	25.10
2/1	3/2	4/2	5/2	6/2	7/4

G.MIYAJIMA,Y.SASAKI,M.SUZUKI
CHEM PHARM BULL   20, 429 (1972)

2385   R 003800 3-METHYLPYRIDINE

```
    4    7
    C    CH3
   ⁄⁄ ＼ ⁄
  5C    C3
  |     ||
  6C    C2
   ＼ ⁄
     N
     1
```

FORMULA C6H7N				MOL WT	93.13
SOLVENT C6H7N					
ORIG ST CS2				TEMP	AMB
149.30	132.60	136.10	122.90	146.60	17.70
2/2	3/1	4/2	5/2	6/2	7/4

P.C.LAUTERBUR
J CHEM PHYS   43, 360 (1965)

2386   R 003800 3-METHYLPYRIDINE

```
    4    7
    C    CH3
   ⁄⁄ ＼ ⁄
  5C    C3
  |     ||
  6C    C2
   ＼ ⁄
     N
     1
```

FORMULA C6H7N				MOL WT	93.13
SOLVENT CDCL3					
ORIG ST C4H12SI				TEMP	AMB
150.06	132.84	136.15	122.93	146.73	18.34
2/2	3/1	4/2	5/2	6/2	7/4

R.J.CUSHLEY,D.NAUGLER,C.ORTIZ
CAN J CHEM   53, 3419 (1975)

2387   R 003800 4-METHYLPYRIDINE

```
    7CH3
     |
     C4
    ⁄⁄ ＼
  5C    C3
  |     ||
  6C    C2
   ＼ ⁄
     N
     1
```

FORMULA C6H7N			MOL WT	93.13
SOLVENT CDCL3				
ORIG ST C4H12SI			TEMP	300
149.65	124.60	146.85	20.80	
2/2	3/2	4/1	7/4	

E.BREITMAIER
UNPUBLISHED   (1974)

2388   R 003800 4-METHYLPYRIDINE

```
    7CH3
     |
     C4
    ⁄⁄ ＼
  5C    C3
  |     ||
  6C    C2
   ＼ ⁄
     N1
```

FORMULA C6H7N			MOL WT	93.13
SOLVENT CDCL3				
ORIG ST C4H12SI			TEMP	AMB
149.33	124.43	146.72	20.89	
2/2	3/2	4/1	7/4	

R.J.CUSHLEY,D.NAUGLER,C.ORTIZ
CAN J CHEM   53, 3419 (1975)

2389   R 003800 2—METHOXYPYRIDINE

```
      4
      C
     ⁄ \
   5C   C3
    |   ||
   6C   C2
    \ ⁄ \ 7
     N   O—CH3
     1
```

FORMULA	C6H7NO			MOL WT	109.13
SOLVENT	C6H7NO				
ORIG ST	C6H6			TEMP	AMB

| 164.90 | 111.10 | 138.30 | 116.70 | 147.40 | 53.20 |
| 2/1 | 3/2 | 4/2 | 5/2 | 6/2 | 7/4 |

G.MIYAJIMA,Y.SASAKI,M.SUZUKI
CHEM PHARM BULL   20, 429 (1972)

2390   R 003800 2—METHYLPYRIDINE N—OXIDE

```
     C4
    ⁄ \
  5C   C3
   |   ||
  6C   C2
   \+⁄ \7
    N1  CH3
    |
    O—
```

FORMULA	C6H7NO			MOL WT	109.13
SOLVENT	CDCL3				
ORIG ST	C4H12SI			TEMP	AMB

| 148.83 | 126.43 | 125.30 | 123.52 | 139.21 | 17.75 |
| 2/1 | 3/2 | 4/2 | 5/2 | 6/2 | 7/4 |

R.J.CUSHLEY,D.NAUGLER,C.ORTIZ
CAN J CHEM   53, 3419 (1975)

2391   R 003800 3—METHYLPYRIDINE N—OXIDE

```
     C4 7CH3
    ⁄ \ ⁄
  5C   C3
   |   ||
  6C   C2
   \+⁄
    N1
    |
    O—
```

FORMULA	C6H7NO			MOL WT	109.13
SOLVENT	CDCL3				
ORIG ST	C4H12SI			TEMP	AMB

| 138.92 | 136.74 | 127.22 | 125.37 | 136.26 | 18.19 |
| 2/2 | 3/1 | 4/2 | 5/2 | 6/2 | 7/4 |

R.J.CUSHLEY,D.NAUGLER,C.ORTIZ
CAN J CHEM   53, 3419 (1975)

2392   R 003800 4—METHYLPYRIDINE—N—OXIDE

```
    7CH3
     |
     C4
    ⁄ \
  5C   C3
   |   ||
  6C   C2
   \ ⁄
   1N+
    |
    O—
```

FORMULA	C6H7NO		MOL WT	109.13
SOLVENT	CDCL3			
ORIG ST	C4H12SI		TEMP	300

| 138.55 | 126.90 | 137.65 | 20.20 |
| 2/2 | 3/2 | 4/1 | 7/4 |

E.BREITMAIER
UNPUBLISHED   (1974)

2393   R 003800 4—METHYLPYRIDINE N—OXIDE

```
    7CH3
     |
     C4
    ⁄ \
  5C   C3
   |   ||
  6C   C2
   \+⁄
    N1
    |
    O—
```

FORMULA	C6H7NO		MOL WT	109.13
SOLVENT	CDCL3			
ORIG ST	C4H12SI		TEMP	AMB

| 138.44 | 126.63 | 137.46 | 20.20 |
| 2/2 | 3/2 | 4/1 | 7/4 |

R.J.CUSHLEY,D.NAUGLER,C.ORTIZ
CAN J CHEM   53, 3419 (1975)

2394   R 003800 1—METHYL—3—PIPERIDEINE

```
      4
      C
     ╱ ╲
   5C   C3
    |    |
   6C   C2
    ╲1╱
     N
     |
     CH3
     7
```

FORMULA   C6H11N   MOL WT   97.16
SOLVENT   CCL4
ORIG ST   C4H12SI   TEMP   AMB

       54.20   125.00   124.30   26.20   51.70   45.90
        2/3   3/2   4/2   5/3   6/3   7/4

E.WENKERT,D.W.COCHRAN,E.W.HAGAMAN,F.M.SCHELL,
N.NEUSS,A.S.KATNER,P.POTIER,C.KAN,M.PLAT,
M.KOCH,H.MEHRI,J.POISSON,N.KUNESCH,Y.ROLLAND
J AM CHEM SOC   95, 4990 (1973)

2395   R 003800 N—METHYLPIPERIDINE

```
      4
      C
     ╱ ╲
   5C   C3
    |    |
   6C   C2
    ╲1╱
     N
     |
     7CH3
```

FORMULA   C6H13N   MOL WT   99.18
SOLVENT   C6H13N
ORIG ST   C6H12   TEMP   AMB

       56.05   26.46   23.65   46.15
        2/3   3/3   4/3   7/4

G.ELLIS,R.G.JONES
J CHEM SOC PERKIN TRANS II   437 (1972)

2396   R 003800 2—METHYLPIPERIDINE

```
   6CH3
    |
   1C—C2
   ╱    ╲
  N      C3
   ╲    ╱
    C—C
    5  4
```

FORMULA   C6H13N   MOL WT   99.18
SOLVENT   CDCL3
ORIG ST   C4H12SI   TEMP   AMB

       52.40   34.90   25.00   26.30   47.30   23.10
        1/2   2/3   3/3   4/3   5/3   6/4

L.F.JOHNSON,W.C.JANKOWSKI
CARBON—13 NMR SPECTRA,JOHN
WILEY AND SONS,NEW YORK   201 (1972)

2397   R 003800 2—METHYLPIPERIDINE

```
      4
      C
     ╱ ╲
   5C   C3
    |    |
   6C   C2
    ╲1╱ ╲
     N   CH3
         7
```

FORMULA   C6H13N   MOL WT   99.18
SOLVENT   C6H13N
ORIG ST   C6H12   TEMP   AMB

       52.05   34.55   24.85   26.05   46.95   22.55
        2/2   3/3   4/3   5/3   6/3   7/4

G.ELLIS,R.G.JONES
J CHEM SOC PERKIN TRANS II   437 (1972)

2398   R 003800 3—METHYLPIPERIDINE

```
      4
      C   CH3
     ╱ ╲ ╱
   5C   C3
    |    |
   6C   C2
    ╲1╱
     N
```

FORMULA   C6H13N   MOL WT   99.18
SOLVENT   C6H13N
ORIG ST   C6H12   TEMP   AMB

       53.65   33.75   31.75   26.56   46.45   19.15
        2/3   3/2   4/3   5/3   6/3   7/4

G.ELLIS,R.G.JONES
J CHEM SOC PERKIN TRANS II   437 (1972)

2399   R 003800   4-METHYLPIPERIDINE

```
  1 2
  C-C
 /   \3 6
N     C-CH3
 \   /
  C-C
  5 4
```

FORMULA	C6H13N			MOL WT	99.18
SOLVENT	CDCL3				
ORIG ST	C4H12SI			TEMP	AMB

46.80	35.70	31.30	35.70	46.80	22.50
1/3	2/3	3/2	4/3	5/3	6/4

L.F.JOHNSON,W.C.JANKOWSKI
CARBON-13 NMR SPECTRA,JOHN
WILEY AND SONS,NEW YORK   199 (1972)

2400   R 003800   4-METHYLPIPERIDINE

```
    7CH3
     |
     C
    /4\
  5C   C3
   |   |
  6C   C2
    \1/
     N
```

FORMULA	C6H13N		MOL WT	99.18
SOLVENT	C6H13N			
ORIG ST	C6H12		TEMP	AMB

46.55	35.45	31.35	22.25
2/3	3/3	4/2	7/4

G.ELLIS,R.G.JONES
J CHEM SOC PERKIN TRANS II   437 (1972)

2401   R 003800   4-HYDROXY-N-METHYLPIPERIDINE

```
    OH
     |
     C
    /4\
  5C   C3
   |   |
  6C   C2
    \1/
     N
     |
    7CH3
```

FORMULA	C6H13NO		MOL WT	115.18
SOLVENT	C6H13NO			
ORIG ST	C6H12		TEMP	AMB

53.05	34.05	30.45	46.05
2/3	3/3	4/2	7/4

G.ELLIS,R.G.JONES
J CHEM SOC PERKIN TRANS II   437 (1972)

2402   R 003800   4-VINYLPYRIDINE

```
     8CH2
      //
    7CH
     |
     C4
    // \
  5C   C3
   |   ||
  6C   C2
    \1/
     N
```

FORMULA	C7H7N			MOL WT	105.14
SOLVENT	C7H7N				
ORIG ST	CS2			TEMP	AMB

149.70	120.20	143.80	134.30	117.70
2/2	3/2	4/1	7/2	8/3

H.L.RETCOFSKY,R.A.FRIEDEL
J PHYS CHEM   71, 3592 (1967)

2403   R 003800   2-ACETYLPYRIDINE

```
     4
     C
    // \
  5C   C3
   |   ||   8
  6C   C2  CH3
    \ / \7/
     1   ||
         O
```

FORMULA	C7H7NO			MOL WT	121.14
SOLVENT	C7H7NO				
ORIG ST	C6H6			TEMP	AMB

153.90	121.40	136.90	127.20	149.30	
2/1	3/2	4/2	5/2	6/2	7/1

G.MIYAJIMA,Y.SASAKI,M.SUZUKI
CHEM PHARM BULL   20, 429 (1972)

2404   R 003800   4-ACETYLPYRIDINE

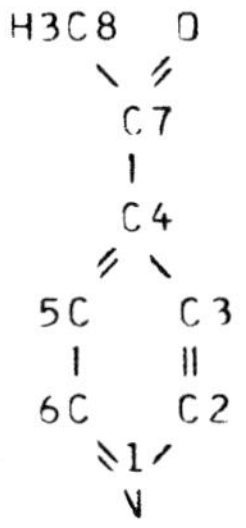

```
 H3C8  O
   \  //
    C7
    |
    C4
   //  \
  5C    C3
  |     ||
  6C    C2
   \1 //
    N
```

FORMULA	C7H7NO			MOL WT	121.14
SOLVENT	C7H7NO				
ORIG ST	CS2			TEMP	AMB

| 150.20 | 120.70 | 141.90 | 196.60 | 25.70 |
| 2/2 | 3/2 | 4/1 | 7/1 | 8/4 |

H.L.RETCOFSKY,R.A.FRIEDEL
J PHYS CHEM   71, 3592 (1967)

2405   R 003800   2,3-DIMETHYLPYRIDINE

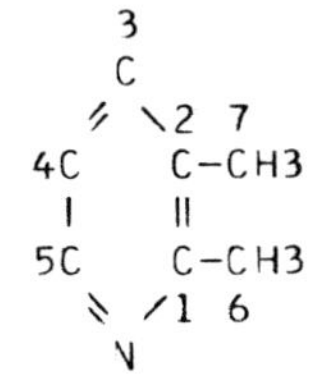

```
        3
        C
      //  \2 7
    4C     C-CH3
    |      ||
    5C     C-CH3
     \  /1 6
      N
```

FORMULA	C7H9N			MOL WT	107.16
SOLVENT	CDCL3				
ORIG ST	C4H12SI			TEMP	AMB

156.90	131.10	136.70	121.00	146.40	22.40
1/1	2/1	3/2	4/2	5/2	6/4
18.90					
7/4					

L.F.JOHNSON,W.C.JANKOWSKI
CARBON-13 NMR SPECTRA,JOHN
WILEY AND SONS,NEW YORK   259 (1972)

2406   R 003800   2,6-DIMETHYLPYRIDINE

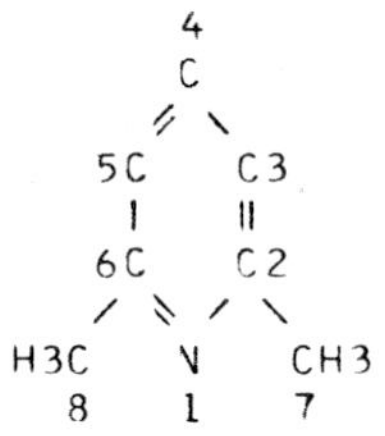

```
      4
      C
    //  \
  5C     C3
  |      ||
  6C     C2
  /  \ / \
H3C    N   CH3
 8     1    7
```

FORMULA	C7H9N			MOL WT	107.16
SOLVENT	C7H9N				
ORIG ST	CS2			TEMP	AMB

| 156.90 | 119.50 | 136.50 | 24.00 |
| 2/1 | 3/2 | 4/2 | 7/4 |

P.C.LAUTERBUR
J CHEM PHYS   43, 360 (1965)

2407   R 003800   2,6-DIMETHYLPYRIDINE

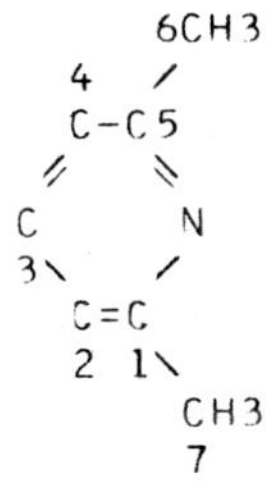

```
      6CH3
   4   /
   C-C5
  //   \
 C      N
 3\    /
  C=C
  2 1\
     CH3
      7
```

FORMULA	C7H9N			MOL WT	107.16
SOLVENT	CDCL3				
ORIG ST	C4H12SI			TEMP	AMB

157.50	120.00	136.30	120.00	157.50	24.40
1/1	2/2	3/2	4/2	5/1	6/4
24.40					
7/4					

L.F.JOHNSON,W.C.JANKOWSKI
CARBON-13 NMR SPECTRA,JOHN
WILEY AND SONS,NEW YORK   252 (1972)

2408   R 003800   3,4-DIMETHYLPYRIDINE

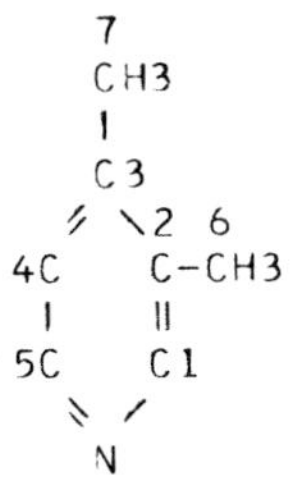

```
      7
     CH3
      |
     C3
    //  \2 6
  4C     C-CH3
  |      ||
  5C     C1
   \  //
    N
```

FORMULA	C7H9N			MOL WT	107.16
SOLVENT	CDCL3				
ORIG ST	C4H12SI			TEMP	AMB

149.90	132.00	145.30	124.40	147.30	16.10
1/2	2/1	3/1	4/2	5/2	6/4
18.80					
7/4					

L.F.JOHNSON,W.C.JANKOWSKI
CARBON-13 NMR SPECTRA,JOHN
WILEY AND SONS,NEW YORK   257 (1972)

2409   R 003800 3,5-DIMETHYLPYRIDINE

```
    8   4   7
  H3C   C   CH3
    \\  //  \\  /
    5C      C3
    |       ||
    6C      C2
     \\    /
       \\ /
        N
        1
```

FORMULA   C7H9N   MOL WT   107.16
SOLVENT   C7H9N
ORIG ST   CS2   TEMP   AMB

147.10	131.70	136.70	17.50
2/2	3/1	4/2	7/4

P.C.LAUTERBUR
J CHEM PHYS   43, 360 (1965)

2410   R 003800 3,5-DIMETHYLPYRIDINE

```
        7
        CH3
    5 4/
    C—C
   //     \\
  N         C
   \\      /3
    C=C
    1 2\
        6CH3
```

FORMULA   C7H9N   MOL WT   107.16
SOLVENT   CDCL3
ORIG ST   C4H12SI   TEMP   AMB

147.30	132.30	136.90	132.30	147.30	18.10
1/2	2/1	3/2	4/1	5/2	6/4
18.10					
7/4					

L.F.JOHNSON,W.C.JANKCWSKI
CARBON-13 NMR SPECTRA,JOHN
WILEY AND SONS,NEW YORK   256 (1972)

2411   R 003800 2-ETHYLPYRIDINE

```
        C3
       //  \\
    4C      C2
    |       ||
    5C      C-CH2-CH3
     \\   /1 6   7
       \\ /
        N
```

FORMULA   C7H9N   MOL WT   107.16
SOLVENT   CDCL3
ORIG ST   C4H12SI   TEMP   AMB

163.40	121.80	136.10	120.70	149.10	31.40
1/1	2/2	3/2	4/2	5/2	6/3
13.80					
7/4					

L.F.JOHNSON,W.C.JANKCWSKI
CARBON-13 NMR SPECTRA,JOHN
WILEY AND SONS,NEW YORK   255 (1972)

2412   R 003800 2-ETHYLPYRIDINE

```
    4
    C
   //  \\
  5C     C3
  |      ||
  6C     C2
   \\  / \\7   8
     N    CH2-CH3
     1
```

FORMULA   C7H9N   MOL WT   107.16
SOLVENT   C7H9N
ORIG ST   C6H6   TEMP   AMB

163.60	122.10	136.30	121.10	149.40
2/1	3/2	4/2	5/2	6/2

G.MIYAJIMA,Y.SASAKI,M.SUZUKI
CHEM PHARM BULL   20, 429 (1972)

2413   R 003800 3-ETHYLPYRIDINE

```
    4   7   8
    C   CH2-CH3
   //  \\ /
  5C     C3
  |      ||
  6C     C2
   \\  /
     N
     1
```

FORMULA   C7H9N   MOL WT   107.16
SOLVENT   C7H9N
ORIG ST   CS2   TEMP   AMB

149.10	138.20	133.70	122.90	146.80	25.60
2/2	3/1	4/2	5/2	6/2	7/3
14.20					
8/4					

H.L.RETCOFSKY,R.A.FRIEDEL
J PHYS CHEM   72, 290 (1968)

2414   R 003800   4-ETHYLPYRIDINE

```
    8CH3
    /
  7CH2
    |
    C4
  // \
 5C   C3
  |   ||
 6C   C2
   \1/
    N
```

FORMULA	C7H9N		MOL WT	107.16
SOLVENT	C7H9N			
ORIG ST	CS2		TEMP	AMB

148.60	121.90	151.10	27.10	12.70
2/2	3/2	4/1	7/3	8/4

H.L.RETCOFSKY,R.A.FRIEDEL
J PHYS CHEM   71, 3592 (1967)

2415   R 003800   4-ETHYLPYRIDINE

```
   5 4
   C=C
  /    \
 N      C-CH2-CH3
  \\   //3 6    7
   C-C
   1 2
```

FORMULA	C7H9N		MOL WT	107.16
SOLVENT	CDCL3			
ORIG ST	C4H12SI		TEMP	AMB

149.70	123.20	152.70	123.20	149.70	28.10
1/2	2/2	3/1	4/2	5/2	6/3
14.20					
7/4					

L.F.JOHNSON,W.C.JANKOWSKI
CARBON-13 NMR SPECTRA,JOHN
WILEY AND SONS,NEW YORK   253 (1972)

2416   R 003800   3-ACETYLPYRIDINE

```
      O
   4  ||
   C  C7
  // \ / \
 5C   C3  CH3
  |   ||   8
 6C   C2
  \  /
   N
   1
```

FORMULA	C7H7NO		MOL WT	121.14
SOLVENT	C7H7NO			
ORIG ST	CS2		TEMP	AMB

148.90	131.30	134.30	123.10	152.70	196.30
2/2	3/1	4/2	5/2	6/2	7/1
25.60					
8/4					

H.L.RETCOFSKY,R.A.FRIEDEL
J PHYS CHEM   72, 290 (1968)

2417   R 003800   METHYL 1,4,5,6-TETRAHYDRONICOTINATE

```
      O
   4  ||
   C  C
  / \ /7\
 5C   C3  O-CH3
  |   ||   8
 6C   C2
   \1/
    N
```

FORMULA	C7H11NO2		MOL WT	141.17
SOLVENT	CHCL3			
ORIG ST	C4H12SI		TEMP	AMB

142.80	95.30	20.60	20.90	40.60	169.00
2/2	3/1	4/3	5/3	6/3	7/1
50.30					
8/4					

E.WENKERT,D.W.COCHRAN,E.W.HAGAMAN,F.M.SCHELL,
N.NEUSS,A.S.KATNER,P.POTIER,C.KAN,M.PLAT,
M.KOCH,H.MEHRI,J.POISSON,N.KUNESCH,Y.ROLLAND
J AM CHEM SOC   95, 4990 (1973)

2418   R 003800   1,2-DIMETHYL-3-PIPERIDEINE

```
    4
    C
   / \\
  5C   C3
  |    |
  6C   C2
   \1/ \
    N   CH3
    |    8
   CH3
    7
```

FORMULA	C7H13N		MOL WT	111.19
SOLVENT	CHCL3			
ORIG ST	C4H12SI		TEMP	AMB

57.20	131.40	124.10	26.00	51.60	43.00
2/2	3/2	4/2	5/3	6/3	7/4
17.20					
8/4					

E.WENKERT,D.W.COCHRAN,E.W.HAGAMAN,F.M.SCHELL,
N.NEUSS,A.S.KATNER,P.POTIER,C.KAN,M.PLAT,
M.KOCH,H.MEHRI,J.POISSON,N.KUNESCH,Y.ROLLAND
J AM CHEM SOC   95, 4990 (1973)

2419   R 003800 1,2-DIMETHYLPIPERIDINE

```
        4
        C
      /   \
    5C     C3
    |      |
    6C     C2
     \1/  \
      N    CH3
      |     8
      7CH3
```

FORMULA	C7H15N			MOL WT	113.20
SOLVENT	C7H15N				
ORIG ST	C6H12			TEMP	AMB

58.85	34.45	24.45	25.95	56.55	42.45
2/2	3/3	4/3	5/3	6/3	7/4
19.75					
8/4					

G.ELLIS,R.G.JONES
J CHEM SOC PERKIN TRANS II   437 (1972)

2420   R 003800 2,6-DIMETHYLPIPERIDINE

```
          7CH3
      4  5/
      C-C
    /      \
   C        N
  3\       /
    C-C
   2 1\
      CH3
       6
```

FORMULA	C7H15N			MOL WT	113.20
SOLVENT	C4H8O2				
ORIG ST	C4H12SI			TEMP	AMB

52.60	34.60	25.40	34.60	52.60	23.20
1/2	2/3	3/3	4/3	5/2	6/4
23.20					
7/4					

L.F.JOHNSON,W.C.JANKOWSKI
CARBON-13 NMR SPECTRA,JOHN
WILEY AND SONS,NEW YORK   266 (1972)

2421   R 003800 2,6-DIMETHYLPIPERIDINE

```
        4
        C
      /   \
    5C     C3
    |      |
    6C     C2
   /  \1/ \
  H3C   N   CH3
   8         7
```

FORMULA	C7H15N			MOL WT	113.20
SOLVENT	C7H15N				
ORIG ST	C6H12			TEMP	AMB

52.05	33.95	24.75	22.79
2/2	3/3	4/3	7/4

G.ELLIS,R.G.JONES
J CHEM SOC PERKIN TRANS II   437 (1972)

2422   R 003800 3,3-DIMETHYLPIPERIDINE

```
      4   7
      C   CH3
     / \3/
   5C    C-CH3
   |    |  8
   6C   C2
    \1/
     N
```

FORMULA	C7H15N			MOL WT	113.20
SOLVENT	CHCL3				
ORIG ST	C4H12SI			TEMP	AMB

58.50	29.70	37.80	23.20	47.80	26.60
2/3	3/1	4/3	5/3	6/3	7/4

E.WENKERT,D.W.COCHRAN,E.W.HAGAMAN,F.M.SCHELL,
N.NEUSS,A.S.KATNER,P.POTIER,C.KAN,M.PLAT,
M.KOCH,H.MEHRI,J.POISSON,N.KUNESCH,Y.ROLLAND
J AM CHEM SOC   95, 4990 (1973)

2423   R 003800 1,2-DIMETHYL-4-PIPERIDEINE

```
        4
        C
      //  \
    5C     C3
    |      |
    6C     C2
     \1/  \
      N    CH3
      |     8
      CH3
       7
```

FORMULA	C7H15N			MOL WT	113.20
SOLVENT	CHCL3				
ORIG ST	C4H12SI			TEMP	AMB

53.70	33.80	124.10	124.90	53.70	41.50
2/2	3/3	4/2	5/2	6/3	7/4
17.20					
8/4					

E.WENKERT,D.W.COCHRAN,E.W.HAGAMAN,F.M.SCHELL,
N.NEUSS,A.S.KATNER,P.POTIER,C.KAN,M.PLAT,
M.KOCH,H.MEHRI,J.POISSON,N.KUNESCH,Y.ROLLAND
J AM CHEM SOC   95, 4990 (1973)

2424   R 003800   N—ETHYLPIPERIDINE

```
        4
        C
      / \
    5C   C3
    |    |
    6C   C2
      \1/
       N
       |    8
      7CH2-CH3
```

FORMULA	C7H15N		MOL WT	113.20
SOLVENT	C7H15N			
ORIG ST	C6H12		TEMP	AMB

53.95	25.85	24.45	52.55	11.95
2/3	3/3	4/3	7/3	8/4

G.ELLIS,R.G.JONES
J CHEM SOC PERKIN TRANS II   437 (1972)

2425   R 003800   2—ETHYLPIPERIDINE

```
        4
        C
      / \
    5C   C3
    |    |
    6C   C2
      \1/ \
       N   CH2-CH3
           7   8
```

FORMULA	C7H15N		MOL WT	113.20
SOLVENT	C7H15N			
ORIG ST	C6H12		TEMP	AMB

58.35	32.45	25.05	26.56	47.85	30.05
2/2	3/3	4/3	5/3	6/3	7/3

9.85
8/4

G.ELLIS,R.G.JONES
J CHEM SOC PERKIN TRANS II   437 (1972)

2426   R 003800   N—(2—HYDROXYETHYL)—PIPERIDINE

```
        4
        C
      / \
    5C   C3
    |    |
    6C   C2
      \1/
       N
       |    8
      7CH2-CH2-OH
```

FORMULA	C7H15NO		MOL WT	129.20
SOLVENT	C7H15NO			
ORIG ST	C6H12		TEMP	AMB

54.55	25.45	24.15	58.15	61.15
2/3	3/3	4/3	7/3	8/4

G.ELLIS,R.G.JONES
J CHEM SOC PERKIN TRANS II   437 (1972)

2427   R 003800   4—ISOPROPYLPYRIDINE

```
  H3C8 9CH3
     \7/
      CH
      |
      C4
    / \
  5C   C3
   |   ||
  6C   C2
    \1/
     N
```

FORMULA	C8H11N		MOL WT	121.18
SOLVENT	C8H11N			
ORIG ST	CS2		TEMP	AMB

149.80	121.50	156.60	32.80	22.40
2/2	3/2	4/1	7/2	8/4

H.L.RETCOFSKY,R.A.FRIEDEL
J PHYS CHEM   71, 3592 (1967)

2428   R 003800   2,4,6—TRIMETHYLPYRIDINE

```
   9CH3
    |
    C4
   / \
 5C   C3
  |   ||
 6C   C2
 / \ / \
H3C  N  CH3
 8   1   7
```

FORMULA	C8H11N		MOL WT	121.18
SOLVENT	C8H11N			
ORIG ST	CS2		TEMP	AMB

156.70	120.60	146.50	23.50	20.50
2/1	3/2	4/1	7/4	9/4

P.C.LAUTERBUR
J CHEM PHYS   43, 360 (1965)

2429   R 003800 METHYL 1-METHYL-1,4-DIHYDRONICOTINATE

```
         O
   4     ||
   C     C
  / \  /8\
5C    C3  O-CH3
 ||   ||    9
6C    C2
  \1/
   N
   |
   CH3
   7
```

FORMULA	C8H11NO2			MOL WT	153.18
SOLVENT	CHCL3				
ORIG ST	C4H12SI			TEMP	AMB

141.00	95.40	21.30	103.10	128.50	36.60
2/2	3/1	4/3	5/2	6/2	7/4
166.50	49.60				
8/1	9/4				

E.WENKERT,D.W.COCHRAN,E.W.HAGAMAN,F.M.SCHELL,
N.NEUSS,A.S.KATNER,P.POTIER,C.KAN,M.PLAT,
M.KOCH,H.MEHRI,J.POISSON,N.KUNESCH,Y.ROLLAND
J AM CHEM SOC   95, 4990 (1973)

2430   R 003800 METHYL 1-METHYL-1,4,5,6-TETRAHYDRONICOTINATE

```
         O
   4     ||
   C     C
  / \  /8\
5C    C3  O-CH3
 |    ||    9
6C    C2
  \1/
   N
   |
   CH3
   7
```

FORMULA	C8H13NO2			MOL WT	155.20
SOLVENT	CHCL3				
ORIG ST	C4H12SI			TEMP	AMB

146.10	93.40	19.30	20.80	47.30	47.30
2/2	3/1	4/3	5/3	6/3	7/4
168.40	49.80				
8/1	9/4				

E.WENKERT,D.W.COCHRAN,E.W.HAGAMAN,F.M.SCHELL,
N.NEUSS,A.S.KATNER,P.POTIER,C.KAN,M.PLAT,
M.KOCH,H.MEHRI,J.POISSON,N.KUNESCH,Y.ROLLAND
J AM CHEM SOC   95, 4990 (1973)

2431   R 003800 METHYL 1-METHYLPIPERIDINE-3-CARBOXYLATE

```
         O
   4     ||
   C     C
  / \  /8\
5C    C3  O-CH3
 |    |     9
6C    C2
  \1/
   N
   |
   CH3
```

FORMULA	C8H15NO2			MOL WT	157.21
SOLVENT	CHCL3				
ORIG ST	C4H12SI			TEMP	AMB

57.20	46.30	23.70	26.10	55.50	41.20
2/3	3/2	4/3	5/3	6/3	7/4
173.30	50.90				
8/1	9/4				

E.WENKERT,D.W.COCHRAN,E.W.HAGAMAN,F.M.SCHELL,
N.NEUSS,A.S.KATNER,P.POTIER,C.KAN,M.PLAT,
M.KOCH,H.MEHRI,J.POISSON,N.KUNESCH,Y.ROLLAND
J AM CHEM SOC   95, 4990 (1973)

2432   R 003800 2-PROPYLPIPERIDINE

```
       3
       C
      / \
   4C    C2
   |    |   7
   5C   C1  CH2
    \ / \ / \
     N   CH2 CH3
         6    8
```

FORMULA	C8H17N			MOL WT	127.23
SOLVENT	C4H8O2				
ORIG ST	C4H12SI			TEMP	AMB

57.10	33.70	25.70	27.40	47.60	40.20
1/2	2/3	3/3	4/3	5/3	6/3
19.40	14.50				
7/3	8/4				

L.F.JOHNSON,W.C.JANKOWSKI
CARBON-13 NMR SPECTRA,JOHN
WILEY AND SONS,NEW YORK   319 (1972)

2433   R 003800 4-TERT-BUTYLPYRIDINE

```
    9CH3
   8 | 10
 H3C-C-CH3
    |7
    C4
   / \
 5C    C3
  |    ||
 6C    C2
  \1/
   N
```

FORMULA	C9H13N			MOL WT	135.21
SOLVENT	C9H13N				
ORIG ST	CS2			TEMP	AMB

149.50	119.90	158.60	33.90	29.80
2/2	3/2	4/1	7/1	8/4

H.L.RETCOFSKY,R.A.FRIEDEL
J PHYS CHEM   71, 3592 (1967)

2434   R 003800 2,2,6,6-TETRAMETHYLPIPERIDINE

```
        4
        C
    9  / \  8
  H3C  C5 3C  CH3
      \ |  | /
      6C    C2
   10/ \ | / \
  H3C    N    CH3
         7
```

FORMULA   C9H19N   MOL WT   141.26
SOLVENT   C9H19N
ORIG ST   C6H12   TEMP   AMB

        48.95   38.15   18.05   31.75
         2/1   3/3   4/3   7/4

G.ELLIS,R.G.JONES
J CHEM SOC PERKIN TRANS II   437 (1972)

2435   R 003800 2,2'-DIPYRIDYL

```
        C4
      / \
   5C     C3
    |     ||
   6C   C2   C
    \ / \ / \ //
     N   C   C
     1   ||  |
         N   C
          \ //
           C
```

FORMULA   C10H8N2   MOL WT   156.19
SOLVENT   C2D6OS
ORIG ST   C4H12SI   TEMP   300

       155.25   124.20   137.25   120.40   149.20
        2/1   3/2   4/2   5/2   6/2

E.BREITMAIER
UNPUBLISHED   (1974)

2436   R 003800 CIS-1,2-DIPYRIDYLETHYLENE

```
          C
        / \
       C   C
    4  |   ||
    C   C   N
  / \  / \ /
 5C    C3  C
  |    ||  |
 6C    C2  C
  \ / \ / \ /
   N   C7  H
   1   |
       H
```

FORMULA   C12H10N2   MOL WT   182.23
SOLVENT   CDCL3
ORIG ST   C4H12SI   TEMP   300

       155.40   123.90   135.50   121.90   149.10   132.70
        2/1   3/2   4/2   5/2   6/2   7/2

G.WITTE,E.BREITMAIER
UNPUBLISHED   (1974)

2437   R 003800 TRANS-1,2-DIPYRIDYLETHYLENE

```
        4
        C
      / /
   5C    C3   H
    |    ||
   6C    C2   C    N
  \ / \ / \ / \ //
   N   C7  C    C
   1   |   ||   |
       H   C    C
            \ //
             C
```

FORMULA   C12H10N2   MOL WT   182.23
SOLVENT   CDCL3
ORIG ST   C4H12SI   TEMP   300

       154.50   122.80   136.20   122.10   149.30   131.30
        2/1   3/2   4/2   5/2   6/2   7/2

G.WITTE,E.BREITMAIER
UNPUBLISHED   (1974)

2438

```
          15C
        ╱  ╲
     14C    C16
      ‖     |
     13N    C17
    1 C    C12
    ╱ ╲   |
  9C  11C  4CH2
   |   ‖  |3  1
  8C  6C--C-C≡CH
   ╲ ╱    |  2
    N     5CH2
    7     |
         C18
        ╱ ╲
     23C   N19
      |    ‖
     22C   C20
       ╲  ╱
        C21
```

R 003800   1'-ETHYNYL-1',1'-DI-ALPHA-PICOLYL-ALPHA-
PICOLINE

FORMULA	C23H17N3			MOL WT	335.41
SOLVENT	CDCL3				
ORIG ST	C4H12SI			TEMP	AMB

78.24	86.88	50.26	50.65	161.70	150.22
1/D	2/S	3/S	4/T	6/S	8/D
123.06	137.26	124.13	159.07	149.93	122.68
9/D	10/D	11/D	12/S	14/D	15/D
136.62	126.10				
16/D	17/D				

W.M.LITCHMAN,A.E.ZUNE,U.HOLLSTEIN
J MAGN RESON   17, 241 (1975)

2439

```
         14C
        ╱ ╲
     13C    C15
      ‖     |
     N12    C16
   9  ╲   ╱
    C     C11      C25
    ╱ ╲10 |       ╱ ╲
  8C   C 2CH2   N24  C26
   |   ‖ 11 3    |    ‖
  7C  5C--C-CH2-C23  C27
   ╲ ╱    |      ╲  ╱
   6N     4CH2    C28
          |
         C17
        ╱ ╲
     22C   N18
      |    ‖
     21C   C19
       ╲  ╱
        C20
```

R 003800   1',1',1'-TRI-ALPHA-PICOLYL-ALPHA-
PICOLINE

FORMULA	C24H22N4			MOL WT	366.47
SOLVENT	CDCL3				
ORIG ST	C4H12SI			TEMP	AMB

50.55	51.05	163.07	150.14	122.79	137.03
1/1	2/3	5/1	7/2	8/2	9/2
124.87	160.10	150.03	122.47	136.48	126.35
10/2	11/1	13/2	14/2	15/2	16/2

W.M.LITCHMAN,A.E.ZUNE,U.HOLLSTEIN
J MAGN RESON   17, 241 (1975)

2440

```
      7  6
      C-C
     ╱    ╲
   8C      C5
    ╲     ╱
     9C=C4
    ╱    ╲
   N      C3
    ╲    ╱
     1C-C2
      |
      BR
```

R 004100   3-BROMOQUINOLINE

FORMULA	C9H6BRN			MOL WT	208.06
SOLVENT	CDCL3				
ORIG ST	C4H12SI			TEMP	AMB

150.90	116.80	136.70	128.60	127.20	126.50
1/2	2/1	3/2	4/1	5/2	6/2
129.30	129.30	145.90			
7/2	8/2	9/1			

L.F.JOHNSON,W.C.JANKOWSKI
CARBON-13 NMR SPECTRA,JOHN
WILEY AND SONS,NEW YORK   330 (1972)

2441

```
       CL
    4  |
    C   C3
   ╱ ╲8╱ ╲
  5C  C   C2
  |   ‖   |
  6C  C   C1
   ╲ ╱9╲ ╱
    C   N
    7
```

R 004100   4-CHLOROQUINOLINE

FORMULA	C9H6CLN			MOL WT	163.61
SOLVENT	CDCL3				
ORIG ST	C4H12SI			TEMP	AMB

149.50	120.90	142.10	126.10	148.90	123.70
1/2	2/2	3/1	8/1	9/1	
127.20	129.60	130.00			

L.F.JOHNSON,W.C.JANKOWSKI
CARBON-13 NMR SPECTRA,JOHN
WILEY AND SONS,NEW YORK   331 (1972)

2442 R 004100 QUINOLINE

```
        4   3
        C   C
      ⁄ \8⁄ \
    5C   C   C2
    |   ||   |
    6C   C   C1
      \ ⁄9\ ⁄
        C   N
        7
```

FORMULA	C9H7N			MOL WT	129.16
SOLVENT	CDCL3				
ORIG ST	C4H12SI			TEMP	AMB

150.00	120.80	135.70	127.60	126.30	129.20
1/2	2/2	3/2	4/2	5/2	6/2
129.20	128.00	148.10			
7/2	8/1	9/1			

L.F.JOHNSON,W.C.JANKOWSKI
CARBON-13 NMR SPECTRA,JOHN
WILEY AND SONS,NEW YORK   335 (1972)

2443 R 004100 2,6-DIMETHYLQUINOLINE

```
        CH3
    7 8⁄11
    C-C
  ⁄     \
 6C      C9
  \     ⁄
   5C=C4
  ⁄     \
 N       C3
  \     ⁄
   1C-C2
  ⁄
 H3C
  10
```

FORMULA	C11H11N			MOL WT	157.22
SOLVENT	CDCL3				
ORIG ST	C4H12SI			TEMP	AMB

157.60	121.60	135.20	126.30	146.40	135.00
1/1	2/2	3/2	4/1	5/1	8/1
25.10	21.10	126.20	128.20	131.30	
10/4	11/4				

L.F.JOHNSON,W.C.JANKOWSKI
CARBON-13 NMR SPECTRA,JOHN
WILEY AND SONS,NEW YORK   418 (1972)

2444 R 004100 2-METHYL-6-METHOXYQUINOLINE

```
          4   3
 H3C-O    C   C
  11  \ ⁄ \8⁄ \
      5C   C   C2
      |   ||   |
      6C   C   C1
        \ ⁄9\ \
         C   N   CH3
         7       10
```

FORMULA	C11H11NO			MOL WT	173.22
SOLVENT	CDCL3				
ORIG ST	C4H12SI			TEMP	AMB

155.90	121.60	134.70	105.10	157.00	121.90
1/1	2/2	3/2	4/2	5/1	6/2
129.90	127.20	143.80	24.80	55.10	
7/2	8/1	9/1	10/4	11/4	

L.F.JOHNSON,W.C.JANKOWSKI
CARBON-13 NMR SPECTRA,JOHN
WILEY AND SONS,NEW YORK   419 (1972)

2445 R 004200 ISOQUINOLINE

```
     4 5
     C-C
   ⁄     \
  3C      C6
   \     ⁄
    2C=C7
   ⁄     \
  1C      C8
   \     ⁄
    N-C9
```

FORMULA	C9H7N			MOL WT	129.16
SOLVENT	CDCL3				
ORIG ST	C4H12SI			TEMP	AMB

152.20	128.50	127.30	127.00	130.10	126.20
1/2	2/1	3/2	4/2	5/2	6/2
135.50	120.20	142.70			
7/1	8/2	9/2			

L.F.JOHNSON,W.C.JANKOWSKI
CARBON-13 NMR SPECTRA,JOHN
WILEY AND SONS,NEW YORK   334 (1972)

2446 R 004200 3-METHYLISOQUINOLINE

```
    4   3   10
    C   C   CH3
  ⁄ \8⁄ \ ⁄
 5C   C   C2
  |   ||   |
 6C   C   N
  \ ⁄9\ ⁄
   C   C
   7   1
```

FORMULA	C10H9N			MOL WT	143.19
SOLVENT	CDCL3				
ORIG ST	C4H12SI			TEMP	AMB

151.80	151.50	118.20	136.40	126.70
1/2	2/1	3/2	8/1	9/1
24.10	125.70	126.00	127.30	130.00
10/4				

L.F.JOHNSON,W.C.JANKOWSKI
CARBON-13 NMR SPECTRA,JOHN
WILEY AND SONS,NEW YORK   369 (1972)

2447

```
      8   1
      C   C
    /  \9/  \
  7C    C    \
   |    |     )2
  6C    N    /
    \  /4\  /
      C   C
      5   3
```

R 004500 HEXAHYDRO-3H-OXAZOLO(3,4-A)PYRIDINE

FORMULA	C7H13NO		MOL WT	127.19
SOLVENT	CDCL3			50.0
ORIG ST	C4H12SI		TEMP	AMB

68.40	86.00	46.90	24.40	21.90	25.10
1/T	3/T	5/T	6/T	7/T	8/T
60.10					
9/D					

Y.TAKEUCHI,P.J.CHIVERS,T.A.CRABB
J CHEM SOC PERKIN TRANS II   51 (1975)

2448

```
    8     1
    C     N
   // \  / \
 7C   C9   C2
  |    ||   |
 6C   10C   C3
   \  / \  //
     N    C
     5    4
```

R 004500 1,5-DIAZANAPHTHALENE   1,5-NAPHTHYRIDINE

FORMULA	C8H6N2	MOL WT	130.15
SOLVENT	C2H3N		
ORIG ST	C6H6	TEMP	AMB

151.70	125.00	140.40	144.20
2/2	3/2	4/2	9/1

A.BOICELLI,R.DANIELI,A.MANGINI,L.LUNAZZI,
G.PLACUCCI
J CHEM SOC PERKIN TRANS II   1024 (1973)

2449

```
    3     1
    N     N
   // \  / \
 7C   C9   C2
  |    ||   |
 6C   10C   C3
   \  / \  //
     C    C
     5    4
```

R 004500 1,8-DIAZANAPHTHALENE   1,8-NAPHTHYRIDINE

FORMULA	C8H6N2	MOL WT	130.15
SOLVENT	C2H3N		
ORIG ST	C6H6	TEMP	AMB

154.00	122.80	139.00	156.40	123.40
2/2	3/2	4/2	9/1	10/1

A.BOICELLI,R.DANIELI,A.MANGINI,L.LUNAZZI,
G.PLACUCCI
J CHEM SOC PERKIN TRANS II   1024 (1973)

2450

```
    8     1
    C     C
   // \  / \
 7C   C9   N2
  |    ||   |
 6N   10C   C3
   \  / \  //
     C    C
     5    4
```

R 004500 2,6-DIAZANAPHTHALENE   2,6-NAPHTHYRIDINE

FORMULA	C8H6N2	MOL WT	130.15
SOLVENT	C2H3N		
ORIG ST	C6H6	TEMP	AMB

152.60	145.40	119.90	131.00
1/2	3/2	4/2	9/1

A.BOICELLI,R.DANIELI,A.MANGINI,L.LUNAZZI,
G.PLACUCCI
J CHEM SOC PERKIN TRANS II   1024 (1973)

2451

```
    8     1
    C     C
   // \  / \
 7N   C9   N2
  |    ||   |
 6C   10C   C3
   \  / \  //
     C    C
     5    4
```

R 004500 2,7-DIAZANAPHTHALENE   2,7-NAPHTHYRIDINE

FORMULA	C8H6N2	MOL WT	130.15
SOLVENT	C2H3N		
ORIG ST	C6H6	TEMP	AMB

153.40	147.20	119.60	124.30	138.50
1/2	3/2	4/2	9/1	10/1

A.BOICELLI,R.DANIELI,A.MANGINI,L.LUNAZZI,
G.PLACUCCI
J CHEM SOC PERKIN TRANS II   1024 (1973)

2452   R 004500 CIS-(6-H,8A-H)-6-METHYLHEXAHYDRO-3H-OXAZOLO-
(3,4-A)PYRIDINE

```
      8   1
      C   C
    / \9/ \
  7C   C   \
   |   |   )2
  6C   N   /
 10/ \ /4\ /
 H3C   C   C
      5   3
```

FORMULA	C8H15NO			MOL WT	141.21
SOLVENT	CDCL3				50.0
ORIG ST	C4H12SI			TEMP	AMB

68.70	87.80	54.40	30.10	27.50	22.50
1/T	3/T	5/T	6/D	7/T	8/T
57.30	19.30				
9/D	10/Q				

Y.TAKEUCHI,P.J.CHIVERS,T.A.CRABB
J CHEM SOC PERKIN TRANS II   51 (1975)

2453   R 004500 CIS-(7-H,8A-H)-7-METHYLHEXAHYDRO-3H-OXAZOLO-
(3,4-A)PYRIDINE

```
  10   8   1
  H3C   C   C
    \ / \9/ \
  7C   C   \
   |   |   )2
  6C   N   /
    \ /4\ /
     C   C
     5   3
```

FORMULA	C8H15NO			MOL WT	141.21
SOLVENT	CDCL3				50.0
ORIG ST	C4H12SI			TEMP	AMB

70.80	85.20	46.80	32.40	30.10	34.80
1/T	3/T	5/T	6/T	7/D	8/T
61.20	18.90				
9/D	10/Q				

Y.TAKEUCHI,P.J.CHIVERS,T.A.CRABB
J CHEM SOC PERKIN TRANS II   51 (1975)

2454   R 004500 CIS-(8-H,8A-H)-8-METHYLHEXAHYDRO-3H-OXAZOLO-
(3,4-A)PYRIDINE

```
      10
      CH3
       |   1
      8C   C
     / \9/ \
    7C   C   \
     |   |   )2
    5C   N   /
      \ /4\ /
       C   C
       5   3
```

FORMULA	C8H15NO			MOL WT	141.21
SOLVENT	CDCL3				50.0
ORIG ST	C4H12SI			TEMP	AMB

61.20	87.50	46.80	24.10	27.20	29.50
1/T	3/T	5/T	6/T	7/T	8/D
63.00	17.50				
9/D	10/Q				

Y.TAKEUCHI,P.J.CHIVERS,T.A.CRABB
J CHEM SOC PERKIN TRANS II   51 (1975)

2455   R 004500 TRANS-(6-H,8A-H)-6-METHYLHEXAHYDRO-3H-OXAZOLO-
(3,4-A)PYRIDINE

```
      8   1
      C   C
    / \9/ \
  7C   C   \
   |   |   )2
  6C   N   /
 10/ \ /4\ /
 H3C   C   C
      5   3
```

FORMULA	C8H15NO			MOL WT	141.21
SOLVENT	CDCL3				50.0
ORIG ST	C4H12SI			TEMP	AMB

71.00	85.30	54.60	30.20	32.50	26.00
1/T	3/T	5/T	6/D	7/T	8/T
61.40	19.10				
9/D	10/Q				

Y.TAKEUCHI,P.J.CHIVERS,T.A.CRABB
J CHEM SOC PERKIN TRANS II   51 (1975)

2456   R 004500 QUINOLIZIDINE

```
      5   4
      C   C
    / \ / \
  6C  C10 C3
   |   |   |
  7C   N   C2
    \ /9\ /
     C   C
     8   1
```

FORMULA	C9H17N			MOL WT	139.24
SOLVENT	CHCL3				
ORIG ST	C4H12SI			TEMP	AMB

56.40	25.60	24.40	33.20	62.90
1/3	2/3	3/3	4/3	10/2

E.WENKERT,J.S.BINDRA,C.-J.CHANG,D.W.COCHRAN,
F.M.SCHELL
ACC CHEM RES   7,   46 (1974)

2457   R 004500   NB—METHYLTETRAHYDROHARMINE

```
      4        10
      C         C
     ⁄ \8     ⁄ \
   5C   C---C3  C11
   |    ||   ||  |
   6C   C    C2  N12
  14 ⁄ \ ⁄9\1⁄ \ ⁄ \
 H3C-O  C   N   C13 CH3
    7   |        |  15
           16CH3
```

FORMULA	C14H18N2O			MOL WT	230.31
SOLVENT	CHCL3				
ORIG ST	C4H12SI			TEMP	AMB

134.80	107.30	118.40	108.60	155.50	95.10
2/1	3/1	4/2	5/2	6/1	7/2
121.70	136.90	20.50	50.80	58.00	55.60
8/1	9/1	10/3	11/3	13/2	14/4
42.20	18.40				
15/4	16/4				

E.WENKERT,J.S.BINDRA,C.—J.CHANG,D.W.COCHRAN,
F.M.SCHELL
ACC CHEM RES   7,   46 (1974)

2458   R 004500   HEXAHYDROINDOLO(2,3-A)QUINOLIZINE

```
      4        10
      C         C
     ⁄ \8     ⁄ \
   5C   C---C3  C11
   |    ||   ||  |
   6C   C    C2  N12
     \ ⁄9\1⁄ \ ⁄ \
      C   N  17C   C13
      7       |     |
           16C     C14
            \ ⁄     ⁄
              C
              15
```

FORMULA	C15H18N2			MOL WT	226.32
SOLVENT	CHCL3				
ORIG ST	C4H12SI			TEMP	AMB

135.40	108.20	118.40	121.40	119.40	111.00
2/1	3/1	4/2	5/2	6/2	7/2
127.80	136.40	21.80	53.70	55.90	25.90
8/1	9/1	10/3	11/3	13/3	14/3
24.50	30.10	30.40			
15/3	16/3	17/2			

E.WENKERT,J.S.BINDRA,C.—J.CHANG,D.W.COCHRAN,
F.M.SCHELL
ACC CHEM RES   7,   46 (1974)

2459   R 004600   NORTROPANE

```
     1   2
     C--C
    ⁄ \   \
  7C    \   \
   |   H-N8  C3
  6C   ⁄   ⁄
    \ ⁄   ⁄
     C--C
     5   4
```

FORMULA	C7H13N		MOL WT	111.19
SOLVENT	CHCL3			
ORIG ST	C4H12SI		TEMP	AMB

54.70	32.90	17.20	29.00
1/2	2/3	3/3	6/3

E.WENKERT,J.S.BINDRA,C.—J.CHANG,D.W.COCHRAN,
F.M.SCHELL
ACC CHEM RES   7,   46 (1974)

2460   R 004600   QUINUCLIDINE

```
       C4
      ⁄|\
     ⁄ | \
   5C  C7 C3
    |  |  |
    |  |  |
   6C  C8 C2
     \ | ⁄
      \|⁄
       N1
```

FORMULA	C7H13N		MOL WT	111.19
SOLVENT	CHCL3			
ORIG ST	C4H12SI		TEMP	AMB

47.60	26.80	20.80
2/3	3/3	4/2

E.WENKERT,J.S.BINDRA,C.—.J.CHANG,D.W.COCHRAN,
F.M.SCHELL
ACC CHEM RES   7,   46 (1974)

2461   R 004600   QUINUCLIDINE

```
      4C
     ⁄|\
    ⁄ | \
   ⁄  |  \
  7C  C3 C5
  |   |  |
  8C  C2 C6
   \  |  ⁄
    \ | ⁄
     \|⁄
      N
      1
```

FORMULA	C7H13N				MOL WT	111.19
SOLVENT	C2D6OS					
ORIG ST	C4H12SI				TEMP	AMB

48.20	27.30	21.50	27.30	48.20	27.30
2/3	3/3	4/2	5/3	6/3	7/3
48.20					
8/3					

F.I.CARROLL,D.SMITH,M.E.WALL,C.G.MORELAND
J MED CHEM   17,   985 (1974)

2462   R 004600 QUINUCLIDINE OXIDE

```
      C4
     /|\
    / | \
  5C  C7 C3
   |  |  |
   |  |  |
  6C  C8 C2
   \  | /
    \ |/
     N+
     |
     O-
```

FORMULA	C7H13NO	MOL WT 127.19
SOLVENT	CHCL3	
ORIG ST	C4H12SI	TEMP AMB

62.70	26.10	19.50
2/3	3/3	4/2

E.WENKERT,J.S.BINDRA,C.—J.CHANG,D.W.COCHRAN,
F.M.SCHELL
ACC CHEM RES   7,   46 (1974)

2463   R 004600 3-QUINUCLIDINOL

```
      4C
     /|\
    / | \
  7C  RC3 C5
   |  |  |
  8C  C2 C6
   \  | /
    \ |/
     N
     1
```

FORMULA	C7H13NO	MOL WT 127.19
SOLVENT	C2D6OS	
ORIG ST	C4H12SI	TEMP AMB

57.90	66.39	28.05	18.97	47.17	24.70
2/3	3/2	4/2	5/3	6/3	7/3
46.15					
8/3					

R —OH   F.I.CARROLL,D.SMITH,M.E.WALL,C.G.MORELAND
         J MED CHEM   17,   985 (1974)

2464   R 004600 TROPIDINE

```
   1   2
   C--C
  / \   \
 7C  9\   \
 1H3C-N8 C3
 6C  /   /
  \ /   /
   C--C
   5   4
```

FORMULA	C8H13N	MOL WT 123.20
SOLVENT	CHCL3	
ORIG ST	C4H12SI	TEMP AMB

58.90	130.80	122.90	29.90	57.80	33.90
1/2	2/3	3/2	4/3	5/2	6/3
31.80	36.60				
7/3	9/4				

E.WENKERT,J.S.BINDRA,C.—J.CHANG,D.W.COCHRAN,
F.M.SCHELL
ACC CHEM RES   7,   46 (1974)

2465   R 004600 TROPINONE

```
   1   2
   C--C
  / \   \
 7C  9\   \3
 1H3C-N8 C=O
 6C  /   /
  \ /   /
   C--C
   5   4
```

FORMULA	C8H13NO	MOL WT 139.20
SOLVENT	CHCL3	
ORIG ST	C4H12SI	TEMP AMB

60.20	47.10	207.80	27.30	37.80
1/2	2/3	3/1	6/3	9/4

E.WENKERT,J.S.BINDRA,C.—J.CHANG,D.W.COCHRAN,
F.M.SCHELL
ACC CHEM RES   7,   46 (1974)

2466   R 004600 TROPANE

```
   1   2
   C--C
  / \   \
 7C  9\   \
 1H3C-N8 C3
 6C  /   /
  \ /   /
   C--C
   5   4
```

FORMULA	C8H15N	MOL WT 125.22
SOLVENT	CHCL3	
ORIG ST	C4H12SI	TEMP AMB

61.20	29.90	15.90	25.60	40.40
1/2	2/3	3/3	6/3	9/4

E.WENKERT,J.S.BINDRA,C.—J.CHANG,D.W.COCHRAN,
F.M.SCHELL
ACC CHEM RES   7,   46 (1974)

2467   R 004600 PSEUDOTROPINE

```
    1   2
    C--C
   /  \  \
 7C  9\  \3
 |H3C-N8 C-OH
 6C   /  /E
   \ /  /
    C--C
    5   4
```

FORMULA	C8H15NO		MOL WT	141.21
SOLVENT	CHCL3			
ORIG ST	C4H12SI		TEMP	AMB

| 60.10 | 38.30 | 62.70 | 26.70 | 39.20 |
| 1/2 | 2/3 | 3/2 | 6/3 | 9/4 |

E.WENKERT,J.S.BINDRA,C.-J.CHANG,D.W.COCHRAN,
F.M.SCHELL
ACC CHEM RES   7,   46 (1974)

2468   R 004600 TROPINE

```
    1   2
    C--C
   /  \  \
 7C  9\  \4
 |H3C-N8 C3
 6C   /  /|
   \ /  / OH
    C--C
```

FORMULA	C8H15NO		MOL WT	141.21
SOLVENT	CHCL3			
ORIG ST	C4H12SI		TEMP	AMB

| 59.80 | 39.10 | 63.60 | 25.70 | 40.00 |
| 1/2 | 2/3 | 3/2 | 6/3 | 9/4 |

E.WENKERT,J.S.BINDRA,C.-J.CHANG,D.W.COCHRAN,
F.M.SCHELL
ACC CHEM RES   7,   46 (1974)

2469   R 004700 1,4-DIOXANE

```
   1  2
   C-C
  /    \
 O      O
  \    /
   C-C
   4  3
```

FORMULA	C4H8O2		MOL WT	88.11
SOLVENT	CDCL3			
ORIG ST	C4H12SI		TEMP	AMB

| 67.40 | 67.40 | 67.40 | 67.40 |
| 1/3 | 2/3 | 3/3 | 4/3 |

N.R.EASTON,JR.,F.A.L.ANET,P.A.BURNS,C.S.FOOTE
J AM CHEM SOC   96, 3945 (1974)

2470   R 004800 1,4-OXATHIANE

```
      1
      O
     / \
   6C   C2
   |    |
   5C   C3
     \ /
      S
      4
```

FORMULA	C4H8OS		MOL WT	104.17
SOLVENT	CDCL3			
ORIG ST	C4H12SI		TEMP	AMB

| 68.50 | 27.00 | 27.00 | 68.50 |
| 2/3 | 3/3 | 4/3 | 5/3 |

W.A.SZAREK,D.M.VYAS,A.-M.SEPULCHRE,S.D.GERO,
G.LUKACS
CAN J CHEM   52, 2041 (1974)

2471   R 004800 1,4-OXATHIANE 4-OXIDE

```
      1
      O
     / \
   6C   C2
   |    |
   5C   C3
     \ /
      S4
      ||
      O
```

FORMULA	C4H8O2S		MOL WT	120.17
SOLVENT	CDCL3			
ORIG ST	C4H12SI		TEMP	AMB

| 59.00 | 46.20 | 46.20 | 59.00 |
| 2/3 | 3/3 | 5/3 | 6/3 |

W.A.SZAREK,D.M.VYAS,A.-M.SEPULCHRE,S.D.GERO,
G.LUKACS
CAN J CHEM   52, 2041 (1974)

2472   R 004800 1,4-OXATHIANE 4,4-DIOXIDE

```
    O1
   / \
 6C   C2
  |   |
 5C   C3
   \ /
    S4
   // \\
  O    O
```

FORMULA	C4H8O3S			MOL WT	136.17
SOLVENT	CDCL3				
ORIG ST	C4H12SI			TEMP	AMB

66.00	52.80	52.80	66.00
2/3	3/3	5/3	6/3

W.A.SZAREK,D.M.VYAS,A.—M.SEPULCHRE,S.D.GERO,
G.LUKACS
CAN J CHEM   52, 2041 (1974)

2473   R 004800 2-METHOXY-1,4-OXATHIANE

```
    1
    O
   / \2    7
 6C   C-O-CH3
  |   |
 5C   C3
   \ /
    S
    4
```

FORMULA	C5H10O2S				MOL WT	134.20
SOLVENT	CDCL3					
ORIG ST	C4H12SI				TEMP	AMB

98.90	30.20	26.00	65.10	55.30
2/2	3/3	5/3	6/3	7/4

W.A.SZAREK,D.M.VYAS,A.—M.SEPULCHRE,S.D.GERO,
G.LUKACS
CAN J CHEM   52, 2041 (1974)

2474   R 004800 3-ACETOXY-1,4-OXATHIANE

```
    1
    O
   / \
 6C   C2
  |   |  7 8
 5C   C-O-C-CH3
   \ /3  ||
    S    O
    4
```

FORMULA	C6H10O3S				MOL WT	162.21
SOLVENT	CDCL3					
ORIG ST	C4H12SI				TEMP	AMB

70.90	67.90	23.30	67.60	169.10	21.00
2/3	3/2	5/3	6/3	7/1	8/4

W.A.SZAREK,D.M.VYAS,A.—M.SEPULCHRE,S.D.GERO,
G.LUKACS
CAN J CHEM   52, 2041 (1974)

2475   R 004800 (2R,6S)-2-METHOXY-6-METHYL-1,4-OXATHIANE

```
      1
      O
  8 6/ \2   7
 H3C-C   C-O-CH3
    |   |
   5C   C3
    \ /
     S
     4
```

FORMULA	C6H12O3S				MOL WT	164.22
SOLVENT	CDCL3					
ORIG ST	C4H12SI				TEMP	AMB

95.10	29.30	31.90	64.30	54.60	21.50
2/2	3/3	5/3	6/2	7/4	8/4

W.A.SZAREK,D.M.VYAS,A.—M.SEPULCHRE,S.D.GERO,
G.LUKACS
CAN J CHEM   52, 2041 (1974)

2476   R 005000 3-HYDROXY-6-OXO-1,6-DIHYDROPYRIDAZINE

```
    O
    ||
    6C
   / \1
 5C   N
 ||   |
 4C   N2
   \ //
    C3
    |
    OH
```

FORMULA	C4H4N2O2			MOL WT	112.09
SOLVENT	C3D7NO				
ORIG ST	C4H12SI			TEMP	233

154.16	128.91	133.87	160.42
3/1	4/2	5/2	6/1

H.P.FRITZ,F.H.KOEHLER,B.LIPPERT
CHEM BER   106, 2918 (1973)

2477 R 005000 3-CHLORO-6-METHOXYPYRIDAZINE

```
        C2  CL
       ╱ ╲ ╱
      3C    C1
      |     ||
      4C    N
   5 ╱ ╲ ╱
  H3C-O   N
```

FORMULA	C5H5CLN2O			MOL WT	144.56
SOLVENT	CDCL3				
ORIG ST	C4H12SI			TEMP	AMB

151.00	130.80	120.10	164.40	55.00
1/1	2/2	3/2	4/1	5/4

L.F.JOHNSON,W.C.JANKOWSKI
CARBON-13 NMR SPECTRA,JOHN
WILEY AND SONS,NEW YORK   105 (1972)

2478 R 005000 3,6-DIMETHYLDIHYDROPYRIDAZINE DIMER

```
        15CH3
     8   |    1  13
      N---C9  N    CH3
     ╱ ╱ ╲ ╱ ╲ ╱
    ╱   ╱  N10 C2
  7N   C12  |   |
   ╲    ╲   C11 C3
    ╲    ╲5╱|╲4╱
     6C---C | C
     ╱      CH3
    CH3     16
    14
```

FORMULA	C11H20N4			MOL WT	208.31
SOLVENT	CDCL3				
ORIG ST	C4H12SI			TEMP	AMB

139.90	23.90	25.00	47.50	146.50	75.40
2/1	3/3	4/3	5/2	6/1	9/1
62.90	24.50	24.70	27.40	22.20	
11/1	13/4	14/4	15/4	16/4	

P.DE MAYO,J.B.STOTHERS,M.C.USSELMAN
CAN J CHEM   50, 612 (1972)

2479 R 005100 4,6-DIMETHYLPYRIMIDINE

```
     7CH3
      |
      C4
    ╱ ╲
  3N    C5
   |    ||
  2C    C6
   ╲ ╱ ╲
    N    CH3
    1    8
```

FORMULA	C6H8N2		MOL WT	108.14
SOLVENT	C6H8N2			
ORIG ST	CS2		TEMP	AMB

158.00	166.00	119.90	23.40
2/2	4/1	5/2	7/4

P.C.LAUTERBUR
J CHEM PHYS   43, 360 (1965)

2480 R 005150 PYRROLO(2,3-D)PYRIMIDINE

```
      4
      C
    ╱ ╲
  3N  8C---C5
   |   ||   ||
  2C  9C    C6
   ╲ ╱ ╲ ╱
    N    N7
    1
```

FORMULA	C6H5N3			MOL WT	119.13
SOLVENT	C2H6OS				
ORIG ST	C4H12SI			TEMP	313

150.88	148.88	99.43	127.24	118.22	151.25
2/2	4/2	5/2	6/2	8/1	9/1

M.-T.CHENON,R.J.PUGMIRE,D.M.GRANT,R.P.PANZICA,
L.B.TOWNSEND
J AM CHEM SOC   97, 4627 (1975)

2481 R 005150 PYRROLO(2,3-D)PYRIMIDIN-4-ONE

```
      O
      ||
      C4
    3╱ ╲
    N  8C---C5
    |   ||   ||
   2C  9C    C6
    ╲ ╱ ╲ ╱
     N    N7
     1
```

FORMULA	C6H5N3O			MOL WT	135.13
SOLVENT	C2H6OS				
ORIG ST	C4H12SI			TEMP	310

143.37	158.72	102.20	120.55	107.84	148.26
2/2	4/1	5/2	6/2	8/1	9/1

M.-T.CHENON,R.J.PUGMIRE,D.M.GRANT,R.P.PANZICA,
L.B.TOWNSEND
J AM CHEM SOC   97, 4627 (1975)

2482   R 005150 PYRROLO(2,3-D)PYRIMIDINE-4-THIONE

```
      S
      ||
      C4
     3/ \
     N  8C---C5
     |  ||   ||
     2C 9C   C6
      \ / \ /
       N   N7
       1
```

FORMULA C6H5N3S   MOL WT 151.19
SOLVENT C2H6OS
ORIG ST C4H12SI   TEMP   302

142.84	176.31	104.40	123.80	119.99	143.68
2/2	4/1	5/2	6/2	8/1	9/1

M.-T.CHENON,R.J.PUGMIRE,D.M.GRANT,R.P.PANZICA,
L.B.TOWNSEND
J AM CHEM SOC   97, 4627 (1975)

2483   R 005150 4-AMINOPYRROLO(2,3-D)PYRIMIDINE

```
      10
      NH2
      |
      C4
     // \
     3N  8C---C5
     |  ||   ||
     2C 9C   C6
      \ / \ /
       N   N7
       1
```

FORMULA C6H6N4   MOL WT 134.14
SOLVENT C2H6OS
ORIG ST C4H12SI   TEMP   305

151.19	157.23	99.20	121.54	102.30	150.00
2/2	4/1	5/2	6/2	8/1	9/1

M.-T.CHENON,R.J.PUGMIRE,D.M.GRANT,R.P.PANZICA,
L.B.TOWNSEND
J AM CHEM SOC   97, 4627 (1975)

2484   R 005150 4-AMINO-7-METHYLPYRROLO(2,3-D)PYRIMIDINE

```
      NH2
      |
      C4
     // \
     3N  8C---C5
     |  ||   ||
     2C 9C   C6
      \ / \ /
       N   N7
       1   |
           CH3
           10
```

FORMULA C7H8N4   MOL WT 148.17
SOLVENT C2D6OS
ORIG ST C4H12SI   TEMP   313

151.60	157.45	98.31	124.96	102.38	149.92
2/2	4/1	5/2	6/2	8/1	9/1
30.67					
10/4					

M.-T.CHENON,R.J.PUGMIRE,D.M.GRANT,R.P.PANZICA,
L.B.TOWNSEND
J AM CHEM SOC   97, 4627 (1975)

2485   R 005150 7-(BETA-D-RIBOFURANOSYL)PYRROLO(2,3-D)
                       PYRIMIDINE

```
      C4
     // \
     3N  8C---C5
     |  ||   ||
     2C 9C   C6
      \ / \ /
       N1   N7
            |
      14    |
   HO-H2C O |
      |/ \|
     13C   C10
      |    |
     12C---C11
      |    |
      OH   OH
```

FORMULA C11H13N3O4   MOL WT 251.24
SOLVENT C2H6OS
ORIG ST C4H12SI   TEMP   310

151.08	149.62	100.57	127.87	119.34	150.79
2/2	4/2	5/2	6/2	8/1	9/1
86.90	74.22	70.80	85.39	61.80	
10/2	11/2	12/2	13/2	14/3	

M.-T.CHENON,R.J.PUGMIRE,D.M.GRANT,R.P.PANZICA,
L.B.TOWNSEND
J AM CHEM SOC   97, 4627 (1975)

2486　　　　　　R 005150　7-(BETA-D-RIBOFURANOSYL)PYRROLO(2,3-D)-
　　　　　　　　　　　　　　PYRIMIDINE-4-THIONE

```
         S
         ‖
         C4
       3⁄ ⧵
      N  8C---C5
      |   ‖   ‖
     2C  9C   C6
      ⧵ ⁄ ⧵ ⁄
       N1   N7
            |
      14    |
   HO-H2C O |
       |⁄ ⧵|
      13C   C10
       |    |
      12C---C11
       |    |
       OH   OH
```

FORMULA　C11H13N3O4S　　　　　　MOL WT　283.31
SOLVENT　C2H6OS
ORIG ST　C4H12SI　　　　　　　　TEMP　　　310

143.50	176.67	104.96	124.29	120.79	143.47
2/2	4/1	5/2	6/2	8/1	9/1
87.14	74.63	70.72	85.43	61.67	
10/2	11/2	12/2	13/2	14/3	

M.-T.CHENON,R.J.PUGMIRE,D.M.GRANT,R.P.PANZICA,
L.B.TOWNSEND
J AM CHEM SOC　　　　　　　　　97, 4627 (1975)

2487　　　　　　R 005150　7-(BETA-D-RIBOFURANOSYL)PYRROLO(2,3-D)-
　　　　　　　　　　　　　　PYRIMIDIN-4-ONE

```
         O
         ‖
         C4
       3⁄ ⧵
      N  8C---C5
      |   ‖   ‖
     2C  9C   C6
      ⧵ ⁄ ⧵ ⁄
       N1   N7
            |
      14    |
   HO-H2C O |
       |⁄ ⧵|
      13C   C10
       |    |
      12C---C11
       |    |
       OH   OH
```

FORMULA　C11H13N3O5　　　　　　MOL WT　267.24
SOLVENT　C2H6OS
ORIG ST　C4H12SI　　　　　　　　TEMP　　　310

144.00	158.53	102.77	121.44	108.65	148.01
2/2	4/1	5/2	6/2	8/1	9/1
87.36	74.56	70.83	85.33	61.88	
10/2	11/2	12/2	13/2	14/3	

M.-T.CHENON,R.J.PUGMIRE,D.M.GRANT,R.P.PANZICA,
L.B.TOWNSEND
J AM CHEM SOC　　　　　　　　　97, 4627 (1975)

2488　　　　　　R 005150　4-AMINO-7-(BETA-D-RIBOFURANOSYL)PYRROLO(2,3-D)-
　　　　　　　　　　　　　　PYRIMIDINE　TUBERCIDIN

```
        NH2
         |
         C4
       ⁄⁄ ⧵
      3N  8C---C5
      |   ‖   ‖
     2C  9C   C6
      ⧵ ⁄ ⧵ ⁄
       N1   N7
            |
      14    |
   HO-H2C O |
       |⁄ ⧵|
      13C   C10
       |    |
      12C---C11
       |    |
       OH   OH
```

FORMULA　C11H14N4O4　　　　　　MOL WT　266.26
SOLVENT　C2H6OS
ORIG ST　C4H12SI　　　　　　　　TEMP　　　305

151.35	157.42	99.80	122.81	103.28	149.55
2/2	4/1	5/2	6/2	8/1	9/1
87.85	73.39	70.77	85.16	61.87	
10/2	11/2	12/2	13/2	14/3	

M.-T.CHENON,R.J.PUGMIRE,D.M.GRANT,R.P.PANZICA,
L.B.TOWNSEND
J AM CHEM SOC　　　　　　　　　97, 4627 (1975)

2489　　　　　　R 005150　4-AMINO-5-CYANO-7-(BETA-D-RIBOFURANOSYL)
　　　　　　　　　　　　　　PYRROLO(2,3-D)PYRIMIDINE　TOYOCAMYCIN

```
        NH2
         |
         C4
       ⁄⁄ ⧵    5 10
      3N  8C---C-C≡N
      |   ‖   ‖
     2C  9C   C6
      ⧵ ⁄ ⧵ ⁄
       N1   N7
            |
      15    |
   HO-H2C O |
       |⁄ ⧵|
      14C   C11
       |    |
      13C---C12
       |    |
       OH   OH
```

FORMULA　C12H13N5O4　　　　　　MOL WT　291.27
SOLVENT　C2H6OS
ORIG ST　C4H12SI　　　　　　　　TEMP　　　310

153.76	157.22	83.27	132.61	101.58	150.32
2/2	4/1	5/1	6/2	8/1	9/1
115.55	88.16	74.56	70.48	85.74	61.49
10/1	11/2	12/2	13/2	14/2	15/3

M.-T.CHENON,R.J.PUGMIRE,D.M.GRANT,R.P.PANZICA,
L.B.TOWNSEND
J AM CHEM SOC　　　　　　　　　97, 4627 (1975)

2490   R 005150 4-AMINO-5-CARBOXAMIDO-7-(BETA-D-RIBOFURANOSYL)-
```
      NH2     O              PYRROLO(2,3-D)PYRIMIDINE  SANGIVAMYCIN
       |      ||
       C4      C-NH2     FORMULA   C12H15N5O5              MOL WT  309.28
      ∥ \    /10         SOLVENT   C2H6OS
     3N  8C---C5         ORIG ST   C4H12SI                TEMP      310
      |   ∥    ∥
     2C  9C   C6          153.07  158.37  111.27  126.09  101.49  151.10
      \ / \ / \ /          2/2     4/1     5/1     6/2     8/1     9/1
      N1   N7             166.72   87.74   74.21   70.91   85.68   62.19
           |               10/1    11/2    12/2    13/2    14/2    15/3
       15   |
    HO-H2C O |
      | /  \|
     14C    C11
      |      |
     13C---C12          M.-T.CHENON,R.J.PUGMIRE,D.M.GRANT,R.P.PANZICA,
      |      |          L.B.TOWNSEND
      OH    OH          J AM CHEM SOC                     97, 4627 (1975)
```

2491   R 005200 5-BROMOURACIL
```
       O
       ||
       C4   BR       FORMULA   C4H3BRN2O2              MOL WT  190.98
      / \  /         SOLVENT   C2D6OS/C2H6OS
     3N   C5         ORIG ST   C4H12SI                TEMP      AMB
      |    ||
     2C    C6         150.25  159.90   94.50  142.05
      ∥ \ /1/         2/1     4/1     5/1     6/2
      O    N          P.D.ELLIS,R.B.DUNLAP,A.R.POLLARD,K.SEIDMAN,
                      A.D.CARDIN
                      J AM CHEM SOC                     95, 4398 (1973)
```

2492   R 005200 5-CHLOROURACIL
```
       O
       ||
       C4   CL       FORMULA   C4H3CLN2O2              MOL WT  146.53
      / \  /         SOLVENT   C2D6OS/C2H6OS
     3N   C5         ORIG ST   C4H12SI                TEMP      AMB
      |    ||
     2C    C6         150.05  159.75  106.00  139.60
      ∥ \ /1/         2/1     4/1     5/1     6/2
      O    N          P.D.ELLIS,R.B.DUNLAP,A.L.POLLARD,K.SEIDMAN,
                      A.D.CARDIN
                      J AM CHEM SOC                     95, 4398 (1973)
```

2493   R 005200 5-FLUOROURACIL
```
       O
       ||
       C4   F        FORMULA   C4H3FN2O2               MOL WT  130.08
      / \  /         SOLVENT   C2D6OS/C2H6OS
     3N   C5         ORIG ST   C4H12SI                TEMP      AMB
      |    ||
     2C    C6         150.00  158.30  139.75  126.15
      ∥ \ /1/         2/1     4/1     5/1     6/2
      O    N          P.D.ELLIS,R.B.DUNLAP,A.L.POLLARD,K.SEIDMAN,
                      A.D.CARDIN
                      J AM CHEM SOC                     95, 4398 (1973)
```

2494   R 005200 5-IODOURACIL

```
      O
      ‖
      C4   J
     / \ /
   3N    C5
    |    ‖
   2C    C6
   ⁄ \1⁄
  O    N
```

FORMULA   C4H3IN2O2   MOL WT   237.99
SOLVENT   C2D6OS/C2H6OS
ORIG ST   C4H12SI   TEMP   AMB

150.75   161.35   67.65   146.90
  2/1   4/1   5/1   6/2

P.D.ELLIS,R.B.DUNLAP,A.R.POLLARD,K.SEIDMAN,
A.D.CARDIN
J AM CHEM SOC   95, 4398 (1973)

2495   R 005200 5-NITROURACIL

```
      O
      ‖
      C4   NO2
     / \ /
   3N    C5
    |    ‖
   2C    C6
   ⁄ \1⁄
  O    N
```

FORMULA   C4H3N3O4   MOL WT   157.09
SOLVENT   C2D6OS/C2H6OS
ORIG ST   C4H12SI   TEMP   AMB

150.40   156.20   126.05   148.30
  2/1   4/1   5/1   6/2

P.D.ELLIS,R.B.DUNLAP,A.L.POLLARD,K.SEIDMAN,
A.D.CARDIN
J AM CHEM SOC   95, 4398 (1973)

2496   R 005200 URACIL

```
      O
      ‖
      C4
     / \
   3N    C5
    |    ‖
   2C    C6
   ⁄ \1⁄
  O    N
```

FORMULA   C4H4N2O2   MOL WT   112.09
SOLVENT   C2D6OS/C2H6OS
ORIG ST   C4H12SI   TEMP   AMB

151.45   164.25   100.30   142.15
  2/1   4/1   5/2   6/2

P.D.ELLIS,R.B.DUNLAP,A.R.POLLARD,K.SEIDMAN,
A.D.CARDIN
J AM CHEM SOC   95, 4398 (1973)

2497   R 005200 5-HYDROXYURACIL

```
      O
      ‖
      C4   OH
     / \ /
   3N    C5
    |    ‖
   2C    C6
   ⁄ \1⁄
  O    N
```

FORMULA   C4H4N2O3   MOL WT   128.09
SOLVENT   C2D6OS/C2H6OS
ORIG ST   C4H12SI   TEMP   AMB

149.70   161.10   131.55   120.40
  2/1   4/1   5/1   6/2

P.D.ELLIS,R.B.DUNLAP,A.L.POLLARD,K.SEIDMAN,
A.D.CARDIN
J AM CHEM SOC   95, 4398 (1973)

2498   R 005200 5-AMINOURACIL

```
      O
      ‖
      C4   NH2
     / \ /
   3N    C5
    |    ‖
   2C    C6
   ⁄ \1⁄
  O    N
```

FORMULA   C4H5N3O2   MOL WT   127.10
SOLVENT   C2D6OS/C2H6OS
ORIG ST   C4H12SI   TEMP   AMB

149.50   161.40   116.40   121.60
  2/1   4/1   5/1   6/2

P.D.ELLIS,R.B.DUNLAP,A.L.POLLARD,K.SEIDMAN,
A.D.CARDIN
J AM CHEM SOC   95, 4398 (1973)

2499   R 005200 5-TRIFLUOROMETHYLURACIL

```
    O
    ||    7
   C4    CF3
   / \  /
  3N    C5
  |     ||
  2C    C6
  // \1/
  O    N
```

FORMULA   C5H3F3N2O2   MOL WT 180.09
SOLVENT   C2D6OS/C2H6OS
ORIG ST   C4H12SI   TEMP   AMB

150.40   159.70   101.85   143.45   122.80
 2/1   4/1   5/1   6/2   7/4

P.D.ELLIS,R.B.DUNLAP,A.R.POLLARD,K.SEIDMAN,
A.D.CARDIN
J AM CHEM SOC   95, 4398 (1973)

2500   R 005200 5-CYANOURACIL

```
    O
    ||    7
   C4    C≡N
   / \  /
  3N    C5
  |     ||
  2C    C6
  // \1/
  O    N
```

FORMULA   C5H3N3O2   MOL WT 137.10
SOLVENT   C2D6OS/C2H6OS
ORIG ST   C4H12SI   TEMP   AMB

149.85   160.90   87.00   151.75   114.50
 2/1   4/1   5/1   6/2   7/1

P.D.ELLIS,R.B.DUNLAP,A.R.POLLARD,K.SEIDMAN,
A.D.CARDIN
J AM CHEM SOC   95, 4398 (1973)

2501   R 005200 5-FORMYLURACIL

```
    O   O
    ||  ||
   C4   C7
   / \ / \
  3N   C5  H
  |    ||
  2C   C6
  // \1/
  O    N
```

FORMULA   C5H4N2O3   MOL WT 140.10
SOLVENT   C2D6OS/C2H6OS
ORIG ST   C4H12SI   TEMP   AMB

150.25   162.30   110.10   149.00   186.15
 2/1   4/1   5/1   6/2   7/2

P.D.ELLIS,R.B.DUNLAP,A.L.POLLARD,K.SEIDMAN,
A.D.CARDIN
J AM CHEM SOC   95, 4398 (1973)

2502   R 005200 URACIL-5-CARBOXYLIC ACID

```
    O   O
    ||  ||
   C4   C7
   / \ / \
  3N   C5  OH
  |    ||
  2C   C6
  // \1/
  O    N
```

FORMULA   C5H4N2O4   MOL WT 156.10
SOLVENT   C2D6OS/C2H6OS
ORIG ST   C4H12SI   TEMP   AMB

150.10   163.00   110.10   149.90   164.95
 2/1   4/1   5/1   6/2   7/1

P.D.ELLIS,R.B.DUNLAP,A.L.POLLARD,K.SEIDMAN,
A.D.CARDIN
J AM CHEM SOC   95, 4398 (1973)

2503   R 005200 5-METHYLURACIL

```
    O
    ||    7
   C4    CH3
   / \  /
  3N    C5
  |     ||
  2C    C6
  // \1/
  O    N
```

FORMULA   C5H6N2O2   MOL WT 126.12
SOLVENT   C2D6OS/C2H6OS
ORIG ST   C4H12SI   TEMP   AMB

151.15   164.60   107.45   137.40
 2/1   4/1   5/1   6/2   7/4

P.D.ELLIS,R.B.DUNLAP,A.L.POLLARD,K.SEIDMAN,
A.D.CARDIN
J AM CHEM SOC   95, 4398 (1973)

2504   R 005200 5-HYDROXYMETHYLURACIL

```
      O
      ‖    7
     C4   CH2-OH
    / \ /
  3N   C5
   |   ‖
  2C   C6
   ⁄ \1⁄
  O   N
```

FORMULA   C5H6N2O3   MOL WT   142.12
SOLVENT   C2D6OS/C2H6OS
ORIG ST   C4H12SI   TEMP   AMB

150.95	163.40	112.50	137.80	55.70
2/1	4/1	5/1	6/2	7/3

P.D.ELLIS,R.B.DUNLAP,A.L.POLLARD,K.SEIDMAN,
A.D.CARDIN
J AM CHEM SOC   95, 4398 (1973)

2505   R 005200 5-METHOXYURACIL

```
      O
      ‖    7
     C4   O-CH3
    / \ /
  3N   C5
   |   ‖
  2C   C6
   ⁄ \1⁄
  O   N
```

FORMULA   C5H6N2O3   MOL WT   142.12
SOLVENT   C2D6OS/C2H6OS
ORIG ST   C4H12SI   TEMP   AMB

150.70	160.80	135.90	122.60	58.20
2/1	4/1	5/1	6/2	7/4

P.D.ELLIS,R.B.DUNLAP,A.L.POLLARD,K.SEIDMAN,
A.D.CARDIN
J AM CHEM SOC   95, 4398 (1973)

2506   R 005200 METHYL URACIL-5-CARBOXYLATE

```
      O   O
      ‖   ‖
     C4   C7
    / \ / \  8
  3N   C5  O-CH3
   |   ‖
  2C   C6
   ⁄ \1⁄
  O   N
```

FORMULA   C6H6N2O4   MOL WT   170.13
SOLVENT   C2D6OS/C2H6OS
ORIG ST   C4H12SI   TEMP   AMB

150.40	159.90	103.00	149.40	163.05	51.50
2/1	4/1	5/1	6/2	7/1	8/4

P.D.ELLIS,R.B.DUNLAP,A.R.POLLARD,K.SEIDMAN,
A.D.CARDIN
J AM CHEM SOC   95, 4398 (1973)

2507   R 005200 1,3-DIMETHYL URACIL

```
    3 4
    C=C
   /   \
 O=C    N-CH3
  2\   / 5
    N-C
   / 1\
 CH3   O
  6
```

FORMULA   C6H8N2O2   MOL WT   140.14
SOLVENT   CDCL3
ORIG ST   C4H12SI   TEMP   AMB

151.80	163.20	100.90	143.20	27.50	36.80
1/1	2/1	3/2	4/2	5/4	6/4

L.F.JOHNSON,W.C.JANKOWSKI
CARBON-13 NMR SPECTRA,JOHN
WILEY AND SONS,NEW YORK   170 (1972)

2508   R 005200 DI-(5-URACILYL)-DISULFIDE

```
     O          O
     ‖          ‖
    C4  S-S    C
   / \ /   \ /
 3N   C5    C   N
  |   ‖     ‖   |
 2C   C6    C   C
  ⁄ \1⁄      \ /
 O   N       N   O
```

FORMULA   C8H6N4O4S2   MOL WT   286.29
SOLVENT   C2D6OS/C2H6OS
ORIG ST   C4H12SI   TEMP   AMB

150.85	161.90	106.45	147.60
2/1	4/1	5/1	6/2

P.D.ELLIS,R.B.DUNLAP,A.L.POLLARD,K.SEIDMAN,
A.D.CARDIN
J AM CHEM SOC   95, 4398 (1973)

2509   R 005400 6-BROMOPURINE

```
      BR
      |
      C6
    ╱ ╲5   7
  1N    C---N
   |    ‖   ‖
  2C    C   C8
    ╲ ╱4╲9╱
     N    N
        3
```

FORMULA   C5H3BRN4   MOL WT   199.01
SOLVENT   C2D6OS
ORIG ST   C4H12SI   TEMP   AMB

151.50   146.65   131.85   152.90   146.00
 2/2   4/1   5/1   6/1   8/2

E.BREITMAIER,W.VOELTER
TETRAHEDRON   30, 3941 (1974)

2510   R 005400 6-CHLOROPURINE

```
      CL
      |
      C6
    ╱ ╲5   7
  1N    C---N
   |    ‖   ‖
  2C    C   C8
    ╲ ╱4╲9╱
     N    N
        3
```

FORMULA   C5H3CLN4   MOL WT   154.56
SOLVENT   C2D6OS
ORIG ST   C4H12SI   TEMP   AMB

151.35   147.80   129.05   154.20   146.10
 2/2   4/1   5/1   6/1   8/2

E.BREITMAIER,W.VOELTER
TETRAHEDRON   30, 3941 (1974)

2511   R 005400 6-IODOPURINE

```
      J
      |
      C6
    ╱ ╲5   7
  1N    C---N
   |    ‖   ‖
  2C    C   C8
    ╲ ╱4╲9╱
     N    N
        3
```

FORMULA   C5H3IN4   MOL WT   246.01
SOLVENT   C2D6OS
ORIG ST   C4H12SI   TEMP   AMB

151.80   149.95   120.30   136.50   145.10
 2/2   4/1   5/1   6/1   8/2

E.BREITMAIER,W.VOELTER
TETRAHEDRON   30, 3941 (1974)

2512   R 005400 PURINE

```
      6
      C
    ╱ ╲5   7
  1N    C---N
   |    ‖   ‖
  2C    C4  C8
    ╲ ╱ ╲9╱
     N    N
        3
```

FORMULA   C5H4N4   MOL WT   120.11
SOLVENT   C2D6OS
ORIG ST   C4H12SI   TEMP   AMB

152.15   154.70   130.45   145.55   146.10
 2/2   4/1   5/1   6/2   8/2

E.BREITMAIER,W.VOELTER
TETRAHEDRON   30, 3941 (1974)

2513   R 005400 HYPOXANTHINE

```
      OH
      |
      C6
    ╱ ╲5   7
  1N    C---N
   |    ‖   ‖
  2C    C   C8
    ╲ ╱4╲9╱
     N    N
        3
```

FORMULA   C5H4N4O   MOL WT   136.11
SOLVENT   C2D6OS
ORIG ST   C4H12SI   TEMP   AMB

150.20   155.50   118.00   158.90   144.80
 2/2   4/1   5/1   6/1   8/2

E.BREITMAIER,W.VOELTER
TETRAHEDRON   30, 3941 (1974)

2514   R 005400   6-MERCAPTOPURINE

```
  10SH
   |
   C6
  ∥ \5   7
 1N    C---N
  |   ∥   ∥
 2C   C   C8
  \ /4\9/
   N   N
   3
```

FORMULA	C5H4N4S		MOL WT	152.18
SOLVENT	C2D6OS			
ORIG ST	C4H12SI		TEMP	AMB

144.80	150.60	128.80	171.25	144.80
2/2	4/1	5/1	6/1	8/2

E.BREITMAIER,W.VOELTER
TETRAHEDRON   30, 3941 (1974)

2515   R 005400   ADENINE

```
  1CNH2
   |
   C6
  ∥ \5   7
 1N    C---N
  |   ∥   ∥
 2C   C   C8
  \ /4\9/
   N   N
   3
```

FORMULA	C5H5N5		MOL WT	135.13
SOLVENT	C2D6OS			
ORIG ST	C4H12SI		TEMP	AMB

152.45	151.35	117.60	155.35	139.40
2/2	4/1	5/1	6/1	8/2

E.BREITMAIER,W.VOELTER
TETRAHEDRON   30, 3941 (1974)

2516   R 005400   6-CYANOPURINE

```
 N≡C10
   |
   C6
  ∥ \5   7
 1N    C---N
  |   ∥   ∥
 2C   C   C8
  \ /4\9/
   N   N
   3
```

FORMULA	C6H3N5		MOL WT	145.12
SOLVENT	C2D6OS			
ORIG ST	C4H12SI		TEMP	AMB

152.25	133.45	127.85	154.95	149.30	114.45
2/2	4/1	5/1	6/1	8/2	10/1

E.BREITMAIER,W.VOELTER
TETRAHEDRON   30, 3941 (1974)

2517   R 005400   6-METHYLPURINE

```
  10CH3
   |
   C6
  ∥ \5   7
 1N    C---N
  |   ∥   ∥
 2C   C   C8
  \ /4\9/
   N   N
   3
```

FORMULA	C6H6N4		MOL WT	134.14
SOLVENT	C2D6OS			
ORIG ST	C4H12SI		TEMP	AMB

151.60	153.85	135.10	155.45	144.45	19.40
2/2	4/1	5/1	6/1	8/2	10/4

E.BREITMAIER,W.VOELTER
TETRAHEDRON   30, 3941 (1974)

2518   R 005400   7-METHYLPURINE

```
   6      10
   C      CH3
  ∥ \5   /
 1N    C---N7
  |   ∥   |
 2C   C   C8
  \ /4\ /
   N   N
   3   9
```

FORMULA	C6H6N4		MOL WT	134.14
SOLVENT	C2H6OS			
ORIG ST	C4H12SI		TEMP	301

152.04	159.88	125.79	140.72	149.78	31.64
2/2	4/1	5/1	6/2	8/2	10/4

M.-T.CHENON,R.J.PUGMIRE,D.M.GRANT,R.P.PANZICA,
L.B.TOWNSEND
J AM CHEM SOC   97, 4627 (1975)

2519   R 005400 9-METHYLPURINE

```
      6
      C
    ∥ \5
  1N    C---N7
  I    ∥    ∥
  2C   C    C8
   \ /4\ /
    N   N9
    3   I
       CH3
       10
```

FORMULA   C6H6N4   MOL WT   134.14
SOLVENT   C2H6OS
ORIG ST   C4H12SI   TEMP   313

151.86   151.38   133.46   147.44   147.44   29.35
 2/2   4/1   5/1   6/2   8/2   10/4

M.-T.CHENON,R.J.PUGMIRE,D.M.GRANT,R.P.PANZICA,
L.B.TOWNSEND
J AM CHEM SOC   97, 4627 (1975)

2520   R 005400 6-METHOXYPURINE

```
    CH3
    /10
    O
    I
    C6
   ∥ \5   7
  1N   C---N
   I   ∥   ∥
  2C   C   C
   \ /4\9/8
   N3  N
```

FORMULA   C6H6N4O   MOL WT   150.14
SOLVENT   C2D6OS
ORIG ST   C4H12SI   TEMP   AMB

151.50   155.15   118.25   159.35   142.75   53.85
 2/2   4/1   5/1   6/1   8/2   10/4

E.BREITMAIER,W.VOELTER
TETRAHEDRON   30, 3941 (1974)

2521   R 005400 1-METHYLHYPOXANTHINE

```
        O
  10    ∥
  H3C   C6
   \ / \5   7
   1N   C---N
    I   ∥   ∥
   2C   C   C8
    \ /4\9/
    N   N
    3
```

FORMULA   C6H6N4O   MOL WT   150.14
SOLVENT   C2H6OS
ORIG ST   C4H12SI   TEMP   310

147.57   153.01   118.26   155.12   140.55   33.10
 2/2   4/1   5/1   6/1   8/2   10/4

M.-T.CHENON,R.J.PUGMIRE,D.M.GRANT,R.P.PANZICA,
L.B.TOWNSEND
J AM CHEM SOC   97, 4636 (1975)

2522   R 005400 7-METHYLHYPOXANTHINE

```
    O
    ∥        10
    C6       CH3
  1/ \5      /
   N   C---N7
   I   ∥   I
  2C   C   C8
   \ /4\ ∥
    N   N
    3   9
```

FORMULA   C6H6N4O   MOL WT   150.14
SOLVENT   C2H6OS
ORIG ST   C4H12SI   TEMP   308

144.37   157.02   115.48   154.63   144.37   33.31
 2/2   4/1   5/1   6/1   8/2   10/4

M.-T.CHENON,R.J.PUGMIRE,D.M.GRANT,R.P.PANZICA,
L.B.TOWNSEND
J AM CHEM SOC   97, 4627 (1975)

2523   R 005400 1-METHYL-6-MERCAPTOPURINE

```
       S
  10   ∥
  H3C  C6
   \ / \5   7
   1N   C---N
    I   ∥   ∥
   2C   C   C8
    \ /4\9/
    N   N
    3
```

FORMULA   C6H6N4S   MOL WT   166.21
SOLVENT   C2H6OS
ORIG ST   C4H12SI   TEMP   310

147.64   149.23   128.25   171.69   145.09   39.81
 2/2   4/1   5/1   6/1   8/2   10/4

M.-T.CHENON,R.J.PUGMIRE,D.M.GRANT,R.P.PANZICA,
L.B.TOWNSEND
J AM CHEM SOC   97, 4636 (1975)

2524   R 005400   6-METHYLMERCAPTOPURINE

```
        CH3
      10/11
       S
       |
       C6
     // \5    7
   1N    C---N
    |    ||  ||
   2C    C   C8
     \  /4\9/
      3N   N
```

FORMULA	C6H6N4S		MOL WT	166.21
SOLVENT	C2D6OS			
ORIG ST	C4H12SI		TEMP	AMB

151.60	158.60	129.35	150.20	143.15	11.35
2/2	4/1	5/1	6/1	8/2	11/4

E.BREITMAIER,W.VOELTER
TETRAHEDRON   30, 3941 (1974)

2525   R 005400   7-METHYLPURINE-6-THIONE

```
       S
       ||       10
       C6     CH3
     1/  \5   /
      N    C---N7
      |    ||  |
     2C    C   C8
       \  /4\ //
        N   N
        3   9
```

FORMULA	C6H6N4S		MOL WT	166.21
SOLVENT	C2H6OS			
ORIG ST	C4H12SI		TEMP	313

144.74	152.68	125.87	170.40	148.33	34.66
2/2	4/1	5/1	6/1	8/2	10/4

M.-T.CHENON,R.J.PUGMIRE,D.M.GRANT,R.P.PANZICA,
L.B.TOWNSEND
J AM CHEM SOC   97, 4627 (1975)

2526   R 005400   7-METHYLADENINE

```
      NH2
       |        10
       C6     CH3
     // \5    /
   1N    C---N7
    |    ||  |
   2C    C   C8
     \  /4\ //
      N   N
      3   9
```

FORMULA	C6H7N5		MOL WT	149.16
SOLVENT	C2H6OS			
ORIG ST	C4H12SI		TEMP	305

152.31	159.82	111.77	151.91	145.94	33.76
2/2	4/1	5/1	6/1	8/2	10/4

M.-T.CHENON,R.J.PUGMIRE,D.M.GRANT,R.P.PANZICA,
L.B.TOWNSEND
J AM CHEM SOC   97, 4627 (1975)

2527   R 005400   9-METHYLADENINE

```
      NH2
       |
       C6
     // \5
   1N    C---N7
    |    ||  ||
   2C    C   C8
     \  /4\ /
      N   N9
      3   |
         10CH3
```

FORMULA	C6H7N5		MOL WT	149.16
SOLVENT	C2H6OS			
ORIG ST	C4H12SI		TEMP	305

152.50	149.94	118.72	155.98	141.47	29.39
2/2	4/1	5/1	6/1	8/2	10/4

M.-T.CHENON,R.J.PUGMIRE,D.M.GRANT,R.P.PANZICA,
L.B.TOWNSEND
J AM CHEM SOC   97, 4627 (1975)

2528   R 005400   6-DIMETHYLAMINOPURINE

```
    H3C   CH3
     10\ /11
        N
        |
        C6
      // \5    7
    1N    C---N
     |    ||  ||
    2C    C   C8
      \  /4\9/
       N3   N
```

FORMULA	C7H9N5		MOL WT	163.18
SOLVENT	C2D6OS			
ORIG ST	C4H12SI		TEMP	AMB

151.80	151.15	118.90	154.30	137.80	37.85
2/2	4/1	5/1	6/1	8/2	10/4

E.BREITMAIER,W.VOELTER
TETRAHEDRON   30, 3941 (1974)

```
2529              R 005400  6-CHLORO-9-(1,4-OXATHIAN-2-YL)-9H-PURINE
           2 1
           C-N              FORMULA   C9H9CLN4OS              MOL WT   256.72
         ⁄⁄  =6             SOLVENT   CDCL3
      3N     C-CL           ORIG ST   C4H12SI                TEMP       AMB
       \   ⁄
        4C=C5                151.50   150.60   131.10   150.10   142.00   82.00
    10    ⁄   \               2/2      4/1      5/1      6/1      8/2     11/2
     ]  9N    \7             31.10    25.90    70.20
    ⁄ \ ⁄ \  ⁄⁄             12/3     14/3     15/3
  15C   C11 \ ⁄⁄
   |    |    C
  14C  C12  8               W.A.SZAREK,D.M.VYAS,A.-M.SEPULCHRE,S.D.GERO,
   \  ⁄                     G.LUKACS
    S13                     CAN J CHEM                        52, 2041 (1974)
```

```
2530              R 005400  6-CHLORO-9-(1,4-OXATHIAN-3-YL)-9H-PURINE
          2 1
          C-N               FORMULA   C9H9CLN4OS              MOL WT   256.72
    10   ⁄⁄  \6             SOLVENT   CDCL3
     O    3N    C-CL        ORIG ST   C4H12SI                TEMP       AMB
    ⁄ \   \   ⁄
  15C   C11 4C=C5            151.50   150.80   131.00   150.80   143.70   71.10
   |    |   ⁄   \             2/2      4/1      5/1      6/1      8/2     11/3
  14C   C--N    \7          49.40    24.10    68.10
   \  ⁄12 9\  ⁄⁄           12/2     14/3     15/3
    S    \ ⁄⁄
    13    C8               W.A.SZAREK,D.M.VYAS,A.-M.SEPULCHRE,S.D.GERO,
                           G.LUKACS
                           CAN J CHEM                        52, 2041 (1974)
```

```
2531              R 005400  9-(BETA-D-RIBOFURANOSYL)PURINE   NEBULARINE
      C6
     ⁄ \5              FORMULA   C10H12N4O4              MOL WT   252.23
   1N   C---N7         SOLVENT   C2H6OS
    |   |   ||         ORIG ST   C4H12SI                TEMP       313
   2C   C   C8
    \ ⁄4\ ⁄            152.22   151.09   134.28   148.32   143.53   87.73
     N3   N9            2/2      4/1      5/1      6/2      8/2     10/2
          |            73.94    70.47    85.86    61.42
   14     |           11/2     12/2     13/2     14/3
 HO-H2C O |
   |⁄ \|
   13C   C10
    |    |
  12C---C11           M.-T.CHENON,R.J.PUGMIRE,D.M.GRANT,R.P.PANZICA,
    |    |            L.B.TOWNSEND
    OH   OH           J AM CHEM SOC                     97, 4627 (1975)
```

```
2532              R 005400  7-(BETA-D-RIBOFURANOSYL)PURINE-6-THIONE
        OH  OH
         |   |        FORMULA   C10H12N4O4S             MOL WT   284.30
    S   11C---C12      SOLVENT   C2H6OS
    ||    |   |        ORIG ST   C4H12SI                TEMP       312
    C6  10C   C13
   1⁄ \5  |\ ⁄|        144.99   153.34   125.32   169.83   144.99   89.14
   N   C---N ] CH2      2/2      4/1      5/1      6/1      8/2     10/2
   |   ||  7|   |14    75.75    68.98    84.68    60.36
   2C   C   C8  OH     11/2     12/2     13/2     14/3
    \ ⁄4\ ⁄⁄
     N   N            M.-T.CHENON,R.J.PUGMIRE,D.M.GRANT,R.P.PANZICA,
     3   9            L.B.TOWNSEND
                      J AM CHEM SOC                     97, 4627 (1975)
```

2533

```
        S
        ||
        C6
      1/  \5
      N   C---N7
      |   ||   ||
     2C    C   C8
       \\  /4\ /
        N3  N9
            |
       14   |
   HO-H2C ] |
       |/ \|
      13C   C10
       |    |
      12C---C11
       |    |
       ]H   OH
```

R 005400   9-(BETA-D-RIBOFURANOSYL)PURINE-6-THIONE

FORMULA   C10H12N4O4S   MOL WT   284.30
SOLVENT   C2H6OS
ORIG ST   C4H12SI   TEMP   313

145.49	144.12	135.61	176.19	141.45	87.93
2/2	4/1	5/1	6/1	8/2	10/2
74.55	70.45	85.92	61.39		
11/2	12/2	13/2	14/3		

M.-T.CHENON,R.J.PUGMIRE,D.M.GRANT,R.P.PANZICA,
L.B.TOWNSEND
J AM CHEM SOC   97, 4627 (1975)

2534

```
       OH   OH
       |    |
    ]  11C---C11
    ||   |    |
    C6  10C   C13
   1/ \5  |\ /|
   N   C---N ] CH2
   |   ||  7|   |14
  2C    C   C8  OH
    \\  /4\ //
     N    N
     3    9
```

R 005400   7-(BETA-D-RIBOFURANOSYL)HYPOXANTHINE

FORMULA   C10H12N4O5   MOL WT   268.23
SOLVENT   C2H6OS
ORIG ST   C4H12SI   TEMP   310

144.80	157.72	114.79	154.14	142.49	89.47
2/2	4/1	5/1	6/1	8/2	10/2
75.15	69.78	85.48	61.06		
11/2	12/2	13/2	14/3		

M.-T.CHENON,R.J.PUGMIRE,D.M.GRANT,R.P.PANZICA,
L.B.TOWNSEND
J AM CHEM SOC   97, 4627 (1975)

2535

```
        ]
        ||
        C6
      1/  \5
      N   C---N7
      |   ||   ||
     2C    C   C8
       \\  /4\ /
        N3  N9
            |
       14   |
   HO-H2C ] |
       |/ \|
      13C   C10
       |    |
      12C---C11
       |    |
       ]H   OH
```

R 005400   9-(BETA-D-RIBOFURANOSYL)HYPOXANTHINE   INOSINE

FORMULA   C10H12N4O5   MOL WT   268.23
SOLVENT   C2H6OS
ORIG ST   C4H12SI   TEMP   310

146.19	148.46	124.63	156.88	139.11	87.82
2/2	4/1	5/1	6/1	8/2	10/2
74.43	70.59	85.91	61.58		
11/2	12/2	13/2	14/3		

M.-T.CHENON,R.J.PUGMIRE,D.M.GRANT,R.P.PANZICA,
L.B.TOWNSEND
J AM CHEM SOC   97, 4627 (1975)

2536

```
       OH   OH
       |    |
    NH2 11C---C12
     |   |    |
    C6  10C   C13
   / \5  |\ /|
  1N   C---N ] CH2
   |   ||  7|   |14
  2C    C   C8  OH
    \\  /4\ //
     N    N
     3    9
```

R 005400   7-(BETA-D-RIBOFURANOSYL)ADENINE

FORMULA   C10H13N5O4   MOL WT   267.25
SOLVENT   C2H6OS
ORIG ST   C4H12SI   TEMP   310

152.85	160.74	110.27	151.73	144.64	89.47
2/2	4/1	5/1	6/1	8/2	10/2
75.05	69.03	86.45	60.54		
11/2	12/2	13/2	14/3		

M.-T.CHENON,R.J.PUGMIRE,D.M.GRANT,R.P.PANZICA,
L.B.TOWNSEND
J AM CHEM SOC   97, 4627 (1975)

2537

```
        NH2
         |
         C6
       ⁄⁄ \5
    1N    C---N7
     |    ||   ||
    2C    C    C8
      \  ⁄4\  ⁄
       N3    N9
              |
       14     |
    HO-H2C  O |
        |⁄  \|
       13C    C10
        |     |
       12C---C11
        |     |
        OH   OH
```

R 005400 9-(BETA-D-RIBOFURANOSYL)ADENINE   ADENOSINE

FORMULA   C10H13N5O4   MOL WT   267.25
SOLVENT   C2H6OS
ORIG ST   C4H12SI   TEMP   310

152.60	149.27	119.57	156.30	140.20	88.24
2/2	4/1	5/1	6/1	8/2	10/2
73.77	70.91	86.16	61.91		
11/2	12/2	13/2	14/3		

M.-T.CHENON,R.J.PUGMIRE,D.M.GRANT,R.P.PANZICA,
L.B.TOWNSEND
J AM CHEM SOC   97, 4627 (1975)

2538

```
        S
   10   ||
   H3C  C6
      \ ⁄ \5
    1N    C---N7
     |    ||   ||
    2C    C    C8
      \  ⁄4\  ⁄
       N3    N9
              |
       15     |
    HO-H2C  O |
        |⁄  \|
       14C    C11
        |     |
       13C---C12
        |     |
        OH   OH
```

R 005400 1-METHYL-9-(BETA-D-RIBOFURANOSYL)PURINE-
6-THIONE

FORMULA   C11H14N4O4S   MOL WT   298.32
SOLVENT   C2H6OS
ORIG ST   C4H12SI   TEMP   312

148.49	142.01	135.78	177.43	141.60	40.42
2/2	4/1	5/1	6/1	8/2	10/4
87.67	74.32	70.28	35.78	61.25	
11/2	12/2	13/2	14/2	15/3	

M.-T.CHENON,R.J.PUGMIRE,D.M.GRANT,R.P.PANZICA,
L.B.TOWNSEND
J AM CHEM SOC   97, 4627 (1975)

2539

```
        CH3
        ⁄10
        S
        |
        C6
       ⁄⁄ \5
    1N    C---N7
     |    ||   ||
    2C    C    C8
      \  ⁄4\  ⁄
       N3    N9
              |
       15     |
    HO-H2C  O |
        |⁄  \|
       14C    C11
        |     |
       13C---C12
        |     |
        OH   OH
```

R 005400 6-METHYLTHIO-9-(BETA-D-RIBOFURANOSYL)PURINE

FORMULA   C11H14N4O4S   MOL WT   298.32
SOLVENT   C2H6OS
ORIG ST   C4H12SI   TEMP   311

151.53	148.01	131.34	160.46	143.06	11.26
2/2	4/1	5/1	6/1	8/2	10/4
88.02	73.96	70.38	85.83	61.37	
11/2	12/2	13/2	14/2	15/3	

M.-T.CHENON,R.J.PUGMIRE,D.M.GRANT,R.P.PANZICA,
L.B.TOWNSEND
J AM CHEM SOC   97, 4627 (1975)

2540

```
        O
   10   ||
   H3C  C6
      \ ⁄ \5
    1N    C---N7
     |    ||   ||
    2C    C    C8
      \  ⁄4\  ⁄
       N3    N9
              |
       15     |
    HO-H2C  O |
        |⁄  \|
       14C    C11
        |     |
       13C---C12
        |     |
        OH   OH
```

R 005400 1-METHYL-9-(BETA-D-RIBOFURANOSYL)HYPOXANTHINE

FORMULA   C11H14N4O5   MOL WT   282.26
SOLVENT   C2H6OS
ORIG ST   C4H12SI   TEMP   313

148.77	147.63	123.68	156.42	139.22	33.56
2/2	4/1	5/1	6/1	8/2	10/4
87.56	74.20	70.41	87.73	61.41	
11/2	12/2	13/2	14/2	15/3	

M.-T.CHENON,R.J.PUGMIRE,D.M.GRANT,R.P.PANZICA,
L.B.TOWNSEND
J AM CHEM SOC   97, 4627 (1975)

```
2541        CH3        R 005400  6-METHOXY-9-(BETA-D-RIBOFURANOSYL)PURINE
           /10
           O                FORMULA   C11H14N4O5              MOL WT   282.26
           |                SOLVENT   C2H6OS
           C6               ORIG ST   C4H12SI                 TEMP        313
          // \5
       1N    C---N7          151.67  151.83  121.24  160.46  142.33   54.03
       |     ||   ||           2/2     4/1     5/1     6/1     8/2    10/4
      2C     C    C8          87.80   73.88   70.59   85.82   61.40
        \  /4\  /             11/2    12/2    13/2    14/2    15/3
        N3   N9
             |
       15    |
    HO-H2C O |
       | / \ |
      14C    C11             M.-T.CHENON,R.J.PUGMIRE,D.M.GRANT,R.P.PANZICA,
       |     |               L.B.TOWNSEND
      13C---C12              J AM CHEM SOC                    97, 4627 (1975)
       |     |
       OH    OH
```

```
2542                R 005500  PIPERAZINE

                          FORMULA   C4H10N2                MOL WT    86.14
          N               SOLVENT   C4H10N2
         /4\              ORIG ST   C6H12                  TEMP       AMB
       5C   C3
        |   |             46.95
       6C   C2             2/3
        \1/
         N
                          G.ELLIS,R.G.JONES
                          J CHEM SOC PERKIN TRANS II         437 (1972)
```

```
2543                R 005500  2-METHYLPYRAZINE
        2 1 5
        C=C-CH3           FORMULA   C5H6N2                 MOL WT    94.12
       /    \             SOLVENT   CDCL3
      N      N            ORIG ST   C4H12SI                TEMP       AMB
       \\   //
        C-C              154.00   21.60  141.80  143.80  144.70
        3 4               1/1      5/4
                          L.F.JOHNSON,W.C.JANKOWSKI
                          CARBON-13 NMR SPECTRA,JOHN
                          WILEY AND SONS,NEW YORK            108 (1972)
```

```
2544                R 005500  N-METHYLPIPERAZINE

                          FORMULA   C5H12N2                MOL WT   100.16
          N               SOLVENT   C5H12N2
         /4\              ORIG ST   C6H12                  TEMP       AMB
       5C   C3
        |   |             56.05   45.95   46.55
       6C   C2             2/3     3/3     7/4
        \1/
         N
         |
        7CH3              G.ELLIS,R.G.JONES
                          J CHEM SOC PERKIN TRANS II         437 (1972)
```

2545   R 005500   2-METHYLPIPERAZINE

```
      N
     /4\
   5C    C3
   |     |
   6C    C2
    \1/ \
     N    CH3
          7
```

FORMULA	C5H12N2				MOL WT	100.16
SOLVENT	C5H12N2					
ORIG ST	C6H12				TEMP	AMB

| 51.45 | 53.55 | 47.05 | 46.05 | 19.65 |
| 2/2 | 3/3 | 5/3 | 6/3 | 7/4 |

G.ELLIS,R.G.JONES
J CHEM SOC PERKIN TRANS II   437 (1972)

2546   R 005500   2,5-DIMETHYLPYRAZINE

```
    8      4
   H3C    N
     \ /
    5C    C3
    |     ||
    6C    C2
     \ /  \
      N    CH3
      1    7
```

FORMULA	C6H8N2		MOL WT	108.14
SOLVENT	C6H8N2			
ORIG ST	CS2		TEMP	AMB

| 150.10 | 142.40 | 20.30 |
| 2/1 | 3/2 | 7/4 |

P.C.LAUTERBUR
J CHEM PHYS   43, 360 (1965)

2547   R 005500   2,5-DIMETHYLPIPERAZINE

```
    8
   H3C    N
     \ /4\
    5C    C3
    |     |
    6C    C2
     \1/ \
      N    CH3
           7
```

FORMULA	C6H14N2		MOL WT	114.19
SOLVENT	C6H14N2			
ORIG ST	C6H12		TEMP	AMB

| 50.75 | 54.55 | 19.25 |
| 2/2 | 3/3 | 7/4 |

G.ELLIS,R.G.JONES
J CHEM SOC PERKIN TRANS II   437 (1972)

2548   R 005500   2,6-DIMETHYLPIPERAZINE

```
       3
       CH3
    2 1/
    C-C
   /     \
   N      N
    \     /
    C-C
    5 4\
       CH3
       6
```

FORMULA	C6H14N2				MOL WT	114.19
SOLVENT	CDCL3					
ORIG ST	C4H12SI				TEMP	AMB

| 52.10 | 53.30 | 19.90 | 52.10 | 53.30 | 19.90 |
| 1/2 | 2/3 | 3/4 | 4/2 | 5/3 | 6/4 |

L.F.JOHNSON,W.C.JANKOWSKI
CARBON-13 NMR SPECTRA,JOHN
WILEY AND SONS,NEW YORK   211 (1972)

2549   R 005500   2,6-DIMETHYLPIPERAZINE

```
      N
     /4\
   5C    C3
   |     |
   5C    C2
    / \1/ \
  H3C   N   CH3
   8         7
```

FORMULA	C6H14N2		MOL WT	114.19
SOLVENT	C6H14N2			
ORIG ST	C6H12		TEMP	AMB

| 52.95 | 51.85 | 19.55 |
| 2/2 | 3/3 | 7/4 |

G.ELLIS,R.G.JONES
J CHEM SOC PERKIN TRANS II   437 (1972)

```
2550                  R 005500  N,N'-DIMETHYLPIPERAZINE
        8CH3
          |
          N                   FORMULA   C6H14N2            MOL WT   114.19
         /4\                  SOLVENT   C6H14N2
       5C   C3                ORIG ST   C6H12              TEMP       AMB
        |   |
       6C   C2                   54.75
         \1/                      2/3
          N
          |
        7CH3                  G.ELLIS,R.G.JONES
                              J CHEM SOC PERKIN TRANS II        437 (1972)
```

```
2551                  R 005500  CIS-2,5-DIMETHYLPIPERAZINE
         5
        H3C                   FORMULA   C6H14N2            MOL WT   114.19
          \1 2               SOLVENT   CDCL3
           C-C                ORIG ST   C4H12SI            TEMP       AMB
          /   \
         N     N              49.30  49.70  49.30  49.70  18.30  18.30
          \   /                1/2    2/3    3/2    4/3    5/4    6/4
           C-C
         4 3\6                L.F.JOHNSON,W.C.JANKOWSKI
            CH3               CARBON-13 NMR SPECTRA,JOHN
                             WILEY AND SONS,NEW YORK          210 (1972)
```

```
2552                  R 005500  N-(2-HYDROXYETHYL)-PIPERAZINE
                              FORMULA   C6H14N2O           MOL WT   130.19
          N                   SOLVENT   C6H14N2O
         /4\                  ORIG ST   C6H12              TEMP       AMB
       5C   C3
        |   |                  54.55  45.45  58.05  61.05
       6C   C2                  2/3    3/3    7/3    8/3
         \1/
          N
          |   8               G.ELLIS,R.G.JONES
        7CH2-CH2-OH           J CHEM SOC PERKIN TRANS II        437 (1972)
```

```
2553                  R 005500  1,2,4-TRIMETHYLPIPERAZINE
        9CH3
          |
          N                   FORMULA   C7H16N2            MOL WT   128.22
         /4\                  SOLVENT   C7H16N2
       5C   C3                ORIG ST   C6H12              TEMP       AMB
        |   |
       6C   C2                 62.55  56.95  55.05  55.35  41.95
         \1/ \                  2/2    3/3    5/3    6/3    7/4
          N   CH3             45.45
          |   8                9/4
        7CH3                  G.ELLIS,R.G.JONES
                              J CHEM SOC PERKIN TRANS II        437 (1972)
```

```
2554                  R 005500  QUINOXALINE
        5C-C
        // 4\                 FORMULA   C8H6N2             MOL WT   130.15
       6C    C3               SOLVENT   CDCL3
         \  /                 ORIG ST   C4H12SI            TEMP       AMB
         7C=C2
         /   \                144.80  142.80  129.60  129.40  129.40  129.60
        N     N                1/2     2/1     3/2     4/2     5/2     6/2
         \   //               142.80  144.80
        8C-C                   7/1     8/2
          1
                              L.F.JOHNSON,W.C.JANKOWSKI
                              CARBON-13 NMR SPECTRA,JOHN
                             WILEY AND SONS,NEW YORK          282 (1972)
```

2555   R 005600 PTERIDINE

```
    4C    N5
    ⁄ \   ⁄ \
  3N   C10 C6
   |    ||   |
  2C    C   C7
    \ ⁄9\ ⁄
    1N    N8
```

FORMULA   C6H4N4   MOL WT   132.13
SOLVENT   C2H6OS
ORIG ST   C4H12SI   TEMP   AMB

164.60   159.50   149.80   154.50   154.50   135.50
 2/2   4/2   6/2   7/2   9/1   10/1

U.EWERS,H.GUENTHER,L.JAENICKE
CHEM BER   106, 3951 (1973)

2556   R 005600 PTERIDINE

```
    4C    N5
    ⁄ \   ⁄ \
  3N   C10 C6
   |    ||   |
  2C    C   C7
    \ ⁄9\ ⁄
    1N    N8
```

FORMULA   C6H4N4   MOL WT   132.13
SOLVENT   CDCL3
ORIG ST   C4H12SI   TEMP   AMB

163.70   159.00   147.80   152.40   152.40   135.00
 2/2   4/2   6/2   7/2   9/1   10/1

U.EWERS,H.GUENTHER,L.JAENICKE
CHEM BER   106, 3951 (1973)

2557   R 005600 6,7-DIOXO-5,6,7,8-TETRAHYDROPTERIDINE

FORMULA   C6H4N4O2   MOL WT   164.12
SOLVENT   C2H6OS
ORIG ST   C4H12SI   TEMP   AMB

```
    4C    N5   O
    ⁄ \   ⁄ \ ⁄
  3N   C10 C6
   |    ||   |
  2C    C   C7
    \ ⁄9\ ⁄ \
    1N    N8   O
```

151.60   141.10   154.20   156.40   145.00   120.60
 2/2   4/2   6/1   7/1   9/1   10/1

U.EWERS,H.GUENTHER,L.JAENICKE
CHEM BER   106, 3951 (1973)

2558   R 005600 2,4-DIAMINOPTERIDINE

```
     NH2
      |
    4C    N5
    ⁄ \   ⁄ \
  3N   C10 C6
   |    ||   |
  2C    C   C7
   ⁄ \ ⁄9\ ⁄
 H2N  1N    N8
```

FORMULA   C6H6N6   MOL WT   162.15
SOLVENT   C2H6OS
ORIG ST   C4H12SI   TEMP   AMB

164.20   164.20   140.60   151.30   160.50   124.40
 2/1   4/1   6/2   7/2   9/1   10/1

U.EWERS,H.GUENTHER,L.JAENICKE
CHEM BER   106, 3951 (1973)
```
```

2559   R 005600 FOLIC ACID

```
        O
        ‖       11
    4C   N5  CH2 H
   ╱ ╲ ╱ ╲ ╱ ╲ ╱
  3N   C10 C6  N
  |    ‖   |   |
  2C   C   C7  C12 O   OH
  ╱ ╲ ╱9╲ ╱   ╱ ╲   ╲ ╱
 H2N  1N  N8  C17 C13 C23
           |   ‖   |
          16C 14C 22CH2
            ╲ ╱   ╱
          15C  21CH2
            |    |
          18C  19CH  OH
           ╱ ╲ ╱ ╲ ╱
          O   N   C20
              |   ‖
              H   O
```

FORMULA	C19H19N7O6			MOL WT	441.41
SOLVENT	C2H6OS				
ORIG ST	C4H12SI			TEMP	AMB

157.40	162.50	150.00	150.00	155.10	129.20
2/1	4/1	6/1	7/2	9/1	10/1
40.50	152.10	112.50	130.30	122.60	167.70
11/3	12/1	13/2	14/2	15/1	18/1
53.00	175.20	27.40	31.70	175.00	
19/2	20/1	21/3	22/3	23/1	

SIGNALS OF 11 AND C2H6SO OVERLAP (40.50)

U.EWERS,H.GUENTHER,L.JAENICKE
CHEM BER   106, 3951 (1973)

2560   R 005600 AMETHOPTERINE

```
   NH2
    |       11  24
    4C   N5  CH2 CH3
   ╱ ╲ ╱ ╲ ╱ ╲ ╱
  3N   C10 C6  N
  |    ‖   |   |
  2C   C   C7  C12 O   OH
  ╱ ╲ ╱9╲ ╱   ╱ ╲   ╲ ╱
 H2N  1N  N8  C17 C13 C23
           |   ‖   |
          16C 14C 22CH2
            ╲ ╱   ╱
          15C  21CH2
            |    |
          18C  19CH  OH
           ╱ ╲ ╱ ╲ ╱
          O   N   C20
              |   ‖
              H   O
```

FORMULA	C20H22N8O5			MOL WT	454.45
SOLVENT	C2H6OS				
ORIG ST	C4H12SI			TEMP	300

164.00	163.50	148.20	150.30	154.60	122.80
2/1	4/1	6/1	7/2	9/1	10/1
40.50	152.10	112.30	130.30	122.50	167.70
11/3	12/1	13/2	14/2	15/1	18/1
53.00	175.40	27.40	31.70	175.00	40.50
19/2	20/1	21/3	22/3	23/1	24/4

SIGNALS OF 11, 24 AND C2H6SO OVERLAP (40.50)

U.EWERS,H.GUENTHER,L.JAENICKE
CHEM BER   106, 3951 (1973)

2561   R 006000 N—ETHYLMORPHOLINE

```
        3 4
        C—C
       ╱   ╲
H3C—CH2—N    O
  6 5  ╲   ╱
        C—C
        2 1
```

FORMULA	C6H13NO			MOL WT	115.18
SOLVENT	CDCL3				
ORIG ST	C4H12SI			TEMP	AMB

53.40	66.90	66.90	53.40	52.70	11.70
1/3	2/3	3/3	4/3	5/3	6/4

L.F.JOHNSON,W.C.JANKOWSKI
CARBON—13 NMR SPECTRA,JOHN
WILEY AND SONS,NEW YORK   202 (1972)

2562   R 006000 2—AMINOPHENOXAZONE—3

```
    9  10   1
    C   N   C   NH2
   ╱ ╲ ╱ ╲ ╱ ╲ ╱
  8C   C13 C14 C2
  |    ‖   |   |
  7C   C12 C11 C3
   ╲ ╱ ╲ ╱ ╲ ╱ ╲
    C   O   C   O
    6   5   4
```

FORMULA	C12H8N2O2			MOL WT	212.21
SOLVENT	C2D6OS				
ORIG ST	C4H12SI			TEMP	300

103.70	149.10	180.40	98.50	116.20	128.10
1/2	2/1	3/1	4/2	6/2	7/2
129.00	125.50	147.50	142.10	133.90	148.50
8/2	9/2	11/1	12/1	13/1	14/1

U.HOLLSTEIN,E.BREITMAIER,G.JUNG
J AM CHEM SOC   96, 8036 (1974)

2563   R 006000 ACTINOCIN

```
   HO   O  HC   O
    \  //    \  //
     C18      C15
     |   10   |
     C9   N   C1  NH2
    // \ / \ / \ /
  8C   C13 C14 C2
    |   ||   |   |
  7C   C12 C11 C3
    \  / \  / \  \
     C6  O5  C4  O
     |            |
   17CH3    16CH3
```

FORMULA C16H12N2O6				MOL WT	328.28
SOLVENT C2D6OS					
ORIG ST C4H12SI				TEMP	300

108.75	151.15	177.05	113.50	128.05	126.35
1/1	2/1	3/1	4/1	6/1	7/2
128.95	129.45	145.00	140.15	129.45	146.10
8/2	9/1	11/1	12/1	13/1	14/1
165.10	6.70	13.80	168.00		
15/1	16/4	17/4	18/1		

U.HOLLSTEIN,E.BREITMAIER,G.JUNG
J AM CHEM SOC   96, 8036 (1974)

2564   R 006500 TRIMETHYL-SYM-TRIAZINE

```
    8CH3
     |
     C4
    // \
   5N   N3
    |    ||
   6C    C2
   / \  / \
 H3C   N   CH3
  9    1    7
```

FORMULA C6H9N3	MOL WT	123.16
SOLVENT CHCL3		
ORIG ST CS2	TEMP	AMB

175.60	25.10
2/1	7/4

P.C.LAUTERBUR
J CHEM PHYS   43, 360 (1965)

2565   R 006700 DIMETHYL-SYM-TETRAZINE

```
    7CH3
     |
     C3
    // \
   4N   N2
    |    ||
   5N   N1
    \   /
     C6
     |
    8CH3
```

FORMULA C4H6N4	MOL WT	110.12
SOLVENT CHCL3		
ORIG ST CS2	TEMP	AMB

166.30	20.50
3/1	7/4

P.C.LAUTERBUR
J CHEM PHYS   43, 360 (1965)

2566   R 006700 3,5-DICHLORO-1-DIISOPROPYLAMINO-1H-1,2,4,6-THIA(4)TRIAZINE

```
        9
      CH3      CL
  10  8/    2 3/
  H3C-HC    N-C
      \  1/    \
       7N-S     N4
       /  \    /
   H3C-HC   6N=C5
   12 11\      \
        CH3     CL
        13
```

FORMULA C8H14CL2N4S	MOL WT	269.20
SOLVENT CDCL3		
ORIG ST C4H12SI	TEMP	298

164.42	48.99	23.26
3/1	8/2	9/4

W.STOREK,W.SCHRAMM,G.VOSS,G.REMBARZ,E.FISCHER
Z CHEM   15, 104 (1975)

2567   R 006700 5-AMINO-3-CHLORO-1-DIISOPROPYLAMINO-1H-1,2,4,6-THIA(4)TRIAZINE

```
        9
      CH3      CL
  10  8/    2 3/
  H3C-HC    N-C
     \7  1/    \
      N-S       N4
      /  \    /
  H3C-HC   6N=C5
  12 11\      \
       CH3     NH2
       13
```

FORMULA C8H16CLN5S	MOL WT	249.77
SOLVENT CDCL3		
ORIG ST C4H12SI	TEMP	298

163.04	118.67	46.41	18.98
3/1	5/1	8/2	9/4

W.STOREK,W.SCHRAMM,G.VOSS,G.REMBARZ,E.FISCHER
Z CHEM   15, 104 (1975)

2568   R 006700   3-CHLORO-1-DIISOPROPYLAMINO-5-METHYLTHIO-1H-1,2,4,6-THIA(4)TRIAZINE

```
          9
         CH3      CL
   10  8/      2 3/
  H3C-HC       N-C
        \7 1/       \
        N-S         N4
       /   \      /
  H3C-HC    6N=C 5
   12 11\        \
        CH3       S-CH3
        13        14 15
```

FORMULA	C9H17CLN4S2		MOL WT	280.84
SOLVENT	CDCL3			
ORIG ST	C4H12SI		TEMP	298

163.38	159.23	46.84	23.33	45.67	20.70
3/1	5/1	8/2	9/4	11/2	15/4

W.STOREK,W.SCHRAMM,G.VOSS,G.REMBARZ,E.FISCHER
Z CHEM   15, 104 (1975)

2569   R 006700   1-DIISOPROPYLAMINO-3,5-DI(METHYLTHIO)-1H-1,2,4,6-THIA(4)TRIAZINE

```
          9          14 15
         CH3         S-CH3
   10  8/      2 3/
  H3C-HC       N-C
        \7 1/       \
        N-S    .   N4
       /   \      /
  H3C-HC    6N=C 5
   12 11\        \
        CH3       S-CH3
        13        16 17
```

FORMULA	C10H20N4S3		MOL WT	292.49
SOLVENT	CDCL3			
ORIG ST	C4H12SI		TEMP	298

174.08	47.30	23.09	12.51
3/1	8/2	9/4	15/4

W.STOREK,W.SCHRAMM,G.VOSS,G.REMBARZ,E.FISCHER
Z CHEM   15, 104 (1975)

2570   R 006700   3-CHLORO-5-N,N-DIETHYLDITHIOCARBAMYL-1-DIISOPROPYLAMINO-1H-1,2,4,6-THIA(4)TRIAZINE

```
          9
         CH3      CL
   10  8/      2 3/
  H3C-HC       N-C
        \ 1/       \
        N-S         N4
       /   \      /
  H3C-HC    6N=C 5         17  18
   12 11\        \        CH2-CH3
        CH3       S-C-N16
        13        14 ||   \
                  S    CH2-CH3
                       19 20
```

FORMULA	C13H24CLN5S3		MOL WT	382.01
SOLVENT	CDCL3			
ORIG ST	C4H12SI		TEMP	298

163.31	173.90	48.64	23.49	23.37	182.17
3/1	5/1	8/2	9/4	12/4	15/1
50.27	13.72	48.46	10.60		
17/3	18/4	19/3	20/4		

W.STOREK,W.SCHRAMM,G.VOSS,G.REMBARZ,E.FISCHER
Z CHEM   15, 104 (1975)

2571   R 006900   CAPROLACTAM

```
          2
          C
        / \   O
       /   \1/
      3C     C
      |      |
      |      |
      4C     N
       \    /
        C-C
        5 6
```

FORMULA	C6H11NO		MOL WT	113.16
SOLVENT	CDCL3			
ORIG ST	C4H12SI		TEMP	AMB

179.50	36.80	23.20	29.70	30.60	42.60
1/1	2/3	3/3	4/3	5/3	6/3

L.F.JOHNSON,W.C.JANKOWSKI
CARBON-13 NMR SPECTRA,JOHN
WILEY AND SONS,NEW YORK   185 (1972)

2572   R 006900   5,14-DIHYDRODIBENZO(B,I)5,9,14,18-TETRAAZA-(14)ANNULENE

```
            C16
          / \\
     15C    C17
   13  ||   |   1
    C 14N  N18 C
   //  \ /    \ / \\
  12C   C22   C19 C2
   |    ||     || |
  11C   C21   C20 C3
   \\  / \    / \ //
    C  9N  N5  C4
   10  |   ||
      8C   C6
       \  /
```

FORMULA	C18H16N4		MOL WT	288.36
SOLVENT	C6D18N3OP			
ORIG ST	C4H12SI		TEMP	300

113.82	123.97	147.13	96.13	137.35
1/2	2/2	6/2	7/2	19/1

E.LORCH,E.BREITMAIER
CHEMIKER-ZTG   99, 87 (1975)

2573

R 006900   7,16-DIMETHYL-5,14-DIHYDRODIBENZO(B,I)-
5,9,14,18-TETRAAZA(14)ANNULENE

```
        24CH3
          |
         C16
        / \
   15C    C17
13  ||   |   1
 C  14N   N18 C
   / \ /   \ / \
12C   C22   C19 C2
 |    ||    ||  |
11C   C21   C20 C3
 \\ / \     / \ //
  C  9N   N5  C
 10  |    ||   4
   8C    C6
    \ /
    C7
     |
   23CH3
```

FORMULA	C20H20N4			MOL WT	316.41
SOLVENT	C6D18N3OP				
ORIG ST	C4H12SI			TEMP	343

113.50	123.54	147.38	102.50	137.46	17.50
1/2	2/2	6/2	7/1	19/1	23/4

E.LORCH,E.BREITMAIER
CHEMIKER-ZTG   99,   87 (1975)

2574

R 006900   7,16-DIETHYL-5,14-DIHYDRODIBENZO(B,I)-
5,9,14,18-TETRAAZA(14)ANNULENE

```
        26CH3
         /
       H2C25
         |
        C16
       / \
  15C    C17
13  ||   |   1
 C  14N   N18 C
   / \ /   \ / \
12C   C22   C19 C2
 |    ||    ||  |
11C   C21   C20 C3
 \\ / \     / \ //
  C  9N   N5  C
 10  |    ||   4
   8C    C6
    \ /
    C7
     |
   23CH2
     /
   H3C24
```

FORMULA	C22H24N4			MOL WT	344.46
SOLVENT	CDCL3				
ORIG ST	C4H12SI			TEMP	303

113.50	124.15	146.70	110.20	137.90	26.40
1/2	2/2	6/2	7/1	19/1	23/3
16.90					
24/4					

E.LORCH,E.BREITMAIER
CHEMIKER-ZTG   99,   87 (1975)

2575

R 006900   7,16-DIBUTYL-5,14-DIHYDRODIBENZO(B,I)-
5,9,14,18-TETRAAZA(14)ANNULENE

```
     28CH2 CH3
     27/ \ /30
     H2C   CH2
      |    29
      C16
     / \
 15C    C17
13 ||   |   1
 C  14N   N18 C
   / \ /   \ / \
12C   C22   C19 C2
 |    ||    ||  |
11C   C21   C20 C3
 \\ / \     / \ //
  C  9N   N5  C
 10  |    ||   4
   8C    C6
    \ /
    C7
     |
   H2C25 CH2
   / \   /23
  H3C H2C
   26   24
```

FORMULA	C22H32N4			MOL WT	352.53
SOLVENT	CDCL3				
ORIG ST	C4H12SI			TEMP	333

113.53	124.21	147.04	108.79	138.18	34.67
1/2	2/2	6/2	7/1	19/1	23/3
33.09	22.09	13.90			
24/3	25/3	26/4			

E.LORCH,E.BREITMAIER
CHEMIKER-ZTG   99,   87 (1975)

2576

R 006900   7,16-DIPROPYL-5,14-DIHYDRODIBENZO(B,I)-
5,9,14,18-TETRAAZA(14)ANNULENE

```
      27CH2
      26/ \
     H2C   CH3
      |    28
      C16
     / \
 15C    C17
13 ||   |   1
 C  14N   N18 C
   / \ /   \ / \
12C   C21   C19 C2
 |    ||    ||  |
11C   C21   C20 C3
 \\ / \     / \ //
  C  9N   N5  C
 10  |    ||   4
   8C    C6
    \ /
    C7
  25 |
  H3C   CH2
     \ /23
     24CH2
```

FORMULA	C24H28N4			MOL WT	372.52
SOLVENT	CDCL3				
ORIG ST	C4H12SI			TEMP	309

113.47	124.15	147.04	108.43	137.99	35.46
1/2	2/2	6/2	7/1	19/1	23/3
25.44	13.42				
24/3	25/4				

E.LORCH,E.BREITMAIER
CHEMIKER-ZTG   99,   87 (1975)

2577

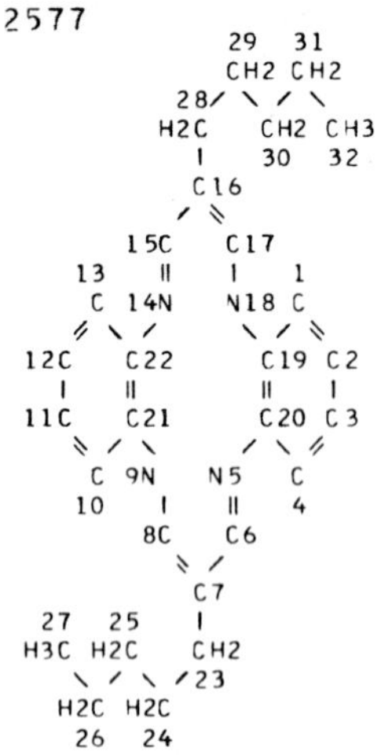

R 006900 7,16-DIPENTYL-5,14-DIHYDRODIBENZO(B,I)
5,9,14,18-TETRAAZA(14)ANNULENE

```
        29   31
        CH2  CH2
      28/ \  / \
    H2C   CH2 CH3
     |     30  32
     C16
     / \\
   15C   C17
  13  ||   |  1
   C 14N   N18 C
   // \ /     \ / \\
 12C   C22   C19 C2
  |    ||     || |
 11C   C21   C20 C3
   \\  / \   / \ /
    C 9N    N5  C
   10 |     ||  4
      8C    C6
       \\  /
        C7
  27 25 |
 H3C H2C  CH2
   \ / \ /23
    H2C H2C
    26  24
```

FORMULA	C28H36N4			MOL WT	428.63
SOLVENT	CDCL3				
ORIG ST	C4H12SI			TEMP	303

113.41	124.09	146.92	108.61	137.88	33.27
1/2	2/2	6/2	7/1	19/1	23/3
32.18	31.21	22.58	14.09		
24/3	25/3	26/3	27/4		

E.LORCH,E.BREITMAIER
CHEMIKER-ZTG   99,   87 (1975)

2578

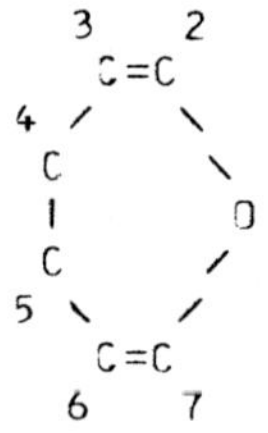

R 006910 4,5-DIHYDROOXEPINE

```
    3   2
     C=C
   4 /    \
    C       \
    |        O
    C       /
   5 \     /
     C=C
     6   7
```

FORMULA	C6H8O		MOL WT	96.13
SOLVENT	CCL4/CDCL3			
ORIG ST	C4H12SI		TEMP	AMB

142.60	107.80	26.80
2/2	3/2	4/3

H.GUENTHER,G.JIKELI
CHEM BER   106, 1863 (1973)

2579

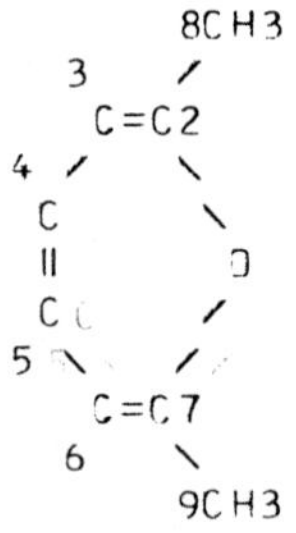

R 006910 2,7-DIMETHYLOXEPINE

```
       8CH3
   3    /
    C=C2
   4 /   \
    C      \
    ||      O
    C      /
   5 \    /
     C=C7
    6   \
        9CH3
```

FORMULA	C8H10O		MOL WT	122.17
SOLVENT	CCL4/CDCL3			
ORIG ST	C4H12SI		TEMP	AMB

150.20	112.30	127.60	21.10
2/1	3/2	4/2	8/4

H.GUENTHER,G.JIKELI
CHEM BER   106, 1863 (1973)

2580

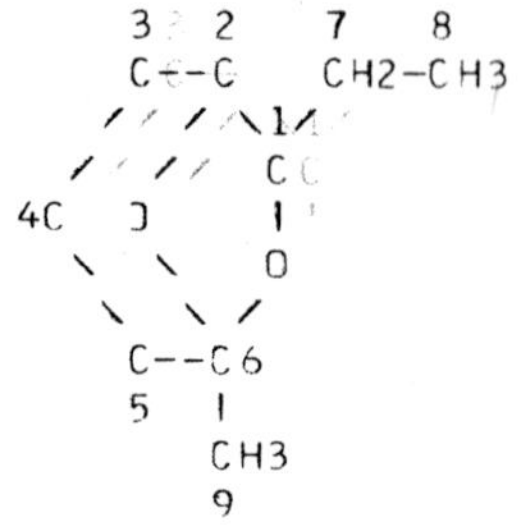

R 006910 EXO-BREVICOMIN

```
   3   2   7   8
    C--C  CH2-CH3
   / / /\14/
  / / / C
 4C  ]  |
  \  \  O
   \ \ /
    C--C6
   5 |
     CH3
     9
```

FORMULA	C9H16O2			MOL WT	156.23
SOLVENT	CDCL3				
ORIG ST	C4H12SI			TEMP	AMB

81.100	78.30	107.60	9.70	25.00	17.30
1/2	2/2	6/1	8/4	9/4	
28.00	28.60	35.00			

L.F.JOHNSON,W.C.JANKOWSKI
CARBON-13 NMR SPECTRA,JOHN
WILEY AND SONS,NEW YORK   359 (1972)

2581

```
      11   4    5
       C    C=C
   10 // \3/    \
     C    C      \
     |    ||      C6
     C    C      //
   9 \\ /2\     //
      C    D-C
      8     7
```

R 006910 2,3-BENZOXEPINE

FORMULA	C10H8O			MOL WT	144.17
SOLVENT	CCL4/CDCL3				
ORIG ST	C4H12SI			TEMP	AMB

156.30	131.30	120.90	124.20	114.60	146.70
2/1	3/1	4/2	5/2	6/2	7/2
132.50	128.80	129.70	126.50		
8/2	9/2	10/2	11/2		

H.GUENTHER,G.JIKELI
CHEM BER   106, 1863 (1973)

2582

```
      8    3    2
       C    C=C
   9 // \4/    \
     C    C      \
     |    ||      D
   10C    C      /
     \\ /5\     /
      C    C=C
      11   6    7
```

R 006910 4,5-BENZOXEPINE

FORMULA	C10H8O			MOL WT	144.17
SOLVENT	CCL4/CDCL3				
CRIG ST	C4H12SI			TEMP	AMB

145.50	112.80	135.70	128.70	127.30
2/2	3/2	4/1	8/2	9/2

H.GUENTHER,G.JIKELI
CHEM BER   106, 1863 (1973)

2583

```
            13CH3
            /
          12CH
     11   4  5/ \
      C    C=C 14CH3
   10 // \3/    \
     C    C      \
     |    ||      C6
     C    C      //
   9 \\ /2\     //
      C    D-C
      8     7
```

R 006910 5-ISOPROPYL-2,3-BENZOXEPINE

FORMULA	C13H14O			MOL WT	186.26
SOLVENT	CCL4/CDCL3				
ORIG ST	C4H12SI			TEMP	AMB

156.00	131.10	120.50	145.20	115.30	146.10
2/1	3/1	4/2	5/1	6/2	7/2
128.90	128.80	128.80	124.10	35.40	
8/2	9/2	10/2	11/2	12/2	

H.GUENTHER,G.JIKELI
CHEM BER   106, 1863 (1973)

2584

```
           12C-D-CH3
      8    3  / D  14
       C    C=C2   D
   9 // \4/    \
     C    C      \
     |    ||      D
     C    C      /
   10 \\ /5\     /
      C    C=C7   D
      11   6  \ // 15
           13C-D-CH3
```

R 006910 2,7-DIMETHOXYCARBONYL-4,5-BENZOXEPINE

FORMULA	C14H12O5			MOL WT	260.25
SOLVENT	CCL4/CDCL3				
ORIG ST	C4H12SI			TEMP	AMB

116.00	124.30	133.70	131.10	129.30	162.00
2/1	3/2	4/1	8/2	9/2	12/1
52.40					
14/4					

H.GUENTHER,G.JIKELI
CHEM BER   106, 1863 (1973)

2585

S 000400 TRI-N-PROPYL BORATE

```
     D-CH2-CH2-CH3
     |  1   2    3
     B
    / \
H3C-CH2-CH2-D   D-CH2-CH2-CH3
 9  8   7        4   5   6
```

FORMULA	C9H21BO3			MOL WT	188.08
SOLVENT	CDCL3				
ORIG ST	C4H12SI			TEMP	AMB

64.90	24.90	10.30	64.90	24.90	10.30
1/3	2/3	3/4	4/3	5/3	6/4
64.90	24.90	10.30			
7/3	8/3	9/4			

L.F.JOHNSON,W.C.JANKOWSKI
CARBON-13 NMR SPECTRA,JOHN
WILEY AND SONS,NEW YORK   365 (1972)

```
2586                    S 000400  TRI-TERT-BUTYL BORATE
      H3C10    6CH3
       11 19      15 7       FORMULA   C12H27BO3          MOL WT   230.16
      H3C-C-O   O-C-CH3      SOLVENT   CDCL3
        |    \ /   |         ORIG ST   C4H12SI            TEMP       AMB
       H3C    B   CH3
        12     |    8        72.00    30.20    30.20    30.20    72.00    30.20
              O                1/1      2/4      3/4      4/4      5/1      6/4
             4 11            30.20    30.20    72.00    30.20    30.20    30.20
          H3C-C-CH3            7/4      8/4      9/1     10/4     11/4     12/4
              | 2
             CH3             L.F.JOHNSON,W.C.JANKOWSKI
              3              CARBON-13 NMR SPECTRA,JOHN
                            WILEY AND SONS,NEW YORK              448 (1972)
```

```
2587                    S 000600  ALLYLTRICHLOROSILANE

             CL            FORMULA   C3H5CL3SI          MOL WT   175.52
      3  2  1  |           SOLVENT   C3H5CL3SI
     CH2=CH-CH2-SI-CL      ORIG ST   C4H12SI            TEMP       AMB
             |
             CL            31.60    128.10   120.60
                            1/3      2/2      3/3

                           Y.K.GRISHIN,N.M.SERGEYEV,Y.A.USTYNYUK
                           ORG MAGN RESON                   4,  377 (1972)
```

```
2588                    S 000600  METHYLALLYLDICHLOROSILANE

             CL            FORMULA   C4H8CL2SI          MOL WT   155.10
      3  2  1  | 4         SOLVENT   C4H8CL2SI
     CH2=CH-CH2-SI-CH3     ORIG ST   C4H12SI            TEMP
             |
             CL            29.60    130.60   118.90    5.20
                            1/3      2/2      3/3      4/4

                           Y.K.GRISHIN,N.M.SERGEYEV,Y.A.USTYNYUK
                           ORG MAGN RESON                   4,  377 (1972)
```

```
2589                    S 000600  DIMETHYLALLYLCHLOROSILANE
           4
           CH3            FORMULA   C5H11CLSI          MOL WT   134.68
      3  2  1  |          SOLVENT   C5H11CLSI
     CH2=CH-CH2-SI-CL     ORIG ST   C4H12SI            TEMP       AMB
             |
             CH3          28.30    133.00   116.30    2.40
             5             1/3      2/2      3/3      4/4

                           Y.K.GRISHIN,N.M.SERGEYEV,Y.A.USTYNYUK
                           ORG MAGN RESON                   4,  377 (1972)
```

```
2590                    S 000600  TRIMETHYLALLYLSILANE
           4
           CH3            FORMULA   C6H14SI            MOL WT   114.26
      3  2  1  | 5        SOLVENT   C6H14SI
     CH2=CH-CH2-SI-CH3    ORIG ST   C4H12SI            TEMP       AMB
             |
             CH3          25.40    135.30   113.50    1.50
             6             1/3      2/2      3/3      4/4

                           Y.K.GRISHIN,N.M.SERGEYEV,Y.A.USTYNYUK
                           ORG MAGN RESON                   4,  377 (1972)
```

2591   S 000600   2,2-DIMETHYL-2-SILAPENTANESULFONIC ACID
SODIUM SALT

```
      6
     CH3          O          FORMULA   C6H15NAO3SSI          MOL WT  158.17
      |           ||-   +    SOLVENT   H2O
  H3C-SI-CH2-CH2-CH2-S=O NA  ORIG ST   C4H12SI               TEMP      AMB
   4  |  3   2   1   ||
     CH3          O            55.30   19.90   16.30  -1.10  -1.10  -1.10
      5                         1/3     2/3     3/3    4/4    5/4    6/4
```

L.F.JOHNSON,W.C.JANKOWSKI
CARBON-13 NMR STECTRA,JOHN
WILEY AND SONS,NEW YORK   220 (1972)

2592   S 000600   2,2,4,4-TETRAMETHYLDISILAPENTANE-2,4

```
   3      5                    FORMULA   C7H20SI2              MOL WT  160.41
  CH3    CH3                   SOLVENT   CDCL3
 2 1   1    6                  ORIG ST   C4H12SI               TEMP      AMB
 H3C-SI-CH2-SI-CH3
    |       |                    4.40    1.30    1.30   1.30   1.30   1.30
   4CH3    7CH3                   1/3     2/4     3/4    4/4    5/4    6/4
                                  1.30
                                  7/4
```

L.F.JOHNSON,W.C.JANKOWSKI
CARBON-13 NMR SPECTRA,JOHN
WILEY AND SONS,NEW YORK   280 (1972)

2593   S 000600   TRIETHOXYETHYLSILANE

```
                               FORMULA   C8H20O3SI             MOL WT  192.33
   4 3                         SOLVENT   CDCL3
 H3C-CH2-O                     ORIG ST   C4H12SI               TEMP      AMB
 6 5    |   1   2
 H3C-CH2-O-SI-CH2-CH3            2.50    6.50   58.40  18.30  58.40  18.30
         |                       1/3     2/4     3/3    4/4    5/3    6/4
 H3C-CH2-O                      58.40   18.30
  8  7                           7/3     8/4
```

L.F.JOHNSON,W.C.JANKOWSKI
CARBON-13 NMR SPECTRA,JOHN
WILEY AND SONS,NEW YORK   329 (1972)

2594   S 000600   TRIALLYLCHLOROSILANE

```
  3   2   1                    FORMULA   C9H15CLSI             MOL WT  186.76
  CH2=CH-CH2                   SOLVENT   C9H15CLSI
      |                        ORIG ST   C4H12SI               TEMP      AMB
 CH2=CH-CH2-SI-CL
      |                         23.20  132.40  116.90
  CH2=CH-CH2                     1/3     2/2     3/3
   9   8   7
```

Y.K.GRISHIN,N.M.SERGEYEV,Y.A.USTYNYUK
ORG MAGN RESON   4, 377 (1972)

2595   S 000600   METHYLTRIALLYLSILANE   3

```
  3   2   1                    FORMULA   C10H18SI              MOL WT  166.34
  CH2=CH-CH2                   SOLVENT   C10H18SI
 6  5  4  | 10                 ORIG ST   C4H12SI               TEMP      AMB
 CH2=CH-CH2-SI-CH3
      |                         21.50  134.80  114.40  +5.70
  CH2=CH-CH2                     1/3     2/2     3/3    10/4
   9   8   7
```

Y.K.GRISHIN,N.M.SERGEYEV,Y.A.USTYNYUK
ORG MAGN RESON   4, 377 (1972)

2596   S 000800 DIMETHYL METHYLPHOSPHONATE

```
                        FORMULA   C3H9O3P            MOL WT   124.08
        O               SOLVENT   CDCL3
        ‖    2          ORIG ST   C4H12SI            TEMP        AMB
   CH3-P-O-CH3
    1   |              9.80      52.10     52.10
        O-CH3         1/4        2/4       3/4
          3
```

L.F.JOHNSON,W.C.JANKOWSKI
CARBON-13 NMR SPECTRA,JOHN
WILEY AND SONS,NEW YORK   42 (1972)

2597   S 000800 DIETHYL ETHYLPHOSPHONATE

```
                        FORMULA   C6H15O3P           MOL WT   166.16
        5   6           SOLVENT   CDCL3
     O  O-CH2-CH3       ORIG ST   C4H12SI            TEMP        AMB
     ‖ /
  H3C-CH2-P           19.00     6.60     61.40    16.50    61.40    16.50
    2  1    \         1/3       2/4      3/3      4/4      5/3      6/4
          O-CH2-CH3
            3   4
```

L.F.JOHNSON,W.C.JANKOWSKI
CARBON-13 NMR SPECTRA,JOHN
WILEY AND SONS,NEW YORK   219 (1972)

2598   S 000800 HEXAMETHYL PHOSPHORAMIDE

```
    4        2
  H3C    O    CH3       FORMULA   C6H18N3OP          MOL WT   179.20
    \    ‖   /          SOLVENT   CDCL3
  H3C-N- P-N-CH3        ORIG ST   C4H12SI            TEMP        AMB
    3    |   1
         N            36.80     36.80    36.80    36.80    36.80    36.80
        / \          1/4       2/4      3/4      4/4      5/4      6/4
     H3C   CH3
      5     6
```

L.F.JOHNSON,W.C.JANKOWSKI
CARBON-13 NMR SPECTRA,JOHN
WILEY AND SONS,NEW YORK   221 (1972)

2599   S 000800 TRIPHENYLPHOSPHITE

```
        15C-C16
        ⁄⁄    \\              FORMULA   C18H15O3P       MOL WT   310.29
     14C      C17            SOLVENT   CDCL3
        \    /               ORIG ST   C4H12SI          TEMP       AMB
        13C=C18
   3 2    |     8 9         151.50    120.60   129.50   124.10   129.50   120.60
   C-C    O    C-C          1/1       2/2      3/2      4/2      5/2      6/2
  4⁄⁄  \\1  |  7⁄⁄  \       151.50    120.60   129.50   124.10   129.50   120.60
   C    C-O-P-O-C    C10    7/1       8/2      9/2      10/2     11/2     12/2
    \5 6⁄    \    /         151.50    120.60   129.50   124.10   129.50   120.60
   C=C       11C=C12        13/1      14/2     15/2     16/2     17/2     18/2
```

L.F.JOHNSON,W.C.JANKOWSKI
CARBON-13 NMR SPECTRA,JOHN
WILEY AND SONS,NEW YORK   476 (1972)

2600   S 000800 TRIPHENYL PHOSPHATE

```
   17 18           2 3
   C-C     O     C-C           FORMULA   C18H15O4P       MOL WT   326.29
   ⁄⁄  \\   ‖   1⁄⁄  \\         SOLVENT   CDCL3
  16C     C-O-P-O-C    C4      ORIG ST   C4H12SI          TEMP       AMB
    \   ⁄13  |   \   /
    C=C      O    C=C        150.40    120.10   129.70   125.50   129.70   120.10
   15 14     |    6 5        1/1       2/2      3/2      4/2      5/2      6/2
             C7             150.40    120.10   129.70   125.50   129.70   120.10
            ⁄⁄ \            7/1       8/2      9/2      10/2     11/2     12/2
        12C    C8          150.40    120.10   129.70   125.50   129.70   120.10
         |    ‖            13/1      14/2     15/2     16/2     17/2     18/2
        11C    C9
          \ ⁄
          C10
```

L.F.JOHNSON,W.C.JANKOWSKI
CARBON-13 NMR SPECTRA,JOHN
WILEY AND SONS,NEW YORK   477 (1972)

2601   S 000800 TRIPHENYLPHOSPHINE

```
      16C-C17
       /    \
    15C       C18        FORMULA   C18H15P           MOL WT  262.29
       \    / 8 9        SOLVENT   CDCL3
      14C=C13 C-C        ORIG ST   C4H12SI           TEMP      AMB
      3 2  |  /  \
     C-C  P-C    C10     137.20  133.60  128.40  128.50  128.40  133.60
    4/  \1/  7\  /        1/1     2/2     3/2     4/2     5/2     6/2
    C    C      C=C       137.20  133.60  128.40  128.50  128.40  133.60
     \  /    12 11        7/1     8/2     9/2    10/2    11/2    12/2
      C=C                 137.20  133.60  128.40  128.50  128.40  133.60
      5 6                 13/1    14/2    15/2    16/2    17/2    18/2
```

L.F.JOHNSON,W.C.JANKOWSKI
CARBON-13 NMR SPECTRA,JOHN
WILEY AND SONS,NEW YORK   478 (1972)

2602   S 000800 TRI-ORTHO-TOLYLPHOSPHINE

```
     21        7          FORMULA   C21H21P           MOL WT  304.38
     CH3      CH3         SOLVENT   CDCL3
   17 |        |          ORIG ST   C4H12SI           TEMP      AMB
     C-C15    20-C3
    /  \15  1/   \        142.50  134.40  128.50  132.90  126.00  129.90
  18C   C--P--C    C4      1/1     2/1     3/2     4/2     5/2     6/2
    \  /  |   \  /        21.00  142.50  134.40  128.50  132.90  126.00
    C=C   C8  6C=C5        7/4     8/1     9/1    10/2    11/1    12/2
    19 20 /  \9           129.90   21.00  142.50  134.40  128.50  132.90
      13C    C-CH3        13/2    14/4    15/1    16/1    17/2    18/2
       |     || 14        126.00  129.90   21.00
      12C    C10          19/2    20/2    21/4
        \  /
         C11
```

L.F.JOHNSON,W.C.JANKOWSKI
CARBON-13 NMR SPECTRA,JOHN
WILEY AND SONS,NEW YORK   489 (1972)

2603   S 000800 TRI-PARA-TOLYLPHOSPHINE

```
      21CH3
        |              FORMULA   C21H21P           MOL WT  304.38
        C16            SOLVENT   CDCL3
       / \\            ORIG ST   C4H12SI           TEMP      AMB
    17C    C15
     ||    |           134.20  133.50  129.10  138.10  129.10  133.50
    18C    C14          1/1     2/2     3/2     4/1     5/2     6/2
   3 2  \ /  8 9       134.20  133.50  129.10  138.10  129.10  133.50
   C-C   C13 C-C        7/1     8/2     9/2    10/1    11/2    12/2
 19 4/  \1 | 7/  \10   134.20  133.50  129.10  138.10  129.10  133.50
 H3C-C    C-P-C    C-CH3 13/1   14/2    15/2    16/1    17/2    18/2
    \  /    \  / 20     21.10   21.10   21.10
    5C=C6   12C=C11     19/4    20/4    21/4
```

L.F.JOHNSON,W.C.JANKOWSKI
CARBON-13 NMR SPECTRA,JOHN
WILEY AND SONS,NEW YORK   488 (1972)

2604   S 001300 SELENOPHENE

```
    4   3
    C---C              FORMULA   C4H4SE            MOL WT  131.04
    ||  ||             SOLVENT   C3H6O
   5C   C2             ORIG ST   CS2               TEMP      AMB
     \ /
      SE               130.20  128.90
      1                 2/2     3/2
```

F.J.WEIGERT,J.D.ROBERTS
J AM CHEM SOC   90, 3543 (1968)

2605

```
   3   2
   C---C
   ||  ||
  4C   C1
    \5/
     C
     |
     HG
     |
    6CH3
```

S 019000 METHYL–CYCLOPENTADIENYL–MERCURY

FORMULA C6H8HG   MOL WT 280.72
SOLVENT C6H8HG
ORIG ST C4H12SI   TEMP   293

117.60   15.60
1/2   6/4

Y.K.GRISHIN,N.M.SERGEYEV,Y.A.USTYNYUK
ORG MAGN RESON   4, 377 (1972)

2606

```
   3   2
   C---C
   ||  ||
  4C   C1
    \5/
     C
  8 |  6
  H3C-GE-CH3
     |
    7CH3
```

S 047000 TRIMETHYL–5–CYCLOPENTADIENYLGERMANE

FORMULA C8H14GE   MOL WT 189.16
SOLVENT C8H14GE
ORIG ST C4H12SI   TEMP   253

133.90   129.80   52.10   +0.90
1/2   2/2   5/2   6/4

Y.K.GRISHIN,N.M.SERGEYEV,Y.A.USTYNYUK
ORG MAGN RESON   4, 377 (1972)

2607

```
   3   2
   C---C
   ||  ||
  4C   C1
    \5/ \6
     C   CH3
  9 |
  H3C-GE-CH3
     |  7
    8CH3
```

S 047000 TRIMETHYL–5–(1–METHYLCYCLOPENTADIENYL)–GERMANE

FORMULA C9H16GE   MOL WT 203.19
SOLVENT C9H16GE
ORIG ST C4H12SI   TEMP   243

144.90   126.70   130.20   131.50   55.60   18.10
1/1   2/2   3/2   4/2   5/2   6/4
+0.70
7/4

Y.K.GRISHIN,N.M.SERGEYEV,Y.A.USTYNYUK
ORG MAGN RESON   4, 377 (1972)

2608

```
        6CH3
   3   2/
   C---C
   ||  ||
  4C   C1
    \5/
     C
  9 |  7
  H3C-GE-CH3
     |
    8CH3
```

S 047000 TRIMETHYL–5–(2–METHYLCYCLOPENTADIENYL)–GERMANE

FORMULA C9H16GE   MOL WT 203.19
SOLVENT C9H16GE
ORIG ST C4H12SI   TEMP   243

128.80   139.90   133.10   134.60   52.80   16.20
1/1   2/1   3/2   4/2   5/2   6/4
+0.70
7/4

Y.K.GRISHIN,N.M.SERGEYEV,Y.A.USTYNYUK
ORG MAGN RESON   4, 377 (1972)

2609

```
   3   2
   C---C
   ||  ||
  4C   C1
    \5/
     C
  8 |  6
  H3C-SN-CH3
     |
    7CH3
```

S 048000 TRIMETHYL–CYCLOPENTADIENYL–TIN

FORMULA C8H14SN   MOL WT 189.16
SOLVENT C8H14SN
ORIG ST C4H12SI   TEMP   293

114.30   +6.6
1/2   6/4

Y.K.GRISHIN,N.M.SERGEYEV,Y.A.USTYNYUK
ORG MAGN RESON   4, 377 (1972)

2610   S 048000 DIBUTYLTIN DIACETATE

```
       O
   6  ||5    4    3    2    1    FORMULA   C12H24O4SN        MOL WT    295.87
  H3C-C-]   CH2-CH2-CH2-CH3 SOLVENT  CDCL3
        \   /                 ORIG ST   C4H12SI            TEMP        AMB
         SN
    12 11 / \              13.50    24.90    26.60    26.30   181.00    20.50
   H3C-C-]   CH2-CH2-CH2-CH3  1/4      2/3      3/3      4/3     5/1      6/4
     ||      10  9   8   7   13.50    24.90    26.60    26.30   181.00    20.50
      O                      7/4      8/3      9/3     10/3    11/1     12/4
```

                   L.F.JOHNSON,W.C.JANKOWSKI
                   CARBON-13 NMR SPECTRA,JOHN
                   WILEY AND SONS,NEW YORK   446 (1972)

2611   S 049000 HEXAPHENYLDILEAD

```
     34C        C10
    ⁄ \       ⁄ \
   33C 35C  C11 C9      FORMULA   C36H30PB2         MOL WT    620.56
    |  ||  |   ||       SOLVENT   CDCL3
   32C 36C  C12 C8      ORIG ST   C4H12SI           TEMP        AMB
  29   \ /   \ / 23
   C-C30 C31  C7 C-C    152.80   137.70   129.50   128.10   129.50   137.70
  ⁄   \251   |  1⁄  \4   1/1      2/2      3/2      4/2      5/2      6/2
 C28   C-PB---PB-C    C 152.80   137.70   129.50   128.10   129.50   137.70
  \   ⁄ |     |  \  ⁄    7/1      8/2      9/2     10/2     11/2     12/2
 27C=C26 C19   C13 C=C  152.80   137.70   129.50   128.10   129.50   137.70
    ⁄ \    ⁄ \  6 5     13/1     14/2     15/2     16/2     17/2     18/2
   24C 20C C18 C14      152.80   137.70   129.50   128.10   129.50   137.70
    |  ||  |   ||       19/1     20/2     21/2     22/2     23/2     24/2
   23C 21C C17 C15      152.80   137.70   129.50   128.10   129.50   137.70
    \ /    \ /          25/1     26/2     27/2     28/2     29/2     30/2
    C22    C16          152.80   137.70   129.50   128.10   129.50   137.70
                        31/1     32/2     33/2     34/2     35/2     36/2
```

                   L.F.JOHNSON,W.C.JANKOWSKI
                   CARBON-13 NMR SPECTRA,JOHN
                   WILEY AND SONS,NEW YORK   498 (1972)

2612   S 071000 DICARBONYL-DICYCLOPENTADIENYL-IRON

```
   C1    2
  ⁄ \   C=O
  C   C  |  C---C       FORMULA   C12H10FEO2         MOL WT    242.06
  ||  || -FE- ||  ||    SOLVENT   C3H6O
  C---C  |   C   C      ORIG ST   C4H12SI           TEMP        333
       C=O   \  ⁄
        2     C1        113.10
                         1/2      2/1
```

                   V.K.GRISHIN,N.M.SERGEYEV,Y.A.USTYNYUK
                   ORG MAGN RESON   4, 377 (1972)

2613   T 013000 BIS(1,3,3-TRIMETHYL-INDOLENIN-2-YL)-3-BROMO-
                        PENTAMETHINIUM TETRAFLUOROBORATE

```
    6  7
    C=C              24C=C23
   ⁄  | 12       29 |  \       FORMULA   C27H30BBRF4N2      MOL WT    549.26
  5C  8C CH3     H3C C25 C22 SOLVENT  C2D6OS
  || 9⁄ \| 11    28 1⁄ \26|| ORIG ST   C4H12SI           TEMP        AMB
 4C-C  1C-CH3    H3C-C18 C-C21
   |   |  14 16   |    |      49.40   175.00   111.80   128.60   125.60   122.40
   3N---N2  CH  CH  C===N+     1/1      2/1      4/2      5/2      6/2      7/2
 10⁄    \ ⁄ \ ⁄ \ ⁄19 20\27  141.50   142.70    31.40    26.40   102.20   149.40
  CH3    CH  C15 CH   H3C      8/1      9/1     10/4     11/4     13/2     14/2
       13  |  17             115.60
          BR          -      15/1
                3F4
```

                   W.GRAHN,C.REICHARDT
                   TETRAHEDRON   32, 125 (1976)

```
2614                    T 013000  BIS(1,3,3-TRIMETHYL-INDOLENIN-2-YL)-3-CHLORO-
   6 7                            PENTAMETHINIUM TETRAFLUOROBORATE
   C=C                24C=C23
  /  | 12          29 |  \    FORMULA   C27H30BCLF4N2            MOL WT   504.81
 5C  8C CH3        H3C C25 C22 SOLVENT  C2D6OS
 || 9/ \| 11       28 I/ \26|| ORIG ST  C4H12SI                 TEMP        AMB
 4C-C  1C-CH3      H3C-C18 C-C21
  |    |   14 16    |   |       49.30   174.80   111.70   128.60   125.60   122.50
  3N---C2  CH  CH  C===N+       1/1     2/1      4/2      5/2      6/2      7/2
 10/       \ / \ / \ /19 20\27  141.50  142.70   31.40    26.50   100.00   147.40
 CH3        CH  C15 CH     H3C  8/1     9/1      10/4     11/4     13/2     14/2
            13  |   17            122.30
            CL              -      15/1
                     BF4    W.GRAHN,C.REICHARDT
                            TETRAHEDRON                          32,   125 (1976)
```

```
2615                    T 013000  BIS(1,3,3-TRIMETHYL-INDOLENIN-2-YL)-3-IODO-
   6 7                            PENTAMETHINIUM TETRAFLUOROBORATE
   C=C                24C=C23
  /  | 12          29 |  \    FORMULA   C27H30BF4IN2            MOL WT   637.20
 5C  8C CH3        H3C C25 C22 SOLVENT  C2D6OS
 || 9  \| 11       28 I/ \26|| ORIG ST  C4H12SI                 TEMP        AMB
 4C-C  1C-CH3      H3C-C18 C-C21
  |    |   14 16    |   |       49.40   175.00   111.80   128.70   125.60   122.60
  3N---C2  CH  CH  C===N+       1/1     2/1      4/2      5/2      6/2      7/2
 10/       \ / \ / \ /19 20\27  141.40  142.70   31.50    26.60   106.70   153.80
 CH3        CH  C15 CH     H3C  8/1     9/1      10/4     11/4     13/2     14/2
            13  |   17            97.80
            J              -       15/1
                     BF4    W.GRAHN,C.REICHARDT
                            TETRAHEDRON                          32,   125 (1976)
```

```
2616                    T 013000  BIS(1,3,3-TRIMETHYL-INDOLENIN-2-YL)-3-NITRO-
   6 7                            PENTAMETHINIUM TETRAFLUOROBORATE
   C=C                24C=C23
  /  | 12          29 |  \    FORMULA   C27H30BF4N3O2           MOL WT   515.36
 5C  8C CH3        H3C C25 C22 SOLVENT  C2D6OS
 || 9/ \| 11       28 I/ \26|| ORIG ST  C4H12SI                 TEMP        AMB
 4C-C  1C-CH3      H3C-C18 C-C21
  |    |   14 16    |   |       50.40   179.40   112.70   128.90   127.30   122.70
  3N---C2  CH  CH  C===N+       1/1     2/1      4/2      5/2      6/2      7/2
 10/       \ / \ / \ /19 20\27  142.20  142.40   32.60    26.80   102.30   141.50
 CH3        CH  C15 CH     H3C  8/1     9/1      10/4     11/4     13/2     14/2
            13  |   17           131.50
            N              -      15/1
           / \             BF4   W.GRAHN,C.REICHARDT
          J   J                  TETRAHEDRON                     32,   125 (1976)
```

```
2617                    T 013000  BIS(1,3,3-TRIMETHYL-INDOLENIN-2-YL)-3-FLUORO-
   6 7                            PENTAMETHINIUM TETRAFLUOROBORATE
   C=C                24C=C23
  /  | 12          29 |  \    FORMULA   C27H30BF5N2             MOL WT   488.36
 5C  8C CH3        H3C C25 C22 SOLVENT  C2D6OS
 || 9/ \| 11       28 I/ \26|| ORIG ST  C4H12SI                 TEMP        AMB
 4C-C  1C-CH3      H3C-C18 C-C21
  |    |   14 16    |   |       49.00   173.30   111.40   128.40   125||30   122.40
  3N---C2  CH  CH  C===N+       1/S     2/S      4/D      5/D      6/D      7/D
 10/       \ / \ / \ /19 20\27  141.30  142.80   31.30    26.50    96.40   135.50
 CH3        CH  C15 CH     H3C  8/S     9/S      10/Q     11/Q     13/D     14/D
            13  |   17           151.40
            F              -      15/D
                     BF4    W.GRAHN,C.REICHARDT
                            TETRAHEDRON                          32,   125 (1976)
```

```
2618                    T 013000  BIS(1,3,3-TRIMETHYL-INDOLENIN-2-YL)-PENTA-
   6 7                            METHINIUM TETRAFLUOROBORATE
   C=C                24C=C23
  /  | 12          29 |  \    FORMULA   C27H31BF4N2             MOL WT   470.37
 5C  8C CH3        H3C C25 C22 SOLVENT  C2D6OS
 || 9/ \| 11       28 I/ \26|| ORIG ST  C4H12SI                 TEMP        AMB
 4C-C  1C-CH3      H3C-C18 C-C21
  |    |   14 16    |   |       48.70   173.30   110.90   128.40   124.70   122.30
  3N---C2  CH  CH  C===N+       1/1     2/1      4/2      5/2      6/2      7/2
 10/       \ / \ / \ /19 20\27  141.10  142.90   30.90    26.90   103.30   154.00
 CH3        CH  CH  CH     H3C  8/1     9/1      10/4     11/4     13/2     14/2
            13  15  17           125.20
                           -      15/2
                     BF4    W.GRAHN,C.REICHARDT
                            TETRAHEDRON                          32,   125 (1976)
```

2619   T 013000   BIS(1,3,3-TRIMETHYL-INDOLENIN-2-YL)-3-CYANO-
PENTAMETHINIUM TETRAFLUOROBORATE

```
   6  7
   C=C                    24C=C23
  /  |  12          29  |  \       FORMULA   C28H30BF4N3            MOL WT   495.38
5C   8C  CH3            H3C C25 C22 SOLVENT  C2D6OS
 || 9/ \| 11        28 |/ \26||     ORIG ST  C4H12SI               TEMP       AMB
4C-C   1C-CH3       H3C-C18 C-C21
   |   |   14 16       |    |        49.90   176.50  112.30  128.70  126.30  122.60
   3N---C2  CH  CH  C===N+            1/1     2/1     4/2     5/2     6/2     7/2
10/       \ / \ / \ /19 20\27       141.80  142.40   31.80   26.40  101.30  153.20
  CH3        CH  C15 CH     H3C       8/1     9/1     10/4    11/4    13/2    14/2
          13  |   17                 98.60   115.80
            30C≡N              -      15/1    30/1
                          BF4
```

W.GRAHN,C.REICHARDT
TETRAHEDRON   32,   125 (1976)

2620   T 013000   BIS(1,3,3-TRIMETHYL-INDOLENIN-2-YL)-3-FORMYL-
PENTAMETHINIUM TETRAFLUOROBORATE

```
   6  7
   C=C                    24C=C23
  /  |  12          29  |  \       FORMULA   C28H31BF4N2            MOL WT   482.38
5C   8C  CH3            H3C C25 C22 SOLVENT  C2D6OS
 || 9/ \| 11        28 |/ \26||     ORIG ST  C4H12SI               TEMP       AMB
4C-C   1C-CH3       H3C-C18 C-C21
   |   |   14 16       |    |        49.90   178.00  112.50  128.70  126.40  122.50
   3N---C2  CH  CH  C===N+            1/1     2/1     4/2     5/2     6/2     7/2
10/       \ / \ / \ /19 20\27       141.90  142.50   32.00   26.90  101.80  151.00
  CH3        CH  C15 CH     H3C       8/1     9/1     10/4    11/4    13/2    14/2
          13  |   17                120.00  190.90
            C30                -      15/1    30/2
           // \              BF4
          O    H
```

W.GRAHN,C.REICHARDT
TETRAHEDRON   32,   125 (1976)

2621   T 013000   BIS(1,3,3-TRIMETHYL-INDOLENIN-2-YL)-3-METHYL-
PENTAMETHINIUM TETRAFLUOROBORATE

```
   6  7
   C=C                    24C=C23
  /  |  12          29  |  \       FORMULA   C28H33BF4N2            MOL WT   484.39
5C   8C  CH3            H3C C25 C22 SOLVENT  C2D6OS
 || 9/ \| 11        28 |/ \26||     ORIG ST  C4H12SI               TEMP       AMB
4C-C   1C-CH3       H3C-C18 C-C21
   |   |   14 16       |    |        48.90   173.30  110.90  128.40  124.80  122.30
   3N---C2  CH  CH  C===N+            1/1     2/1     4/2     5/2     6/2     7/2
10/       \ / \ / \ /19 20\27       141.20  142.90   31.00   26.80  100.10  153.80
  CH3        CH  C15 CH     H3C       8/1     9/1     10/4    11/4    13/2    14/2
          13  |   17                128.70   21.60
            CH3                -      15/1    30/4
            30                BF4
```

W.GRAHN,C.REICHARDT
TETRAHEDRON   32,   125 (1976)

2622   T 013000   BIS(1,3,3-TRIMETHYL-INDOLENIN-2-YL)-3-METHOXY-
PENTAMETHINIUM TETRAFLUOROBORATE

```
   6  7              24 23
   C=C                C=C          FORMULA   C28H33BF4N2O          MOL WT   500.39
  /  |  12          29  |  \       SOLVENT  C2D6OS
5C   8C  CH3            H3C C25 C22 ORIG ST  C4H12SI               TEMP       AMB
 || 9/ \| 11        28 |/ \26||
4C-C   1C-CH3       H3C-C18 C-C21   48.90   173.00  111.10  128.50  124.90  122.40
   |   |   14 16       |    |        1/1     2/1     4/2     5/2     6/2     7/2
   3N---C2  CH  CH  C===N+          141.20  142.90   31.10   26.70   97.90  143.40
10/       \ / \ / \ /19 20\27        8/1     9/1     10/4    11/4    13/2    14/2
  CH3        CH  C15 CH     H3C     150.20   59.50
          13  |   17                 15/1    30/4
              O                -
         30/              BF4
        H3C
```

W.GRAHN,C.REICHARDT
TETRAHEDRON   32,   125 (1976)

```
2623                      T 013000  BIS(1,3,3,5-TETRAMETHYL-INDOLENIN-2-YL)-
   CH3                      H3C     3-NITRO-PENTAMETHINIUM TETRAFLUOROBORATE
13\  7                     /31
   6C=C               25C=C24   FORMULA   C29H34BF4N3O2          MOL WT   543.42
  /  | 12              30 |  \  SOLVENT   C2D6OS
5C   8C CH3           H3C C26 C23ORIG ST  C4H12SI               TEMP        AMB
 || 9% \| 11            29 1/ \27||
4C-C   1C-CH3    H3C-C19 C-C22     50.30   178.50   112.90   129.30   137.20   123.20
  |    |    15 17    |    |         1/1     2/1      4/2      5/2      6/1      7/2
   3N---C2   CH   CH   C===N+      142.30   140.10    32.70    26.90    21.10   102.20
10/      \ / \ / \ /20  21\28       8/1     9/1     10/4     11/4     13/4     14/2
   CH3       CH   C    CH    H3C    141.00   131.00
          14  |16  18             15/2     16/1
                     N                   -
                  // \               BF4
                 O    O                     W.GRAHN,C.REICHARDT
                                            TETRAHEDRON               32,  125  (1976)
```

```
2624                      T 013000  BIS(1,3,3,5-TETRAMETHYL-INDOLENIN-2-YL)-PENTA-
   CH3                      H3C     METHINIUM TETRAFLUOROBORATE
13\  7                     /31
   6C=C               25C=C24   FORMULA   C29H35BF4N2           MOL WT   498.42
  /  | 12              30 |  \  SOLVENT   C2D6OS
5C   8C CH3           H3C C26 C23ORIG ST  C4H12SI               TEMP        AMB
 || 9% \| 11            29 1/ \27||
4C-C   1C-CH3    H3C-C19 C-C22     48.60   172.50   110.40   128.50   134.10   122.90
  |    |    15 17    |    |         1/1     2/1      4/2      5/2      6/1      7/2
   3N---C2   CH   CH   C===N+      141.10   140.60    31.00    26.90    20.70   102.90
10/      \ / \ / \ /20  21\28       8/1     9/1     10/4     11/4     13/4     14/2
   CH3       CH   CH   CH    H3C   153.20   124.40
          14    16   18            15/2     16/2
                                     -
                  BF4                       W.GRAHN,C.REICHARDT
                                            TETRAHEDRON               32,  125  (1976)
```

```
2625                      T 013000  BIS(1,3,3,5-TETRAMETHYL-INDOLENIN-2-YL)-
   CH3                      H3C     3-CYANO-PENTAMETHINIUM TETRAFLUOROBORATE
13\  7                     /
   6C=C               25C=C24   FORMULA   C30H34BF4N3           MOL WT   523.43
  /  | 12              30 |  \  SOLVENT   C2D6OS
5C   8C CH3           H3C C26 C23ORIG ST  C4H12SI               TEMP        AMB
 || 9% \| 11            29 1/ \27||
4C-C   1C-CH3    H3C-C19 C-C22     49.70   175.70   111.90   128.90   136.00   123.10
  |    |    15 17    |    |         1/1     2/1      4/2      5/2      6/1      7/2
   3N---C2   CH   CH   C===N+      141.70   140.20    31.80    26.50    20.80   100.90
10/      \ / \ / \ /20  21\28       8/1     9/1     10/4     11/4     13/4     14/2
   CH3       CH   C16 CH    H3C    152.40    97.90   115.80
          14   |    18            15/2     16/1     31/1
             31C≡N                   -
                     BF4                    W.GRAHN,C.REICHARDT
                                            TETRAHEDRON               32,  125  (1976)
```

```
2626                      T 013000  BIS(1,3,3-TRIMETHYL-INDOLENIN-2-YL)-3-CYCLO-
   6  7                             PROPYL-PENTAMETHINIUM TETRAFLUOROBORATE
   C=C               24C=C23
  /  | 12              29 |  \  FORMULA   C30H35BF4N2           MOL WT   510.43
5C   8C CH3           H3C C25 C22SOLVENT  C2D6OS
 || 9% \| 11            28 1/ \26|| ORIG ST  C4H12SI            TEMP        AMB
4C-C   1C-CH3    H3C-C18 C-C21
  |    |    14 16    |    |         48.80   173.50   110.80   128.50   124.90   122.40
   3N---C2   CH   CH   C===N+       1/1     2/1      4/2      5/2      6/2      7/2
10/      \ / \ / \ /19  21\27      141.20   142.90    31.00    27.00   100.80   154.60
   CH3       CH   C15 CH    H3C     8/1     9/1     10/4     11/4     13/2     14/2
          13  |    17              132.40     7.10     6.70
             C30           -       15/1     30/2     31/3
            / \          BF4
          32C---C31                         W.GRAHN,C.REICHARDT
                                            TETRAHEDRON               32,  125  (1976)
```

```
2627                    T 013000  BIS(1,3,3,5-TETRAMETHYL-INDOLENIN-2-YL)-
  13                          31  3-METHOXY-PENTAMETHINIUM TETRAFLUOROBORATE
 CH3                         H3C
   \  7              25  /       FORMULA  C30H37BF4N2O          MOL WT   528.45
    6C=C              C=C 24     SOLVENT  C2D6OS
   /  | 12           30 |  \     ORIG ST  C4H12SI               TEMP        AMB
 5C  8C CH3        H3C C26 C23
  || 9⁄ \| 11      29 |/ \27||   48.80   172.20   110.60   128.70   134.30   123.00
 4C-C  1C-CH3     H3C-C19 C-C22    1/1     2/1      4/2      5/2      6/1      7/2
  |    |  15 17      |    |       141.20   140.60    31.10    26.70    20.70    97.40
   3N---C2  CH  CH  C===N+         8/1     9/1      10/4     11/4     13/4     14/2
 10/     \ / \ / \ /20 21\28      142.70   149.70    59.30
 CH3       CH  C16 CH    H3C       15/2    16/1     32/4
          14  |  18
              O              _
            32/           BF4      W.GRAHN,C.REICHARDT
           H3C                     TETRAHEDRON               32,   125 (1976)
```

```
2628                    T 013000  BIS(1,3,3,-TRIMETHYL-INDOLENIN-2-YL)-3-CYCLO-
  6 7                             BUTYL-PENTAMETHINIUM TETRAFLUOROBORATE
 C=C  .              24C=C 23
   /  | 12           29 |  \     FORMULA  C31H37BF4N2          MOL WT   524.46
 5C  8C CH3        H3C C25 C22   SOLVENT  C2D6OS
  || 9⁄ \| 11      28 |/ \26||   ORIG ST  C4H12SI              TEMP        AMB
 4C-C  1C-CH3     H3C-C18 C-C21
  |    |  14 16      |    |        48.70   173.10   111.10   128.60   124.90   122.40
   3N---C2  CH  CH  C===N+         1/1     2/1      4/2      5/2      6/2      7/2
 10/     \ / \ / \ /19 20\27      141.10   142.90    31.20    27.10   100.50   151.30
 CH3       CH  C15 CH    H3C       8/1     9/1      10/4     11/4     13/2     14/2
          13  |  17             135.10    35.00    29.20    18.50
              C30                 15/1    30/2     31/3     32/3
            /  \              _
          33C   C31           BF4
            \  /
             C32                   W.GRAHN,C.REICHARDT
                                   TETRAHEDRON               32,   125 (1976)
```

```
2629                    T 013000  BIS(1,3,3-TRIMETHYL-INDOLENIN-2-YL)-3-PHENYL-
  6 7                             PENTAMETHINIUM TETRAFLUOROBORATE
 C=C                24C=C 23
   /  | 12           29 |  \     FORMULA  C33H35BF4N2          MOL WT   546.46
 5C  8C CH3        H3C C25 C22   SOLVENT  C2D6OS
  || 9⁄ \| 11      28 |/ \26||   ORIG ST  C4H12SI              TEMP        AMB
 4C-C  1C-CH3     H3C-C18 C-C21
  |    |  14 16      |    |        49.00   173.50   111.10   128.40   125.00   122.40
   3N---C2  CH  CH  C===N+         1/1     2/1      4/2      5/2      6/2      7/2
 10/     \ / \ / \ /19 20\27      141.10   142.70    30.70    26.80   100.80   152.90
 CH3       CH  C15 C     H3C       8/1     9/1      10/4     11/4     13/2     14/2
          13  |  17             135.40   134.80   129.30   130.00   128.10
              C30                 15/1    30/1     31/2     32/2     33/2
            /  \              _
          35C   C31           BF4
            ||   |
          34C   C32
            \  ⁄
             C33                   W.GRAHN,C.REICHARDT
                                   TETRAHEDRON               32,   125 (1976)
```

```
2630                    T 013000  BIS(1,3,3-TRIMETHYL-INDOLENIN-2-YL)-3-PHENYL-
                                  AZO-PENTAMETHINIUM TETRAFLUOROBORATE
  6 7
 C=C                24C=C 23
   /  | 12           29 |  \     FORMULA  C33H35BF4N4          MOL WT   574.48
 5C  8C CH3        H3C C25 C22   SOLVENT  C2D6OS
  || 9⁄ \| 11      28 |/ \26||   ORIG ST  C4H12SI              TEMP        AMB
 4C-C  1C-CH3     H3C-C18 C-C21
  |    |  14 16      |    |        49.70   177.40   112.30   128.60   126.10   122.60
   3N---C2  CH  CH  C===N+         1/1     2/1      4/2      5/2      6/2      7/2
 10/     \ / \ / \ /19 20\27      142.00   142.60    31.80    26.90   101.60   146.00
 CH3       CH  C15 CH    H3C       8/1     9/1      10/4     11/4     13/2     14/2
          13  |  17             136.70   153.20   122.10   129.60   131.20
              N                   15/1    30/1     31/2     32/2     33/2
              ||              _
              N              BF4
              |
              C30
            /  \
          35C   C31
            ||   |
          34C   C32
            \  ||
             C33                   W.GRAHN,C.REICHARDT
                                   TETRAHEDRON               32,   125 (1976)
```

```
2631                 T 013000  BIS(1,3,3-TRIMETHYL-INDOLENIN-2-YL)-3-CYCLO-
    6  7                        HEXYL-PENTAMETHINIUM TETRAFLUOROBORATE
    C=C              24C=C23
   /   |  12          29  |  \    FORMULA   C33H41BF4N2          MOL WT  552.51
 5C   8C  CH3        H3C  C25 C22 SOLVENT   C2D6OS
 || 9/ \| 11         28 |/ \26||  ORIG ST   C4H12SI              TEMP      AMB
 4C-C  1C-CH3        H3C-C18 C-C21
   |   |   14 16      |    |   |       48.90  173.20  111.20  128.60  125.00  122.40
   3N---C2  CH  CH  C===N+             1/1     2/1     4/2     5/2     6/2     7/2
 10/     \ / \ / \ /19 20\27         141.20  143.00   31.30   27.30  100.00  149.90
  CH3    CH C15 CH    H3C             8/1     9/1    10/4    11/4    13/2    14/2
         13  |  17                   135.10   36.90   30.90   26.40   25.80
            C30            -          15/1    30/2    31/3    32/3    33/3
           /  \         BF4
         35C   C31
          |    |
         34C   C32
           \  /                    W.GRAHN,C.REICHARDT
            C33                    TETRAHEDRON                   32,  125 (1976)
```

```
2632               U 002000  CARVONE

                              FORMULA   C10H14O              MOL WT  150.22
    10      5 6                SOLVENT   CDCL3
   H3C      C-C                ORIG ST   C4H12SI              TEMP      AMB
     \8 4/    \\ 7
      C-C      C-CH3          135.30  198.60   42.50   43.10   31.20  144.20
     /   \    /1               1/1     2/1     3/3     4/2     5/3     6/2
   H2C   3C-C2                 15.60  146.60  110.40   20.40
    9        \\               7/4     8/1     9/3    10/4
              O
                             L.F.JOHNSON,W.C.JANKOWSKI
                             CARBON-13 NMR SPECTRA,JOHN
                             WILEY AND SONS,NEW YORK           386 (1972)
```

```
2633               U 002000  VERBENONE
         10
         CH3                  FORMULA   C10H14O              MOL WT  150.22
          | 2                 SOLVENT   CDCL3
         3C===C               ORIG ST   C4H12SI              TEMP      AMB
        /  9 \
       CH3    \              203.00  121.00  169.70   49.60   40.60   57.50
      / 7/     \1             1/1     2/2     3/1     4/2     5/3     6/2
   4C---C-CH3  C=O            53.60   26.40   23.30   21.90
      \  \8   /               7/1     8/4     9/4    10/4
       \   \ /
        \   \ /
         C---C              L.F.JOHNSON,W.C.JANKOWSKI
         5   6              CARBON-13 NMR SPECTRA,JOHN
                           WILEY AND SONS,NEW YORK           387 (1972)
```

```
2634               U 002000  CAMPHENE
         10   9
        H3C   CH3            FORMULA   C10H16               MOL WT  136.24
          \  /               SOLVENT   CDCL3
         3C---C2             ORIG ST   C4H12SI              TEMP      AMB
        /  \    \
       /    \    \          165.90   41.70   48.20   23.80   28.90   47.00
      /      \    \1 8       1/1     2/1     3/2     4/3     5/3     6/2
   4C        7C   C=CH2      37.40   99.10   29.40   25.80
     \        |   /          7/3     8/3    10/4    9/4
      \       | /
       \      |/
        5C---C6            L.F.JOHNSON,W.C.JANKOWSKI
                           CARBON-13 NMR SPECTRA,JOHN
                           WILEY AND SONS,NEW YORK           393 (1972)
```

2635   U 002000 3-CARENE

```
      6    9
      C   CH3        FORMULA  C10H16              MOL WT  136.24
  10 1/  \5  |       SOLVENT  CDCL3
 H3C-C    C--C-CH3   ORIG ST  C4H12SI             TEMP       AMB
    ||    | /7 8
    ||    |/          131.20  119.50   24.90   18.70   16.80   20.80
   2C    C4            1/1     2/2      3/3      4/2     5/2     6/3
     \  /              16.70   28.40   13.10   23.60
      C                7/1      8/4     9/4     10/4
      3
```

```
              L.F.JOHNSON,W.C.JANKOWSKI
              CARBON-13 NMR STECTRA,JOHN
              WILEY AND SONS,NEW YORK              394 (1972)
```

2636   U 002000 ALPHA-PINENE

```
        7
       CH3          FORMULA  C10H16              MOL WT  136.24
      /             SOLVENT  C4H8O2
   6 /              ORIG ST  C4H12SI             TEMP       AMB
   C---C1
  / \ / \
 /  9\8   \C2        144.20  116.10   31.50   41.00   31.50   47.30
 5C H3C-CR C2         1/1     2/2      3/3      4/2     5/3     6/2
  \   /   /           20.80   38.10   26.40   22.80
   \ /   /            7/4      8/1     9/4     10/4
   4C---C3
```

```
            10    L.F.JOHNSON,W.C.JANKOWSKI
          R -CH3   CARBON-13 NMR SPECTRA,JOHN
                   WILEY AND SONS,NEW YORK         396 (1972)
```

2637   U 002000 BETA-PINENE

```
                    FORMULA  C10H16              MOL WT  136.24
   3    2           SOLVENT  CDCL3
   C---C            ORIG ST  C4H12SI             TEMP       AMB
  / \ / \
 /   \   \          151.80   51.90   27.00   40.50   23.60   23.60
/  7/9   \1          1/1     2/2      3/3      4/2     5/3     6/3
4C---C-CH3  C=CH2    40.50   21.80   26.10  106.00
 \    \    / 10      7/1      8/4     9/4     10/3
  \   CH3 /
   \   8 /          L.F.JOHNSON,W.C.JANKOWSKI
    C---C           CARBON-13 NMR SPECTRA,JOHN
    5   6           WILEY AND SONS,NEW YORK         395 (1972)
```

2638   U 002000 CAMPHOR

```
     10CH3          FORMULA  C10H16O             MOL WT  152.24
       /            SOLVENT  CDCL3
   3C---C2          ORIG ST  C4H12SI             TEMP       AMB
  / \ / \
 /   \   \          218.40   57.40   29.90   27.00   43.20   43.10
/   /9   \1          1/1     2/1      3/3      4/3     5/2     6/3
4C  7C-CH3  C=0     46.60   19.10   19.70    9.20
 \  |\    /          7/1      8/4     9/4     10/4
  \ | CH3 /
   \| 8  /          L.F.JOHNSON,W.C.JANKOWSKI
   5C---C6          CARBON-13 NMR SPECTRA,JOHN
                    WILEY AND SONS,NEW YORK         397 (1972)
```

2639   U 002000 FENCHONE

```
   8
   CH3  O           FORMULA  C10H16O             MOL WT  152.24
   ||  //           SOLVENT  CDCL3
   C--C2  10         ORIG ST  C4H12SI             TEMP       AMB
  /|   \  CH3
 6/ |    \3/         53.90  222.30   47.20   45.30   24.90   31.80
 C  C7    C           1/1     2/1      3/1      4/2     5/3     6/3
  \  \   / \9        41.60   14.50   23.30   21.60
   \  \ /  CH3        7/3      8/4     9/4     10/4
    5C--C4
```

```
              L.F.JOHNSON,W.C.JANKOWSKI
              CARBON-13 NMR SPECTRA,JOHN
              WILEY AND SONS,NEW YORK              398 (1972)
```

2640   U 002000 PULEGONE

```
                    FORMULA  C10H16O              MOL WT  152.24
                    SOLVENT  CDCL3
           O        ORIG ST  C4H12SI              TEMP        AMB
           ||    9
        3C-C2   CH3        131.70  199.10  50.70  31.50  32.80  28.50
       4/   \1  /           1/1     2/1    3/3    4/2    5/3    6/3
     H3C-C     C=C         141.50  22.90   22.00  21.70
      10 \   / 7\           7/1     8/4    9/4    10/4
         C-C     CH3
         5 6     8     L.F.JOHNSON,W.C.JANKOWSKI
                         CARBON-13 NMR SPECTRA,JOHN
                         WILEY AND SONS,NEW YORK        399 (1972)
```

2641   U 002000 LIMONENE

```
      2 3      9        FORMULA  C10H18              MOL WT  138.25
      C-C     CH2       SOLVENT  CDCL3
    7 1/  \4 8/         ORIG ST  C4H12SI             TEMP        AMB
  H3C-C     C-C
      \    /   \10      133.50  120.70  30.90  41.20  28.00  30.70
      6C-C5    CH3       1/1     2/2    3/3    4/2    5/3    6/3
                         23.40  149.90  108.40  20.70
                          7/4     8/1    9/3    10/4

                    L.F.JOHNSON,W.C.JANKOWSKI
                    CARBON-13 NMR SPECTRA,JOHN
                    WILEY AND SONS,NEW YORK         400 (1972)
```

2642   U 002000 CINEOLE

```
     O
     |\              FORMULA  C10H18O             MOL WT  154.25
     |  \            SOLVENT  CDCL3
     |    \          ORIG ST  C4H12SI             TEMP        AMB
     |  5 6\    8
     | C-C  \  CH3   32.90  22.80   31.50  73.50  31.50  22.80
     |/    \ \ /      1/2    2/3    3/3    4/2    5/3    6/3
  10 |/      \ \ /   69.60  28.80   28.80  27.50
  H3C-C       C-C     7/1    8/4    9/4    10/4
     4\    /1 7\
       C-C     CH3  L.F.JOHNSON,W.C.JANKOWSKI
       3 2     9    CARBON-13 NMR SPECTRA,JOHN
                    WILEY AND SONS,NEW YORK        406 (1972)
```

2643   U 002000 CITRONELLAL

```
                O       FORMULA  C10H18O             MOL WT  154.25
         10    ||       SOLVENT  CDCL3
         CH3  C8        ORIG ST  C4H12SI             TEMP        AMB
  1     4  5  |  / \
 H3C   CH2-CH2-CH-CH2 H   17.60  131.50  124.10  25.40  37.00  27.80
  \2 3/      6  7        1/4     2/1    3/2    4/3    5/3    6/2
   C=C                   51.00  202.20  25.60  19.80
  /   \   H               7/3     8/2    9/4    10/4
 H3C
  9               L.F.JOHNSON,W.C.JANKOWSKI
                  CARBON-13 NMR SPECTRA,JOHN
                  WILEY AND SONS,NEW YORK           403 (1972)
```

2644   U 002000 GERANIOL

```
   8        10    1    FORMULA  C10H18O             MOL WT  154.25
  H3C   H CH3  CH2-OH  SOLVENT  C4H8O2
    \7 6/  \3 2/       ORIG ST  C4H12SI             TEMP        AMB
    C=C     C=C
   9/  \5 4/   \       58.60  125.30  136.90  39.80  26.80  124.40
  H3C  H2C-CH2   H      1/3     2/2    3/1    4/3    5/3    6/2
                       131.10  25.60  17.60  16.10
                        7/1     8/4    9/4    10/4

                  L.F.JOHNSON,W.C.JANKOWSKI
                  CARBON-13 NMR SPECTRA,JOHN
                  WILEY AND SONS,NEW YORK           402 (1972)
```

2645   U 002000 ISOPULEGOL

```
           OH
      3  2/      8
       C-C      CH2
   10 4/    \1 7/
  H3C-C        C-C
      \   /   \9
       C-C      CH3
       5  6
```

FORMULA   C10H18O   MOL WT 154.25
SOLVENT   CDCL3
ORIG ST   C4H12SI   TEMP   AMB

54.10	70.40	42.90	31.50	34.40	29.90
1/2	2/2	3/3	4/2	5/3	6/3
146.70	112.40	19.30	22.20		
7/1	8/3	9/4	10/4		

L.F.JOHNSON,W.C.JANKOWSKI
CARBON-13 NMR SPECTRA,JOHN
WILEY AND SONS,NEW YORK   404 (1972)

2646   U 002000 LINALOOL

FORMULA   C10H18O   MOL WT 154.25
SOLVENT   CDCL3
ORIG ST   C4H12SI   TEMP   AMB

```
              OH
   1      4    5  | 9
  H3C   CH2-CH2-C-CH3
    \2 3/       6\
     C=C          C=CH2
    /    \       /7 8
  H3C     H     H
  10
```

17.60	131.50	124.50	27.70	42.20	73.30
1/4	2/1	3/2	4/3	5/3	6/1
145.10	111.50	22.80	25.60		
7/2	8/3	9/4	10/4		

L.F.JOHNSON,W.C.JANKOWSKI
CARBON-13 NMR SPECTRA,JOHN
WILEY AND SONS,NEW YORK   407 (1972)

2647   U 002000 MENTHONE OXIME

```
          NOH
      3  2/      8
       C-C      CH3
   10 4/    \1 7/
  H3C-C        C-CH
      \   /   \9
       C-C      CH3
       5  6
```

FORMULA   C10H19NO   MOL WT 169.27
SOLVENT   CDCL3
ORIG ST   C4H12SI   TEMP   AMB

48.70	160.80	32.80	32.30	31.90	26.80
1/2	2/1	3/3	4/2	5/3	6/3
26.30	21.70	19.00	21.40		
7/2	8/4	9/4	10/4		

L.F.JOHNSON,W.C.JANKOWSKI
CARBON-13 NMR SPECTRA,JOHN
WILEY AND SONS,NEW YORK   408 (1972)

2648   U 002000 MENTHOL

```
          OH
      3  2/      8
       C-C      CH3
   10 4/    \1 7/
  H3C-C        C-C4
      \   /   \9
       C-C      CH3
       5  6
```

FORMULA   C10H20O   MOL WT 156.27
SOLVENT   CDCL3
ORIG ST   C4H12SI   TEMP   AMB

50.10	71.30	45.10	31.70	34.60	23.20
1/2	2/2	3/3	4/2	5/3	6/3
25.80	16.10	21.00	22.20		
7/2	8/4	9/4	10/4		

L.F.JOHNSON,W.C.JANKOWSKI
CARBON-13 NMR SPECTRA,JOHN
WILEY AND SONS,NEW YORK   410 (1972)

2649   U 002000 MARRUBIIN

```
          O  \15
       14/   \
        C     C
     12 ||    ||
     H2C-C---C
     20 11/  13  16
     H3C CH2  H
       |   |
     1C  |  C-O CH3
     /  \|/9\  /17
    2C   C10 C8
    |    |   |
    3C   C5  C7
     \4/  \  /
    H3C-C    C6
    18 |     |
       C---O
      /19
     O
```

FORMULA   C20H28O4   MOL WT 332.44
SOLVENT   CDCL3
ORIG ST   C4H12SI   TEMP   AMB

35.30	18.20	28.60	43.90	44.90	76.60
1/3	2/3	3/3	4/1	5/2	6/2
31.60	32.30	75.70	39.80	28.30	21.00
7/3	8/2	9/1	10/1	11/3	12/3
125.50	111.00	143.20	138.80	16.60	23.00
13/1	14/2	15/2	16/2	17/4	18/4
184.30	22.30				
19/1	20/4				

S.O.ALMQVIST,C.R.ENZELL,F.W.WEHRLI
ACTA CHEM SCAND   29, 695 (1975)

2650   U 002000 2-OXOMANOYL OXIDE

```
          16
      12C   CH3
    20 /  \ /
   H3C C11 C13
      |  |  |\
    O 1C 19C  O CH=CH2
     \ / \ / \R/  14 15
    2C 10C   C8
    |   |    |
    3C  C5   C7      17
     \4/ \  /    R -CH3
      C    C6
     / \
   H3C   CH3
   19    18
```

FORMULA   C20H32O2   MOL WT 304.48
SOLVENT   CDCL3
ORIG ST   C4H12SI   TEMP   AMB

54.60	211.00	56.40	38.70	55.60	20.10
1/3	2/1	3/3	4/1	5/2	6/3
42.60	74.60	55.10	42.20	15.50	35.50
7/3	8/1	9/2	10/1	11/3	12/3
73.20	147.50	110.40	28.50	25.00	33.40
13/1	14/2	15/3	16/4	17/4	18/4
22.90	16.50				
19/4	20/4				

S.O.ALMQVIST,C.R.ENZELL,F.W.WEHRLI
ACTA CHEM SCAND   29, 695 (1975)

2651   U 002000 3-OXOMANOYL OXIDE

```
          16
      12C   CH3
    20 /  \ /
   H3C C11 C13
      |  |  |\
    1C 19C  O CH=CH2
    /  \|/ \R/  14 15
   2C 10C   C8
   |   |    |
   3C  C5   C7      17
  // \4/ \  /    R -CH3
  O   C    C6
     / \
   H3C   CH3
   19    18
```

FORMULA   C20H32O2   MOL WT 304.48
SOLVENT   CDCL3
ORIG ST   C4H12SI   TEMP   AMB

37.80	33.80	217.30	47.30	54.90	20.80
1/3	2/3	3/1	4/1	5/2	6/3
42.50	74.40	54.90	36.50	15.60	35.70
7/3	8/1	9/2	10/1	11/3	12/3
73.40	147.90	110.30	28.50	24.80	26.60
13/1	14/2	15/3	16/4	17/4	18/4
20.80	15.00				
19/4	20/4				

S.O.ALMQVIST,C.R.ENZELL,F.W.WEHRLI
ACTA CHEM SCAND   29, 695 (1975)

2652   U 002000 MANOOL

```
        12
      H2C   OH
   20 11/  \ /16
   H3C CH2 C-CH3
     | |  13\14 15
    1C 19C    CH=CH2
    /  \|/  \8
   2C 10C   C=CH2
   |   |    |  17
   3C  C5   C7
    \4/ \  /
     C    C6
    / \
  H3C   CH3
  19    18
```

FORMULA   C20H34O   MOL WT 290.49
SOLVENT   CDCL3
ORIG ST   C4H12SI   TEMP   AMB

39.00	19.30	42.20	33.30	55.60	24.30
1/3	2/3	3/3	4/1	5/2	6/3
38.30	148.60	57.20	39.70	17.60	41.40
7/3	8/1	9/2	10/1	11/3	12/3
73.50	145.10	111.40	27.90	106.30	33.60
13/1	14/2	15/3	16/4	17/3	18/4
21.60	14.40				
19/4	20/4				

S.O.ALMQVIST,C.R.ENZELL,F.W.WEHRLI
ACTA CHEM SCAND   29, 695 (1975)

```
2653              U 002000 MANOYL OXIDE

              16              FORMULA   C20H34O                    MOL WT  290.49
        12C     CH3           SOLVENT   CDCL3
      20   / \ /              ORIG ST   C4H12SI                    TEMP         AMB
      H3C  C11 C13
       |    |    |  |\          39.00    18.60    42.10    33.20    56.40    19.90
      1C  19C    O CH=CH2        1/3      2/3      3/3      4/1      5/2      6/3
      / \|/ \R/ 14 15          43.20    74.80    55.70    36.90    15.40    35.80
     2C  10C    C8               7/3      8/1      9/2     10/1     11/3     12/3
      |    |    |       17      73.00   147.80   110.10    28.50    25.50    33.40
      3C   C5   C7    R -CH3     13/1     14/2     15/3     16/4     17/4     18/4
       \4/ \  /                 21.30    15.30
        C    C6                  19/4     20/4
       / \
     H3C    CH3                S.O.ALMQVIST,C.R.ENZELL,F.W.WEHRLI
      19    18                 ACTA CHEM SCAND              29,  695 (1975)
```

```
2654              U 002000 2-ALPHA-HYDROXYMANOYL OXIDE

              16              FORMULA   C20H34O2                   MOL WT  306.49
        12C     CH3           SOLVENT   CDCL3
      20   / \ /              ORIG ST   C4H12SI                    TEMP         AMB
      H3C  C11 C13
       |    |    |  |\          48.10    64.80    51.10    34.70    54.80    19.50
     H C  19C    O CH=CH2        1/3      2/2      3/3      4/1      5/2      6/3
     1/\|/ \R/ 14 15           43.00    74.50    55.80    38.50    15.40    35.30
     2C  10C    C8               7/3      8/1      9/2     10/1     11/3     12/3
     /|    |    |       17      73.00   147.10   109.70    28.60    25.40    33.30
     HO C3  C5  C7  R -CH3      13/1     14/2     15/3     16/4     17/4     18/4
       \4/ \  /                 22.10    16.30
        C    C6                  19/4     20/4
       / \
     H3C    CH3                S.O.ALMQVIST,C.R.ENZELL,F.W.WEHRLI
      19    18                 ACTA CHEM SCAND              29,  695 (1975)
```

```
2655              U 002000 DIHYDROMANOYL OXIDE

              16              FORMULA   C20H36O                    MOL WT  292.51
        12C     CH3           SOLVENT   CDCL3
      20   / \ /              ORIG ST   C4H12SI                    TEMP         AMB
      H3C  C11 C13
       |    |    |  |\          39.20    18.70    42.20    33.30    56.50    19.90
      1C  19C    O CH2-CH3       1/3      2/3      3/3      4/1      5/2      6/3
      / \|/ \R/ 14  15         43.10    74.50    58.30    36.80    15.40    35.80
     2C  10C    C8               7/3      8/1      9/2     10/1     11/3     12/3
      |    |    |       17      72.80    37.90     8.00    27.20    24.80    33.40
      3C   C5   C7    R -CH3     13/1     14/3     15/4     16/4     17/4     18/4
       \4/ \  /                 21.30    15.70
        C    C6                  19/4     20/4
       / \
     H3C    CH3                S.O.ALMQVIST,C.R.ENZELL,F.W.WEHRLI
      19    18                 ACTA CHEM SCAND              29,  695 (1975)
```

```
2656              U 002000 LABD-13E-EN-8-OL

        12   14                FORMULA   C20H36O                   MOL WT  292.51
        H2C    CH              SOLVENT   CDCL3
      20 11/ \ / \15           ORIG ST   C4H12SI                   TEMP         AMB
      H3C CH2 C13 CH3
       |    |    |               39.80    18.50    42.10    33.30    56.20    20.60
      1C  19C    CH3             1/3      2/3      3/3      4/1      5/2      6/3
      / \|/ \R 16      17      43.30    74.10    61.60    39.30    23.90    44.60
     2C  10C    C8    R -CH3     7/3      8/1      9/2     10/1     11/3     12/3
      |    |    | |\           137.00   118.30    13.30    15.90    23.90    33.40
      3C   C5  7C OH            13/1     14/2     15/4     16/4     17/4     18/4
       \4/ \  /                 21.50    15.50
        C    C6                  19/4     20/4
       / \
     H3C    CH3                S.O.ALMQVIST,C.R.ENZELL,F.W.WEHRLI
      19    18                 ACTA CHEM SCAND              29,  695 (1975)
```

2657   U 002000 SCLAREOL
```
          12
        H2C   OH        FORMULA  C20H36O2              MOL WT  308.51
     20 11/ \ /16       SOLVENT  CDCL3
     H3C CH2 C-CH3      ORIG ST  C4H12SI               TEMP      AMB
      |  |  13\
      1C 19C    CH=CH2      39.50   1H.80   41.90   33.10   55.90   20.40
     / \|/ \8  14 15        1/3     2/3     3/3     4/1     5/2     6/3
    2C 10C    C-CH3         43.90   74.30   61.40   39.00   18.40   44.80
     |   |    | 17          7/3     8/1     9/2    10/1    11/3    12/3
    3C  C5 7C OH           73.00  146.20  110.30   26.20   24.00   33.30
     \4/ \ /               13/1    14/2    15/3    16/4    17/4    18/4
      C   C6               21.40   15.30
     / \                   19/4    20/4
   H3C   CH3
   19    18               S.O.ALMQVIST,C.R.ENZELL,F.W.WEHRLI
                          ACTA CHEM SCAND           29,  695 (1975)
```

2658   U 002000 SQUALENE
```
    1       26      8
  H3C   H H3C    CH2        FORMULA  C30H50              MOL WT  410.73
    \   \ \   / /           SOLVENT  C4H8O2
    C=C       C=C   \       ORIG ST  C4H12SI             TEMP      AMB
   /2 3\4   5/6 7\   \
  H3C    CH2-CH2  H  9CH2    25.50  130.70  124.60   27.00   40.00  134.80
   25                  |      1/4     2/1     3/2     4/3     5/3     6/1
      16CH2   H   H   |     124.60   27.00   40.00  134.80  124.60   28.50
      / \   /   \   |        7/2     8/3     9/3    10/1    11/2    12/3
     / 15C=C14  11C=C10      28.50  124.60  134.80   40.00   27.00  124.60
     / 28/  \13 12/  /27    13/3    14/2    15/1    16/3    17/3    18/2
   17CH2 CH3  CH2-CH2 CH3   134.80   40.00   27.00  124.60  130.70   25.50
    |                       19/1    20/3    21/3    22/2    23/1    24/4
    | 29CH3 H    CH3         25.50   15.90   15.90   15.90   15.90   17.50
    | /    \  /24           25/4    26/4    27/4    28/4    29/4    30/4
  18C=C19  22C=C23
   /   \20/  /   \          L.F.JOHNSON,W.C.JANKOWSKI
  H     CH2-CH2   CH3       CARBON-13 NMR SPECTRA,JOHN
        21      30          WILEY AND SONS,NEW YORK        497 (1972)
```

2659   U 003000 3-HYDROXYESTRA-1,3,5(10)-TRIEN-16,17-DIONE
```
     18CH3 O
        | /              FORMULA  C18H20O3              MOL WT  284.36
     12C | C17 O         SOLVENT  C4H8O2
    / \|/ \ /            ORIG ST  C4H12SI               TEMP      AMB
   11C   C13 C16
    |   |   |            126.91  113.81  156.06  115.98  138.20   38.31
   1C   C9  C---C15       1/2     2/2     3/1     4/2     5/1     6/3
  / \ / \ / \ /14        27.43   29.96   44.45  131.50   26.33   31.64
  2C  C10 C8              7/3     8/2     9/2    10/1    11/3    12/3
   |   ||  |             48.71   43.20   36.09  204.58  204.58   13.73
  3C  C5  C7             13/1    14/2    15/3    16/1    17/1    18/4
  / \ / \ /
 HO  4C  C6              T.A.WITTSTRUCK,K.I.H.WILLIAMS
                         J ORG CHEM                38,  1542 (1973)
```

2660   U 003000 ESTRA-1,3,5(10)-TRIEN-17-ONE
```
    H3C18 O
       | /               FORMULA  C18H22O               MOL WT  254.38
    12C | C17            SOLVENT  C4H8O2
   / \|/ \               ORIG ST  C4H12SI               TEMP      AMB
   11C   C  C16
    |   |   |            126.31  126.69  126.69  129.86  137.49   39.06
   1C   C9  C---C15       1/2     2/2     3/2     4/2     5/1     6/3
  / \ / \ / \ /14        27.33   30.04   45.43  141.10   26.31   32.56
  2C  C10 C8              7/3     8/2     9/2    10/1    11/3    12/3
   |   ||  |             48.37   51.26   22.16   35.93  218.92   13.92
  3C  C5  C7             13/1    14/2    15/3    16/3    17/1    18/4
   \ / \ /
    4C  C6               T.A.WITTSTRUCK,K.I.H.WILLIAMS
                         J ORG CHEM                38,  1542 (1973)
```

2661 — U 003000 3-HYDROXYESTRA-1,3,5(10)-TRIEN-17-ONE (ESTRONE)

```
        18CH3 O
          |  //
       12C | C17
       / \|/ \
     11C   C13 C16
      |    |   |
    1C   C9  C---C15
    // \ / \ /14
  2C   C10 C8
   |    ||  |
  3C   C5  C7
  / \ / \ /
HO  4C   C6
```

FORMULA	C18H22O2			MOL WT	270.37
SOLVENT	C4H8O2				
ORIG ST	C4H12SI			TEMP	AMB

126.90	113.45	155.76	115.86	138.20	39.33
1/2	2/2	3/1	4/2	5/1	6/3
27.41	30.19	44.95	131.92	26.44	32.51
7/3	8/2	9/2	10/1	11/3	12/3
48.33	51.11	22.17	35.86	219.98	13.92
13/1	14/2	15/3	16/3	17/1	18/4

T.A.WITTSTRUCK,K.I.H.WILLIAMS
J ORG CHEM   38, 1542 (1973)

2662 — U 003000 3,16-ALPHA-DIHYDROXYESTRA-1,3,5(10)-TRIEN-17-ONE

```
        18CH3 O
          |  //
       12C | C17 OH
       / \|/ \ /
     11C   C13 C16
      |    |   |
    1C   C9  C---C15
    // \ / \ /14
  2C   C10 C8
   |    ||  |
  3C   C5  C7
  / \ / \ /
HO  4C   C6
```

FORMULA	C18H22O3			MOL WT	286.37
SOLVENT	C2D6OS				
ORIG ST	C4H12SI			TEMP	300

126.15	112.85	155.15	115.00	137.15	38.00
1/2	2/2	3/1	4/2	5/1	6/3
26.10	29.15	43.50	130.00	25.35	31.50
7/3	8/2	9/2	10/1	11/3	12/3
47.05	47.35	37.75	70.35	219.00	14.25
13/1	14/2	15/3	16/2	17/1	18/4

E.BREITMAIER,W.VOELTER
UNPUBLISHED   (1974)

2663 — U 003000 ESTRA-1,3,5(10)-TRIENE

```
        18
        CH3
       12C | C17
       / \|/ \
     11C 13C   C16
      |   |    |
    1C   C9  C---C15
    // \ / \ /14
  2C   C10 C8
   |    ||  |
  3C   C5  C7
  \ / \ /
    4C   C6
```

FORMULA	C18H24			MOL WT	240.39
SOLVENT	C4H8O2				
ORIG ST	C4H12SI			TEMP	AMB

125.96	126.22	126.22	129.73	137.40	39.48
1/2	2/2	3/2	4/2	5/1	6/3
26.87	29.95	45.12	141.47	28.56	40.89
7/3	8/2	9/2	10/1	11/3	12/3
41.44	54.19	25.54	20.88	39.24	17.56
13/1	14/2	15/3	16/3	17/3	18/4

T.A.WITTSTRUCK,K.I.H.WILLIAMS
J ORG CHEM   38, 1542 (1973)

2664 — U 003000 3-HYDROXYESTRA-1,3,5(10)-TRIENE

```
        18
        CH3
       12C | C17
       / \|/ \
     11C 13C   C16
      |   |    |
    1C   C9  C---C15
    // \ / \ /14
  2C   C10 C8
   |    ||  |
  3C   C5  C7
  / \ / \ /
HO  4C   C6
```

FORMULA	C18H24O			MOL WT	256.39
SOLVENT	C4H8O2				
ORIG ST	C4H12SI			TEMP	AMB

126.88	113.40	155.59	115.76	138.42	39.94
1/2	2/2	3/1	4/2	5/1	6/3
27.25	30.10	44.64	132.44	28.67	40.96
7/3	8/2	9/2	10/1	11/3	12/3
41.47	54.05	25.56	20.90	39.31	17.64
13/1	14/2	15/3	16/3	17/3	18/4

T.A.WITTSTRUCK,K.I.H.WILLIAMS
J ORG CHEM   38, 1542 (1973)

2665 — U 003000 17-BETA-HYDROXYESTRA-1,3,5(10)-TRIENE

```
      H3C18 OH
         |  /
       12C | C17
       / \|/ \
     11C 13C   C16
      |   |    |
    1C   C9  C---C15
    // \ / \ /14
  2C   C10 C8
   |    ||  |
  3C   C5  C7
   \ / \ /
    4C   C6
```

FORMULA	C18H24O			MOL WT	256.39
SOLVENT	C4H8O2				
ORIG ST	C4H12SI			TEMP	AMB

126.12	126.44	126.44	129.76	137.50	39.69
1/2	2/2	3/2	4/2	5/1	6/3
26.98	30.23	45.53	141.39	28.18	37.88
7/3	8/2	9/2	10/1	11/3	12/3
44.08	51.25	23.90	31.25	81.86	11.61
13/1	14/2	15/3	16/3	17/2	18/4

T.A.WITTSTRUCK,K.I.H.WILLIAMS
J ORG CHEM   38, 1542 (1973)

2666   U 003000 3,17—ALPHA—DIHYDROXYESTRA—1,3,5(10)—TRIENE

```
        18CH3 JH
          |  /
       12C | C17           FORMULA   C18H24O2              MOL WT   272.39
      / \|/ \             SOLVENT   C4H8O2
     11C   C13 C16         ORIG ST   C4H12SI               TEMP        AMB
      |    |   |
    1C   C9  C---C15        127.17   113.71   155.73   116.05   138.66   40.13
    / \ / \ /14             1/2      2/2      3/1      4/2      5/1      6/3
   2C   C10 C8              27.05    30.37    44.59    132.64   28.92    33.14
    |    ||  |              7/3      8/2      9/2      10/1     11/3     12/3
   3C   C5  C7              46.22    48.39    24.88    32.36    79.70    17.53
   / \ / \ /                13/1     14/2     15/3     16/3     17/2     18/4
  HO   4C   C6
                           T.A.WITTSTRUCK,K.I.H.WILLIAMS
                           J ORG CHEM                       38, 1542 (1973)
```

2667   U 003000 3,17—BETA—DIHYDROXYESTRA—1,3,5(10)—TRIENE

```
        18CH3 JH
          |  /
       12C | C17           FORMULA   C18H24O2              MOL WT   272.39
      / \|/ \             SOLVENT   C4H8O2
     11C   C13 C16         ORIG ST   C4H12SI               TEMP        AMB
      |    |   |
    1C   C9  C---C15        126.90   113.51   155.64   115.92   138.43   39.78
    / \ / \ /14             1/2      2/2      3/1      4/2      5/1      6/3
   2C   C10 C8              27.10    30.18    44.79    132.28   28.01    37.61
    |    ||  |              7/3      8/2      9/2      10/1     11/3     12/3
   3C   C5  C7              43.88    50.77    23.66    30.99    81.93    11.47
   / \ / \ /                13/1     14/2     15/3     16/3     17/2     18/4
  HO   4C   C6
                           T.A.WITTSTRUCK,K.I.H.WILLIAMS
                           J ORG CHEM                       38, 1542 (1973)
```

2668   U 003000 ESTRONE METHYL ETHER

```
           O             FORMULA   C19H24O2              MOL WT   284.40
      12   ||            SOLVENT   CDCL3
        C   C17           ORIG ST   C4H12SI               TEMP        AMB
      / \R/ \
     11C   C13 C16         126.10   111.40   157.50   113.80   137.50   29.60
    |    |   |   |         1/2      2/2      3/1      4/2      5/1      6/3
    C   9C  C---C15        26.50    38.30    43.90    131.90   25.90    31.50
   / \ / \ /14             7/3      8/2      9/2      10/1     11/3     12/3
  2C   C10 C8              47.80    50.30    21.50    35.70    220.20   55.00
   |    ||  |              13/1     14/2     15/3     16/3     17/1     19/4
  3C   C   C7              13.70
  / \  /5\ /               18/4
 H3C-O  C   C
  19   4   6       18     L.F.JOHNSON,W.C.JANKOWSKI
               R -CH3     CARBON—13 NMR SPECTRA,JOHN
                          WILEY AND SONS,NEW YORK            481 (1972)
```

2669   U 003000 16—ALPHA—METHYL—3,6—ALPHA,7—ALPHA—TRIHYDROXY—
                                  ESTRA—1,3,5(10)—TRIEN—17—ONE

```
        18CH3 J
          |  / 19
       12C | C17 CH3       FORMULA   C19H24O4              MOL WT   316.40
      / \|/ \ /           SOLVENT   C2D6OS
     11C   C13 C16         ORIG ST   C4H12SI               TEMP        300
      |    |   |
    1C   C9  C---C15        126.00   103.60   143.60   108.10   118.90   65.90
    / \ / \ /14             1/2      2/2      3/1      4/2      5/1      6/2
   2C   C10 C8              62.60    37.60    38.40    130.60   24.90    32.80
    |    ||  |              7/2      8/2      9/2      10/1     11/3     12/3
   3C   C5  C7              47.40    44.50    32.90    43.10    203.40   13.20
   / \ / \ / \              13/1     14/2     15/3     16/3     17/1     18/4
  HO   4C   C6  OH          19.20
         |                  19/4
        OH
                           E.BREITMAIER,W.VOELTER
                           UNPUBLISHED                          (1974)
```

2670

```
           18CH3 O
             |  //
         12C | C17
         / \|/ \
       11C    C13 C16
         |     |   |
    1C   C9   C---C15
     //  \  /  \  /14
    2C    C10  C8
     |     ||   |
    3C    C5   C7
    / \  / \  / \
   O  4C    C6
   |
   CH3
   19
```

U 003000 3-METHOXYESTRA-1,3,5(10)-TRIEN-17-ONE

FORMULA C19H26O2 MOL WT 286.42
SOLVENT C4H8O2
ORIG ST C4H12SI TEMP AMB

127.32	112.49	158.92	114.74	138.43	39.36
1/2	2/2	3/1	4/2	5/1	6/3
27.30	30.26	45.00	133.18	26.57	32.43
7/3	8/2	9/2	10/1	11/3	12/3
48.32	51.10	22.04	35.87	219.08	13.76
13/1	14/2	15/3	16/3	17/1	18/4
55.28					
19/4					

T.A.WITTSTRUCK,K.I.H.WILLIAMS
J ORG CHEM 38, 1542 (1973)

2671

```
           18CH3 OH
             |  /
         12C | C17
         / \|/ \
       11C    C13 C16
         |     |   |
    1C   C9   C---C15
     //  \  /  \  /14
    2C    C10  C8
     |     ||   |
    3C    C5   C7
    / \  / \  / \
   O  4C    C6
   |
   CH3
   19
```

U 003000 3-METHOXY-16-BETA-HYDROXESTRA-1,3,5(10)-TRIENE

FORMULA C19H26O2 MOL WT 286.42
SOLVENT C4H8O2
ORIG ST C4H12SI TEMP AMB

126.95	112.34	158.70	114.57	138.60	39.78
1/2	2/2	3/1	4/2	5/1	6/3
27.03	30.34	44.71	133.57	28.71	40.98
7/3	8/2	9/2	10/1	11/3	12/3
39.41	52.22	37.53	71.31	53.73	19.21
13/1	14/2	15/3	16/3	17/2	18/4
55.05					
19/4					

T.A.WITTSTRUCK,K.I.H.WILLIAMS
J ORG CHEM 38, 1542 (1973)

2672

```
           18CH3 OH
            |17/19
        12C | C-CH3
        / \|/ \
       11C    C13 C16
         |     |   |
    1C   C9   C---C15
     //  \  /  \  /14
    2C    C10  C8
     |     ||   |
    3C    C5   C7
    / \  / \  / \
   HO  4C    C6
```

U 003000 17-ALPHA-METHYL-3,17-BETA-DIHYDROXYESTRA-1,3,5(10)-TRIENE

FORMULA C19H26O2 MOL WT 286.42
SOLVENT C2D6OS
ORIG ST C4H12SI TEMP 300

126.00	112.85	155.05	115.10	137.25	38.65
1/2	2/2	3/1	4/2	5/1	6/3
26.35	29.35	43.60	130.65	27.40	39.60
7/3	8/2	9/2	10/1	11/3	12/3
45.65	49.40	22.85	31.60	79.95	14.25
13/1	14/2	15/3	16/3	17/1	18/4
26.35					
19/4					

E.BREITMAIER,W.VOELTER
UNPUBLISHED (1974)

2673

```
              O
         12   ||
          C   C17
       R1 / \R/ \
        | C11 C13 C16
        | | |   |   |
        C 19C   C---C15
       / \|/ \  /14
      2C   C10 C8
       |    |   |
      3C    C   C7
      // \ //5\ /   18
     O   C   C   R -CH3
         4   6
              19
        R1 -CH3
```

U 003000 TESTOSTERONE

FORMULA C19H28O2 MOL WT 288.43
SOLVENT CDCL3
ORIG ST C4H12SI TEMP AMB

35.60	33.80	199.40	123.60	171.40	32.70
1/3	2/3	3/1	4/2	5/1	6/3
31.50	35.60	53.90	38.60	20.60	36.40
7/3	8/2	9/2	10/1	11/3	12/3
42.70	50.40	23.20	30.10	81.20	17.30
13/1	14/2	15/3	16/3	17/1	19/4
11.00					
18/4					

L.F.JOHNSON,W.C.JANKOWSKI
CARBON-13 NMR SPECTRA,JOHN
WILEY AND SONS,NEW YORK 483 (1972)

2674

```
                    U 003000  17-ALPHA-ETHYNYL-3,17-BETA-DIHYDROXYESTRA-
      H3C18 OH               1,3,5(10)-TRIENE
        |17/19
      12C | C-C≡CH           FORMULA   C20H24O2              MOL WT  296.41
      / \|/ \19              SOLVENT   C2D6OS
     11C   C13 C16           ORIG ST   C4H12SI               TEMP      300
      |    |   |
    1C   C9  C---C15         126.15  112.85  155.05  115.00  137.25   38.75
    ⁄ \ / \ /14               1/2     2/2     3/1     4/2     5/1      6/3
   2C   C10 C8               26.10   29.25   43.35  130.45   27.10    38.75
    |   ||   |                7/3     8/2     9/2    10/1     11/3     12/3
   3C   C5  C7               46.85   49.10   22.55   32.70   78.35    12.85
   / \ / \ /                 13/1    14/2    15/3    16/3     17/1     18/4
  HO  4C   C6                89.10   75.10
                             19/1    20/2
```

E.BREITMAIER,W.VOELTER
UNPUBLISHED   (1974)

2675

```
                  U 030000  3-IODO-17-BETA-ACETOXYESTRA-1,3,5(10)-TRIENE
        18  19 20
      H3C O-C-CH3           FORMULA   C20H25IO2             MOL WT  424.33
        |  |  \             SOLVENT   C4H8O2
     12C | C17 O            ORIG ST   C4H12SI               TEMP       AMB
      / \|/ \
     11C   C13 C16          128.51  135.55   91.63  138.57  140.36   38.90
      |    |   |             1/2     2/2      3/1     4/2     5/1      6/3
    1C   C9  C---C15         27.62   29.55   44.81  141.02   26.41    37.57
    ⁄ \ / \ /14               7/3     8/2      9/2    10/1     11/3     12/3
   2C   C10 C8               43.54   50.54   23.76   28.16   83.07    12.37
    |   ||   |               13/1    14/2    15/3    16/3     17/2     18/4
   3C   C5  C7                       20.69
   / \ / \ /                 19/1    20/4
  J   4C   C6
```

T.A.WITTSTRUCK,K.I.H.WILLIAMS
J ORG CHEM   38, 1542 (1973)

2676

```
                    U 0030000  17-ALPHA-VINYL-3,17-BETA-DIHYDROXYESTRA-
      18CH3 OH                 1,3,5(10)-TRIENE
        |17/19
      12C | C-CH=CH2         FORMULA   C20H26O2             MOL WT  298.43
      / \|/ \ 20             SOLVENT   C2D6OS
     11C   C13 C16           ORIG ST   C4H12SI              TEMP       300
      |    |   |
    1C   C9  C---C15         126.15  111.45  157.10  113.50  137.45   38.75
    ⁄ \ / \ /14               1/2     2/2     3/1     4/2     5/1      6/3
   2C   C10 C8               26.10   29.35   43.35  132.30   27.10    39.60
    |   ||   |                7/3     8/2     9/2    10/1     11/3     12/3
   3C   C5  C7               46.30   48.35   23.10   31.95   82.75    14.15
   / \ / \ /                 13/1    14/2    15/3    16/3     17/1     18/4
  HO  4C   C6               144.35  111.45
                             19/2    20/3
```

E.BREITMAIER,W.VOELTER
UNPUBLISHED   (1974)

2677

```
                    U 003000  21-HYDROXYPROGESTERONE
          O
      19   \20              FORMULA   C21H30O3             MOL WT  330.47
     H3C   C-CH2-OH         SOLVENT   C2D6OS/CDCL3
      |   /   21            ORIG ST   C4H12SI              TEMP       303
     12C | C17
   19  / \|/ \              35.00   33.40  197.50  123.00  170.40   31.80
   H3C C11 C13 C16           1/3     2/3     3/1     4/2     5/1      6/3
    |   |   |   |           31.50   34.80   52.90   38.00   20.40    37.60
   1C | C9  C---C15          7/3     8/2     9/2    10/1     11/3     12/3
   / \|/ \ /14              43.50   55.30   22.30   23.90   57.40    13.10
   2C   C10 C8              13/1    14/2    15/3    16/3     17/2     18/4
    |   |   |               16.70  209.60   68.60
   3C   C5  C7              19/4    20/1    21/3
   ⁄ \ / \ /
  O   4C   C6
```

N.S.BHACCA,D.D.GIANNINI,W.S.JANKOWSKI,M.E.WOLFF
J AM CHEM SOC   95, 8421 (1973)

2678   U 003000   11-ALPHA-HYDROXYPROGESTERONE

```
          O
   18    \20
   H3C    C-CH3          FORMULA   C21H30O3              MOL WT   330.47
    |   /  21            SOLVENT   C2D6OS/CDCL3
   HO 12C | C17          ORIG ST   C4H12SI               TEMP        303
   19\E/ \|/ \
   H3C C11 C13 C16           33.80    36.90   198.40   123.50   171.30    32.80
    |   |   |   |             1/3      2/3      3/1      4/2      5/1      6/3
   1C | C9  C---C15          31.70    34.20    58.10    39.50    67.10    49.40
   / \|/ \ /14               7/3      8/2      9/2     10/1     11/2     12/3
  2C   C10 C8               43.50    54.80    22.80    23.70    62.30    14.10
   |    |   |               13/1     14/2     15/3     16/3     17/2     18/4
  3C   C5  C7               17.90   208.00    30.80
  // \ // \ /               19/4     20/1     21/4
  O  4C   C6
                        N.S.BHACCA,D.D.GIANNINI,W.S.JANKOWSKI,M.E.WOLFF
                        J AM CHEM SOC                   95, 8421 (1973)
```

2679   U 003000   11-BETA-HYDROXYPROGESTERONE

```
          O
   18    \20
   H3C    C-CH3          FORMULA   C21H30O3              MOL WT   330.47
    |   /  21            SOLVENT   C2D6OS/CDCL3
   HO 12C | C17          ORIG ST   C4H12SI               TEMP        303
   19\A/ \|/ \
   H3C C11 C13 C16           34.00    33.40   197.80   121.50   172.10    32.40
    |   |   |   |             1/3      2/3      3/1      4/2      5/1      6/3
   1C | C9  C---C15          31.30    31.00    55.60    38.80    66.20    46.80
   / \|/ \ /14               7/3      8/2      9/2     10/1     11/2     12/3
  2C   C10 C8               42.70    56.10    21.80    23.90    62.90    15.30
   |    |   |               13/1     14/2     15/3     16/3     17/2     18/4
  3C   C5  C7               20.30   208.10    30.90
  // \ // \ /               19/4     20/1     21/4
  O  4C   C6
                        N.S.BHACCA,D.D.GIANNINI,W.S.JANKOWSKI,M.E.WOLFF
                        J AM CHEM SOC                   95, 8421 (1973)
```

2680   U 003000   17-ALPHA-HYDROXYPROGESTERONE

```
          O
   18    \20
   H3C    C-CH3          FORMULA   C21H30O3              MOL WT   330.47
     |17/  21            SOLVENT   C2D6OS/CDCL3
   12C | C-OH            ORIG ST   C4H12SI               TEMP        303
  19  / \|/ \
   H3C C11 C13 C16           35.10    33.50   198.20   123.10   171.20    31.90
    |   |   |   |             1/3      2/3      3/1      4/2      5/1      6/3
   1C | C9  C---C15          30.40    35.10    53.10    38.20    20.30    32.30
   / \|/ \ /14               7/3      8/2      9/2     10/1     11/3     12/3
  2C   C10 C8               46.30    49.90    23.10    32.10    89.20    14.50
   |    |   |               13/1     14/2     15/3     16/3     17/1     18/4
  3C   C5  C7               17.00   210.30    26.60
  // \ // \ /               19/4     20/1     21/4
  O  4C   C6
                        N.S.BHACCA,D.D.GIANNINI,W.S.JANKOWSKI,M.E.WOLFF
                        J AM CHEM SOC                   95, 8421 (1973)
```

2681   U 003000   CORTEXOLONE

```
          O
   18    \20
   H3C    C-CH2-OH       FORMULA   C21H30O4              MOL WT   346.47
     |17/  21            SOLVENT   C2D6OS/CDCL3
   12C | C-OH            ORIG ST   C4H12SI               TEMP        303
  19  / \|/ \
   H3C C11 C13 C16           35.20    33.40   197.40   123.00   170.40    31.90
    |   |   |   |             1/3      2/3      3/1      4/2      5/1      6/3
   1C | C9  C---C15          30.10    35.20    53.00    38.10    20.30    33.50
   / \|/ \ /14               7/3      8/2      9/2     10/1     11/3     12/3
  2C   C10 C8               47.00    49.90    23.20    32.10    88.40    14.50
   |    |   |               13/1     14/2     15/3     16/3     17/1     18/4
  3C   C5  C7               17.00   211.30    65.90
  // \ // \ /               19/4     20/1     21/3
  O  4C   C6
                        N.S.BHACCA,D.D.GIANNINI,W.S.JANKOWSKI,M.E.WOLFF
                        J AM CHEM SOC                   95, 8421 (1973)
```

2682 U 003000 CORTICOSTERONE

```
        O
   18   20
   H3C  C-CH3
    |  /   21
 HO 12C | C17
  19\ / \|/ \
  H3C C11 C13 C16
   |   |   |   |
   1C  | C9  C---C15
  / \|/ \ /14
 2C  C10 C8
  |   |   |
 3C  C5  C7
 // \ / \ /
 O  4C  C6
```

FORMULA	C21H30O4			MOL WT	346.47
SOLVENT	C2D6OS/CDCL3				
ORIG ST	C4H12SI			TEMP	303

34.20	33.50	197.70	121.50	171.90	32.60
1/3	2/3	3/1	4/2	5/1	6/3
31.50	31.20	55.70	38.90	66.30	46.80
7/3	8/2	9/2	10/1	11/2	12/3
43.20	57.10	22.00	24.20	58.20	15.70
13/1	14/2	15/3	16/3	17/2	18/4
20.50	209.80	68.60			
19/4	20/1	21/4			

N.S.BHACCA,D.D.GIANNINI,W.S.JANKOWSKI,M.E.WOLFF
J AM CHEM SOC   95, 8421 (1973)

2683 U 003000 11-BETA-17-ALPHA-DIHYDROXYPROGESTERONE

```
        O
   18   20
   H3C  C-CH3
     |17/   21
 HO 12C | C-OH
  19\ / \|/ \
  H3C C11 C13 C16
   |   |   |   |
   1C  | C9  C---C15
  / \|/ \ /14
 2C  C10 C8
  |   |   |
 3C  C5  C7
 // \ / \ /
 O  4C  C6
```

FORMULA	C21H30O4			MOL WT	346.47
SOLVENT	C2D6OS/CDCL3				
ORIG ST	C4H12SI			TEMP	303

34.00	33.40	197.90	121.50	172.20	32.70
1/3	2/3	3/1	4/2	5/1	6/3
31.40	31.10	55.60	38.80	66.50	39.30
7/3	8/2	9/2	10/1	11/2	12/3
45.60	51.40	23.20	31.90	89.10	16.80
13/1	14/2	15/3	16/3	17/1	18/3
20.40	209.80	26.50			
19/4	20/1	21/4			

N.S.BHACCA,D.D.GIANNINI,W.S.JANKOWSKI,M.E.WOLFF
J AM CHEM SOC   95, 8421 (1973)

2684 U 003000 CORTISOL

```
        O
   18   20
   H3C   C-CH2-OH
     |17/   21
 HO 12C | C-OH
  19\ / \|/ \
  H3C C11 C13 C16
   |   |   |   |
   1C  | C9  C---C15
  / \|/ \ /14
 2C  C10 C8
  |   |   |
 3C  C5  C7
 // \ / \ /
 O  4C  C6
      C
```

FORMULA	C21H30O5			MOL WT	362.47
SOLVENT	C2D6OS/CDCL3				
ORIG ST	C4H12SI			TEMP	303

34.00	33.40	197.70	121.40	172.00	32.70
1/3	2/3	3/1	4/2	5/1	6/3
31.30	31.10	55.50	38.80	66.40	39.00
7/3	8/2	9/2	10/1	11/2	12/3
46.20	51.50	23.30	32.90	88.30	16.80
13/1	14/2	15/3	16/3	17/1	18/4
20.40	211.40	65.80			
19/4	20/1	21/3			

N.S.BHACCA,D.D.GIANNINI,W.S.JANKOWSKI,M.E.WOLFF
J AM CHEM SOC   95, 8421 (1973)

2685 U 003000 5-BETA-PREGNAN-3-ALPHA-OL-20-ONE

```
        21
       H3C
  12 18   20
   C CH3   C=O
 19 11/ \| 17/
  H3C C 13C---C
   |   |   |   |
   1C  | C9  C14 C
  / \|/ \ / \ /16
 2C  C10 C8  C
  |   |   |  15
 3C  C5  C7
 / \4/|\ /
 HO  C  C6
```

FORMULA	C21H34O2			MOL WT	318.50
SOLVENT	CH2CL2				
ORIG ST	C4H12SI			TEMP	AMB

35.70	30.80	71.60	36.70	42.40	27.50
1/3	2/3	3/2	4/3	5/2	6/3
26.70	36.10	40.70	34.90	21.20	39.50
7/3	8/2	9/2	10/1	11/3	12/3
44.40	56.90	23.20	24.70	64.00	13.40
13/1	14/2	15/2	16/3	17/2	18/4
23.40	209.70	31.50			
19/4	20/1	21/4			

D.LEIBFRITZ,J.D.ROBERTS
J AM CHEM SOC   95, 4996 (1973)

2686 U 003000 3,17-ALPHA-DIACETOXYESTRA-1,3,5(10)-TRIENE

```
          18   21 22
          H3C  D-C-CH3
               |  |  \
          12C  |  C17 D
          / \|/ \
       11C    C13 C16
        |    |    |
      1C    C9   C---C15
      ∥ \  / \  /14
    2C    C10 C8
    |    ∥    |
   3C    C5   C7
   / \ / \ / \/
  D   4C   C6
19I 20
D=C-CH3
```

FORMULA	C22H2804			MOL WT	356.47
SOLVENT	C4H802				
ORIG ST	C4H12SI			TEMP	AMB
126.91	119.52	149.76	122.28	138.45	39.49
1/2	2/2	3/1	4/2	5/1	6/3
28.49	30.06	44.55	138.70	26.57	32.64
7/3	8/2	9/2	10/1	11/3	12/3
45.57	49.96	24.82	30.48	82.22	16.77
13/1	14/2	15/3	16/3	17/2	18/4
	20.70		20.70		
19/1	20/4	21/1	22/4		

T.A.WITTSTRUCK,K.I.H.WILLIAMS
J ORG CHEM   38, 1542 (1973)

2687 U 003000 3,17-BETA-DIACETOXYESTRA-1,3,5(10)-TRIENE

```
          18   21 22
          H3C  D-C-CH3
               |  |  \
          12C  |  C17 D
          / \|/ \
       11C    C13 C16
        |    |    |
      1C    C9   C---C15
      ∥ \  / \  /14
    2C    C10 C8
    |    ∥    |
   3C    C5   C7
   / \ / \ / \/
  D   4C   C6
19I 20
D=C-CH3
```

FORMULA	C22H2804			MOL WT	356.47
SOLVENT	C4H802				
ORIG ST	C4H12SI			TEMP	AMB
127.02	119.56	149.78	122.33	138.46	39.14
1/2	2/2	3/1	4/2	5/1	6/3
27.82	30.05	44.92	138.64	26.74	37.70
7/3	8/2	9/2	10/1	11/3	12/3
43.66	51.66	23.79	28.18	82.96	12.35
13/1	14/2	15/3	16/3	17/2	18/4
	20.78		20.78		
19/1	20/4	21/1	22/4		

T.A.WITTSTRUCK,K.I.H.WILLIAMS
J ORG CHEM   38, 1542 (1973)

2688 U 003000 DIANHYDROGITOXIGENIN

```
               O
        22    ∥
          C-C23
         ∥  |
          C20 |
     18 |\  |
     H3C | C-O
         | | 21
     12C | C17
    19 / \|/ \16
   H3C C11 C13 C
    |   |   |   |
   1C  |  C 14C===C
   / \|/9\ /    15
  2C   C10 C8
  |    |   |
 3C   C5   C7
 / \ / \ / \/
HD  4C   C6
```

FORMULA	C23H3003			MOL WT	354.49
SOLVENT	CDCL3/CD4C				
ORIG ST	C4H12SI			TEMP	AMB
30.70	27.90	66.70	33.50	36.80	26.60
1/3	2/3	3/2	4/3	5/2	6/3
24.00	36.70	45.10	36.20	21.40	37.70
7/3	8/2	9/2	10/1	11/3	12/3
54.20	146.30	108.30	135.80	158.00	20.10
13/1	14/1	15/2	16/2	17/1	18/4
24.00	173.50	72.10	119.50	176.80	
19/4	20/1	21/3	22/2	23/1	

K.TORI,H.ISHII,Z.W.WOLKOWSKI,C.CHACHATY,
M.SANGARE,F.PIRIOU,G.LUKACS
TETRAHEDRON LETT   1077 (1973)

2689 U 003000 STROPHANTHIDIN

```
        22    ∥
          C-C23
         ∥  |
          C20 |
     18 |\  |
     H3C | C-O
     12 | | 21
  D   H C | C17
   \ / / \|/ \16
   19C C11 C13 C
    |   |   |   |
   1C 19C 14C---C
   / \|/8/|  \  15
  2C   C10 C OH
  |    |   |
 3C   C5   C7
 / \ / \|\ /
HD  4C | C6
        DH
```

FORMULA	C23H3206			MOL WT	404.51
SOLVENT	CDCL3/CD4C				
ORIG ST	C4H12SI			TEMP	AMB
24.80	27.40	67.20	38.10	75.30	37.00
1/3	2/3	3/2	4/3	5/1	6/3
18.10	42.20	40.20	55.80	22.80	40.20
7/3	8/2	9/2	10/1	11/3	12/3
50.10	85.30	32.20	27.50	51.40	16.20
13/1	14/1	15/3	16/3	17/2	18/4
195.70	177.20	74.80	117.80	176.60	
19/1	20/1	21/3	22/2	23/1	

K.TORI,H.ISHII,Z.W.WOLKOWSKI,C.CHACHATY,
M.SANGARE,F.PIRIOU,G.LUKACS
TETRAHEDRON LETT   1077 (1973)

2690   U 003000 DIGITOXIGENIN

```
                    O   FORMULA   C23H34O4                MOL WT   374.53
            22  //      SOLVENT   CDCL3/CD4O
            C-C23       ORIG ST   C4H12SI               TEMP       AMB
             //  |
            C20 |          30.00   28.00   66.80   33.50   35.90   27.10
        18  |\  |           1/3     2/3     3/2     4/3     5/2     6/3
       H3C  | C-O          21.60   41.90   35.80   35.80   21.70   40.40
        |  | 21             7/3     8/2     9/2    10/1    11/3    12/3
      12C  | C17           50.30   85.60   33.00   27.30   51.50   16.10
     19   / \|/ \16        13/1    14/1    15/3    16/3    17/2    18/4
    H3C C11 C13 C          23.90  177.10   74.50  117.40  176.30
     |   |   |   |         19/4    20/1    21/3    22/2    23/1
    1C  |9C 14C---C
    / \|/ \8/|   15
   2C   C10 C OH
    |    |   |
   3C   C5  C7              K.TORI,H.ISHII,Z.W.WOLKOWSKI,C.CHACHATY,
  / \  / \ /                M.SANGARE,F.PIRIOU,G.LUKACS
 HO   4C   C6               TETRAHEDRON LETT                1077 (1973)
```

2691   U 003000 DIGOXIGENIN

```
                    O  FORMULA   C23H34O5                MOL WT   390.52
            22  //     SOLVENT   CDCL3/CD4O
            C-C23      ORIG ST   C4H12SI               TEMP       AMB
             //  |
            C20 |          30.00   27.90   66.60   33.30   36.40   26.90
        18  |\  |           1/3     2/3     3/2     4/3     5/2     6/3
    HO H3C  | C-O          21.90   41.30   32.60   35.50   30.00   74.80
     \  |  | 21             7/3     8/2     9/2    10/1    11/3    12/2
      12C  | C17           56.40   85.80   33.00   27.90   46.10    9.40
     19   / \|/ \16        13/1    14/1    15/3    16/3    17/2    18/4
    H3C C11 C13 C          23.80  177.10   74.60  117.00  176.30
     |   |   |   |         19/4    20/1    21/3    22/2    23/1
    1C  |9C 14C---C
    / \|/ \8/|   15
   2C   C10 C OH
    |    |   |
   3C   C5  C7              K.TORI,H.ISHII,Z.W.WOLKOWSKI,C.CHACHATY,
  / \  / \ /                M.SANGARE,F.PIRIOU,G.LUKACS
 HO   4C   C6               TETRAHEDRON LETT                1077 (1973)
```

2692   U 003000 GITOXIGENIN

```
                    O   FORMULA   C23H34O5                MOL WT   390.52
            22  //      SOLVENT   CDCL3/CD4O
            C-C23       ORIG ST   C4H12SI               TEMP       AMB
             //  |
            C20 |          30.00   28.00   66.80   33.50   36.40   27.00
        18  |\  |           1/3     2/3     3/2     4/3     5/2     6/3
       H3C  | C-O          21.40   41.80   35.80   35.80   21.90   41.20
        |  | 21             7/3     8/2     9/2    10/1    11/3    12/3
      12C  | C17 OH        50.40   85.20   42.60   72.80   58.80   16.90
     19   / \|/ \ /        13/1    14/1    15/3    16/2    17/2    18/4
    H3C C11 C13 C16        23.90  171.80   76.70  119.60  175.30
     |   |   |   |         19/4    20/1    21/3    22/2    23/1
    1C  |9C 14C---C
    / \|/ \8/|   15
   2C   C10 C OH
    |    |   |
   3C   C5  C7              K.TORI,H.ISHII,Z.W.WOLKOWSKI,C.CHACHATY,
  / \  / \ /                M.SANGARE,F.PIRIOU,G.LUKACS
 HO   4C   C6               TETRAHEDRON LETT                1077 (1973)
```

2693   U 003000 3-ALPHA-ACETOXY-5-BETA-PREGNAN-20-ONE

```
         H3C          FORMULA   C23H36O3               MOL WT   360.54
     12 18 21\20      SOLVENT   CH2CL2
     C CH3   C=O      ORIG ST   C4H12SI              TEMP       AMB
  19 11/ \|  17/
   H3C C 13C---C          35.40   27.00   74.60   32.60   42.30  165.50
    |   |   |   |          1/3     2/3     3/2     4/3     5/2     6/3
   1C  | C9  C14 C        26.70   36.20   40.80   35.00   21.20   39.60
   / \|/ \ \ / \/16        7/3     8/2     9/2    10/1    11/3    12/3
  2C   C10 C8  C          44.60   57.10   23.60   24.80   64.30   13.30
   |    |   |              13/1    14/2    15/3    16/3    17/2    18/4
  3C   C5  C7   15        23.30  209.80   31.80  171.00   21.80
 O  / \4/|\ /              19/4    20/1    21/4    22/1    23/4
 \22/ \4/|\ /
  C-O   C   C6
 23/                       D.LEIBFRITZ,J.D.ROBERTS
 CH3                       J AM CHEM SOC                 95, 4996 (1973)
```

2694   U 003000 1',2',3',4'-TETRAHYDRO-17-BETA-HYDROXY-2'-
                        THIOXOANDROSTA-2,4-DIENO(2,3-G')(PYRIDO(2',3"-
18CH3 OH DD)PYRIMIDIN)-4'-ONE
     | /
    12C | C17   FORMULA   C24H29N3O2S   MOL WT   423.58
    19 / \|/ \ SOLVENT   C2D6OS
   O   H3C C11 C13 C16ORIG ST C4H12SI   TEMP   300
   || 23 | | | |
   21C   C 1C | C9 C---C15   41.45 109.30 150.70 121.40 157.65 31.40
   / \ / \ /\|/ \ /14   1/3   2/1   3/1   4/2   5/1   6/3
   HN   C22 C2 C10 C8   30.45   35.40   53.30   37.75   20.95 36.45
   | || | | | 7/3   8/2   9/2   10/1   11/3   12/3
   20C   C24 C3 C5 C7   42.50   50.05   23.20   30.00   79.95 11.20
   / \ / \ / \ / \ / \ /   13/1   14/2   15/3   16/3   17/2   18/4
   S   N   N 4C   C6   17.50 175.35 161.40 125.90 133.70 159.70
                                 19/4   20/1   21/1   22/1   23/2   24/1

              H.PECH,G.BOUCHON,E.BREITMAIER
              CHEM BER   107, 1389 (1974)

2695   U 003000 1',2',3',4'-TETRAHYDRO-17-BETA-HYDROXYANDROSTA-
                        2,4-DIENO(2,3-G')(PYRIDO(2',3'-D')PYRIMIDINE)-
18CH3 OH 2',4'-DIONE
     | /
    12C | C17   FORMULA   C24H29N3O3   MOL WT   407.52
    19 / \|/ \ SOLVENT   C2D6OS
   O   H3C C11 C13 C16ORIG ST C4H12SI   TEMP   300
   || 23 | | | |
   21C   C 1C | C9 C---C15   40.55 107.05 150.50 121.40 157.20 31.40
   / \ / \ /\|/ \ /14   1/3   2/1   3/1   4/2   5/1   6/3
   HN   C22 C2 C10 C8   30.50   35.50   53.40   38.70   21.05 36.45
   | || | | | 7/3   8/2   9/2   10/1   11/3   12/3
   20C   C24 C3 C5 C7   42.50   50.05   23.20   30.00   79.95 11.20
   / \ / \ / \ / \ / \ /   13/1   14/2   15/3   16/3   17/2   18/4
   O   N   N 4C   C6   17.50 151.35 162.25 123.85 134.10 160.45
                                 19/4   20/1   21/1   22/1   23/2   24/1

              H.PECH,G.BOUCHON,E.BREITMAIER
              CHEM BER   107, 1389 (1974)

2696   U 003000 LITHOCHOLIC ACID

         21   FORMULA   C24H40O3   MOL WT   376.58
        H3C   SOLVENT   CH4O
   12 18   \20 22   ORIG ST   C4H12SI   TEMP   AMB
    C CH3   CH-CH2
   19 11/ \| 17/   \ 35.60   30.40   71.80   36.50   42.50   27.50
   H3C C 13C---C   23CH2 1/3   2/3   3/2   4/3   5/2   6/3
   | | | | | 26.70   36.10   40.60   34.80   21.10 40.50
   1C | C9 C14 C   C24   7/3   8/2   9/2   10/1   11/3   12/3
   / \|/ \ / \ /16   / \\ 43.00   56.70   24.40   28.40   56.50 11.80
   2C   C10 C8 C   HO O 13/1   14/2   15/3   16/3   17/2   18/4
   | | | 15   23.20   35.60   18.10   31.20   31.10 174.50
   3C   C5 C7   19/4   20/2   21/4   22/3   23/3   24/1
   . \4/ \ /
   HO C C6   D.LEIBFRITZ,J.D.ROBERTS
                         J AM CHEM SOC   95, 4996 (1973)

2697   U 003000 CHENODESOXYCHOLIC ACID

         21   FORMULA   C24H40O4   MOL WT   392.58
        H3C   SOLVENT   CH4O
   12 18   \20 22   ORIG ST   C4H12SI   TEMP   AMB
    C CH3   CH-CH2
   19 11/ \| 17/   \ 35.80   31.00   71.40   39.70   42.10   35.40
   H3C C 13C---C   23CH2 1/3   2/3   3/2   4/3   5/2   6/3
   | | | | | 67.80   39.90   33.30   35.50   21.10 39.90
   1C | C9 C14 C   C24   7/2   8/2   9/2   10/1   11/1   12/3
   / \|/ \ / \ /15   / \\ 43.00   50.00   24.00   28.50   56.30 11.90
   2C   C10 C8 C   HO O 13/1   14/2   15/3   16/3   17/2   18/4
   | | | 15   23.00   35.80   18.30   31.20   31.10 174.60
   3C   C5 C7   19/4   20/2   21/4   22/3   23/3   24/1
   . \4/ \ / .
   HO C C6 OH   D.LEIBFRITZ,J.D.ROBERTS
                         J AM CHEM SOC   95, 4996 (1973)

2698   U 003000 DESOXYCHOLIC ACID

```
              21                FORMULA   C24H40O4              MOL WT   392.58
            H3C                 SOLVENT   CH4O
        • 18     \20 22         ORIG ST   C4H12SI              TEMP         AMB
        12C  CH3   CH-CH2
      19 11/ \|  17/      \        35.70   30.50   71.70   36.10   42.70   27.60
      H3C  C 13C---C     23CH2      1/3     2/3     3/2     4/3     5/2     6/3
        |  |  |    |       |       26.60   36.40   33.90   34.50   29.00   72.80
      1C  |  C9   C14 C        C24   7/3     8/2     9/2    10/1    11/3    12/2
      /  \|/ \  / \  /15    / \    46.80   48.40   24.20   26.50   47.40   12.90
    2C   C10 C8   C      HO   O    13/1    14/2    15/3    16/3    17/2    18/4
     |    |   |    15             23.40   35.70   17.50   31.30   31.20  177.20
    3C   C5  C7                   19/4    20/2    21/4    22/3    23/3    24/1
    • \4/|\ /
   HO   C   C6                   D.LEIBFRITZ,J.D.ROBERTS
                                 J AM CHEM SOC                95, 4996 (1973)
```

2699   U 003000 CHOLIC ACID

```
              21                FORMULA   C24H40O5              MOL WT   408.58
          OH  H3C               SOLVENT   CH4O
        • 18     \20 22         ORIG ST   C4H12SI              TEMP         AMB
        12C  CH3   CH-CH2
      19 11/ \|  17/      \        35.70   30.60   71.60   39.60   42.00   35.10
      H3C  C 13C---C     23CH2      1/3     2/3     3/2     4/3     5/2     6/3
        |  |  |    |       |       68.00   39.90   26.70   35.10   28.50   72.50
      1C  |  C9   C14 C        C24   7/2     8/2     9/2    10/1    11/3    12/2
      /  \|/ \  / \  /15    / \    46.80   42.00   23.50   28.00   47.40   12.80
    2C   C10 C8   C      HO   O    13/1    14/2    15/3    16/3    17/2    18/4
     |    |   |    15             22.80   35.70   17.60   31.30   31.20  174.50
    3C   C5  C7                   19/4    20/2    21/4    22/3    23/3    24/1
    • \4/|\ /
   HO   C   C6   OH              D.LEIBFRITZ,J.D.ROBERTS
                                 J AM CHEM SOC                95, 4996 (1973)
```

2700   U 003000 CHENODESOXYCHOLANOL

```
              21                FORMULA   C24H42O3              MOL WT   378.60
            H3C                 SOLVENT   CH2CL2
        12 18     \20 22        ORIG ST   C4H12SI              TEMP        AMB
         C  CH3   CH-CH2
      19 11/ \|  17/      \        35.80   31.30   72.30   39.70   42.10   35.40
      H3C  C 13C---C     23CH2      1/3     2/3     3/2     4/3     5/2     6/3
        |  |  |    |       |       68.80   40.00   33.30   35.50   20.90   39.90
      1C  |  C9   C14 C        24CH2   7/2     8/2     9/2    10/1    11/3    12/3
      /  \|/ \  / \  /16     |       43.00   50.90   24.00   28.50   56.50   11.90
    2C   C10 C8   C         OH      13/1    14/2    15/3    16/3    17/2    18/4
     |    |   |    15             23.00   35.90   18.60   32.30   29.50   62.90
    3C   C5  C7                   19/4    20/2    21/4    22/3    23/3    24/3
    • \4/|\ /
   HO   C   C6   OH              D.LEIBFRITZ,J.D.ROBERTS
                                 J AM CHEM SOC                95, 4996 (1973)
```

2701   U 003000 DESOXYCHOLANOL

```
              21                FORMULA   C24H42O3              MOL WT   378.60
          OH  H3C               SOLVENT   CH2CL2
        • 18     \20 22         ORIG ST   C4H12SI              TEMP        AMB
        12C  CH3   CH-CH2
      19 11/ \|  17/      \        35.70   30.50   71.40   36.30   42.70   27.70
      H3C  C 13C---C     23CH2      1/3     2/3     3/2     4/3     5/2     6/3
        |  |  |    |       |       26.60   36.30   33.90   34.50   29.50   73.10
      1C  |  C9   C14 C        24CH2   7/3     8/2     9/2    10/1    11/3    12/2
      /  \|/ \  / \  /16     |       46.90   48.40   24.20   26.50   47.60   12.90
    2C   C10 C8   C         OH      13/1    14/2    15/3    16/3    17/2    18/4
     |    |   |    15             23.40   35.90   18.00   32.10   29.50   63.80
    3C   C5  C7                   19/4    20/2    21/4    22/3    23/3    24/3
    • \4/|\ /
   HO   C   C6                   D.LEIBFRITZ,J.D.ROBERTS
                                 J AM CHEM SOC                95, 4996 (1973)
```

2702   U 003000   MONOANHYDROGITOXIGENIN-3-ACETATE

```
              O   FORMULA   C25H34O5              MOL WT   414.55
         22  //   SOLVENT   CDCL3/CD4O
         C-C23    ORIG ST   C4H12SI               TEMP       AMB
        //  |
        C20 |
    18  |\  |
    H3C | C-O
    |  | 21
  12C | C17
  19  / \|/ \16
  H3C C11 C13 C
   |  |  |  |
  1C 19C 14C---C
  / \|/ \8/|  15
 O  2C  C10 C OH
 25 ||  |   |  |
 H3C-C24 C3 C5  C7
   \ /  \ / \ /
    O   4C   C6
```

30.80	25.40	71.30	30.80	37.30	26.60
1/3	2/3	3/2	4/3	5/2	6/3
20.20	41.20	36.80	35.40	21.30	40.60
7/3	8/2	9/2	10/1	11/3	12/3
52.60	85.70	38.80	133.80	161.20	16.60
13/1	14/1	15/3	16/2	17/1	18/4
24.10	172.80	72.60	111.70	176.30	171.40
19/4	20/1	21/3	22/2	23/1	24/1
21.30					
25/4					

K.TORI,H.ISHII,Z.W.WOLKOWSKI,C.CHACHATY,
M.SANGARE,F.PIRIOU,G.LUKACS
TETRAHEDRON LETT   1077 (1973)

2703   U 003000   DIGITOXIGENIN-3-ACETATE

```
              O   FORMULA   C25H36O5              MOL WT   416.56
         22  //   SOLVENT   CDCL3/CD4O
         C-C23    ORIG ST   C4H12SI               TEMP       AMB
        //  |
        C20 |
    18  |\  |
    H3C | C-O
    |  | 21
  12C | C17
  19  / \|/ \16
  H3C C11 C13 C
   |  |  |  |
  1C 19C 14C---C
  / \|/ \8/|  15
 O  2C  C10 C OH
 25 ||  |   |  |
 H3C-C24 C3 C5  C7
   \ /  \ / \ /
    O   4C   C6
```

30.80	25.40	71.40	30.80	37.40	26.80
1/3	2/3	3/2	4/3	5/2	6/3
21.60	41.80	36.10	35.80	21.60	40.30
7/3	8/2	9/2	10/1	11/3	12/3
50.30	85.60	33.00	27.30	51.50	16.00
13/1	14/1	15/3	16/3	17/2	18/4
23.90	177.10	74.70	117.40	176.30	171.40
19/4	20/1	21/3	22/2	23/1	24/1
21.30					
25/4					

K.TORI,H.ISHII,Z.W.WOLKOWSKI,C.CHACHATY,
M.SANGARE,F.PIRIOU,G.LUKACS
TETRAHEDRON LETT   1077 (1973)

2704   U 003000   DIGITOXIGENIN-3-ACETATE   ISOMERIC FORM

```
              O   FORMULA   C25H36O5              MOL WT   416.56
         22  //   SOLVENT   CDCL3/CD4O
         C-C23    ORIG ST   C4H12SI               TEMP       AMB
        //  |
        C20 |
    18  .\  |
    H3C .  C-O
    |  .  21
  12C | C17
  19  / \|/ \16
  H3C C11 C13 C
   |  |  |  |
  1C 19C 14C---C
  / \|/ \8/|  15
 O  2C  C10 C OH
 25 ||  |   |  |
 H3C-C24 C3 C5  C7
   \ /  \ / \ /
    O   4C   C6
```

30.80	25.30	71.30	30.80	37.40	26.80
1/3	2/3	3/2	4/3	5/2	6/3
20.60	41.50	36.20	35.50	21.20	31.30
7/3	8/2	9/2	10/1	11/3	12/3
49.50	86.10	31.30	24.80	48.90	18.50
13/1	14/1	15/3	16/3	17/2	18/4
24.00	173.60	74.80	116.60	175.80	171.40
19/4	20/1	21/3	22/2	23/1	24/1
21.30					
25/4					

K.TORI,H.ISHII,Z.W.WOLKOWSKI,C.CHACHATY,
M.SANGARE,F.PIRIOU,G.LUKACS
TETRAHEDRON LETT   1077 (1973)

2705   U 003000   GITOXIGENIN-16-ACETATE

```
              O   FORMULA   C25H36O6              MOL WT   432.56
         22  //   SOLVENT   CDCL3/CD4O
         C-C23    ORIG ST   C4H12SI               TEMP       AMB
        //  |
        C20 |
    18  |\  |
    H3C | C-O     O
    |  | 21
  12C | C17 O-C
  19  / \|/ \ / 24\
  H3C C11 C13 C16    CH3
   |  |  |  |    25
  1C 19C 14C---C
  / \|/ \8/|  15
 2C  C10 C OH
 |   |  |  |
 3C  C5  C7
 / \ / \ /
HO 4C   C6
```

30.00	27.90	66.80	33.40	36.40	26.90
1/3	2/3	3/2	4/3	5/2	6/3
21.20	41.80	35.90	35.60	21.30	41.00
7/3	8/2	9/2	10/1	11/3	12/3
50.70	84.10	39.50	75.00	56.80	16.10
13/1	14/1	15/3	16/2	17/2	18/4
23.90	171.50	76.80	121.30	175.80	171.40
19/4	20/1	21/3	22/2	23/1	24/1
21.30					
25/4					

K.TORI,H.ISHII,Z.W.WOLKOWSKI,C.CHACHATY,
M.SANGARE,F.PIRIOU,G.LUKACS
TETRAHEDRON LETT   1077 (1973)

2706   U 003000   METHYL CHOLANATE

```
                21
               H3C
        12 18    \20 22
         C CH3   CH-CH2
      19 11/ \I  17/      \
      H3C C 13C---C      23CH2
        I  I  I    I       I
       1C I C9  C14 C       C24
       / \I/ \ / \ /16    / \
     2C   C10 C8  C      O    O
      I    I   I    15    I
     3C   C5  C7          CH3
      \ /I\ /             25
      C4   C6
```

FORMULA	C25H42O2			MOL WT	374.61
SOLVENT	CH2CL2				
ORIG ST	C4H12SI			TEMP	AMB
57.90	21.60	27.80	27.40	44.10	27.50
1/3	2/3	3/3	4/3	5/2	6/3
26.90	36.20	40.80	35.60	21.10	40.60
7/3	8/2	9/2	10/1	11/3	12/3
43.00	56.90	24.40	28.40	56.40	12.10
13/1	14/2	15/3	16/3	17/2	18/4
24.20	35.60	18.30	31.20	31.10	175.00
19/4	20/2	21/4	22/3	23/3	24/1
141.50					
25/4					

D.LEIBFRITZ,J.D.ROBERTS
J AM CHEM SOC   95, 4996 (1973)

2707   U 003000   METHYL LITHOCHOLATE

```
                21
               H3C
        12 18    \20 22
         C CH3   CH-CH2
      19 11/ \I  17/      \
      H3C C 13C---C      23CH2
        I  I  I    I       I
       1C I C9  C14 C       C24
       / \I/ \ / \ /15    / \
     2C   C10 C8  C      O    O
      I    I   I    15    I
     3C   C5  C7          CH3
      . \4/I\ /           25
     HO  C  C6
```

FORMULA	C25H42O3			MOL WT	390.61
SOLVENT	CH2CL2				
ORIG ST	C4H12SI			TEMP	AMB
35.60	30.90	71.90	36.80	42.50	27.50
1/3	2/3	3/2	4/3	5/2	6/3
26.70	36.10	40.70	34.80	21.10	40.50
7/3	8/2	9/2	10/1	11/3	12/3
43.00	56.80	24.40	28.40	56.30	11.40
13/1	14/2	15/3	16/3	17/2	18/4
23.40	35.60	18.30	31.30	31.00	175.10
19/4	20/2	21/4	22/3	23/3	24/1
51.40					
25/4					

D.LEIBFRITZ,J.D.ROBERTS
J AM CHEM SOC   95, 4996 (1973)

2708   U 003000   METHYL CHENODESOXYCHOLATE

```
                21
               H3C
        12 18    \20 22
         C CH3   CH-CH2
      19 11/ \I  17/      \
      H3C C 13C---C      23CH2
        I  I  I    I       I
       1C I C9  C14 C       C24
       / \I/ \ / \ /16    / \
     2C   C10 C8  C      O    O
      I    I   I    15    I
     3C   C5  C7          CH3
      . \4/I\ / .         25
     HO  C  C6  OH
```

FORMULA	C25H42O4			MOL WT	406.61
SOLVENT	CH2CL2				
ORIG ST	C4H12SI			TEMP	AMB
35.80	31.30	72.10	39.90	42.10	35.50
1/3	2/3	3/2	4/3	5/2	6/3
68.60	39.90	33.30	35.50	21.10	39.90
7/2	8/2	9/2	10/1	11/3	12/3
43.00	50.90	24.00	28.50	56.30	12.10
13/1	14/2	15/3	16/3	17/2	18/4
23.20	35.80	18.40	31.30	31.00	175.20
19/4	20/2	21/4	22/3	23/3	24/1
51.80					
25/4					

D.LEIBFRITZ,J.D.ROBERTS
J AM CHEM SOC   95, 4996 (1973)

2709   U 003000   METHYL DESOXYCHOLATE

```
                21
         OH   H3C
        . 18    \20 22
        12C CH3   CH-CH2
      19 11/ \I  17/      \
      H3C C 13C---C      23CH2
        I  I  I    I       I
       1C I C9  C14 C       24C
       / \I/ \ / \ /16    / \
     2C   C10 C8  C      O    O
      I    I   I    15    I
     3C   C5  C7          CH3
      . \4/I\ /           25
     HO  C  C6
```

FORMULA	C25H42O4			MOL WT	406.61
SOLVENT	CH2CL2				
ORIG ST	C4H12SI			TEMP	AMB
35.70	30.70	71.80	36.40	42.60	27.60
1/3	2/3	3/2	4/3	5/2	6/3
26.60	36.40	33.90	34.50	29.00	73.30
7/3	8/2	9/2	10/1	11/3	12/2
46.80	48.40	24.20	26.50	47.40	12.90
13/1	14/2	15/3	16/3	17/2	18/4
23.40	35.70	17.50	31.40	31.20	175.20
19/4	20/2	21/4	22/3	23/3	24/1
51.80					
25/4					

D.LEIBFRITZ,J.D.ROBERTS
J AM CHEM SOC   95, 4996 (1973)

2710   U 003000 METHYL HYODESOXYCHOLATE

```
          21
         H3C
    12 18   \20 22
     C CH3   CH-CH2
  19 11/ \|  17/    \
  H3C C 13C---C    23CH2
   |   |  |   |      |
  1C  | C9  C14 C        C24
  / \|/  \ / \ /16   / \
 2C  C10 C8  C     O   O
  |   |   |  15      |
 3C   C5  C7        CH3
  · \ /|\ /         25
 HO  C4  C6
       ·
       OH
```

36.30	29.70	71.80	35.30	48.80	68.10
1/3	2/3	3/2	4/3	5/2	6/2
30.60	35.30	40.20	36.10	21.00	40.20
7/3	8/2	9/2	10/1	11/3	12/3
43.10	56.30	24.50	28.40	56.30	12.10
13/1	14/2	15/3	16/3	17/2	18/4
169.20	35.80	18.30	31.30	31.10	175.10
19/4	20/2	21/4	22/3	23/3	24/1
141.40					
25/4					

FORMULA C25H42O4   MOL WT 406.61
SOLVENT CH2CL2
ORIG ST C4H12SI   TEMP   AMB

D.LEIBFRITZ,J.D.ROBERTS
J AM CHEM SOC   95, 4996 (1973)

2711   U 003000 METHYL CHOLATE

```
          21
      OH  H3C
      · 18  \20 22
    12C CH3   CH-CH2
  19 11/ \|  17/    \
  H3C C 13C---C    23CH2
   |   |  |   |      |
  1C  | C9  C14 C        C 24
  / \|/  \ / \ /16   / \
 2C  C10 C8  C     O   O
  |   |   |  15      |
 3C   C5  C7        CH3
  · \4/|\ /  ·       25
 HO   C  C6  OH
```

35.70	30.90	72.20	39.90	42.00	35.10
1/3	2/3	3/2	4/3	5/2	6/3
68.90	39.90	26.70	35.10	28.50	73.40
7/2	8/2	9/2	10/1	11/3	12/2
46.80	42.00	23.50	28.00	47.40	12.80
13/1	14/2	15/3	16/3	17/2	18/4
22.80	35.70	17.60	31.30	31.20	175.30
19/4	20/2	21/4	22/3	23/3	24/1
51.80					
25/4					

FORMULA C25H42O5   MOL WT 422.61
SOLVENT CH2CL2
ORIG ST C4H12SI   TEMP   AMB

D.LEIBFRITZ,J.D.ROBERTS
J AM CHEM SOC   95, 4996 (1973)

2712   U 003000 1'-ETHYL-17-BETA-ACETOXYANDROSTA-2,4-DIENO-(2,3-G')(1H-PYRAZOLO(3',4'-B')PYRIDINE)

```
    H3C  O
   21\ //
      C20
    18 |
    H3C O
    | |
    12C | C17
  19 / \|/ \
   H3C C11 C13 C16
 24  | | | |
  C  1C | C9  C---C15
 23/ \ / \|/ \ /14
22C---C  C2 C10 C8
  ||  ||  |  |  |
  N   C25 C3  C5 C7
   \ / \ / \ // \ /
    N   N  4C  C6
    |
  26CH2
    |
  27CH3
```

41.65	113.85	152.80	123.45	156.10	31.60
1/3	2/1	3/1	4/2	5/1	6/3
30.85	35.70	53.40	39.05	21.15	36.80
7/3	8/2	9/2	10/1	11/3	12/3
42.50	50.30	23.50	27.40	82.45	12.00
13/1	14/2	15/3	16/3	17/2	18/4
17.40	170.90	21.15	131.00	122.90	127.40
19/4	20/1	21/4	22/2	23/1	24/2
149.45	42.50	15.00			
25/1	26/3	27/4			

FORMULA C27H35N3O2   MOL WT 433.60
SOLVENT CDCL3
ORIG ST C4H12SI   TEMP   300

J.HAEUFEL,H.PECH,E.BREITMAIER
CHEMIKER-ZTG   97, 658 (1973)

2713   U 003000 1',3'-DIMETHYL-17-BETA-ACETOXYANDROSTA-2,4-DIENO(2,3-G')(1H-PYRAZOLO(3',4'-B')PYRIDINE)

```
    H3C  O
   21\ //
      C20
    18 |
    H3C O
    | |
    12C | C17
  19 / \|/ \
   H3C C11 C13 C16
 27   24  | | | |
 H3C  C  1C | C9  C---C15
   \ 23/ \ / \|/ \ /14
  22C---C  C2 C10 C8
   ||  ||  |  |  |
    N   C25 C3  C5 C7
     \ / \ // \ // \ /
      N   N  4C  C6
      |
    26CH3
```

41.95	112.65	152.25	122.90	155.60	31.05
1/3	2/1	3/1	4/2	5/1	6/3
30.30	35.15	52.85	38.50	20.60	36.25
7/3	8/2	9/2	10/1	11/3	12/3
41.95	49.75	23.00	26.85	81.90	11.45
13/1	14/2	15/3	16/3	17/2	18/4
16.95	170.25	20.60	138.80	121.40	126.45
19/4	20/1	21/4	22/1	23/1	24/2
150.20	32.90	11.85			
25/1	26/4	27/4			

FORMULA C27H35N3O2   MOL WT 433.60
SOLVENT CDCL3
ORIG ST C4H12SI   TEMP   300

J.HAEUFEL,H.PECH,E.BREITMAIER
CHEMIKER-ZTG   97, 658 (1973)

2714

U 003000 1',3'-DIMETHYL-3-BETA-ACETOXY-5-ALPHA-ANDROST-16-ENO(16,17-G')(1H-PYRAZOLO(3',4'-B')PYRIDINE)

```
        CH3
        26\
         N---N
         |   ||
        25C   C22
    18  / \  / \
   H3C  N   C23 CH3
    |   |   ||   27
   12C  |  C17 C24
  19  / \|/  \ / \
  H3C C11 C13 C16
   |   |   |   |
  1C   |  C9  C---C15
   / \ |/ \  /14
  2C   C10 C8
   |    |   |
  3C   C5  C7
   / \  / \ / \
  O   C4  C6
  | 21
  O=C-CH3
  20
```

FORMULA C27H37N3O2   MOL WT 435.61
SOLVENT CDCL3
ORIG ST C4H12SI   TEMP   300

36.60	33.75	73.50	27.40	44.80	28.40
1/3	2/3	3/2	4/3	5/2	6/3
31.50	34.55	54.70	35.70	20.80	34.00
7/3	8/2	9/2	10/1	11/3	12/3
45.55	56.00	29.80	113.30	173.90	17.70
13/1	14/2	15/3	16/1	17/1	18/4
12.30	170.45	21.35	139.50	128.95	124.30
19/4	20/1	21/4	22/1	23/1	24/2
151.05	33.45	12.30			
25/1	26/4	27/4			

J.HAEUFEL,H.PECH,E.BREITMAIER
CHEMIKER-ZTG   97, 658 (1973)

2715

U 003000 1'-ETHYL-3-BETA-ACETOXY-5-ALPHA-ANDROST-16-ENO-(16,17-G')(1H-PYRAZOLO(3',4'-B')PYRIDINE)

```
     CH3-CH2
     27  26\
         N---N
         |   ||
        25C   C22
    18  / \ /
   H3C  N   C23
    |   |   ||
   12C  |  C17 C24
  19  / \|/  \ / \
  H3C C11 C13 C16
   |   |   |   |
  1C   |  C9  C---C15
   / \ |/ \  /14
  2C   C10 C8
   |    |   |
  3C   C5  C7
   / \  / \ / \
  O   C4  C6
  | 21
  O=C-CH3
  20
```

FORMULA C27H37N3O2   MOL WT 435.61
SOLVENT CDCL3
ORIG ST C4H12SI   TEMP   300

36.05	33.25	72.95	26.95	44.25	27.95
1/3	2/3	3/2	4/3	5/2	6/3
30.95	34.00	54.15	35.30	20.40	33.45
7/3	8/2	9/2	10/1	11/3	12/3
45.10	55.45	29.35	113.50	173.30	17.15
13/1	14/2	15/3	16/1	17/1	18/4
11.75	169.95	20.95	130.65	129.45	124.40
19/4	20/1	21/4	22/2	23/1	24/2
149.10	41.30	14.45			
25/1	26/3	27/4			

J.HAEUFEL,H.PECH,E.BREITMAIER
CHEMIKER-ZTG   97, 658 (1973)

2716

U 003000 GITOXIGENIN-3,16-DIACETATE

```
              O
        22   //
         C-C23
        // |
        C20 |
   18  |\  |
  H3C  | C-O   O
    |  | 21   //
  12C  | C17 O-C 26
 19  / \|/ \ /  \
 H3C C11 C13 C16 H3C
  |  |   |   |    27
  C  |9C 14C---C
 / \|/ \8/|  \ 15
 2C   C10 C OH
  |    |   |
 3C   C5  C7
  \24 / \ / \ /
  C-O  4C   C6
 25/
 CH3
```

FORMULA C27H38O7   MOL WT 474.60
SOLVENT CDCL3/CD4O
ORIG ST C4H12SI   TEMP   AMB

30.70	25.20	71.10	30.70	37.20	26.60
1/3	2/3	3/2	4/3	5/2	6/3
20.90	41.60	35.80	35.40	21.40	40.90
7/3	8/2	9/2	10/1	11/3	12/3
50.50	83.80	39.30	74.70	56.60	16.10
13/1	14/1	15/3	16/2	17/2	18/4
23.80	171.50	76.50	121.10	175.40	171.40
19/4	20/1	21/3	22/2	23/1	24/1
21.30					
25/4					

K.TORI,H.ISHII,Z.W.WOLKOWSKI,C.CHACHATY,
M.SANGARE,F.PIRIOU,G.LUKACS
TETRAHEDRON LETT   1077 (1973)

2717

U 003000 METHYL 3-ACETYLLITHOCHOLATE

```
       H3C
   12 18 21\20 22
    C CH3  CH-CH2
 19  11/ \| 17/
  H3C C 13C---C   23CH2
   |  |  |   |    |
  1C  | C9  C14 C  C24
 / \ |/ \ / \ /16 / \
 2C  C10 C8  C  O  O
  |   |   |  15  |
 3C  C5  C7      CH3
  . \4/|\ /      25
 O   C   C6
  \
 26C-CH3
 // 27
 O
```

FORMULA C27H45O4   MOL WT 433.66
SOLVENT CH2CL2
ORIG ST C4H12SI   TEMP   AMB

35.10	26.40	74.40	32.20	42.00	27.30
1/3	2/3	3/2	4/3	5/2	6/3
26.70	36.10	40.60	34.80	21.10	40.50
7/3	8/2	9/2	10/1	11/3	12/3
43.00	56.70	24.40	28.40	56.50	11.70
13/1	14/2	15/3	16/3	17/2	18/4
23.00	35.60	18.00	31.30	31.10	175.20
19/4	20/2	21/4	22/3	23/3	24/1
51.60	169.40	21.00			
25/4	26/1	27/4			

D.LEIBFRITZ,J.D.ROBERTS
J AM CHEM SOC   95, 4996 (1973)

2718   U 003000 CHOLESTEROL

```
        27    21      FORMULA   C27H46O              MOL WT   386.67
        H3C   CH2     SOLVENT   C5D5N
     18  \  /  \      ORIG ST   C4H12SI              TEMP       AMB
     H3C  HC20 CH2
       |   /    22\      37.70   32.10   71.00   43.20   141.70  120.90
    12C | C17    CH2      1/3     2/3     3/2     4/3     5/1     6/2
   19 / \|/  \    23\     32.30   32.10   50.40   36.70   21.30   28.40
   H3C C11 C13 C6   24CH   7/3     8/2     9/2    10/1    11/3    12/3
     |   |   |   |    / \   42.40   56.90   24.40   40.00   56.40   11.90
   1C  | C9  C---C  H3C H3C  13/1    14/2    15/3    16/3    17/2    18/4
   / \|/ \ /14 15  25  26   19.50   36.00   36.40   24.10   39.70   28.10
  2C   C10 C8             19/4    20/2    21/3    22/3    23/3    24/2
   |    |   |             22.80   22.60   18.90
  3C    C   C7            25/4    26/4    27/4
  / \  /5\ /
 HO   C   C          L.F.JOHNSON,W.C.JANKOWSKI
      4   6          CARBON-13 NMR SPECTRA,JOHN
                     WILEY AND SONS,NEW YORK        494 (1972)
```

2719   U 003000 CHOLESTANE

```
        21          FORMULA   C27H48               MOL WT   372.68
        H3C         SOLVENT   CH2CL2
    12 18  \20 22   ORIG ST   C4H12SI              TEMP       AMB
     C CH3  CH-CH2
   19 11/ \|  17/    \   39.20   22.70   27.30   29.20   47.60   29.20
   H3C C 13C---C   23CH2   1/3     2/3     3/3     4/3     5/2     6/3
     |   |   |   |    /    32.60   36.00   55.30   36.70   21.30   40.70
   C11  C9  C14  C  H2C24   7/3     8/2     9/2    10/1    11/3    12/3
   / \|/ \ / \ / \ /16  \25  43.00   57.10   24.60   28.70   56.80   12.60
  2C   C10 C8   C       CH   13/1    14/2    15/3    16/3    17/2    18/4
   |    |   |   15      / \   12.50   36.30   19.10   36.70   24.40   40.00
  3C   C5  C7        H3C H3C  19/4    20/2    21/4    22/3    23/3    24/3
   \4/.\ /           26  27   28.50   23.10   23.80
    C   C6                    25/2    26/4    27/4

                     D.LEIBFRITZ,J.D.ROBERTS
                     J AM CHEM SOC               95, 4996 (1973)
```

2720   U 003000 COPROSTANE

```
        21          FORMULA   C27H48               MOL WT   372.68
        H3C         SOLVENT   CH2CL2
    12 18  \20 22   ORIG ST   C4H12SI              TEMP       AMB
     C CH3  CH-CH2
   19 11/ \|  17/    \   37.90   21.70   27.50   27.90   44.20   27.60
   H3C C 13C---C   23CH2   1/3     2/3     3/3     4/3     5/2     6/3
     |   |   |   |    /    26.90   36.20   40.70   35.70   21.10   40.60
   1C  | C9  C14  C  H2C24   7/3     8/2     9/2    10/1    11/3    12/3
   / \|/ \ / \ / \ /15  \25  43.00   57.00   24.50   28.60   56.80   12.10
  2C   C10 C8   C       CH   13/1    14/2    15/3    16/3    17/2    18/4
   |    |   |   15      / \   24.40   36.10   18.80   36.50   24.10   39.80
  3C   C5  C7        H3C H3C  19/4    20/2    21/4    22/3    23/3    24/3
   \4/|\ /           26  27   28.30   22.80   22.60
    C   C6                    25/2    26/4    27/4

                     D.LEIBFRITZ,J.D.ROBERTS
                     J AM CHEM SOC               95, 4996 (1973)
```

2721   U 003000 CHLOLESTANOL

```
        21          FORMULA   C27H48O              MOL WT   388.68
        H3C         SOLVENT   CH2CL2
    12 18  \20 22   ORIG ST   C4H12SI              TEMP       AMB
     C CH3  CH-CH2
   19 11/ \|  17/    \   37.40   31.90   71.60   38.70   45.30   29.60
   H3C C 13C---C   23CH2   1/3     2/3     3/2     4/3     5/2     6/3
     |   |   |   |    /    32.50   35.90   54.80   35.90   21.40   40.50
   C11  C9  C14  C  H2C24   7/3     8/2     9/2    10/1    11/3    12/3
   / \|/ \ / \ / \ /16  \25  43.00   56.90   24.60   28.70   56.90   12.60
  2C   C10 C8   C       CH   13/1    14/2    15/3    16/3    17/2    18/4
   |    |   |   15      / \   180.40  36.20   19.00   36.60   24.30   40.00
  3C   C5  C7        H3C H3C  19/4    20/2    21/4    22/3    23/3    24/3
   / \4/.\ /         26  27   164.30  23.10   22.90
  HO   C   C6                 25/2    26/4    27/4

                     D.LEIBFRITZ,J.D.ROBERTS
                     J AM CHEM SOC               95, 4996 (1973)
```

2722   U 003000   EPICOPROSTANOL

```
            21
           H3C
   12 18      \20 22
    C  CH3   CH-CH2
  19 11/  \|   17/    \
  H3C C 13C---C    23CH2
   |  |   |   |    |       /
   C11| C9  C14 C    H2C24
  / \|/ \ / \ / \ /15     \25
 2C   C10 C8   C          CH
  |   |   |   15        / \
 3C   C5  C7         H3C H3C
 / \4/|\ /           26   27
HO   C   C6
```

35.80	31.10	72.20	37.00	42.70	27.70
1/3	2/3	3/2	4/3	5/2	6/3
26.90	36.20	40.80	35.00	21.30	40.70
7/3	8/2	9/2	10/1	11/3	12/3
43.10	57.00	24.70	28.70	57.00	12.00
13/1	14/2	15/3	16/3	17/2	18/4
23.70	36.20	19.00	36.60	24.50	39.90
19/4	20/2	21/4	22/3	23/3	24/3
28.40	23.10	22.90			
25/2	26/4	27/4			

FORMULA   C27H48O   MOL WT   388.68
SOLVENT   CH2CL2
ORIG ST   C4H12SI   TEMP   AMB

D.LEIBFRITZ,J.D.ROBERTS
J AM CHEM SOC   95, 4996 (1973)

2723   U 003000   METHYL 3,7-DIACETYLCHENODESOXYCHOLATE

```
            H3C
   12 18 21\20 22
    C  CH3   CH-CH2
  19 11/  \|   17/    \
  H3C C 13C---C    23CH2
   |  |   |   |    |       |
  1C  | C9  C14 C         C24
  / \|/ \ / \ / \ /15    / \\
 2C   C10 C8   C        O   O
  |   |   |   15        |
 3C   C5  C7   O        CH3
 . \4/|\ /  .    //      25
 O   C   C6  O-C
  \        28\29
26C-CH3         CH3
 // 27
 O
```

35.30	31.60	74.30	35.20	41.20	31.20
1/3	2/3	3/2	4/3	5/2	6/3
71.20	38.10	34.80	35.80	21.10	39.90
7/2	8/2	9/2	10/1	11/3	12/3
43.00	51.20	23.90	28.30	56.30	12.10
13/1	14/2	15/3	16/3	17/2	18/4
23.00	35.70	17.30	31.20	31.00	175.30
19/4	20/2	21/4	22/3	23/3	24/1
51.80	169.20	20.90			
25/4	26/1	27/4			

FORMULA   C29H46O6   MOL WT   490.69
SOLVENT   CH2CL2
ORIG ST   C4H12SI   TEMP   AMB

D.LEIBFRITZ,J.D.ROBERTS
J AM CHEM SOC   95, 4996 (1973)

2724   U 003000   METHYL 3,12-DIACETYLDESOXYCHOLATE

```
      O
  29 ||       21
 H3C-C-O   H3C
  28 . 18     \20 22
    12C CH3   CH-CH2
  19 11/  \|   17/    \
  H3C C 13C---C    23CH2
   |  |   |   |    |       |
  1C  | C9  C14 C         C24
  / \|/ \ / \ / \ /16    / \\
 2C   C10 C8   C        O   O
  |   |   |   15        |
 3C   C5  C7           CH3
 . \4/|\ /             25
 O   C   C6
  \
26C-CH3
 \\ 27
 O
```

35.30	26.70	74.20	32.20	43.10	27.60
1/3	2/3	3/2	4/3	5/2	6/3
26.60	34.90	36.40	34.50	27.30	75.80
7/3	8/2	9/2	10/1	11/3	12/2
44.70	50.10	23.80	26.50	48.40	12.80
13/1	14/2	15/3	16/3	17/2	18/4
23.50	35.70	17.50	31.40	31.60	175.20
19/4	20/2	21/4	22/3	23/3	24/1
51.80	165.00	20.90			
25/4	26/1	27/4			

FORMULA   C29H46O6   MOL WT   490.69
SOLVENT   CH2CL2
ORIG ST   C4H12SI   TEMP   AMB

D.LEIBFRITZ,J.D.ROBERTS
J AM CHEM SOC   95, 4996 (1973)

2725   U 003000   1'-PHENYL-17-BETA-ACETOXYANDROSTA-2,4-DIENO-
                           (2,3-G')(1H-PYRAZOLO(3',4'-B')PYRIDINE)

```
       H3C  O
     21\  //
         C20
    18 |
   H3C  O
    |  |
    12C | C17
  19  / \|/ \
  H3C C11 C13 C16
 24   |  |  |  |
  C  1C | C9  C---C15
 23/ \ / \|/ \ /14
22C---C  C2 C10 C8
 ||  ||  |  |  |
  N  C25 C3 C5  C7
  \ / \ / / \ / \ /
  N   N  4C  C6
  |
  C26
 // \
31C   C27
 |   ||
30C   C28
 \ //
  C29
```

42.50	115.55	153.55	123.55	156.75	31.70
1/3	2/1	3/1	4/2	5/1	6/3
30.95	35.80	53.50	39.15	21.25	36.90
7/3	8/2	9/2	10/1	11/3	12/3
42.50	50.40	23.65	27.50	82.65	11.85
13/1	14/2	15/3	16/3	17/2	18/4
17.50	171.10	21.25	133.35	124.10	127.65
19/4	20/1	21/4	22/3	23/1	24/2
149.95	139.95	120.85	128.95	125.50	
25/1	26/1	27/2	28/2	29/2	

FORMULA   C31H35N3O2   MOL WT   481.64
SOLVENT   CDCL3
ORIG ST   C4H12SI   TEMP   300

J.HAEUFEL,H.PECH,E.BREITMAIER
CHEMIKER-ZTG   97, 658 (1973)

2726 — U 003000 1'-PHENYL-3-BETA-ACETOXY-5-ALPHA-ANDROST-16-ENO(16,17-G')(1H-PYRAZOLO(3',4'-B')PYRIDINE)

```
           C30
        /  \
     29C    C31
      |      ||
     28C    C26
      \  /  \ /
       C27  N---N
        |      ||
       25C    C22
     18 / \ / \
    H3C   N    C23
     |  |       \
    12C  |  C17 C24
   19 / \|/ \ /
   H3C C11 C13 C16
    |   |   |    |
   1C  | C9  C---C15
  / \|/ \ / \14
 2C    C10 C8
  |     |    |
 3C    C5   C7
 / \  / \  /
O    C4  C6
|   21
O=C-CH3
  20
```

FORMULA C31H37N3O2   MOL WT 483.66
SOLVENT CDCL3
ORIG ST C4H12SI   TEMP   300

36.45	33.55	73.50	27.40	45.55	28.25
1/3	2/3	3/2	4/3	5/2	6/3
31.30	34.40	54.60	35.60	20.70	33.90
7/3	8/2	9/2	10/1	11/3	12/3
45.65	55.80	29.65	115.75	174.35	17.70
13/1	14/2	15/3	16/1	17/1	18/4
12.10	170.45	21.35	133.45	131.00	128.80
19/4	20/1	21/4	22/2	23/1	24/2
149.95	140.15	120.30	128.80	125.15	
25/1	26/1	27/2	28/2	29/2	

J.HAEUFEL,H.PECH,E.BREITMAIER
CHEMIKER-ZTG   97, 658 (1973)

2727 — U 003000 METHYL 3,7,12-TRIACETYLCHOLATE

```
        O
  31 ||30    21
 H3C-C-O   H3C
   .  18  \20 22
  12C CH3  CH-CH2
 19 11/ \| 17/ \
 H3C  C 13C---C  23CH2
   |  |   |      |
  1C  | C9 C14 C   C24
 / \|/ \ / \ /16  / \
2C   C10 C8  C    O   O
 |    |   |  15    |
3C   C5  C7   CH3 CH3
 . \4/|\ /  .  /29  25
O   C   C6  O-C 28
 \              O
26C-CH3        O
 /  27
O
```

FORMULA C31H48O8   MOL WT 548.72
SOLVENT CH2CL2
ORIG ST C4H12SI   TEMP   AMB

35.40	27.40	74.30	35.50	41.50	31.60
1/3	2/3	3/2	4/3	5/2	6/3
71.20	38.00	29.60	35.10	27.10	75.60
7/2	8/2	9/2	10/1	11/3	12/2
45.70	43.90	23.50	28.00	48.40	12.80
13/1	14/2	15/3	16/3	17/2	18/4
23.10	35.60	18.50	31.40	31.20	175.40
19/4	20/2	21/4	22/3	23/3	24/1
51.80	169.30	20.80			
25/4	26/1	27/4			

D.LEIBFRITZ,J.D.ROBERTS
J AM CHEM SOC   95, 4996 (1973)

2728 — U 003000 1'-PHENYL-3'-METHYL-17-BETA-ACETOXYANDROSTA-2,-4-DIENO(2,3-G')(1H-PYRAZOLO(3',4'-B')PYRIDINE)

```
     H3C   O
    21\  //
        C20
    18 |
    H3C O
     | |
    12C | C17
   19 / \|/ \
   H3C C11 C13 C16
32     24  |   |   |    |
H3C   C  1C | C9  C---C15
   \  23/ \|/ \ / \14
 22C---C  C2  C10 C8
    ||    |    |   |
    N   C25 C3  C5  C7
   \ / \ / \ / \ / \
    N   N   4C  C6
     |
    C26
    / \
  31C   C27
   |     ||
  30C   C28
    \ /
    C29
```

FORMULA C32H37N3O2   MOL WT 495.67
SOLVENT CDCL3
ORIG ST C4H12SI   TEMP   300

42.30	115.10	153.00	123.55	156.00	31.50
1/3	2/1	3/1	4/2	5/1	6/3
30.65	35.60	53.30	38.85	21.05	36.70
7/3	8/2	9/2	10/1	11/3	12/3
42.30	50.05	23.40	27.40	82.45	12.00
13/1	14/2	15/3	16/3	17/2	18/4
17.35	170.80	21.05	140.00	123.00	126.90
19/4	20/1	21/4	22/1	23/1	24/2
150.40	141.75	119.85	128.70	124.50	12.30
25/1	26/1	27/2	28/2	29/2	32/4

J.HAEUFEL,H.PECH,E.BREITMAIER
CHEMIKER-ZTG   97, 658 (1973)

2729 — U 003000 1'-PHENYL-3'-METHYL-3-BETA-ACETOXY-5-ALPHA-ANDROST-16-ENO(16,17-G')(1H-PYRAZOLO(3',4'-B')-PYRIDINE)

```
           C30
        / \
     29C    C31
      |      ||
     28C    C26
      \  /  \ /
       C27  N---N
        |      ||
       25C    C22
     18 / \ / \
    H3C   N    C23 CH3
     |  |       ||  32
    12C  |  C17 C24
   19 / \|/ \ /
   H3C C11 C13 C16
    |   |   |    |
   1C  | C9  C---C15
  / \|/ \ / \14
 2C    C10 C8
  |     |    |
 3C    C5   C7
 / \  / \  /
O    C4  C6
|   21
O=C-CH3
  20
```

FORMULA C32H39N3O2   MOL WT 497.69
SOLVENT CDCL3
ORIG ST C4H12SI   TEMP   300

36.35	33.55	73.35	27.30	44.55	28.25
1/3	2/3	3/2	4/3	5/2	6/3
31.20	34.30	54.50	35.60	20.70	33.90
7/3	8/2	9/2	10/1	11/3	12/3
45.55	55.80	29.65	115.45	174.00	17.60
13/1	14/2	15/3	16/1	17/1	18/4
12.30	170.35	21.35	140.25	130.00	128.70
19/4	20/1	21/4	22/1	23/1	24/2
150.50	141.90	119.75	128.70	124.50	12.30
25/1	26/1	27/2	28/2	29/2	32/4

J.HAEUFEL,H.PECH,E.BREITMAIER
CHEMIKER-ZTG   97, 658 (1973)

2730   U 004000 GLYCINE

```
          O
   + 2  1/
 H3N-CH2-C
        \  -
          O
```

FORMULA   C2H5NO2   MOL WT   75.07
SOLVENT   D2O
ORIG ST   C4H12SI   TEMP   303

173.50   42.50
1/1   2/3

W.VOELTER,G.JUNG,E.BREITMAIER,E.BAYER
Z NATURFORSCH   26B, 213 (1971)

2731   U 004000 GLYCINE

FORMULA   C2H5NO2   MOL WT   75.07
SOLVENT   D2O
ORIG ST   C4H12SI   TEMP   300

```
          O
    2  1/
 H2N-CH2-C
        \
          OH
```

173.50   42.50
1/1   2/3

W.VOELTER,S.FUCHS,R.H.SEUFFER,K.ZECH
MONATSH CHEM   105, 1110 (1974)

2732   U 004000 GLYCINAMIDE HYDROCHLORIDE

```
  HCL     O
  . 2  1/
 H2N-CH2-C
        \
          NH2
```

FORMULA   C2H6ON2   MOL WT   74.08
SOLVENT   D2O
ORIG ST   C4H12SI   TEMP   303

174.90   50.80
1/1   2/3

W.VOELTER,G.JUNG,E.BREITMAIER,E.BAYER
Z NATURFORSCH   26B, 213 (1971)

2733   U 004000 GLYCINAMIDE HYDROCHLORIDE

FORMULA   C2H6N2O   MOL WT   74.08
SOLVENT   C2D6OS
ORIG ST   C4H12SI   TEMP   300

```
          O
    2  1/
 H2N-CH2-C
        \
          NH2
          .
          HCL
```

168.00   40.25
1/1   2/3

W.VOELTER,S.FUCHS,R.H.SEUFFER,K.ZECH
MONATSH CHEM   105, 1110 (1974)

2734   U 004000 L—ALANINE

```
          O
   + 2  1/
 H3N-CH-C
    |    \  -
   3CH3   O
```

FORMULA   C3H7NO2   MOL WT   89.09
SOLVENT   D2O
ORIG ST   C4H12SI   TEMP   303

176.80   51.60   17.30
1/1   2/2   3/4

W.VOELTER,G.JUNG,E.BREITMAIER,E.BAYER
Z NATURFORSCH   26B, 213 (1971)

2735   U 004000 ALANINE

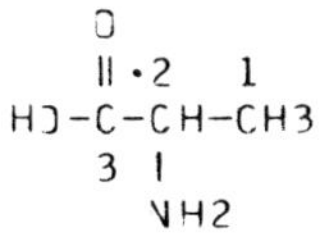

```
        O
        ‖·2  1
   HO-C-CH-CH3
     3 |
        NH2
```

FORMULA	C3H7NO2		MOL WT	89.09
SOLVENT	H2O			
ORIG ST	C4H12SI		TEMP	AMB

| 17.20 | 51.50 | 176.50 |
| 1/4 | 2/2 | 3/1 |

L.F.JOHNSON,W.C.JANKOWSKI
CARBON-13 NMR SPECTRA,JOHN
WILEY AND SONS,NEW YORK   36 (1972)

2736   U 004000 BETA—ALANINE

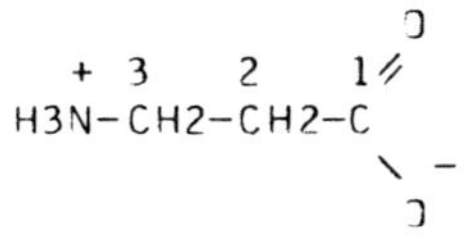

```
                  O
    + 3   2  1⁄
   H3N-CH2-CH2-C
               \ -
                 O
```

FORMULA	C3H7NO2		MOL WT	89.09
SOLVENT	D2O			
ORIG ST	C4H12SI		TEMP	303

| 183.40 | 35.70 | 38.45 |
| 1/1 | 2/3 | 3/3 |

W.VOELTER,G.JUNG,E.BREITMAIER,E.BAYER
Z NATURFORSCH   26B, 213 (1971)

2737   U 004000 N—METHYLGLYCINE

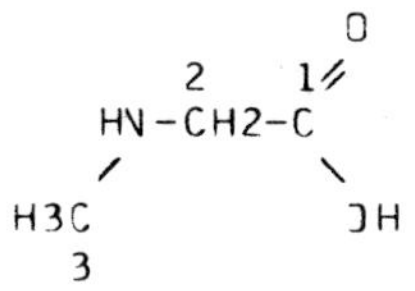

```
            O
      2   1⁄
   HN-CH2-C
   /       \
 H3C        OH
  3
```

FORMULA	C3H7NO2		MOL WT	89.09
SOLVENT	D2O			
ORIG ST	C4H12SI		TEMP	303

| 170.00 | 50.20 | 78.45 |
| 1/1 | 2/3 | 3/4 |

W.VOELTER,G.JUNG,E.BREITMAIER,E.BAYER
Z NATURFORSCH   26B, 213 (1971)

2738   U 004000 CYSTEINE

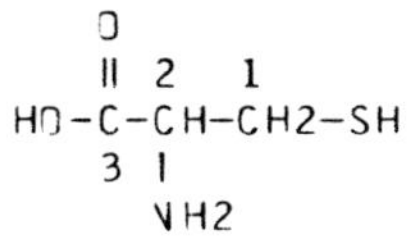

```
       O
       ‖ 2  1
   HO-C-CH-CH2-SH
     3 |
        NH2
```

FORMULA	C3H7NO2S		MOL WT	121.16
SOLVENT	H2O			
ORIG ST	C4H12SI		TEMP	AMB

| 28.20 | 58.40 | 176.80 |
| 1/3 | 2/2 | 3/1 |

L.F.JOHNSON,W.C.JANKOWSKI
CARBON-13 NMR SPECTRA,JOHN
WILEY AND SONS,NEW YORK   38 (1972)

2739   U 004000 L—CYSTEINE HYDROCHLORIDE

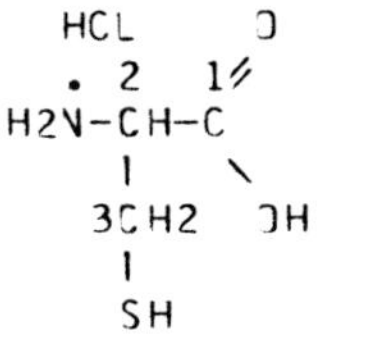

```
   HCL      O
   · 2   1⁄
   H2N-CH-C
      |     \
    3CH2    OH
      |
      SH
```

FORMULA	C3H7NO2S		MOL WT	121.16
SOLVENT	D2O			
ORIG ST	C4H12SI		TEMP	303

| 171.95 | 57.45 | 27.40 |
| 1/1 | 2/2 | 3/3 |

W.VOELTER,G.JUNG,E.BREITMAIER,E.BAYER
Z NATURFORSCH   26B, 213 (1971)

2740　　　　　　　U 004000　L—CYSTEIC ACID

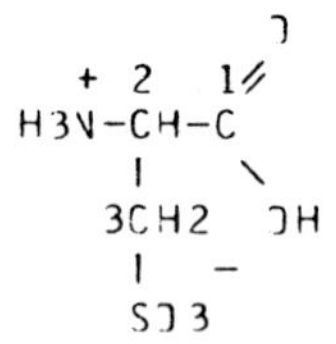

```
          O
    + 2  1//
   H3N-CH-C
       |    \
      3CH2  OH
       |    -
       SO3
```

FORMULA　C3H7NO5S　　　MOL WT　169.16
SOLVENT　D2O
ORIG ST　C4H12SI　　　　TEMP　　　303

175.10　　51.70　　52.50
1/1　　　2/2　　　3/3

W.VOELTER,G.JUNG,E.BREITMAIER,E.BAYER
Z NATURFORSCH　　　　　26B, 213 (1971)

2741　　　　　　　U 004000　L—SERINE

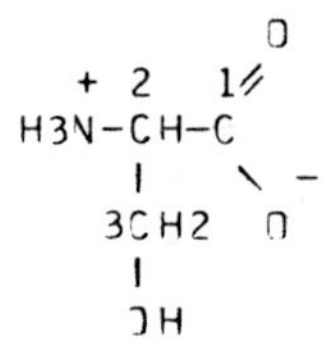

```
          O
    + 2  1//
   H3N-CH-C
       |    \ -
      3CH2  O
       |
       OH
```

FORMULA　C3H7NO3　　　MOL WT　105.09
SOLVENT　D2O
ORIG ST　C4H12SI　　　　TEMP　　　303

173.10　　57.40　　61.30
1/1　　　2/2　　　3/3

W.VOELTER,G.JUNG,E.BREITMAIER,E.BAYER
Z NATURFORSCH　　　　　26B, 213 (1971)

2742　　　　　　　U 004000　L—ASPARTIC ACID

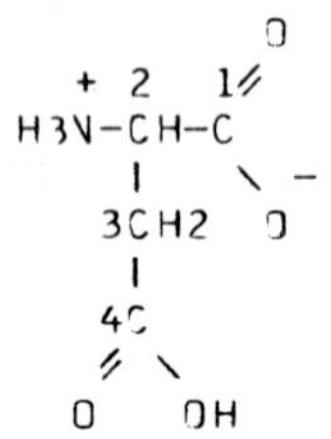

```
          O
    + 2  1//
   H3N-CH-C
       |    \ -
      3CH2  O
       |
       4C
      // \
     O    OH
```

FORMULA　C4H7NO4　　　MOL WT　133.10
SOLVENT　D2O
ORIG ST　C4H12SI　　　　TEMP　　　303

175.50　　53.20　　37.60　　178.80
1/1　　　2/2　　　3/3　　　4/1

W.VOELTER,G.JUNG,E.BREITMAIER,E.BAYER
Z NATURFORSCH　　　　　26B, 213 (1971)

2743　　　　　　　U 004000　N—GLYCYLGLYCINE

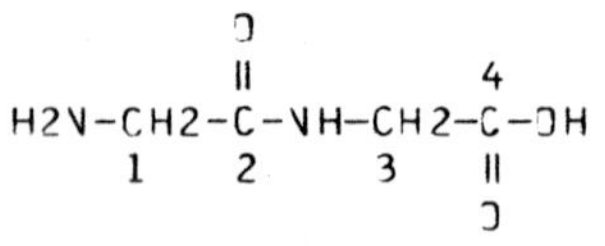

```
      O
      ||              4
  H2N-CH2-C-NH-CH2-C-OH
   1   2   3     ||
               O
```

FORMULA　C4H8N2O3　　　MOL WT　132.12
SOLVENT　H2O
ORIG ST　C4H12SI　　　　TEMP　　　AMB

44.20　　167.90　　41.50　　177.10
1/3　　　2/1　　　3/3　　　4/1

L.F.JOHNSON,W.C.JANKOWSKI
CARBON-13 NMR SPECTRA,JOHN
WILEY AND SONS,NEW YORK　　　72 (1972)

2744　　　　　　　U 004000　L—ASPARAGINE

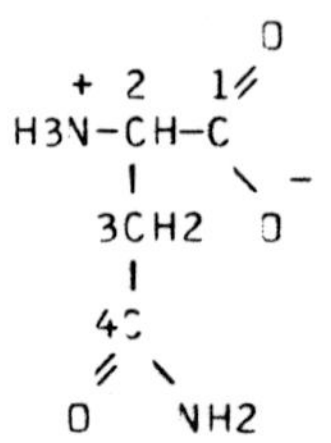

```
          O
    + 2  1//
   H3N-CH-C
       |    \ -
      3CH2  O
       |
       4C
      // \
     O    NH2
```

FORMULA　C4H8N2O3　　　MOL WT　132.12
SOLVENT　D2O/DCL
ORIG ST　C4H12SI　　　　TEMP　　　303

174.20　　52.65　　36.50　　174.45
1/1　　　2/2　　　3/3　　　4/1

W.VOELTER,G.JUNG,E.BREITMAIER,E.BAYER
Z NATURFORSCH　　　　　26B, 213 (1971)

2745   U 004000 L—THREONINE

```
            O              FORMULA   C4H9NO3                MOL WT  119.12
    + 2  1//             SOLVENT   D2O
  H3N—CH—C               ORIG ST   C4H12SI                 TEMP      303
     13   \ -
      HC—OH O             174.00    61.50    67.10    20.50
       |                   1/1       2/2      3/2      4/4
      4CH3
```

W.VOELTER,G.JUNG,E.BREITMAIER,E.BAYER
Z NATURFORSCH   26B, 213 (1971)

2746   U 004000 THREONINE

```
                         FORMULA   C4H9NO3                MOL WT  119.12
   O   OH                SOLVENT   H2O
   || 2  |               ORIG ST   C4H12SI                TEMP      AMB
  C—CH—CH—CH3
 //1  |  3  4             181.40    62.70    70.40    20.00
HO    NH2                  1/1       2/2      3/2      4/4
```

L.F.JOHNSON,W.C.JANKOWSKI
CARBON-13 NMR SPECTRA,JOHN
WILEY AND SONS,NEW YORK   87 (1972)

2747   U 004000 L—CYSTEINE METHYL ESTER HYDROCHLORIDE

```
  HCL     O             FORMULA   C4H9NO2S               MOL WT  135.19
  . 2  1//             SOLVENT   C2D6OS
 H2N—CH—C               ORIG ST   C4H12SI                TEMP      303
     |   \ 4
   3CH2  O—CH3           173.75    57.20    26.50    56.10
     |                    1/1       2/2      3/3      4/4
     SH
```

W.VOELTER,G.JUNG,E.BREITMAIER,E.BAYER
Z NATURFORSCH   26B, 213 (1971)

2748   U 004000 L—CYSTEINE METHYL ESTER HYDROCHLORIDE

```
  HCL     O             FORMULA   C4H9NO2S               MOL WT  135.19
  . 2  1//             SOLVENT   D2O
 H2N—CH—C               ORIG ST   C4H12SI                TEMP      303
     |   \ 4
   3CH2  O—CH3           174.20    57.35    25.95    56.90
     |                    1/1       2/2      3/3      4/4
     SH
```

W.VOELTER,G.JUNG,E.BREITMAIER,E.BAYER
Z NATURFORSCH   26B, 213 (1971)

2749   U 004000 PYROGLUTAMIC ACID

```
   3    2                FORMULA   C5H7NO3                MOL WT  129.12
   C———C   O             SOLVENT   H2O
   |   |  //            ORIG ST   C4H12SI                TEMP      300
  4C   C—C
  || \ //1 5\            58.35    25.35    29.65   181.20   180.20
  O    N     OH           1/2      2/3      3/3      4/1      5/1
```

W.VOELTER,S.FUCHS,R.H.SEUFFER,K.ZECH
MONATSH CHEM   105, 1110 (1974)

2750   U 004000 MONOSODIUM GLUTAMATE

                              FORMULA   C5H8NNAO4   MOL WT   169.11
                              SOLVENT   H2O
        O NH2   O   ORIG ST   C4H12SI   TEMP   AMB
        ‖ | ‖   - +
     HO-C-CH-CH2-CH2-C-O NA   182.00   34.30   27.80   55.50   175.40
        5 4 3   2   1   1/1   2/3   3/3   4/2   5/1

                              L.F.JOHNSON,W.C.JANKOWSKI
                              CARBON—13 NMR SPECTRA,JOHN
                              WILEY AND SONS,NEW YORK   111 (1972)

2751   U 004000 PROLINE

                O   FORMULA   C5H9NO2   MOL WT   115.13
          1 5∥   SOLVENT   D2O
        N---C-C   ORIG ST   C4H12SI   TEMP   300
        | | \
       4C   C 2 OH   61.60   29.70   24.40   46.50   174.60
         \ /   1/2   2/3   3/3   4/3   5/1
          C
          3   W.VOELTER,S.FUCHS,R.H.SEUFFER,K.ZECH
                              MONATSH CHEM   105, 1110 (1974)

2752   U 004000 L—PROLINE

                O   FORMULA   C5H9NO2   MOL WT   115.13
        +   2 ∥   SOLVENT   D2O
     H2N---C-C1   ORIG ST   C4H12SI   TEMP   303
        | | \ -
       5C   C3   O   174.60   61.60   29.70   24.40   46.50
         \ /   1/1   2/2   3/3   4/3   5/3
          C4
                              W.VOELTER,G.JUNG,E.BREITMAIER,E.BAYER
                              Z NATURFORSCH   26B, 213 (1971)

2753   U 004000 PROLINE

        O OH   FORMULA   C5H9NO2   MOL WT   115.13
         \ /   SOLVENT   H2O
          C1   ORIG ST   C4H12SI   TEMP   AMB
          |
          C2   174.20   61.10   28.70   23.60   46.00
         / \   1/1   2/2   3/3   4/3   5/3
       HN   C3
        | | D.E.DORMAN,F.A.BOVEY
       5C---C4   J ORG CHEM   38, 2379 (1973)

2754   U 004000 PROLINE

                              FORMULA   C5H9NO2   MOL WT   115.13
       3C---C2   SOLVENT   H2O
        | | ORIG ST   C4H12SI   TEMP   AMB
       4C   C1 OH
         \ / \ /   62.10   29.80   24.60   47.00   175.30
          N   C5   1/2   2/3   3/3   4/3   5/1
               ‖
               O   L.F.JOHNSON,W.C.JANKOWSKI
                              CARBON—13 NMR SPECTRA,JOHN
                              WILEY AND SONS,NEW YORK   117 (1972)

2755   U 004000 HYDROXYPROLINE

```
   HO
     \
      3C---C2
      |    |
     4C   C1  OH
      \  / \ /
       N  5C
       |  ||
          O
```

FORMULA   C5H9NO3   MOL WT   131.13
SOLVENT   H2O
ORIG ST   C4H12SI   TEMP   AMB

60.70	38.20	70.90	53.90	174.90
1/2	2/3	3/2	4/3	5/1

L.F.JOHNSON,W.C.JANKOWSKI
CARBON-13 NMR SPECTRA,JOHN
WILEY AND SONS,NEW YORK   118 (1972)

2756   U 004000 N-ACETYL-L-CYSTEINE

```
   O         O
   \\4   2  1//
    C-NH-CH-C
  5//      |   \
 H3C    3CH2   OH
         |
         SH
```

FORMULA   C5H9NO3S   MOL WT   163.20
SOLVENT   C2H6OS
ORIG ST   C4H12SI   TEMP   303

179.40	57.20	27.30	178.60	23.75
1/1	2/2	3/3	4/1	5/4

W.VOELTER,G.JUNG,E.BREITMAIER,E.BAYER
Z NATURFORSCH   26B, 213 (1971)

2757   U 004000 GLUTAMIC ACID

```
  O            NH2  O
   \           |   //
    C-CH2-CH2-CH-C
   /5  4   3   2  1\
  HO                OH
```

FORMULA   C5H9NO4   MOL WT   147.13
SOLVENT   D2O
ORIG ST   C4H12SI   TEMP   300

175.60	55.70	28.10	34.50	182.30
1/1	2/2	3/3	4/3	5/1

W.VOELTER,S.FUCHS,R.H.SEUFFER,K.ZECH
MONATSH CHEM   105, 1110 (1974)

2758   U 004000 L-GLUTAMIC ACID

```
          O
    + 2  1//
  H3N-CH-C
      |   \ -
  O  3CH2  O
   \  |
    C-CH2
   /5  4
  HO
```

FORMULA   C5H9NO4   MOL WT   147.13
SOLVENT   D2O
ORIG ST   C4H12SI   TEMP   303

175.60	55.70	28.10	34.50	182.30
1/1	2/2	3/3	4/3	5/1

W.VOELTER,G.JUNG,E.BREITMAIER,E.BAYER
Z NATURFORSCH   26B, 213 (1971)

2759   U 004000 PROLINAMIDE

```
          O
    1  5//
  N---C-C
  |   |   \
 4C   C2  NH2
  \  /
   C
   3
```

FORMULA   C5H10N2O   MOL WT   114.15
SOLVENT   D2O
ORIG ST   C4H12SI   TEMP   300

59.90	30.55	25.15	46.40	180.05
1/2	2/3	3/3	4/3	5/1

W.VOELTER,S.FUCHS,R.H.SEUFFER,K.ZECH
MONATSH CHEM   105, 1110 (1974)

2760   U 004000 GLUTAMINE

```
                        FORMULA   C5H10N2O3           MOL WT  146.15
                        SOLVENT   D2O/DCL
     NH2      NH2  O     ORIG ST   C4H12SI             TEMP      300
      I        I   //
   O=C-CH2-CH2-CH-C       179.50   52.20   28.30   33.00   174.45
    5  4   3   2  1\        1/1     2/2     3/3     4/3     5/1
                    OH
                        W.VOELTER,S.FUCHS,R.H.SEUFFER,K.ZECH
                        MONATSH CHEM                  105, 1110 (1974)
```

2761   U 004000 L—GLUTAMINE

```
               O        FORMULA   C5H10N2O3           MOL WT  146.15
     + 2   1//           SOLVENT   D2O/DCL
   H3N-CH-C             ORIG ST   C4H12SI             TEMP      303
      I     \ -
    3CH2    O            179.50   55.20   28.30   33.00   174.45
      I                    1/1     2/2     3/3     4/3     5/1
    4CH2
     I 5
   O=C-NH2              W.VOELTER,G.JUNG,E.BREITMAIER,E.BAYER
                        Z NATURFORSCH                 26B, 213 (1971)
```

2762   U 004000 L—VALINE

```
               O        FORMULA   C5H11NO2            MOL WT  117.15
     + 2   1//           SOLVENT   D2O
   H3N-CH-C             ORIG ST   C4H12SI             TEMP      303
      I     \ -
    3HC    O             175.30   61.60   30.20   17.90   19.10
    5/  \4                 1/1     2/2     3/2     4/4     5/4
   H3C   CH3
                        W.VOELTER,G.JUNG,E.BREITMAIER,E.BAYER
                        Z NATURFORSCH                 26B, 213 (1971)
```

2763   U 004000 METHYL 2—AMINO—2—METHYLPROPANOATE

```
                        FORMULA   C5H11NO2            MOL WT  117.15
    CH3 O               SOLVENT   CDCL3
   3 I  //              ORIG ST   C4H12SI             TEMP      300
  H3C-C-C1
    I2  \ 4              177.15   53.50   26.55   50.82
    NH2 O-CH3             1/1     2/1     3/4     4/4

                        G.JUNG,E.BREITMAIER
                        UNPUBLISHED                         (1974)
```

2764   U 004000 METHIONINE

```
                        FORMULA   C5H11NO2S           MOL WT  149.21
                        SOLVENT   H2O
    O                   ORIG ST   C4H12SI             TEMP      AMB
    II 2  3   4    5
  HO-C-CH-CH2-CH2-S-CH3   182.70   56.10   30.60   34.60   15.00
    1 I                    1/1     2/2     3/3     4/3     5/4
    NH2
                        L.F.JOHNSON,W.C.JANKOWSKI
                        CARBON-13 NMR SPECTRA,JOHN
                        WILEY AND SONS,NEW YORK         135 (1972)
```

2765   U 004000 L—METHIONINE

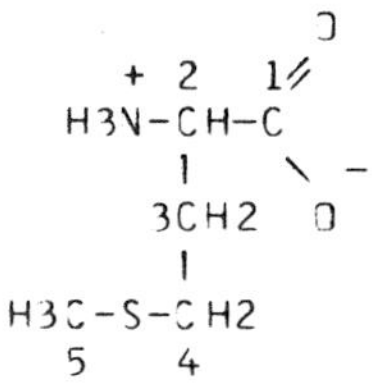

```
         O
   + 2  1//
  H3N-CH-C
    |     \  -
   3CH2   O
    |
  H3C-S-CH2
    5    4
```

FORMULA C5H11NO2S   MOL WT 149.21
SOLVENT D2O
ORIG ST C4H12SI   TEMP   303

175.30   55.30   31.00   30.10   15.20
1/1   2/2   3/3   4/3   5/4

W.VOELTER,G.JUNG,E.BREITMAIER,E.BAYER
Z NATURFORSCH   26B, 213 (1971)

2766   U 004000 BETAINE HYDROCHLORIDE

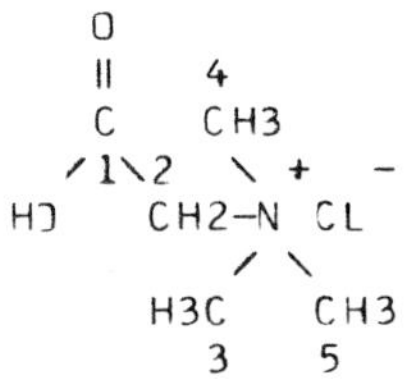

```
       O
       ||   4
       C    CH3
     //1\2 \ + -
   HO   CH2-N CL
       / \
     H3C   CH3
       3     5
```

FORMULA C5H12CLNO2   MOL WT 153.61
SOLVENT H2O
ORIG ST C4H12SI   TEMP   AMB

167.60   64.60   54.80   54.80   54.80
1/1   2/3   3/4   4/4   5/4

L.F.JOHNSON,W.C.JANKOWSKI
CARBON—13 NMR SPECTRA,JOHN
WILEY AND SONS,NEW YORK   134 (1972)

2767   U 004000 L—ORNITHINE HYDROCHLORIDE

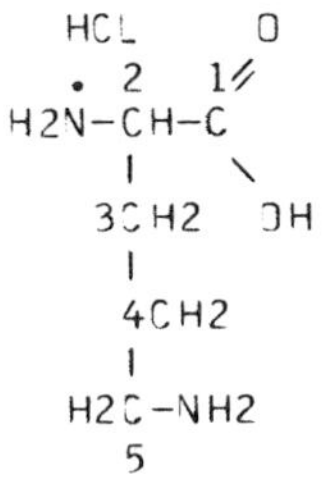

```
  HCL      O
  . 2  1//
 H2N-CH-C
    |     \
   3CH2   OH
    |
   4CH2
    |
  H2C-NH2
    5
```

FORMULA C5H12N2O2   MOL WT 132.16
SOLVENT D2O
ORIG ST C4H12SI   TEMP   303

179.40   56.60   29.20   24.50   41.10
1/1   2/2   3/3   4/3   5/3

W.VOELTER,G.JUNG,E.BREITMAIER,E.BAYER
Z NATURFORSCH   26B, 213 (1971)

2768   U 004000 CIS—N—FORMYLPROLINE

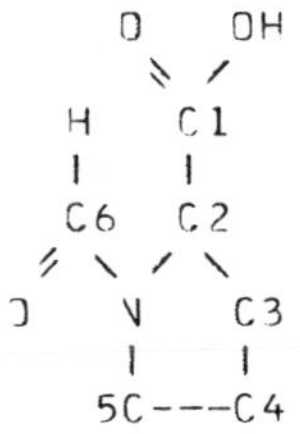

```
   O  OH
    \ /
   H  C1
   |  |
  C6  C2
  // \ / \
 O   N   C3
     |   |
    5C---C4
```

FORMULA C6H9NO3   MOL WT 143.14
SOLVENT H2O
ORIG ST C4H12SI   TEMP   AMB

175.90   59.40   29.10   22.20   44.00   164.20
1/1   2/2   3/3   4/3   5/3   6/2

D.E.DORMAN,F.A.BOVEY
J ORG CHEM   38, 2379 (1973)

2769   U 004000 TRANS—N—FORMYLPROLINE

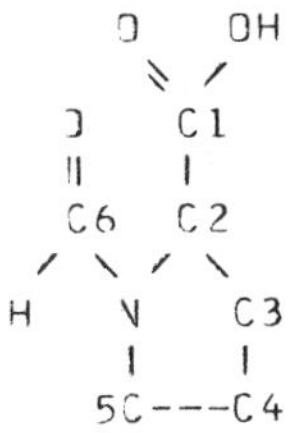

```
   O  OH
    \ /
   O  C1
   ||  |
  C6  C2
  / \ / \
 H   N   C3
     |   |
    5C---C4
```

FORMULA C6H9NO3   MOL WT 143.14
SOLVENT H2O
ORIG ST C4H12SI   TEMP   AMB

175.20   56.70   29.10   23.40   47.00   163.10
1/1   2/2   3/3   4/3   5/3   6/2

D.E.DORMAN,F.A.BOVEY
J ORG CHEM   38, 2379 (1973)

2770   U 004000 L—HISTIDINE

```
                 O
      4 3    2  1/
   HN---C-CH2-CH-C
    |   ||    |  \ -
   C6   C5   NH3  O
    \  /       +
     N
```

FORMULA	C6H9N3O2			MOL WT	155.16
SOLVENT	D2O				
ORIG ST	C4H12SI			TEMP	303

174.90	58.80	29.00	118.20	137.20
1/1	2/2	3/3	4/1	5/2

W.VOELTER,G.JUNG,E.BREITMAIER,E.BAYER
Z NATURFORSCH   26B, 213 (1971)

2771   U 004000 HISTIDINE HYDROCHLORIDE

FORMULA	C6H9N3O2				MOL WT	155.16
SOLVENT	H2O					
ORIG ST	C4H12SI				TEMP	AMB

```
      O
      || 5   4    3
   HO-C-CH-CH2-C===C2
      6 |     |  |
       NH2    N  N
        .      \ /
       HCL     1C
```

128.40	118.60	135.00	26.70	54.50	173.30
1/2	2/2	3/1	4/3	5/2	6/1

L.F.JOHNSON,W.C.JANKOWSKI
CARBON-13 NMR SPECTRA,JOHN
WILEY AND SONS,NEW YORK   174 (1972)

2772   U 004000 GLYCYLSARKOSINE METHYL ESTER

FORMULA	C6H12N2O3			MOL WT	160.17
SOLVENT	CDCL3				
ORIG ST	C4H12SI			TEMP	300

```
         O       O
    5    ||   2  1/
   H2N-CH2-C-N-CH2-C
      4  |      \ 6
       3CH3      O-CH3
```

165.40	51.50	34.85	163.35	44.55	54.15
1/1	2/3	3/4	4/1	5/3	6/4

U.HOLLSTEIN,E.BREITMAIER,G.JUNG
J AM CHEM SOC   96, 8036 (1974)

2773   U 004000 CYSTINE

```
   O                       O
   ||                      ||
   6C                      3C
  / \5 4       1    2 / \
 HO    CH-CH2-S-S-CH2-CH   OH
        |            |
        NH2         NH2
```

FORMULA	C6H12N2O4S2			MOL WT	240.30
SOLVENT	H2O				
ORIG ST	C4H12SI			TEMP	AMB

44.60	55.80	181.20	44.60	55.80	181.20
1/3	2/2	3/1	4/3	5/2	6/1

L.F.JOHNSON,W.C.JANKOWSKI
CARBON-13 NMR SPECTRA,JOHN
WILEY AND SONS,NEW YORK   189 (1972)

2774   U 004000 L—CYSTINE HYDROCHLORIDE

```
    HCL          HCL
     .            .
   H2N          NH2
     \5 6     3 2/
      HC-CH2 H2C-CH
     4/   \   /  \1
   O=C     S-S    C=O
     |            |
    HO            OH
```

FORMULA	C6H12N2O4S2			MOL WT	240.30
SOLVENT	D2O/DCL				
ORIG ST	C4H12SI			TEMP	303

175.60	54.70	39.00	175.60	54.70	39.00
1/1	2/2	3/3	4/1	5/2	6/3

W.VOELTER,G.JUNG,E.BREITMAIER,E.BAYER
Z NATURFORSCH   26B, 213 (1971)

2775   U 004000 ETHYL 2-AMINO-2-METHYLPROPANOATE

```
     CH3 O
      3 |  /
   H3C-C-C 1
      |2 \  4   5
     NH2 O-CH2-CH3
```

FORMULA	C6H13NO2			MOL WT	131.18
SOLVENT	CDCL3				
ORIG ST	C4H12SI			TEMP	300

176.85	53.50	26.65	59.75	13.15
1/1	2/1	3/4	4/3	5/4

G.JUNG,E.BREITMAIER
UNPUBLISHED   (1974)

2776   U 004000 LEUCINE

```
              5
    O        CH3
    || 2 3 4|
  HO-C-CH-CH2-CH
    1 |         |
     NH2       CH3
               6
```

FORMULA	C6H13NO2			MOL WT	131.18
SOLVENT	H2O				
ORIG ST	C4H12SI			TEMP	AMB

173.30	52.50	39.80	24.80	22.00	22.50
1/1	2/2	3/3	4/2	5/4	6/4

L.F.JOHNSON,W.C.JANKOWSKI
CARBON-13 NMR SPECTRA,JOHN
WILEY AND SONS,NEW YORK   204 (1972)

2777   U 004000 L-LEUCINE

```
           O
    + 2  1/
  H3N-CH-C
     |    \ -
    3CH2  O
     |
    4CH
    6/ \5
   H3C   CH3
```

FORMULA	C6H13NO2			MOL WT	131.18
SOLVENT	D2O				
ORIG ST	C4H12SI			TEMP	303

176.60	54.70	41.00	25.40	22.10	23.20
1/1	2/2	3/3	4/2	5/4	6/4

W.VOELTER,G.JUNG,E.BREITMAIER,E.BAYER
Z NATURFORSCH   26B, 213 (1971)

2778   U 004000 L-ISOLEUCINE

```
           O
    + 2  1/
  H3N-CH-C
     |    \ -
    3CH   O
    4/ \6
   H2C   CH3
     |
    H3C5
```

FORMULA	C6H13NO2			MOL WT	131.18
SOLVENT	D2O				
ORIG ST	C4H12SI			TEMP	303

175.20	60.90	39.70	25.70	12.50	15.90
1/1	2/2	3/2	4/3	5/4	6/4

W.VOELTER,G.JUNG,E.BREITMAIER,E.BAYER
Z NATURFORSCH   26B, 213 (1971)

2779   U 004000 LEUCINE

```
              5
    O        CH3
    || 2 3 4/
  HO-C-CH-CH2-CH
    1 |         \
     NH2       CH3
               6
```

FORMULA	C6H13NO2			MOL WT	131.18
SOLVENT	H2O				
ORIG ST	C4H12SI			TEMP	AMB

184.70	55.50	45.20	25.20	22.30	23.50
1/1	2/2	3/3	4/2	5/4	6/4

L.F.JOHNSON,W.C.JANKOWSKI
CARBON-13 NMR SPECTRA,JOHN
WILEY AND SONS,NEW YORK   203 (1972)

2780 — U 004000 TRANS-N-ACETYLPROLINE METHYL ESTER

```
            8
      ⊃   O-CH3
       \ /
      ⊃   C1
      ‖
      C6  C2
     7/ \ / \
   H3C  N   C3
       |   |
       5C---C4
```

FORMULA	C8H13NO3			MOL WT	171.20
SOLVENT	H2O				
ORIG ST	C4H12SI			TEMP	AMB

174.60	58.70	28.90	24.10	48.10	172.50
1/1	2/2	3/3	4/3	5/3	6/1
21.50	52.60				
7/4	8/4				

D.E.DORMAN,F.A.BOVEY
J ORG CHEM   38, 2379 (1973)

2781 — U 004000 L-LYSINE

```
            O
    + 2  1⁄
  H3N-CH-C
      |   \ -
     3CH2  ⊃
      |
     4CH2
      |
     5CH2  NH2
       \ /
      6CH2
```

FORMULA	C6H14N2O2			MOL WT	146.19
SOLVENT	D2O				
ORIG ST	C4H12SI			TEMP	303

175.40	55.30	27.20	22.40	30.70	40.00
1/1	2/2	3/3	4/3	5/3	6/3

W.VOELTER,G.JUNG,E.BREITMAIER,E.BAYER
Z NATURFORSCH   26B, 213 (1971)

2782 — U 004000 LYSINE HYDROCHLORIDE

```
    ⊃
    ‖ 2  3    4    5    6
 HO-C-CH-CH2-CH2-CH2-CH2-NH2
    1 |                    •
      NH2              HCL
```

FORMULA	C6H14N2O2			MOL WT	146.19
SOLVENT	H2O				
ORIG ST	C4H12SI			TEMP	AMB

175.30	55.20	27.30	22.30	30.80	40.20
1/1	2/2	3/3	4/3	5/3	6/3

L.F.JOHNSON,W.C.JANKOWSKI
CARBON-13 NMR SPECTRA,JOHN
WILEY AND SONS,NEW YORK   212 (1972)

2783 — U 004000 ARGININE

```
  HN
    \
    6C-NH2
    /
  HN         NH2  ⊃
    \          |  ⁄⁄
     CH2-CH2-CH2-CH-C
     5   4   3   2  1\
                     ⊃H
```

FORMULA	C6H14N4O2			MOL WT	174.20
SOLVENT	D2O				
ORIG ST	C4H12SI			TEMP	300

175.20	55.10	28.50	24.90	41.50	157.50
1/1	2/2	3/3	4/3	5/3	6/1

W.VOELTER,S.FUCHS,R.H.SEUFFER,K.ZECH
MONATSH CHEM   105, 1110 (1974)

2784 — U 004000 ARGININE HYDROCHLORIDE

```
    ⊃            H
    ‖            |
    C1           N   NH
   ⁄\2  3   4   5 ⁄\ ⁄⁄
 HO  CH-CH2-CH2-CH2   C6
     |              |
     NH2           NH2

   HCL
```

FORMULA	C6H14N4O2			MOL WT	174.20
SOLVENT	H2O				
ORIG ST	C4H12SI			TEMP	AMB

175.00	55.30	28.40	24.80	41.50	157.70
1/1	2/2	3/3	4/3	5/3	6/1

L.F.JOHNSON,W.C.JANKOWSKI
CARBON-13 NMR SPECTRA,JOHN
WILEY AND SONS,NEW YORK   213 (1972)

2785 U 004000 L-ARGININE

```
         O
  + 2  1//
 H3N-CH-C
   |     \ -
  3CH2   O
   |
  4CH2    NH
   |    6//
 H2C-NH-C
   5     \
          NH2
```

FORMULA	C6H14N4O2		MOL WT	174.20
SOLVENT	D2O			
ORIG ST	C4H12SI		TEMP	303

175.20	55.10	28.50	24.90	41.50	157.50
1/1	2/2	3/3	4/3	5/3	6/1

W.VOELTER,G.JUNG,E.BREITMAIER,E.BAYER
Z NATURFORSCH   26B, 213 (1971)

2786 U 004000 CIS-ACETYLPROLINE

```
            O
  7        //
 H3C       5C
   \      / \
   6C-N---C1  OH
  // |    |
 O  4C    C2
     \   /
       C
       3
```

FORMULA	C7H11NO3		MOL WT	157.17
SOLVENT	C2D6OS			
ORIG ST	C4H12SI		TEMP	300

59.55	30.95	22.55	45.95	173.95	168.45
1/2	2/3	3/3	4/3	5/1	6/1
22.25					
7/4					

W.VOELTER,S.FUCHS,R.H.SEUFFER,K.ZECH
MONATSH CHEM   105, 1110 (1974)

2787 U 004000 CIS-N-ACETYLPROLINE

```
    O  OH
  7  \ /
 H3C  C1
   |  |
   C6 C2
  // \ / \
 O   N   C3
     |   |
     5C---C4
```

FORMULA	C7H11NO3		MOL WT	157.17
SOLVENT	H2O			
ORIG ST	C4H12SI		TEMP	AMB

176.10	60.70	30.60	22.30	46.40	172.80
1/1	2/2	3/3	4/3	5/3	6/1
21.10					
7/4					

D.E.DORMAN,F.A.BOVEY
J ORG CHEM   38, 2379 (1973)

2788 U 004000 TRANS-ACETYLPROLINE

```
           O
          //
  O        5C
   \\     / \
   6C-N---C1  OH
   / |    |
 H3C 4C   C2
   7  \  /
        C
        3
```

FORMULA	C7H11NO3		MOL WT	157.17
SOLVENT	C2D6OS			
ORIG ST	C4H12SI		TEMP	300

58.25	29.25	24.50	47.35	173.60	168.75
1/2	2/3	3/3	4/3	5/1	6/1
22.25					
7/4					

W.VOELTER,S.FUCHS,R.H.SEUFFER,K.ZECH
MONATSH CHEM   105, 1110 (1974)

2789 U 004000 TRANS-N-ACETYLPROLINE

```
    O  OH
     \ /
  O   C1
  ||  |
  C6  C2
  7/ \ / \
 H3C  N   C3
      |   |
     5C---C4
```

FORMULA	C7H11NO3		MOL WT	157.17
SOLVENT	H2O			
ORIG ST	C4H12SI		TEMP	AMB

176.00	58.80	29.20	24.00	48.20	172.50
1/1	2/2	3/3	4/3	5/3	6/1
21.10					
7/4					

D.E.DORMAN,F.A.BOVEY
J ORG CHEM   38, 2379 (1973)

2790　　U 004000　CIS-N-ACETYLPROLINAMIDE

```
    O   NH2
     \ /
H3C7  C1
  |   |
  C6  C2
 ⁄ \ ⁄ \
O   N   C3
    |   |
   5C---C4
```

FORMULA	C7H12N2O2			MOL WT	156.19
SOLVENT	H2O				
ORIG ST	C4H12SI			TEMP	AMB

177.00	61.00	31.40	22.20	46.70	172.70
1/1	2/2	3/3	4/3	5/3	6/1
21.20					
7/4					

D.E.DORMAN,F.A.BOVEY
J ORG CHEM　　　　　　　　　　38, 2379 (1973)

2791　　U 004000　TRANS-N-ACETYLPROLINAMIDE

```
    O   NH2
     \ /
  O   C1
  ||  |
  C6  C2
 7⁄ \ ⁄ \
H3C   N   C3
      |   |
     5C---C4
```

FORMULA	C7H12N2O2			MOL WT	156.19
SOLVENT	H2O				
ORIG ST	C4H12SI			TEMP	AMB

177.00	59.50	29.80	23.80	48.30	172.70
1/1	2/2	3/3	4/3	5/3	6/1
21.00					
7/4					

D.E.DORMAN,F.A.BOVEY
J ORG CHEM　　　　　　　　　　38, 2379 (1973)

2792　　U 004000　CIS-N-GLYCYLPROLINE

```
H2N O   OH
  | \ ⁄
H2C7  C1
  |   |
  C6  C2
 ⁄ \ ⁄ \
O   N   C3
    |   |
   5C---C4
```

FORMULA	C7H12N2O3			MOL WT	172.19
SOLVENT	H2O				
ORIG ST	C4H12SI			TEMP	AMB

178.00	61.40	31.30	22.10	47.00	165.30
1/1	2/2	3/3	4/3	5/3	6/1
40.20					
7/3					

D.E.DORMAN,F.A.BOVEY
J ORG CHEM　　　　　　　　　　38, 2379 (1973)

2793　　U 004000　TRANS-N-GLYCYLPROLINE

```
     O   OH
      \ ⁄
   O   C1
   ||  |
   C6  C2
  7⁄ \ ⁄ \
H2N-H2C   N   C3
       |   |
      5C---C4
```

FORMULA	C7H12N2O3			MOL WT	172.19
SOLVENT	H2O				
ORIG ST	C4H12SI			TEMP	AMB

178.70	61.70	29.30	24.00	46.40	164.70
1/1	2/2	3/3	4/3	5/3	6/1
40.40					
7/3					

D.E.DORMAN,F.A.BOVEY
J ORG CHEM　　　　　　　　　　38, 2379 (1973)

2794　　U 004000　N-ACETYL-L-GLUTAMINE

```
    O          O
  7 ||   2  1 ⁄
H3C-C-NH-CH-C
  6    |    \
      3CH2  OH
       |
      4CH2
       |
     O=C-NH2
       5
```

FORMULA	C7H12N2O4			MOL WT	188.18
SOLVENT	C2D6OS				
ORIG ST	C4H12SI			TEMP	303

175.55	54.60	29.15	33.70	179.85	179.20
1/1	2/2	3/3	4/3	5/1	6/1
24.40					
7/4					

W.VOELTER,G.JUNG,E.BREITMAIER,E.BAYER
Z NATURFORSCH　　　　　　　26B, 213 (1971)

2795   U 004000   N—ACETYL—D,L—VALINE

```
      O   ·      O
       \6    2  1/
        C—NH—CH—C
      7/    |     \
    H3C   3CH   OH
        4/   \5
      H3C    CH3
```

FORMULA	C7H13NO3			MOL WT	159.19
SOLVENT	CDCL3				
ORIG ST	C4H12SI			TEMP	300

173.60	57.70	30.20	19.50	18.30	170.35
1/1	2/2	3/2	4/4	5/4	6/1
22.65					
7/4					

U.HOLLSTEIN,E.BREITMAIER,G.JUNG
J AM CHEM SOC   96, 8036 (1974)

2796   U 004000   N—TERT—BUTYLOXYCARBONYLGLYCINE

```
      6
      CH3          O
    5 |4  3   2  1/
  H3C—C—O—C—NH—CH2—C
      |   ||          \
    7CH3  O           OH
```

FORMULA	C7H13NO4			MOL WT	175.19
SOLVENT	C2D6OS				
ORIG ST	C4H12SI			TEMP	303

172.10	41.90	156.25	78.05	27.95	27.95
1/1	2/3	3/1	4/1	5/4	6/4
27.95					
7/4					

W.VOELTER,G.JUNG,E.BREITMAIER,E.BAYER
Z NATURFORSCH   26B, 213 (1971)

2797   U 004000   TERT—BUTYLOXYCARBONYLGLYCINE

```
    5
  H3C    O          O
    6\4  ||   2   1/
  H3C—C—O—C—NH—CH2—C
    /     3           \
  H3C               OH
  7
```

FORMULA	C7H13NO4			MOL WT	175.19
SOLVENT	C2D6OS				
ORIG ST	C4H12SI			TEMP	300

172.10	42.10	156.10	78.35	28.40	28.40
1/1	2/3	3/1	4/1	5/4	6/4
28.40					
7/4					

W.VOELTER,S.FUCHS,R.H.SEUFFER,K.ZECH
MONATSH CHEM   105, 1110 (1974)

2798   U 004000   TERT—BUTYLOXYCARBONYLGLYCINAMIDE

```
    5
  H3C    O          O
    6\4  ||   2   1/
  H3C—C—O—C—NH—CH2—C
    /     3           \
  H3C               NH2
  7
```

FORMULA	C7H14N2O3			MOL WT	174.20
SOLVENT	C2D6OS				
ORIG ST	C4H12SI			TEMP	300

171.75	43.25	156.00	78.20	28.25	28.25
1/1	2/3	3/1	4/1	5/4	6/4
28.25					
7/4					

W.VOELTER,S.FUCHS,R.H.SEUFFER,K.ZECH
MONATSH CHEM   105, 1110 (1974)

2799   U 004000   GLYCYL—L—THREONINE METHYL ESTER

```
      O          O
    6 ||   2   1/
  H2N—CH2—C—NH—CH—C
    5      |     \ 7
         3CH   O—CH3
         4/  \
        H3C    OH
```

FORMULA	C7H14N2O4			MOL WT	190.20
SOLVENT	CDCL3				
ORIG ST	C4H12SI			TEMP	300

171.65	58.05	67.85	19.95	159.55	43.90
1/1	2/2	3/2	4/4	5/1	6/3
52.55					
7/4					

U.HOLLSTEIN,E.BREITMAIER,G.JUNG
J AM CHEM SOC   96, 8036 (1974)

2800　U 004000 ARGININE METHYL ESTER DIHYDROCHLORIDE

```
HN
  \
  6C-NH2      HCL
  /    •
HN   HCL      NH2  O
  \       |      //
   CH2-CH2-CH2-CH-C
   5    4    3    2  1\
                      O-CH3
                      7
```

FORMULA C7H16N4O2　　　　MOL WT 188.23
SOLVENT C2D6OS
ORIG ST C4H12SI　　　　　TEMP 300

| 169.80 | 53.10 | 27.20 | 24.40 | 40.25 | 157.00 |
| 1/1 | 2/2 | 3/3 | 4/3 | 5/3 | 6/1 |
| 51.70 |
| 7/4 |

W.VOELTER,S.FUCHS,R.H.SEUFFER,K.ZECH
MONATSH CHEM　　　　105, 1110 (1974)

2801　U 004000 CIS-N-ACETYLPROLINE METHYL ESTER

```
        8
    O   O-CH3
     \ /
H3C7  C1
  |    |
  C6   C2
 //\ / \
O  N    C3
   |    |
   5C---C4
```

FORMULA C8H13NO3　　　　MOL WT 171.20
SOLVENT H2O
ORIG ST C4H12SI　　　　　TEMP AMB

174.60	60.40	30.70	22.10	46.40	172.50
1/1	2/2	3/3	4/3	5/3	6/1
21.00	52.60
7/4	8/4

D.E.DORMAN,F.A.BOVEY
J ORG CHEM　　　　38, 2379 (1973)

2802　U 004000 N-ACETYL-L-VALINE METHYL ESTER

```
    O          O
    \7    2    //
     C-NH-CH-C
   8/    |   1\  6
  H3C    3CH   O-CH3
         5/ \4
        H3C   CH3
```

FORMULA C8H15NO3　　　　MOL WT 173.21
SOLVENT C2H6OS
ORIG ST C4H12SI　　　　　TEMP 303

| 176.10 | 59.30 | 31.90 | 21.00 | 20.50 | 55.40 |
| 1/1 | 2/2 | 3/2 | 4/4 | 5/4 | 6/4 |

W.VOELTER,G.JUNG,E.BREITMAIER,E.BAYER
Z NATURFORSCH　　　　26B, 213 (1971)

2803　U 004000 N-TERT-BUTYLOXYCARBONYL-L-ALANINE

```
   6
  H3C    O        O
   7\5   ||   2  1//
  H3C-C-O-C-NH-CH-C
   8/   4   |    \
  H3C      3CH3  OH
```

FORMULA C8H15NO4　　　　MOL WT 189.21
SOLVENT C2D6OS
ORIG ST C4H12SI　　　　　TEMP 303

175.10	48.95	17.30	155.90	78.10	28.05
1/1	2/2	3/4	4/1	5/1	6/4
28.05	28.05
7/4	8/4

W.VOELTER,G.JUNG,E.BREITMAIER,E.BAYER
Z NATURFORSCH　　　　26B, 213 (1971)

2804　U 004000 N-TERT-BUTYLOXYCARBONYL-SARKOSINE CIS-TRANS ISOMER

```
   6
  H3C    O        O
   7\5   ||   2  1/
  H3C-C-O-C-N-CH2-C
   8/   4  |      \
  H3C     3CH3    OH
```

FORMULA C8H15NO4　　　　MOL WT 189.21
SOLVENT CDCL3
ORIG ST C4H12SI　　　　　TEMP 300

| 173.60 | 50.80 | 35.60 | 156.60 | 80.70 | 28.25 |
| 1/1 | 2/3 | 3/4 | 4/1 | 5/1 | 6/4 |

U.HOLLSTEIN,E.BREITMAIER,G.JUNG
J AM CHEM SOC　　　　96, 8036 (1974)

2805   U 004000   N-TERT-BUTYLOXYCARBONYL-SARKOSINE   CIS-TRANS ISOMER

```
        6
  H3C      O        O
    7\5    ||   2  1/
  H3C-C-O-C-N-CH2-C
    8/    4 |      \
  H3C      3CH3    OH
```

FORMULA	C8H15NO4			MOL WT	189.21
SOLVENT	CDCL3				
ORIG ST	C4H12SI			TEMP	300

173.60	50.10	35.60	156.00	80.70	28.25
1/1	2/3	3/4	4/1	5/1	6/4

U.HOLLSTEIN,E.BREITMAIER,G.JUNG
J AM CHEM SOC   96, 8036 (1974)

2806   U 004000   L-CYSTINE DIMETHYL ESTER HYDROCHLORIDE

```
    HCL           HCL
     •             •
   H2N           NH2
     \5 6     3 2/
     HC-CH2 H2C-CH
    4/    \  /   \1
  O=C      S-S    C=O
    |              |
  H3C-O          O-CH3
    8              7
```

FORMULA	C8H16N2O4S2			MOL WT	268.36
SOLVENT	C2D6OS				
ORIG ST	C4H12SI			TEMP	303

174.00	53.15	42.85	56.25	174.00	53.15
1/1	2/2	3/3	7/4	4/1	5/2
42.85	56.25				
6/3	8/4				

W.VOELTER,G.JUNG,E.BREITMAIER,E.BAYER
Z NATURFORSCH   26B, 213 (1971)

2807   U 004000   L-CYSTINE DIMETHYL ESTER HYDROCHLORIDE

```
    HCL           HCL
     •             •
   H2N           NH2
     \5 6     3 2/
     HC-CH2 H2C-CH
    4/    \  /   \1
  O=C      S-S    C=O
    |              |
  H3C-O          O-CH3
    8              7
```

FORMULA	C8H16N2O4S2			MOL WT	268.36
SOLVENT	D2O				
ORIG ST	C4H12SI			TEMP	303

174.55	54.55	38.45	57.10	174.55	54.55
1/1	2/2	3/3	7/4	4/1	5/2
38.45	57.10				
6/3	8/4				

W.VOELTER,G.JUNG,E.BREITMAIER,E.BAYER
Z NATURFORSCH   26B, 213 (1971)

2808   U 004000   D,L-HOMOCYSTINE

```
    +               +
  H3N             NH3
    \6 7      3 2/
    CH-CH2 H2C-CH
   5/   |     |   \1
  O=C  8CH2 H2C4  C=O
    |    \    /    |
    O     S-S      O
    —              —
```

FORMULA	C8H16N2O4S2			MOL WT	268.36
SOLVENT	C2D6OS				
ORIG ST	C4H12SI			TEMP	303

173.80	54.20	35.30	31.80	173.80	54.20
1/1	2/2	3/3	4/3	5/1	6/2
35.30	31.80				
7/3	8/3				

W.VOELTER,G.JUNG,E.BREITMAIER,E.BAYER
Z NATURFORSCH   26B, 213 (1971)

2809   U 004000   TERT-BUTYL 2-AMINO-2-METHYLPROPANOATE

```
    CH3 O
   3 |  / 5
  H3C-C-C1  CH3
    | 2 \  1 4
   NH2 O-C-CH3
        |
       CH3
```

FORMULA	C8H17NO2			MOL WT	159.23
SOLVENT	CDCL3				
ORIG ST	C4H12SI			TEMP	300

175.85	53.75	26.45	78.85	26.65
1/1	2/1	3/4	4/1	5/4

G.JUNG,E.BREITMAIER
UNPUBLISHED   (1974)

2810 U 004000 L-PHENYLALANINE

```
      O
    - ||          5  6
   O-C1          C-C
      \2  3  4/    \7
       CH-CH2-C     C
      +/        \9 8/
    H3N          C=C
```

FORMULA	C9H11NO2			MOL WT	165.19
SOLVENT	D2O				
ORIG ST	C4H12SI			TEMP	303
175.00	57.30	37.50		131.10	130.70
1/1	2/2	3/3	4/1	5/2	6/2
129.50					
7/2					

W.VOELTER,G.JUNG,E.BREITMAIER,E.BAYER
Z NATURFORSCH   26B, 213 (1971)

2811 U 004000 TYROSINE

```
     8  9
     C=C       NH2  O
    /    \      |  /
  HO-C    C-CH2-CH-C
    7\    /4 3  2  1\
     C-C              OH
     6  5
```

FORMULA	C9H11NO3			MOL WT	181.19
SOLVENT	D2O/DCL				
ORIG ST	C4H12SI			TEMP	300
175.00	57.30	37.50	130.50	117.50	156.30
1/1	2/2	3/3	5/2	6/2	7/1
117.50	130.50				
8/2	9/2				

W.VOELTER,S.FUCHS,R.H.SEUFFER,K.ZECH
MONATSH CHEM   105, 1110 (1974)

2812 U 004000 L-TYROSINE

```
      O
    - ||          5  6
   O-C1          C-C
      \2  3  4/    \7
       CH-CH2-C     C-OH
      +/        \9 8/
    H3N          C=C
```

FORMULA	C9H11NO3			MOL WT	181.19
SOLVENT	D2O				
ORIG ST	C4H12SI			TEMP	303
175.00	57.30	37.50		130.50	117.50
1/1	2/2	3/3	4/1	5/2	6/2
156.30	117.50	130.50			
7/1	8/2	9/2			

W.VOELTER,G.JUNG,E.BREITMAIER,E.BAYER
Z NATURFORSCH   26B, 213 (1971)

2813 U 004000 CARNOSINE

```
                 O
    9  8  7   2  1/
  H2N-CH2-CH2-C-NH-CH-C
              ||   |   \
              O   3CH2  OH
                   |
                  4C
                 // \
                6C   N
                 |   ||
                 N---C5
```

FORMULA	C9H14N3O3			MOL WT	212.23
SOLVENT	D2O				
ORIG ST	C4H12SI			TEMP	303
179.40	56.00	31.80	135.10	137.70	119.60
1/1	2/2	3/3	4/1	5/2	6/2
173.30	35.50	38.90			
7/1	8/3	9/3			

W.VOELTER,G.JUNG,E.BREITMAIER,E.BAYER
Z NATURFORSCH   26B, 213 (1971)

2814 U 004000 N-ACETYL-L-LEUCINE METHYL ESTER

```
   O        O
    \7   2  1/
     C-NH-CH-C
    8/    |    \ 7
  H3C   3CH2  O-CH3
          |
         4CH
        6/ \5
      H3C   CH3
```

FORMULA	C9H17NO3			MOL WT	187.24
SOLVENT	C2H6OS				
ORIG ST	C4H12SI			TEMP	303
176.40	51.90	43.00	26.00	25.00	23.00
1/1	2/2	3/3	4/2	5/4	6/4

W.VOELTER,G.JUNG,E.BREITMAIER,E.BAYER
Z NATURFORSCH   26B, 213 (1971)

2815   U 004000   N—TERT—BUTYLOXYCARBONYL—L—THREONINE

```
      7
 H3C      O           O
   8\6   ||    2  1/
 H3C-C-O-C-NH-CH-C
    9/   5    |    \
 H3C        3CH    OH
         4/  \
      H3C    OH
```

FORMULA	C9H17NO5		MOL WT	219.24
SOLVENT	CDCL3			
ORIG ST	C4H12SI		TEMP	300

174.70	58.90	68.10	19.40	156.75	80.25
1/1	2/2	3/2	4/4	5/1	6/1
28.40					
7/4					

U.HOLLSTEIN,E.BREITMAIER,G.JUNG
J AM CHEM SOC   96, 8036 (1974)

2816   U 004000   N—BENZYLOXYCARBONYLGLYCINE

```
  9 10
  C=C                      O
 8/   \5 4    3    2   1/
 C    C-CH2-O-C-NH-CH2-C
  \7 6/      ||         \
  C-C        O          OH
```

FORMULA	C10H11NO4		MOL WT	209.20
SOLVENT	C2D6OS			
ORIG ST	C4H12SI		TEMP	303

171.00	42.60	156.00	65.80	136.65	127.20
1/1	2/3	3/1	4/3	5/1	6/2
127.20	127.95	127.20	127.20		
7/2	8/2	9/2	10/2		

W.VOELTER,G.JUNG,E.BREITMAIER,E.BAYER
Z NATURFORSCH   26B, 213 (1971)

2817   U 004000   TERT—BUTYLOXYCARBONYLPYROGLUTAMIC ACID

```
      O   OH
       \\ /
        C5
  10     |
 H3C   O  1C-C2
   9\  ||  /  |
 H3C-C-O-C-N  |
    /7  6  \  |
 H3C        C-C3
   8      //4
         O
```

FORMULA	C10H15NO5		MOL WT	229.23
SOLVENT	C2D6OS			
ORIG ST	C4H12SI		TEMP	300

58.60	20.95	30.95	173.15	173.15	149.10
1/2	2/3	3/3	4/1	5/1	6/1
82.10	27.50	27.50	27.50		
7/1	8/4	9/4	10/4		

W.VOELTER,S.FUCHS,R.H.SEUFFER,K.ZECH
MONATSH CHEM   105, 1110 (1974)

2818   U 004000   N,N—DIACETYL—L—CYSTINE

```
   10            5
 O  CH3      H3C  O
  \ /          \ //
   C9           C4
   |  7 8    3 2 |
 HN-HC-CH2 H2C-CH-NH
   |  \  /  |
   C6  S-S  C1
  // \      / \\
 O   OH   HO   O
```

FORMULA	C10H16N2O6S2		MOL WT	324.38
SOLVENT	D2O			
ORIG ST	C4H12SI		TEMP	303

178.85	54.35	41.10	179.30	23.65	178.85
1/1	2/2	3/3	4/1	5/4	6/1
54.35	41.10	179.30	23.65		
7/2	8/3	9/1	10/4		

W.VOELTER,G.JUNG,E.BREITMAIER,E.BAYER
Z NATURFORSCH   26B, 213 (1971)

2819   U 004000   CIS—TERT—BUTYLOXYCARBONYLPROLINE

```
            O
  10       //
 H3C      5C
   9\  6  / \
 H3C-C-O-C-N---C1  OH
   8/7 || |    |
 H3C    O C4  C2
     \ /
      C
      3
```

FORMULA	C10H17NO4		MOL WT	215.25
SOLVENT	CDCL3			
ORIG ST	C4H12SI		TEMP	300

58.80	30.75	23.50	46.20	178.35	153.95
1/2	2/3	3/3	4/3	5/1	6/1
80.25	28.25	28.25	28.25		
7/1	8/4	9/4	10/4		

W.VOELTER,S.FUCHS,R.H.SEUFFER,K.ZECH
MONATSH CHEM   105, 1110 (1974)

2820   U 004000   TRANS-TERT-BUTYLOXYCARBONYLPROLINE

```
                          O
    10                    ∥
  H3C        O     5C
    9\       ∥    /  \
  H3C-C-O-C-N---C1   OH
    /7   6 |     |
  H3C      4C    C2
    8        \  /
              C
              3
```

FORMULA	C10H17NO4			MOL WT	215.25
SOLVENT	CDCL3				
ORIG ST	C4H12SI			TEMP	300

58.80	29.35	24.15	46.70	176.60	155.35
1/2	2/3	3/3	4/3	5/1	6/1
80.25	28.25	28.25	28.25		
7/1	8/4	9/4	10/4		

W.VOELTER,S.FUCHS,R.H.SEUFFER,K.ZECH
MONATSH CHEM   105, 1110 (1974)

2821   U 004000   N-TERT-BUTYLOXYCARBONYL-L-PROLINE CIS ISOMER

```
        HO   O
         \  ∥
          C1
    8     |
  H3C     C2
    9\7  6  / \
  H3C-C-O-C-N   C3
   10/  ∥ |     |
  H3C   O C---C4
        5
```

FORMULA	C10H17NO4			MOL WT	215.25
SOLVENT	CDCL3				
ORIG ST	C4H12SI			TEMP	300

176.30	58.50	30.50	23.30	46.40	154.85
1/1	2/2	3/3	4/3	5/3	6/1
80.15	27.95				
7/1	8/4				

U.HOLLSTEIN,E.BREITMAIER,G.JUNG
J AM CHEM SOC   96, 8036 (1974)

2822   U 004000   N-TERT-BUTYLOXYCARBONYL-L-PROLINE TRANS ISOMER

```
        HO   O
         \  ∥
          C1
    8     |
  H3C    O  C2
    9\7  ∥ / \
  H3C-C-O-C-N   C3
   10/   6|    |
  H3C     5C---C4
```

FORMULA	C10H17NO4			MOL WT	215.25
SOLVENT	CDCL3				
ORIG ST	C4H12SI			TEMP	300

177.05	58.70	29.40	24.00	46.10	153.85
1/1	2/2	3/3	4/3	5/3	6/1
80.15	27.95				
7/1	8/4				

U.HOLLSTEIN,E.BREITMAIER,G.JUNG
J AM CHEM SOC   96, 8036 (1974)

2823   U 004000   CIS-TERT-BUTYLOXYCARBONYLPROLINAMIDE

```
                     O
    10               ∥
  H3C        6     5C
    9\       /  \
  H3C-C-O-C-C---C1   NH2
    /7   ∥ |     |
  H3C   O C4    C2
    8        \  /
              C
              3
```

FORMULA	C10H18N2O3			MOL WT	214.27
SOLVENT	D2O				
ORIG ST	C4H12SI			TEMP	300

60.40	31.10	23.50	46.70	178.70	155.70
1/2	2/3	3/3	4/3	5/1	6/1
81.80	27.60	27.60	27.60		
7/1	8/4	9/4	10/4		

W.VOELTER,S.FUCHS,R.H.SEUFFER,K.ZECH
MONATSH CHEM   105, 1110 (1974)

2824   U 004000   TRANS-TERT-BUTYLOXYCARBONYLPROLINAMIDE

```
                     O
    10               ∥
  H3C      O    5C
    9\     ∥   /  \
  H3C-C-O-C-N---C1   NH2
    /7   6 |     |
  H3C      4C    C2
    8        \  /
              C
              3
```

FORMULA	C10H18N2O3			MOL WT	214.27
SOLVENT	D2O				
ORIG ST	C4H12SI			TEMP	300

60.00	30.45	23.95	47.25	178.15	155.70
1/2	2/3	3/3	4/3	5/1	6/1
81.80	27.60	27.60	27.60		
7/1	8/4	9/4	10/4		

W.VOELTER,S.FUCHS,R.H.SEUFFER,K.ZECH
MONATSH CHEM   105, 1110 (1974)

2825   U 004000   TERT—BUTYLOXYCARBONYLGLUTAMINE

```
                8
        ]           CH3       FORMULA   C10H18N2O5              MOL WT   246.27
         \\    7/9             SOLVENT   C2D6OS
          6C-]-C-CH3           ORIG ST   C4H12SI                TEMP      300
           |      \
 NH2          NH   CH3         173.90    53.20    26.65    31.50   173.90   155.60
    \         |  1  10         1/1       2/2      3/3      4/3     5/1      6/1
   5C-CH2-CH2-CH-C=]           78.10     28.15    28.15    28.15
   //  4   3    2  1           7/1       8/4      9/4      10/4
  ]              OH
```

W.VOELTER,S.FUCHS,R.H.SEUFFER,K.ZECH
MONATSH CHEM   105, 1110 (1974)

2826   U 004000   N—TERT—BUTYLOXYCARBONYL—L—GLUTAMINE

```
     8                        FORMULA   C10H18N2O5              MOL WT   246.27
   H3C      ]        0        SOLVENT   C2D6OS
     9\7     ||    2  1//      ORIG ST   C4H12SI                TEMP      303
   H3C-C-]-C-NH-CH-C
    10/   6   |    \           172.70    53.05    26.85    31.45   172.70   154.40
   H3C          3CH2 OH        1/1       2/2      3/3      4/3     5/1      6/1
                |              77.60     28.15    28.15    28.15
              4CH2             7/1       8/4      9/4      10/4
                |
             ]=C-NH2           W.VOELTER,G.JUNG,E.BREITMAIER,E.BAYER
                5              Z NATURFORSCH                26B, 213 (1971)
```

2827   U 004000   N—TERT—BUTYLOXYCARBONYL—L—VALINE

```
     8                        FORMULA   C10H19NO4               MOL WT   217.27
   H3C      ]        0        SOLVENT   C2D6OS
     9\7     \\    2  1//      ORIG ST   C4H12SI                TEMP      303
   H3C-C-]-C-NH-CH-C
    10/   6   |    \           172.40    58.95    29.85    18.20   19.25   151.50
   H3C          3CH  OH        1/1       2/2      3/2      4/4     5/4     6/1
             5/  \4            78.00     28.25    28.25    28.25
          H3C    CH3           7/1       8/4      9/4      10/4
```

W.VOELTER,G.JUNG,E.BREITMAIER,E.BAYER
Z NATURFORSCH   26B, 213 (1971)

2828   U 004000   N—TERT—BUTYLOXYCARBONYL—L—VALINE   CIS—TRANS
                            ISOMER

```
     8                        FORMULA   C10H19NO4               MOL WT   217.27
   H3C      ]        0        SOLVENT   CDCL3
     9\7     ||    2  1//      ORIG ST   C4H12SI                TEMP      300
   H3C-C-]-C-NH-CH-C
    10/   6   |    \           176.60    60.50    31.40    18.30   17.80   157.85
   H3C          3CH  OH        1/1       2/2      3/2      4/4     5/4     6/1
             4/  \5            80.00     28.60
          H3C    CH3           7/1       8/4
```

U.HOLLSTEIN,E.BREITMAIER,G.JUNG
J AM CHEM SOC   96, 8036 (1974)

2829   U 004000   N—TERT—BUTYLOXYCARBONYL—L—VALINE   CIS—TRANS
                            ISOMER

```
     8                        FORMULA   C10H19NO4               MOL WT   217.27
   H3C      ]        0        SOLVENT   CDCL3
     9\7     ||    2  1//      ORIG ST   C4H12SI                TEMP      300
   H3C-C-]-C-NH-CH-C
    10/   6   |    \           176.60    58.80    31.40    18.30   17.80   156.35
   H3C          3CH  OH        1/1       2/2      3/2      4/4     5/4     6/1
             4/  \5            80.25     28.60
          H3C    CH3           7/1       8/4
```

U.HOLLSTEIN,E.BREITMAIER,G.JUNG
J AM CHEM SOC   96, 8036 (1974)

2830   U 004000   PYROGLUTAMIC ACID 2,4,5-TRICHLOROPHENYL ESTER

```
    3   2
    C---C   O   CL
    |   |   ∥
  4C   C-C   7C=C8
   ∥  \  /1 5\  /  \9
   O   N   O-C   C-CL
          6\  /
          C-C10
         11  \
             CL
```

FORMULA	C11H7CL3NO3		MOL WT	307.54
SOLVENT	C2D6OS			
ORIG ST	C4H12SI		TEMP	300

54.80	24.60	28.90	177.25	170.70	145.65
1/2	2/3	3/3	4/1	5/1	
131.30	130.75	130.00	125.90		

W.VOELTER,S.FUCHS,R.H.SEUFFER,K.ZECH
MONATSH CHEM   105, 1110 (1974)

2831   U 004000   L-TRYPTOPHANE

```
              O
      3   2  1/
  C6      CH2-CH-C
  / \10  /   |   \ -
 C7  C---C5  NH3  O
 |   ∥   ∥    +
 C8  C11 C4
  \ / \ /
   C9   N
```

FORMULA	C11H12N2O2		MOL WT	204.23
SOLVENT	D2O/DCL			
ORIG ST	C4H12SI		TEMP	303

174.70	56.00	28.20	127.60	114.25	124.50
1/1	2/2	3/3	4/2	5/1	6/2
121.80	120.50	114.25	128.90	138.60	
7/2	8/2	9/2	10/1	11/1	

W.VOELTER,G.JUNG,E.BREITMAIER,E.BAYER
Z NATURFORSCH   26B, 213 (1971)

2832   U 004000   BENZYL 2-AMINO-2-METHYLPROPANOATE

```
  CH3 O
  3 | /
 H3C-C-C1
  |2 \  4
 NH2 O-CH2 C
      \ / \
      5C   C
      |   ∥
      6C   C8
       \ /
        C7
```

FORMULA	C11H13NO2		MOL WT	191.23
SOLVENT	CDCL3			
ORIG ST	C4H12SI		TEMP	300

176.75	53.85	26.85	65.60	135.30	127.00
1/1	2/1	3/4	4/3	5/1	6/2
127.65	127.00				
7/2	8/2				

G.JUNG,E.BREITMAIER
UNPUBLISHED   (1974)

2833   U 004000   N-BENZYLOXYCARBONYL-D,L-ALANINE

```
 10 11
  C=C          O        O
 9/  \6 5  ∥   2  1/
 C    C-CH2-O-C-NH-CH-C
  \8 7/        4  |   \
   C-C          3CH3  OH
```

FORMULA	C11H13NO4		MOL WT	223.23
SOLVENT	C2D6OS			
ORIG ST	C4H12SI		TEMP	303

174.00	49.55	17.55	155.15	65.50	136.55
1/1	2/2	3/4	4/1	5/3	6/1
127.20	127.20	127.75	127.20	127.20	
7/2	8/2	9/2	10/2	11/2	

W.VOELTER,G.JUNG,E.BREITMAIER,E.BAYER
Z NATURFORSCH   26B, 213 (1971)

2834   U 004000   TRYPTOPHANAMIDE HYDROCHLORIDE

```
   10
    C              O
   / \11  4 3   2  1/
  9C    C---C-CH2-CH-C
  |   ∥   ∥    |   \
 8C    C   C5  NH2  NH2
  \ /6\ /       .
   C   N       HCL
   7
```

FORMULA	C11H13N3O		MOL WT	203.25
SOLVENT	C2D6OS			
ORIG ST	C4H12SI		TEMP	300

170.25	52.65	27.10	107.15	124.85	136.25
1/1	1/2	3/3	4/1	5/2	6/1
127.20	111.45	118.35	121.05	118.80	
7/2	8/2	9/2	10/2	11/1	

W.VOELTER,S.FUCHS,R.H.SEUFFER,K.ZECH
MONATSH CHEM   105, 1110 (1974)

2835   U 004000   N-TERT-BUTYLOXYCARBONYL-6-AMINOCAPROIC ACID

```
                    O     FORMULA   C11H21NO4              MOL WT   231.29
   6  5    4    3    2    1/   SOLVENT   C2D6OS
 H2C-CH2-CH2-CH2-CH2-C    ORIG ST   C4H12SI               TEMP       303
    \                    \
    HN       9CH3         OH    173.25   40.35   29.40   28.25   33.70   58.10
    |       8/10               1/1      2/3     3/3     4/3     5/3     6/2
  O=C-O-C-CH3                  154.60   76.95   28.25
   7      \11              W.VOELTER,G.JUNG,E.BREITMAIER,E.BAYER
          CH3              Z NATURFORSCH            26B, 213 (1971)
```

2836   U 004000   N-TERT-BUTYLOXYCARBONYL-L-LEUCINE

```
   9                      FORMULA   C11H21NO4              MOL WT   231.29
 H3C      O        O      SOLVENT   C2D6OS
  10\8  ||     2   1/     ORIG ST   C4H12SI               TEMP       303
 H3C-C-O-C-NH-CH-C
  11/   7   |    \        173.55   52.00   40.50   24.50   21.30   22.90
 H3C        3CH2  OH      1/1      2/2     3/3     4/2     5/4     6/4
            |             154.70   77.60   28.25   28.25   28.25
            4CH           7/1      8/1     9/4     10/4    11/4
            6/  \5
         H3C    CH3       W.VOELTER,G.JUNG,E.BREITMAIER,E.BAYER
                         Z NATURFORSCH            26B, 213 (1971)
```

2837   U 004000   N-TERT-BUTYLOXYCARBONYL-L-ISOLEUCINE

```
   9                      FORMULA   C11H21NO4              MOL WT   231.29
 H3C      O        O      SOLVENT   C2D6OS
  10\8  ||     2   1/     ORIG ST   C4H12SI               TEMP       303
 H3C-C-O-C-NH-CH-C
  11/   7   |3   \        173.90   58.30   40.55   24.80   11.75   15.95
 H3C        CH   OH       1/1      2/2     3/2     4/3     5/4     6/4
           4/  \6         154.95   78.00   28.55   28.55   28.55
         H2C    CH3       7/1      8/1     9/4     10/4    11/4
           |
         H3C5             W.VOELTER,G.JUNG,E.BREITMAIER,E.BAYER
                         Z NATURFORSCH            26B, 213 (1971)
```

2838   U 004000   N-TERT-BUTYLOXYCARBONYL-L-METHYLVALINE   CIS-
                    TRANS ISOMER

```
   9        6             FORMULA   C11H21NO4              MOL WT   231.29
 H3C      O CH3    O      SOLVENT   CDCL3
  10\8  || |  2   1/      ORIG ST   C4H12SI               TEMP       300
 H3C-C-O-C-N-CH-C
  11/   7   |    \        175.75   64.95   27.50   19.75   18.90   31.30
 H3C        3CH   OH      1/1      2/2     3/2     4/4     5/4     6/4
           4/  \5         155.70   80.60   28.15
         H3C    CH3       7/1      8/1     9/4

         U.HOLLSTEIN,E.BREITMAIER,G.JUNG
         J AM CHEM SOC              96, 8036 (1974)
```

2839   U 004000   N-TERT-BUTYLOXYCARBONYL-L-METHYLVALINE   CIS-
                    TRANS ISOMER

```
   9        6             FORMULA   C11H21NO4              MOL WT   231.29
 H3C      O CH3    O      SOLVENT   CDCL3
  10\8  || |  2   1/      ORIG ST   C4H12SI               TEMP       300
 H3C-C-O-C-N-CH-C
  11/   7   |    \        175.75   64.20   27.50   19.75   18.90   31.05
 H3C        3CH   OH      1/1      2/2     3/2     4/4     5/4     6/4
           4/  \5         155.70   80.60   28.15
         H3C    CH3       7/1      8/1     9/4

         U.HOLLSTEIN,E.BREITMAIER,G.JUNG
         J AM CHEM SOC              96, 8036 (1974)
```

2840 — U 004000 — TERT-BUTYLOXYCARBONYL-NG-NITROARGININE

```
                9
 HN          ]      CH3      FORMULA  C11H21N5O6          MOL WT  319.32
   \          \\   8/10      SOLVENT  C2D6OS
   6C-NH-NO2   7C-O-C-CH3     ORIG ST  C4H12SI              TEMP      300
   /           |    \
 HN           NH     CH3      174.25   53.40   28.35   25.25   40.55  159.55
   \           |  |  11        1/1      2/2     3/3     4/3     5/3    6/1
   CH2-CH2-CH2-CH-C=]        155.80   78.35   28.40   28.40   28.40
    5   4   3   2  |            7/1      8/1     9/4    10/4    11/4
                 OH
```

W.VOELTER,S.FUCHS,R.H.SEUFFER,K.ZECH
MONATSH CHEM   105, 1110 (1974)

2841 — U 004000 — N-TERT-BUTYLOXYCARBONYL-L-NITROARGININE

```
      9
 H3C       ]       ]         FORMULA  C11H21N4O6          MOL WT  305.31
  10\8   ||     2  1/         SOLVENT  C2D6OS
 H3C-C-]-C-NH-CH-C            ORIG ST  C4H12SI              TEMP      303
  11/   7    |    \
 H3C       3CH2  OH           173.00   53.30   28.35   25.45   40.25  158.55
            |                  1/1      2/2     3/3     4/3     5/3    6/1
          4CH2 NH2            154.85   78.00   28.35
            |   |              7/1      8/1     9/4
          H2C5  C=N-NO2
            \  /6
             NH                W.VOELTER,G.JUNG,E.BREITMAIER,E.BAYER
                               Z NATURFORSCH               26B, 213 (1971)
```

2842 — U 004000 — TRYPTOPHAN METHYL ESTER HYDROCHLORIDE

```
    10
     C              ]         FORMULA  C12H14N2O2          MOL WT  218.26
   / \11  4  3   2  1/        SOLVENT  C2D6OS
 9C    C---C-CH2-CH-C          ORIG ST  C4H12SI              TEMP      300
  |    ||   ||      |  \
 8C    C    C5    NH2  ]-CH3   170.15   53.00   26.45  106.70  125.05  136.25
  \  /6\  /          .   12     1/1      2/2     3/3     4/1     5/2    6/1
    C    N          HCL        127.00  111.65  118.05  121.25  118.55   45.30
    7                           7/2      8/2     9/2    10/2    11/1    12/4
```

W.VOELTER,S.FUCHS,R.H.SEUFFER,K.ZECH
MONATSH CHEM   105, 1110 (1974)

2843 — U 004000 — N-(PARA-AMINOBENZOYL)GLUTAMIC ACID

```
       ]   OH
        \ /                   FORMULA  C12H14N2O5          MOL WT  266.26
    ]    C9                    SOLVENT  C2H6OS
    ||   |  11                ORIG ST  C4H12SI              TEMP      AMB
  C6  C7 8CH  CH2 ]H
  // \ / \ / \ / \ /          121.90  130.30  113.80  153.10  167.90   53.00
 5C   ]1  N   CH2 C12          1/1      2/2     3/2     4/1     7/1     8/2
  |   ||  |  10  ||           175.20   27.40   31.70  175.10
 4C   C2  H      ]             9/1     10/3    11/3    12/1
  / \ /
 H2N   C3                     U.EWERS,H.GUENTHER,L.JAENICKE
                              CHEM BER                    106, 3951 (1973)
```

2844 — U 004000 — N-ACETYLPHENYLALANINE METHYL ESTER

```
          6  7               FORMULA  C12H15NO3           MOL WT  221.26
    ]       C-C               SOLVENT  CDCL3
    ||      / \\              ORIG ST  C4H12SI              TEMP      AMB
   2C    5C    C8
  1 / \3  / \  /              52.00   170.20   53.50   37.70  136.30  128.40
 H3C-]   CH-CH2 C=]            1/4      2/1     3/2     4/3     5/1    6/2
         |  4  10 9           129.10  126.80  129.10  128.40  172.30   22.50
         N   CH3               7/2      8/2     9/2    10/2    11/1    12/4
        / \ /12
       H   ]11
           ||
           ]
```

L.F.JOHNSON,W.C.JANKOWSKI
CARBON-13 NMR SPECTRA,JOHN
WILEY AND SONS,NEW YORK   435 (1972)

2845 — U 004000 GLUTAMIC ACID GAMMA-BENZYL ESTER

```
  9  8                              FORMULA   C12H15NO4        MOL WT  237.26
   C=C        O        NH2          SOLVENT   C2D6OS
 10/   \7 6   || 4  3 2|            ORIG ST   C4H12SI          TEMP      300
  C    C-CH2-O-C-CH2-CH2-CH
   \   //      5        |
   C-C                 1C
  11 12               / \\
                    HO   O
                         CH
```

177.40	55.15	24.80	29.35	174.70	63.25
1/1	2/2	3/3	4/3	5/1	6/3
142.75	128.30	126.65			
7/1					

W.VOELTER,S.FUCHS,R.H.SEUFFER,K.ZECH
MONATSH CHEM   105, 1110 (1974)

2846 — U 004000 CIS-N-TERT-BUTYLOXYCARBONYL-GLYCYLPROLINE

```
        O
  10    ||                          FORMULA   C12H20N2O5       MOL WT  272.30
 H3C    C8                          SOLVENT   H2O
  11\9  / \                         ORIG ST   C4H12SI          TEMP      AMB
 H3C-C-O  HN  O   OH
  11/     |   \ /
 H3C      H2C7 C1
          |    |
          C6   C2
         // \ / \
        O   N   C3
            |   |
           5C---C4
```

174.40	59.10	31.00	21.90	46.50	169.30
1/1	2/2	3/3	4/3	5/3	6/1
42.50	157.00	79.40	27.40		
7/3	8/1	9/1	10/4		

D.E.DORMAN,F.A.BOVEY
J ORG CHEM   38, 2379 (1973)

2847 — U 004000 TRANS-N-TERT-BUTYLOXYCARBONYL-GLYCYLPROLINE

```
            O   OH              FORMULA   C12H20N2O5       MOL WT  272.30
             \ /               SOLVENT   H2O
   O    O   C1                 ORIG ST   C4H12SI          TEMP      AMB
 10   ||   ||  |
 H3C    C8   C6  C2
 11\9  / \   / \ / \
 H3C-C-O  HN-H2C   N    C3
 12/     |   |
 H3C       5C---C4
```

174.40	59.10	28.70	24.30	45.90	168.80
1/1	2/2	3/3	4/3	5/3	6/1
42.50	157.00	79.40	27.40		
7/3	8/1	9/1	10/4		

D.E.DORMAN,F.A.BOVEY
J ORG CHEM   38, 2379 (1973)

2848 — U 004000 BENZYLOXYCARBONYLPYROGLUTAMIC ACID

```
            O   OH              FORMULA   C13H13NO5        MOL WT  263.25
             \ /               SOLVENT   C2D6OS
  10 9        C5               ORIG ST   C4H12SI          TEMP      300
   C=C        |
  11/   \8 7  |
   C    C-CH2-O  1C-C2
   \   //       | / |
   C-C       O=C-N   |
  12 13      6       |
                    C-C3
                   //4
                   O
```

58.50	21.25	30.85	173.15	173.15	150.75
1/2	2/3	3/3	4/1	5/1	6/1
67.35	135.60	128.50	128.20	127.65	
7/3					

W.VOELTER,S.FUCHS,R.H.SEUFFER,K.ZECH
MONATSH CHEM   105, 1110 (1974)

2849 — U 004000 CIS-BENZYLOXYCARBONYLPROLINE

```
        O  FORMULA   C13H15NO4        MOL WT  249.27
 12 13      //  SOLVENT   CDCL3
   C=C        5C  ORIG ST   C4H12SI.          TEMP      300
  / \      6    / \
 11C    C-CH2-O-C-N---C1  OH
   \  //8 7   || |   |
   C-C        O C4  C2
  10 9         \ /
                C
                3
```

58.25	30.45	23.00	46.05	176.60	154.30
1/2	2/3	3/3	4/3	5/1	6/1
66.90	135.95	128.05			
7/3					

W.VOELTER,S.FUCHS,R.H.SEUFFER,K.ZECH
MONATSH CHEM   105, 1110 (1974)

2850   U 004000   TRANS—BENZYLOXYCARBONYLPROLINE

```
                          O   FORMULA   C13H15NO4              MOL WT   249.27
   12 13                  ∥   SOLVENT   CDCL3
   C=C          O     5C      ORIG ST   C4H12SI                TEMP       300
      ∥               / \
  11C     C-CH2-O-C-N---C1  OH   58.80    29.25    23.85    46.60   176.20   155.05
    \   ∥8 7   6 |   |             1/2      2/3      3/3      4/3     5/1      6/1
     C-C          4C    C2      66.90   127.40   127.10
    10 9               \ /       7/3
                        C
                        3      W.VOELTER,S.FUCHS,R.H.SEUFFER,K.ZECH
                               MONATSH CHEM                   105, 1110 (1974)
```

2851   U 004000   BENZYLOXYCARBONYLGLUTAMINE

```
           O            FORMULA   C13H16N2O5              MOL WT   280.28
           ∥    7   8   SOLVENT   C2D6OS
          6C-O-CH2-C    ORIG ST   C4H12SI                TEMP       300
         /      / \
  NH2      HN  13C   C9   174.15   53.85    26.85    31.70   174.15   156.55
   |       |   ∥   |       1/1      2/2      3/3      4/3     5/1      6/1
  5C-CH2-CH2-CH  12C   C10   65.80   137.25   128.70   128.05
  ∥ 4   3   2\   \ ∥        7/3      8/1
  O         1C=O    C
           /        11
         HO             W.VOELTER,S.FUCHS,R.H.SEUFFER,K.ZECH
                        MONATSH CHEM                   105, 1110 (1974)
```

2852   U 004000   PARA—TOLUYL—L—THREONINE METHYL ESTER

```
    10 11              FORMULA   C13H17NO4              MOL WT   251.28
    C=C   O       O    SOLVENT   CDCL3
  12 9/  \6 ∥   2 1/   ORIG ST   C4H12SI               TEMP       300
  H3C-C     C-C-NH-CH-C
      \  ∥ 5  |   /  \  13   171.00   57.85    67.35    19.65   168.10   130.25
      C-C    3CH  O-CH3       1/1      2/2      3/2      4/4     5/1      6/1
      8  7    4/ \           128.70   126.90   141.90   20.95    52.00
           H3C   OH           7/2      8/2      9/1     12/4     13/4

                      U.HOLLSTEIN,E.BREITMAIER,G.JUNG
                      J AM CHEM SOC                    96, 8036 (1974)
```

2853   U 004000   N—TERT—BUTYLOXYCARBONYL—L—PHENYLALANINE

```
      O   OH            FORMULA   C14H19NO4              MOL WT   265.31
  12   \ /       5 6    SOLVENT   C2D6OS
  H3C   O   C1   C-C    ORIG ST   C4H12SI               TEMP       303
  13\11 ∥  |  3  4/  \7
  H3C-C-O-C-NH-CH-CH2-C    C   172.70   55.20    36.90   137.50   128.60   127.55
  14/   10   2      \9 8/        1/1      2/2      3/3      4/1     5/2      6/2
  H3C                  C=C    125.85   154.70    78.00    28.35   127.55   128.60
                               7/2     10/1     11/1     12/4     8/2      9/2
                               28.35    28.35
                               13/4     14/4

                      W.VOELTER,G.JUNG,E.BREITMAIER,E.BAYER
                      Z NATURFORSCH                   26B, 213 (1971)
```

2854   U 004000   BENZYLOXYCARBONYLARGININE

```
       13C=C14
      /   \  8       FORMULA   C14H20N4O4              MOL WT   308.34
   12C      C-CH2    SOLVENT   C2D6OS
     \     ∥9 |      ORIG ST   C4H12SI                TEMP       300
  HN    C-C    O
    \   11 10  |      176.20   55.65    29.90    25.35    40.55   157.65
   6C-NH2      7C=O    1/1      2/2      3/3      4/3     5/3      6/1
    /          |      155.80   65.30   137.45   127.65   128.50   127.65
  HN           NH   O   7/1      8/3      9/1     10/2     11/2     12/2
    \          |  ∥   128.50   127.65
     CH2-CH2-CH2-CH-C   13/2     14/2
      5   4   3   2  1\
                      OH   W.VOELTER,S.FUCHS,R.H.SEUFFER,K.ZECH
                           MONATSH CHEM              105, 1110 (1974)
```

2855 U 004000 N—TERT—BUTYLOXYCARBONYL—L—GLUTAMIC ACID TERT—
BUTYL ESTER

```
H3C      0           0
 12\11   ||    2  1/
H3C-C-0-C-NH-CH-C
 13/    10    |     \
H3C           3CH2  0H
 14           |
              4CH2  7CH3
              |    6/8
              0=C-0-C-CH3
              5     \9
                    CH3
```

FORMULA   C14H25N06
SOLVENT   C2D60S
ORIG ST   C4H12SI

MOL WT   303.36

TEMP   303

170.35	52.45	26.45	31.35	172.50	79.20
1/1	2/2	3/3	4/3	5/1	6/1
27.70	154.50	77.60	28.15	27.70	27.70
7/4	10/1	11/1	12/4	8/4	9/4
28.15	28.15				
13/4	14/4				

W.VOELTER,G.JUNG,E.BREITMAIER,E.BAYER
Z NATURFORSCH   26B, 213 (1971)

2856 U 004000 TERT—BUTYLOXYCARBONYLTRYPTOPHAN

```
  10
   C              0
  / \11  4  3   2  1/
 9C   C---C-CH2-CH-C
 |    ||  ||    |    \
8C    C   C5   NH    0H
 \  /6 \  /          |
   C    N      0=C12 CH3
   7           |   /14
           0-C-CH3
          13\15
            CH3
            16
```

FORMULA   C16H20N204
SOLVENT   C2D60S
ORIG ST   C4H12SI

MOL WT   416.11

TEMP   300

174.25	54.80	26.95	110.35	123.85	136.25
1/1	2/2	3/3	4/1	5/2	6/1
127.40	111.55	118.45	121.05	118.70	155.60
7/2	8/2	9/2	10/2	11/1	12/1
78.20	28.40	28.40	28.40		
13/1	14/4	15/4	16/4		

W.VOELTER,S.FUCHS,R.H.SEUFFER,K.ZECH
MONATSH CHEM   105, 1110 (1974)

2857 U 004000 N—TERT—BUTYLOXYCARBONYL—0—BENZYL—L—ASPARTIC
ACID

```
 14
H3C      0 H        0
 15\13   || |   2  1/
H3C-C-0-C-N-CH-C
 16/   12    |    \
H3C           3CH2  0H
              |          7  8
              C4        C-C
             / \ 5   6/    \\
            0   0-CH2-C      C9
                      \     /
                       C=C
                      11  10
```

FORMULA   C16H21N06
SOLVENT   C2D60S
ORIG ST   C4H12SI

MOL WT   323.35

TEMP   303

168.85	49.85	36.05	171.30	65.25	135.35
1/1	2/2	3/3	4/1	5/3	6/1
127.00	126.80	127.45	154.10	77.80	28.05
7/2	8/2	9/2	12/1	13/1	14/3

W.VOELTER,G.JUNG,E.BREITMAIER,E.BAYER
Z NATURFORSCH   26B, 213 (1971)

2858 U 004000 TERT—BUTYLOXYCARBONYLTRYPTOPHANAMIDE

```
  10
   C              0
  / \11  4  3   2  1/
 9C   C---C-CH2-CH-C
 |    ||  ||    |    \
8C    C   C5   NH    NH2
 \  /6 \  /          |
   C    N      0=C12 CH3
   7           |   /14
           0-C-CH3
          13\15
            CH3
            16
```

FORMULA   C16H21N303
SOLVENT   C2D60S
ORIG ST   C4H12SI

MOL WT   303.36

TEMP   300

174.90	55.25	28.40	110.60	124.30	136.25
1/1	2/2	3/3	4/1	5/2	6/1
127.65	111.45	118.45	123.95	121.05	155.45
7/2	8/2	9/2	10/2	11/1	12/1
78.20	28.40	28.40	28.40		
13/1	14/4	15/4	16/4		

W.VOELTER,S.FUCHS,R.H.SEUFFER,K.ZECH
MONATSH CHEM   105, 1110 (1974)

2859 U 004000 N—TERT—BUTYLOXYCARBONYL—0—BENZYL—L—THREONINE

```
 14
H3C      0          0
 15\13   ||    2  1/
H3C-C-0-C-NH-CH-C
 16/   12    |     \
H3C           3CH  0H
             / \        7  8
            H3C  0    C=C
             4   \5 6/    \9
               H2C-C       C
                  \\       /
                   C-C
                  11 10
```

FORMULA   C16H23N05
SOLVENT   C2D60S
ORIG ST   C4H12SI

MOL WT   309.37

TEMP   303

171.40	58.20	70.30	16.70	74.70	138.05
1/1	2/2	3/2	4/4	5/3	6/1
126.80	126.80	127.55	155.15	78.30	28.35
7/2	8/2	9/2	12/1	13/1	14/4
126.80	126.80	28.35	28.35		
10/2	11/2	15/4	16/4		

W.VOELTER,G.JUNG,E.BREITMAIER,E.BAYER
Z NATURFORSCH   26B, 213 (1971)

2860   U 004000 N,N—DI—TERT—BUTYLOXYCARBONYL—L—CYSTINE

```
   14  15           6  7
  H3C  CH3        H3C  CH3
    \ |             | /
    13C-CH3      H3C-C 5
     | 16        8 |
     O   O        O   O
      \ /          \ //
    12C              C4
     | 10 11     3 2 |
   HN-HC-CH2   H2C-CH-NH
     |   / |        |
     C9  S-S    C1
    // \       / \
   O   OH   HO   O
```

FORMULA   C16H28N2O8S2   MOL WT   440.54
SOLVENT   C2D6OS
ORIG ST   C4H12SI   TEMP   303

178.20	55.80	42.75	160.60	81.75	30.35
1/1	2/2	3/3	4/1	5/1	6/4
30.35	30.35	178.20	55.80	42.75	160.60
7/4	8/4	9/1	10/2	11/3	12/1
81.75	30.35	30.35	30.35		
13/1	14/4	15/4	16/4		

W.VOELTER,G.JUNG,E.BREITMAIER,E.BAYER
Z NATURFORSCH   26B, 213 (1971)

2861   U 004000 N—TERT—BUTYLOXYCARBONYL—O—BENZYL—L—GLUTAMIC
                         ACID

```
   15
  H3C     O H      O
 16\14  || | 2  1/
 H3C-C-O-C-N-CH-C
 17/  13   |   \
 H3C      3CH2  OH
           |
          4CH2
           |
          C5        8 9
        // \  6  7/  \
       O   O-CH2-C     C
              \    /10
              C=C
             12 11
```

FORMULA   C17H23N6   MOL WT   337.38
SOLVENT   C2D6OS
ORIG ST   C4H12SI   TEMP   303

171.00	52.65	26.45	30.15	172.50	65.05
1/1	2/2	3/3	4/3	5/1	6/3
135.35	127.00	126.80	127.45	154.40	77.60
7/1	8/2	9/2	10/2	13/1	14/1
28.05					
15/4					

W.VOELTER,G.JUNG,E.BREITMAIER,E.BAYER
Z NATURFORSCH   26B, 213 (1971)

2862   U 004000 N—TERT—BUTYLOXYCARBONYL—VALYLVALINE ISOPROPYL
                         ESTER

```
            18CH3
            17 |
           H3C-CH         O
   13         16\        O
  H3C    O      O
 14\12  ||  7  ||   2  1/
 H3C-C-O-C-NH-CH-C-NH-CH-C
 15/  11    | 6    |    \
 H3C      8CH    3CH    O
        9/ \10  4/ \5
      H3C   CH3 CH3 CH3
```

FORMULA   C18H34N2O5   MOL WT   358.48
SOLVENT   CDCL3
ORIG ST   C4H12SI   TEMP   300

172.30	60.00	31.00	19.30	17.90	171.30
1/1	2/2	3/2	4/4	5/4	6/1
57.30	28.50	19.00	18.40	156.00	79.20
7/2	8/2	9/4	10/4	11/1	12/1
28.40	68.60	21.80			
13/4	16/2	17/4			

U.HOLLSTEIN,E.BREITMAIER,G.JUNG
J AM CHEM SOC   96, 8036 (1974)

2863   U 004000 TERT—BUTYLOXYCARBONYL—O—BENZYLTYROSINE

```
  15  16           6  5
   C=C             C=C
 14/  \ 10    7/    \4  3
 C     C-CH2-O-C      C--CH2
  \   //11        \  //    |
  C-C              C-C     |
 13  12         19  8 9    |
         H3C           2CH
       20\18 17    /  \
       H3C-C-O-C-NH 1C=O
        /    ||        |
       H3C   O        OH
        21
```

FORMULA   C21H25NO5   MOL WT   371.44
SOLVENT   C2D6OS
ORIG ST   C4H12SI   TEMP   300

173.80	55.55	35.70	69.25	157.10	78.10
1/1	2/2	3/3	10/3	17/1	18/1
28.25	28.25	28.25	155.60	137.35	130.25
19/4	20/4	21/4			
127.65	125.50	114.70			

W.VOELTER,S.FUCHS,R.H.SEUFFER,K.ZECH
MONATSH CHEM   105, 1110 (1974)

2864 — U 004000 N-TERT-BUTYLOXYCARBONYL-O-BENZYL-L-TYROSINE

```
H3C         O           O
 19\18      ||    2   1//
H3C-C-O-C-NH-CH-C
 20/        17   |      \
H3C               3CH2  OH
 21               |
                  C4
                 //  \
              C9   C5
              |    ||
              C8   C6
              \\  /      12  13
               C7        C=C
               | 10 11/      \14
            O-CH2-C          C
                    \\      //
                     C-C
                    16  15
```

FORMULA	C21H25NO5			MOL WT	371.44
SOLVENT	C2D6OS				
ORIG ST	C4H12SI			TEMP	303

172.40	55.10	35.65	136.30	129.25	127.45
1/1	2/2	3/3	4/1	5/2	6/2
126.80	126.60	113.65	68.90	136.30	129.25
7/1	8/2	9/2	10/3	11/1	12/2
127.45	126.80	126.60	113.65	154.30	77.60
13/2	14/2	15/2	16/2	17/1	18/1
27.95					
19/4					

W.VOELTER,G.JUNG,E.BREITMAIER,E.BAYER
Z NATURFORSCH   26B, 213 (1971)

2865 — U 004000 BENZYLOXYCARBONYL-NG-TOSYLARGININE

```
     16C=C17        12C
      15/    21     //  \11
  O    C    C-CH3  13C    C
  \\  / \   //18    |     ||
   S    C-C        14C    C
  // \ 20 19        \\  //10
  O    N             C9
        \\           7  |
      6C-NH2       O=C-O-CH2
      /             |   8
   HN              NH   O
     \             |   //
      CH2-CH2-CH2-CH-C-OH
      5    4    3   2  1
```

FORMULA	C21H26N4O6S			MOL WT	462.53
SOLVENT	C2D6OS				
ORIG ST	C4H12SI			TEMP	300

173.90	53.85	28.25	25.90	40.55	156.90
1/1	2/2	3/3	4/3	5/3	6/1
156.35	65.70	21.05	142.00	141.25	137.15
7/1	8/3	21/4			
129.25	128.50	125.80	127.85		

W.VOELTER,S.FUCHS,R.H.SEUFFER,K.ZECH
MONATSH CHEM   105, 1110 (1974)

2866 — U 004000 BENZYLOXYCARBONYL-O-BENZYLTYROSINE

```
  15  16           6   5
   C=C              C=C
  /    \ 10    7/    \4  3
 C    C-CH2-O-C      C--CH2
14\  //11          \\ // |
   C-C               C-C  |
  13 12             8  9  2CH
       21 20            /  \
        C=C         O-C-NH 1C=O
       /   \19 /   ||17   |
    22C     C-CH2 O       OH
       \\  //  18
        23C-C24
```

FORMULA	C24H23NO5			MOL WT	405.45
SOLVENT	C2D6OS				
ORIG ST	C4H12SI			TEMP	300

173.60	55.90	35.80	69.30	157.20	65.40
1/1	2/2	3/3	10/3	17/1	18/3
156.25	137.55	130.25	128.50	127.75	114.60

W.VOELTER,S.FUCHS,R.H.SEUFFER,K.ZECH
MONATSH CHEM   105, 1110 (1974)

2867 — U 004000 N-TERT-BUTYLOXYCARBONYL-THREONYL-S-BENZYL-CYSTEINYL-ISOLEUCINE METHYL ESTER

```
    H3C4        O
     \3  2  1//
      HC-CH-C
   6 5/  |    \ 14
   H3C-CH2 NH   O-CH3
   21  15    \7
    C H2C-S   C=O
   // \ \    \9 8/
  20C  C16 H2C-CH
   |   ||     |
  19C  C17    NH
   \  /       \10
    C18 HO     C=O
      \  11/     24
      12CH-CH    CH3
      13/  |   23/25
      H3C  HN  O-C-CH3
            \  /  \26
             C22   CH3
             ||
             O
```

FORMULA	C26H41N3O7S			MOL WT	539.70
SOLVENT	CDCL3				
ORIG ST	C4H12SI			TEMP	300

171.85	52.65	37.55	15.45	25.15	11.45
1/1	2/2	3/2	4/4	5/3	6/4
171.00	56.85	28.25	170.35	59.00	67.50
7/1	8/2	9/3	10/1	11/2	12/2
18.45	52.00	56.85	137.90	128.50	128.95
13/4	14/4	15/3	16/1	17/2	18/2
127.10	156.00	79.95	28.25		
19/2	22/1	23/1	24/4		

U.HOLLSTEIN,E.BREITMAIER,G.JUNG
J AM CHEM SOC   96, 8036 (1974)

2868 — U 004000 TERT-BUTYLOXYCARBONYL-O-BENZYLTYROSINE PARA-NITROPHENYL ESTER

```
   15 16            6  5
   C=C              C=C
  /    \ 10    7 /     \ 4   3
 C    C-CH2-O-C       C--CH2
14 \  / 11        \  //     |
  C-C    H3C      8C-C9     |
  13 12     19 \ 18          |
        H3C-C-O-C--NH---CH
         20 /    || 17     21
        H3C      O          |
         21            23   |
              24C=C    1C \
             /    \ 22 /    \
        O2N-C      C-O    O
           25 \    //
            26C-C27
```

FORMULA C27H28N2O7
SOLVENT C2D6OS
ORIG ST C4H12SI

MOL WT 492.53
TEMP 300

170.70	56.00	35.40	69.25	157.30	78.85
1/1	2/2	3/3	10/3	17/1	18/1
28.25	28.25	28.25	155.70	155.35	145.20
19/4	20/4	21/4			
137.35	130.45	129.25	128.50	127.65	125.50
122.90	114.70				

W.VOELTER,S.FUCHS,R.H.SEUFFER,K.ZECH
MONATSH CHEM   105, 1110 (1974)

2869 — U 004000 BENZYLOXYCARBONYL-(4,4-DIMETHOXYBENZHYDRYL)-GLUTAMINE

```
    19 20          27 26
    C=C            C=C
  18 /  \  14  /     \ 25
 O-C    C-CH-C       C-O
  /   // 15 |  22 /     \
CH3  C-C    |    C-C    CH3
21   17 16  NH  23 24    28
          /   O=C-O-CH2
        5 /     6 |   7 \
      O=C       NH    C8
        |  3   21   /   \
       CH2-CH2-CH  9C    C13
       4      /     ||    |
            C1   10C    C12
          // \      \  //
          O  OH     C11
```

FORMULA C28H30N2O7
SOLVENT C4H8O2/H2O
ORIG ST C4H12SI

MOL WT 506.56
TEMP 300

173.90	53.60	27.10	31.95	170.80	156.35
1/1	2/2	3/3	4/3	5/1	6/1
65.60	48.75	158.30	55.05	158.30	55.05
7/3	14/2	18/1	21/4	25/1	28/4
137.15	134.95	128.50	127.75	113.70	

W.VOELTER,S.FUCHS,R.H.SEUFFER,K.ZECH
MONATSH CHEM   105, 1110 (1974)

2870 — U 004000 ACTINOMYCIN D

```
        O            O
    64 /   60 59    \ 63
   ---C    H3C CH3   C---
  /    \  56 /  \ 55  /    \
 |    54CH-CH   HC-HC 53    |
 | 62 /    \   /     \ 61   |
 |H3C-N    H3C CH3    N-CH3 |
 |    \    58  57    /      |
 |   52C=O       O=C 51     |
 |    /              \      |
 |  H2C48            47CH2  |
 |    \ 50        49 /      |
 |     N-CH3   H3C-N        |
 |   46 /          \ 45     |
 |  O=C  C40    39C  C=O    |
 |   \ / \      / \ /       |
 |   38C 42C  C41 C37       |
 |    | 44 |   | 43 |       |
 |    N---C    C---N        |
 |   36 /          \ 35     |
 |  O=C  H3C    CH3  C=O     |
 |   \ 28 / 34  33 \ 27 /   |
 |    HC-CH    HC-CH        |
 |   / 30 \     / 29 \      |
 |  HN   H3C   CH3   NH     |
 |   \ 32    31 /           |
 O  26C=O       O=C25     O
  \ 20 /            \ 19 /
 22CH-CH   O   O   HC-HC 21
   \   \   \\ // /  /
 H3C  HN-C18 17C-NH  CH3
  24  | 10  |  23
     9C    N   C1   NH2
    // \ / \ / \ //
   8C   C13 C14 C2
   |    ||  |   |
   7C   C12 C11 C3
    \ / \ / \ / \
    6C   O   C4   O
    | 5       |
   H3C        CH3
   16         15
```

FORMULA C62H86N12O16
SOLVENT CDCL3
ORIG ST C4H12SI

MOL WT 1255.45
TEMP 300

101.90	147.85	179.20	113.60	127.55	125.95
1/S	2/S	3/S	4/S	6/S	7/D
130.30	129.25	145.25	140.55	132.80	146.05
8/D	9/S	11/S	12/S	13/S	14/S
7.80	14.95	166.10	168.65	71.65	71.50
15/Q	16/Q	17/S	18/S	19/D	20/D
75.10	75.10	21.70	21.60	169.05	169.05
21/D	22/D	23/Q	24/Q	25/S	26/S
56.50	56.35	31.90	31.60	19.10	19.10
27/D	28/D	29/D	30/D	31/Q	32/Q
17.85	17.40	173.45	173.45	55.45	55.10
33/Q	34/Q	35/S	36/S	37/D	38/D
31.40	31.10	23.10	22.90	47.70	47.40
39/T	40/T	41/T	42/T	43/T	44/T
167.65	167.55	51.50	51.50	39.30	39.20
45/S	46/S	47/T	48/T	49/Q	50/Q
166.60	166.45	59.00	58.85	27.05	27.05
51/S	52/S	53/D	54/D	55/D	56/D
19.30	19.30	19.30	19.30	34.90	34.90
57/Q	58/Q	59/Q	60/Q	61/Q	62/Q
173.85	173.25				
63/S	64/S				

U.HOLLSTEIN,E.BREITMAIER,G.JUNG
J AM CHEM SOC   96, 8036 (1974)

2871   U 005000 METHYL ALPHA-D-ERYTHROFURANOSIDE

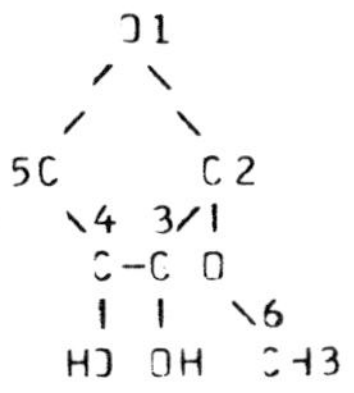

```
        O1
       / \
      /   \
    5C      C2
     \4 3/ |
     C-C O
     | |  \6
    HO OH  CH3
```

FORMULA C5H10O4   MOL WT 134.13
SOLVENT D2O
ORIG ST C4H12SI   TEMP   AMB

103.60   72.80   69.90   73.60   56.70
 2/2   3/2   4/2   5/3   6/4

R.G.S.RITCHIE,N.CYR,B.KORSCH,H.J.KOCH,
A.S.PERLIN
CAN J CHEM   53, 1424 (1975)

2872   U 005000 METHYL BETA-D-ERYTHROFURANOSIDE

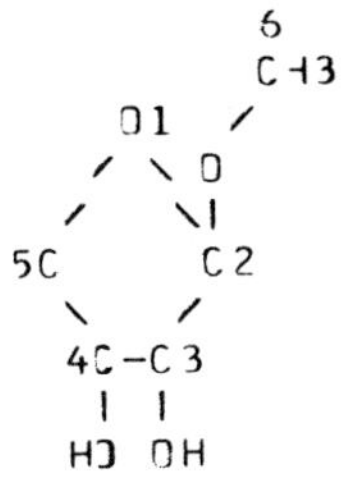

```
        6
       CH3
    O1 /
    / \ O
   /   \|
  5C     C2
   \    /
   4C-C3
   | |
   HO OH
```

FORMULA C5H10O4   MOL WT 134.13
SOLVENT D2O
ORIG ST C4H12SI   TEMP   AMB

109.60   76.40   71.40   72.60   56.60
 2/2   3/2   4/2   5/3   6/4

R.G.S.RITCHIE,N.CYR,B.KORSCH,H.J.KOCH,
A.S.PERLIN
CAN J CHEM   53, 1424 (1975)

2873   U 005000 METHYL ALPHA-L-THREOFURANOSIDE

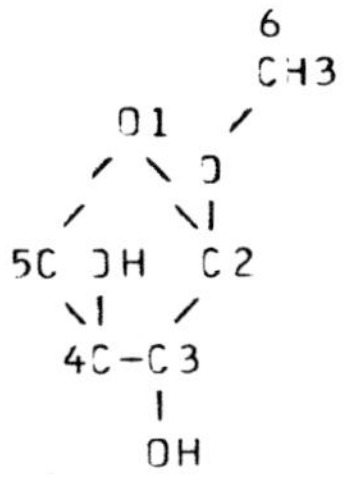

```
        6
       CH3
    O1 /
    / \ O
   /   \|
  5C OH C2
   \|  /
   4C-C3
    |
    OH
```

FORMULA C5H10O4   MOL WT 134.13
SOLVENT D2O
ORIG ST C4H12SI   TEMP   AMB

109.40   80.50   76.40   73.70   55.50
 2/2   3/2   4/2   5/3   6/4

R.G.S.RITCHIE,N.CYR,B.KORSCH,H.J.KOCH,
A.S.PERLIN
CAN J CHEM   53, 1424 (1975)

2874   U 005000 METHYL BETA-L-THREOFURANOSIDE

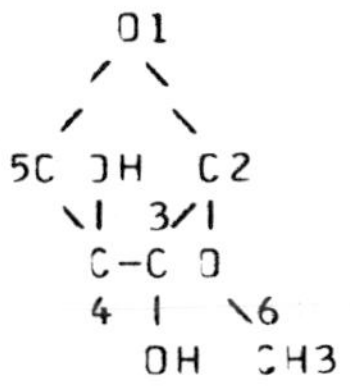

```
      O1
     / \
    /   \
  5C OH  C2
   \| 3/ |
   C-C O
   4 |  \6
    OH  CH3
```

FORMULA C5H10O4   MOL WT 134.13
SOLVENT D2O
ORIG ST C4H12SI   TEMP   AMB

103.80   77.40   75.80   72.00   56.20
 2/2   3/2   4/2   5/3   6/4

R.G.S.RITCHIE,N.CYR,B.KORSCH,H.J.KOCH,
A.S.PERLIN
CAN J CHEM   53, 1424 (1975)

2875   U 005000 ALPHA-XYLOSE

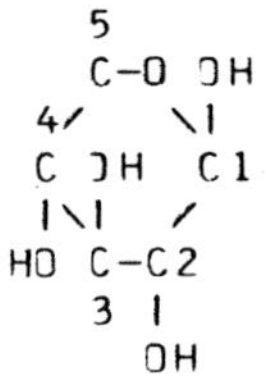

```
    5
   C-O OH
  4/  \|
  C OH C1
  |\| /
 HO C-C2
   3 |
    OH
```

FORMULA C5H10O5   MOL WT 150.13
SOLVENT H2O
ORIG ST C4H12SI   TEMP   AMB

 93.00   73.60   72.30   70.20   61.80
 1/2   2/2   3/2   4/2   5/3

L.F.JOHNSON,W.C.JANKOWSKI
CARBON-13 NMR SPECTRA,JOHN
WILEY AND SONS,NEW YORK   128 (1972)

2876   U 005000 BETA-XYLOSE

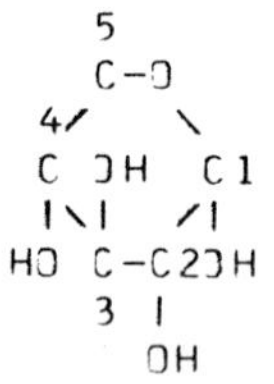

```
          5
        C-O
      4/   \
    C OH  C1
    |\|  /|
  HO C-C2OH
     3 |
       OH
```

FORMULA	C5H10O5			MOL WT	150.13
SOLVENT	H2O				
ORIG ST	C4H12SI			TEMP	AMB

| 97.40 | 76.60 | 74.80 | 70.00 | 66.00 |
| 1/2 | 2/2 | 3/2 | 4/2 | 5/3 |

L.F.JOHNSON,W.C.JANKOWSKI
CARBON-13 NMR SPECTRA,JOHN
WILEY AND SONS,NEW YORK   128 (1972)

2877   U 005000 ASCORBIC ACID

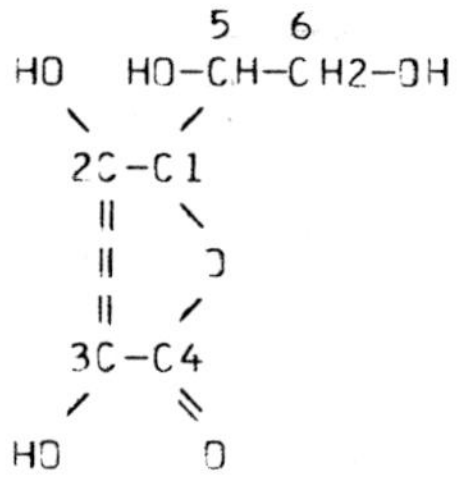

```
       5   6
  HO  HO-CH-CH2-OH
    \   /
    2C-C1
    ||  \
    ||   O
    ||  /
    3C-C4
    /   \
  HO     O
```

FORMULA	C6H8O6			MOL WT	176.13
SOLVENT	H2O				
ORIG ST	C4H12SI			TEMP	AMB

| 77.10 | 156.30 | 118.80 | 174.00 | 69.90 | 63.20 |
| 1/2 | 2/1 | 3/1 | 4/1 | 5/2 | 6/3 |

L.F.JOHNSON,W.C.JANKOWSKI
CARBON-13 NMR SPECTRA,JOHN
WILEY AND SONS,NEW YORK   171 (1972)

2878   U 005000 BETA-D-ALLOPYRANOSE

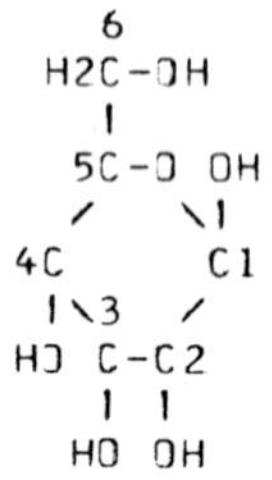

```
      6
  H2C-OH
    |
  5C-O OH
   /  \|
  4C    C1
  |\3  /
 HO C-C2
   | |
  HO OH
```

FORMULA	C6H12O6			MOL WT	180.16
SOLVENT	C2D6OS				
ORIG ST	CS2			TEMP	AMB

| 92.60 | 70.60 | 70.20 | 66.40 | 72.80 | 60.30 |
| 1/2 | 2/2 | 3/2 | 4/2 | 5/2 | 6/3 |

W.A.SZAREK,D.M.VYAS,S.D.GERO,G.LUKACS
CAN J CHEM   52, 3394 (1974)

2879   U 005000 METHYL ALPHA-L-ARABINOFURANOSIDE

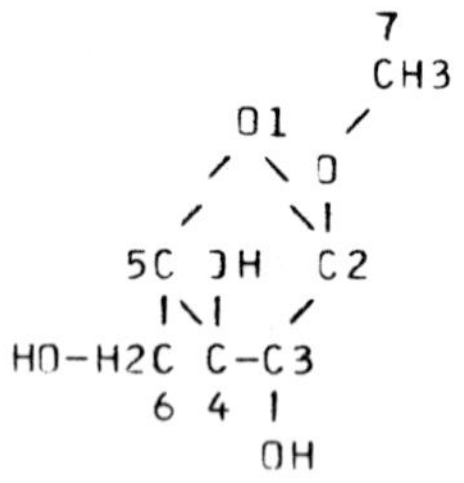

```
       7
      CH3
   O1 /
  / \ O
 /   \|
5C OH C2
|\| /
HO-H2C C-C3
 6 4 |
     OH
```

FORMULA	C6H12O5			MOL WT	164.16
SOLVENT	D2O				
ORIG ST	C4H12SI			TEMP	AMB

| 109.20 | 81.80 | 77.50 | 84.90 | 62.40 | 56.00 |
| 2/2 | 3/2 | 4/2 | 5/2 | 6/3 | 7/4 |

R.G.S.RITCHIE,N.CYR,B.KORSCH,H.J.KOCH,
A.S.PERLIN
CAN J CHEM   53, 1424 (1975)

2880   U 005000 METHYL BETA-L-ARABINOFURANOSIDE

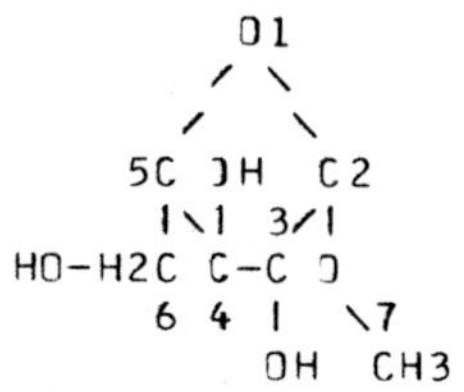

```
     O1
    / \
   /   \
  5C OH C2
  |\| 3/|
 HO-H2C C-C O
  6 4 |  \7
     OH  CH3
```

FORMULA	C6H12O5			MOL WT	164.16
SOLVENT	D2O				
ORIG ST	C4H12SI			TEMP	AMB

| 103.10 | 77.40 | 75.70 | 82.90 | 62.40 | 56.30 |
| 2/2 | 3/2 | 4/2 | 5/2 | 6/3 | 7/4 |

R.G.S.RITCHIE,N.CYR,B.KORSCH,H.J.KOCH,
A.S.PERLIN
CAN J CHEM   53, 1424 (1975)

2881   U 005000   METHYL ALPHA-D-LYXOFURANOSIDE

```
        6  01
    HO-H2C / \
        I/    \
       5C     C2
        \3 R/I R -OH
         C-C 0
         4 3 \7
            CH3
```

FORMULA	C6H12O5			MOL WT	164.16
SOLVENT	D2O				
ORIG ST	C4H12SI			TEMP	AMB

109.20	77.00	72.20	81.40	61.50	56.90
2/2	3/2	4/2	5/2	6/3	7/4

R.G.S.RITCHIE,N.CYR,B.KORSCH,H.J.KOCH,
A.S.PERLIN
CAN J CHEM   53, 1424 (1975)

2882   U 005000   METHYL BETA-D-LYXOFURANOSIDE

```
          7
         CH3
     6  01 /
  HO-H2C / \ 0
     I/    \I
    5C     C2
     \3 R/  3 -OH
     4C-C3
```

FORMULA	C6H12O5			MOL WT	164.16
SOLVENT	D2O				
ORIG ST	C4H12SI			TEMP	AMB

103.30	73.20	71.00	82.10	62.70	56.70
2/2	3/2	4/2	5/2	6/3	7/4

R.G.S.RITCHIE,N.CYR,B.KORSCH,H.J.KOCH,
A.S.PERLIN
CAN J CHEM   53, 1424 (1975)

2883   U 005000   ALPHA-RHAMNOSE

```
       6
      CH3
      I
     5C---0
     /     \
  4C 0H H0 C1
   I\I  I/ \
  HO C---C  0H
     3   2
```

FORMULA	C6H12O5			MOL WT	164.16
SOLVENT	H2O				
ORIG ST	C4H12SI			TEMP	AMB

94.30	72.20	73.70	72.70	72.70	17.70
1/2	2/2	3/2	4/2	5/2	6/4

L.F.JOHNSON,W.C.JANKOWSKI
CARBON-13 NMR SPECTRA,JOHN
WILEY AND SONS,NEW YORK   195 (1972)

2884   U 005000   BETA-RHAMNOSE

```
       6
      CH3
      I
     5C---0   OH
     /    \ /
  4C 0H H0 C1
   I\I  I/
  H0 C---C
     3   2
```

FORMULA	C6H12O5			MOL WT	164.16
SOLVENT	H2O				
ORIG ST	C4H12SI			TEMP	AMB

94.80	71.70	70.90	73.10	69.00	17.70
1/2	2/2	3/2	4/2	5/2	6/4

L.F.JOHNSON,W.C.JANKOWSKI
CARBON-13 NMR SPECTRA,JOHN
WILEY AND SONS,NEW YORK   195 (1972)

2885   U 005000   METHYL ALPHA-D-RIBOFURANOSIDE

```
      6  01
   HO-H2C / \
     I/    \
    5C     C2
     \4 3/I
      C-C 0
      I I \7
     H0 0H CH3
```

FORMULA	C6H12O5			MOL WT	164.16
SOLVENT	D2O				
ORIG ST	C4H12SI			TEMP	AMB

103.10	71.10	69.80	84.60	61.90	55.50
2/2	3/2	4/2	5/2	6/3	7/4

R.G.S.RITCHIE,N.CYR,B.KORSCH,H.J.KOCH,
A.S.PERLIN
CAN J CHEM   53, 1424 (1975)

2886   U 005000   METHYL BETA-D-RIBOFURANOSIDE

```
          7
        CH3        FORMULA   C6H12O5              MOL WT  164.16
    6   O1   /      SOLVENT   D2O
  HO-H2C  /  \  O   ORIG ST   C4H12SI             TEMP       AMB
     I/    \I
    5C      C2        108.00   74.30   70.90   83.00   62.90   55.30
     \     /          2/2      3/2     4/2     5/2     6/3     7/4
      4C-C3
      I  I           R.G.S.RITCHIE,N.CYR,B.KORSCH,H.J.KOCH,
     HO OH           A.S.PERLIN
                     CAN J CHEM                  53, 1424 (1975)
```

2887   U 005000   METHYL ALPHA-D-XYLOFURANOSIDE

```
                   FORMULA   C6H12O5              MOL WT  164.16
    6   O1         SOLVENT   D2O
  HO-H2C  /  \     ORIG ST   C4H12SI             TEMP       AMB
     I/    \
    5C  OH  C2       103.00   77.80   76.20   79.30   61.60   56.70
     \I  3/I         2/2      3/2     4/2     5/2     6/3     7/4
      C-C  O
      4 I   \7      R.G.S.RITCHIE,N.CYR,B.KORSCH,H.J.KOCH,
        OH  CH3     A.S.PERLIN
                    CAN J CHEM                   53, 1424 (1975)
```

2888   U 005000   METHYL BETA-D-XYLOFURANOSIDE

```
         CH3       FORMULA   C6H12O5              MOL WT  164.16
    6   O1  /7      SOLVENT   D2O
  HO-H2C  /  \  O   ORIG ST   C4H12SI             TEMP       AMB
     I/    \I
    5C  OH  C2       109.70   81.00   76.00   83.60   62.20   56.40
     \I  /           2/2      3/2     4/2     5/2     6/3     7/4
      4C-C3
       I            R.G.S.RITCHIE,N.CYR,B.KORSCH,H.J.KOCH,
       OH           A.S.PERLIN
                    CAN J CHEM                   53, 1424 (1975)
```

2889   U 005000   ALPHA-GALACTOSE

```
       6
    HO-CH2          FORMULA   C6H12O6              MOL WT  180.16
      I            SOLVENT   H2O
   HO C-O          ORIG ST   C4H12SI             TEMP       AMB
    I/5  \I
   4C  OH  C         93.20    70.10   69.30   70.20   71.30   62.10
    \I  2/I          1/2      2/2     3/2     4/2     5/2     6/3
      C-C  OH
      3  I          L.F.JOHNSON,W.C.JANKOWSKI
      HO           CARBON-13 NMR STECTRA,JOHN
                   WILEY AND SONS,NEW YORK            198 (1972)
```

2890   U 005000   BETA-GALACTOSE

```
       6
    HO-CH2          FORMULA   C6H12O6              MOL WT  180.16
      I            SOLVENT   H2O
   HO C-O OH        ORIG ST   C4H12SI             TEMP       AMB
    I/5  \I
   4C  OH  C         97.30    72.90   73.70   69.60   75.90   61.90
    \I  2/I          1/2      2/2     3/2     4/2     5/2     6/3
      C-C
      3  I          L.F.JOHNSON,W.C.JANKOWSKI
      OH           CARBON-13 NMR STECTRA,JOHN
                   WILEY AND SONS,NEW YORK            198 (1972)
```

```
2891                   U 005000  ALPHA-D-GALACTOSE
        6
      H2C-OH            FORMULA   C6H12O6              MOL WT   180.16
        |               SOLVENT   D2O
     HO C-O             ORIG ST   C4H12SI              TEMP     AMB
      1/5  \
     4C OH  C1            92.35    69.40    68.40    69.20    70.50    61.25
      \| 2/|              1/2      2/2      3/2      4/2      5/2      6/3
       C-C OH
       3 |              W.VOELTER,E.BREITMAIER,E.B.RATHBONE,A.M.STEPHEN
        OH              TETRAHEDRON                   29, 3845 (1973)
```

```
2892                   U 005000  BETA-D-GALACTOSE
        6
      H2C-OH            FORMULA   C6H12O6              MOL WT   180.16
        |               SOLVENT   D2O
     HO C-O OH          ORIG ST   C4H12SI              TEMP     AMB
      1/5  \|
     4C OH  C1            96.50    71.90    72.85    68.80    75.20    61.05
      \|  /               1/2      2/2      3/2      4/2      5/2      6/3
       3C-C 2
         |              W.VOELTER,E.BREITMAIER,E.B.RATHBONE,A.M.STEPHEN
        OH              TETRAHEDRON                   29, 3845 (1973)
```

```
2893                   U 005000  ALPHA-GLUCOSE
         6
       CH2-OH           FORMULA   C6H12O6              MOL WT   180.16
         |              SOLVENT   H2O
       5C-O             ORIG ST   C4H12SI              TEMP     AMB
      /    \
     4C OH  C1            92.80    72.30    73.60    70.40    72.30    61.60
     / \|  / \            1/2      2/2      3/2      4/2      5/2      6/3
    HO   C-C2  OH
         3 |            L.F.JOHNSON,W.C.JANKOWSKI
          OH            CARBON-13 NMR SPECTRA,JOHN
                        WILEY AND SONS,NEW YORK            197 (1972)
```

```
2894                   U 005000  BETA-GLUCOSE
         6
       CH2-OH           FORMULA   C6H12O6              MOL WT   180.16
         |              SOLVENT   H2O
       5C-O   OH        ORIG ST   C4H12SI              TEMP     AMB
      /    \ /
     4C OH  C1            96.70    74.90    76.70    70.40    76.50    61.60
     / \|  /              1/2      2/2      3/2      4/2      5/2      6/3
    HO   C-C2
         3 |            L.F.JOHNSON,W.C.JANKOWSKI
          OH            CARBON-13 NMR SPECTRA,JOHN
                        WILEY AND SONS,NEW YORK            197 (1972)
```

```
2895                   U 005000  ALPHA-GLUCOSAMINE HYDROCHLORIDE
         6
       CH2-OH           FORMULA   C6H13NO5             MOL WT   179.17
         |              SOLVENT   H2O
       5C-O             ORIG ST   C4H12SI              TEMP     AMB
      /    \
     4C OH  C1            90.00    55.30    70.50    70.50    72.40    61.30
     / \|  / \            1/2      2/2      3/2      4/2      5/2      6/3
    HO  3C-C2  OH
         |              L.F.JOHNSON,W.C.JANKOWSKI
         |              CARBON-13 NMR SPECTRA,JOHN
        NH2.HCL         WILEY AND SONS,NEW YORK            205 (1972)
```

```
2896              U 005000  BETA-GLUCOSAMINE HYDROCHLORIDE
         6
        CH2-OH                FORMULA   C6H13NO5              MOL WT   179.17
         |                    SOLVENT   H2O
       5C-O    OH             ORIG ST   C4H12SI              TEMP       AMB
       /   \  /
     4C OH  C1               93.60   57.80   72.90   70.50   76.90   61.30
     / \|  /                  1/2     2/2     3/2     4/2     5/2     6/3
   HO   3C-C2
         |                    L.F.JOHNSON,W.C.JANKOWSKI
         |                    CARBON-13 NMR SPECTRA,JOHN
       NH2.HCL               WILEY AND SONS,NEW YORK          205 (1972)
```

```
2897              U 005000  GALACTITOL
         1
       H2C-OH                 FORMULA   C6H14O6              MOL WT   182.17
        |2                    SOLVENT   D2O
      H-C-OH                  ORIG ST   C4H12SI              TEMP       AMB
        |3
     HO-C-H                  63.25   69.25   70.15
        |4                    1/3     2/2     3/2
     HO-C-H
        |5
      H-C-OH
        |6
       H2C-OH                 W.VOELTER,E.BREITMAIER,E.B.RATHBONE,A.M.STEPHEN
                              TETRAHEDRON                 29, 3845 (1973)
```

```
2898              U 005000  METHYL 3,4,6-TRICHLORO-3,4,6-TRIDEOXY-ALPHA-
                            D-ALLOPYRANOSIDE
         6
       H2C-CL                 FORMULA   C7H11CL3O3           MOL WT   249.52
        |                     SOLVENT   C2D6OS
       5C-O                   ORIG ST   CS2                  TEMP       AMB
       /   \
     4C     C1               98.20   64.80   63.40   55.00   67.20   44.10
     / \3 2/|                 1/2     2/2     3/2     4/2     5/2     6/3
   CL  C-C O                 54.20
     /  | \7                  7/4
   CL  OH  CH3
                              W.A.SZAREK,D.M.VYAS,S.D.GERO,G.LUKACS
                              CAN J CHEM                  52, 3394 (1974)
```

```
2899              U 005000  METHYL 4,6-DICHLORO-4,6-DIDEOXY-ALPHA-
                            D-GALACTOPYRANOSIDE
         6
       H2C-CL                 FORMULA   C7H12CL2O4           MOL WT   231.08
        |                     SOLVENT   C2D6OS
     CL C-O                   ORIG ST   CS2                  TEMP       AMB
      |/5 \1
     4C OH  C                98.80   66.50   66.50   63.60   68.10   43.10
      \| 2/|                  1/2     2/2     3/2     4/2     5/2     6/3
      3C-C O                 53.60
       |  \7                  7/4
      OH  CH3
                              W.A.SZAREK,D.M.VYAS,S.D.GERO,G.LUKACS
                              CAN J CHEM                  52, 3394 (1974)
```

```
2900              U 005000  METHYL 4,6-DICHLORO-4,6-DIDEOXY-BETA-
                            D-GALACTOPYRANOSIDE
        6      7
      H2C-CL  CH3             FORMULA   C7H12CL2O4           MOL WT   231.08
       |    /                 SOLVENT   C2D6OS
     CL C-O O                 ORIG ST   CS2                  TEMP       AMB
      |/5 \1
     4C OH  C1               103.20   68.60   70.00   62.40   71.50   42.40
      \|  /                   1/2     2/2     3/2     4/2     5/2     6/3
      3C-C2                  55.20
       |                      7/4
      OH
                              W.A.SZAREK,D.M.VYAS,S.D.GERO,G.LUKACS
                              CAN J CHEM                  52, 3394 (1974)
```

2901   U 005000 METHYL 3,6-DICHLORO-3,6-DIDEOXY-BETA-
D-ALLOPYRANOSIDE

```
    6        7
  H2C-CL   CH3
    |      /
    5C-O O
   /    \|
  4C      C1
  |\3   /
  HO C-C2
   /  |
  CL  OH
```

FORMULA	C7H12CL2O4			MOL WT	231.08
SOLVENT	C2D6OS				
ORIG ST	CS2			TEMP	AMB

99.90	67.40	67.40	65.50	71.50	43.90
1/2	2/2	3/2	4/2	5/2	6/3
54.70					
7/4					

W.A.SZAREK,D.M.VYAS,S.D.GERO,G.LUKACS
CAN J CHEM   52, 3394 (1974)

2902   U 005000 METHYL 6-CHLORO-4,6-DIDECXY-ALPHA-D-XYLO-
HEXOPYRANOSIDE

```
    6
  H2C-CL
    |
    5C-O
   /   \1
  4C OH  C
   \| 2/|
   3C-C O
     |  \7
    OH  CH3
```

FORMULA	C7H13CLO4			MOL WT	196.63
SOLVENT	C2D6OS				
ORIG ST	CS2			TEMP	AMB

99.10	72.40	65.20	35.30	66.10	46.00
1/2	2/2	3/2	4/3	5/2	6/3
53.20					
7/4					

W.A.SZAREK,D.M.VYAS,S.D.GERO,G.LUKACS
CAN J CHEM   52, 3394 (1974)

2903   U 005000 METHYL 4-CHLORO-4-DEOXY-ALPHA-D-GALACTO-
PYRANOSIDE

```
    6
  H2C-OH
    |
  CL C-O
  1/5 \
  4C OH  C1
   \| 2/|
   3C-C O
     |  \7
    OH  CH3
```

FORMULA	C7H13CLO5			MOL WT	212.63
SOLVENT	C2D6OS				
ORIG ST	CS2			TEMP	AMB

98.60	66.80	66.80	63.60	68.20	59.70
1/2	2/2	3/2	4/3	5/2	6/3
53.30					
7/4					

W.A.SZAREK,D.M.VYAS,S.D.GERO,G.LUKACS
CAN J CHEM   52, 3394 (1974)

2904   U 005000 METHYL 2-CHLORO-2-DEOXY-BETA-D-GLUCOPYRANOSIDE

```
    6       7
  H2C-OH  CH3
    |     /
    5C-O O
   /   \|
  4C OH  C1
  |\|  /
  HO C-C2
   3 |
     CL
```

FORMULA	C7H13CLO5			MOL WT	212.63
SOLVENT	C2D6OS				
ORIG ST	CS2			TEMP	AMB

101.20	62.00	75.50	69.60	75.50	59.40
1/2	2/2	3/2	4/2	5/2	6/3
54.80					
7/4					

W.A.SZAREK,D.M.VYAS,S.D.GERO,G.LUKACS
CAN J CHEM   52, 3394 (1974)

2905   U 005000 METHYL ALPHA-D-ALLOFURANOSIDE

```
     7
  HO-CH2
   |6  O1
  HO-CH/ \
    |/    \
    5C     C2
    \4 3/ |
    C-C O
    | |  \8
   HO OH  CH3
```

FORMULA	C7H14O6			MOL WT	194.19
SOLVENT	D2O				
ORIG ST	C4H12SI			TEMP	AMB

103.80	72.30	69.90	85.90	72.70	63.50
2/2	3/2	4/2	5/2	6/2	7/3
56.60					
8/4					

R.G.S.RITCHIE,N.CYR,B.KORSCH,H.J.KOCH,
A.S.PERLIN
CAN J CHEM   53, 1424 (1975)

2906   U 005000   METHYL BETA-D-ALLOFURANOSIDE

```
      7          8
  HO-CH2       CH3
     16 01  /
  HO-CH/  \ O
     |/   \|
    5C     C2
      \   /
      4C-C3
      |  |
     HO  OH
```

FORMULA C7H14O6				MOL WT	194.19
SOLVENT D2O					
ORIG ST C4H12SI			TEMP		AMB
109.00	75.60	72.70	83.40	73.80	63.90
2/2	3/2	4/2	5/2	6/2	7/3
56.40					
8/4					

R.G.S.RITCHIE,N.CYR,B.KORSCH,H.J.KOCH,
A.S.PERLIN
CAN J CHEM   53, 1424 (1975)

2907   U 005000   METHYL ALPHA-D-GALACTOFURANOSIDE

```
        01
       / \
      /   \
    5C OH C2
    6|\| 3/|
  HO-HC C-C O
    | 4 |  \8
  HO-H2C7 OH CH3
```

FORMULA C7H14O6				MOL WT	194.19
SOLVENT D2O					
ORIG ST C4H12SI			TEMP		AMB
103.80	78.20	76.20	83.10	74.50	64.10
2/2	3/2	4/2	5/2	6/2	7/3
57.20					
8/4					

R.G.S.RITCHIE,N.CYR,B.KORSCH,H.J.KOCH,
A.S.PERLIN
CAN J CHEM   53, 1424 (1975)

2908   U 005000   METHYL BETA-D-GALACTOFURANOSIDE

```
           8
          CH3
     01  /
    / \ O
   /   \|
 5C OH C2
 6|\|  /
HO-HC C-C3
  | 4 |
HO-H2C7 OH
```

FORMULA C7H14O6				MOL WT	194.19
SOLVENT D2O					
ORIG ST C4H12SI			TEMP		AMB
109.90	81.30	78.40	84.70	71.70	63.60
2/2	3/2	4/2	5/2	6/2	7/3
55.60					
8/4					

R.G.S.RITCHIE,N.CYR,B.KORSCH,H.J.KOCH,
A.S.PERLIN
CAN J CHEM   53, 1424 (1975)

2909   U 005000   METHYL ALPHA-D-GALACTOPYRANOSIDE

```
      6
   H2C-OH
      |
   HO C-O
    |/5 \1
   4C OH  C
    \| 2/|
    3C-C O
    |  \7
    OH  CH3
```

FORMULA C7H14O6				MOL WT	194.19
SOLVENT C2D6OS					
ORIG ST CS2			TEMP		AMB
98.80	67.70	68.50	67.30	69.80	59.60
1/2	2/2	3/2	4/2	5/2	6/3
53.30					
7/4					

W.A.SZAREK,D.M.VYAS,S.D.GERO,G.LUKACS
CAN J CHEM   52, 3394 (1974)

2910   U 005000   METHYL ALPHA-D-GALACTOPYRANOSIDE

```
      6
   H2C-OH
      |
   HO C-O
    |/5 \
   4C OH  C1
    \| 2/|
    3C-C O
    |  \7
    OH  CH3
```

FORMULA C7H14O6				MOL WT	194.19
SOLVENT D2O					
ORIG ST C4H12SI			TEMP		AMB
99.50	69.60	68.30	69.30	70.80	61.30
1/2	2/2	3/2	4/2	5/2	6/3
55.15					
7/4					

W.VOELTER,E.BREITMAIER,E.B.RATHBONE,A.M.STEPHEN
TETRAHEDRON   29, 3845 (1973)

2911   U 005000 METHYL BETA-D-GALACTOPYRANOSIDE

```
    6       7
 H2C-OH   CH3
    |     /
  HO C-O O
   |/5  \|
  4C OH  C1
    \|  /
    3C-C2
     |
     OH
```

FORMULA C7H14O6				MOL WT	194.19
SOLVENT D2O					
ORIG ST C4H12SI				TEMP	AMB
103.90	70.80	72.85	68.75	75.20	61.05
1/2	2/2	3/2	4/2	5/2	6/3
57.30					
7/4					

W.VOELTER,E.BREITMAIER,E.B.RATHBONE,A.M.STEPHEN
TETRAHEDRON   29, 3845 (1973)

2912   U 005000 3-O-METHYL-D-GALACTITOL

```
      1
   H2C-OH
    | 2
   H-C-OH
 7    | 3
 H3C-O-C-H
    | 4
  HO-C-H
    | 5
   H-C-OH
    | 6
   H2C-OH
```

FORMULA C7H14O6				MOL WT	194.19
SOLVENT D2O					
ORIG ST C4H12SI				TEMP	AMB
69.20	70.15	79.40	70.80	68.75	63.25
1/3	2/2	3/2	4/2	5/2	6/3
60.20					
7/4					

W.VOELTER,E.BREITMAIER,E.B.RATHBONE,A.M.STEPHEN
TETRAHEDRON   29, 3845 (1973)

2913   U 005000 METHYL ALPHA-D-GLUCOFURANOSIDE

```
     7
  HO-CH2
    | O1
  HO-HC6/ \
   |/    \
   5C OH  C2
   \| 3/|
    C-C O
   4 |  \8
    OH  CH3
```

FORMULA C7H14O6				MOL WT	194.19
SOLVENT D2O					
ORIG ST C4H12SI				TEMP	AMB
104.00	77.70	76.60	78.80	70.70	64.20
2/2	3/2	4/2	5/2	6/2	7/3
57.00					
8/4					

R.G.S.RITCHIE,N.CYR,B.KORSCH,H.J.KOCH,
A.S.PERLIN
CAN J CHEM   53, 1424 (1975)

2914   U 005000 METHYL BETA-D-GLUCOFURANOSIDE

```
    7         8
 HO-CH2     CH3
    | O1   /
 HO-HC6/ \ O
   |/   \|
   5C OH  C2
   \|  /
   4C-C3
    |
    OH
```

FORMULA C7H14O6				MOL WT	194.19
SOLVENT D2O					
ORIG ST C4H12SI				TEMP	AMB
110.00	80.60	75.80	82.30	70.70	64.70
2/2	3/2	4/2	5/2	6/2	7/3
56.30					
8/4					

R.G.S.RITCHIE,N.CYR,B.KORSCH,H.J.KOCH,
A.S.PERLIN
CAN J CHEM   53, 1424 (1975)

2915   U 005000 METHYL BETA-D-GLUCOPYRANOSIDE

```
    6      7
 H2C-OH   CH3
   |     /
  5C-O O
  /    \|
 4C OH  C1
  |\|  /
 HO C-C2
   3 |
     OH
```

FORMULA C7H14O6				MOL WT	194.19
SOLVENT C2D6OS					
ORIG ST CS2				TEMP	AMB
102.70	72.20	75.60	68.90	75.60	60.00
1/2	2/2	3/2	4/2	5/2	6/3
54.80					
7/4					

W.A.SZAREK,D.M.VYAS,S.D.GERO,G.LUKACS
CAN J CHEM   52, 3394 (1974)

```
2916              U 005000  METHYL ALPHA-D-MANNOFURANOSIDE
        7
     HO-CH2            FORMULA   C7H14O6            MOL WT   194.19
       | O1            SOLVENT   D2O
     HO-HC6/  \        ORIG ST   C4H12SI            TEMP        AMB
       |/   \
       5C    C2        109.70   77.90   72.50   80.50   70.60   64.50
       \R R/|  R -OH     2/2     3/2     4/2     5/2     6/2     7/3
        C-C O          57.20
        4 3  \8        8/4
           CH3
                       R.G.S.RITCHIE,N.CYR,B.KORSCH,H.J.KOCH,
                       A.S.PERLIN
                       CAN J CHEM                 53, 1424 (1975)
```

```
2917              U 005000  METHYL BETA-D-MANNOFURANOSIDE
       7        8
     HO-CH2    CH3      FORMULA   C7H14O6            MOL WT   194.19
       | O1  /          SOLVENT   D2O
     HO-HC6/  \ O       ORIG ST   C4H12SI            TEMP        AMB
       |/   \|
       5C    C2         103.60   73.10   71.20   80.70   71.00   64.40
       \R R/  R -OH       2/2     3/2     4/2     5/2     6/2     7/3
        4C-C3           56.80
                        8/4

                       R.G.S.RITCHIE,N.CYR,B.KORSCH,H.J.KOCH,
                       A.S.PERLIN
                       CAN J CHEM                 53, 1424 (1975)
```

```
2918              U 005000  3,4-DI-O-METHYL-GALACTITOL
        1
     H2C-OH            FORMULA   C8H16O6            MOL WT   208.21
      |2               SOLVENT   D2O
     H-C-OH            ORIG ST   C4H12SI            TEMP        AMB
     7  |3
   H3C-O-C-H           62.90   70.70   78.85   60.30
      |4               1/3     2/2     3/2     7/4
   H3C-O-C-H
     8  |5
     H-C-OH
      |6               W.VOELTER,E.BREITMAIER,E.B.RATHBONE,A.M.STEPHEN
     H2C-OH            TETRAHEDRON                29, 3845 (1973)
```

```
2919              U 005000  METHYL 3-O-METHYL-BETA-D-GALACTOPYRANOSIDE
       6    7
     H2C-OH  CH3       FORMULA   C8H16O6            MOL WT   208.21
       |   /           SOLVENT   D2O
     HO C-O O          ORIG ST   C4H12SI            TEMP        AMB
      |/5  \|
      4C    C1    8    103.90   69.80   82.00   64.20   75.10   61.20
      \R  /  R -O-CH3    1/2     2/2     3/2     4/2     5/2     6/3
       3C-C2           57.30   56.20
        |              7/4     8/4
        OH
                       W.VOELTER,E.BREITMAIER,E.B.RATHBONE,A.M.STEPHEN
                       TETRAHEDRON                29, 3845 (1973)
```

```
2920              U 005000  METHYL 2,6-DI-O-METHYL-ALPHA-D-GALACTO-
          9CH3                        PYRANOSIDE
       6 /
     H2C-O             FORMULA   C9H18O6            MOL WT   222.24
       |               SOLVENT   D2O
     HO C-O            ORIG ST   C4H12SI            TEMP        AMB
      |/5  \
      4C OH C1         96.80   77.45   68.60   69.50   73.15   71.85
      \| 2/|             1/2     2/2     3/2     4/2     5/2     6/3
       C-C O           55.05   57.70   58.50
       3 |  \7         7/4     8/4     9/4
       O   CH3
        \8             W.VOELTER,E.BREITMAIER,E.B.RATHBONE,A.M.STEPHEN
        CH3            TETRAHEDRON                29, 3845 (1973)
```

```
2921                      U 005000  METHYL 2,3,4,6-TETRA-O-METHYL-BETA-
            11                      D-GALACTOPYRANOSIDE
            CH3
    10   6  /
    H3C H2C-O  7CH3        FORMULA  C11H2206              MOL WT  250.29
       \  I    /           SOLVENT  D2O
        O C-O O            ORIG ST  C4H12SI               TEMP      AMB
        I/5  \I            103.35   79.75   85.20   73.05   74.90   71.00
    4C      C1      9       1/2      2/2     3/2     4/2     5/2     6/3
      \R  /  R -O-CH3      57.10    57.10   58.50   60.20   60.95
       3C-C2                7/4      8/4     9/4    10/4    11/4
         I
         O
          \8                        W.VOELTER,E.BREITMAIER,E.B.RATHBONE,A.M.STEPHEN
          CH3                       TETRAHEDRON                29, 3845 (1973)
```

```
2922                      U 005000  SUCROSE

         6                 FORMULA  C12H22O11             MOL WT  342.30
         CH2-OH            SOLVENT  H2O
         I                 ORIG ST  C4H12SI               TEMP      AMB
        5C-O        O   OH
       /   \      / \  I   92.90    70.10   61.10  104.40   77.40   82.20
    4C OH  C1    /   \ 11CH2  1/2    4/2     6/3     7/1     8/2     9/2
    / \I  / \  7/     \ /  74.90    62.30   63.20   71.90   73.20   73.50
   HO  3C-C2  O-C   HO C10  10/2    11/3    12/3
      I     / \    I/
     HO  H2C  8C---C9
        /12  I                      L.F.JOHNSON,W.C.JANKOWSKI
       HO    OH                     CARBON-13 NMR SPECTRA,JOHN
                                    WILEY AND SONS,NEW YORK         443 (1972)
```

```
2923                      U 005000  KASUGAMYCIN HYDROCHLORIDE
         12
    HCL   CH3      HO  OH  FORMULA  C14H25N3O9            MOL WT  379.37
     .    I11      I  I    SOLVENT  H2O
    NH    C-O     2C-C3    ORIG ST  C4H12SI               TEMP      AMB
    II  10/   \7  1/   \
    13C   C H2N C   C OH  C4   81.80    97.00   49.50   26.80   50.80   17.50
    / \  / \  I/ \ / \I 5/I    1/2      7/2     8/2     9/3    10/2    12/4
  HO-C14 N   C-C   O    C-C OH   158.10  159.80   68.40   69.90   71.30   72.30
    II   I   9 8      6 I        13/1     14/1
    O    H           OH          72.80    73.90


                                  L.F.JOHNSON,W.C.JANKOWSKI
                                  CARBON-13 NMR SPECTRA,JOHN
                                  WILEY AND SONS,NEW YORK          462 (1972)
```

```
2924                      U 005000  METHYL-4,6-O-BENZYLIDENE-2,3-DIDEOXY-
            6                       2-ETHYLAMINO-3-NITRO-ALPHA-GLUCOPYRANOSIDE
   12  13    O-CH2
   C-C      /  I          FORMULA  C16H22N2O6            MOL WT  338.36
  //  \8 7/ 5C-O          SOLVENT  CDCL3
 11C    C-C  /  \         ORIG ST  C4H12SI               TEMP      AMB
   \   /  \4C NO2 C1
   C=C    \I\I  / \       98.25    60.08   87.70   78.41   62.06   68.89
  10  9    O C-C2  O       1/2      2/2     3/2     4/2     5/2     6/3
          3 I    I        101.35   136.27  128.97  128.02  125.87   55.48
          HN 14CH3         7/2      8/1     9/2    10/2    11/2    14/4
            \             41.59    15.63
           H2C-CH3        15/3     16/4
           15  16                  N.GURUDATA,F.J.M.RAJABALEE
                                   CAN J CHEM                 51, 1797 (1973)
```

```
2925                      U 005000  METHYL-4,6-O-BENZYLIDENE-2,3-DIDEOXY-
            6      14                2-ETHYLAMINO-3-NITRO-BETA-GLUCOPYRANOSIDE
   12  13    O-CH2  CH3
   C-C      /  I    I     FORMULA  C16H22N2O6            MOL WT  338.36
  //  \8 7/ 5C-O    O     SOLVENT  CDCL3
 11C    C-C  4/  \  /     ORIG ST  C4H12SI               TEMP      AMB
   \   /  \ C NO2 C1
   C=C    \I\I  /        104.52   61.43   89.13   77.94   66.35   68.57
  10  9    O C-C2         1/2      2/2     3/2     4/2     5/2     6/3
          3 I            101.27   136.27  129.05  128.10  125.87   59.29
          HN              7/2      8/1     9/2    10/2    11/2    14/4
            \            41.27    15.56
           H2C-CH3        15/3     16/4
           15  16                  N.GURUDATA,F.J.M.RAJABALEE
                                   CAN J CHEM                 51, 1797 (1973)
```

```
2926              U 005000  METHYL-4,6-O-BENZYLIDENE-2,3-DIDEOXY-
              6                        3-NITRO-2-PYRROLIDYL-ALPHA-GLUCOPYRANOSIDE
    12 13      O-CH2
     C-C      / |              FORMULA  C18H24N2O6            MOL WT  364.40
    //  =8 7/  5C-O            SOLVENT  CDCL3
  11C     C-C  4/   \          ORIG ST  C4H12SI               TEMP     AMB
    \  /    \ C NO2 C1
     C=C     \I\I  / \           99.52   61.98   84.52   79.05   61.98   68.97
    10 9     O C-C2 O            1/2     2/2     3/2     4/2     5/2     6/3
             3  |   |          101.43  136.35  128.97  128.02  125.95   54.76
               N  14CH3          7/2     8/1     9/2    10/2    11/2    14/4
              / \               48.33   24.05
          18C   C15             15/3    16/3
           |    |
          17C---C16            N.GURUDATA,F.J.M.RAJABALEE
                               CAN J CHEM                    51, 1797 (1973)
```

```
2927              U 005000  METHYL-4,6-O-BENZYLIDENE-2,3-DIDEOXY-
              6      14                3-NITRO-2-PYRROLIDYL-BETA-GLUCOPYRANOSIDE
    12 13      O-CH2  CH3
     C-C      / |     |        FORMULA  C18H24N2O6            MOL WT  364.40
    //  \8 7/  5C-O   O         SOLVENT  CDCL3
  11C     C-C  4/   \ /         ORIG ST  C4H12SI               TEMP     AMB
    \  /    \ C NO2 C1
     C=C     \I\I  / \          102.62   62.30   87.14   78.33   66.27   68.65
    10 9     O C-C2             1/2     2/2     3/2     4/2     5/2     6/3
             3  |             101.11  136.27  128.97  128.02  125.87   56.51
               N                7/2     8/1     9/2    10/2    11/2    14/4
              / \               48.17   23.97
          18C   C15             15/3    16/3
           |    |
          17C---C16            N.GURUDATA,F.J.M.RAJABALEE
                               CAN J CHEM                    51, 1797 (1973)
```

```
2928              U 005000  METHYL-4,6-O-BENZYLIDENE-2,3-DIDEOXY-
              6                        3-NITRO-2-PIPERIDYL-ALPHA-GLUCOPYRANOSIDE
    12 13      O-CH2
     C-C      / |              FORMULA  C19H26N2O6            MOL WT  378.43
    //  \8 7/  5C-O            SOLVENT  CDCL3
  11C     C-C  4/   \          ORIG ST  C4H12SI               TEMP     AMB
    \  /    \ C NO2 C1
     C=C     \I\I  / \           98.97   66.98   83.65   79.29   62.14   69.05
    10 9     O C-C2 O            1/2     2/2     3/2     4/2     5/2     6/3
             3  |   |          101.43  136.43  128.97  128.02  125.95   54.68
               N   CH3           7/2     8/1     9/2    10/2    11/2    14/4
              / \  14           50.87   26.90   24.68
          19C   C15             15/3    16/3    17/3
           |    |
          18C   C16
            \  /                N.GURUDATA,F.J.M.RAJABALEE
             C17               CAN J CHEM                    51, 1797 (1973)
```

```
2929              U 005000  METHYL-4,6-O-BENZYLIDENE-2,3-DIDEOXY-
              6      14                3-NITRO-2-PIPERIDYL-BETA-GLUCOPYRANOSIDE
    12 13      O-CH2  CH3
     C-C      / |     |        FORMULA  C19H26N2O6            MOL WT  378.43
    //  \8 7/  5C-O   O         SOLVENT  CDCL3
  11C     C-C  4/   \ /         ORIG ST  C4H12SI               TEMP     AMB
    \  /    \ C NO2 C1
     C=C     \I\I  /           102.78   67.78   86.59   78.33   66.35   68.65
    10 9     O C-C2             1/2     2/2     3/2     4/2     5/2     6/3
             3  |             101.19  136.27  128.97  128.02  125.87   56.51
               N                7/2     8/1     9/2    10/2    11/2    14/4
              / \               57.03   26.75   24.52
          19C   C15             15/3    16/3    17/3
           |    |
          18C   C16
            \  /                N.GURUDATA,F.J.M.RAJABALEE
             C17               CAN J CHEM                    51, 1797 (1973)
```

```
2930              U 005000  RICKAMYCIN
H2N-CH2
   6|                   FORMULA   C19H37N5O7           MOL WT  447.54
   5C-O      NH2        SOLVENT   H2O
  // \1  | 11          ORIG ST   C4H12SI              TEMP      AMB
 4C     C 12C-C NH2
  \  2/1 7/   \1        96.70    25.60   100.70   150.20   43.50   85.40
   3C-C O-C OH  C10      1/2      3/3     4/2      5/1     6/3     7/2
    |    \| /            75.40    87.80   36.40   101.40   70.20   64.30
   H2N    8C-C9          8/3      9/2    11/3     13/2    14/2    15/2
        |   17 19        73.30    68.50   37.90    22.60   47.50   50.30
        O O-C CH3       16/1     17/3    18/4     19/4
        |/   \|          51.80
       13C OH  C16
        \ |15/|
        14C-C OH        L.F.JOHNSON,W.C.JANKOWSKI
        18 |            CARBON-13 NMR SPECTRA,JOHN
       H3C-NH           WILEY AND SONS,NEW YORK            484 (1972)
```

```
2931              U 005000  SUCROSE OCTAACETATE
       6
     CH2-OR                FORMULA   C28H38O19          MOL WT  678.60
       |                   SOLVENT   CDCL3
     5C-O                  ORIG ST   C4H12SI            TEMP      AMB
    /    \
  4C  OR   C1       O   OR  90.00    61.80   104.00   63.60   62.80   20.50
  / \|  / \    / \121|      1/2      6/3      7/1    11/3    12/3    14/4
 RO  C-C2  O   /   \ CH2    68.20    68.50    69.60   70.30   75.10   75.80
    3 |   \ /    \|         79.20   169.30   169.40  169.70  169.90  170.20
      OR  7C      C10
       / \  RO /           170.40
     RO-CH2 \  |/
    O       11  8C-9
    ||            |
  R -C-CH3        OR       L.F.JOHNSON,W.C.JANKOWSKI
   13 14                   CARBON-13 NMR SPECTRA,JOHN
                           WILEY AND SONS,NEW YORK            495 (1972)
```

```
2932              U 006000  ARECOLINE
        O
    4   ||                 FORMULA   C8H13NO2           MOL WT  155.20
    C   C8                 SOLVENT   CCL4
   / \ / \ 9               ORIG ST   C4H12SI            TEMP      AMB
  5C   C3  O-CH3
   |   |                   52.50   128.90   136.20   26.20   50.50   45.30
  6C   C2                   2/3      3/1     4/2      5/3     6/3     7/4
   \1/                    164.60    50.50
    N                       8/1      9/4
    |
   7CH3                    E.WENKERT,J.S.BINDRA,C.-J.CHANG,D.W.COCHRAN,
                           F.M.SCHELL
                           ACC CHEM RES                 7,   46 (1974)
```

```
2933              U 006000  INDOLIZIDINE
       4
       C                   FORMULA   C8H15N             MOL WT  125.22
     / \8                  SOLVENT   CHCL3
   5C   C---C3             ORIG ST   C4H12SI            TEMP      AMB
    |   |   |
   6C   N   C2             53.90    20.30    30.10   30.70   24.20   25.10
    \ /9\ /                 1/3      2/3      3/3     4/3     5/3     6/3
     C   C                 52.70    64.10
     7   1                  7/3      8/2

                           E.WENKERT,J.S.BINDRA,C.-J.CHANG,D.W.COCHRAN,
                           F.M.SCHELL
                           ACC CHEM RES                 7,   46 (1974)
```

2934 U 006000 CONIINE

```
        4
        C
      ╱ ╲
    5C   C3
   8  │   │
   CH2 C6  C2
   ╱ ╲ ╱ ╲1╱
  CH3 CH2 N
   9   7
```

FORMULA	C8H16N			MOL WT	126.22
SOLVENT	C4H8O2				
ORIG ST	C4H12SI			TEMP	AMB

46.50	26.10	24.60	32.40	56.10	39.10
2/3	3/3	4/3	5/3	6/2	7/3
18.10	13.10				
8/3	9/4				

E.WENKERT,J.S.BINDRA,C.—J.CHANG,D.W.COCHRAN,
F.M.SCHELL
ACC CHEM RES   7,   46 (1974)

2935 U 006000 NICOTINE

```
  3C    7C---C8
  ╱ ╲2  │   │
 4C   C--C   C9
 │    ║  6╲ ╱
 5C   C1  N
  ╲ ╱    │
   N     CH3
        10
```

FORMULA	C10H14N2			MOL WT	162.24
SOLVENT	C4H8O2				
ORIG ST	C4H12SI			TEMP	AMB

149.60	138.90	134.20	123.20	148.60	68.70
1/2	2/1	3/2	4/2	5/2	6/2
35.60	22.80	56.80	40.40		
7/3	8/3	9/3	10/4		

L.F.JOHNSON,W.C.JANKOWSKI
CARBON-13 NMR SPECTRA,JOHN
WILEY AND SONS,NEW YORK   381 (1972)

2936 U 006000 GRAMINE

```
    4      10   11
    C      CH2  CH3
   ╱ ╲8   ╱ ╲  ╱
  5C   C---C3  N
  │    ║   ║   │
  6C   C   C2  CH3
   ╲ ╱9╲1╱  12
    C   N
    7   │
        H
```

FORMULA	C11H14N2			MOL WT	174.25
SOLVENT	CHCL3				
ORIG ST	C4H12SI			TEMP	AMB

123.90	112.20	119.00	121.60	119.20	111.10
2/2	3/1	4/2	5/2	6/2	7/2
127.80	136.20	54.30	45.10		
8/1	9/1	10/3	11/4		

E.WENKERT,J.S.BINDRA,C.—J.CHANG,D.W.COCHRAN,
F.M.SCHELL
ACC CHEM RES   7,   46 (1974)

2937 U 006000 N,N—DIMETHYLTRYPTAMINE

```
           12
           CH3
    4    10  │
    C    CH2 N
   ╱ ╲8  ╱ ╲ ╱
  5C   C---C3 CH2 CH3
  │    ║   ║  11  13
  6C   C   C2
   ╲ ╱9╲1╱
    C   N
    7
```

FORMULA	C12H16N2			MOL WT	188.27
SOLVENT	CHCL3				
ORIG ST	C4H12SI			TEMP	AMB

128.20	113.20	118.60	121.50	118.80	111.10
2/2	3/1	4/2	5/2	6/2	7/2
127.30	136.40	23.40	60.30	45.20	
8/1	9/1	10/3	11/3	12/4	

E.WENKERT,J.S.BINDRA,C.—J.CHANG,D.W.COCHRAN,
F.M.SCHELL
ACC CHEM RES   7,   46 (1974)

2938 U 006000 OLIVETOL DIMETHYLETHER

```
  H3C-O
   7  ╲1 2
      C=C
     ╱   ╲3 9
    C     C-CH2
   6╲  ╱   ╲
      C-C   10CH2
  8 ╱5 4   ╱
 H3C-O   11CH2
           ╲
          12CH2
           ╱
         13CH3
```

FORMULA	C13H20O2			MOL WT	208.30
SOLVENT	CCL4				
ORIG ST	CS2			TEMP	AMB

157.70	106.00	144.00	97.30	54.40	36.30
1/1	2/2	3/1	6/2	7/4	9/3
31.60	31.10	22.60	14.00		
10/3	11/3	12/3	13/4		

E.WENKERT,D.W.COCHRAN,F.M.SCHELL,R.A.ARCHER,
K.MATSUMOTO
EXPERIENTIA   28, 250 (1972)

2939 — U 006000 SPIRO-(INDOLIZIDINE-3.3'-OXINDOLE) ISOMER 1

```
        11  13
        C   C
   4   / \ / \
   C  10C  N12 C14
  ⁄ \8  |   |   |
5C   C---C---C   C15
 |   ||  13 17\ ⁄
6C   C   C2    C16
 \ ⁄9\1⁄ \
  C   N   O
  7
```

FORMULA	C15H18N2O			MOL WT	242.32
SOLVENT	CHCL3				
ORIG ST	C4H12SI			TEMP	AMB

182.60	56.60	122.90	122.40	128.00	109.80
2/1	3/1	4/2	5/2	6/2	7/2
134.10	141.70	34.40	55.30	53.80	24.80
8/1	9/1	10/3	11/3	13/3	14/3
24.30	25.60	75.40			
15/3	16/3	17/2			

E.WENKERT,J.S.BINDRA,C.-J.CHANG,D.W.COCHRAN,
F.M.SCHELL
ACC CHEM RES   7, 46 (1974)

2940 — U 006000 SPIRO-(INDOLIZIDINE-3.3'-OXINDOLE) ISOMER 2

```
        11  13
        C   C
   4   / \ / \
   C  10C  N12 C14
  ⁄ \8  |   |   |
5C   C---C---C   C15
 |   ||  13 17\ ⁄
6C   C   C2    C16
 \ ⁄9\1⁄ \
  C   N   O
  7
```

FORMULA	C15H18N2O			MOL WT	242.32
SOLVENT	CHCL3				
ORIG ST	C4H12SI			TEMP	AMB

182.70	57.30	125.10	122.40	127.40	109.70
2/1	3/1	4/2	5/2	6/2	7/2
134.40	140.80	34.70	54.30	53.60	25.20
8/1	9/1	10/3	11/3	13/3	14/3
23.80	26.20	72.10			
15/3	6/3	17/2			

E.WENKERT,J.S.BINDRA,C.-J.CHANG,D.W.COCHRAN,
F.M.SCHELL
ACC CHEM RES   7, 46 (1974)

2941 — U 006000 AGROCLAVINE

```
 H3C   C7  CH3
  18\  ⁄ \6⁄17
    8C   N  H
    ||   | ⁄
    9C   C5
     \ ⁄ \
     10C   C4
      |    |
      C11  C3
     ⁄ \  ⁄ \
   12C   C16 \
    |    ||   C2
   13C   C15 ⁄
     \ ⁄ \ ⁄
      C   N1
      14
```

FORMULA	C16H18N2			MOL WT	238.34
SOLVENT	C5D5N				
ORIG ST	C4H12SI			TEMP	AMB

118.30	111.20	26.40	63.60	60.20	131.90
2/2	3/1	4/3	5/2	7/3	8/1
119.40	40.80	131.90	112.		
9/2	10/2	11/1	12/2	13/2	14/2
134.00	126.60	40.20	19.90		
15/1	16/1	17/4	18/4		

N.J.BACH,H.E.BOAZ,E.C.KORNFELD,C.-J.CHANG,
H.G.FLOSS,E.W.HAGAMAN,E.WENKERT
J ORG CHEM   39, 1272 (1974)

2942 — U 006000 SETOCLAVINE

```
  HO   C7  CH3
   \ ⁄ \6⁄17
 H3C-C8   N
  18 |    |
     C9   C5
     \ ⁄ \
     C10  C4
      |    |
      C11  C3
     ⁄ \  ⁄ \
   12C   C16 \
    |    ||   C2
   13C   C15 ⁄
     \ ⁄ \ ⁄
      C   N1
      14
```

FORMULA	C16H18N2O			MOL WT	254.33
SOLVENT	CDCL3/CD4O				
ORIG ST	C4H12SI			TEMP	303

118.45	110.90	27.30	63.45	66.70	77.10
2/2	3/1	4/3	5/2	7/3	8/1
127.15	136.05	127.80	112.70	123.45	110.00
9/2	10/1	11/1	12/2	13/2	14/2
134.10	126.70	43.46	24.90		
15/1	16/1	17/4	18/4		

A.WALTER,E.BREITMAIER
UNPUBLISHED

2943   U 006000 FESTUCLAVINE

```
  H3C   C7  CH3
   18\  /  \6/17
      C8   N
      |    |
      C9   C5
      \  /  \
      10C    C4
       |     |
      C11   C3
      ⁄ \  / ⩘
   12C    C16 ⩘
    |    ||    C2
   13C    C15 ⁄
     \ ⁄  \ ⁄
       C    N1
       14
```

FORMULA	C16H20N2			MOL WT	240.35
SOLVENT	CDCL3				
ORIG ST	C4H12SI			TEMP	AMB
117.70	110.50	26.60	66.70	65.00	30.20
2/2	3/1	4/3	5/2	7/3	8/2
36.20	40.40	132.70	112.00	122.00	108.30
9/3	10/2	11/1	12/2	13/2	14/2
133.10	125.90	42.70	19.30		
15/1	16/1	17/4	18/4		

N.J.BACH,H.E.BOAZ,E.C.KORNFELD,C.-J.CHANG,
H.G.FLOSS,E.W.HAGAMAN,E.WENKERT
J ORG CHEM   39, 1272 (1974)

2944   U 006000 CHANOCLAVINE

```
        OH
   18⁄   7   17
  H2C    CH3 CH3
   \  /  6⁄
    8C   HN  H
    ||   |⁄
    9C    C5
    \  ⁄  \
     C10  C4
      |    |
     C11  C3
     ⁄ \  / ⩘
   12C    C16 ⩘
    |    ||    C2
   13C    C15 ⁄
     \ ⁄  \ ⁄
       C    N1
       14
```

FORMULA	C16H20N2O			MOL WT	256.35
SOLVENT	CDCL3/CD4O				
ORIG ST	C4H12SI			TEMP	303
119.35	110.45	26.10	68.30	14.30	139.80
2/2	3/1	4/3	5/2	7/4	8/1
126.20	33.55	131.70	116.15	123.10	109.65
9/2	10/2	11/1	12/2	13/2	14/2
134.80	126.75	43.15	62.00		
15/1	16/1	17/4	18/3		

A.WALTER,E.BREITMAIER
UNPUBLISHED

2945   U 006000 FUMIGACLAVINE B

```
  H3C   C7  CH3
   18\  /  \6/17
      C8   N
      |    |
      C9   C5
     /  \  / \
   HO    C10 C4
         |    |
        C11  C3
        ⁄ \  / ⩘
   12C    C16 ⩘
    |    ||    C2
   13C    C15 ⁄
     \ ⁄  \ ⁄
       C    N1
       14
```

FORMULA	C16H20N2O			MOL WT	256.35
SOLVENT	C5D5N				
ORIG ST	C4H12SI			TEMP	AMB
117.90	110.60	26.60	60.70	56.90	35.80
2/1	3/1	4/3	5/2	7/3	8/2
68.10	41.40	130.80	112.90	122.00	108.00
9/2	10/2	11/1	12/2	13/2	14/2
134.00	122.90	42.90	16.50		
15/1	16/1	17/4	18/4		

N.J.BACH,H.E.BOAZ,E.C.KORNFELD,C.-J.CHANG,
H.G.FLOSS,E.W.HAGAMAN,E.WENKERT
J ORG CHEM   39, 1272 (1974)

2946   U 006000 SCOPOLAMINE

```
        H-O
   1   2        \
   C--C    O   12CH2
  ⁄  \  \    ⩘ 11⁄
 7C    \  \  10C-CH
 ⁄|   9\  \3 ⁄    \
 O |  H3C-N8 C-O   13C-C14
  \|    ⁄   ⁄    ⁄  ⩘
  6C   ⁄   ⁄    C18 15C
   \ ⁄  ⁄        \  ⁄
    C--C        17C=C16
    5   4
```

FORMULA	C16H21NO4			MOL WT	291.35
SOLVENT	CHCL3				
ORIG ST	C4H12SI			TEMP	AMB
58.20	31.70	66.60	55.90	43.40	171.70
1/2	2/3	3/2	6/2	9/4	10/1
54.50	63.70	135.90	128.50	127.90	127.40
11/2	12/3	13/1	14/2	15/2	16/2

E.WENKERT,J.S.BINDRA,C.-J.CHANG,D.W.COCHRAN,
F.M.SCHELL
ACC CHEM RES   7,   46 (1974)

2947   U 006000 METHYL LYSERGATE

```
    19CH3
      /
     O
     ‖           17
   18C   C7   CH3
    ⁄  \  \6⁄
  J   C8   N
      |    |
     C9   C5
      \  ⁄ ⁄
     10C   C4
      |    |
      C11 C3
     ⁄ \  ⁄ \
   12C   C16 \
    |    ‖     C2
   13C   C15 ⁄
    \  ⁄ \  ⁄
      C   N1
      14
```

FORMULA	C17H18N2O2			MOL WT	282.35
SOLVENT	CDCL3				
ORIG ST	C4H12SI			TEMP	AMB
118.20	110.20	26.90	62.60	54.60	41.80
2/2	3/1	4/3	5/2	7/3	8/2
117.60	136.00	127.60	112.00	122.90	109.40
9/2	10/1	11/1	12/2	13/2	14/2
133.70	125.90	43.40	172.40	51.90	
15/1	16/1	17/4	18/1	19/4	

N.J.BACH,H.E.BOAZ,E.C.KORNFELD,C.-J.CHANG,
H.G.FLOSS,E.W.HAGAMAN,E.WENKERT
J ORG CHEM   39, 1272 (1974)

2948   U 006000 METHYL 9,10-DIHYDROLYSERGATE

```
    19CH3
      /
     O
     ‖           17
   18C   C7   CH3
    ⁄  \  \6⁄
  O   C8   N
      |    |
     C9   C5
      \  ⁄ ⁄
     C10 C4
      |    |
      C11 C3
     ⁄ \  ⁄ \
   12C   C16 \
    |    ‖     C2
   13C   C15 ⁄
    \  ⁄ \  ⁄
      C   N1
      14
```

FORMULA	C17H20N2O2			MOL WT	284.36
SOLVENT	C2D6OS				
ORIG ST	C4H12SI			TEMP	AMB
118.40	109.90	26.40	66.40	58.30	39.30
2/2	3/1	4/3	5/2	7/3	8/2
30.30	40.70	132.00	112.00	122.00	108.70
9/3	10/2	11/1	12/2	13/2	14/2
133.20	125.80	42.40	173.60	51.50	
15/1	16/1	17/1	18/1	19/4	

N.J.BACH,H.E.BOAZ,E.C.KORNFELD,C.-J.CHANG,
H.G.FLOSS,E.W.HAGAMAN,E.WENKERT
J ORG CHEM   39, 1272 (1974)

2949   U 006000 ELYMOCLAVINE ACETATE

```
      J
     ⁄⁄
    C19
    ⁄ \20
   O   CH3
   |         17
  H2C   C7   CH3
  18\  ⁄ \6⁄
   8C   N
   ‖    |
  9C   C5
    \  ⁄ ⁄
   10C   C4
    |    |
    C11 C3
   ⁄ \  ⁄ \
 12C   C16 \
  |    ‖     C2
 13C   C15 ⁄
  \  ⁄ \  ⁄
    C   N1
    14
```

FORMULA	C18H20N2O2			MOL WT	296.37
SOLVENT	CDCL3				
ORIG ST	C4H12SI			TEMP	AMB
117.90	111.30	26.40	63.40	56.80	130.90
2/2	3/1	4/3	5/2	7/3	8/1
124.80	40.50	131.30	112.20	122.60	108.70
9/2	10/2	11/1	12/2	13/2	14/2
133.40	126.10	40.50	66.20	170.70	20.60
15/1	16/1	17/4	18/3	19/1	20/4

N.J.BACH,H.E.BOAZ,E.C.KORNFELD,C.-J.CHANG,
H.G.FLOSS,E.W.HAGAMAN,E.WENKERT
J ORG CHEM   39, 1272 (1974)

2950   U 006000 ERGOBASINE

```
  20CH2-OH
      |
    HC19
    ⁄ \21
  HN   CH3
   |
  18C   C7  CH3
  ⁄ \  ⁄ \6⁄17
 J   C8   N
     |    |
    C9   C5
     \  ⁄ ⁄
    10C   C4
     |    |
     C11 C3
    ⁄ \  ⁄ \
  12C   C16 \
   |    ‖     C2
  13C   C15 ⁄
   \  ⁄ \  ⁄
     C   N1
     14
```

FORMULA	C19H23N3O2			MOL WT	325.41
SOLVENT	C2D6OS				
ORIG ST	C4H12SI			TEMP	AMB
119.10	108.90	26.80	62.60	55.50	42.80
2/2	3/1	4/3	5/2	7/3	8/2
120.10	135.00	127.40	111.00	122.40	109.00
9/2	10/1	11/1	12/2	13/2	14/2
133.70	125.80	43.40	171.20	46.40	64.40
15/1	16/1	17/4	18/1	19/2	20/3
17.40					
21/4					

N.J.BACH,H.E.BOAZ,E.C.KORNFELD,C.-J.CHANG,
H.G.FLOSS,E.W.HAGAMAN,E.WENKERT
J ORG CHEM   39, 1272 (1974)

2951   U 006000   ERGOBASININE

```
   20CH2-OH
     |
    HC19
    / \21
  HN   CH3
   |       17
  18C  C7 CH3
  / \ / \6/
 D   C8  N
     |   |
    C9  C5
     \ / \
   10C   C4
    |   |
   C11 C3
   / \ / \
 12C   C16 \
  |   ||   C2
 13C  C15 /
  \ / \ /
   C   N1
   14
```

FORMULA	C19H23N3O2			MOL WT	325.41
SOLVENT	C2D6OS				
ORIG ST	C4H12SI			TEMP	AMB

119.00	108.90	26.90	62.00	54.00	42.20
2/2	3/1	4/3	5/2	7/3	8/2
119.00	136.10	127.60	111.00	122.10	109.80
9/2	10/1	11/1	12/2	13/2	14/2
133.70	125.70	43.60	172.10	46.20	64.30
15/1	16/1	17/4	18/1	19/2	20/3
17.20					
21/4					

N.J.BACH,H.E.BOAZ,E.C.KORNFELD,C.-J.CHANG,
H.G.FLOSS,E.W.HAGAMAN,E.WENKERT
J ORG CHEM   39, 1272 (1974)

2952   U 006000   CORYNANTHEIDINE DERIVATIVE

```
    4      10
    C      C
   / \8   / \
  5C  C---C3 C11
  |   ||  ||  |
  6C  C   2C  N12
   \ /9\1/ \ / \
    C  N 17C   C13
    7    |   |   19
        16C   C14 CH3
         \ / \ /
         15C  CH2
          |  18
         20CH2 O
           \ /
           21C
            |
            H
```

FORMULA	C19H24N2O			MOL WT	296.42
SOLVENT	CHCL3				
ORIG ST	C4H12SI			TEMP	AMB

135.40	108.60	118.30	121.40	119.70	110.90
2/1	3/1	4/2	5/2	6/2	7/2
127.80	136.50	22.00	53.60	57.90	40.20
8/1	9/1	10/3	11/3	13/3	14/2
34.60	32.10	60.20	18.60	12.50	47.90
15/2	16/3	17/2	18/3	19/4	20/3
202.00					
21/2					

E.WENKERT,J.S.BINDRA,C.-J.CHANG,D.W.COCHRAN,
F.M.SCHELL
ACC CHEM RES   7,   46 (1974)

2953   U 006000   RHYNCOPHYLLAL

```
      11  13  18
      C   C   CH2
   4   \ / \ / \
   C  10C   N  C14 CH3
  / \8  |   |   |  19
 5C  C---C---C  C15
 |   ||  13 17\ / \
 6C  C   C2  16C 20CH2
  \ /9\1/ \     \ / \
   C  N   O     21C
   7               / \
             H   O
```

FORMULA	C19H24N2O2			MOL WT	312.42
SOLVENT	CHCL3				
ORIG ST	C4H12SI			TEMP	AMB

181.70	56.10	123.00	122.60	128.00	109.70
2/1	3/1	4/2	5/2	6/2	7/2
145.60	141.30	34.90	54.70	57.60	41.10
8/1	9/1	10/3	11/3	13/3	14/2
35.40	32.00	74.50	23.80	10.80	47.90
15/2	16/3	17/2	18/3	19/4	20/3
202.20					
21/2					

E.WENKERT,J.S.BINDRA,C.-J.CHANG,D.W.COCHRAN,
F.M.SCHELL
ACC CHEM RES   7,   46 (1974)

2954   U 006000   GELSEDINE

```
   20C-CH2-CH3
   / \19 18
  /    \
 N      \
  \  16C-C15
   \ / \ / \
   5C 17C  \
  9  |   |  C14
  C  6C  O /
  / \8 \7 1/
10C  C---C-C3
 |   ||  |
11C  C13 C2
 \ / \ / \
  C   N   O
 12  | 21
    O-CH3
```

FORMULA	C19H24N2O3			MOL WT	328.41
SOLVENT	CCL4				
ORIG ST	CS2			TEMP	AMB

174.80	74.70	65.70	34.10	53.10	132.10
2/1	3/2	5/2	6/3	7/1	8/1
125.60	123.80	128.20	107.30	138.40	21.60
9/2	10/2	11/2	12/2	13/1	14/3
34.90	42.10	64.00	12.10	21.60	59.80
15/2	16/2	17/3	18/4	19/3	20/2
63.50					
21/4					

E.WENKERT,C.-J.CHANG,D.W.COCHRAN,R.PELLICCIARI
EXPERIENTIA   28,   377 (1972)

2955 — U 006000 GELSEMINE

```
        17    5
         C   C-------
        /·\ /·\ /  22  |
       /  16C    N-CH3  |
      /    |    |       |
     O    15C   C21     |
      \  14/·\ ·/  18   |
       \  C  20C-CH=CH2 |
      9  \|    | 19     |
        C   C3  C-------
       // \8  \ /6
    10C    C---C7
     |    ||   |
    11C    C13 C2
     \\ / ·\ / \\
        C   N   O
        12
```

FORMULA	C20H22N2O2			MOL WT	322.41
SOLVENT	CCL4				
ORIG ST	CS2			TEMP	AMB

179.40	69.60	72.10	40.60	54.10	132.20
2/1	3/2	5/2	6/2	7/1	8/1
128.10	121.80	128.40	109.10	140.70	23.00
9/2	10/2	11/2	12/2	13/1	14/3
38.20	36.00	61.50	112.30	138.90	54.10
15/2	16/2	17/3	18/2	19/3	20/1
66.30	50.80				
21/3	22/4				

E.WENKERT,C.-J.CHANG,D.W.COCHRAN,R.PELLICCIARI
EXPERIENTIA   28, 377 (1972)

2956 — U 006000 QUINIDINE

```
          15C   19
          /·\    CH
         / · \ / \\
      14C  C16C20 CH2
       |   |   |  18
       |   |   |
      HO  H C3 C17C21
       \·|/·\  | /
          C2   \·|/
      22  9  |   \\
    H3C-O  C   C7  4
       \ // \8·/
      10C   C   C6
       |   ||   |
      11C    C13 C5
       \\ / \·1/
         12C   N
```

FORMULA	C20H24N2O2			MOL WT	324.43
SOLVENT	CHCL3				
ORIG ST	C4H12SI			TEMP	AMB

71.70	60.70	147.30	121.30	148.30	126.80
2/2	3/2	5/2	6/2	7/1	8/1
101.80	157.70	118.50	131.30	144.00	21.20
9/2	10/1	11/2	12/2	13/1	14/3
28.40	26.70	49.50	114.50	140.60	40.10
15/2	16/3	17/3	18/3	19/2	20/2
50.00	55.60				
21/3	22/4				

E.WENKERT,J.S.BINDRA,C.-J.CHANG,D.W.COCHRAN,
F.M.SCHELL
ACC CHEM RES   7,   46 (1974)

2957 — U 006000 QUINIDINE

```
          15C
   21 22      /·\
  R -CH=CH2  / · \
           / · \
     18C  RC14 C16
       |   |   |
      HO 19C   C13 C17
       \ / \ · | /
      20CH  \ · |/
    11  5  |   \·|/
   H3C-O  C  4C   N
      \ // \9/· \\  12
     6C   C   C3
       |   ||   |
      7C 10C   C2
       \\ / \ / \\
        8C   1N
```

FORMULA	C20H24N2O2			MOL WT	324.43
SOLVENT	C2D6OS				
ORIG ST	C4H12SI			TEMP	AMB

147.52	120.97	149.46	102.50	156.83	119.81
2/2	3/2	4/1	5/2	6/1	7/2
131.16	127.10	143.95	55.47	49.20	39.89
8/2	9/1	10/1	11/4	13/3	14/2
27.94	26.37	48.56	23.28	60.61	70.91
15/2	16/3	17/3	18/3	19/2	20/2
141.37	114.41				
21/2	22/3				

F.I.CARROLL,D.SMITH,M.E.WALL,C.G.MORELAND
J MED CHEM   17, 985 (1974)

2958 — U 006000 QUININE

```
        18
       O-CH3
   13  /
   C-C14
  12/  \\
  C    C15
   \  /
 17C=C16    OH
  / \\9 /
 N    C-C8
  \\ //  \·1 2
11C-C10   C-C
   /·6 7\
  N-C-C-C3
   \  /
   5C-C4
     \
      CH=CH2
      19 20
```

FORMULA	C20H24N2O2			MOL WT	324.43
SOLVENT	CDCL3				
ORIG ST	C4H12SI			TEMP	AMB

60.10	27.70	27.90	40.00	57.00	43.20
1/2	2/3	3/2	4/2	5/3	6/3
21.60	71.80	143.50	118.40	147.20	131.10
7/3	8/2	9/1	10/2	11/2	12/2
121.20	157.60	101.60	126.60	143.90	55.60
13/2	14/1	15/2	16/1	17/1	18/4
141.90	114.20				
19/2	20/3				

L.F.JOHNSON,W.C.JANKOWSKI
CARBON-13 NMR SPECTRA,JOHN
WILEY AND SONS,NEW YORK   486 (1972)

2959 U 006000 QUININE

```
        15C    19
        /I\    CH      FORMULA   C20H24N2O2           MOL WT   324.43
       / I \ / \       SOLVENT   CHCL3
    14C  C16C20 CH2    ORIG ST   C4H12SI              TEMP        AMB
     I   I   I   18
     I   I   I           71.40    60.00   147.00   121.20   148.60   126.60
    HO  H C3 C17C21       2/2      3/2      5/2      6/2      7/1      8/1
     \I/ \ I /          ·101.70   158.50   118.50   130.80   143.90    20.70
      C2   \I/            9/2     10/1     11/2     12/2     13/1     14/3
  22    9   I    N        27.70    27.70    43.10   113.70   141.70    39.90
 H3C-O   C   C7   4       15/2     16/3     17/3     18/3     19/1     20/2
   \ / \8/ \             56.90    55.50
   10C   C   C6           21/3    22/4
    I    II   I
   11C   C13 C5        E.WENKERT,J.S.BINDRA,C.-J.CHANG,D.W.COCHRAN,
    \ / \1/            F.M.SCHELL
    12C    N           ACC CHEM RES                    7,   46 (1974)
```

2960 U 006000 3-HYDROXYQUINIDINE

```
              15C
      21 22   /I\       FORMULA   C20H24N2O3           MOL WT   340.43
  R -CH=CH2  / I \      SOLVENT   C2D6OS
            / I14\      ORIG ST   C4H12SI              TEMP        AMB
  R1 -OH  18C RCR1 C16
    I    I   I          147.87   121.56   149.37   102.46   157.37   119.16
   HO 19C  C13 C17       2/2      3/2      4/1      5/2      6/1      7/2
    \ / \  I /          131.41   127.00   144.00    55.86    57.08    71.00
    20CH  \ I /          8/2      9/1     10/1     11/4     13/3     14/1
  11   5    I    \I/     33.62    20.69    49.25    24.11    59.29    71.10
 H3C-O   C   4C   N      15/2     16/3     17/3     18/3     19/2     20/2
   \ / \9/ \    12      144.24   112.65
   6C    C   C3          21/2    22/3
    I    II   I
   7C 10C    C2
    \ / \ /
     8C   IN          F.I.CARROLL,D.SMITH,M.E.WALL,C.G.MORELAND
                      J MED CHEM                      17,  985 (1974)
```

2961 U 006000 QUINIDINE NB-OXIDE

```
        15C    19
        /I\    CH      FORMULA   C20H24N2O3           MOL WT   340.43
       / I \ / \       SOLVENT   CHCL3
    14C  C16C20 CH2    ORIG ST   C4H12SI              TEMP        AMB
     I   I   I   18
     I   I   I           72.90    62.80   147.40   122.10   147.40   125.80
    HO  H C3 C17C21       2/2      3/2      5/2      6/2      7/1      8/1
     \I/ \ I /           100.40   157.70   119.20   130.90   143.70    20.20
      C2   \I/            9/2     10/1     11/2     12/2     13/1     14/3
  22    9   I    N+       27.70    26.60    63.70   111.60   147.40    41.30
 H3C-O   C   C7   I       15/2     16/3     17/3     18/3     19/2     20/2
   \ / \8/ \    O-       65.40    54.60
   10C   C   C6           21/3    22/4
    I    II   I
   11C   C13 C5        E.WENKERT,J.S.BINDRA,C.-J.CHANG,D.W.COCHRAN,
    \ / \1/            F.M.SCHELL
    12C    N           ACC CHEM RES                    7,   46 (1974)
```

2962 U 006000 QUININE NB-OXIDE

```
        15C    19
        /I\    CH      FORMULA   C20H24N2O3           MOL WT   340.43
       / I \ / \       SOLVENT   CHCL3
    14C  C16C20 CH2    ORIG ST   C4H12SI              TEMP        AMB
     I   I   I   18
     I   I   I           73.20    62.90   147.20   122.00   147.20   126.00
    HO  H C3 C17C21       2/2      3/2      5/2      6/2      7/1      8/1
     \I/ \ I /           100.60   157.80   119.00   131.00   143.90    20.20
      C2   \I/            9/2     10/1     11/2     12/2     13/1     14/3
  22    9   I    N+       27.20    27.20    58.90   116.40   138.30    40.90
 H3C-O   C   C7   I       15/2     16/3     17/3     18/3     19/2     20/2
   \ / \8/ \    O-       70.80    54.90
   10C   C   C6           21/3    22/4
    I    II   I
   11C   C13 C5        E.WENKERT,J.S.BINDRA,C.-J.CHANG,D.W.COCHRAN,
    \ / \1/            F.M.SCHELL
    12C    N           ACC CHEM RES                    7,   46 (1974)
```

2963   U 006000   TABERSONINE

```
        3
        C
    5  / \
    C---N  C14
  9  |   |   ||
  C  6C  C21 C15
  // \8  \7/ \ /
10C   C---C  C20
  |   ||  |\
11C  C13 C 17C C12-CH3
  \ / \1/2\ / 19  18
   C   N  C16
  12        |
          C22
         // \ 23
       J   J-CH3
```

FORMULA	C20H24N2O2			MOL WT	324.43
SOLVENT	CHCL3				
ORIG ST	C4H12SI			TEMP	AMB

166.70	50.30	50.80	44.30	55.00	137.80
2/1	3/3	5/3	6/3	7/1	8/1
121.40	120.50	127.60	109.20	143.10	124.80
9/2	10/2	11/2	12/2	13/1	14/2
132.90	92.20	26.70	7.30	28.40	41.20
15/2	16/1	17/3	18/4	19/3	20/1
69.90	168.80	50.80			
21/2	22/1	23/4			

E.WENKERT,D.W.COCHRAN,E.W.HAGAMAN,F.M.SCHELL,
N.NEUSS,A.S.KATNER,P.POTIER,C.KAN,M.PLAT,
M.KOCH,H.MEHRI,J.POISSON,N.KUNESCH,Y.ROLLAND
J AM CHEM SOC   95, 4990 (1973)

2964   U 006000   2-QUINIDINONE

```
        15C
        /|\
   21 22 / | \
R -CH=CH2 / | \
      18C  RC14 C16
       |   |   |
      HJ 19C  C13 C17
       \ / \  | /
      20CH  \ | /
  11   5  |  \|/
 H3C-J  C  4C  N
   \ // \9/ \  12
   6C  C  C3
    |  ||  |
   7C 10C  C2
    \ / \ / \
    8C  1N  J
```

FORMULA	C20H24N2O3			MOL WT	340.43
SOLVENT	C2D6OS				
ORIG ST	C4H12SI			TEMP	AMB

161.74	118.67	153.70	106.96	154.04	119.06
2/1	3/2	5/1	5/2	6/1	7/2
119.06	117.20	133.56	55.67	49.40	39.89
8/2	9/1	10/1	11/4	13/3	14/2
28.04	26.47	48.56	22.74	59.78	71.49
15/2	16/3	17/3	18/3	19/3	20/3
141.40	114.60				
21/2	22/4				

F.I.CARROLL,D.SMITH,M.E.WALL,C.G.MORELAND
J MED CHEM   17, 985 (1974)

2965   U 006000   LYSERGIC ACID N,N-DIETHYLAMIDE

```
  19   21
  H2C  CH2
  / \ / \
H3C20 N 22CH3
    |   17
  C18 C7  CH3
  // \ / \6/
  J  C8  N
     |   |
    C9  C5
     \ / \
     C10 C4
      |   |
     C11 C3
     // \ /
   12C  C16 \
    |   ||  C2
  13C  C15 /
    \ / \ /
     C   N1
    14
```

FORMULA	C20H25N3O			MOL WT	323.44
SOLVENT	CD4O				
ORIG ST	C4H12SI			TEMP	303

118.40	110.85	27.45	63.30	56.16	40.35
2/2	3/1	4/3	5/2	7/3	8/2
119.70	136.30	128.10	112.75	123.40	109.70
9/2	10/1	11/1	12/2	13/2	14/2
134.10	126.35	43.95	171.80	40.00	13.25
15/1	16/1	17/4	18/1	19/3	20/4
42.05	14.95				
21/3	22/4				

A.WALTER,E.BREITMAIER
UNPUBLISHED

2966   U 006000   18,19-DIHYDROQUINIDINE

```
     15C   19
     /|\   C12
    / | \ / \
   14C  C16C20 CH3
    |   |   |  18
    |   |   |
   HJ  H C3 C17C21
    \|/ \ | /
    C2  \|/
 22  9  |  N
H3C-J  C  C7  4
  \ // \8/ \
  10C  C  C6
   |   ||  |
  11C  C13 C5
   \ / \1/
   12C  N
```

FORMULA	C20H26N2O2			MOL WT	326.44
SOLVENT	CHCL3				
ORIG ST	C4H12SI			TEMP	AMB

71.50	59.80	147.20	121.40	148.80	126.60
2/2	3/2	5/2	6/2	7/1	8/1
101.60	157.60	118.60	131.00	144.00	20.60
9/2	10/1	11/2	12/2	13/1	14/3
26.00	26.70	51.10	12.00	25.10	37.40
15/2	16/3	17/3	18/4	19/3	20/2
50.10	55.50				
21/3	22/4				

E.WENKERT,J.S.BINDRA,C.-J.CHANG,D.W.COCHRAN,
F.M.SCHELL
ACC CHEM RES   7, 46 (1974)

2967

```
                  U 006000  18,19-DIHYDROQUININE
         15C    19
         /I\    CH2      FORMULA   C20H26N2O2          MOL WT   326.44
        / I \  / \       SOLVENT   CHCL3
      14C  C16C20 CH3    ORIG ST   C4H12SI             TEMP      AMB
       I   I   I  18
       I   I   I          71.20    59.60   146.80   120.70   148.70   126.40
      HD  H C3 C17C21      2/2      3/2     5/2      6/2      7/1      8/1
       \I/ \  I /         101.60   157.40  118.20   130.60   143.60   20.60
         C2   \I/          9/2     10/1    11/2     12/2     13/1     14/3
   22    9   I    \N       25.20    28.00   42.90    11.60    27.30    37.30
 H3C-D    C   C7   4       15/2     16/3    17/3     18/4     19/3     20/2
   \ / \8/ \             58.20    55.30
    10C   C   C6          21/3     22/4
     I    II  I
     11C   C13 C5         E.WENKERT,J.S.BINDRA,C.-J.CHANG,D.W.COCHRAN,
      \ / \1/             F.M.SCHELL
       12C   N            ACC CHEM RES                     7,   46 (1974)
```

2968

```
               U 006000  VINCADIFFORMINE
          3
          C              FORMULA   C20H26N2O2          MOL WT   326.44
     5   / \             SOLVENT   CHCL3
     C---N   C14         ORIG ST   C4H12SI             TEMP      AMB
   9  I   I   I
   C  6C   C21 C15        167.80   50.30   50.80    44.30    55.00   .138.00
  / \8 \7/ \ /             2/1      3/3     5/3      6/3      7/1      8/1
 10C   C---C   C20        121.00   120.50  127.40   109.30   143.40   22.20
  I    II   I\             9/2     10/2    11/2     12/2     13/1     14/3
 11C   C13 C 17C CH2-CH3   32.90    92.80   25.60     7.30    29.30    38.20
  \ / \1/2\ / 19  18      15/3     16/1    17/3     18/4     19/3     20/1
   C   V   C16            72.70   169.20   50.90
  12       I             21/2    22/1     23/4
          C22
        / \ 23           E.WENKERT,D.W.COCHRAN,E.W.HAGAMAN,F.M.SCHELL,
     D   O-CH3           N.NEUSS,A.S.KATNER,P.POTIER,C.KAN,M.PLAT,
                         M.KOCH,H.MEHRI,J.POISSON,N.KUNESCH,Y.ROLLAND
                         J AM CHEM SOC                    95, 4990 (1973)
```

2969

```
                  U 006000  18,19-DIHYDROQUININE NB-OXIDE
         15C    19
         /I\    CH2      FORMULA   C20H26N2O3          MOL WT   342.44
        / I \  / \       SOLVENT   CHCL3
      14C  C16C20 CH3    ORIG ST   C4H12SI             TEMP      AMB
       I   I   I  18
       I   I   I          73.20    63.20   147.10   121.90   147.20   125.90
      HD  H C3 C17C21      2/2      3/2     5/2      6/2      7/1      8/1
       \I/ \  I /         100.80   157.80  119.10   131.10   143.80   20.20
         C2   \I/          9/2     10/1    11/2     12/2     13/1     14/3
   22    9   I    N+       25.20    27.70   59.00    11.40    27.30    39.30
 H3C-D    C   C7   I       15/2     16/3    17/3     18/4     19/3     20/2
   \ / \8/ \   O-         73.20    55.10
    10C   C   C6          21/3     22/4
     I    II  I
     11C   C13 C5         E.WENKERT,J.S.BINDRA,C.-J.CHANG,D.W.COCHRAN,
      \ / \1/             F.M.SCHELL
       12C   N            ACC CHEM RES                     7,   46 (1974)
```

2970

```
                  U 006000  18,19-DIHYDROQUINIDINE NB-OXIDE
         15C   '19
         /I\    CH2      FORMULA   C20H26N2O3          MOL WT   342.44
        / I \  / \       SOLVENT   CHCL3
      14C  C16C20 CH3    ORIG ST   C4H12SI             TEMP      AMB
       I   I   I  18
       I   I   I          72.70    62.60   147.60   121.70   147.60   125.70
      HD  H C3 C17C21      2/2      3/2     5/2      6/2      7/1      8/1
       \I/ \  I /         100.00   157.40  119.10   130.80   143.50   19.80
         C2   \I/          9/2     10/1    11/2     12/2     13/1     14/3
   22    9   I    N+       25.30    27.00   65.50    11.60    24.90    38.80
 H3C-D    C   C7   I       15/2     16/3    17/3     18/4     19/3     20/2
   \ / \8/ \   D-         65.50    54.60
    10C   C   C6          21/3     22/4
     I    II  I
     11C   C13 C5         E.WENKERT,J.S.BINDRA,C.-J.CHANG,D.W.COCHRAN,
      \ / \1/             F.M.SCHELL
       12C   N            ACC CHEM RES                     7,   46 (1974)
```

2971   U 006000   GELSEVIRINE

```
        17    5
        C    C--------
       / \  / \  22  |
      /  16C   N-CH3  |
     /     |     |    |
    O    15C    C21   |
     \  14/ \  /  18  |
      \  C  20C-CH=CH2|
    9  \|   |  19     |
     C   C3  C--------
    // \8  \ /6
  10C    C---C7
   |     ||   |
  11C   C13 C2
    \ / \ / \
     C   N   O
    12   | 23
        O-CH3
```

FORMULA	C21H24N2O3		MOL WT	352.44
SOLVENT	CCL4			
ORIG ST	CS2		TEMP	AMB

173.20	69.60	72.50	40.70	52.50	128.30
2/1	3/2	5/2	6/2	7/1	8/1
128.30	122.80	128.30	107.40	139.80	23.30
9/2	10/2	11/2	12/2	13/1	14/3
38.20	36.30	61.60	113.20	138.50	54.30
15/2	16/2	17/3	18/3	19/2	20/1
66.30	50.80	63.20			
21/3	22/4	23/4			

E.WENKERT,C.-J.CHANG,D.W.COCHRAN,R.PELLICCIARI
EXPERIENTIA   28,   377 (1972)

2972   U 060000   N-A-METHYLGELSEMINE

```
        17    5
        C    C--------
       / \  / \  22  |
      /  16C   N-CH3  |
     /     |     |    |
    O    15C    C21   |
     \  14/ \  /  18  |
      \  C    C-CH=CH2|
    9  \|   |  19     |
     C   C3  C--------
    // \8  \ /6
  10C    C---C7
   |     ||   |
  11C   C13 C2
    \ / \ / /
     C   N   O
    12   |
        23CH3
```

FORMULA	C21H24N2O2		MOL WT	336.44
SOLVENT	CCL4			
ORIG ST	CS2		TEMP	AMB

176.90	69.60	72.20	40.70	53.80	131.40
2/1	3/2	5/2	6/2	7/1	8/1
128.10	121.90	128.10	107.60	143.30	22.80
9/2	10/2	11/2	12/2	13/1	14/3
38.20	35.80	61.50	111.90	139.10	54.00
15/2	16/2	17/3	18/3	19/2	20/1
66.30	50.60	26.10			
21/3	22/4	23/4			

E.WENKERT,C.-J.CHANG,D.W.COCHRAN,R.PELLICCIARI
EXPERIENTIA   28,   377 (1972)

2973   U 006000   CIS-DELTA-8-TETRAHYDROCANNABINOL

```
   22
  H3C    HO      \1 2
   \  10      \1 2
    9C-C      C=C
   // \   /  \3 15
   C   13C-C14   C-CH2
  8\  /   \  /  \
   C-C12 11C-C   16CH2
  7 \6 5/  4    /
    C-O     17CH2
   / \       \
  H3C   CH3   18CH2
  20   21      /
           19CH3
```

FORMULA	C21H30O2		MOL WT	314.47
SOLVENT	CCL4			
ORIG ST	CS2		TEMP	AMB

154.70	109.60	142.90	102.60	75.80	28.50
1/1	2/2	3/1	4/2	6/1	7/3
120.30	134.30	39.10	154.90	46.70	27.60
8/2	9/1	10/3	11/1	12/2	13/2
116.20	33.50	32.10	31.00	22.60	14.20
14/1	15/3	16/3	17/3	18/3	19/4
22.60	18.30	23.60			
20/4	21/4	22/4			

E.WENKERT,D.W.COCHRAN,F.M.SCHELL,R.A.ARCHER,
K.MATSUMOTO
EXPERIENTIA   28,   250 (1972)

2974   U 006000   TRANS-DELTA-8-TETRAHYDROCANNABINOL

```
   22
  H3C    HO      \1 2
   \  10      \1 2
    9C-C      C=C
   // \   /  \3 15
   C   13C-C14   C-CH2
  8\  /   \  /  \
   C-C12 11C-C   16CH2
  7 \6 5/  4    /
    C-O     17CH2
   / \       \
  H3C   CH3   18CH2
  20   21      /
           19CH3
```

FORMULA	C21H30O2		MOL WT	314.47
SOLVENT	CCL4			
ORIG ST	CS2		TEMP	AMB

154.20	110.10	141.50	107.30	76.20	31.60
1/1	2/2	3/1	4/2	6/1	7/3
118.70	134.30	36.10	154.60	45.00	27.90
8/2	9/1	10/3	11/1	12/2	13/2
110.30	35.60	31.60	30.60	22.60	14.10
14/1	15/3	16/3	17/3	18/3	19/4
18.50	27.60	23.50			
20/4	21/4	22/4			

E.WENKERT,D.W.COCHRAN,F.M.SCHELL,R.A.ARCHER,
K.MATSUMOTO
EXPERIENTIA   28,   250 (1972)

2975   U 006000 DELTA-9-TETRAHYDROCANNABINOL

```
    22
H3C        HO
   \ 10    \1 2
   9C=C      C=C
   /   \    / \3 15
  C  13C-C14  C-CH2
 8\  /  \  /  \
  C-C12 11C-C  16CH2
  7  \6 5/  4   /
      C-)    17CH2
     / \       \
  H3C   CH3  18CH2
   20   21      /
            19CH3
```

FORMULA C21H30O2				MOL WT	314.47
SOLVENT CCL4					
ORIG ST CS2				TEMP	AMB
154.00	109.60	141.40	107.40	76.60	25.10
1/1	2/2	3/1	4/2	6/1	7/3
31.50	131.90	124.70	154.00	45.70	33.80
8/3	9/1	10/2	11/1	12/2	13/2
108.90	35.50	31.50	30.50	22.50	14.10
14/1	15/3	16/3	17/3	18/3	19/4
19.30	27.60	23.30			
20/4	21/4	22/4			

E.WENKERT,D.W.COCHRAN,F.M.SCHELL,R.A.ARCHER,
K.MATSUMOTO
EXPERIENTIA   28,   250 (1972)

2976   U 006000 7,8,9,10-TETRAHYDROCANNABINOL

```
    22
H3C        HO
   \ 10    \1 2
   9C-C      C=C
   /   \   /  \3 15
  C  13C-C14  C-CH2
 8\  /  \  /  \
  C-C12 11C-C  16CH2
  7  \6 5/  4   /
      C-)    17CH2
   20/ \21     \
  H3C   CH3  18CH2
                /
            19CH3
```

FORMULA C21H30O2				MOL WT	314.47
SOLVENT CCL4					
ORIG ST CS2				TEMP	AMB
152.30	109.50	142.70	109.50	77.30	25.50
1/1	2/2	3/1	4/2	6/1	7/3
30.70	28.90	36.80	153.90	124.00	131.70
8/3	9/2	10/3	11/1	12/1	13/1
110.70	35.50	31.50	30.50	22.50	14.10
14/1	15/3	16/3	17/3	18/3	19/4
20/4	21/4	22/4			

E.WENKERT,D.W.COCHRAN,F.M.SCHELL,R.A.ARCHER,
K.MATSUMOTO
EXPERIENTIA   28,   250 (1972)

2977   U 006000 CORYNANTHEINE

```
     4      10
     C       C
    / \8    / \
  5C   C---C3  C11
   |    ||   ||  |
  6C    C   2C  N12
   \  /9\1/  \ / \
    C   N 17C   C13
    7   |   |   |  19
       16C   C14 CH2
        \  / \  /
       15C    CH
       24  |  18
     H3C-) 20C   22CH3
        \ /  \  /
       23C    CH-)
         ||  21
         O
```

FORMULA C22H26N2O3				MOL WT	366.46
SOLVENT CHCL3					
ORIG ST C4H12SI				TEMP	AMB
135.20	107.50	117.90	120.90	119.00	110.80
2/1	3/1	4/2	5/2	6/2	7/2
127.40	136.20	21.80	52.60	61.30	42.40
8/1	9/1	10/3	11/3	13/3	14/2
38.80	33.10	59.90	139.20	115.40	111.70
15/2	16/3	17/2	18/2	19/3	20/1
159.80	61.30	168.90	51.10		
21/2	22/4	23/1	24/4		

E.WENKERT,J.S.BINDRA,C.-J.CHANG,D.W.COCHRAN,
F.M.SCHELL
ACC CHEM RES   7,   46 (1974)

2978   U 006000 VANDRIKIDINE

```
         3
         C
    5   / \
   C---N  C14
  9 |   |   ||
  C  6C  C21 C15
 / \8 \7/ \ /
10C  C---C  C20
 |   ||  |  |\
11C  C13 C 17C  CH-CH3
 / \ / \1/2\ / 19\ 18
 O   C  N  C16  OH
 | 12       |
CH3      C22
24     / \ 23
     )   )-CH3
```

FORMULA C22H26N2O4				MOL WT	382.46
SOLVENT CDCL3					
ORIG ST C4H12SI				TEMP	AMB
166.00	49.90	50.80	44.20	54.80	130.40
2/1	3/3	5/3	6/3	7/1	8/1
122.00	105.20	159.90	96.60	144.00	127.60
9/2	10/2	11/1	12/2	13/1	14/2
129.60	90.80	28.10	17.70	67.90	46.00
15/2	16/1	17/3	18/4	19/2	20/1
66.30	168.30	50.80	55.30		
21/2	22/1	23/4	24/4		

E.WENKERT,D.W.COCHRAN,E.W.HAGAMAN,F.M.SCHELL,
N.NEUSS,A.S.KATNER,P.POTIER,C.KAN,M.PLAT,
M.KOCH,H.MEHRI,J.POISSON,N.KUNESCH,Y.ROLLAND
J AM CHEM SOC   95, 4990 (1973)

2979   U 006000 VANDRIKINE

```
        3
        C
     5  / \
     C---N  C14
     9  | |   |
     C  6C 21C 15C-O
   / \8 \7/ \ / |
 10C  C---C  C20 C18
   |   ||  |  |\ /
 11C  C13 C 17C C19
  / \ / \ /2\ /
 O   C  N  C16
 |  12
 CH3        C22
24      / \ 23
       J  O-CH3
```

FORMULA	C22H26N2O4			MOL WT	382.46
SOLVENT	CHCL3				
ORIG ST	C4H12SI			TEMP	AMB
167.40	45.70	51.20	45.10	54.20	130.50
2/1	3/3	5/3	6/3	7/1	8/1
121.50	104.80	159.80	96.50	144.10	27.40
9/2	10/2	11/1	12/2	13/1	14/3
79.80	93.90	26.60	64.70	34.60	46.40
15/2	16/1	17/3	18/3	19/3	20/1
68.70	168.50	50.80	55.20		
21/2	22/1	23/4	24/4		

E.WENKERT,D.W.COCHRAN,E.W.HAGAMAN,F.M.SCHELL,
N.NEUSS,A.S.KATNER,P.POTIER,C.KAN,M.PLAT,
M.KOCH,H.MEHRI,J.POISSON,N.KUNESCH,Y.ROLLAND
J AM CHEM SOC   95, 4990 (1973)

2980   U 006000 CORYNANTHEIDINE

```
   4     10
   C      C
 / \8   / \
5C   C---C3  C11
 |   ||   ||  |
6C   C  2C   C12
 \ /9\1/ \ / \
   C  N 17C   C13
   7  |   |  19
     16C   C14 CH3
       \ / \ /
      15C   CH2
    24  |  18
   H3C-O 20C   22CH3
       \ / \  /
      23C   C4-O
       || 21
        O
```

FORMULA	C22H28N2O3			MOL WT	368.48
SOLVENT	CHCL3				
ORIG ST	C4H12SI			TEMP	AMB
136.00	107.90	117.90	121.00	119.20	110.90
2/1	3/1	4/2	5/2	6/2	7/2
127.70	136.20	21.90	53.40	57.90	40.00
8/1	9/1	10/3	11/3	13/3	14/2
40.80	29.80	61.20	19.10	12.80	111.80
15/2	16/3	17/2	18/3	19/4	20/1
160.70	61.20	169.50	51.20		
21/2	22/4	23/1	24/4		

E.WENKERT,J.S.BINDRA,C.-J.CHANG,D.W.COCHRAN,
F.M.SCHELL
ACC CHEM RES   7,   46 (1974)

2981   U 006000 DIHYDROCORYNANTHEINE

```
   4     10
   C      C
 / \8   / \
5C   C---C3  C11
 |   ||   ||  |
6C   C  2C   V12
 \ /9\1/ \ / \
   C  V 17C   C13
   7  |   |  19
     16C   C14 CH3
       \ / \ /
      15C   CH2
    24  |  18
   H3C-O 20C   22CH3
       \ / \  /
      23C   C4-O
       || 21
        O
```

FORMULA	C22H28N2O3			MOL WT	368.48
SOLVENT	CHCL3				
ORIG ST	C4H12SI			TEMP	AMB
135.20	107.50	117.90	120.90	119.00	110.80
2/1	3/1	4/2	5/2	6/2	7/2
127.40	136.20	21.90	53.10	61.30	39.30
8/1	9/1	10/3	11/3	13/3	14/2
38.70	33.80	60.20	24.40	11.30	111.70
15/2	16/3	17/2	18/3	19/4	20/1
159.80	61.30	168.90	51.10		
21/2	22/4	23/1	24/4		

E.WENKERT,J.S.BINDRA,C.-J.CHANG,D.W.COCHRAN,
F.M.SCHELL
ACC CHEM RES   7,   46 (1974)

2982   U 006000 ISORHYNCOPHYLLINE

```
      11  13  18
       C   C H C H2
   4  / \ / \|/ \
   C  10C  N  C14 CH3
 / \8  |  |  |  19
5C   C---C---C  C15
 |   ||  |3 17/ \
6C   C  C2  16C 20C   H
 \ /9\1/ \  \  / \ /
   C  N  O  O=C23 C21
   7  |   |   |
        O   O
       /     \
      CH3    CH3
      24      22
```

FORMULA	C22H28N2O4			MOL WT	384.48
SOLVENT	CHCL3				
ORIG ST	C4H12SI			TEMP	AMB
182.40	57.00	125.20	122.10	127.40	109.60
2/1	3/1	4/2	5/2	6/2	7/2
134.20	140.70	36.50	54.20	58.20	38.30
8/1	9/1	10/3	11/3	13/3	14/2
38.30	30.10	75.20	24.30	11.20	113.00
15/2	16/3	17/2	18/3	19/4	20/1
159.50	61.20	168.40	50.90		
21/2	22/4	23/1	24/4		

E.WENKERT,J.S.BINDRA,C.-J.CHANG,D.W.COCHRAN,
F.M.SCHELL
ACC CHEM RES   7,   46 (1974)

2983　U 006000　RHYNCOPHYLLINE

```
        11  13   18
         C   C   CH2        FORMULA   C22H28N2O4              MOL WT   384.48
    4   / \ / \ / \         SOLVENT   CHCL3
    C  10C   N  G14 CH3     ORIG ST   C4H12SI                 TEMP        AMB
   / \8   |   |   | 19
 5C   C---C---C   C15
  |   II  13 17\ / \
 6C   C   C2  16C 20C   H
  \ /9\1/   \   / \ /
   C   N   O   O=C23 C21
   7           |   |
               O   O
              / \   \
            CH3    CH3
            24     22
```

182.20	56.20	122.80	122.40	127.80	109.70
2/1	3/1	4/2	5/2	6/2	7/2
134.10	141.50	34.80	55.10	58.20	39.30
8/1	9/1	10/3	11/3	13/3	14/2
38.00	29.20	75.30	24.20	11.20	112.40
15/2	16/3	17/2	18/3	19	
159.60	61.20	168.80	51.00		
21/2	22/4	23/1	24/4		

E.WENKERT,J.S.BINDRA,C.-J.CHANG,D.W.COCHRAN,
F.M.SCHELL
ACC CHEM RES　　　　　　　　　7,　46 (1974)

2984　U 006000　7,8,9,10-TETRAHYDROCANNABINOL METHYLETHER

```
          23CH3
  22        /            FORMULA   C22H32O2               MOL WT   328.50
 H3C       O             SOLVENT   CCL4
   \  10    \1 2         ORIG ST   CS2                    TEMP        AMB
  9C-C     C=C
   / \ / \   \3 15
   C  13C-C14  C-CH2
  8\  / \ / \ / \
    C-C12 11C-C   16CH2
   7 \6 5/  4     /
     C-O      17CH2
     / \        \
    H3C  CH3   18CH2
     20  21     /
             19CH3
```

156.10	104.30	142.00	109.80	76.70	25.40
1/1	2/2	3/1	4/2	6/1	7/3
30.50	28.90	36.70	153.50	124.00	131.50
8/3	9/2	10/3	11/1	12/1	13/1
104.30	36.00	31.40	30.50	22.40	14.00
14/1	15/3	16/3	17/3	18/3	19/4
25.40	23.70	21.80	54.90		
20/4	21/4	22/4	23/4		

E.WENKERT,D.W.COCHRAN,F.M.SCHELL,R.A.ARCHER,
K.MATSUMOTO
EXPERIENTIA　　　　　　　　　28,　250 (1972)

2985　U 006000　HAZUNTININE

```
             C
 24      5   / \           FORMULA   C23H28N2O5            MOL WT   412.49
 CH3     C---N  14C-O      SOLVENT   CDCL3
  | 9    |    |   |/       ORIG ST   C4H12SI               TEMP        AMB
  O  C  6C  21C 15C
   \ /8 \7/ \ / \
 10C   C---C   C20
   |   II  |   I\
 11C  13C   C  17C CH2-CH3
  / \ / \ / /2\ /  19 18
  O   C   N   C16
  |  12       |
 CH3         C22
 25        / \ 23
         O   O-CH3
```

166.00	49.20	51.40	43.60	54.80	128.70
2/1	3/2	5/3	6/3	7/1	8/1
103.50	149.30	143.50	95.60	137.00	51.80
9/2	10/1	11/1	12/2	13/1	14/2
57.00	90.70	23.30	7.00	26.30	36.80
15/2	16/1	17/3	18/4	19/3	20/1
70.80	168.80	50.70	55.90	55.90	
21/2	22/1	23/4	24/4	25/4	

E.WENKERT,D.W.COCHRAN,E.W.HAGAMAN,F.M.SCHELL,
N.NEUSS,A.S.KATNER,P.POTIER,C.KAN,M.PLAT,
M.KOCH,H.MEHRI,J.POISSON,N.KUNESCH,Y.ROLLAND
J AM CHEM SOC　　　　　　　95, 4990 (1973)

2986　U 006000　DELTA-9-TETRAHYDROCANNABINOL ACETATE

```
       O
        \ 24
       23C-CH3           FORMULA   C23H32O4               MOL WT   372.51
 22      /               SOLVENT   CCL4
 H3C    O                ORIG ST   CS2                    TEMP        AMB
   \ 10  \1 2
  9C=C    C=C
   / \   / \3 15
   C  13C-C14  C-CH2
  8\  / \ / \ / \
    C-C12 11C-C   16CH2
   7 \6 5/  4     /
     C-O      17CH2
    / \         \
  H3C  CH3    18CH2
   20  21      /
           19CH3
```

148.20	114.90	141.60	113.50	76.60	24.70
1/1	2/2	3/1	4/2	6/1	7/3
30.80	133.50	123.60	154.30	45.30	33.90
8/3	9/1	10/2	11/1	12/2	13/2
114.60	34.80	31.20	30.00	22.20	13.80
14/1	15/3	16/3	17/3	18/3	19/4
19.10	27.20	23.10	166.60	20.50	
20/4	21/4	22/4	23/1	24/4	

E.WENKERT,D.W.COCHRAN,F.M.SCHELL,R.A.ARCHER,
K.MATSUMOTO
EXPERIENTIA　　　　　　　　28,　250 (1972)

2987   U 006000 LYSERGIC ACID N,N-DIETHYLAMIDE TARTRATE

```
   19   21
  H2C    CH2
  ╱ ╲ ╱ ╲
H3C20  N 22CH3        O   OH
   |        17   ‖   |25
   C18 C7  CH3  C23 CH
   ╱ ╲ ╱ ╲+╱  - ╱ ╲ ╱ ╲26
 O   C8  N6   O 24CH  C=O
   |   |          |   |
   C9  C5        OH  OH
    ╲ ╱ ╲
    C10 C4
     |   |
    C11 C3
    ╱ ╲ ╱ ╲
  12C    C16 ╲
   |     ‖    C2
  13C    C15 ╱
    ╲ ╱ ╲ ╱
      C   N1
      14
```

FORMULA	C24H31N3O7			MOL WT	473.53
SOLVENT	CD4O				
ORIG ST	C4H12SI			TEMP	303
118.95	107.90	25.80	63.45	55.10	38.25
2/2	3/1	4/3	5/2	7/3	8/2
120.20	134.85	126.50	112.75	123.50	111.35
9/2	10/1	11/1	12/2	13/2	14/2
134.60	126.35	42.55	171.45	42.45	13.10
15/1	16/1	17/4	18/1	19/3	20/4
43.15	14.95	177.40	73.70		
21/3	22/4	23/1	24/2		

A.WALTER,E.BREITMAIER
UNPUBLISHED

2988   U 006000 VINDOLINE

```
        3
        C
     5    ╱ ╲
     C---N  C14
  9  |   |   ‖
   C  6C  C21 C15
  ╱ ╲8  ╲7╱ ╲ ╱
 10C  C---C  C20
  |  ‖  |  |╲19 18
 11C 13C  C 17C CH2-CH3
 ╱ ╲ ╱ ╲ ╱2╲ ╱ ╲ 26
O   C  N 16C  O-C-CH3
 |  12 |  ╱ ╲  ‖27
CH3  H3C OH  C=O  O
24    25   ╱22
         O
         |
       23CH3
```

FORMULA	C25H32N2O6			MOL WT	456.54
SOLVENT	CDCL3				
ORIG ST	C4H12SI			TEMP	AMB
83.20	50.90	51.90	43.60	52.60	124.90
2/2	3/3	5/3	6/3	7/1	8/1
122.40	104.50	161.10	95.60	153.60	123.90
9/2	10/2	11/2	12/2	13/1	14/2
130.20	79.50	76.20	7.50	30.60	42.80
15/2	16/1	17/2	18/4	19/3	20/1
67.00	170.40	51.90	55.10	38.00	171.70
21/2	22/1	23/4	24/4	25/4	26/1
20.80					
27/4					

E.WENKERT,D.W.COCHRAN,E.W.HAGAMAN,F.M.SCHELL,
N.NEUSS,A.S.KATNER,P.POTIER,C.KAN,M.PLAT,
M.KOCH,H.MEHRI,J.POISSON,N.KUNESCH,Y.ROLLAND
J AM CHEM SOC   95, 4990 (1973)

2989   U 006000 DIHYDROVINDOLINE

```
        3
        C
     5    ╱ ╲
     C---N  C14
  9  |   |   |
   C  6C  C21 C15
  ╱ ╲8  ╲7╱ ╲ ╱
 10C  C---C  C20
  |  ‖  |  |╲19 18
 11C 13C  C 17C CH2-CH3
 ╱ ╲ ╱ ╲ ╱2╲ ╱ ╲ 26
O   C  N 16C  O-C-CH3
 |  12 |  ╱ ╲  ‖27
CH3  H3C OH  C=O  O
24    25   ╱22
         O
         |
       23CH3
```

FORMULA	C25H34N2O6			MOL WT	458.56
SOLVENT	CDCL3				
ORIG ST	C4H12SI			TEMP	AMB
83.50	52.40	51.40	43.40	52.40	125.20
2/2	3/3	5/3	6/3	7/1	8/1
122.70	104.10	160.60	95.60	154.00	22.40
9/2	10/2	11/1	12/2	13/1	14/3
33.00	78.30	75.60	7.80	29.90	40.00
15/3	16/1	17/2	18/4	19/3	20/1
72.30	170.00	52.00	55.00	37.70	172.30
21/2	22/1	23/4	24/4	25/4	26/1
20.70					
27/4					

E.WENKERT,D.W.COCHRAN,E.W.HAGAMAN,F.M.SCHELL,
N.NEUSS,A.S.KATNER,P.POTIER,C.KAN,M.PLAT,
M.KOCH,H.MEHRI,J.POISSON,N.KUNESCH,Y.ROLLAND
J AM CHEM SOC   95, 4990 (1973)

```
2990              U 006000 ERGOKRYPTINE
    27C---C26
     |    |            FORMULA   C32H41N5O5        MOL WT  575.71
    28C   N25 O        SOLVENT   CDCL3
     \ / \ // 37      ORIG ST   C4H12SI           TEMP     303
    29C   C24 CH3
     |    |   |35      119.10   110.80    26.55    59.30    53.35    40.10
   HO-C30 C23 CH        2/2      3/1      4/3      5/2      7/3      8/2
  32   / \ / \ /\36    119.10   139.20   129.85   112.20   123.55   110.15
 H3C   O   N22 CH2 CH3   9/2     10/1     11/1     12/2     13/2     14/2
   \311  |    34        134.00   126.20    43.55   176.35    89.80   166.30
   HC-C---C21           15/1     16/1     17/4     18/1     20/1     21/1
 33/ /20     \          48.20   165.50    46.00    21.65    25.15    64.55
 H3C NH       O         23/2     24/1     26/3     27/3     28/3     29/2
    |19                103.65    34.35    16.95    15.40    44.40    22.65
   C18 C7  CH3          30/1     31/2     32/4     33/4     34/3     35/2
    / \ / \6/17         22.15    22.65
   O   C8  N            36/4     37/4
       |   |
       C9  C5
        \ / \
        C10 C4
         |   |
        C11 C3
        / \ / \
      12C   C16 \
       |    ||   C2
      13C   C15 /
       \ / \ /
        C   N1          A.WALTER,E.BREITMAIER
        14              UNPUBLISHED
```

```
2991              U 006000 ERGOKRYPTININE
    27C---C26
     |    |            FORMULA   C32H41N5O5        MOL WT  575.71
    28C   N25 O        SOLVENT   C2D6OS
     \ / \ // 36       ORIG ST   C4H12SI           TEMP     AMB
     C29 C24 CH3
      |   |   |35      119.40   108.20    26.70    61.90    53.70    42.20
    HO-C30 C23 CH       2/2      3/1      4/3      5/2      7/3      8/2
  32   / \ / \ / \37   117.60   136.70   126.70   111.50   122.20   110.20
 H3C   O   N22 CH2 CH3   9/2     10/1     11/1     12/2     13/2     14/2
   \31120 |    34       133.60   125.80    42.60   175.80    89.10   164.80
   HC-C---C21           15/1     16/1     17/4     18/1     20/1     21/1
 33/ /        \         52.30   164.80    45.50    21.40    25.90    63.40
 H3C NH        O        23/2     24/1     26/3     27/3     28/3     29/2
    |19                102.80    33.80    16.40    15.30    42.60    25.00
   C18 C7  CH3          30/1     31/2     32/4     33/4     34/3     35/2
    / \ / \6/17         22.20
   O   C8  N            36/4
       |   |
       9C  C5
        \ / \
        C10 C4
         |   |
        C11 C3
        / \ / \
      12C   C16 \
       |    ||   C2
      13C   C15 /
       \ / \ /
        C   N1          N.J.BACH,H.E.BOAZ,E.C.KORNFELD,C.-J.CHANG,
        14              H.G.FLOSS,E.W.HAGAMAN,E.WENKERT
                        J ORG CHEM              39, 1272 (1974)
```

```
2992                  U 006000  DIHYDROERGOTAMINE
   27C---C26
    |    |                    FORMULA   C33H37N5O5              MOL WT   583.69
   28C   N25 O   C35          SOLVENT   COCL3
    \  / \  /  / \            ORIG ST   C4H12SI                TEMP     303
     C29 C24 C34 C36
      |30 |   ||  |           118.70   108.85    26.65    66.55    58.80    41.85
     HO-C 23C   C33 C37        2/2      3/1       4/3      5/2      7/3      8/2
     /  \  /  \  /  \          30.65    39.50    127.90   112.10   122.15   110.05
     O   N22 CH2 C38           9/3      10/2      11/1     12/2     13/2     14/2
   31 |   |    32            133.35   126.00     42.70   175.95    85.90   166.05
   H3C-C---C21                15/1     16/1      17/4     18/1     20/1     21/1
     /20    \                  56.30   164.45     45.95    21.80    26.00    64.00
   HN19     O'                 23/2     24/1      26/3     27/3     28/3     29/2
     |      17                102.80    23.95     38.70   138.85   129.90   127.85
   C18 C7  CH3                 30/1     31/4      32/3     33/1     34/2     35/2
   /  \ / \6/                 127.85
  O   C8   N                   36/2
      |    |
      C9   C5
       \  / \
       C10 C4
        |   |
       C11 C3
       / \ / \
    12C    C16 \
     |     ||   C2
    13C    C15 /
      \  / \  /
       C    N1
       14
                            A.WALTER,E.BREITMAIER
                            UNPUBLISHED
```

```
2993                  U 006000  ERGOTAMINE
   27C---C26
    |    |      35          FORMULA   C33H35N5O5              MOL WT   581.68
   28C   N25 O   C          SOLVENT   C2D6OS
    \  / \  /  / \          ORIG ST   C4H12SI                TEMP     AMB
     C29 C24 C34 C36
      |30 |   ||  |          119.40   108.80    26.60    62.40    55.10    42.50
     HO-C 23C   C33 C37       2/2      3/1       4/3      5/2      7/3      8/2
     /  \  /  \  /  \         118.30   136.00   127.10   111.00   122.20   110.20
     O   N22 CH2 C38          9/2      10/1      11/1     12/2     13/2     14/2
   31 |   |    32           133.80   125.90     43.40   174.30    85.90   165.80
   H3C-C---C21               15/1     16/1      17/4     18/1     20/1     21/1
     /20    \                 56.10   164.20     45.80    21.70    25.90    63.90
   HN19     O                 23/2     24/1      26/3     27/3     28/3     29/2
     |                       102.80    23.60     38.70   138.70   129.90   127.70
   C18 C7  CH3                30/1     31/4      32/3     33/1     34/2     35/2
   /  \ / \6/17              127.40
  O   C8   N                  36/2
      |    |
      C9   C5
       \  / \
      10C    C4
        |   |
       C11 C3
       / \ / \
    12C    C16 \
     |     ||   C2
    13C    C15 /
      \  / \  /
       C    N1
       14
                            N.J.BACH,H.E.BOAZ,E.C.KORNFELD,C.-J.CHANG,
                            H.G.FLOSS,E.W.HAGAMAN,E.WENKERT
                            J ORG CHEM                    39, 1272 (1974)
```

```
2994              U 006000 ERGOTAMININE
   27C---C26
    |    |
   28C   N25 O   C35         FORMULA   C33H35N5O5          MOL WT  581.68
    \  / \ //   / \          SOLVENT   C2D6OS
     C29 C24 C34 C36         ORIG ST   C4H12SI             TEMP     AMB
      |   |   ||   |
    HO-C30 C23 C33 C37        119.70  109.00   26.90   61.70   53.00   41.80
     /  \ / \ / \ //           2/2     3/1     4/3     5/2     7/3     8/2
    O   N22 CH2 C38           118.10  137.10  127.90  111.40  122.40  110.30
   31 |   |    32              9/2     10/1    11/1    12/2    13/2    14/2
   H3C-C---C21                133.80  126.10   42.50  175.30   85.70  165.90
     /20    \                  15/1    16/1    17/4    18/1    20/1    21/1
   HN19     O                  56.10  164.50   45.70   21.80   25.90   63.90
    |                          23/2    24/1    26/3    27/3    28/3    29/2
   18C   C7  CH3              102.90   23.80   38.70  138.90  129.90  127.90
   / \  / \ /6/17              30/1    31/4    32/3    33/1    34/2    35/2
  O   C8  N                   126.10
      |   |                    36/2
      C9  C5
      \ / \
      C10 C4
       |   |
      C11 C3
     / \ / \
   12C   C16 \
    |    ||    C2
   13C   C15 /
    \ / \ /
     C   N1                   N.J.BACH,H.E.BOAZ,E.C.KORNFELD,C.-J.CHANG,
     14                       H.G.FLOSS,E.W.HAGAMAN,E.WENKERT
                              J ORG CHEM              39, 1272 (1974)
```

```
2995              U 006000 ERGOCRISTINE
   27C---C26
    |    |
   28C   N25 O   C37          FORMULA   C35H39N5O5          MOL WT  609.73
    \  / \ //   / \           SOLVENT   CDCL3
     C29 C24 C36 C38          ORIG ST   C4H12SI             TEMP     303
      |30 |   ||   |
    HO-C 23C   C35 C39        118.80  110.80   26.55   64.45   59.45   41.05
   32   /  \ / \ / \ //         2/2     3/1     4/3     5/2     7/3     8/2
   H3C  O   N22 CH2 C40        119.30  139.40  129.75  112.00  123.40  110.25
    \31|    |      34           9/2     10/1    11/1    12/2    13/2    14/2
    HC-C---C21                 134.05  126.35   44.55  176.25   90.05  165.85
   33/ /20   \                  15/1    16/1    17/4    18/1    20/1    21/1
   H3C NH     O                 56.80  165.55   46.25   21.80   22.35   64.45
      |19                       23/2    24/1    26/3    27/3    28/3    29/2
    C18 C7  CH3                103.85   34.40   16.85   15.40   39.70  139.10
   / \ / \ /6/17                30/1    31/2    32/4    33/4    34/3    35/1
  O   C8  N                    130.15  128.05  126.35
      |   |                     36/2    37/2    38/2
      C9  C5
      \ / \
      C10 C4
       |   |
      C11 C3
     / \ / \
   12C   C16 \
    |    ||    C2
   13C   C15 /
    \ / \ /
     C   N1                    A.WALTER,E.BREITMAIER
     14                        UNPUBLISHED
```

2996   U 006000 DIHYDROERGOCRISTINE
```
     27C---C26
      |    |                  FORMULA   C35H41N5O5          MOL WT   611.75
     28C   N25 O   C37        SOLVENT   CDCL3
       \ / \ / // / \         ORIG ST   C4H12SI             TEMP       303
       C29 C24 C36 C38
       |30 |    ||   |        117.95   108.90    27.00    64.30    59.25    44.15
     HO-C 23C    C35 C39       2/2      3/1      4/3      5/2      7/3      8/2
   32  / \ / \ / \ / \        30.60    40.15   132.30   113.35   123.35   111.70
  H3C   O  N22 CH2 C40         9/3     10/2     11/1     12/2     13/2     14/2
    \31|   |     34           133.55   126.45    43.05   175.65    89.95   165.20
     HC-C---C21               15/1     16/1     17/4     18/1     20/1     21/1
  33/  /20      \             57.15   165.20    46.25    22.30    26.55    66.80
  H3C  NH        O            23/2     24/1     26/3     27/3     28/3     29/2
      |19                     104.00    34.30    17.00    15.55    39.55   138.65
     C18 C7   CH3             30/1     31/2     32/4     33/4     34/3     35/1
     / \ / \ /6/17            130.20   128.10   126.45
    O   C8   N                36/2     37/2     38/2
        |    |
        C9   C5
          \ / \
          C10 C4
           |   |
          C11 C3
         / \ / \
      12C   C16 \
       |    ||    C2
      13C   C15 /
        \ / \ /
         C    N1
         14
```
                              A.WALTER,E.BREITMAIER
                              UNPUBLISHED

2997   U 007000 FORMYCIN B
```
          O
          ||                 FORMULA   C10H12N4O5          MOL WT   268.23
         2C   N              SOLVENT   C2H6OS
        / \4/ \              ORIG ST   C4H12SI             TEMP       AMB
       N   C   \
       |   ||   N            153.70   143.10    77.60    74.90    72.10    85.60
      3C   C   //             2/1      3/2      6/2      7/2      8/2      9/2
        \ /5\ //
         N   C1              62.50   128.50   136.50   144.60
     10   |               10/3
  HO-H2C  O |
      \ / \|
      9C   C6
       |   |
      8C---C7
       |   |                 L.F.JOHNSON,W.C.JANKOWSKI
      OH  OH                 CARBON-13 NMR SPECTRA,JOHN
                             WILEY AND SONS,NEW YORK        375 (1972)
```

2998   U 007000 ADENOSINE 5'-DIPHOSPHATE TRISODIUM SALT
```
          NH2
           |                 FORMULA   C10H12N5NA3O10P2    MOL WT   493.15
           C6                SOLVENT   D2O
         // \                ORIG ST   C4H12SI             TEMP       298
        1N  5C---N7
         |   ||   ||         152.60   148.50   118.30   155.10   139.95    87.55
    +   2C  4C   C8           2/D      4/S      5/S      6/S      8/D      10/D
  3NA    \ / \9/             70.15    74.70    83.75    64.90
    -   -  3N   N            11/D     12/D     13/D     14/T
    O   O   14  O  |
    -   |   |     |
  O-P-O-P-O-H2C  / \ |
    ||  ||   |/   \|
    O   O   C13   C10
              \   /
             12C-C11
              |   |          E.BREITMAIER,W.VOELTER
             HO  OH          EUR J BIOCHEM                 31, 234 (1972)
```

2999 U 007000 9-ALPHA-ARABINOFURANOSYL ADENINE

```
              O
        14   / \
     HO-CH2 /   \
        \ /      \
        13C    HO C10
          \     |/|
          12C---C |
          |  11   |
          HO       |
                   |
              N    N
            //  \4/  \\
           2C    C    \
            |    ||     C 8
            N    C     //
            \\  /5\   //
             6C    N
              |
             NH2
```

FORMULA	C10H13N5O4			MOL WT	267.25
SOLVENT	C2D6OS				
ORIG ST	C4H12SI			TEMP	AMB

151.80	148.50	118.40	155.30	139.40	87.80
2/2	4/1	5/1	6/1	8/2	10/2
78.80	74.70	84.70	60.50		
11/2	12/2	13/2	14/3		

L.F.JOHNSON,W.C.JANKOWSKI
CARBON-13 NMR SPECTRA,JOHN
WILEY AND SONS,NEW YORK   377 (1972)

3000 U 007000 GUANOSINE

```
        OH
        |
       6C    N
      //  \5/  \\
      N    C    \
      |    ||     C 8
      2C    C    /
     /  \  /4\  /
   H2N    N    N
                |
          O     |
         14   / \ |
      HO-CH2 /   \ |
         \  /     \|
         13C      C 10
           \      /
           12C---C11
           |     |
           OH    OH
```

FORMULA	C10H13N5O5			MOL WT	283.25
SOLVENT	C2D6OS				
ORIG ST	C4H12SI			TEMP	AMB

153.80	151.50	116.70	157.10	136.10	86.70
2/1	4/1	.5/1	6/1	8/2	10/2
73.90	70.60	85.50	61.60		
11/2	12/2	13/2	14/3		

L.F.JOHNSON,W.C.JANKOWSKI
CARBON-13 NMR SPECTRA,JOHN
WILEY AND SONS,NEW YORK   383 (1972)

3001 U 007000 ADENOSINE-5-TRIPHOSPHATE

```
            NH2
            |
            C6  N7
           //  \5/  \\
          1N    C    \
           |    ||     C 8
          2C    C    /
           \\  /4\  /
            3N    N9
                   |
            O      |
    2+     14   / \ |
   2NA         /   \ |
           \  /     \|
   2-      13C      C 10
  O9P3O-CH2 \      /
           12C---C11
            |     |
            OH    OH
```

FORMULA	C10H14N5NA2O13P3			MOL WT	551.15
SOLVENT	H2O				
ORIG ST	C4H12SI			TEMP	AMB

146.20	150.60	118.80	148.70	143.00	88.80
2/2	4/1	5/1	6/1	8/2	10/2
75.60	70.90	84.80	66.10		
11/2	12/2	13/2	14/3		

L.F.JOHNSON,W.C.JANKOWSKI
CARBON-13 NMR SPECTRA,JOHN
WILEY AND SONS,NEW YORK   383 (1972)

3002 U 007000 RIBOFLAVIN 5'-MONOPHOSPHATE DISODIUM SALT

```
                O
   15   6   5   ||
   H3C   C   N   C4
     \ / \ / \ / \3
     7C   C12 C14 NH
      |    ||   ||   |
     8C   C11 C13 C2
   16/ \ / \ / \ / \
   H3C   N10 N1   O
        9    |
           17CH2
            |18
           HC-OH
            |19
           HC-OH
        +  - |20
   2NA  O   HC-OH
    -  |  |  |
     O-P-O-CH2
       ||   21
       O
```

FORMULA	C17H19N4NA2O9P			MOL WT	500.32
SOLVENT	D2O				
ORIG ST	C4H12SI			TEMP	298

157.55	160.25	130.05	133.50	134.15	117.20 P
2/S	4/S	6/D	7/S	8/S	9/D
131.35	139.55	150.85	149.55	18.70	20.85
11/S	12/S	13/S	14/S	15/Q	16/Q
47.60	69.40	72.65	71.55	65.85	
17/T	18/D	19/D	20/D	21/T	

E.BREITMAIER,W.VOELTER
EUR J BIOCHEM   31, 234 (1972)

3003   U 007000 FLAVIN-ADENINE DINUCLEOTIDE DISODIUM SALT

```
        O   O
  21    ||  ||   35
 H2C-O-P-O-P-O-CH2
 120  |   |   |
 HO-CH O   O  34C
  119  - -   / \33
 HO-CH  +  /   C-OH
  118  2NA O   |
 HO-CH      \   C-OH
  |          \ /32
 17CH2      31C
 16 9  |  |     | 24
 CH3 C  \10 N  O  N30 N
   \ / \ / \ / \   / \ / \23
  8C   C11 C13 C2  C 29 C25 C
  |     ||  |  |   ||  ||  |
  7C    C12 C14 N  N---C26 N
 15/ \  \ / \ / /3  28  \ //22
 CH3 C   \  C4      27C
   6  5   ||         |
          O         NH2
```

FORMULA	C27H31N9NA2O15P2			MOL WT	829.53
SOLVENT	D2O				
ORIG ST	C4H12SI			TEMP	298

157.45	160.60	130.15	133.30	133.80	116.75
2/S	4/S	6/D	7/S	8/S	9/D
131.35	139.20	150.65	149.80	18.70	20.85
11/S	12/S	13/S	14/S	15/Q	16/Q
47.60	69.40	72.65	71.45	67.80	152.40
17/T	18/D	19/D	20/D	21/T	23/D
148.15	117.75	154.65	139.20	87.45	70.25
25/S	26/S	27/S	29/D	31/D	32/D
75.15	83.85	65.40			
33/D	34/D	35/T			

E.BREITMAIER,W.VOELTER
EUR J BIOCHEM   31, 234 (1972)

3004   U 008000 COUMARIN

```
        7
        C   O   O
      // \8/ \  //
   6C    C    C1
   |    ||    |
   5C    C   C2
     \  /9\  //
      C    C
      4    3
```

FORMULA	C9H6O2			MOL WT	146.15
SOLVENT	CDCL3				
ORIG ST	C4H12SI			TEMP	AMB

160.40	143.40	116.50	116.40	153.80	118.70
1/1	3/2	5/2	7/2	8/1	9/1
124.30	127.90	131.70			

L.F.JOHNSON,W.C.JANKOWSKI
CARBON-13 NMR SPECTRA,JOHN
WILEY AND SONS,NEW YORK   333 (1972)

3005   U 008000 AZULENE

```
   5   4
   C==C   C3
  /    \9/ \
 /      C   \
6C      |    C2
 \\    10C   //
  \     / \ //
   C--C    C
   7  8    1
```

FORMULA	C10H8			MOL WT	128.18
SOLVENT	CDCL3				
ORIG ST	C4H12SI			TEMP	AMB

117.90	137.00	136.40	122.60	136.90	140.10
1/2	2/2	4/2	5/2	6/2	9/1
117.90	122.60	136.40	140.10		
3/2	7/2	8/2	10/1		

L.F.JOHNSON,W.C.JANKOWSKI
CARBON-13 NMR SPECTRA,JOHN
WILEY AND SONS,NEW YORK   367 (1972)

3006   U 008000 METHYL 8-DEOXYNONACTINATE DIASTEREOMERS

```
    4   5
 O   C---C
 ||  |   |   8
 C1  C   C   CH2
 / \2/3\ /6\7/ \
O   CH O   CH2 CH3
|   |          9
CH3 CH3
11  10
```

FORMULA	C11H20O3			MOL WT	200.28
SOLVENT	CDCL3				
ORIG ST	C4H12SI			TEMP	AMB

175.50	175.30	45.30	80.20	80.10	29.20
1/1	1/1	2/2	3/2	3/2	4/3
28.50	31.00	79.50	38.30	19.40	14.20
4/3	5/3	6/2	7/3	8/3	9/4
14.20	13.30	51.50			
10/4	10/4	11/4			

E.PRETSCH,M.VASAK,W.SIMON
HELV CHIM ACTA   55, 1098 (1972)

3007 U 008000 NONACTINE

```
       4   5   9
     O   C---C  CH3
     ||  |   |   |
     C1  C   C   CH
   (/ \2/3\ /6\7/8\ /)
      CH O   CH2 O   4
      |
      CH3
      10
```

FORMULA	C40H64O12			MOL WT	736.95
SOLVENT	CDCL3				
ORIG ST	C4H12SI			TEMP	AMB
174.30	45.20	80.10	28.20	31.50	76.40
1/1	2/2	3/2	4/3	5/3	6/2
42.40	69.10	20.60	12.80		
7/3	8/2	9/4	10/4		

E.PRETSCH,M.VASAK,W.SIMON
HELV CHIM ACTA   55, 1098 (1972)

3008 U 009000 THIAMINE HYDROCHLORIDE

```
         HCL
  12        •        10  11
  H3C   N   NH2 S   CH2-CH2
    \ / \4/  / \ /    |
    1C   C  6C   C8    OH
    |    ||   ||   ||
    N    C   N---C
    \ /3\ / + 7\
     C    CH2   - CH3
     2    5   CL   9
```

FORMULA	C12H17CLN4OS			MOL WT	300.81
SOLVENT	H2O				
ORIG ST	C4H12SI			TEMP	AMB
164.00	155.50	107.00	163.80	50.70	145.60
1/1	2/2	3/1	4/1	5/3	6/2
143.60	137.30	12.10	30.20	61.20	22.00
7/1	8/1	9/4	10/3	11/3	12/4

L.F.JOHNSON,W.C.JANKOWSKI
CARBON-13 NMR SPECTRA,JOHN
WILEY AND SONS,NEW YORK   437 (1972)

3009 U 009000 THIAMINE HYDROCHLORIDE

```
    NH2
    |  7        - 14
    C4  CH2 CL  CH3
    // \ / \10  /
  3N   C5 +N---C11
  |    ||  ||   || 16
  2C   C6 9C   C12 CH2
 13/ \ /     \ / \ /
 H3C   N1    8S 15CH2 OH
       •
       HCL
```

FORMULA	C12H18CLN4OS			MOL WT	301.82
SOLVENT	CH4O/H2O				
ORIG ST	C4H12SI			TEMP	AMB
163.78	164.18	106.50	146.44	50.96	155.28
2/1	4/1	3/1	6/2	7/3	9/10
143.62	137.30	21.93	12.20	30.34	61.04
11/1	12/1	13/4	14/4	15/3	16/3

R.E.ECHOLS,G.C.LEVY
J ORG CHEM   39, 1321 (1974)

3010 U 009000 RIBOFLAVINE

```
       OH OH OH OH
    13  |  |  |  |
    CH2-CH-CH-CH-CH2
  11  |  15 16 17  14
  H3C   C   N   N   O
    \ / \6/ \7\ \ //
     2C   C   C   C10
     |    ||   |   |
     3C   C   C   N
   \ / /5\ /8\ /
   H3C   C   N   C9
    12   4       ||
                 O
```

FORMULA	C17H20N4O6			MOL WT	376.37
SOLVENT	C2D6OS				
ORIG ST	C4H12SI			TEMP	AMB
117.20	130.50	18.60	20.60	47.10	63.20
1/2	4/2	11/4	12/4	13/3	14/3
68.70	72.60	73.50	131.90	133.80	135.50
136.50	145.80	150.60	155.20	159.70	

L.F.JOHNSON,W.C.JANKOWSKI
CARBON-13 NMR SPECTRA,JOHN
WILEY AND SONS,NEW YORK   475 (1972)

3011 U 009000 DELTA-TOCOPHEROL

```
       5    4
 HO    C    C
   \ / \ / \    27      24
   6C    C9  3C  CH3    CH3
    |    ||   |/   12    |14
   7C  10C   2C   CH2   CH
    \ / \ |/ \ / \ / \15
     C8   O   CH2 CH2 CH2
     |     11  13   |
     CH3            16CH2
     28      21  19   |
           H2C  H2C  17CH2
       23 22/ \ / \ /
       H3C-HC  H2C  18CH
          |   20    |
         26CH3    25CH3
```

FORMULA	C27H46O2			MOL WT	402.67
SOLVENT	CDCL3				
ORIG ST	C4H12SI			TEMP	AMB

75.50	31.40	22.70	112.70	147.50	115.80
2/1	3/3	4/3	5/2	6/1	7/2
127.10	121.10	145.80	40.00	21.00	37.50
8/1	9/1	10/1	11/3	12/3	13/3
32.70	37.50	24.50	37.50	32.70	37.50
14/2	15/3	16/3	17/3	18/2	19/3
24.80	39.40	28.00	22.70	19.70	19.70
20/3	21/3	22/2	23/4	24/4	25/4
22.70	24.00	16.00			
26/4	27/4	28/4			

M.MATSUO,S.URANO
TETRAHEDRON   32, 229 (1976)

3012 U 009000 BETA-TOCOPHEROL

```
      28
      CH3
       |    4
 HO   C5   C
   \ / \ • / \    27      24
   6C    C9  3C  CH3    CH3
    |    ||   |/   12    |14
   7C  10C   2C   CH2   CH
    \ / \ |/ \ / \ / \15
     C9   O   CH2 CH2 CH2
     |     11  13   |
     CH3            16CH2
     29      21  19   |
           H2C  H2C  17CH2
       23 22/ \ / \ /
       H3C-HC  H2C  18CH
          |   20    |
         26CH3    25CH3
```

FORMULA	C28H48O2			MOL WT	416.69
SOLVENT	CDCL3				
ORIG ST	C4H12SI			TEMP	AMB

74.40	31.50	20.80	119.20	145.50	115.40
2/1	3/3	4/3	5/1	6/1	7/2
123.80	120.10	145.70	39.80	21.00	37.40
8/1	9/1	10/1	11/3	12/3	13/3
32.70	37.40	24.50	37.40	32.70	37.40
14/2	15/3	16/3	17/3	18/2	19/3
24.80	39.40	28.00	22.70	19.70	19.70
20/3	21/3	22/2	23/4	24/4	25/4
22.70	23.80	11.00	15.80		
26/4	27/4	28/4	29/4		

M.MATSUO,S.URANO
TETRAHEDRON   32, 229 (1976)

3013 U 009000 GAMMA-TOCOPHEROL

```
       5    4
 HO    C    C
   \ / \ / \    27      24
   6C    C9  3C  CH3    CH3
    |    ||   |/   12    |14
   7C  10C   2C   CH2   CH
  28/ \ / \ |/ \ / \ / \15
 H3C   C8   O   CH2 CH2 CH2
    |     11  13   |
    CH3            16CH2
    29      21  19   |
          H2C  H2C  17CH2
      23 22/ \ / \ /
      H3C-HC  H2C  18CH
         |   20    |
        26CH3    25CH3
```

FORMULA	C28H48O2			MOL WT	416.69
SOLVENT	CDCL3				
ORIG ST	C4H12SI			TEMP	AMB

75.30	31.40	22.30	112.00	146.00	121.50
2/1	3/3	4/3	5/2	6/1	7/1
125.50	118.00	145.50	40.00	21.00	37.50
8/1	9/1	10/1	11/3	12/3	13/3
32.80	37.50	24.40	37.50	32.80	37.50
14/2	15/3	16/3	17/3	18/2	19/3
24.80	39.40	27.90	22.60	19.70	19.70
20/3	21/3	22/2	23/4	24/4	25/4
22.60	24.00	11.90	11.90		
26/4	27/4	28/4	29/4		

M.MATSUO,S.URANO
TETRAHEDRON   32, 229 (1976)

3014 U 009000 ALPHA-TOCOPHEROL

```
      28
      CH3
       |    4
 HO   C5   C
   \ / \ / \    27      24
   6C    9C  3C  CH3    CH3
    |    ||   |/   12    |14
   7C  10C   2C   CH2   CH
  29/ \ / \ |/ \ / \ / \15
 H3C   C8   O   CH2 CH2 CH2
    |     11  13   |
    CH3            16CH2
    30      21  19   |
          H2C  H2C  17CH2
      23 22/ \ / \ /
      H3C-HC  H2C  18CH
         |   20    |
        26CH3    25CH3
```

FORMULA	C29H50O2			MOL WT	430.72
SOLVENT	CDCL3				
ORIG ST	C4H12SI			TEMP	AMB

74.30	31.60	20.80	118.50	114.40	121.00
2/1	3/3	4/3	5/1	6/1	7/1
122.30	117.00	145.40	39.80	21.00	37.58
8/1	9/1	10/1	11/3	12/3	13/3
32.70	37.50	24.50	37.50	32.70	37.50
14/2	15/3	16/3	17/3	18/2	19/3
24.80	39.40	28.00	22.60	19.70	19.70
20/3	21/3	22/2	23/4	24/4	25/4
22.60	23.80	11.20	12.10	11.80	
26/4	27/4	28/4	29/4	30/4	

M.MATSUO,S.URANO
TETRAHEDRON   32, 229 (1976)

3015　U 009000　VITAMIN E

```
   HO  CH3
    |   |
      2C-C3
    1/    \4
H3C-C       C-CH3
    \    /
     6C=C5
    /     \
  7C       )
    \  9/ 26      25
   8C-C-CH3      CH3
  10/  11   12   |13 14 15
    CH2-CH2-CH2-C-CH2-CH2
     21          17 |
H3C-CH-CH2-CH2-C-2-C-CH2
  22 |  20   19  18 |  16
    CH3            CH3
     23             24
```

FORMULA	C29H50O2		MOL WT	430.72	
SOLVENT	CDCL3				
ORIG ST	C4H12SI		TEMP	AMB	
145.50	144.40	20.70	31.50	78.20	39.80
2/1	5/1	7/3	8/3	9/1	10/3
21.00	37.40	32.70	37.40	24.40	37.40
11/3	12/3	13/2	14/3	15/3	16/3
32.70	37.40	24.80	39.40	27.90	22.60
17/2	18/3	19/3	20/3	21/2	22/4
22.60	19.60	19.60	23.70	11.10	11.70
23/4	24/4	25/4	26/4		
12.10	117.10	118.50	121.10	122.40	

L.F.JOHNSON,W.C.JANKOWSKI
CARBON—13 NMR STECTRA,JOHN
WILEY AND SONS,NEW YORK　　496 (1972)

3016　U 012010　PENICILLIN G POTASSIUM

```
    3 2
   C-C        O
  /   \1 7   ||
 4C      C-CH2-C    H
  \    /     8\  /
   C=C         N    )
   5 6         |  //       +
        9C--C13  K
         |  |          -
      10C--N    O
         |  \12 /
         |    C-C16
         |   /   \
       C--C11    O
        14/ \15
       H3C    CH3
```

FORMULA	C16H17KN2O4S		MOL WT	372.49	
SOLVENT	H2O				
ORIG ST	C4H12SI		TEMP	AMB	
135.30	130.20	129.60	128.00	129.60	130.20
1/1	2/2	3/2	4/2	5/2	6/2
43.00	58.90	67.60	65.30	74.00	31.80
7/3	9/2	10/2	11/1	12/2	14/4
27.50	174.00	174.70	175.30		
15/4					

L.F.JOHNSON,W.C.JANKOWSKI
CARBON—13 NMR SPECTRA,JOHN
WILEY AND SONS,NEW YORK　　471 (1972)

3017　U 012040　TETRACYCLINE HYDROCHLORIDE

```
  OH  O  OH  O    NH2.HCL
   |  ||  |  ||   /
  1C 17C 15C   C13 C22
  / \ / \ / \3/ \ / \
 2C  C18 C16 C14 C12 O
  |  ||  |   |   ||
 3C   C   C   C   C11
  \ /5\ /7\ /9\ / \
   4C  6C  8C  C10 OH
    / \        |
  HO    CH3   N
    19      / \
       /H3C   CH3
      20    21   R -OH
```

FORMULA	C22H24N2O8		MOL WT	444.45	
SOLVENT	H2O				
ORIG ST	C4H12SI		TEMP	AMB	
116.80	138.40	118.50	146.80	70.20	42.20
2/2	3/2	4/2	5/1	6/1	7/2
26.90	35.30	70.60	74.20	22.30	43.20
8/3	9/2	10/2	14/1	19/4	20/4
43.20	97.20	107.10	114.80	161.90	173.30
21/4					
173.80	187.10	193.70			

L.F.JOHNSON,W.C.JANKOWSKI
CARBON—13 NMR SPECTRA,JOHN
WILEY AND SONS,NEW YORK　　491 (1972)